Plasma Modeling (Second Edition)

Methods and applications

Online at: https://doi.org/10.1088/978-0-7503-3559-1

IOP Series in Plasma Physics

Series Editor

Richard Dendy

Culham Centre for Fusion Energy and the University of Warwick, UK

About the series

The IOP Plasma Physics ebook series aims at comprehensive coverage of the physics and applications of natural and laboratory plasmas, across all temperature regimes. Books in the series range from graduate and upper-level undergraduate textbooks, research monographs and reviews.

The conceptual areas of plasma physics addressed in the series include:

- Equilibrium, stability and control
- Waves: fundamental properties, emission, and absorption
- Nonlinear phenomena and turbulence
- Transport theory and phenomenology
- Laser-plasma interactions
- Non-thermal and suprathermal particle populations
- Beams and non-neutral plasmas
- High energy density physics
- Plasma-solid interactions, dusty, complex and non-ideal plasmas
- Diagnostic measurements and techniques for data analysis

The fields of application include:

- Nuclear fusion through magnetic and inertial confinement
- Solar-terrestrial and astrophysical plasma environments and phenomena
- Advanced radiation sources
- Materials processing and functionalisation
- Propulsion, combustion and bulk materials management
- Interaction of plasma with living matter and liquids
- Biological, medical and environmental systems
- Low temperature plasmas, glow discharges and vacuum arcs
- Plasma chemistry and reaction mechanisms
- Plasma production by novel means

A full list of titles published in this series can be found here: https://iopscience.iop.org/bookListInfo/iop-plasma-physics-series.

Plasma Modeling (Second Edition)

Methods and applications

Edited by
Gianpiero Colonna
Consiglio Nazionale delle Ricerche (CNR), Istituto per la Scienza e la Tecnologia dei Plasmi, Sezione di Bari, Italy

Antonio D'Angola
Università della Basilicata, Scuola di Ingegneria, Potenza, Italy
and
Consiglio Nazionale delle Ricerche, Istituto per la Scienza e la Tecnologia dei Plasmi, Sezione di Bari, Italy

IOP Publishing, Bristol, UK

ISBN 978-0-7503-3559-1 (ebook)
ISBN 978-0-7503-3557-7 (print)
ISBN 978-0-7503-3560-7 (myPrint)
ISBN 978-0-7503-3558-4 (mobi)

DOI 10.1088/978-0-7503-3559-1

Version: 20221101

IOP ebooks

British Library Cataloguing-in-Publication Data: A catalogue record for this book is available from the British Library.

Published by IOP Publishing, wholly owned by The Institute of Physics, London

IOP Publishing, No.2 The Distillery, Glassfields, Avon Street, Bristol, BS2 0GR, UK

US Office: IOP Publishing, Inc., 190 North Independence Mall West, Suite 601, Philadelphia, PA 19106, USA

To Mariachiara, Gerardina, Michelangelo and Maria Teresa, our future.

Contents

Preface

In these last years, research activities in the field of plasma physics and chemistry have grown exponentially, not only focusing on fundamental aspects, but also on industrial applications. Due to their inherent complexity, both experiments and numerical modeling are necessary to fully characterize plasma systems. The mostly used numerical simulation techniques for plasma modeling include fluid dynamic, kinetic and hybrid models. These simulation models are significantly different in principles, strengths and limitations. The kinetic models, such as PIC-MCC, are used for non-equilibrium systems, and the fluid models are employed for faster computation, reducing the accuracy. Finally, hybrid models balance between precision and efficiency.

The Golden Age of plasma modeling was in the 1980s–1990s, when several research groups all over the world developed theoretical and numerical approaches to investigate the plasma behaviors. Nowadays, plasma modeling is mainly addressed to technological applications and due to the complexity of systems, commercial multipurpose codes are often used. These numerical codes are based on a fluid dynamics approach, describing realistic discharge geometries. *COMSOL Multiphysics*, *ANSYS Fluent* and *STAR-CCM+* are examples of very popular commercial general-purpose software platforms, based on advanced numerical methods, for modeling and simulating physics-based problems. They are suitable to design and develop optimized products by researchers and engineers working for leading technical enterprises, research labs, and universities.

In strong non-equilibrium cases it is necessary to calculate the energy distribution function of particles. In fact, kinetic modeling is recurrent in gas discharge investigations due to the non-equilibrium energy distribution of charged particles, not representable by the Maxwell distribution function, showing abrupt change in the slope or dumps and bumps, affecting the macroscopic properties of the plasma. In this contest, a critical aspect is to determine the actual distribution function of free electrons, solving the Boltzmann equation. The electron energy distribution directly affects plasma composition, inducing non-equilibrium also in internal states, which, on their side, influence the electron gas, creating a strong synergy between the different plasma components. *BOLSIG+*, originally developed in the LAPLACE laboratory in Toulouse, is an example a freeware non-commercial code for the numerical solution of the Boltzmann equation for free electrons in weakly ionized gases in the framework of the two-term approximation.

Several non commercial codes have been developed by research groups to simulate one- or two-dimensional problems and suitable plasma applications. It is worth mentioning *nonPDPSIM*, developed by the Kushner research group, which is a two-dimensional computational platform that solves, by using the hybrid approach, transport equations for charged and neutral species, Poisson's equation for the electric potential, the electron energy conservation equation for the electron temperature, radiation transport and Navier–Stokes equations for the neutral gas flow, coupled with a Monte Carlo solver for the electron energy distribution function. The two-dimensional platform has three major modules addressing plasma dynamics, fluid dynamics and surface kinetics and is implemented on an unstructured mesh to enable resolution of a large dynamic range in space.

The aim of this book is to present a review of the different approaches that can be adopted to model a plasma, giving a comprehensive bibliography for students and researchers that want to start a project in plasma modeling, orienting the reader toward the choice of the suitable numerical method. Chapters have been written by experienced researchers in the field, who have developed advanced theoretical models and high performance numerical codes. The theoretical models used in commercial or freeware codes are widely described along this book, guiding students, researchers and engineers to select the proper approach for the problem under investigation. The book describes theoretical and numerical approaches that nowadays the plasma community adopt to investigate plasmas and for this reason helps the reader in creating a customized in-house code especially when commercial or freeware codes cannot be used.

Another fundamental aspect of plasma modeling, missing in the first edition, resides in the input data, specific for each physical system. The need to calculate plasma thermodynamic as well as dynamical data (rates and cross sections) is highlighted. Very accurate quantum and semiclassical methods and dedicated experiments are available to determine these quantities with very high precision. However, these methods are time consuming and therefore approximate methods have been developed, in order to obtain complete databases with an adequate accuracy and in a relatively small computational time. In this second edition the book is divided into four parts: (I) Kinetic theory; (II) Fluid and hybrid models; (III) Applications; (IV) Data for plasma modeling. In part I, chapter 1 gives a general overview on the Boltzmann equation, methods of solutions and applications. Chapter 2 presents the general theory of the two-term expansion of the Boltzmann equation, while chapter 3 introduces the multi-term expansion, including also non uniform plasmas. Chapter 4 describes the particle-in-cell algorithm while chapter 5 presents the Ergodic method, suitable for multiscale problems in plasma physics.

In part II, chapter 6 presents the theory of the fluid model for a plasma, focusing on magnetic reconnection application, while chapter 7 describes the magneto-hydrodynamics modeling of plasmas. Chapter 8 shows the drift-diffusion approach and suitable numerical methods adopted to solve equations. Chapter 9 describes the self-consistent model, which couples the state-to-state with the free electron kinetics. In this edition, a section dedicated to polyatomic molecules (CO_2) has been added. In chapter 10 a 2D fluid dynamic solver coupled to the state-to-state kinetics is presented. The code is implemented using MPI-CUDA and the strategies adopted to improve the efficiency have been widely described. Chapter 11 discusses hybrid models, considering electrons as a fluid and following ions as particles.

In part III, some applications are presented. In particular, chapter 12 introduces some applications of the self-consistent model describing plasmas formed by diatomic molecules in discharges and high enthalpy flows. Chapter 13 is focused on the non-equilibrium in high pressure discharges, also considering the problem of fast gas heating. Chapters 14 and 15 are dedicated to the modeling of microarcs and RF plasma jets. Chapter 16 discusses the role of radiation heating for different space vehicles entering planetary atmospheres and chapter 17 discusses the effects of plasma on flying vehicles and how the flow characteristics are affected by the presence of an ionized gas. Chapter 18 treats the complex problem of dust formation in low pressure plasmas. Chapters 19 and 20 are, respectively, dedicated to the plasma confinement in fusion devices and to the problem of validation and verification of numerical codes, a fundamental aspect for modelling activities.

Finally, in part IV some methods dedicated to the data calculations have been presented. Chapter 21 is dedicated to the calculation of thermodynamic and transport properties of a complex mixture, presenting simplified approaches to calculate the internal partition function of atoms and molecules and the phenomenological potential approach to determine the transport cross sections from species parameters, depending on the molecular properties such as polarization. In chapter 22 accurate and approximate methods to determine electron–molecule cross sections are described. In chapter 23 semi-empirical models to determine the rate of internal transitions induced by heavy particle collisions and the state-to-state chemical processes extending Arrhenius macroscopic rates are presented. The chapter discusses also the popular QCT (quasi-classical-trajectory) method that allows one to calculate the probability of a transition with a random sampling of the initial conditions and following particle trajectories on accurate potential energy surfaces.

In the book the authors tried to find the right compromise, synthesizing, in a single chapter subjects that are worthy of dedicated books. Moreover, the editors, due to the limited space, could not dedicate discussions on some numerical approaches, such as Monte Carlo and molecular dynamics, and some applications, such as waves in plasma, self-organized plasmas, probe theory and so on. The extended reference section in each chapter will help in finding additional material for more in-depth analysis.

Gianpiero Colonna and Antonio D'Angola

Acknowledgements

The Editors want to thank all the contributing authors for their efforts and invaluable work. Without their co-operation this book would not have been published. We want to thank Professor Mario Capitelli and Professor Gianni Coppa for their precious and patient guidance and for the enthusiastic encouragements during our research activities.

For chapter 1: The author wants to thank Dr Annarita Laricchiuta for useful suggestions, in particular on the section devoted to cross-sections.

For chapter 3: The author would like to thank their colleagues Dr Markus M Becker and Dr Florian Sigeneger for their helpful discussions and kind assistance in preparing this chapter.

For chapter 4: The work presented in chapter 4 is supported by the European Commission via the Deep-ER project (http://www.deep-er.eu), by the Onderzoekfonds KU Leuven (Research Fund KU Leuven, GOA scheme and Space Weaves RUN project), by the US Air Force EOARD Project FA2386-14-1-0052 and by the Interuniversity Attraction Poles Programme of the Belgian Science Policy Oce (IAPP7/08CHARM).

For chapter 8: The author acknowledges various discussions through the years on weakly-ionized plasmas with Sergey Macheret (Purdue University) and Mikhail Shneider (Princeton University) that have improved this work.

For chapter 14: This work is based on developments and results in the framework of the project 'Theoretical description and modelling of low-current arcs at small gap distances' funded by the Deutsche Forschungsgemeinschaft (Project number 390828847).
We thank our colleagues Dr Detlef Loffhagen for useful comments and suggestions on the manuscript and Prof. Dr Dirk Uhrlandt for the encouragement to go into the fascinating field of microarcs.

For chapter 15: The author thanks his colleagues Dr Detlef Loffhagen, Dr Jan Schäfer, Dr Markus Becker and Dr Margarita Baeva for their fruitful discussions and support.

For chapter 16: This work was performed in the framework of the Program of Fundamental Research of the Russian Academy of Sciences and was supported by the Russian Foundation of Basic Research grant #16-01-00379.

For chapter 18: The authors wish to thank S Ratynskaia, M Rosenberg and P Tolias for useful comments on the manuscript.

This work was funded by the US Department of Energy Office of Science, Office of Fusion Energy Sciences, under the auspices of the National Nuclear Security Administration of the U.S. Department of Energy by Los Alamos National Laboratory, operated by Los Alamos National Security LLC under Contract No. DE-AC52-06NA25396.

For chapter 20: The simulations presented herein were carried out at the Swiss National Supercomputing Centre (CSCS), under project ID s346, and at the Computational Simulation Centre of the International Fusion Energy Research Centre (IFERCCSC), Aomori, Japan, under the Broader Approach collaboration between Euratom and Japan, implemented by Fusion for Energy and JAEA. This work was carried out within the framework of the EUROfusion Consortium. It was supported by the Swiss National Science Foundation and received funding from the European Unions Horizon 2020 research and innovation programme under grant agreement number 633053. The views and opinions expressed herein do not necessarily reflect those of the European Commission.

Editor biographies

Gianpiero Colonna

Gianpiero Colonna graduated in Physics in 1991 with a thesis on the stationary solution of the Boltzmann equation, a PhD in Chemistry in 1995 with a thesis on the state-to-state kinetics in the boundary layer of hypersonic flows, under the supervision of Mario Capitelli, and a post-doctoral position at University of British Columbia in Vancouver, under the supervision of Bernie Shizgal. After a period as Researcher at the Pharmacy Faculty of the University of Bari, he became Researcher at the National Research Council (CNR). Presently, he is Research Director in the same institution. His research activities were focused on plasma modeling, on state-to-state self-consistent kinetics in gas discharges and hypersonic flows, thermodynamic and transport properties of plasmas, modeling plasma plume produced by ns laser pulses. He is author of more than two hundreds papers, conference proceedings and book chapters. He also co-authored two books of the series Fundamental Aspects of Plasma Chemical Physics and for IOP co-edited the book *Meteoroid Entry Physics*.

Antonio D'Angola

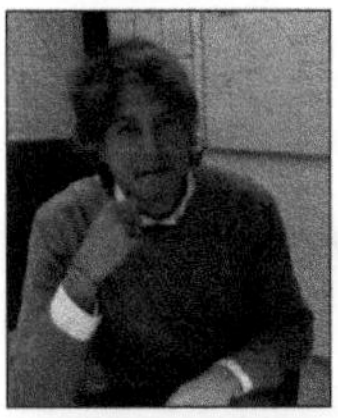

Antonio D'Angola received the Master Degree in Nuclear Engineering in 1996 and the PhD in Energetics in 2000 both at Politecnico di Torino, Italy. Since February 2018 he has been Associate Professor at the University of Basilicata, Italy, in Nuclear Reactor Physics. In 2000 he has been Visiting Researcher at the Plasma Technology Research Centre (CRTP) of Sherbrooke, Quebec. His scientific activity and areas of interest are: numerical methods for the simulation of plasmas by using particle-in-cell and Monte Carlo methods, the calculation of thermodynamic and transport properties of ionized plasmas for industrial and aerospace applications, investigations of non-neutral plasmas for ultra-high vacuum systems and laser–plasma interaction for medical applications. He is Associate Researcher at the National Research Council (CNR). He is author of about one hundred papers and conference proceedings. He also co-authored a book of the series Fundamental Aspects of Plasma Chemical Physics.

List of contributors

Igor V Adamovich
adamovich.1@osu.edu
The Ohio State University, Department of Mechanical and Aerospace Engineering, Columbus, OH, USA

Jayr Amorim
jayr.de.amorim@gmail.com
Departamento de Física, Instituto Tecnológico de Aeronáutica, 12228-900, Sáo José dos Campos, Brazil

Margarita Baeva
baeva@inp-greifswald.de
Department of Plasma Modelling, Leibniz Institute for Plasma Science and Technology (INP), Greifswald, Germany

Markus M Becker
markus.becker@inp-greifswald.de
Department of Plasma Modelling, Leibniz Institute for Plasma Science and Technology (INP), Greifswald, Germany

Francesco Bonelli
francesco.bonelli@poliba.it
Polytechnic University of Bari, Department of Mechanics, Mathematics and Management Bari, Italy

Dario Borgogno
dario.borgogno@cnr.it
ISC-CNR and Politecnico di Torino Dipartimento Energia Torino, Italy

Mario Capitelli
mario.capitelli@ba.istp.cnr.it
Consiglio Nazionale delle Ricerche, Institute for Plasma Science and Technology, Bari section, Italy

Roberto Celiberto
roberto.celiberto@poliba.it
Polytechnic University of Bari, DICATECh and Consiglio Nazionale delle Ricerche, Institute for Plasma Science and Technology, Bari section, Italy

Gianpiero Colonna
gianpiero.colonna@cnr.it
Consiglio Nazionale delle Ricerche, Institute for Plasma Science and Technology, Bari section, Italy

Gianni Coppa
gianni.coppa@polito.it
DET, Department of Electronics and Telecommunications, Politecnico di Torino Torino, Italy

Andrea Cristofolini
andrea.cristofolini@unibo.it
Department of Electrical, Electronic, and Information Engineering 'Guglielmo Marconi' ALMA MATER STUDIORUM—Università di Bologna Bologna, Italy

Antonio D'Angola
antonio.dangola@unibas.it
Scuola di Ingegneria, Università della Basilicata Potenza, Italy
and
Consiglio Nazionale delle Ricerche, Institute for Plasma Science and Technology, Bari section, Italy

Gian Luca Delzanno
delzanno@lanl.gov
Los Alamos National Laboratory, Los Alamos, NM, United States of America

Fabrizio Esposito
fabrizio.esposito@cnr.it
Consiglio Nazionale delle Ricerche, Institute for Plasma Science and Technology, Bari section, Italy

Ambrogio Fasoli
ambrogio.fasoli@epfl.ch
École Polytechnique Fédérale de Lausanne, Swiss Plasma Center Lausanne, Switzerland

Ivo Furno
ivo.furno@epfl.ch
École Polytechnique Fédérale de Lausanne, Swiss Plasma Center Lausanne, Switzerland

Jimena D Gorfinkiel
jimena.gorfinkiel@open.ac.uk
The Open University-School of Physical Sciences Milton Keynes, United Kingdom

Daniela Grasso
daniela.grasso@polito.it
ISC-CNR and Politecnico di Torino Dipartimento Energia Torino, Italy

Vasco Guerra
vguerra@tecnico.ulisboa.pt
Instituto de Plasmas e Fusão Nuclear, Instituto Superior Técnico Lisboa, Portugal

Giovanni Lapenta
giovanni.lapenta@wis.kuleuven.be
KU Leuven, Departement Wiskunde,
Afdeling Plasma-astrofysica Leuven, Belgium

Annarita Laricchiuta
annarita.laricchiuta@cnr.it
Consiglio Nazionale delle Ricerche, Institute for Plasma Science and Technology, Bari Section, Bari, Italy

Enzo Lazzaro
enzo.lazzaro@istp.cnr.it
Consiglio Nazionale delle Ricerche, Institute for Plasma Science and Technology, Milano, Italy

Detlef Loffhagen
loffhagen@inp-greifswald.de
Department of Plasma Modeling, Leibniz Institute for Plasma Science and Technology (INP), Greifswald, Germany

Sergey O Macheret
macheret@purdue.edu
Purdue University, School of Aeronautics and Astronautics
West Lafayette, IN, USA

Pierpaolo Minelli
pierpaolo.minelli@cnr.it
Consiglio Nazionale delle Ricerche, Institute for Plasma Science and Technology, Bari section, Italy

Davide Ninni
davide.ninni@poliba.it
Polytechnic University of Bari, Department of Mechanics, Mathematics and Management Bari, Italy

Bernard Parent
bparent@arizona.edu
Aerospace and Mechanical Engineering, University of Arizona Tucson, USA

Giuseppe Pascazio
giuseppe.pascazio@poliba.it
Polytechnic University of Bari, Department of Mechanics, Mathematics and Management Bari, Italy

Fabio Peano
fabio.peano@gmail.com

Federico Peinetti
federico.peinetti@gmail.com

Lucia Daniela Pietanza
luciadaniela.pietanza@cnr.it
Consiglio Nazionale delle Ricerche, Institute for Plasma Science and Technology, Bari section, Italy

Jonathan Poggie
jpoggie@purdue.edu
School of Aeronautics and Astronautics, Purdue University
West Lafayette, IN, USA

Paolo Ricci
paolo.ricci@epfl.ch
École Polytechnique Fédérale de Lausanne, Swiss Plasma Center CH-1015
Lausanne, Switzerland

Marco A. Ridenti
marcoridenti@gmail.com
Departamento de Física, Instituto Tecnológico de Aeronáutica, 12228-900
Sáo José dos Campos, Brazil

Fabio Riva
fabio.riva@epfl.ch
École Polytechnique Fédérale de Lausanne, Swiss Plasma Center Lausanne, Switzerland

Florian Sigeneger
sigeneger@inp-greifswald.de
Department of Plasma Modeling, Leibniz Institute for Plasma Science and Technology (INP), Greifswald, Germany

Luis O Silva
luis.silva@ist.utl.pt
GoLP, Instituto Superior Técnico Lisboa, Portugal

Sergey T Surzhikov
surg@ipmnet.ru
Institute for Problems in Mechanics, Russian Academy of Sciences Moscow, Russia

Francesco Taccogna
francesco.taccogna@cnr.it
Consiglio Nazionale delle Ricerche, Institute for Plasma Science and Technology, Bari section, Italy

Xianzhu Tang
xtang@lanl.gov
Los Alamos National Laboratory, Los Alamos, NM, USA

Jonathan Tennyson
j.tennyson@ucl.ac.uk
University College London, Department of Physics and Astronomy London, UK

Christian Theiler
christian.theiler@epfl.ch
École Polytechnique Fédérale de Lausanne, Swiss Plasma Center Lausanne, Switzerland

Part I

Kinetic theory

IOP Publishing

Plasma Modeling (Second Edition)
Methods and applications
Gianpiero Colonna and Antonio D'Angola

Chapter 1

Boltzmann and Vlasov equations in plasma physics

Gianpiero Colonna

When asking researchers in physics and chemistry what is the scientific discovery at the origin of modern science, most of the answers will be *Einstein's relativity* or *quantum physics*. In our opinion, this role should be assigned to the *Boltzmann kinetic equation* and its application to the kinetic theory of gases. The same opinion was expressed by Max Planck in his Nobel lecture [1]:

> '...my trying to give you the story of the origin of the quantum theory in broad outlines and to couple with this, a picture in a small frame, of the development of this theory up to now, and its present-day significance for physics. ...For this reason, I busied myself, from then on, that is, from the day of its establishment, with the task of elucidating a true physical character for the formula, and this problem led me automatically to a consideration of the connection between entropy and probability, that is, Boltzmann's trend of ideas; until after some weeks of the most strenuous work of my life, light came into the darkness, and a new undreamed-of perspective opened up before me.'

The Boltzmann equation introduced in physics the idea of probability, used a few years later in quantum physics. Also, the concept of time changed, relating its direction to that of entropy exchange [2], breaking the symmetry between the past and the future [3], even if the laws of mechanics are invariant for the time inversion. Despite its importance, students in physics and chemistry often confuse the Boltzmann equation with the Boltzmann distribution.

The Boltzmann equation finds many applications in modern science [4, 5], from fluid dynamics to material science:

doi:10.1088/978-0-7503-3559-1ch1

- the Chapman–Enskog expansion is the method to calculate transport properties of gases [6];
- numerical methods, such as the Bhatnagar–Gross–Krook (BGK) [7] approximation or *lattice Boltzmann method* [8, 9], are commonly used to determine the flow field of complex fluids;
- spectroscopical signature of gases as well as radiative heating is estimated by solving the Boltzmann equation for photons (see chapter 16);
- thermal and electrical properties of metals and semiconductor materials can be estimated by solving the Boltzmann equation for electrons and phonons, adapted to account for quantum statistics [10];
- mathematical aspects, such as existence and unicity of solution and the accuracy of approximate methods [11].

The Boltzmann equation is one of the most powerful tools to investigate the plasma state, from the electron kinetics in weakly ionized gases [12] to fusion [13] and astrophysical [14, 15] plasmas. In this chapter we will present the theoretical foundation of the Boltzmann and Vlasov equations, giving an overview of their applications.

1.1 Fundamentals

Let us consider the single-particle phase space Φ, characterized by the sextuplet $(\vec{r}, \vec{v})$ of particle position and velocity. The state of the particle corresponds to a single point in this space, while a system with $\mathcal{N}$ particles is represented by a set of $\mathcal{N}$ points. In order to give a pictorial view, we limit to a 1D space coordinate and one velocity component. This representation can also be considered the projection of the entire volume in the 2D plane (x, v_x). To determine the time evolution of the volume in the Φ space, the motion of each particle, according to the classical mechanic laws, must be followed

$$\begin{aligned} \frac{d\vec{r}_i}{dt} &= \vec{v}_i \\ \frac{d\vec{v}_i}{dt} &= \vec{a}_i \end{aligned} \tag{1.1}$$

The ensemble of points changes its shape, as can be observed in figure 1.1(left) for a 2D example, because each particle moves independently from the others. The acceleration $\vec{a}_i$ originates from two contributions

$$\vec{a}_i = \vec{a}_{\text{ext}}(\vec{r}_i) + \sum_{k \neq i}^{\mathcal{N}} \frac{\vec{F}_{ik}}{m_i} \tag{1.2}$$

one, $\vec{a}_{\text{ext}}$, is due to an external field, and the other one $\vec{F}_{ik}/m_i$, where m_i is the particle mass, is due to intermolecular forces $\vec{F}_{ik}$. This approach, successfully applied to predict the orbit of planets in the solar system, in principle allows one to determine exactly the evolution of any N-particle system, such as a matter in gas phase.

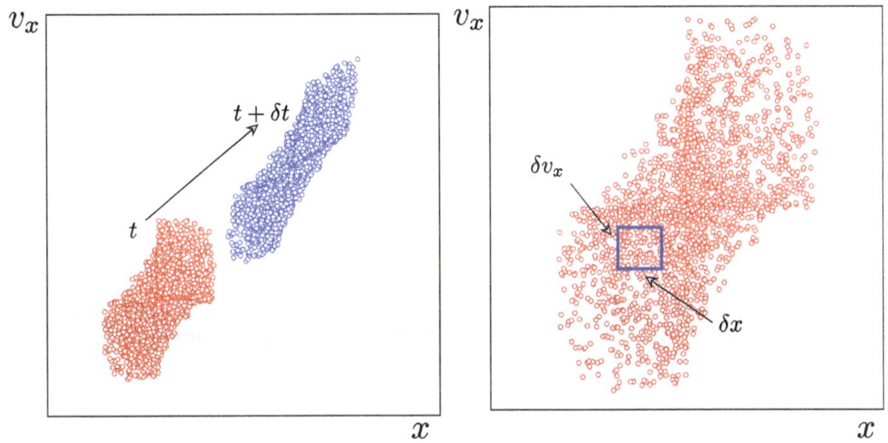

Figure 1.1. Projection of the phase space $(\vec{r}, \vec{v})$ on the 2D plane (x, v_x) of an $\mathcal{N}$ particle system. (left) Evolution of the sampling points with $a_x(x) = a_{x0} + a_{x1}x^2$, (right) starting distribution with the projection of a sampling volume $(\delta\vec{x}, \delta\vec{v}_x)$.

In practice, the number of molecules in a macroscopic system ($\mathcal{N} \approx 10^{23}$) for any computer available nowadays, and the uncertainty in the initial conditions[1] make such a deterministic procedure unfeasible, for any numerical simulation, even for modern high-performance computers.

An alternative approach is to describe the evolution of an N-particle system by using the laws of statistics, developing the theory of *physical kinetics*. Let us consider the following assumptions:

1. An infinitesimal sampling volume, such as that depicted in figure 1.1(right), contains a very large number of particles, so that a *macroscopical* infinitesimal volume still contains a large number of particles: the gas can be considered as a continuous medium.
2. The mean particle distance, $\bar{d} = \sqrt[3]{\frac{V}{\mathcal{N}}}$, with V the physical volume, is larger than the thermal de Broglie length, $\lambda_{\mathrm{th},i} = \sqrt{\frac{2\pi\hbar}{m_i kT}}$: quantum effects are negligible and the classical statistics and laws of motion can be used.
3. The interparticle interaction distance $\bar{r}_i$ is much smaller than the mean free path λ_{mfp}: particle interactions can be described as binary collisions.

As a first step, let us limit it to a system of free particles moving in an external field, without collisions. In an infinitesimal hypercube identified by the vertices $(\vec{r} - \frac{1}{2}\delta\vec{r}, \vec{v} - \frac{1}{2}\delta\vec{v})$ and $(\vec{r} + \frac{1}{2}\delta\vec{r}, \vec{v} + \frac{1}{2}\delta\vec{v})$, see figure 1.1(right), whose volume in the phase space is

[1] The uncertainty of initial position and velocity is assured by *Heisenberg's principle.*

$$\delta\Gamma = \delta x \delta y \delta z \delta v_x \delta v_y \delta v_z = \delta V \delta^3 v \tag{1.3}$$

where δV is the physical volume, the number of particles in the infinitesimal hypercube can be written as

$$\delta\mathcal{N} = f(\vec{r}, \vec{v})\delta\Gamma \tag{1.4}$$

which defines the *distribution function* as the number of particles per unit volume in the phase space. For real systems, n_s different species can be present, each one with its own internal structure, that, in the framework of quantum physics, can be represented by discrete levels, characterized by a set of quantum numbers ℓ.[2] This consideration brings us to a generalized definition of distribution function, considering equation (1.4) for particles of sth species in the ℓ th level, i.e.

$$\delta\mathcal{N}_{s,\ell}(t) = f_s(\vec{r}, \vec{v}, \ell, t)\delta\Gamma \tag{1.5}$$

1.1.1 The convection operator

Let us now evaluate at which rate the number of particles inside the sampling volume changes. It corresponds to the flux of particles crossing the surface of the hypercube. For sake of simplicity, the discussion is restricted to the two-dimensional problem (2D) depicted in figure 1.1(right), with the square sampling volume, where the particle flux is the sum of four contributions

$$\frac{\partial\delta\mathcal{N}_{s,\ell}}{\partial t} = \Phi = \Phi_{\text{left}} + \Phi_{\text{right}} + \Phi_{\text{bottom}} + \Phi_{\text{top}} \tag{1.6}$$

two through the vertical sides (left and right), due to particles with the fixed velocity changing position

$$\begin{aligned}
\Phi_{\text{left}} &= +\int_{v_x-\frac{1}{2}\delta v_x}^{v_x+\frac{1}{2}\delta v_x} w_x f_s\left(x - \frac{1}{2}\delta x, w_x, \ell, t\right) dw_x \approx v_x f_s\left(x - \frac{1}{2}\delta x, v_x, \ell, t\right)\delta v_x \\
\Phi_{\text{right}} &= -\int_{v_x-\frac{1}{2}\delta v_x}^{v_x+\frac{1}{2}\delta v_x} w_x f_s\left(x + \frac{1}{2}\delta x, w_x, \ell, t\right) dw_x \approx v_x f_s\left(x + \frac{1}{2}\delta x, v_x, \ell, t\right)\delta v_x
\end{aligned} \tag{1.7}$$

and two through the horizontal sides (bottom and top), due to particles accelerated in the same positions

[2] As an example, for atomic species we have the quadruplet (n, l, m, s) corresponding to principal, angular, magnetic and spin quantum numbers and for molecules by (n, J, v, s) corresponding to electronic state, rotational, vibrational and total spin quantum numbers. However, internal levels can be ordered by energy and identified by a single integer number.

$$\begin{aligned}\Phi_{\text{bottom}} &= +\int_{x-\frac{1}{2}\delta x}^{x+\frac{1}{2}\delta x} a_x(x')f_s(x', v_x - \frac{1}{2}\delta v_x, \ell, t)dx' \approx a_x(x)f_s(x, v_x - \frac{1}{2}\delta v_x, \ell, t)\delta x \\ \Phi_{\text{top}} &= -\int_{x-\frac{1}{2}\delta x}^{x+\frac{1}{2}\delta x} a_x(x')f_s(x', v_x + \frac{1}{2}\delta v_x, \ell, t)dx' \approx a_x(x)f_s(x, v_x + \frac{1}{2}\delta v_x, \ell, t)\delta x\end{aligned} \tag{1.8}$$

Combining the terms in equations (1.7) and (1.8), it is

$$\begin{aligned}\Phi_{\text{left}} + \Phi_{\text{right}} &= v_x\delta v_x\left[f_s(x - \frac{1}{2}\delta x, v_x, \ell, t) - f_s(x + \frac{1}{2}\delta x, v_x, \ell, t)\right] \\ \Phi_{\text{bottom}} + \Phi_{\text{top}} &= a_x(x)\delta x\left[f_s(x, v_x - \frac{1}{2}\delta v_x, \ell, t) - f_s(x, v_x + \frac{1}{2}\delta v_x, \ell, t)\right]\end{aligned} \tag{1.9}$$

Considering that the sampling volume in the phase space is small, the parentheses at the right-side in equation (1.9) can be approximated with the derivatives

$$\begin{aligned}f_s(x - \frac{1}{2}\delta x, v_x, \ell, t) - f_s(x + \frac{1}{2}\delta x, v_x, \ell, t) &\approx -\frac{\partial f_s(x, v_x, \ell, t)}{\partial x}\delta x \\ f_s(x, v_x - \frac{1}{2}\delta v_x, \ell, t) - f_s(x, v_x + \frac{1}{2}\delta v_x, \ell, t) &\approx -\frac{\partial f_s(x, v_x, \ell, t)}{\partial v_x}\delta v_x\end{aligned} \tag{1.10}$$

and equations (1.6), (1.7), and (1.8) give the total flux as

$$\frac{\partial \delta \mathcal{N}_{s,\ell}}{\partial t} = \Phi = -\delta x\delta v_x\left[v_x\frac{\partial f_s(x, v_x, \ell, t)}{\partial x} + a_x(x)\frac{\partial f_s(x, v_x, \ell, t)}{\partial v_x}\right] \tag{1.11}$$

To extend this result to the whole phase space, we should note that in the multidimensional space:

- the integrals in equations (1.7) and 1.8 must be performed over all the variables but the one perpendicular to the surface, i.e. $\delta x\delta v_x \rightarrow \delta\Gamma$;
- two contributions must be considered for each independent direction, i.e. equation (1.9) should be written also for the y and z directions;
- acceleration depends on the position in the phase space, i.e. $a_x(x) \rightarrow [a_x(\vec{r}), a_y(\vec{r}), a_z(\vec{r})]$.[3]

These assumptions, together with equation (1.5), give

$$\frac{\partial \delta \mathcal{N}_{s,\ell}}{\partial t} = \delta\Gamma\frac{\partial f_s(\vec{r}, \vec{v}, \ell, t)}{\partial t} = -\delta\Gamma[\vec{v} \cdot \vec{\nabla}_r + \vec{a}(\vec{r}) \cdot \vec{\nabla}_v]f_s(\vec{r}, \vec{v}, \ell, t) \tag{1.12}$$

[3] The acceleration can also depend on the velocity as happens in the presence of the magnetic field for charged species as discussed in chapter 2.

where

$$\vec{v}\cdot\vec{\nabla}_r \equiv v_x\frac{\partial}{\partial x} + v_y\frac{\partial}{\partial y} + v_z\frac{\partial}{\partial z}$$
$$\vec{a}(\vec{r})\cdot\vec{\nabla}_v \equiv a_x(\vec{r})\frac{\partial}{\partial v_x} + a_y(\vec{r})\frac{\partial}{\partial v_y} + a_z(\vec{r})\frac{\partial}{\partial v_z} \tag{1.13}$$

The square parentheses in equation (1.12) can be considered as a convection operator, because it describes the evolution of the distribution function due to the free motion of particles subjected to an external force.

1.1.2 The collisional operator

Let us now describe the contribution of collisions to the evolution of the distribution in the phase space. If the gas density is sufficiently small only binary encounters could be considered. If the mean free path is much larger than the interaction distance, these encounters can be treated as collisions. In a collision at given velocities, the final state is completely determined by the deflection angle, and, in the case of internal excitation and chemical reactions, by the energy necessary to activate the considered process. The uncertainty in assigning position and speed to the colliding particles makes it necessary to describe the collisions statistically. The cross-section of a process is the measure of the probability of a collision output. Let us define the following quantities:

- the differential cross-section $\frac{d\sigma}{d\Omega}$ is the ratio between the number of particles deflected in the unit time in a given solid angle $d\Omega$ and the incoming particle flux per unit time and per unit surface. It is a function of the module of the relative velocity $|\vec{v}_A - \vec{v}_B|$, initial and final levels and the solid angle.
- the total cross-section σ is the integral of the differential cross-sections over the whole solid angle. For elastic collisions this quantity is not relevant, including also the contribution of collision with very small deflection angles. On the other hand, it is often used in inelastic collisions, whose angular dependence of differential cross-section is assumed to be isotropic.

Including the collisional operator, equation (1.14) becomes the Boltzmann equation in the usual form

$$\frac{\partial f_s(\vec{r}, \vec{v}, \ell, t)}{\partial t} + [\vec{v}\cdot\vec{\nabla}_r + \vec{a}(\vec{r})\cdot\vec{\nabla}_v]f_s(\vec{r}, \vec{v}, \ell, t) = \left(\frac{\delta f_s}{\delta t}\right)_{\text{coll}} \tag{1.14}$$

Let us construct the collisional operator analyzing the binary encounter sketched in figure 1.2. In a sampling volume given by equation (1.3), the number of particles A with velocity $\vec{v}_A$ flowing through a unitary surface is given by[4]

[4] For sake of simplicity, let us consider implicitly the variables of the distribution substituting $f_{A/B}(\vec{r}, \vec{v}_{A/B}, \ell_{A/B}, t) \rightarrow f_{A/B}$ and $f_{A/B}(\vec{r}, \vec{v}'_{A/B}, \ell'_{A/B}, t) \rightarrow f'_{A/B}$.

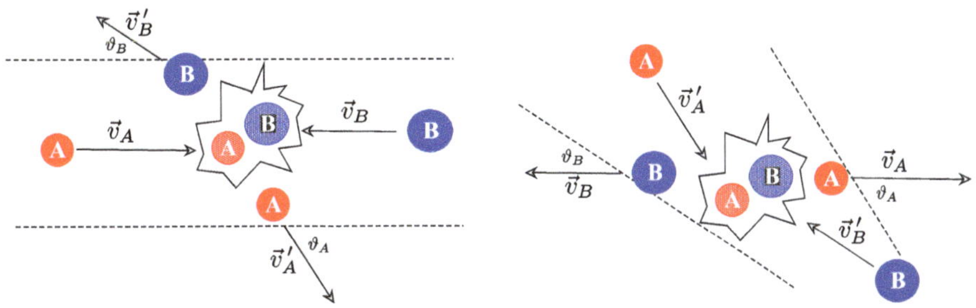

Figure 1.2. Scheme of a binary elastic collision (left) with deflection angle and the picture of the reverse process (right) obtained by reverting time and space. The dashed lines are the directions of the velocities before the collisions and the θs are the corresponding deflection angles.

$$\Phi_A(\vec{v}_A, \vec{v}_B) = f_A \left|\vec{v}_A - \vec{v}_B\right| \delta^3 v_A = f_A v_{AB} \delta^3 v_A \tag{1.15}$$

where $v_{AB} = |\vec{v}_A - \vec{v}_B|$ is the relative velocity, and, from the definition of differential cross-section[5], the number of particles A deflected by the angle ϑ_A in the unit time by a single particle B (see figure 1.3) is

$$\begin{aligned} \left[\frac{d\delta \mathcal{N}_A(\vartheta_A)}{dt}\right]_1 &= \Phi_A(\vec{v}_A, \vec{v}_B)\frac{d\sigma_{AB}}{d\Omega}(\vartheta_A)\delta\Omega_A \\ &= f_A v_{AB}\delta^3 v_A \frac{d\sigma_{AB}}{d\Omega}(\vartheta_A) d\Omega_A \end{aligned} \tag{1.16}$$

being

$$d\Omega_A = \sin\vartheta_A d\vartheta_A d\phi \tag{1.17}$$

with ϕ the angle of the deflection plane[6]. The velocities after the collision $\vec{v}_A'$ and $\vec{v}_B'$, as well as ϑ_B, are univocally determined by ϑ_A and ϕ, because energy and momentum conservation laws can be used to determine all the other components. Equation (1.16), gives the number of particles undergoing the considered collision with a single B particle. To have the total number of particles deflected in the given direction we have to multiply this quantity by the number of scattering centers, i.e.

$$\frac{\partial\delta \mathcal{N}_A(\vartheta_A)}{\partial t} = \left[\frac{\partial\delta \mathcal{N}_A(\vartheta_A)}{\partial t}\right]_1 f_B \delta^3 v_B \delta V = f_A f_B v_{AB} \frac{d\sigma_{AB}}{d\Omega}(\vartheta_A)\delta^3 v_A \delta^3 v_B \delta V d\Omega_A \tag{1.18}$$

It should be noted that equation (1.18) must be symmetric with respect to exchange between A and B, because this expression gives the same rate for the number of particles B deflected of the angle θ_B, therefore

[5] Let us write explicitly only the dependence of cross-section on the deflection angle, using the symbol $\frac{d\sigma'}{d\Omega}$ for the reverse process.

[6] The directions of the velocities after the collision, if not parallel, univocally define the deflection plane. Therefore, the same angle ϕ can be used for both particles.

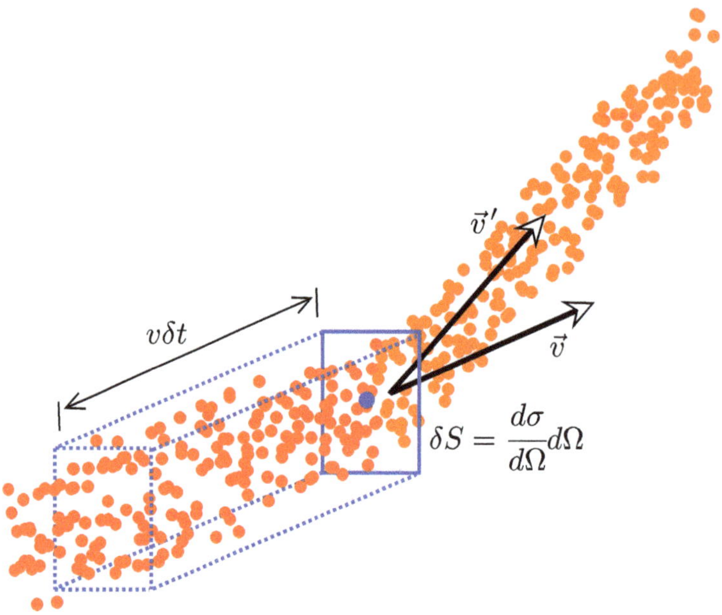

Figure 1.3. The number of particles, all with the same velocity $\vec{v}$ relative to the scattering center (blue dot), flowing through the surface δS in the time interval δt are those contained in the volume $\delta V = v\delta t\delta S$. Considering δS the differential cross-section, all these particles are deflected in the same solid angle $d\Omega$ with velocity $\vec{v}'$.

$$\frac{d\sigma_{AB}}{d\Omega}(\vartheta_A)d\Omega_A = \frac{d\sigma_{AB}}{d\Omega}(\vartheta_B)d\Omega_B \tag{1.19}$$

The variation expressed by equation (1.18) is the loss rate for particles A and B in the state before the collision. At the same time it must be considered the gain rates coming from the reverse process, as illustrated in figure 1.2(right), given by

$$\left[\frac{\partial \delta \mathcal{N}_A(\vartheta_A)}{\partial t}\right]_r = f'_A f'_B v'_{AB} \delta^3 v'_A \delta^3 v'_B \frac{d\sigma'_{AB}}{d\Omega}(\vartheta_A) d\Omega_A \delta V \tag{1.20}$$

noting that the deflection angles are the same as in the direct process. However, to keep the parity with respect to space-time inversion, the probability of a process must be equal to that of its reverse, i.e.

$$v'_{AB}\frac{d\sigma'_{AB}}{d\Omega}(\vartheta_A) = v_{AB}\frac{d\sigma_{AB}}{d\Omega}(\vartheta_A) \tag{1.21}$$

This equation is known as a *micro-reversibility* principle. This assumption can be demonstrated for collisions with spherical symmetry, but in general it is not proved. For strange geometries, like those of large organic molecules, the micro-reversibility principle is not valid. In any case, in this book the micro-reversibility principle is considered valid. The micro-reversibility principle has a corresponding law from the macroscopic point of view, the *detailed balance* principle. The global variation of the distribution function of f_A is obtained by summing over all the collision partners B, integrating over all its initial states and over all the final states of A, obtaining the final expression for the collisional operator

$$\left(\frac{df_A}{dt}\right)_{\text{coll}} = \sum_{\ell'_A} \sum_{B,\ell_B,\ell'_B} \int v_{AB} \frac{d\sigma_{AB}}{d\Omega}(\vartheta_A)(f'_A f'_B - f_A f_B) d^3v_B d\Omega_A \tag{1.22}$$

that has been divided by the volume $\delta\Gamma_A = \delta V \delta^3 v_A$ to have the equation for the distribution function.

1.1.3 Boltzmann's *H*-theorem

It is straightforward to demonstrate that in equilibrium, i.e. in the presence of Boltzmann and Maxwell distributions, at the same temperature, the collisional operator is null, reproducing an important result already obtained in the framework of statistical physics. Another important result obtained for the first time by Boltzmann is the demonstration of the growth of entropy for an isolated rarefied gas, namely, the *H-theorem*. Let us define the functional

$$H(t) = \iiint d^3v f \ln f \tag{1.23}$$

and its time derivative[7] is given by

$$\frac{\partial H(t)}{\partial t} = \int d^3v \frac{\partial f}{\partial t}(\ln f + 1) \tag{1.24}$$

Substituting the Boltzmann equation (see equation (1.14)) in this expression and noticing that the convection operator in equation (1.12) gives null integral[8], only the contribution of the collision integral remains. Distinguishing the colliders as in the previous section it is

$$\frac{\partial H(t)}{\partial t} = \int d^3v_A(\ln f_A + 1) v_{AB} \frac{d\sigma_{AB}}{d\Omega}(\vartheta_A)(f'_A f'_B - f_A f_B) d^3v_A d^3v_B d\Omega_A \tag{1.25}$$

The argument of the integral in equation (1.25) is invariant by exchanging A and B and averaging the two expressions it is

$$\frac{\partial H(t)}{\partial t} = \frac{1}{2} \int d^3v_A[\ln(f_A f_B) + 2] v_{AB} \frac{d\sigma_{AB}}{d\Omega}(\vartheta_A)(f'_A f'_B - f_A f_B) d^3v_A d^3v_B d\Omega_A \tag{1.26}$$

As a consequence of the detailed balance principle, the integral must be the same when initial and final states are exchanged

[7] It should be noted [16, 17] that the entropy is defined as the integral of $(f - f \ln f)$. The following demonstration can be applied straightforwardly to entropy [16], showing that the time derivative of S has opposite sign with respect to that of H.

[8] This assertion can be easily demonstrated, applying Gauss's theorem, transforming the volume integral of the derivatives in a surface integral of the function. At infinite the distribution function is null, therefore the surface integral gives null contribution [16].

$$\frac{\partial H(t)}{\partial t} = \frac{1}{2}\int d^3v_A[\ln(f'_A f'_B) + 2]v_{AB}\frac{d\sigma_{AB}}{d\Omega}(\vartheta_A)(f_A f_B - f'_A f'_B)d^3v_A d^3v_B d\Omega_A \quad (1.27)$$

and averaging equations (1.26) and (1.27) we have the final expression

$$\frac{\partial H(t)}{\partial t} = \frac{1}{4}\int d^3v_A[\ln(f_A f_B) - \ln(f'_A f'_B)]v_{AB}\frac{d\sigma_{AB}}{d\Omega}(\vartheta_A)(f'_A f'_B - f_A f_B) \quad (1.28)$$
$$d^3v_A d^3v_B d\Omega_A$$

The sign of the integrand depends on the product

$$\left[\ln(f_A f_B) - \ln(f'_A f'_B)\right](f'_A f'_B - f_A f_B) \quad (1.29)$$

which is always negative or null when the initial and final distributions are the same, conditions verified at the equilibrium. The integral of a negative function is always negative, it is

$$\frac{\partial H(t)}{\partial t} \leqslant 0 \quad (1.30)$$

demonstrating the H-theorem[9].

1.1.4 Vlasov equation

For ionized gases, the previous theory could not be applied, with the Coulomb interaction being a long-range force. As a consequence charge–charge interaction cannot be treated as a collision. If the gas is weakly ionized, the charge–charge interactions are negligible with respect to collisions with neutrals, but for highly ionized gases, the Coulomb forces gives the dominant contribution, and some adjustments are necessary for the Boltzmann equation to be applied to plasmas.

Let us consider a test particle s with charge q_s, position $\vec{r}_s$ in a plasma in stationary conditions. The s particle will feel the electric field produced by all the other charges

$$\vec{E}(\vec{r}_s, t) = \frac{1}{4\pi\epsilon_0}\sum_{i\neq s}^{\mathcal{N}}\frac{q_i\, \vec{r}_{s,i}(t)}{|\vec{r}_{s,i}(t)|^3} \quad (1.31)$$

where $\vec{r}_{s,i} = \vec{r}_i(t) - \vec{r}_s$ is the position of i relative to s. We can separate two contributions to $\vec{E}(\vec{r}_s, t)$, a constant mean field

$$\vec{E}_0(\vec{r}_s) = \frac{1}{\delta t}\int_t^{t+\delta t}\vec{E}(\vec{r}_s, t)dt \quad (1.32)$$

where δt is a time interval, and $\vec{E}_r(\vec{r}_s, t)$, due to fluctuations, is a function of time with null mean

[9] The demonstration given here refers to homogeneous systems. For non-homogeneous systems see reference [5].

$$\frac{1}{\delta t}\int_{t}^{t+\delta t}\vec{E}_r(\vec{r}_s, t) = \frac{1}{\delta t}\int_{t}^{t+\delta t}[\vec{E}(\vec{r}_s, t) - \vec{E}_0(\vec{r}_s)]dt \approx \vec{0} \tag{1.33}$$

The time interval δt must be a macroscopic time, i.e. $\delta t \gg 1/\tau_c$ where τ_c is the mean time between collisions. In principle, $\vec{E}_r(\vec{r}_s, t)$ is a continuous function of time, with rapid oscillations with mean time amplitude of the order of τ_c. This means that if we sample this field periodically, at successive equal time intervals of amplitude δt, much larger than τ_c, i.e. $\vec{E}_r(\vec{r}_s, t)$, $\vec{E}_r(\vec{r}_s, t + 2\delta t)$, ..., $\vec{E}_r(\vec{r}_s, t + n\delta t)$, we will have a random series. If the system is not in a stationary condition, the above assumptions remain valid if the sampling time δt is small with respect to the evolution time of whole system, i.e. all the macroscopic quantities are almost constant in this time interval. Similar arguments can be used for the magnetic field. This property of the local field suggests that the mean field contributes to the acceleration in the Boltzmann equation, while $\vec{E}_r$ must contribute to the collision integral, due to its random evolution. The mean field does not produce entropy, a consequence of the H-theorem, differently from the random field.

In fully ionized gases it could be demonstrated that the entropy variation due to collisions is much slower than other macroscopic quantities such as charge density ρ_c

$$\left|\frac{1}{S}\frac{\partial S}{\partial t}\right| \ll \left|\frac{1}{\rho_c}\frac{\partial \rho_c}{\partial t}\right| \tag{1.34}$$

therefore the collision integral can be considered null. The Boltzmann equation in equation (1.14) becomes

$$\frac{\partial f_s}{\partial t} + \left[\vec{v}\cdot\vec{\nabla}_r + \frac{q_s}{m_s}(\vec{E}_0 + \vec{v}\wedge\vec{B}_0)\cdot\vec{\nabla}_v\right]f_s = 0 \tag{1.35}$$

where the local fields are related only to local macroscopic quantities such as charge density and current

$$\begin{cases} \rho_c = \sum_s q_s \int f_s d^3v \\ \vec{j}_c = \sum_s q_s \int \vec{v} f_s d^3v \end{cases} \tag{1.36}$$

through the Maxwell equations

$$\begin{cases} \vec{\nabla}\cdot\vec{E}_0 = \dfrac{\rho_c}{\epsilon_0} \\ \vec{\nabla}\cdot\vec{B}_0 = 0 \\ \vec{\nabla}\wedge\vec{E}_0 = -\dfrac{\partial\vec{B}_0}{\partial t} \\ \vec{\nabla}\wedge\vec{B}_0 = \dfrac{\vec{j}_c}{\epsilon_0 c^2} + \dfrac{1}{c^2}\dfrac{\partial E_0}{\partial t} \end{cases} \tag{1.37}$$

This system is known as the *Vlasov equation.* In most practical applications the time derivatives of the fields in the Maxwell equations are negligible, and the quasi-static fields are considered. The validity of the Vlasov equation stands on two conditions:

- The mean particle distance must be smaller than the Debye length

$$\lambda_D = \sqrt{\frac{\epsilon_0 kTV}{\sum_s q_s^2 N_s}} \tag{1.38}$$

i.e. the number of particle inside the Debye sphere must be statistically significant [16];

- The collision frequency is much smaller than the plasma frequency

$$\omega_p = \sqrt{\sum_s \frac{q_s^2 N_s}{\epsilon_0 m_s V}} \tag{1.39}$$

in order to neglect the collision integral.

1.2 Cross sections

The collisional term in the Boltzmann equation results from the dynamics of binary collisions in the plasma promoting elastic, inelastic and reactive processes that govern the temporal evolution of the kinetics at a microscopic level. The construction of a reliable database for the elementary probabilities represents even nowadays a challenge, due to the requirements of accuracy, consistency and completeness, issues significantly affecting the predictive capability of kinetic models, not only in the estimation of macroscopic plasma parameters, but also in shedding light on relevant collisional mechanisms [18]. Modern approaches in quantum molecular dynamics are fully exploited aiming at the highest accuracy, however, the wide spectrum of theoretical methods including semiclassical, quasi-classical, classical and semi-empirical, are actually suitable for specific classes of processes, offering a favorable balance between the reliability of dynamic information and the computational load characterizing the derivation of complete dataset, especially in the state-to-state kinetics, where the dependence of the probability on the quantum state of colliders is also considered.

Elastic collisions usually dominate the kinetics and fully determine the transport properties of thermal equilibrium plasmas [6]. Transport cross-sections (see chapter 21), i.e. momentum transfer and higher ℓ moments of elastic cross-section, and their thermal average, i.e. the collision integrals, $\Omega_{ij}^{\ell,s}(T)$ of the Chapman–Enskog theory [19], can be obtained by quantum [20] or semiclassical WKB estimation of phase-shifts or by direct classical trajectory integration of the interaction potential [21]. The interaction potential, from accurate electronic structure calculations or modeled either with full-range analytical forms [6] or with a phenomenological potential [22], is assumed isotropic, so dependent only on the approaching line of colliding partners, also for interactions involving molecules. Some examples do exist in literature of classical trajectory calculations on *ab initio* surfaces for molecular collisions [23], considering anisotropic potentials, however, for small molecules, rapidly rotating, an isotropic angle-averaged potential can be assumed [24]. Resonant charge-exchange in ion–parent-atom interactions is the only non-elastic process considered in transport calculations, due to the significant contribution in the ℓ-odd collision integrals [6]. A rigorous quantum coupled-channel approach [20] is required, however, it is worth mentioning the asymptotic approach [25, 26], based on the idea of expanding the cross-section in terms of the small parameter connected to the geometry of the collision.

In the case of electron elastic scattering, the quantum nature of the interactions is not negligible and manifests itself in the features of the energy profile of the cross-section, as the low-energy peaks, due to resonances, or the *Ramsauer minimum* [27]. Therefore, the quantum approach is required and the collision integrals are evaluated by integration of theoretical [28] or experimental differential elastic electron-scattering cross-sections.

Inelastic and reactive channels, considered negligible in transport, except for quantum effects at low-temperatures, play instead a key role in non-equilibrium plasmas, chemical species undergoing excitation, dissociation and ionization processes induced by electron impact or collision with heavy particles with increasing threshold energies.

Quantum approaches (R-matrix [29], close-coupling, Schwinger multichannel approach [30], local complex potential [31]) represent the best theoretical framework for the investigation of the dynamics in electron–molecule scattering (see chapter 22), entailing non-adiabatic effects as the resonant electron-capture with formation of long-living negative ion intermediates or the vibronic coupling. However, first-order approaches, Born, Born–Bethe and Born–Ochkur [32], accurate for high collision energies, semiclassical methods [33–35], assuming the impinging electron moving classically in linear trajectories and the quantum molecular target, and the classical Gryzinskii method [36] have been successfully used in the past. More recently, novel simplified approaches have been proposed in literature as the BE*f*-scaling method [37], based on a rescaling procedure of the Born cross-section for dipole-allowed vibronic excitations with the experimental optical oscillator strengths, the similarity approach [38], or the BED (binary-encounter-dipole) model for ionization [39, 40]. Simplified approaches are still nowadays attractive, demonstrating in many cases a striking agreement with experiments and representing very promising tools for more complex molecular systems.

For the dynamics of heavy-particle collisions (see chapter 23), leading to energy transfer among internal and translation degrees of freedom of colliding partners, rearrangement reactions or dissociation, the quasi-classical trajectory (QCT) method [18, 41, 42] is by now well-assessed, proved to be well-comparing with and complementing the quantum approaches [43], provided an accurate potential energy surface (PES) and a robust trajectory statistics, allowing the derivation of state-specific rate coefficients through the pseudo-quantization of energy in reactant and product channels, and able to fully exploit the advantages of high-performance code parallelization [44]. Furthermore the QCT approach can follow the dynamics in collisional systems characterized by non-adiabatic coupling among different electronic states, with trajectories hopping [45] from ground to excited potential energy surfaces. The mechanism of vibrational energy transfer in atom–molecule and molecule–molecule collisions has been also investigated in the framework of the semiclassical approach [46], the force-harmonic oscillator (FHO) [47] and the widely used SSH method [48].

A number of different open-access web-databases are available [49–52], where dynamical information has been collected, validated and shared with the modeling community, including also state-specific cross-sections and rate coefficients of elementary processes involving ground and excited chemical species, with resolution on the electronic, vibrational and rotational degrees of freedom [53, 54].

1.3 Solution of the Boltzmann equation

The first approach for an analytical solution of the Boltzmann equation was the expansion series named *Chapman–Enskog* theory [55] which writes the distribution function as

$$f = f_0 + \sum_{n=1}^{\infty} \epsilon^n f_n \tag{1.40}$$

where ϵ is a small parameter, and stopping the series at the first order. The first term f_0 is the Maxwell distribution, that gives null collision integral. Therefore, the transport coefficients are related to the perturbation f_1. This approach allows one to calculate transport properties relatively to the specific cross-sections. Also, the perturbation is expanded in a polynomial series [6], allowing different degrees of approximation. The Chapman–Enskog theory is valid for small deviations from the Maxwell distributions. Chapman–Enskog theory leads to the Navier–Stokes equations. However, in extremely rarefied gases, Navier–Stokes equations fail and more accurate approaches are required. An improved model was introduced by Grad [56] extending the moments equations to thirteen terms.

Even if it is very difficult to solve the complete Boltzmann equation due to the large dimension of the phase space and to the nonlinearity of the collisional operator, different numerical methods are inspired to it. In recent years the *lattice Boltzmann method* (LBM) [8, 57] is finding growing interest to simulate complex fluids. This method writes the Boltzmann equation in a discrete space $\mathcal{L}$ with a finite number of possible velocities (see figure 1.4). The time must be discrete too in order to have that any advancement in the position is still a point of the lattice, i.e.

$$x_i \in \mathcal{L} \Rightarrow x_i + \vec{v}_j \delta t \in \mathcal{L} \tag{1.41}$$

An interesting feature of LBM is to be a solver for Navier–Stokes equations [8, 57], a characteristic inherited by the original Boltzmann equation. Contrarily to the Navier–Stokes equations, LBM has a linear convective operator. In most of the cases, also the collisional operator is linearized, using BGK approach [9]. This method is practically suitable for investigating complex systems, such as incompressible flows through porous media [58] and multi-phase/multi-component flows [59].

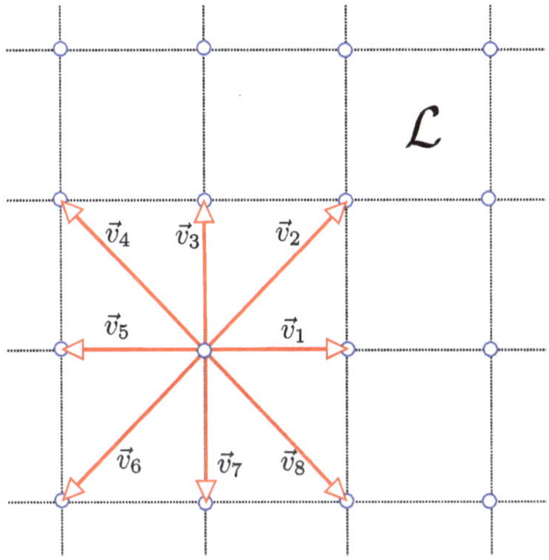

Figure 1.4. Example of 2D lattice with discrete velocities.

Even if the LBM can be derived directly from the Boltzmann equation in BGK approximation [7] under small Mach number expansion [60], the two equations are substantially different. Whereas the Boltzmann equation is valid for rarefied gases, LBM is mainly used for incompressible flows and dense matter. Moreover, the H-theorem is not granted in LBM.

Another interesting approach based on the Boltzmann equation is *direct Monte Carlo simulation*, developed by G Bird in the early 1960s [61, 62]. The method has been developed to investigate rarefied flows in very low density, where the Navier–Stokes equations fail, being based on the idea that the free motion and the collisions can be decoupled. DSMC considers the atomic structure of the matter. The gas is represented by a large number of pseudo-particles, each one representing a large number of real particles in the same position and with the same velocity. In the convection phase all the particles move according to the classical laws of motion for a given time interval, δt, small with respect to the mean collision time. Figure 1.1(left) can also be considered a pictorial view of this system. The computational domain is divided in a grid, (see figure 1.1(right) as an example of grid elements), with the purpose of calculating statistical means such as density, velocity, temperature and so on. In each grid element, according to the total cross-section, the number of collisions is determined and a Monte Carlo approach is used to sample the collisions. Random numbers are generated to decide which particles will collide, what is the result of the collision, changing the state of the particles accordingly. Convective and collision phases are repeated one after the other. With this method it is possible to calculate directly the distribution function, with a sensitivity depending on the number of particles. DSMC has been used to model hypersonic entry in planetary atmosphere [63] also for real vehicles such as the space shuttle [64, 65].

For fluid dynamic applications, simplified cross-section models are used, such as hard spheres [66], variable hard spheres [67] and variable soft spheres [68], since the collision probability is available in closed form. More advanced models include cross-sections obtained from quantum or semiclassical calculations (see section 1.2), including also state-to-state kinetics [69–71], as described in chapter 9. The inclusion of a detailed cross-section dataset in DSMC calculation is a complex problem, not only because of the large amount of required memory to store the data and the time needed to calculate the probability of each event, but also because these data are not easily available. A novel approach, named DSMC-QCT [72, 73], does not go through the cross-section evaluation, but follows the motion of particle on the multidimensional PES. Once the colliding particles have been chosen, random numbers are generated to select the collisional parameters. The interaction is then followed in the center-of-mass frame of reference, moving the particles on the given PES, until the interparticle distance is large enough. The final state obtained in this way determines the new state of the colliders.

DSMC is not adequate for modeling plasma even though some attempts were done in literature using different time-steps for heavy particles and electrons [74]. Other approaches entail the coupling of PIC (see chapter 4) with DSMC to model plasma in rarefied flows [75].

The Boltzmann equation has also been applied to calculate the evolution of non-equilibrium distributions of electrons and phonons in metals heated by femtosecond laser pulses [10, 76–82]. The laser pulse is absorbed by the electron gas and the energy is redistributed by electron–electron collisions. At longer times the energy is transferred to the solid lattice with the production of phonons.

In a metal, electrons partially occupy the conduction band, following the Fermi statistic. In this case we generalize the definition of distribution as (see chapter 21)

$$n_s = \frac{\mathcal{N}_s}{V} = \int_0^\infty g_s f_s^{\mathrm{FD}}(\varepsilon)\mathcal{D}_s(\varepsilon)d\varepsilon \tag{1.42}$$

where g_s is the spin multiplicity of the particle s (= 2 for electrons) and $\mathcal{D}_s$ is the density of states. The equilibrium state is described by the Fermi–Dirac distribution

$$f_s^{\mathrm{FD}}(\varepsilon) = (e^{\beta(\varepsilon - E_F)} + 1)^{-1} \tag{1.43}$$

where E_F is the Fermi energy[10], defined as

$$\mathcal{N}_s = \int_0^{E_F} \mathcal{D}_s(\varepsilon)d\varepsilon \tag{1.44}$$

Even at absolute temperature $T = 0$ the electrons are not in the same quantum state because of the Pauli exclusion principle, but all allowed states below the Fermi energy are fully occupied and the distribution function is a step function

$$\begin{aligned} \varepsilon \leqslant E_F &\Rightarrow f_s^{\mathrm{FD}}(\varepsilon) = 1 \\ \varepsilon > E_F &\Rightarrow f_s^{\mathrm{FD}}(\varepsilon) = 0 \end{aligned} \tag{1.45}$$

[10] Equation (1.43) is only an approximation, valid only at low temperature.

This character of the Fermion distribution should reflected on the Boltzmann equation, and in particular, to account for the Pauli exclusion principle, the product of the distributions in the collision integral (see equation (1.22)) becomes

$$f_A f_B \rightarrow f_A(1 - f_A) f_B(1 - f_B) \tag{1.46}$$

which ensures that the distribution does not become larger than one.

1.4 Plasma modeling numerical codes

To conclude this chapter, we will give a brief description of the numerical codes commonly used by researchers for modeling plasmas, using the theoretical approaches described in the following chapters. The mostly used numerical simulation techniques for plasma modeling include fluid dynamic, kinetic and hybrid models. These simulation models are significantly different in principles, strengths and limitations. The kinetic models, such as PIC-MCC (see chapter 4), are used for non-equilibrium systems, and the fluid models are employed for faster computation, reducing the accuracy. Finally, hybrid models balance between precision and efficiency.

COMSOL Multiphysics [83] is a general-purpose software platform, based on advanced numerical methods, for modeling and simulating physics-based problems. It is suitable to design and develop optimized products by researchers and engineers working for leading technical enterprises, research labs, and universities. The *COMSOL* platform is characterized by an environment designed for cross-disciplinary product development with a unified workflow for electrical, mechanical, fluid, and chemical applications. The plasma module, available in the electrical suite, is a partial differential equation solver including drift-diffusion approximations for ions, the quasi-neutrality assumption for electrons movements, reduced Maxwell equations for electromagnetic fields, electron energy equations for electron temperatures and the Navier–Stokes equation for neutral background gas. Users could conveniently formulate physical equations according to their requirements, and do not have to spend much time in coding the numerical algorithm, especially in complex geometries. As an example, by using the plasma module, low-temperature plasma sources and systems (direct current discharges, inductively-coupled plasmas, and microwave plasmas) can be simulated and scientists can gain insight into the physics of discharges and gauge the performance of existing or future designs. The only drawback of the solver is the lack of kinetic information. In fact, *COMSOL* does not include the Boltzmann equation solver and in many applications kinetic effects must be considered.

ANSYS Fluent [84] is a computational fluid dynamics software tool used by researchers and engineers to optimize a product's performance. *ANSYS Fluent* includes well-validated physical modeling capabilities to deliver fast, accurate results across the widest range of CFD and multiphysics applications. *ANSYS Fluent* offers many great advantages in the modelling of plasmas. In fact, *ANSYS Fluent* can solve very complex fluid dynamic problems by using a large set of numerical algorithms to solve energy, momentum, mass and many other kinds of equations. Moreover, chemical reactions, continuity of species, drift-diffusion models and

Maxwell equations can be coupled and solved by taking advantage of a very important capability of such software: the user defined scalar (UDS) variables. *ANSYS Fluent* can solve the transport equation for an arbitrary UDS in the same way that it solves the transport equation for a scalar such as species mass fraction. Extra scalar transport equations may be needed in certain types of combustion applications or, for example, in plasma applications. *ANSYS Fluent* allows us to define additional scalar transport equations in your model.

By using this capability, for an arbitrary scalar Φ_k, *ANSYS Fluent* solves the equation:

$$\frac{\partial \rho \Phi_k}{\partial t} + \frac{\partial}{\partial x_i}\left(\rho u_i \Phi_k - \Gamma_k \frac{\partial \Phi_k}{\partial x_i}\right) = S_{\Phi_k} \qquad k = 1, \ldots, N \tag{1.47}$$

where Γ_k and S_{Φ_k} are the diffusion coefficient and source term supplied by the user for each one of the N scalar equations, u_i represent the velocity field and x_i the spatial coordinates.

In the case of a steady state problem and in the absence of convective fluxes $\frac{\partial \rho \Phi_k}{\partial t} = 0$ and $u_i = 0$. Furthermore, imposing $\Gamma_k = 0$, equation (1.47) becomes:

$$-\nabla^2 \Phi_k = S_{\Phi_k} \qquad k = 1, \ldots, N \tag{1.48}$$

ANSYS Fluent can solve UDS equations both in fluid and solid regions, and is able to impose continuity condition along a boundary that separates a fluid zone from a solid one. This capability allows one to solve a large number of conditions.

Both *COMSOL* and *ANSYS Fluent* are CFD codes based on the theoretical general models deeply described in chapters 6 and 7 and applied for aerospace applications (chapters 10 and 16).

BOLSIG+ [85] is a non-commercial code for the numerical solution of the Boltzmann equation for free electrons in weakly ionized gases in the framework of the two-term approximation (see chapter 2). Developed by Gerjan Hagelaar [86] of the LAPLACE laboratory in Toulouse in France, *BOLSIG+* was first released in 2005 as a replacement for the earlier *BOLSIG* solver developed by Leanne Pitchford, Jean-Pierre Boeuf and Lowell Morgan, based on the old code *ELENDIF* by Morgan and Penetrante [87].

BOLSIG+ uses data from the web-accessed database LXcat [49], a selection of complete sets of electron–neutral cross-sections sets (see section 1.2) for a wide range of gases, compiled by different contributors and with varying levels of detail.

The main application of *BOLSIG+* is to obtain electron transport coefficients and rates for electron-induced processes starting from cross-section data. These quantities, which can be used as input for fluid models (see chapters 6, 7, and 16), are obtained from the non-Maxwellian electron energy distribution calculated for a given gas composition and under stationary conditions[11]. *BOLSIG+* is solved under the following assumptions:

[11] This procedure, known also as cold gas approximation, should be compared with the self-consistent approach (see chapter 9).

- electric field and collision probabilities are constant in space and time and there are no boundaries for the system;
- constant gas composition
- electron production/loss due to ionization/attachment is assumed to result in exponential growth/decay of the electron number density.

Under the above assumptions, the Boltzmann equation reduces to a convection–diffusion equation with non-local source term in energy space, which is then discretized by an exponential scheme and solved for the electron energy distribution function by a standard matrix inversion technique (see section 2.8).

Another freely available Boltzmann solver, *LOKI* [88], has been recently release by the Portuguese group. This code works under the same approximations as *BOLSIG+*.

There are other non-commercial codes developed by research groups and generalized for a category of plasma applications. Typically, these codes can simulate only 1D or 2D, for a limited range of plasma applications. Among them, one of the best and popular codes is *nonPDPSIM* [89], developed by Kushner and his research group. *nonPDPSIM* is a 2D computational platform that solves transport equations for charged and neutral species, Poisson's equation for the electric potential, the electron energy conservation equation for the electron temperature, radiation transport and Navier–Stokes equations for the neutral gas flow. The 2D platform has three major modules addressing plasma dynamics, fluid dynamics and surface kinetics and is implemented on an unstructured mesh to enable resolution of a large dynamic range in space. *CFD-GEOM* or *SKYMESH2*, a mesh generation program, has been incorporated into the code to allow for unstructured meshes. The model is coupled with a thermodynamics module providing local thermodynamic equilibrium (LTE) properties. To speed up the simulation, a fully implicit scheme and sparse matrix techniques are used. The source function for electron impact reactions and secondary electron emission, and electron transport coefficients needed by this segment of the code are computed assuming the local field approximation, solving the electron energy equation and/or using a Monte Carlo Simulation. The continuity equations for neutral species are solved explicitly.

References

[1] Planck M 1920 The Nobel Lecture (http://www.nobelprize.org/nobel_prizes/physics/laureates/1918/planck-lecture.html)

[2] Bobylev A V and Cercignani C 1999 On the rate of entropy production for the Boltzmann equation *J. Stat. Phys* **94** 603–18

[3] Eddington A 2012 *The Nature of the Physical World: Gifford Lectures (1927)* (Cambridge: Cambridge University Press)

[4] Cercignani C 1988 *The Boltzmann Equation and Its Applications* (Applied Mathematical Sciences vol 67) (New York: Springer) https://doi.org/10.1007/978-1-4612-1039-9

[5] Cercignani C 1995 Aerodynamical applications of the Boltzmann equation *Riv. Nuovo Cim.* **18** 1–40

[6] Capitelli M, Bruno D and Laricchiuta A 2013 *Fundamental Aspects of Plasma Chemical Physics: Transport* (Springer Series on Atomic, Optical, and Plasma Physics vol 74) (New York: Springer) https://doi.org/10.1007/978-1-4419-8172-1

[7] Bhatnagar P L, Gross E P and Krook M 1954 A model for collision processes in gases. I. Small amplitude processes in charged and neutral one-component systems *Phys. Rev.* **94** 511

[8] Chen S and Doolen G D 1998 Lattice Boltzmann method for fluid flows *Annu. Rev. Fluid Mech.* **30** 329–64

[9] Qian Y, d'Humières D and Lallemand P 1992 Lattice BGK models for Navier–Stokes equation *Europhys. Lett.* **17** 479

[10] Greenwood D 1958 The Boltzmann equation in the theory of electrical conduction in metals *Proc. Phys. Soc.* **71** 585

[11] Villani C 2002 A review of mathematical topics in collisional kinetic theory *Handbook of Mathematical Fluid Dynamics* 1 (Amsterdam: Elsevier) pp 71–305

[12] Rockwood S D 1973 Elastic and inelastic cross sections for electron-Hg scattering from Hg transport data *Phys. Rev.* A **8** 2348–60

[13] Brizard A J 1995 Nonlinear gyrokinetic Vlasov equation for toroidally rotating axisymmetric tokamaks *Phys. Plasmas* **2** 459–71

[14] Hénon M 1982 Vlasov equation *Astron. Astrophys.* **114** 211

[15] Maruca B A, Kasper J C and Gary S P 2012 Instability-driven limits on helium temperature anisotropy in the solar wind: observations and linear Vlasov analysis *Astrophys. J.* **748** 137

[16] Landau L D, Lifshitz E and Pitaevskii L 1993 Physical kinetics *Course of Theoretical Physics, Volume 10* (Oxford: Pergamon)

[17] Capitelli M, Colonna G and D'Angola A 2011 *Fundamental Aspects of Plasma Chemical Physics: Thermodynamics* (Springer Series on Atomic, Optical, and Plasma Physics vol 66) (New York: Springer) https://doi.org/10.1007/978-1-4419-8185-1

[18] Capitelli M, Celiberto R, Colonna G, Esposito F, Gorse C, Hassouni K, Laricchiuta A and Longo S 2016 *Fundamental Aspects of Plasma Chemical Physics: Kinetics* (Springer Series on Atomic, Optical, and Plasma Physics vol 85) (New York: Springer) https://doi.org/10.1007/978-1-4419-8185-1

[19] Hirschfelder J O, Curtiss C F and Bird R B 1966 *Molecular Theory of Gases and Liquids* (New York: Wiley)

[20] Krstic` P S and Schultz D R 1999 Elastic scattering and charge transfer in slow collisions: isotopes of H and H^+ colliding with isotopes of H and with He *J. Phys. B: At. Mol. Phys.* **32** 3485–509

[21] Colonna G 2008 A general numerical algorithm for classical collision integral calculation *Comput. Phys. Commun.* **178** 809–16

[22] Capitelli M, Cappelletti D, Colonna G, Gorse C, Laricchiuta A, Liuti G, Longo S and Pirani F 2007 On the possibility of using model potentials for collision integral calculations of interest for planetary atmospheres *Chem. Phys.* **338** 62–8

[23] Viehland L A, Dickinson A S and Maclagan R G A R 1996 Transport coefficients for NO^+ ions in helium gas: a test of the NO^+-He interaction potential *Chem. Phys.* **211** 1–15

[24] Stallcop J R, Partridge H and Levin E 2001 Effective potential energies and transport cross sections for atom-molecule interactions of nitrogen and oxygen *Phys. Rev.* A **64** 042722

[25] Nikitin E E and Smirnov B M 1978 Quasiresonant processes in slow collisions *Sov. Phys. - Usp.* **21** 95–116

[26] Smirnov B M 2001 Atomic structure and the resonant charge exchange process *Phys.-Usp.* **44** 221–53

[27] Gryzinski M 1970 Ramsauer effect as a result of the dynamic structure of the atomic shell *Phys. Rev. Lett.* **24** 45–7

[28] Bray I and Stelbovics A T 2003 Convergent close-coupling approach to electron-atom collisions *Many-Particle Quantum Dynamics in Atomic and Molecular Fragmentation* (Springer Series on Atomic, Optical, and Plasma Physics vol 35) (New York: Springer) pp 121–35

[29] Tennyson J 2010 Electron-molecule collision calculations using the R-matrix method Phys *Phys. Rep.* **491** 29–76

[30] Gianturco F A and Huo W M (ed) 1995 *Computational Methods for Electron–Molecule Collisions* (Springer Series on Atomic Optical, and Plasma Physics) (New York: Springer) https://doi.org/10.1007/978-1-4757-9797-8

[31] Wadehra J M 1986 Vibrational excitation and dissociative attachment *Nonequilibrium Vibrational Kinetics* (Topics in Current Physics vol 39) (New York: Springer) pp 191–232 https://doi.org/10.1007/978-3-642-48615-9_7

[32] Chung S and Lin C C 1972 Excitation of the electronic states of the nitrogen molecule by electron impact *Phys. Rev.* A **6** 988–1002

[33] Hazi A U 1981 Impact-parameter method for electronic excitation of molecules by electron impact *Phys. Rev.* A **23** 2232–40

[34] Redmon M J, Garrett B C, Redmon L T and McCurdy C W 1985 Improved impact-parameter method for electronic excitation and dissociation of diatomic molecules by electron impact *Phys. Rev.* A **32** 3354–65

[35] Celiberto R and Rescigno T N 1993 Dependence of electron-impact excitation cross sections on the initial vibrational quantum number in H_2 and D_2 molecules: $X^1\Sigma_g^+ \rightarrow B^1\Sigma_u^+$ and $^1\Sigma_g^+ \rightarrow C^1\Pi_u$ transitions *Phys. Rev.* A **47** 1939–45

[36] Gryziński M 1965 Classical theory of atomic collisions. I. Theory of inelastic collisions *Phys. Rev.* **138** A336–58

[37] Kim Y-K 2007 Scaled Born cross sections for excitations of H_2 by electron impact *J. Chem. Phys.* **126** 064305

[38] Adamson S, Astapenko V, Deminskii M, Eletskii A, Potapkin B, Sukhanov L and Zaitsevskii A 2007 Electron impact excitation of molecules: calculation of the cross section using the similarity function method and ab initio data for electronic structure *Chem. Phys. Lett.* **436** 308–13

[39] Kim Y-K and Rudd M E 1994 Binary-encounter-dipole model for electron-impact ionization *Phys. Rev.* A **50** 3954–67

[40] Huo W M 2001 Convergent series representation for the generalized oscillator strength of electron-impact ionization and an improved binary-encounter-dipole model *Phys. Rev.* A **64** 042719

[41] Blais N C and Truhlar D G 1976 Monte Carlo trajectory study of Ar+H_2 collisions. I. Potential energy surface and cross sections for dissociation, recombination, and inelastic scattering *J. Chem. Phys.* **65** 5335–56

[42] Blais N C, Zhao M, Mladenovic M, Truhlar D G, Schwenke D W, Sun Y and Kouri D J 1989 Comparison of quasiclassical trajectory calculations to accurate quantum mechanics for state-to-state partial cross sections at low total angular momentum for the reaction D+H $_2\rightarrow$ HD+H *J. Chem. Phys.* **91** 1038–42

[43] Esposito F, Coppola C M and Fazio D D 2015 Complementarity between quantum and classical mechanics in chemical modeling. The H + HeH $^+\rightarrow H_2^+$ + He reaction: a rigorous test for reaction dynamics methods *J. Phys. Chem.* A **119** 12615–26

[44] Lombardi A, Faginas-Lago N and Laganà A 2014 *Grid Calculation Tools for Massive Applications of Collision Dynamics Simulations: Carbon Dioxide Energy Transfer* (Lecture Notes in Computer Science) (Cham: Springer) pp 627–39 https://doi.org/10.1007/978-3-319-09144-0_43

[45] Tully J C 1976 *Dynamics of Molecular Collisions* ed W H Miller (New York: Plenum)

[46] Billing G 1987 Rate constants for vibrational transitions in diatom-diatom collisions *Comput. Phys. Commun.* **44** 121–36

[47] Adamovich I V, Macheret S O, Rich J W and Treanor C E 1998 Vibrational energy transfer rates using a forced harmonic oscillator model *J. Thermophys. Heat Transfer* **12** 57–65

[48] Schwartz R N, Slawsky Z I and Herzfeld K F 1952 Calculation of vibrational relaxation times in gases *J. Chem. Phys.* **20** 1591–9

[49] LXcat 2016 Plasma Data Exchange Project Database (http://fr.lxcat.net/home/)

[50] ALADDIN 2016 Numerical Database Maintained by the IAEA Nuclear Data Section A+M Data Unit (https://www-amdis.iaea.org/ALADDIN/)

[51] GASPAR 2016 GAS and PlasmA Radiation Database (http://esther.ist.utl.pt/gaspar/)

[52] NIFS 2016 National Institute for Fusion Science database (http://dbshino.nifs.ac.jp/)

[53] Phys4Entry 2016 (http://phys4entrydb.ba.imip.cnr.it/Phys4EntryDB/)

[54] STELLAR 2016 (http://esther.ist.utl.pt/pages/stellar.html)

[55] Chapman S and Cowling T G 1970 The mathematical theory of non-uniform gases: an account of the kinetic theory of viscosity *Thermal Conduction and Diffusion in Gases* (Cambridge: Cambridge University Press)

[56] Grad H 1949 On the kinetic theory of rarefied gases *Commun. Pure Appl. Math.* **2** 331–407

[57] Succi S 2001 *The Lattice Boltzmann Equation: For Fluid Dynamics and Beyond* (Oxford: Oxford University Press) https://doi.org/10.1016/S0997-7546(02)00005-5

[58] Guo Z and Zhao T 2002 Lattice Boltzmann model for incompressible flows through porous media *Phys. Rev.* E **66** 036304

[59] Shan X and Chen H 1993 Lattice Boltzmann model for simulating flows with multiple phases and components *Phys. Rev.* E **47** 1815

[60] Abe T 1997 Derivation of the lattice Boltzmann method by means of the discrete ordinate method for the Boltzmann equation *J. Comput. Phys.* **131** 241–6

[61] Bird G A 1963 Approach to translational equilibrium in a rigid sphere gas *Phys. Fluids* **6** 1518

[62] Bird G A 1994 *Molecular Gas Dynamics and the Direct Simulation of Gas Flows* (Oxford: Clarendon)

[63] Bondar Y A, Markelov G, Gimelshein S and Ivanov M 2005 *DSMC Study of Shock-Detachment Process in Hypersonic Chemically Reacting Flow* Tech. Rep., DTIC Document

[64] Bird G 1990 *Application of the Direct Simulation Monte Carlo Method to the Full Shuttle Geometry* (Reston, VA: American Institute of Aeronautics and Astronautics) pp 90–1692 https://dx.doi.org/10.2514/6.1990-1692

[65] Gallis M A 2005 *DSMC Simulations in Support of the STS-107 Accident Investigation* Tech. Rep., DTIC Document https://apps.dtic.mil/sti/pdfs/ADA446030.pdf

[66] Ohwada T 1993 Structure of normal shock waves: direct numerical analysis of the Boltzmann equation for hard-sphere molecules *Phys. Fluids* A **5** 217–34

[67] Hash D, Hassan H and Moss J N 1994 Direct simulation of diatomic gases using the generalized hard sphere model *J. Thermophys. Heat Transfer* **8** 758–64

[68] Fan J 2002 A generalized soft-sphere model for Monte Carlo simulation *Phys. Fluids* **14** 4399–405

[69] Willauer D L and Varghese P L 1993 Direct simulation of rotational relaxation using state-to-state cross sections *J. Thermophys. Heat Transfer* **7** 49–54

[70] Bruno D, Capitelli M, Esposito F, Longo S and Minelli P 2002 Direct simulation of non-equilibrium kinetics under shock conditions in nitrogen *Chem. Phys. Lett.* **360** 31–7

[71] Kim J G and Boyd I D 2014 Monte Carlo simulation of nitrogen dissociation based on state-resolved cross sections *Phys. Fluids* **26** 012006

[72] Fujita K 2007 Assessment of molecular internal relaxation and dissociation by DSMC-QCT analysis *39th AIAA Thermohysics Conference (Miami, FL)* pp 2007–4345 https://doi.org/10.2514/6.2007-4345

[73] Valentini P, Schwartzentruber T E, Bender J D, Nompelis I and Candler G V 2015 Direct simulation of rovibrational excitation and dissociation in molecular nitrogen using an ab initio potential energy surface *Phys. Fluids* **27** 086102

[74] Economou D J, Bartel T J, Wise R S and Lymberopoulos D P 1995 Two-dimensional direct simulation Monte Carlo (DSMC) of reactive neutral and ion flow in a high density plasma reactor *IEEE Trans. Plasma Sci.* **23** 581–90

[75] Serikov V V, Kawamoto S and Nanbu K 1999 Particle-in-cell plus direct simulation Monte Carlo (PIC-DSMC) approach for self-consistent plasma-gas simulations *IEEE Trans. Plasma Sci.* **27** 1389–98

[76] Pietanza L, Colonna G and Capitelli M 2005 Electron and phonon dynamics in laser short pulses-heated metals *Appl. Surf. Sci.* **248** 103–7

[77] Pietanza L, Colonna G, Longo S and Capitelli M 2004 Electron and phonon distribution relaxation in metal films under a femtosecond laser pulse *Thin Solid Films* **453** 506–12

[78] Pietanza L, Colonna G, Longo S and Capitelli M 2004 Electron and phonon relaxation in metal films perturbed by a femtosecond laser pulse *Appl. Phys.* A **79** 1047–50

[79] Pietanza L D, Colonna G, Longo S and Capitelli M 2007 Non-equilibrium electron and phonon dynamics in metals under femtosecond laser pulses *Eur. Phys. J.* D **45** 369–89

[80] Shivanian E, Abbasbandy S and Alhuthali M S 2014 Exact analytical solution to the Poisson–Boltzmann equation for semiconductor devices *Eur. Phys. J. Plus* **129** 1–8

[81] Majorana A 1991 Space homogeneous solutions of the Boltzmann equation describing electron-phonon interactions in semiconductors *Transp. Theory Stat. Phys.* **20** 261–79

[82] Apostolova T, Perlado J and Rivera A 2015 Femtosecond laser irradiation induced-high electronic excitation in band gap materials: a quantum-kinetic model based on Boltzmann equation *Nucl. Instrum. Methods Phys. Res.* B **352** 167–70

[83] COMSOL MULTIPHYSICS (http://www.comsol.it/plasma-module)

[84] ANSIS FLUENT (http://www.ansys.com/Products/Fluids/ANSYS-Fluent)

[85] Hagelaar G and Pitchford L BOLSIG+: Electron Boltzmann Equation Solver (http://www.bolsig.laplace.univ-tlse.fr/)

[86] Hagelaar G and Pitchford L 2005 Solving the Boltzmann equation to obtain electron transport coefficients and rate coefficients for fluid models *Plasma Sources Sci. Technol.* **14** 722

[87] Morgan W and Penetrante B 1990 ELENDIF: a time-dependent Boltzmann solver for partially ionized plasmas *Comput. Phys. Commun.* **58** 127–52

[88] Tejero-del Caz A, Guerra V, Gonçalves D, da Silva M L, Marques L, Pinhao N, Pintassilgo C and Alves L 2019 The LisbOn KInetics Boltzmann solver *Plasma Sources Sci. Technol.* **28** 043001

[89] Bhoj A N and Kushner M J 2007 Continuous processing of polymers in repetitively pulsed atmospheric pressure discharges with moving surfaces and gas flow *J. Phys. D: Appl. Phys.* **40** 6953

Chapter 2

Two-term Boltzmann Equation

Gianpiero Colonna and Antonio D'Angola

Modeling gas discharges and non-equilibrium plasmas requires the knowledge of the rate coefficients of processes induced by electron impact and transport properties of free electrons. In collision-dominated plasmas, usually known as cold plasmas, the Boltzmann distribution is not adequate to describe the statistic of free electrons [1]. A gas subjected to the action of an electric field is an open system, where the energy is injected into the gas by means of electrons. The mechanisms of energy transfer from free electrons to atoms and molecules are very complex, determining the shape of the electron distribution, usually departing from the Maxwellian. In order to determine the electron energy distribution function (EEDF) the Boltzmann equation (see equation (1.14)) is commonly used. If the applied field it is not too large, the two-term, or $P1$, approximation is commonly considered, usually written in the following form

$$\frac{\partial n_0(\varepsilon)}{\partial t} = -\frac{\partial J_{el}(\varepsilon)}{\partial \varepsilon} - \frac{\partial J_f(\varepsilon)}{\partial \varepsilon} - \frac{\partial J_{ee}(\varepsilon)}{\partial \varepsilon} - \frac{\partial J_{ne}(\varepsilon)}{\partial \varepsilon} + S_{in}(\varepsilon) + S_{sup}(\varepsilon) + S_{ion}(\varepsilon) + S_{rec}(\varepsilon) + S_{chem}(\varepsilon) \tag{2.1}$$

corresponding to the first-order expansion in spherical harmonics, where J's are fluxes in the energy space corresponding to elastic collisions (el), external fields (f), electron–electron collisions (ee), the cooling term due to nozzle expansion (ne) and S's are source terms due to inelastic (in), superelastic (sup), ionization (ion) and chemical processes in general ($chem$). For large values of the electric fields further terms are necessary in the spherical-harmonic expansion [2–8] (see chapter 3 for details). An alternative approach is the *test particle Monte Carlo* (TP-MC) [9], where collisions of a single particle in a bath of heavy species are randomly sampled. It can

doi:10.1088/978-0-7503-3559-1ch2

be demonstrated that TP-MC is equivalent to solving the stationary, homogeneous Boltzmann equation with expansion in spherical harmonics including infinite terms.

The two-term approximation is a good compromise between computational time and accuracy, therefore it is often used to model the electron energy distribution in stationary [1, 10, 11] and time dependent [12–16] approaches. Another important application is the self-consistent coupling of the Boltzmann equation with state-to-state chemical kinetics [12, 17, 18] (see chapter 9), determining at the same time the distribution function of internal states of atoms and molecules and gas composition.

To put in evidence the role of elementary processes, such as superelastic [11, 19] and electron–electron [20, 21] collisions, on the shape of the EEDF, the *P*1 approximation represents one of the most used approaches, providing at the same time the fractional power going into different channels [22]. The *P*1 approximation was also applied to calculating the EEDF in non-homogenous conditions [23–27] or to investigate MHD effects [28, 29].

The macroscopic properties of the electron gas are determined by the effective shape of the EEDF, which is strictly related to the energy dependence of the inelastic cross-sections. This behavior was the origin of the swarm data approach [30–32] to determine the cross-sections of electron collisions with different species [26, 27, 33–40].

It is worth noting that an on-line solver named *BOLSIG+* [41] can be used free of charge, through the LXCat web site [42], with a wide cross-section database.

In this chapter we will present the two-term approximation of the Boltzmann equation in the general form, including MHD effects, space gradients and flow conditions.

2.1 Two-term distribution

The two-term approximation decomposes the distribution function as

$$f(\vec{r}, \vec{v}, t) = f_0(\vec{r}, v, t) + \frac{\vec{v}}{v} \cdot \vec{f}_1(\vec{r}, v, t) \tag{2.2}$$

where f_0 and $\vec{f}_1$ represent the isotropic and anisotropic contributions to the velocity distribution. Let us take as angle variables the director cosines

$$\varphi_i = \frac{v_i}{v} = \cos \vartheta_i \tag{2.3}$$

For functions of the type

$$F(\vec{v}) = G(v)\Phi(\vec{\varphi}) \tag{2.4}$$

the integrals in the velocity space can be transformed as

$$\iiint_{-\infty}^{\infty} F(\vec{v})d^3v = \frac{\pi}{2}\int_0^{\infty} v^2 dv G(v) \iiint_{-1}^{1} \Phi(\vec{\varphi})d^3\varphi \tag{2.5}$$

It should be noted that equation (2.5) is not general, but it is valid in the rest of this chapter as demonstrated in appendix 2.9. The integrals over the φ variables are of the kinds

$$\int_{-1}^{1} \varphi_i^{a-1} d\varphi_i = 2\frac{a|2}{a} \tag{2.6}$$

$$\int_{-1}^{1} \varphi_i^{a-1} \varphi_j d\varphi_i = 0 \quad \text{for } i \neq j \tag{2.7}$$

where the symbol $a|2$ is the modulus function (0 for even numbers and 1 for odd numbers). The isotropic (f_0) and the anisotropic ($\vec{f}_1$) parts of the distribution can be obtained as

$$f_0(\vec{r}, v, t) = \frac{1}{8} \iiint_{-1}^{1} f(\vec{r}, \vec{v}, t) d^3\varphi \tag{2.8}$$

$$\vec{f}_1(\vec{r}, v, t) = \frac{3}{8} \iiint_{-1}^{1} \vec{\varphi} f(\vec{r}, \vec{v}, t) d^3\varphi \tag{2.9}$$

The first three moments of the distribution (particle density, mean velocity and mean energy in eV) in the two-term approximation are calculated as

$$N_e = \iiint_v f(\vec{r}, \vec{v}, t) d^3v = 4\pi \int_0^\infty v^2 f_0(\vec{r}, v, t) dv \tag{2.10}$$

$$\vec{v}_d = \frac{1}{N_e} \iiint_v \vec{v} f(\vec{r}, \vec{v}, t) d^3v = \frac{4\pi}{3N_e} \int_0^\infty v^3 \vec{f}_1(\vec{r}, v, t) dv \tag{2.11}$$

$$\varepsilon_{\text{med}} = \frac{m_e}{2eN_e} \iiint_v v^2 f(\vec{r}, \vec{v}, t) d^3v = \frac{4\pi m_e}{eN_e} \int_0^\infty v^4 f_0(\vec{r}, v, t) dv \tag{2.12}$$

It should be noted that the scalar properties, such as density and mean energy, are obtained from f_0 while the vectorial properties, such as the mean velocity, are calculated from $\vec{f}_1$.

2.2 Differential equations

For free electrons in magnetized plasma, the acceleration $\vec{a}$ in equation (1.14) is given by

$$\vec{a} = \vec{\mathcal{A}} + \vec{v} \wedge \vec{\mathcal{R}} \tag{2.13}$$

where

$$\vec{\mathcal{A}} = -\frac{e}{m_e}\vec{E} + \vec{g} \tag{2.14}$$

$$\vec{\mathcal{R}} = -\frac{e}{m_e}\vec{B} \tag{2.15}$$

with $\vec{g}$ the gravity acceleration, usually negligible in weakly ionized gas applications.

To determine the differential equation for f_0 we must integrate all the terms in equation (1.14) over φ's, and, substituting in equation (2.2), we obtain

$$\iiint_{-1}^{1} \frac{\partial f}{\partial t} d^3\varphi = 8\frac{\partial f_0}{\partial t} \tag{2.16}$$

$$\iiint_{-1}^{1} \vec{v} \cdot \vec{\nabla}_r f d^3\varphi = \frac{8v}{3}\vec{\nabla}_r \cdot \vec{f}_1 \tag{2.17}$$

$$\vec{\nabla}_v f = \frac{\vec{v}}{v}\frac{\partial f_0}{\partial v} + \frac{\vec{f}_1}{v} + \frac{\vec{v}}{v}\vec{v} \cdot \frac{\partial}{\partial v}\frac{\vec{f}_1}{v} = \vec{\varphi}\frac{\partial f_0}{\partial v} + \frac{\vec{f}_1}{v} + \vec{\varphi}v\vec{\varphi} \cdot \frac{\partial}{\partial v}\frac{\vec{f}_1}{v} \tag{2.18}$$

$$\iiint_{-1}^{1} \vec{A} \cdot \vec{\nabla}_v f d^3\varphi = \vec{A} \cdot \left(8\frac{\vec{f}_1}{v} + \frac{8v}{3}\frac{\partial}{\partial v}\frac{\vec{f}_1}{v}\right) = \frac{8\vec{A}}{3v^2} \cdot \frac{\partial v^2\vec{f}_1}{\partial v} \tag{2.19}$$

$$\iiint_{-1}^{1} \vec{v} \wedge \vec{\mathcal{R}} \cdot \vec{\nabla}_v f d^3\varphi = 0 \tag{2.20}$$

$$S_0 = \frac{1}{8}\iiint_{-1}^{1} \left(\frac{\delta f}{\delta t}\right)_{\text{coll}} d^3\varphi \tag{2.21}$$

The calculation of S_0 is described in section 2.7 (see also references [43–45]).

The equation for $\vec{f}_1$ comes from the integration of equation (1.14) multiplied by $\vec{\varphi}$

$$\iiint_{-1}^{1} \vec{\varphi}\frac{\partial f}{\partial t} d^3\varphi = \frac{8}{3}\frac{\partial \vec{f}_1}{\partial t} \tag{2.22}$$

$$\iiint_{-1}^{1} \vec{\varphi}\vec{v} \cdot \vec{\nabla}_r f d^3\varphi = \frac{8}{3}v\vec{\nabla}_r f_0 \tag{2.23}$$

$$\iiint_{-1}^{1} \vec{\varphi}\vec{A} \cdot \vec{\nabla}_v f d^3\varphi = \frac{8}{3}\vec{A}\frac{\partial f_0}{\partial v} \tag{2.24}$$

$$\iiint_{-1}^{1} \vec{\varphi}\vec{v} \wedge \vec{\mathcal{R}} \cdot \vec{\nabla}_v f d^3\varphi = \frac{8}{3}\vec{\mathcal{R}} \wedge \vec{f}_1 \tag{2.25}$$

$$\vec{S}_1 = \frac{3}{8}\iiint_{-1}^{1} \vec{\varphi}\left(\frac{\delta f}{\delta t}\right)_{\text{coll}} d^3\varphi \tag{2.26}$$

The term $\vec{S}_1$ is given by (see section 2.7.1)

$$\vec{S}_1 = -\nu_e\vec{f}_1 \tag{2.27}$$

where ν_e is the total elastic collision frequency given in equation (2.134). Other processes like inelastic, superelastic and electron–electron collisions must be taken into account in the collision frequency (see reference [43]), but their contribution is usually negligible.

From basic algebra, the vector product can be written in matrix form as

$$\vec{a} \wedge \vec{b} = \begin{pmatrix} 0 & -a_z & a_y \\ a_z & 0 & -a_x \\ -a_y & a_x & 0 \end{pmatrix} \begin{pmatrix} b_x \\ b_y \\ b_z \end{pmatrix} \tag{2.28}$$

Combining magnetic field (equation (2.25)) and elastic collision (equation (2.27)) contributions

$$\vec{\mathcal{R}} \wedge \vec{f_1} + \nu_e \vec{f_1} = \begin{pmatrix} \nu_e & -\mathcal{R}_z & \mathcal{R}_y \\ \mathcal{R}_z & \nu_e & -\mathcal{R}_x \\ -\mathcal{R}_y & \mathcal{R}_x & \nu_e \end{pmatrix} \begin{pmatrix} f_{1x} \\ f_{1y} \\ f_{1z} \end{pmatrix} = \hat{\Omega} \vec{f_1} \tag{2.29}$$

and space gradient (equation (2.23)) and electric field (equation (2.24)) terms

$$\vec{Q}_0 = -\left(v \vec{\nabla}_r f_0 + \vec{\mathcal{A}} \frac{\partial f_0}{\partial v} \right) \tag{2.30}$$

depending only on the isotropic distribution, we obtain the equation for the anisotropic part. The final result is the following system of equations

$$\begin{cases} \dfrac{\partial f_0}{\partial t} + \dfrac{v}{3} \vec{\nabla}_r \cdot \vec{f_1} + \dfrac{\vec{\mathcal{A}}}{3v^2} \cdot \dfrac{\partial v^2 \vec{f_1}}{\partial v} = S_0 \\ \dfrac{\partial \vec{f_1}}{\partial t} + \hat{\Omega} \vec{f_1} = \vec{Q}_0 \end{cases} \tag{2.31}$$

In the equation for the anisotropic distribution, $\hat{\Omega}\vec{f_1}$ is the dissipation term while $\vec{Q}_0$ is the driving force.

2.3 Quasi-stationary approximation

Except in cases where high-frequency electric fields are applied, $\vec{f_1}$ relaxes more rapidly than f_0, and the quasi-stationary approximation ($\frac{\partial \vec{f_1}}{\partial t} \approx 0$) can be used, calculating the anisotropic distribution by solving the algebraic equation

$$\hat{\Omega} \vec{f_1} = \vec{Q}_0 \tag{2.32}$$

with the analytical solution

$$\vec{f_1} = \hat{\Omega}^{-1} \vec{Q}_0 = \hat{\omega} \vec{Q}_0 \tag{2.33}$$

where

$$\hat{\omega} = \frac{1}{\nu_e(\nu_e^2 + \mathcal{R}^2)} \begin{pmatrix} \nu_e^2 + \mathcal{R}_x^2 & \mathcal{R}_x\mathcal{R}_y + \nu_e\mathcal{R}_z & \mathcal{R}_x\mathcal{R}_z - \nu_e\mathcal{R}_y \\ \mathcal{R}_x\mathcal{R}_y - \nu_e\mathcal{R}_z & \nu_e^2 + \mathcal{R}_y^2 & \mathcal{R}_y\mathcal{R}_z + \nu_e\mathcal{R}_x \\ \mathcal{R}_x\mathcal{R}_z + \nu_e\mathcal{R}_y & \mathcal{R}_y\mathcal{R}_z - \nu_e\mathcal{R}_x & \nu_e^2 + \mathcal{R}_z^2 \end{pmatrix} \tag{2.34}$$

Substituting equation (2.33) in equation (2.31) we have

$$\frac{\partial f_0}{\partial t} + \frac{v}{3}\vec{\nabla}_r \cdot \hat{\omega}\vec{Q}_0 + \frac{\vec{\mathcal{A}}}{3v^2} \cdot \frac{\partial v^2 \hat{\omega}\vec{Q}_0}{\partial v} = S_0 \tag{2.35}$$

For homogeneous systems all the space gradients vanish, giving the equation commonly found for the two-term approximation

$$\frac{\partial f_0}{\partial t} - \frac{1}{3v^2}\frac{\partial}{\partial v}\vec{\mathcal{A}} \cdot \hat{\omega}\vec{\mathcal{A}}v^2\frac{\partial f_0}{\partial v} = S_0 \tag{2.36}$$

2.4 Rapidly varying oscillating field

For rapidly oscillating fields, equation (2.31) cannot consider the anisotropic part as stationary. In the general case, such a system should be solved in time also for $\vec{f}_1$. In the case where the electric field time evolution is given by a harmonic function, it is possible to find a simple solution [46]. This problem has been investigated in chapter 13, in the absence of magnetic field. Here we will give a general description of the problem, allowing one also to consider *electron cyclotron resonance*.

Using the phasor notation for a traveling wave we can write

$$\vec{\mathcal{A}} = \tilde{\vec{\mathcal{A}}} e^{i(\omega_f t - \vec{k}\cdot\vec{x})} \tag{2.37}$$

where $\tilde{\mathcal{A}}$ is in general a complex number. Obviously, we are only interested in the real part of the result, an operation that will be done at the end of the demonstration. A generalization of the problem for a generic periodic field can be done by Fourier series.

In the simplest case of uniform field and homogeneous plasma, we can neglect the space dependence and the gradients from the equation obtaining for equations (2.30) and (2.31)

$$\vec{Q}_0 = -\vec{\mathcal{A}}\frac{\partial f_0}{\partial v} \tag{2.38}$$

$$\begin{cases} \dfrac{\partial f_0}{\partial t} + \dfrac{\vec{\mathcal{A}}}{3v^2} \cdot \dfrac{\partial v^2 \vec{f}_1}{\partial v} = S_0 \\ \dfrac{\partial \vec{f}_1}{\partial t} + \hat{\Omega}\vec{f}_1 = \vec{Q}_0 \end{cases} \tag{2.39}$$

As a first approximation, $\vec{f_1}$ reproduces the same dependence on time as the field, therefore we have

$$\vec{f_1} = \vec{\tilde{f_1}} e^{i\omega_f t} \tag{2.40}$$

Substituting this expression in equation (2.39) we have

$$\begin{cases} \dfrac{\partial f_0}{\partial t} + \dfrac{Re(e^{i\omega_f t}\vec{\tilde{A}})}{3v^2} \cdot \dfrac{\partial v^2 Re(e^{i\omega_f t}\vec{\tilde{f_1}})}{\partial v} = S_0 \\ i\omega_f \vec{\tilde{f_1}} e^{i\omega_f t} + \hat{\Omega}\vec{\tilde{f_1}} e^{i\omega_f t} = -\vec{\tilde{A}} e^{i\omega_f t} \dfrac{\partial f_0}{\partial v} \end{cases} \tag{2.41}$$

$$\begin{cases} (i\omega_f \hat{I} + \hat{\Omega})\vec{\tilde{f_1}} = -\vec{\tilde{A}} \dfrac{\partial f_0}{\partial v} \\ \dfrac{\partial f_0}{\partial t} - \dfrac{1}{3v^2} \dfrac{\partial}{\partial v} Re(e^{i\omega_f t}\vec{\tilde{A}}) Re(\hat{\omega}' e^{i\omega_f t}\vec{\tilde{A}}) v^2 \dfrac{\partial f_0}{\partial v} = S_0 \end{cases} \tag{2.42}$$

where

$$\hat{\omega}' = (i\omega_f \hat{I} + \hat{\Omega})^{-1} \tag{2.43}$$

The matrix $(i\omega_f \hat{I} + \hat{\Omega})$ has the same structure as $\hat{\Omega}$ and $\hat{\omega}'$ is obtained from equation (2.34) by the substitution $\nu_e \to i\omega_f + \nu_e$. Equation (2.42) for f_0 can be written in the general form as

$$\frac{\partial f_0}{\partial t} - \frac{1}{3v^2} \frac{\partial}{\partial v} \mathcal{F}(v, t)\, v^2 \frac{\partial f_0}{\partial v} = S_0 \tag{2.44}$$

where the function $\mathcal{F}(v, t)$ depends on the wave parameters. An important aspect is that, in the presence of high-frequency fields, f_0 shows small oscillations around the mean value. As a consequence, an equation for the mean value of f_0 calculated on a period τ of the applied field, can be obtained by the substitution

$$\mathcal{F}(v, t) \to <\mathcal{F}(v, t)> = \frac{1}{\tau} \int_0^{\tau} \mathcal{F}(v, t) dt \tag{2.45}$$

In the following we will consider some cases.

2.4.1 Case $B = 0$

If the magnetic filed is neglected, $B = 0$,

$$\frac{\partial f_0}{\partial t} - \frac{1}{3v^2} \frac{\partial}{\partial v} Re(e^{i\omega_f t}\vec{\tilde{A}}) Re\left(\frac{e^{i\omega_f t}\vec{\tilde{A}}}{\nu_e + i\omega_f}\right) v^2 \frac{\partial f_0}{\partial v} = S_0$$

$$\frac{\partial f_0}{\partial t} - \frac{1}{3v^2}\frac{\partial}{\partial v} Re(e^{i\omega_f t}\tilde{\vec{\mathcal{A}}}) \frac{Re[e^{i\omega_f t}\tilde{\vec{\mathcal{A}}}(\nu_e - i\omega_f)]}{\nu_e^2 + \omega_f^2} v^2 \frac{\partial f_0}{\partial v} = S_0$$

Writing for the components $\tilde{\vec{\mathcal{A}}} = \vec{a}e^{i\varphi}$, with $\vec{a}$ real vector and φ real number, we have

$$\mathcal{F}(v, t) = \frac{\vec{a}^2}{\nu_e^2 + \omega_f^2}\cos(\omega_f t + \varphi)[\nu_e \cos(\omega_f t + \varphi) + \omega_f \sin(\omega_f t + \varphi)] \tag{2.46}$$

and after some algebra

$$\mathcal{F}(v, t) = \frac{\alpha^2}{\nu_e^2 + \omega_f^2}\left\{\frac{\nu_e}{2} + \frac{\nu_e}{2}\cos[2(\omega_f t + \varphi)] + \frac{\omega_f}{2}\sin[2(\omega_f t + \varphi)]\right\} \tag{2.47}$$

Assuming

$$\tan(\delta\varphi) = \frac{\omega_f}{\nu_e} \tag{2.48}$$

we have[1]

$$\mathcal{F}(v, t) = \frac{\alpha^2 \nu_e}{2(\nu_e^2 + \omega_f^2)\cos(\delta\varphi)}\{\cos(\delta\varphi) + \cos[2(\omega_f t + \varphi) - \delta\varphi]\} \tag{2.49}$$

Equations (2.46)–(2.49) are very general and can be used to determine the evolution of f_0 under the action of a sinusoidal field. However, if the frequency is very high, usually, the evolution of f_0 in the period of oscillations is not followed, as we are interested in the mean effects. The only explicit dependence on time is in $\mathcal{F}$, then it can be substituted by the integral over the period

$$<\mathcal{F}(v, t)>_{0\leqslant t\leqslant 2\pi/\omega_f} = \frac{\nu_e}{\nu_e^2 + \omega_f^2}\frac{\vec{a}^2}{2} \tag{2.50}$$

2.4.2 Generalization to independent frequencies

In the case where the frequencies of the applied field are different in the three directions, the equation must be written in tensor form

$$\begin{cases} \vec{\mathcal{A}} = \hat{\Gamma}\tilde{\vec{\mathcal{A}}} \\ \tilde{\vec{\mathcal{A}}} = \hat{\Phi}\vec{a} \end{cases} \tag{2.51}$$

where $\tilde{\vec{\mathcal{A}}}$ are the field amplitudes while $\vec{a}$ is a real vector. In this case we can assume

$$\vec{f}_1 = \hat{\Gamma}\tilde{\vec{f}}_1 \tag{2.52}$$

[1] Remember that $\cos a \cos b + \sin a \sin b = \cos(a - b)$.

where

$$\hat{\Gamma} = \begin{pmatrix} e^{i\omega_{f1}t} & 0 & 0 \\ 0 & e^{i\omega_{f2}t} & 0 \\ 0 & 0 & e^{i\omega_{f3}t} \end{pmatrix} \tag{2.53}$$

is the evolution operator,

$$\hat{\Phi} = \begin{pmatrix} e^{i\varphi_1} & 0 & 0 \\ 0 & e^{i\varphi_2} & 0 \\ 0 & 0 & e^{i\varphi_3} \end{pmatrix} \tag{2.54}$$

is the phase matrix and

$$\hat{\omega}_f = \begin{pmatrix} \omega_{f1} & 0 & 0 \\ 0 & \omega_{f2} & 0 \\ 0 & 0 & \omega_{f3} \end{pmatrix} \tag{2.55}$$

is the phase matrix, being ω_{fk} and φ_k real numbers. The time derivative of the matrix $\hat{\Gamma}$ can be easily calculated

$$\frac{\partial \hat{\Gamma}}{\partial t} = \begin{pmatrix} i\omega_{f1}e^{i\omega_{f1}t} & 0 & 0 \\ 0 & i\omega_{f2}e^{i\omega_{f2}t} & 0 \\ 0 & 0 & i\omega_{f3}e^{i\omega_{f3}t} \end{pmatrix} = i\hat{\omega}_f\hat{\Gamma} \tag{2.56}$$

and equation (2.42) becomes

$$\begin{cases} (i\hat{\omega}_f + \hat{\Omega})\hat{\Gamma}\tilde{\vec{f}}_1 = -\hat{\Gamma}\hat{\Phi}\vec{\alpha}\dfrac{\partial f_0}{\partial v} \\ \dfrac{\partial f_0}{\partial t} + \dfrac{1}{3v^2}\dfrac{\partial}{\partial v}v^2 Re(\hat{\Gamma}\hat{\Phi}\vec{\alpha})Re(\hat{\Gamma}\tilde{\vec{f}}_1) = S_0 \end{cases}$$

$$\begin{cases} \tilde{\vec{f}}_1 = -\hat{\Gamma}^{-1}\hat{\omega}'\hat{\Gamma}\tilde{\vec{\mathcal{A}}}\dfrac{\partial f_0}{\partial v} \\ \dfrac{\partial f_0}{\partial t} - \dfrac{1}{3v^2}\dfrac{\partial}{\partial v}Re(\hat{\Gamma}\hat{\Phi}\vec{\alpha})Re(\hat{\omega}'\hat{\Gamma}\hat{\Phi}\vec{\alpha})v^2\dfrac{\partial f_0}{\partial v} = S_0 \end{cases} \tag{2.57}$$

where

$$\hat{\omega}' = \left(i\hat{\omega}_f + \hat{\Omega}\right)^{-1} \tag{2.58}$$

Looking at the elements we have

$$\begin{aligned} \{\hat{\Gamma}\hat{\Phi}\}_{ij} &= \delta_{ij}e^{i(\omega_{fi}t+\varphi_i)} \\ \{\hat{\omega}'\hat{\Gamma}\hat{\Phi}\}_{il} &= \hat{\omega}'_{ij}e^{i(\omega_{fj}t+\varphi_j)} \end{aligned} \tag{2.59}$$

where δ_{ij} is the Kronecker delta. Separating the real and imaginary part $\hat{\omega}' = \hat{\omega}^R + i\hat{\omega}^I$, the real part of the two terms is given by

$$\begin{aligned} Re\{\hat{\Gamma}\hat{\Phi}\}_{ij} &= \delta_{ij}cos(\omega_{fj}t + \varphi_j) \\ Re\{\hat{\omega}'\hat{\Gamma}\hat{\Phi}\}_{ij} &= \hat{\omega}^R_{ij}cos(\omega_{fj}t + \varphi_j) - \hat{\omega}^I_{ij}sin(\omega_{fj}t + \varphi_j) \end{aligned} \tag{2.60}$$

resulting in

$$Re(\hat{\Gamma}\hat{\Phi}\vec{\alpha})Re(\hat{\omega}'\hat{\Gamma}\hat{\Phi}\vec{\alpha}) = \sum_{i,j} \hat{\Theta}_{ij}\alpha_i\alpha_j \tag{2.61}$$

where

$$\begin{aligned} \hat{\Theta}_{ij} &= \cos(\omega_{fi}t + \varphi_i)[\hat{\omega}^R_{ij}\cos(\omega_{fj}t + \varphi_j) - \hat{\omega}^I_{ij}\sin(\omega_{fj}t + \varphi_j)] \\ &= \hat{\omega}^R_{ij}\cos(\omega_{fi}t + \varphi_i)\cos(\omega_{fj}t + \varphi_j) - \hat{\omega}^I_{ij}\cos(\omega_{fi}t + \varphi_i)\sin(\omega_{fj}t + \varphi_j) \end{aligned} \tag{2.62}$$

If $\vec{B} = 0$ we have

$$\begin{aligned} \mathcal{F}(v, t) = \sum_{k=1}^{3} \frac{\alpha_k^2}{\nu_e^2 + \omega_{fk}^2} \\ \cos(\omega_{fk}t + \varphi_k)[\nu_e \cos(\omega_{fk}t + \varphi_k) + \omega_{fk} \sin(\omega_{fk}t + \varphi_k)]\} \end{aligned} \tag{2.63}$$

$$\mathcal{F}(v, t) = \sum_{k=1}^{3} \frac{\alpha_k^2}{\nu_e^2 + \omega_{fk}^2} \left\{ \frac{\nu_e}{2}(1 + \cos[2(\omega_{fk}t + \varphi_k)] + \frac{\omega_{fk}}{2} \sin[2(\omega_f t + \varphi_k)] \right\} \tag{2.64}$$

and the mean

$$<\mathcal{F}(v, t)>_{0 \leqslant t \leqslant t_{\max}} = \frac{\nu_e}{2} \sum_{k=1}^{3} \frac{\alpha_k^2}{\nu_e^2 + \omega_{fk}^2} \tag{2.65}$$

2.4.3 Matrices for single frequency

In the general case, finding $\hat{\omega}'$ is cumbersome and direct solution of equation (2.57) must be calculated. For our actual needs it is sufficient that all electric field components have the same frequency ω_f but independent phases, to account for different polarization of the radiation, obtaining $\hat{\Gamma} = \omega_f \hat{I}$.

Remembering the Werner formulas $\cos(a)\cos(b) = 1/2[\cos(a + b) + \cos(a - b)]$ and $\sin(a)\cos(b) = 1/2[\sin(a + b) + \sin(a - b)]$ we have

$$\cos(\omega_f t + \varphi_i)\cos(\omega_f t + \varphi_j) = \frac{\cos(\varphi_i - \varphi_j) + \cos(2\omega_f t + \varphi_i + \varphi_j)}{2}$$

$$\cos(\omega_f t + \varphi_i)\sin(\omega_f t + \varphi_j) = \frac{\sin(\varphi_i - \varphi_j) + \sin(2\omega_f t + \varphi_i + \varphi_j)}{2}$$

obtaining for equation (2.62)

$$\hat{\Theta}_{ij} = \hat{\omega}_{ij}^{R} \frac{\cos(\varphi_i - \varphi_j) + \cos(2\omega_f t + \varphi_i + \varphi_j)}{2} - \hat{\omega}_{ij}^{I} \frac{\sin(\varphi_i - \varphi_j) + \sin(2\omega_f t + \varphi_i + \varphi_j)}{2} = \hat{\omega}_{ij}^{R} K_{ij}^{c}(t) - \hat{\omega}_{ij}^{I} K_{ij}^{s}(t) \quad (2.66)$$

calculating the mean over a period, we have that the sin and cos functions give null value, and therefore

$$<\hat{\Theta}_{ij}> \hat{\omega}_{ij}^{R} \frac{\cos(\varphi_i - \varphi_j)}{2} - \hat{\omega}_{ij}^{I} \frac{\sin(\varphi_i - \varphi_j)}{2} = \hat{\omega}_{ij}^{R} < K_{ij}^{c}(t) > -\hat{\omega}_{ij}^{I} < K_{ij}^{s}(t)> \quad (2.67)$$

In this case the matrix $\hat{\omega}'$ becomes

$$\hat{\omega}' = \frac{\tilde{N}}{\tilde{D}} \begin{pmatrix} (\nu_e + i\omega_f)^2 + \mathcal{R}_x^2 & \mathcal{R}_x\mathcal{R}_y + (\nu_e + i\omega_f)\mathcal{R}_z & \mathcal{R}_x\mathcal{R}_z - (\nu_e + i\omega_f)\mathcal{R}_y \\ \mathcal{R}_x\mathcal{R}_y - (\nu_e + i\omega_f)\mathcal{R}_z & (\nu_e + i\omega_f)^2 + \mathcal{R}_y^2 & \mathcal{R}_y\mathcal{R}_z + (\nu_e + i\omega_f)\mathcal{R}_x \\ \mathcal{R}_x\mathcal{R}_z + (\nu_e + i\omega_f)\mathcal{R}_y & \mathcal{R}_y\mathcal{R}_z - (\nu_e + i\omega_f)\mathcal{R}_x & (\nu_e + i\omega_f)^2 + \mathcal{R}_z^2 \end{pmatrix}$$

where

$$\begin{aligned} \tilde{D} &= (\nu_e^2 + \omega_f^2)[(\nu_e^2 - \omega_f^2 + \mathcal{R}^2)^2 + 4\nu_e^2\omega_f^2] \\ &= (\nu_e^2 + \omega_f^2)[\nu_e^2 + (R - \omega_f)^2][\nu_e^2 + (R + \omega_f)^2] \\ \tilde{N} &= (\nu_e - i\omega)(\nu_e^2 + \mathcal{R}^2 - \omega_f^2 - 2i\nu_e\omega_f) \\ &= \nu_e(\nu_e^2 + \mathcal{R}^2 - 3\omega_f^2) - i\omega_f(3\nu_e^2 + \mathcal{R}^2 - \omega_f^2) = \tilde{N}_R - i\tilde{N}_I \end{aligned} \quad (2.68)$$

with

$$\begin{aligned} \tilde{N}_R &= \nu_e(\nu_e^2 + \mathcal{R}^2 - 3\omega_f^2) \\ \tilde{N}_I &= \omega_f(3\nu_e^2 + \mathcal{R}^2 - \omega_f^2) \end{aligned} \quad (2.69)$$

The matrix can be decomposed as the sum of three matrices

$$\hat{\omega}' = \frac{\tilde{N}}{\tilde{D}}[\hat{M}_R + i\omega_f\hat{M}_I] = \frac{\tilde{N}}{\tilde{D}}[\hat{M}_0 - \omega_f^2\hat{I} + i\omega_f\hat{M}_I] \quad (2.70)$$

where $\hat{M}_0$ is the same matrix as in equation (2.34)

$$\hat{M}_0 = \begin{pmatrix} \nu_e^2 + \mathcal{R}_x^2 & \mathcal{R}_x\mathcal{R}_y + \nu_e\mathcal{R}_z & \mathcal{R}_x\mathcal{R}_z - \nu_e\mathcal{R}_y \\ \mathcal{R}_x\mathcal{R}_y - \nu_e\mathcal{R}_z & \nu_e^2 + \mathcal{R}_y^2 & \mathcal{R}_y\mathcal{R}_z + \nu_e\mathcal{R}_x \\ \mathcal{R}_x\mathcal{R}_z + \nu_e\mathcal{R}_y & \mathcal{R}_y\mathcal{R}_z - \nu_e\mathcal{R}_x & \nu_e^2 + \mathcal{R}_z^2 \end{pmatrix} \quad (2.71)$$

and

$$\hat{M}_i = \begin{pmatrix} 2\nu_e & \mathcal{R}_z & -\mathcal{R}_y \\ -\mathcal{R}_z & 2\nu_e & \mathcal{R}_x \\ \mathcal{R}_y & -\mathcal{R}_x & 2\nu_e \end{pmatrix} \tag{2.72}$$

$$\begin{aligned} \hat{\omega}^R &= \frac{1}{\tilde{D}}(\tilde{N}_R\hat{M}_R + \tilde{N}_I\omega_f\hat{M}_I) \\ \hat{\omega}^I &= \frac{1}{\tilde{D}}(\tilde{N}_R\omega_f\hat{M}_I - \tilde{N}_I\hat{M}_R) \end{aligned} \tag{2.73}$$

$$\hat{\Theta}_{ij} = \frac{1}{\tilde{D}}\left[\hat{M}_{R,ij}\left(K_{ij}^c(t)\tilde{N}_R + K_{ij}^s(t)\tilde{N}_I\right) + \omega_f\hat{M}_I\left(K_{ij}^c(t)\tilde{N}_I - K_{ij}^s(t)\tilde{N}_R\right)\right] \tag{2.74}$$

2.4.4 Some considerations

Comparing equation (2.36) obtained for static field with the above results, the following conclusions are reached for rapidly oscillating field:

- in the time dependent approach, the equations can be obtained from the static field substituting $\vec{A}\hat{\omega}\vec{A} \rightarrow \vec{\alpha}\hat{\Theta}\vec{\alpha}$
- calculating the mean over a period, the equations can be obtained from the static field substituting $\vec{A}\hat{\omega}\vec{A} \rightarrow \vec{\alpha} < \hat{\Theta} > \vec{\alpha}$

2.4.5 Power absorbed by electrons

The general definition of the power density absorbed by the electrons is

$$W = \vec{E} \cdot \vec{J} \tag{2.75}$$

The current density can be calculated as $q_e N_e \vec{v}_d$, where the drift velocity is given by equation (2.11). In the case of equation (2.57) where ($\vec{f}_1 = \hat{\omega}'\hat{\Gamma}\hat{\Phi}\vec{\alpha}$) is valid, the drift velocity is given by

$$\vec{v}_d = \frac{m_e}{q_e}\hat{\mu}_e\vec{\alpha} \tag{2.76}$$

where $\hat{\mu}_e$ is the tensorial mobility

$$\hat{\mu}_e = \frac{4\pi}{3N_e}\int_0^\infty v^3 Re(\hat{\omega}'\hat{\Gamma}\hat{\Phi})\frac{\partial f_0}{\partial v}dv \tag{2.77}$$

and, where $\vec{E} = \frac{m_e}{q_e}Re(\hat{\Gamma}\hat{\Phi})\vec{\alpha}$,

$$W = \frac{m_e^2}{q_e}[Re(\hat{\Gamma}\hat{\Phi})\vec{\alpha}] \cdot (\hat{\mu}_e\vec{\alpha}) = \frac{m_e^2}{q_e}\sum_{i,j}\alpha_i\alpha_j\frac{4\pi}{3}\int_0^\infty v^3\hat{\Theta}_{ij}\frac{\partial f_0}{\partial v}dv \tag{2.78}$$

2.4.6 Mean magnetic dipole moment

The force applied to electrons in the presence of the gradient of the magnetic field is given by

$$\vec{F}_{\mu} = -\mu(\vec{v})\vec{\nabla}B \tag{2.79}$$

where [44, 47][2]

$$\mu(\vec{v}) = \frac{m_e v_{\perp}^2}{2B} \tag{2.80}$$

is the magnetic dipole moment and

$$\vec{v}_{\perp} = \vec{v} - \vec{v}_{\parallel} = \vec{v} - (\vec{v} \cdot \vec{b})\vec{b} \tag{2.81}$$

is the component of the velocity perpendicular to the magnetic field, being $\vec{b} = \frac{\vec{B}}{B}$ the versor in the direction of the magnetic field. On the fluid dynamic approach, the force per unit volume is given by

$$\vec{F} = -N_e\bar{\mu}\vec{\nabla}B \tag{2.82}$$

where $\bar{\mu}$ is the mean magnetic moment calculated as

$$\bar{\mu} = \iiint_{-\infty}^{\infty} \mu(\vec{v})f(\vec{r}, \vec{v}, t)d^3v = \frac{m_e}{2BN_e}\iiint_{-\infty}^{\infty} v_{\perp}^2 f(\vec{r}, \vec{v}, t)d^3v = \frac{m_e}{2B} <v_{\perp}^2> \tag{2.83}$$

To calculate the integral, we need the module of the perpendicular velocity (equation (2.81)) given by

$$v_{\perp}^2 = v^2 + (\vec{v} \cdot \vec{b})^2 b^2 - 2(\vec{v} \cdot \vec{b})^2 = v^2 - (\vec{v} \cdot \vec{b})^2 \tag{2.84}$$

with $\vec{b}$ a unitary vector, and, in the two-term approximation (see equations (2.2) and 2.3), it is

$$(\vec{v} \cdot \vec{b})^2 = v^2(\varphi_x^2 b_x^2 + \varphi_y^2 b_y^2 + \varphi_z^2 b_z^2 + 2\varphi_x b_x \varphi_y b_y + 2\varphi_x b_x \varphi_z b_z + 2\varphi_y b_y \varphi_z b_z) \tag{2.85}$$

Considering the general property of the integral in equation (2.5)

$$\begin{aligned} <v_{\perp}^2> =& \frac{2\varepsilon_{\text{med}}}{m_e} - \frac{\pi}{2N_e}\int_0^{\infty} v^4 f_0(v)dv \iiint_{-1}^{1} \Phi(\vec{\varphi})d^3\varphi \\ &- \frac{\pi}{2N_e}\int_0^{\infty} v^4 \vec{f}_1(v)dv \iiint_{-1}^{1} \vec{\varphi}\Phi(\vec{\varphi})d^3\varphi \end{aligned} \tag{2.86}$$

where here $\Phi(\vec{\varphi})$ is the parenthesis in equation (2.85). Considering the properties of the integrals given in equation (2.6), the second integral is null because it contains

[2] The force and the magnetic dipole moment are expressed as a function of $H = \frac{B}{\mu_0}$ (μ_0 is the vacuum magnetic permeability). The expression of the force, containing μ_0 in both numerator and denominator, the result is the same if we consider B or H.

only terms φ_i^3 or $\varphi_i^2\varphi_{j\neq i}$. In the second integrals there survive only the square terms, being the integral of $\varphi_i\varphi_{j\neq i}$ null, resulting in

$$\iiint_{-1}^{1} \Phi(\vec{\varphi})d^3\varphi = \sum_{i=x,y,z} b_i^2 \iiint_{-1}^{1} \varphi_i^2 d^3\varphi = \frac{8}{3}\sum_{i=x,y,z} b_i^2 = \frac{8}{3} \tag{2.87}$$

and as a consequence the second integral becomes (see equation (2.12))

$$<v_\perp^2> = \frac{2\varepsilon_{\text{med}}}{m_e} - \frac{4\pi}{3N_e}\int_0^\infty v^4 f_0(v)dv = \frac{2\varepsilon_{\text{med}}}{m_e} - \frac{2\varepsilon_{\text{med}}}{3m_e} = \frac{4\varepsilon_{\text{med}}}{3m_e} \tag{2.88}$$

giving for the magnetic dipole moment

$$\bar{\mu} = \frac{2\varepsilon_{\text{med}}}{3B} \tag{2.89}$$

2.4.7 Perpendicular energy equation

The results in equation (2.89) are the consequence of the equipartition of energy. It should be considered as a first-order approximation, a consequence of the two-term expansion. To avoid higher order approaches, to consider that in ECR the energy is absorbed mainly in the perpendicular direction, to evaluate $\bar{\varepsilon}_\perp = \frac{1}{2}m_e v_\perp^2$, we should write a continuity equation. This equation should balance the energy pumped by ECR (w_{ecr}) with the energy transferred by elastic collisions $\left(w_{\text{el}} = -(\varepsilon_\perp - \frac{2}{3}\bar{\varepsilon})\nu_{\text{el}}\right)$, energy lost by inelastic collisions $\left(-\frac{\varepsilon_\perp}{\bar{\varepsilon}}w_{\text{in}}\right)$ and that gained by superelastic collisions $\left(\frac{2}{3}w_{\text{sup}}\right)$, giving the equation

$$\frac{d\varepsilon_\perp}{dt} = w_{\text{ecr}} - \left(\varepsilon_\perp - \frac{2}{3}\bar{\varepsilon}\right)\nu_{\text{el}} - \frac{\varepsilon_\perp}{\bar{\varepsilon}}w_{\text{in}} + \frac{2}{3}w_{\text{sup}} \tag{2.90}$$

where

$$w_{\text{in/sup}} = \sum_{s,t} \varepsilon_{s,t}^{\star}\nu_{s,t}^{\text{in/sup}} \tag{2.91}$$

where s, t are the species and transition indexes. The value $\varepsilon_\perp = \frac{2}{3}\bar{\varepsilon}$ should be considered as the initial and equilibrium conditions.

2.5 Electrons in flow

The differential equations in the previous section have been obtained for a gas at rest. To determine the corrections due to flowing gas, we can suppose that equations (2.35) and 2.36 are valid in a reference frame that moves together with the flow at speed $\vec{u}$ in the laboratory. The particle velocity in this new frame ($\vec{v}'$) is related to the velocity in the laboratory system ($\vec{v}$) by $\vec{v}' = \vec{v} - \vec{u}$. The distribution function in the two reference systems must be equal,

$$f(\vec{v}) = f'(\vec{v}') = f_0(\vec{v}' + \vec{u}) + \frac{\vec{v}' + \vec{u}}{|\vec{v}' + \vec{u}|} \cdot \vec{f}_1(\vec{v}' + \vec{u}) \tag{2.92}$$

and assuming that the flow speed is much lower than the thermal velocity ($\vec{v}' \approx \vec{v}$), we can expand the terms in equation (2.92) in a Taylor series truncated at the first order,

$$\begin{aligned} f_0(\vec{v}' + \vec{u}) &\approx f_0(v') + \vec{u} \cdot \vec{\varphi}' \frac{\partial f_0(v')}{\partial v} \\ \frac{\vec{v}' + \vec{u}}{|\vec{v}' + \vec{u}|} &\approx \vec{\varphi}' \\ \vec{f}_1(\vec{v}' + \vec{u}) &\approx \vec{f}_1(v') + \vec{u} \cdot \vec{\varphi}' \frac{\partial \vec{f}_1(v')}{\partial v} \end{aligned} \tag{2.93}$$

that applied to the definitions of f_0 and $\vec{f}_1$ gives

$$\begin{aligned} f_0'(v') &= \iiint_{-1}^{1} f'(\vec{v}') d^3\varphi' = f_0(v) + \frac{1}{3}\vec{u} \cdot \frac{\partial \vec{f}_1(v)}{\partial v} \\ \vec{f}_1'(v') &= \iiint_{-1}^{1} \vec{\varphi}' f'(\vec{v}') d^3\varphi' = \vec{f}_1(v) + \vec{u} \frac{\partial f_0(v)}{\partial v} \end{aligned} \tag{2.94}$$

These equations can be used to determine the corrections to the Boltzmann equation for a flowing gas. It must be pointed out that equation (2.26) for $\vec{S}_1$, is valid in the reference frame where the gas is at rest, therefore

$$\vec{S}_1 = -\nu_e \vec{f}_1' = -\nu_e \vec{u} \frac{\partial f_0(v)}{\partial v} - \nu_e \vec{f}_1 \tag{2.95}$$

from which it follows that equation (2.35) is still valid if $\vec{Q}_0$ is given by

$$\vec{Q}_0 = -\left[v \vec{\nabla}_r f_0 + (\vec{\mathcal{A}} + \nu_e \vec{u}) \frac{\partial f_0}{\partial v} \right] \tag{2.96}$$

and equation (2.36) becomes

$$\frac{\partial f_0}{\partial t} - \frac{1}{3v^2} \frac{\partial}{\partial v} \vec{\mathcal{A}} \cdot \hat{\omega} (\vec{\mathcal{A}} + \nu_e \vec{u}) v^2 \frac{\partial f_0}{\partial v} = S_0 \tag{2.97}$$

It should be noted that the term S_0 must be calculated considering $f'(\vec{v}')$. Neglecting electron–electron collisions, the collision integral is a linear function of the isotropic distribution

$$S_0 = \hat{S}_{LC}[f_0'(v')] = \hat{S}_{LC}\left[f_0(v) + \frac{1}{3}\vec{u} \cdot \frac{\partial \vec{f}_1(v)}{\partial v} \right] = \hat{S}_{LC}\left[f_0(v) + \frac{1}{3}\vec{u} \cdot \frac{\partial \hat{\omega} \vec{Q}_0}{\partial v} \right] \tag{2.98}$$

A similar approach can be used to calculate the contribution of electron–electron collisions to $\vec{S}_1$ [43], considering the transformation in a reference frame that moves with the electron current ($\vec{f}_1' = \vec{0}$). In any case, the contribution of the flow to the collision integrals S_0 and S_1 is a second-order correction.

To couple electron kinetics with a fluid dynamic model, it is necessary to determine the interaction between the two fluids. Essentially, the two systems exchange, between each other, both energy and momentum. The energy balance is already considered in S_0, providing one includes the corresponding terms in the fluid dynamic equations. On the other hand, the momentum exchange is usually neglected. To estimate the momentum exchange between electrons and heavy particles we can start from the relaxation equation in the absence of the applied field and space gradients

$$\frac{\partial \vec{f}_1}{\partial t} = -\nu_e \vec{u} \frac{\partial f_0}{\partial v} - \nu_e \vec{f}_1 \tag{2.99}$$

obtaining the source of free electron momentum due to drag from heavy particles

$$\begin{aligned} \left(\frac{\partial \rho_e \vec{u}_e}{\partial t}\right)_{\text{drag}} &= N_e m_e \frac{\partial \vec{v}_v}{\partial t} = \frac{4\pi m_e}{3} \int_0^\infty v^3 \frac{\partial \vec{f}_1}{\partial t} dv = \\ &= -\frac{4\pi m_e}{3} \int_0^\infty v^3 \nu_e \left(\vec{u} \frac{\partial f_0}{\partial v} + \vec{f}_1\right) dv \end{aligned} \tag{2.100}$$

2.6 Electron energy distribution

The two-term Boltzmann equation is usually written for the electron energy distribution defined as

$$n_0(\varepsilon) = 4\pi \sqrt{\frac{2e^3 \varepsilon}{m_e^3}} f_0(v) \tag{2.101}$$

$$\vec{n}_1(\varepsilon) = \frac{8\pi e^2 \varepsilon}{3m_e^2} \vec{f}_1(v) \tag{2.102}$$

where

$$\varepsilon = \frac{m_e}{2e} v^2$$

is the electron energy in eV. This change of variable transforms the velocity derivative and differential as

$$\frac{\partial}{\partial v} = \sqrt{\frac{2m_e \varepsilon}{e}} \frac{\partial}{\partial \varepsilon} \tag{2.103}$$

$$dv = \sqrt{\frac{e}{2m_e\varepsilon}}\,d\varepsilon \tag{2.104}$$

obtaining for equations (2.10)–(2.12)

$$N_e = \int_0^\infty n_0(\varepsilon)d\varepsilon \tag{2.105}$$

$$\vec{v}_d = \frac{1}{N_e}\int_0^\infty \vec{n}_1(\varepsilon)d\varepsilon \tag{2.106}$$

$$\varepsilon_{\text{med}} = \frac{1}{N_e}\int_0^\infty n_0(\varepsilon)\varepsilon d\varepsilon \tag{2.107}$$

The resulting differential equations (see equation (2.31)) become

$$\begin{cases} \dfrac{\partial n_0}{\partial t} + \vec{\nabla}_r \cdot \vec{n}_1 + \dfrac{m_e\vec{A}}{e}\cdot\dfrac{\partial \vec{n}_1}{\partial \varepsilon} = S_0' \\ \dfrac{\partial \vec{n}_1}{\partial t} + \dfrac{2e\varepsilon}{3m_e}\vec{\nabla}_r n_0 + \dfrac{2\varepsilon}{3}(\vec{A} + \nu_e\vec{u})\left(\dfrac{\partial n_0}{\partial \varepsilon} - \dfrac{n_0}{2\varepsilon}\right) + \hat{\Omega}\vec{n}_1 = \vec{0} \end{cases} \tag{2.108}$$

where

$$S_0' = 4\pi\sqrt{\frac{2e^3\varepsilon}{m_e^3}}\,S_0 \tag{2.109}$$

and in quasi-stationary approximation we have

$$\vec{n}_1 = -\frac{2e\varepsilon}{3m_e}\hat{\omega}\left[\vec{\nabla}_r n_0 + \frac{m_e}{e}(\vec{A} + \nu_e\vec{u})\left(\frac{\partial n_0}{\partial \varepsilon} - \frac{n_0}{2\varepsilon}\right)\right] \tag{2.110}$$

2.6.1 Current anisotropy

To clarify the role of the matrix $\hat{\omega}$ let us consider the configuration with the magnetic field oriented in the z direction, $\vec{B} = (0, 0, B)$. For this condition, and defining

$$\begin{aligned} \beta &= \frac{eB}{\nu_e m_e} \\ \alpha &= \frac{\beta}{\nu_e(1 + \beta^2)} \\ \gamma &= \frac{1}{\nu_e(1 + \beta^2)} \end{aligned} \tag{2.111}$$

we have

$$\hat{\omega} = \frac{1}{\nu_e(1+\beta^2)}\begin{pmatrix} 1 & \beta & 0 \\ -\beta & 1 & 0 \\ 0 & 0 & 1+\beta^2 \end{pmatrix} = \begin{pmatrix} \gamma & \alpha & 0 \\ -\alpha & \gamma & 0 \\ 0 & 0 & 1 \end{pmatrix} \tag{2.112}$$

Referring to the expression of $\vec{n}_1$ in equation (2.110) and considering that the current density is

$$\vec{j} = N_e e \vec{v}_d = N_e e \int_0^\infty \vec{n}_1 d\varepsilon \tag{2.113}$$

for $\vec{E} = (0,E,0)$, the Hall current is oriented in the x direction and its value is given by[3]

$$j_{\text{Hall}} = -\frac{2N_e e^2}{3m_e}\int_0^\infty \varepsilon\sqrt{\varepsilon}\frac{\partial}{\partial\varepsilon}\frac{n}{\sqrt{\varepsilon}}\alpha E d\varepsilon \tag{2.114}$$

2.6.2 Transport properties

Transport properties of the electron gas can be calculated directly from the EEDF starting from equation (2.110), remembering that the mean electron velocity $\vec{v}_d$ is calculated from the anisotropic distribution $\vec{n}_1$ (see equation (2.106)). We can distinguish three contributions, due to diffusion, drift and drag[3]

$$\vec{v}_{\text{diff}} = \frac{1}{N_e}\hat{D}_e\vec{\nabla}_r N_e = -\frac{2e}{3m_e N_e}\int_0^\infty \varepsilon\hat{\omega}n_0 d\varepsilon\vec{\nabla}_r N_e \tag{2.115}$$

$$\vec{v}_{\text{drift}} = \hat{\mu}_e\vec{E} = -\frac{2e}{3m_e N_e}\vec{E}\int_0^\infty \varepsilon\hat{\omega}\frac{\partial}{\partial\varepsilon}\frac{n}{\sqrt{\varepsilon}}d\varepsilon \tag{2.116}$$

$$\vec{v}_{\text{drag}} = \hat{C}_{\text{drag}}\vec{u} = -\frac{2}{3N_e}\vec{u}\int_0^\infty \nu_e\varepsilon\hat{\omega}\frac{\partial}{\partial\varepsilon}\frac{n}{\sqrt{\varepsilon}}d\varepsilon \tag{2.117}$$

giving the transport coefficients[4]

$$\hat{D}_e = -\frac{2e}{3m_e N_e}\int_0^\infty \varepsilon\hat{\omega}n_0 d\varepsilon \tag{2.118}$$

$$\hat{\mu}_e = -\frac{2e}{3m_e N_e}\int_0^\infty \varepsilon\hat{\omega}\frac{\partial}{\partial\varepsilon}\frac{n}{\sqrt{\varepsilon}}d\varepsilon \tag{2.119}$$

$$\hat{C}_{\text{drag}} = -\frac{2}{3N_e}\int_0^\infty \nu_e\varepsilon\hat{\omega}\frac{\partial}{\partial\varepsilon}\frac{n}{\sqrt{\varepsilon}}d\varepsilon \tag{2.120}$$

---

[3] It is straightforward that $\frac{\partial}{\partial\varepsilon}\frac{n}{\sqrt{\varepsilon}} = \left(\frac{\partial n_0}{\partial\varepsilon} - \frac{n_0}{2\varepsilon}\right)$.

[4] In the absence of magnetic field, the transport coefficients become scalar quantities, since for a generic tensorial property for $\vec{B} = 0$ it is $\hat{G} = G_0\hat{I}$ where $\hat{I}$ is the identity matrix.

In the case of equilibrium distribution, the diffusion coefficient and electron mobility are related through the Einstein relation to the electron temperature

$$\frac{kT_e}{e} = \frac{D_e}{\mu_e} \tag{2.121}$$

In the case of non-Maxwellian distribution, the electron temperature is not defined, and equation (2.121) is used to define the *characteristic energy*, $\varepsilon_{\mathrm{car}} = D_e/\mu_e$, often assumed as a measure of the electron temperature. For a Maxwellian distribution, in general, we have $\frac{3}{2}\varepsilon_{\mathrm{car}} = \varepsilon_{\mathrm{med}}$, a relation not valid in non-equilibrium.

2.6.3 Nozzle flow

To obtain the two-term approximation of the Boltzmann equation for quasi 1-D nozzle flows, analogously to Euler equations, the space gradients must be substituted with

$$\vec{\nabla}_r f \equiv \begin{pmatrix} \dfrac{1}{A}\dfrac{\partial Af}{\partial x} \\ 0 \\ 0 \end{pmatrix} \tag{2.122}$$

which makes the Boltzmann equation very complex, containing mixed derivatives in space and energy. Moreover, it has been observed that, during the expansion, the electron distribution cools down mainly due to collisions, and therefore the contribution of the expansion is usually negligible. However, a correction term can be added estimating the energy variation related to the increase of the mean velocity due to the expansion. It can be easily verified that the mean electron energy variation corresponds to that at of electron kinematical energy. In fact

$$\delta\varepsilon_{\mathrm{med}} = \delta\int_0^\infty \varepsilon n_0(\varepsilon)d\varepsilon = \int_0^\infty \varepsilon\delta n_0(\varepsilon)d\varepsilon = \frac{m_e}{e}u\delta u\int_0^\infty \varepsilon\frac{\partial n_0}{\partial\varepsilon}d\varepsilon \tag{2.123}$$

and integrating by parts

$$\int_0^\infty \varepsilon\frac{\partial n_0}{\partial\varepsilon}d\varepsilon = \varepsilon n_0(\varepsilon)|_0^\infty - \int_0^\infty n_0(\varepsilon)d\varepsilon = 0 - 0 - 1 = -1 \tag{2.124}$$

we have

$$\delta\varepsilon_{\mathrm{med}} = -\frac{m_e}{e}u\delta u \tag{2.125}$$

which is the variation of electron kinematic energy per particle. As a consequence, the correction term introduced in reference [29] to account for the mean energy variation by expansion or compression is $\frac{\partial J_{ne}}{\partial\varepsilon}$ where

$$J_{ne}(\varepsilon) = \frac{m_e}{e}n_0(\varepsilon)u\frac{\partial u}{\partial t} \tag{2.126}$$

2.7 The collision integral

Let us now derive the expression of the collision integral for the two-term Boltzmann equation starting from the general definition in equation (1.22). A heavy species s, in collisions with electrons, does not change its velocity, due to the large mass difference. Therefore, the integral with respect to the velocity of the s particle can be calculated separately, resulting in its density N_s. Only the dependence on internal distribution $f_{s,\ell}$ will be considered[5], giving

$$\left(\frac{\delta f}{\delta t}\right)_{\text{coll}} = \sum_{s,\ell,\ell'} N_s v \int [f(\vec{v}', \vec{r}, t) f_{s,\ell'} - f(\vec{v}, \vec{r}, t) f_{s,\ell}] \frac{d\sigma_p}{d\Omega} d\Omega \tag{2.127}$$

Three types of processes can be distinguished:

1. elastic collisions, with $\ell = \ell'$, where the internal state of s does not change;
2. inelastic collisions, with $\ell' > \ell$, where the final state has higher energy than the initial one;
3. superelastic, or second kind, collisions, with $\ell > \ell'$, where the electron gains the energy lost by s.

Considering the two-term approximation we can separate the collision integral in two contributions

$$\left(\frac{\delta f}{\delta t}\right)_{\text{coll}} = \delta f_0(v) + \delta f_1(v) \tag{2.128}$$

each one calculated using equation (2.127) with the corresponding term. Trivially, for the properties of integrals, $S_0 \propto \delta f_0(v)$ and $S_1 \propto \delta f_1(v)$, in fact, from equations (2.21) and 2.26 we have

$$S_0 = \sum_{s,\ell,\ell'} N_s v \int [f_0(v', \vec{r}, t) f_{s,\ell'} - f_0(v, \vec{r}, t) f_{s,\ell}] \frac{d\sigma_p}{d\Omega} d\Omega \tag{2.129}$$

$$\vec{S}_1 = \frac{3}{8} \iiint_{-1}^{1} d^3\varphi \vec{\varphi} \sum_{s,\ell,\ell'} N_s v \int [\vec{\varphi}' \cdot \vec{f}_1(v', \vec{r}, t) f_{s,\ell'} - \vec{\varphi} \cdot \vec{f}_1(v, \vec{r}, t) f_{s,\ell}] \frac{d\sigma_p}{d\Omega} d\Omega \tag{2.130}$$

$\vec{\varphi}'$ differs by a quantity that depends on the deflection angle and plane. Under the assumption that all the deflection planes are equivalent[6] we can limit the calculation in the z direction considering that $\varphi_z = \cos\vartheta$ and $\varphi_z' = \cos(\vartheta + \vartheta_d) = \cos\vartheta\cos\vartheta_d - \sin\vartheta\sin\vartheta_d$. In equation (2.130) the integrals with mixed terms in different directions as well as those containing the $\sin\vartheta$ are null, therefore only the terms with φ_z^2 give effective contribution, and therefore

[5] Using this notation $\sum_\ell f_{s,\ell} = 1$.

[6] This assumption is not valid only for very large molecules with complex geometry.

$$\vec{S}_1 = \sum_{s,\ell,\ell'} N_s v \int [\cos\vartheta_d \vec{f}_1(v', \vec{r}, t) f_{s,\ell'} - \vec{f}_1(v, \vec{r}, t) f_{s,\ell}] \frac{d\sigma_p}{d\Omega} d\Omega \tag{2.131}$$

Contrarily to S_0, only elastic collisions contribute significantly to $\vec{S}_1$ (see equation (2.27)), since inelastic and superelastic processes have smaller cross-sections.

2.7.1 Elastic collisions with heavy species

From equation (2.131) it is very simple to obtain equation (2.27). Considering that in elastic collisions $v' \approx v$ (only direction changes) and that the internal distribution is a common factor[7] we obtain

$$\begin{aligned}\vec{S}_1 &= -\sum_s N_s v \vec{f}_1(v, \vec{r}, t) \int_\Omega (1 - \cos\vartheta_d) \frac{d\sigma_p}{d\Omega} d\Omega \\ &= -\sum_s N_s v \vec{f}_1(v, \vec{r}, t) \sigma_s^{\mathrm{mt}}(v) = -\nu_e \vec{f}_1(v, \vec{r}, t)\end{aligned} \tag{2.132}$$

where

$$\sigma_s^{\mathrm{mt}}(v) = \int_\Omega (1 - \cos\vartheta_d) \frac{d\sigma_p}{d\Omega} d\Omega \tag{2.133}$$

is the *momentum transfer* cross-section and

$$\nu_e = \sum_s N_s v \sigma_s^{\mathrm{mt}}(v) \tag{2.134}$$

is the elastic collision frequency. When evaluating $\vec{S}_1$, the small variation in the velocity module is negligible with respect to the change of direction. On the other hand, the energy lost by an electron in an elastic collision must be considered, because assuming $v' = v$ gives a null contribution to the S_0. To solve this problem we have to obtain a Fokker–Planck equation by expanding in series the distribution with respect to a weakly varying variable. The distribution of the heavy species cannot be neglected here, therefore we will insert in equation (2.129) f_s again, considered isotropic and not affected by the collision, giving[8]

$$S_{0,s}^{\mathrm{el}} = N_s \int_0^{4\pi} v \frac{d\sigma_p}{d\Omega} d\Omega \int_0^\infty f_s(v_s, \ell)[f_0(v') - f_0(v)] d^3v_s \tag{2.135}$$

and expanding in series around v up to second order and performing the integrations

[7] This assumption is not really valid, because the elastic cross-sections depend on the atomic or molecular level [48, 49]. However, because of the small population of excited levels, the major contribution comes from the ground state. Only in extreme conditions, the dependence of elastic cross-sections on the excited state gives an appreciable contribution.

[8] For sake of compactness, from now on we will consider implicitly the dependence of the distribution on position and time.

$$S^{\rm el}_{0,\,s,\,\ell} = N_s \frac{\partial}{\partial v}\left(A_{s,\ell} f_0(v) + B_{s,\ell} \frac{\partial f_0(v)}{\partial v} \right) \tag{2.136}$$

It must be noted that, being the distribution a function of the module of the velocity, the probability should be expressed in terms of $f_0 v^2$ [47] (see equation (2.10)), therefore the velocity derivative in the Fokker–Planck equation must be substituted as $\frac{\partial}{\partial v} \rightarrow \frac{1}{v^2}\frac{\partial v^2}{\partial v}$. The coefficients $A_{s,\ell}$ and $B_{s,\ell}$, representing the integrals, are not independent and can be related considering the equilibrium condition[9], i.e. a Boltzmann distribution, giving $A_{s,\ell} = B_{s,\ell} m_e v/kT$. Following these considerations, equation (2.136) becomes

$$S^{\rm el}_{0,\,s,\,\ell} = \frac{N_s}{v^2}\frac{\partial}{\partial v} v^2 B_{s,\ell}\left(\frac{m_e v}{kT} f_0(v) + \frac{\partial f_0(v)}{\partial v} \right) \tag{2.137}$$

Let as now calculate explicitly the coefficient B

$$B_{s,\ell} = \frac{1}{2}\int_0^{4\pi} v\frac{d\sigma_p}{d\Omega} d\Omega \int_0^{\infty} f_s(v_s, \ell)(\Delta v)^2 d^3 v_s \tag{2.138}$$

where $\Delta v = v' - v$. This quantity can be calculated considering that in elastic collisions, the amplitude of the relative velocity is the same before and after the collision, giving

$$\begin{aligned} &(\vec{v} - \vec{v}_s)^2 = (\vec{v}\,' - \vec{v}_s')^2 \Rightarrow v^2 + v_s^2 - 2\vec{v}\cdot\vec{v}_s = v'^2 + v_s'^2 - 2\vec{v}\,'\cdot\vec{v}_s' \\ &\overset{v_s \approx v_s'}{\Longrightarrow} v'^2 - v^2 \overset{v\approx v'}{\approx} 2v\Delta v \approx 2\vec{v}_s\cdot(\vec{v} - \vec{v}\,') \Rightarrow \Delta v \approx \vec{v}_s\cdot(\vec{v} - \vec{v}\,')/v \end{aligned} \tag{2.139}$$

$(\Delta v)^2$ contains terms $\propto v_{s,i}v_{s,j}$, and, being the distribution isotropic, only those terms with $i = j$ give a mean different from zero, i.e. $<v^2_{s,\,i}> = <v_s^2>/3 = kT/m_s$, and equation (2.138) becomes

$$\begin{aligned} B_{s,\ell} &= \frac{kT}{2m_s} f_{s,\ell}\int_0^{4\pi} v\frac{d\sigma_p}{d\Omega} d\Omega (v^2 + v'^2 - 2\vec{v}\cdot\vec{v}\,')/v^2 \\ &\overset{v\approx v'}{\approx} \frac{kT}{m_s} f_{s,\ell}\int_0^{4\pi} v\frac{d\sigma_p}{d\Omega} d\Omega (1 - \cos\vartheta_d) = \frac{kT}{m_s} f_{s,\ell}\, v\sigma^{\rm mt}_{s,\,\ell} = \frac{kT}{m_s} f_{s,\ell}\,\nu_s^{\rm mt} \end{aligned} \tag{2.140}$$

If we assume that the momentum transfer is almost independent on the internal level we can eliminate in equation (2.140) the internal distribution considering a single coefficient for the whole species.

To calculate the source term for the electron energy distribution $n_0(\varepsilon)$ in equation (2.108) we have to consider the equations in section 2.6 giving

---

[9] The equilibrium condition is obtained considering the parenthesis in equation (2.137) null when substituting the Boltzmann distribution.

$$
\begin{aligned}
S_{0,s}^{\prime el} &= 4\pi\sqrt{\frac{2e^3\varepsilon}{m_e^3}\frac{N_s}{v^2}}\sqrt{\frac{2m_e\varepsilon}{e}}\frac{\partial}{\partial\varepsilon}v^2 B_s \\
&\times\left(\frac{m_e}{kT}\sqrt{\frac{2e\varepsilon}{m_e}}\frac{1}{4\pi}\sqrt{\frac{m_e^3}{2e^3\varepsilon}}n_0(\varepsilon) + \sqrt{\frac{2m_e\varepsilon}{e}}\frac{\partial}{\partial\varepsilon}\frac{1}{4\pi}\sqrt{\frac{m_e^3}{2e^3\varepsilon}}n_0(\varepsilon)\right) \\
&= \frac{N_s}{\varepsilon}\frac{m_e}{2e}\frac{2e\varepsilon}{m_e}\frac{\partial}{\partial\varepsilon}\varepsilon\frac{2e}{m_e}B_s\frac{m_e^2}{e^2}\left(\frac{e}{kT}n_0(\varepsilon) + \sqrt{\varepsilon}\frac{\partial}{\partial\varepsilon}\frac{n_0(\varepsilon)}{\sqrt{\varepsilon}}\right) \\
&= N_s\frac{2m_e}{e}\frac{\partial}{\partial\varepsilon}\frac{kT}{m_s}\nu_s^{\mathrm{mt}}\varepsilon\sqrt{\varepsilon}\left(\frac{e}{kT}\frac{n_0(\varepsilon)}{\sqrt{\varepsilon}} + \frac{\partial}{\partial\varepsilon}\frac{n_0(\varepsilon)}{\sqrt{\varepsilon}}\right) \\
&= 2N_s\frac{m_e}{m_s e}\frac{\partial}{\partial\varepsilon}\nu_s^{\mathrm{mt}}\left(e\varepsilon n_0(\varepsilon) + kT\varepsilon\sqrt{\varepsilon}\frac{\partial}{\partial\varepsilon}\frac{n_0(\varepsilon)}{\sqrt{\varepsilon}}\right) \\
&= 2N_s\frac{m_e}{m_s e}\frac{\partial}{\partial\varepsilon}\nu_s^{\mathrm{mt}}(\varepsilon)\left[n_0(\varepsilon)\left(e\varepsilon - \frac{kT}{2}\right) + \varepsilon kT\frac{\partial n_0(\varepsilon)}{\partial\varepsilon}\right]
\end{aligned}
\tag{2.141}
$$

therefore, in equation (2.1) we have that the flux due to elastic collisions is given by

$$
J_{\mathrm{el}} = -2\frac{m_e}{e}\bar{\nu}\left[n_0(\varepsilon)\left(e\varepsilon - \frac{kT}{2}\right) + \varepsilon kT\frac{\partial n_0(\varepsilon)}{\partial\varepsilon}\right] \tag{2.142}
$$

where $\bar{\nu} = \sum_s N_s\nu_s^{\mathrm{mt}}/m_s$.

2.7.2 Electron–electron collisions

The procedure to derive the Fokker–Planck equation and the collisional operator in equation (2.142) for electron–molecule elastic collisions can be adapted to electron–electron collisions obtaining for the corresponding flux term J_{ee}

$$
J_{ee} = \alpha_{ee}\left\{P(\varepsilon)\left[\frac{n(\varepsilon)}{\varepsilon} - \frac{\partial n(\varepsilon)}{\partial\varepsilon}\right] - Q(\varepsilon)n(\varepsilon)\right\} \tag{2.143}
$$

where

$$
\alpha_{ee} = \frac{2\pi}{3}(K_e e)^2\sqrt{\frac{2q_e}{m_e}}N_e \ln\Lambda_{ee} = 2.582\,1\cdot 10^{-12}N_e\ln\Lambda_{ee} \tag{2.144}
$$

with

$$
\Lambda_{ee} = \sqrt{\frac{\varepsilon_{\mathrm{car}}}{4\pi K_e e N_e}}\cdot\frac{\varepsilon_{\mathrm{med}}}{K_e e} = 5.152\,7\cdot 10^{12}\sqrt{\frac{\varepsilon_{\mathrm{car}}\,\varepsilon_{\mathrm{med}}^2}{N_e}} \tag{2.145}
$$

$K_e = 1/(4\pi\varepsilon_0)$ is the Coulomb constant being ε_0 the vacuum dielectric constant. The numerical contants in equations (2.144), (2.145) are obtained for all the quantities in I.S. units apart from for the electron energy which is in eV.

The coefficients P and Q are given by

$$P(\varepsilon) = \frac{2}{\sqrt{\varepsilon}} \int_0^{\varepsilon} \xi n_0(\xi) d\xi + 2\varepsilon \int_{\varepsilon}^{\infty} \frac{n_0(\xi)}{\sqrt{\xi}} d\xi = P^a(\varepsilon) + P^b(\varepsilon) \tag{2.146}$$

$$Q(\varepsilon) = \frac{3}{\sqrt{\varepsilon}} \int_0^{\varepsilon} n_0(\xi) d\xi \tag{2.147}$$

The following recursion rules can be easily obtained

$$P^a(\varepsilon) = \sqrt{\frac{\varepsilon - \Delta}{\varepsilon}} P^a(\varepsilon - \Delta) + \frac{2}{\sqrt{\varepsilon}} \int_{\varepsilon-\Delta}^{\varepsilon} \xi n_0(\xi) d\xi \tag{2.148}$$

$$P^b(\varepsilon) = \frac{\varepsilon}{\varepsilon + \Delta} P^b(\varepsilon + \Delta) + 2\varepsilon \int_{\varepsilon}^{\varepsilon+\Delta} \frac{n_0(\xi)}{\sqrt{\xi}} d\xi \tag{2.149}$$

$$Q(\varepsilon) = \sqrt{\frac{\varepsilon - \Delta}{\varepsilon}} Q(\varepsilon - \Delta) + \frac{3}{\sqrt{\varepsilon}} \int_{\varepsilon-\Delta}^{\varepsilon} n_0(\xi) d\xi \tag{2.150}$$

which can help in building the operator, as will be discussed later.

2.7.3 Inelastic and superelastic collisions

Inelastic collisions are endothermal processes where electrons give energy to heavy species, resulting in excitation or chemical reactions. Superelastic collisions are the reverse processes, where chemical or internal energy is given back to the electrons. The energy transferred ($\varepsilon^\star$), known as threshold energy, is constant, independently of the energy of the impinging electron[10]. In the class of inelastic and superelastic processes are included:

$e(\varepsilon) + A(i) \rightleftharpoons e(\varepsilon - \varepsilon_{i,j}^\star) + A(j > i)$	internal transitions
$e(\varepsilon) + A(i) \rightleftharpoons e(\varepsilon - \varepsilon_{i,\,\text{ion}}^\star) + A^+$	ionization and three body recombination
$e(\varepsilon) + M^+ \rightleftharpoons$ products $(+h\nu)$	attachment and radiative recombination
$e(\varepsilon) + A(i) \rightleftharpoons e(\varepsilon - \varepsilon^\star)$ + products $(+h\nu)$	other chemical processes

The presence of the photons, in practice, means that the reverse process must be neglected[11]. A common assumption on inelastic and superelastic collisions is that their cross-sections do not depend on the deflection angle, i.e. all the directions are equiprobable after the collision, i.e.

$$\frac{d\sigma_p}{d\Omega} = \frac{\sigma_p}{4\pi}. \tag{2.151}$$

[10] In some chemical processes the threshold energy is not constant, depending on the interaction mechanism. Nevertheless, it is usually considered constant, equal to the minimum energy required by the transition.

[11] To properly calculate the rate of reverse processes in the presence of photons a radiation transport equation must be considered as described in chapter 9.

The contribution of these processes are usually neglected in $\vec{S}_1$ even if they should be included. In fact, from equations (2.131) and 2.151 the contribution of a single process is[12]

$$\vec{S}_{1,p} = -N_s f_{s,\ell} v \sigma_p \vec{f}_1 = -\nu_p \vec{f}_1 \tag{2.152}$$

being the integral of $\cos \vartheta_d$ null. Therefore, the collision frequency in equation (2.27) should be the sum of the frequencies of all the processes, including elastic, inelastic and superelastic collisions.

Let us calculate the source term $S'_{0,\,p}$,[13] for inelastic processes staring from equations (2.101), (2.109), and (2.129) and under the hypothesis in equation (2.151) of isotropic collisions

$$S'_{0,\,p}(\varepsilon) = N_s v \left[\sqrt{\frac{\varepsilon}{\varepsilon'}} n_0(\varepsilon') f_{s,\ell'} - n_0(\varepsilon) f_{s,\ell} \right] \sigma_p \tag{2.153}$$

neglecting the change in velocity of the heavy particle, we can say that $\varepsilon' = \varepsilon - \varepsilon^\star$. The negative term corresponds to the inelastic process

$$S^{\text{in}}_{0,\,p}(\varepsilon) = -N_s v n_0(\varepsilon) f_{s,\ell} \sigma_p \tag{2.154}$$

i.e. to electrons passing from energy ε to ε' and the positive term correspond to superelastic collisions

$$S^{\text{sup}}_{0,\,p}(\varepsilon) = N_s v \sqrt{\frac{\varepsilon}{\varepsilon'}} n_0(\varepsilon') f_{s,\ell'} \sigma_p \tag{2.155}$$

i.e. to electrons passing from energy[14] ε' to ε. To determine the cross-section $\sigma'_p(\varepsilon')$ for the superelastic process let us use the general definition of the source term

$$S^{\text{sup}}_{0,\,p}(\varepsilon) = N_s v' n_0(\varepsilon') f_{s,\ell'} \sigma'_p \tag{2.156}$$

and comparing with equation (2.155) it gives

$$v' \sigma'_p = v \sqrt{\frac{\varepsilon}{\varepsilon'}} \sigma_p \Rightarrow \sigma'_p = \frac{\varepsilon}{\varepsilon'} \sigma_p \tag{2.157}$$

Equation (2.157) is valid only for elementary processes, i.e. between states with the same statistical weight. In the general case we can obtain the cross-section of the superelastic process by applying the detailed balance principle, i.e. the collision integral is null for Boltzmann and Maxwell distributions at the same temperature

---

[12] Under assumption of isotropic collisions the dependence on ϑ_d of the first term in equation (2.131) is given by the cosine and therefore its integral is null, leaving only the second term.

[13] p represents the set of initial and final parameter of the considered process.

[14] Obviously, for each of these terms there is another one with opposite sign, $S^{\text{in/sup}}_{0,\,p}(\varepsilon')$, assuring the conservation of the total number of electrons.

$$
\begin{aligned}
v'\sigma_p'(\varepsilon')n_0^{\text{eq}}(\varepsilon')f_{s,\ell'}^{\text{eq}} &= vn_0^{\text{eq}}(\varepsilon)f_{s,\ell}^{\text{eq}}\sigma_p \\
\sqrt{\varepsilon'}\,\sigma_p'(\varepsilon')\sqrt{\varepsilon'}\,e^{-\varepsilon'/kT}g_{s,\ell'}e^{-\varepsilon_{s,\ell'}/kT} &= \sqrt{\varepsilon}\,\sigma_p(\varepsilon)\sqrt{\varepsilon}\,e^{-\varepsilon/kT}g_{s,\ell}e^{-\varepsilon_{s,\ell}/kT} \\
\varepsilon'\sigma_p'(\varepsilon')g_{s,\ell'} &= \varepsilon\sigma_p(\varepsilon)g_{s,\ell}e^{-\frac{\varepsilon-\varepsilon'+\varepsilon_{s,\ell}-\varepsilon_{s,\ell'}}{kT}}
\end{aligned}
\tag{2.158}
$$

where $g_{s,\ell}$ is the statistical weight of the ℓ th level. For the energy conservation we have $\varepsilon - \varepsilon' = \varepsilon^\star = \varepsilon_{s,\ell'} - \varepsilon_{s,\ell}$, obtaining

$$
\sigma_p'(\varepsilon - \varepsilon^\star) = \frac{g_{s,\ell}\varepsilon}{g_{s,\ell'}\varepsilon'}\sigma_p(\varepsilon) = \frac{g_{s,\ell}\varepsilon}{g_{s,\ell'}(\varepsilon - \varepsilon^\star)}\sigma_p(\varepsilon)
\tag{2.159}
$$

2.7.4 Chemical processes

The contribution of chemical processes induced by electron impact is given by equation (2.154), as for inelastic processes. For the reverse processes, in principle, the Boltzmann collision integral could not be used because it accounts for binary collisions only. However, using the detailed balance principle we can estimate the *superelastic* source term also for these processes. The general form for a reaction is given by

$$
e(\varepsilon) + X_\ell \rightleftharpoons e(\varepsilon - \varepsilon^\star) + \sum_s c_s P_{s,\ell_s} + c_e e(\varepsilon_e)
\tag{2.160}
$$

The stoichiometric coefficient c_e of the electron is positive for ionization and negative for the attachment. Now we need to define some quantities derived from the thermodynamics (see chapter 21):

$$
\begin{aligned}
\Delta H_r &= \sum_s c_s H_s^f - H_X^f && \text{reaction enthalpy} \\
\mathcal{Q}_s &= \sum_\ell g_{s,\ell_s}e^{-\varepsilon_{s,\ell}/kT} && \text{partition function of } s\text{th species} \\
\mathcal{Q}_e &= \int_0^\infty \sqrt{\varepsilon}\,e^{-\varepsilon/kT}d\varepsilon && \text{partition function of electrons} \\
K_{\text{eq}} &= N_{X,\text{eq}}^{-1}N_{e,\text{eq}}^{c_e}\prod_s N_{s_{\text{eq}}}^{c_s} && \text{global equilibrium constant}
\end{aligned}
\tag{2.161}
$$

Usually, we can calculate the threshold energy applying the energy conservation law

$$
\varepsilon_t^\star = \Delta H_r + \sum_s c_s\varepsilon_{s,\ell_s} - \varepsilon_{X,\ell} + \max(0, c_e)\varepsilon_e
\tag{2.162}
$$

where the max function excludes the electron in the case of attachment. However, depending on the mechanism of reaction, sometime the threshold energy is higher than the value in equation (2.162) and some of the energy is transferred to the translation of heavy particles [50], therefore we use the subscript t to indicate that equation (2.162) defines the theoretical threshold.

Now let us apply the detailed balance to the reaction in equation (2.160) considering $\varepsilon' = \varepsilon - \varepsilon^\star$ and $\nu(\varepsilon) = v(\varepsilon)\sigma(\varepsilon)$

$$\frac{N_{e,\text{eq}}}{\mathcal{Q}_e}\sqrt{\varepsilon}\, e^{-\varepsilon/kT}\frac{N_{X,\text{eq}}}{\mathcal{Q}_X} g_{X,\ell} e^{-\varepsilon_{X,\ell}/kT}\nu$$
$$= \frac{N_{e,\text{eq}}}{\mathcal{Q}_e}\sqrt{\varepsilon'}\, e^{-\varepsilon'/kT}\frac{N_{e,\text{ eq}}^{c_e}}{\mathcal{Q}_e^{c_e}}\sqrt{\varepsilon_e^{c_e}}\, e^{-c_e\varepsilon_e/kT}\prod_s \frac{N_{s,\text{ eq}}^{c_s}}{\mathcal{Q}_s^{c_s}} g_{s,\ell_s} e^{-c_s\varepsilon_{s,\ell_s}/kT}\nu'$$

$$\nu = \nu'\sqrt{\frac{\varepsilon'}{\varepsilon}}\ \frac{N_{e,\text{ eq}}^{c_e}}{N_{X,\text{ eq}}}\prod_s N_S^{c_s}\ \frac{\mathcal{Q}_X}{\mathcal{Q}_e^{c_e}\prod_s \mathcal{Q}_S^{c_s}}\ \frac{\sqrt{\varepsilon_e^{c_e}}}{g_{X,\,\ell}}\prod_s g_{s,\ell_s}^{c_s}\, e^{(\varepsilon-\varepsilon'+\varepsilon_{X,\,\ell}-c_e\varepsilon_e-\sum_s c_s\varepsilon_{s,\,\ell})/kT} =$$

$$\begin{cases} \nu'\sqrt{\dfrac{\varepsilon'}{\varepsilon}}\ \dfrac{K_{\text{eq}}}{K_Q}\, K_g\, e^{(\varepsilon^\star-\varepsilon_t^\star+\Delta H_r)/kT} & \text{for all kinds of processes but attachment} \\ \nu'\sqrt{\dfrac{\varepsilon'}{\varepsilon}}\ \dfrac{K_{\text{eq}}}{K_Q}\, K_g\, e^{(\varepsilon-\varepsilon_t^\star+\Delta H_r)/kT} & \text{for attachment} \end{cases} \tag{2.163}$$

where

$$K_Q = \frac{\mathcal{Q}_e^{c_e}}{\mathcal{Q}_X}\prod_s \mathcal{Q}_S^{c_s} \qquad\qquad K_g = \frac{\sqrt{\varepsilon_e^{c_e}}}{g_{X,\ell}}\prod_s g_{s,\ell_s}^{c_s}$$

It should be noted that $\sqrt{\varepsilon}$ can be considered the statistical weight for electrons of energy ε.

2.8 The numerical solution

Once all the terms in equation (2.1) have been defined, we can proceed to its numerical solution. The first step is to write the equation in discrete form in the energy space. First and second derivative terms come from the J's. Following Rockwood [45], finite difference approach was used, with central difference for the first derivatives. This procedure is unstable when the electric field is very small, a problem solved by Elliot and Greene in reference [51]. Alternately upwind schemes can be used, which apply properly forward or backward discretization according to the derivative coefficient, i.e.

$$\frac{\partial y}{\partial t} = u(x,\,t)\frac{\partial y}{\partial x} \Rightarrow \frac{\partial y_i}{\partial t} = \begin{cases} u(x_{i+\frac{1}{2}},\,t)\dfrac{y_{i+1}-y_i}{x_{i+1}-x_i} & \text{if } u(x_{i+\frac{1}{2}},\,t) \geqslant 0 \\ u(x_{i-\frac{1}{2}},\,t)\dfrac{y_i-y_{i-1}}{x_i-x_{i-1}} & \text{if } u(x_{i-\frac{1}{2}},\,t) < 0 \end{cases} \tag{2.164}$$

so that y_i coefficient is always negative (see figure 2.1 for the symbols).

The discretization of S's terms does not present particular difficulties, except when some cross-sections contain very sharp peaks as in the resonant vibrational excitation [52]. In this case, sampling the cross-sections as $\sigma_i = \sigma(\varepsilon_i)$ gives values strongly dependent on the grid size. On the other hand, taking the mean value inside each mesh interval as (see equation (2.156))

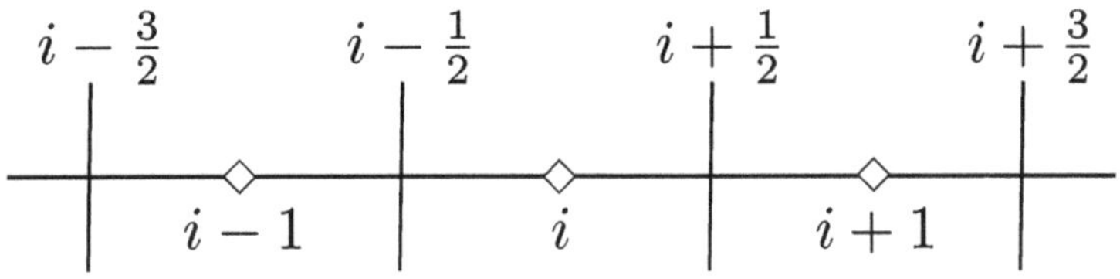

Figure 2.1. Scheme of an uniform grid. The half-integer indexes refer to the interval boundaries.

$$v_i\sigma_i = \frac{1}{\varepsilon_{i+\frac{1}{2}} - \varepsilon_{i-\frac{1}{2}}} \int_{\varepsilon-\frac{1}{2}}^{\varepsilon+\frac{1}{2}} v(\varepsilon)\sigma(\varepsilon)d\varepsilon \tag{2.165}$$

reduces the grid-dependence.

The discretized equation can be expressed in a general matrix form

$$\frac{d\vec{n}}{dt} = [\hat{C} + \hat{T}(\vec{n})]\vec{n} \tag{2.166}$$

where $\vec{n} = \{n_0(\varepsilon_i)\}_{i=1,\ldots,N_\varepsilon}$, or in finite volume discretization, the mean value in the ith interval. The matrix $\hat{C}$ includes all the linear terms and has a diagonal structure

$$\hat{C} = \begin{pmatrix} a_1 & b_1 & 0 & 0 & \cdots & d_1 & 0 & 0 \\ c_2 & a_2 & b_2 & 0 & & 0 & d_2 & 0 \\ 0 & c_3 & a_3 & b_3 & & 0 & 0 & d_3 \\ & \vdots & & & & & \vdots & \\ s_{N_\varepsilon-3} & 0 & 0 & 0 & \cdots & 0 & 0 & 0 \\ 0 & s_{N_\varepsilon-2} & 0 & 0 & c_{N_\varepsilon-2} & a_{N_\varepsilon-2} & b_{N_\varepsilon-2} & 0 \\ 0 & 0 & s_{N_\varepsilon-1} & 0 & 0 & c_{N_\varepsilon-1} & a_{N_\varepsilon-1} & b_{N_\varepsilon-1} \\ 0 & 0 & 0 & s_{N_\varepsilon} & 0 & 0 & c_{N_\varepsilon} & a_{N_\varepsilon} \end{pmatrix} \tag{2.167}$$

The first upper and lower diagonals (b and c) come from the flux terms, while each inelastic process contributes to an upper diagonal (d) and the corresponding superelastic process to the symmetric lower diagonal (s). The main diagonal (a) fulfills the conservation condition, i.e.

$$\sum_{j=1}^{N_\varepsilon} C_{ji} = 0 \tag{2.168}$$

The nonlinear matrix $\hat{T}$, due to electron–electron collisions, has a tridiagonal structure, being generated by the derivative of J_{ee}, and depends linearly on the distribution. Also matrix $\hat{T}$ must fulfill the conservation of total electrons in equation (2.168).

To calculate the stationary solution, the following nonlinear equation must be solved

$$[\hat{C} + \hat{T}(\vec{n})]\vec{n} = 0 \tag{2.169}$$

through iterative methods such as Newton's algorithm, which require the Jacobian matrix, or solving the equation staring from the solution of the linear problem

$$\hat{C}\vec{n}^{\,0} = 0 \tag{2.170}$$

and the successive approximations as

$$[\hat{C} + \hat{T}(\vec{n}^{\,p})]\vec{n}^{\,p+1} = 0 \tag{2.171}$$

equations (2.170) and 2.171 have ∞^1 solutions, because the determinant is null. This problem can be easily solved, substituting the first equation with the conservation law

$$\sum_{i=1}^{N_e} \Delta_i n_i = N_e \tag{2.172}$$

with Δ_i being the amplitude of the ith interval.

The stationary solution is commonly used coupled with the chemical kinetics, to reduce computational time, as the electron evolution is usually faster than level and species concentrations [10]. It should be noted that the EEDF is also a function of the composition and of internal level distributions.

Some general features of EEDF can be evinced in figure 2.2 where the stationary EEDF has been reported for different conditions, listed in table 2.1 , to explore the role of inelastic, superelastic and electron–electron collisions. Inelastic collisions (curves *a*, *b*) induce an abrupt change in the slope at about 20 eV, the threshold energy of the excitation of metastable state. When superelastic collisions are activated by the He metastable state (curves *c*, *d*), some plateaux appear. Electron–electron collisions (curves *b*, *d*) reshape the distribution towards a Maxwellian, with the net result of increasing the distribution tails (curve *b*), hiding the plateaux in the presence of superelastic collisions (curve *d*).

Curve *e* has been calculated by adding to helium some molecular species, with vibrational structure and electronically excited states, in order to investigate the

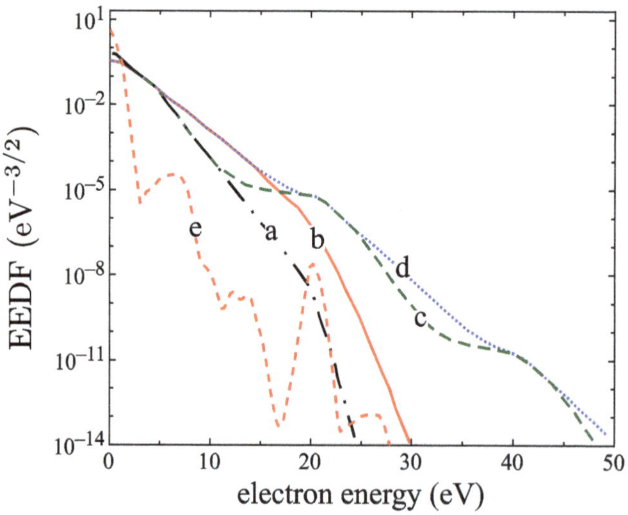

Figure 2.2. EEDF calculated in the conditions listed in table 2.1. Data from references [1, 11].

Table 2.1. Label the curves in figure 2.2. The case with ionization degree $N_e/N = 0$ correspond to the case where e–e collisions are neglected. He ⋆ and $N_2^\star$ are metastable states, with null values indicating that superelastic collisions are neglected. Conditions from references [1, 11].

curve	%He	He ⋆/He	$\%N_2$	$N_2^\star/N_2$	$\%CO_2$	%CO	N_e/N	E/N(Td)
a	100	0	0	0	0	0	0	4
b	100	0	0	0	0	0	10^{-3}	4
c	100	10^{-4}	0	0	0	0	0	4
d	100	10^{-4}	0	0	0	0	10^{-3}	4
e	72	10^{-6}	18	10^{-4}	2.5	7.5	0	5

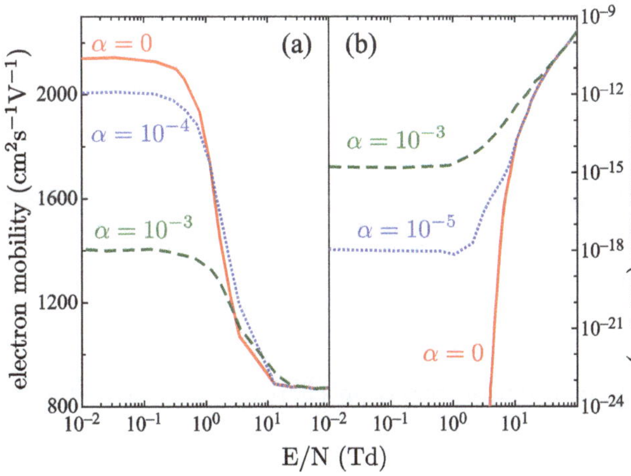

Figure 2.3. Electron mobility (a) and ionization rates (b) in pure helium mixture, with He ⋆/He = $N_e/N = \alpha$, as a function of the reduced electric field. Data from reference [1].

sensitivity of the EEDF to the mixture. The presence of vibrational states keeps the electron distribution cooler than in pure helium, even if the electric field is slightly higher, and, due to lower mean elastic cross-section, in place of plateaux we can observe complex peak structures, generated by the combination of inelastic and superelastic collisions of electrons with different species.

The response of the distribution to changes in external parameters is reflected in transport properties and global rate coefficients. As an example in figure 2.3 electron mobility (a) and ionization rate (b) are reported for different excitation and ionization degrees, represented by α, as a function of the reduced electric field. At high field, all the curves converge, independently of α. At low and intermediate fields the differences are relevant, especially in the case of the ionization degree, where even small excitation increases the ionization rate of orders of magnitude.

In their famous paper [45], Rockwood proposed a time integration based on Euler method, with the linear part taken implicitly, i.e. with backward discretization, and the electron–electron matrix explicitly, i.e. with forward discretization

$$(\hat{I} - \Delta t\hat{C})\vec{n}(t + \Delta t) = \{\hat{I} + \Delta t\hat{T}[\vec{n}(t)]\}\vec{n}(t) \tag{2.173}$$

The advantage of equation (2.173) is that LU decomposition can be applied only once at the beginning. Moreover, in this scheme, the e–e operator implicitly conserves the electron energy. Problems arise when the electron density is high and the time step needed for having a stable explicit integration becomes very small. Taking all the terms implicitly as

$$\{\hat{I} - \Delta t\hat{C} - \Delta t\hat{T}[\vec{n}(t)]\}\vec{n}(t + \Delta t) = \vec{n}(t) \tag{2.174}$$

makes the procedure stable, but the matrix $\hat{T}$ does not conserve the electron energy. To avoid this problem iterative procedure can be used calculating successive approximation of the EEDF, $\vec{n}_z(t + \Delta t)$, as

$$\{\hat{I} - \Delta t\hat{C} - \Delta t\hat{T}[\vec{n}_z(t + \Delta t)]\}\vec{n}_{z+1}(t + \Delta t) = \vec{n}(t) \tag{2.175}$$

To reduce computational time, an *operator-splitting* approach can be used

$$(\hat{I} - \Delta t\hat{C})\{\hat{I} - \Delta t\hat{T}[\vec{n}(t + \Delta t)]\}\vec{n}(t + \Delta t) = \vec{n}(t) \tag{2.176}$$

solving firstly the linear part, calculating the zeroth order approximation

$$(\hat{I} - \Delta t\hat{C})\vec{n}_0 = \vec{n}(t) \tag{2.177}$$

and then iterating the nonlinear matrix for calculating the successive approximations

$$[\hat{I} - \Delta t\hat{T}(\vec{n}_{z-1})]\vec{n}_z = \vec{n}_0 \tag{2.178}$$

until the convergence criterion based on energy conservation by electron–electron collisions

$$\left|\frac{\mathcal{E}_{z_c} - \mathcal{E}_0}{\mathcal{E}_0}\right| < \epsilon_{\max} \tag{2.179}$$

is fulfilled, where $\mathcal{E}_z$ is the mean energy of the zth approximation, and the new distribution is given by $\vec{n}(t + \Delta t) = \vec{n}_{z_c}$. This procedure assures the energy conservation for electron–electron collisions, with good efficiency[15], applying LU decomposition to $\hat{C}$ only once, while $\hat{T}$, being a tridiagonal matrix, can be solved very rapidly, with a CPU time $\propto N_\varepsilon$. Therefore, the critical aspect is the calculation of the $\hat{T}$ matrix.

The traditional approach proposed by Rockwood [45], discretizes the derivative of J_{ee} in the ith interval as

[15] There is another aspect that should be considered for the solution efficiency: the system of equations is very stiff, presenting timescales ranging on many orders of magnitude. A step adapting procedure, such as that proposed in reference [53], must be used to have reasonable total computational time. Alternately, higher order methods for stiff differential equations, such as implicit Runge–Kutta [54], can be adopted. There is also a class of algorithm that does not need the evaluation of the Jacobian such as the method proposed by Verwer [55] and that by Formaggia [56] having higher accuracy. A conservative fast explicit method has also been proposed for stiff equations by Martiradonna [57].

$$\left(\frac{dn_i}{dt}\right)_{ee} = -\frac{J^+_{ee}(i) - J^-_{ee}(i)}{\Delta\varepsilon} \tag{2.180}$$

where $J^+(i)$ and $J^-(i)$ are evaluated at the upper $\left(i + \frac{1}{2}\right)$) and lower $\left(i - \frac{1}{2}\right)$ limits of the ith interval, with the trivial relation

$$J^-_{ee}(i) = J^+_{ee}(i - 1). \tag{2.181}$$

The explicit expression of the of $J^+_{ee}(i)$ is given by

$$J^+_{ee}(i) = \alpha\left[\left(\frac{n_{i+\frac{1}{2}}}{4\varepsilon_{i+\frac{1}{2}}} - \frac{n_{i+1} - n_i}{\Delta\varepsilon}\right)P_{i+\frac{1}{2}} - Q_{i+\frac{1}{2}}n_{i+\frac{1}{2}}\right] \tag{2.182}$$

where the population distribution at the upper limit of the ith interval is calculated as the average of the distribution at i and $i + 1$

$$n_{i+\frac{1}{2}} = \frac{n_i + n_{i+1}}{2} \tag{2.183}$$

As a consequence, the right side of equation (2.180) depends only on n_{i-1},n_i and n_{i+1}

$$\left(\frac{dn_i}{dt}\right)_{ee} = a_{i-1}n_{i-1} - (a_i + b_i)n_i + b_{i+1}n_{i+1} \tag{2.184}$$

confirming that $\hat{T}$ is a tridiagonal matrix. Comparing equations (2.180) and 2.182, a_i and b_i are given by

$$a_i = \frac{\alpha}{\Delta\varepsilon}\left[u_i^+P_{i+\frac{1}{2}} - \frac{Q_{i+\frac{1}{2}}}{2}\right] \tag{2.185}$$

$$b_{i+1} = \frac{\alpha}{\Delta\varepsilon}\left[u_i^-P_{i+\frac{1}{2}} - \frac{Q_{i+\frac{1}{2}}}{2}\right] \tag{2.186}$$

where

$$u_i^\pm = \left(\frac{1}{\Delta\varepsilon} \pm \frac{1}{4\varepsilon_{i+\frac{1}{2}}}\right) \tag{2.187}$$

From definitions of P and Q, it follows that a_i and b_i depend linearly on the EEDF

$$\vec{a} = \hat{A} \cdot \vec{n} \tag{2.188}$$

$$\vec{b} = \hat{B} \cdot \vec{n} \tag{2.189}$$

The matrices $\hat{A}$ and $\hat{B}$ are not independent because electron–electron collisions must conserve also the electron energy and, from equation (2.107), the following relation holds

$$\begin{aligned}\frac{d\varepsilon_{\text{med}}}{dt} &= \sum_{i=1}^{N_\varepsilon} \varepsilon_i \left(\frac{dn_i}{dt}\right)_{ee} \Delta\varepsilon = \sum_{i=1}^{N_\varepsilon} \varepsilon_i [a_i - b_i] n_n \Delta\varepsilon = \\ &= \sum_{i=1}^{N_\varepsilon} \sum_{k=1}^{N_\varepsilon} [A_{i,k} - B_{i,k}] n_i n_k \varepsilon_i \Delta\varepsilon = 0\end{aligned} \tag{2.190}$$

With $\vec{a} \neq \vec{b}$, equation (2.190) is verified if the matrix $\hat{A} - \hat{B}$ is antisymmetric, i.e.

$$\hat{B} = \hat{A}^{+} \tag{2.191}$$

where $\hat{A}^{+}$ is the transpose of $\hat{A}$.

Generally, vanishing fluxes J are imposed as boundary conditions at both sides of the energy interval

$$J_{ee}^{-}(1) = J_{ee}^{+}(N_\varepsilon) = 0 \tag{2.192}$$

which means

$$b_1 = a_{N_\varepsilon} = 0 \tag{2.193}$$

obtaining

$$A_{N_\varepsilon k} = B_{1k} = A_{1k} = B_{1N_\varepsilon} = 0 \quad (k = 1, .., N_\varepsilon) \tag{2.194}$$

To calculate $P_{i+\frac{1}{2}}$ and $Q_{i+\frac{1}{2}}$, Rockwood [45] uses the expressions

$$P_{i+\frac{1}{2}} = \frac{P_i + P_{i+1}}{2}; \; Q_{i+\frac{1}{2}} = \frac{Q_i + Q_{i+1}}{2} \tag{2.195}$$

From definitions of P and Q, it seems that the direct evaluation of these quantities at the boundaries of the domain is more straightforward than equation (2.195). However, evaluating $\hat{T}$ using equations (2.188) and 2.189 requires a CPU time $\propto K^2$, with $\hat{A}$ and $\hat{B}$ being full matrices.

An alternate efficient algorithm to calculate the matrix $\hat{T}$ [58] avoids the explicit calculations of matrices $\hat{A}$ and $\hat{B}$ (see equations (2.188) and 2.189), reducing the computational time, to calculate a_k and b_k. Starting from the recursive relations in equations (2.148)–(2.150)

$$\begin{aligned}P_{i+1}^{a} &= \sqrt{\frac{\varepsilon_i}{\varepsilon_{i+1}}} P_i^{a} + \frac{\Delta\varepsilon}{\sqrt{\varepsilon i + 1}} (\varepsilon_i n_i + \varepsilon_{i+1} n_{i+1}) \\ P_{i+1}^{b} &= \frac{\varepsilon_{i+1}}{\varepsilon_i} P_i^{b} - \varepsilon_{i+1} \Delta\varepsilon \left(\frac{n_i}{\sqrt{\varepsilon_i}} + \frac{n_{i+1}}{\sqrt{\varepsilon_{i+1}}}\right) \\ Q_{i+1} &= \sqrt{\frac{\varepsilon_i}{\varepsilon_{i+1}}} Q_i + \frac{3\Delta\varepsilon}{2\sqrt{\varepsilon_{i+1}}} (n_i + n_{i+1})\end{aligned} \tag{2.196}$$

saving the computational time required to recalculate the terms with $k < i$ for each energy mesh point, as done in equations (2.188) and 2.189. Equation (2.196) can be written in compact form as

$$\begin{aligned}\hat{A}_a\vec{P}^a &= \hat{L}_a\vec{n}\\ \hat{A}_b\vec{P}^b &= \hat{L}_b\vec{n}\\ \hat{A}_Q\vec{Q} &= \hat{L}_Q\vec{n}\end{aligned} \tag{2.197}$$

where $\hat{A}_Q = \hat{A}_a$. The matrices $\hat{A}$'s, $\hat{L}_Q$ and $\hat{L}_a$ are bi-diagonal given by

$$\begin{aligned}\hat{A}_{Q,i,i} &= \hat{A}_{b,i,i} = 1\\ \hat{A}_{Q,i,i-1} &= \sqrt{\frac{i-1}{i}}\\ \hat{A}_{b,i,i-1} &= \frac{i}{i-1}\end{aligned} \tag{2.198}$$

$$\begin{aligned}\hat{L}_{Q,i,i} &= \hat{L}_{Q,i,i-1} = \frac{3}{2}\left(\frac{\Delta\varepsilon}{i}\right)^{\frac{1}{2}}\\ \hat{L}_{a,i,i} &= \Delta\varepsilon^{\frac{3}{2}}\sqrt{i}\\ \hat{L}_{a,i,i-1} &= \Delta\varepsilon^{\frac{3}{2}}\frac{i-1}{\sqrt{i}}\end{aligned} \tag{2.199}$$

while $\hat{L}_2$ has also the first row different from zero

$$\begin{aligned}\hat{L}_{b,i,i} &= -\sqrt{i}\,\Delta\varepsilon^{\frac{3}{2}}i > 1\\ \hat{L}_{b,i,i-1} &= -\Delta\varepsilon^{\frac{3}{2}}\frac{i}{\sqrt{i-1}}\\ \hat{L}_{b,1,i} &= 2\Delta\varepsilon^{\frac{3}{2}}\frac{1}{\sqrt{i}}\end{aligned} \tag{2.200}$$

The elements a_k of the matrix $\hat{T}$ (see equation (2.184)) are calculated as a function of $\vec{P}$ and $\vec{Q}$ by considering equation (2.185)

$$\vec{a} = -\frac{\alpha}{2\Delta\varepsilon}\left[\hat{M}_P\vec{P} + \hat{M}_Q\vec{Q}\right] = \hat{A}\vec{n} \tag{2.201}$$

where $\hat{M}_P$ and $\hat{M}_Q$ are defined as follows (see equation (2.187))

$$\begin{aligned}\hat{M}_{P,i,i} &= \hat{M}_{P,i,i+1} = -u_i^+\\ \hat{M}_{Q,i,i} &= \hat{M}_{Q,i,i+1} = \frac{1}{2}\end{aligned} \tag{2.202}$$

All the matrices are bi-diagonal except $\hat{L}_2$ (which has also the first row different from zero). In the proposed numerical scheme, matrices $\hat{A}$ and $\hat{B}$ are not explicitly calculated and vectors $\vec{a}$ and $\vec{b}$ are obtained directly following equations (2.197)–(2.204). Boundary conditions represented by equations (2.193) and 2.194 are introduced in the numerical procedure requiring that

$$\begin{aligned} &\hat{M}_{P,N_\varepsilon,N_\varepsilon} = \hat{M}_{Q,N_\varepsilon,N_\varepsilon} = 0 \\ &\hat{L}_{Q,l,1} = \hat{L}_{a,l,1} = \hat{L}_{b,l,1} = 0 l = 1, \ldots, N_\varepsilon. \end{aligned} \tag{2.203}$$

To calculate the b_i we have considered the relation (see equation (2.191)) that imposes explicitly the energy conservation. The resulting equations consider the transpose of the matrices already defined (see equations (2.198), (2.200), and (2.202))

$$\vec{b} = -\frac{\alpha}{2\Delta\varepsilon}\left[\hat{L}_a^{+}\vec{x}_a + \hat{L}_b^{+}\vec{x}_b + \hat{L}_Q^{+}\vec{x}_Q\right] = \hat{B}\vec{n} \tag{2.204}$$

where vectors $\vec{x}_a$, $\vec{x}_b$ and $\vec{x}_Q$ are obtained solving the following linear systems

$$\begin{aligned} \hat{A}_a^{+}\vec{x}_a &= \hat{M}_P^{+}\vec{n} \\ \hat{A}_b^{+}\vec{x}_b &= \hat{M}_P^{+}\vec{n} \\ \hat{A}_Q^{+}\vec{x}_Q &= \hat{M}_Q^{+}\vec{n} \end{aligned} \tag{2.205}$$

By using equations (2.197) and 2.201, matrices $\hat{A}$ and $\hat{B}$ can be formally written as a function of the sparse matrices

$$\hat{A} = -\frac{\alpha}{2\Delta\varepsilon}\left[\hat{M}_P(\hat{A}_a^{-1}\hat{L}_a + \hat{A}_b^{-1}\hat{L}_b) + \hat{M}_Q\hat{A}_Q^{-1}\hat{L}_Q\right] \tag{2.206}$$

$$\hat{B} = \hat{A}^{+} = -\frac{\alpha}{2\Delta\varepsilon}\left[(\hat{L}_a^{+}(\hat{A}_a^{-1})^{+} + \hat{L}_b^{+}(\hat{A}_b^{-1})^{+})\hat{M}_P^{+} + \hat{L}_Q^{+}(\hat{A}_Q^{-1})^{+}\hat{M}_Q^{+}\right] \tag{2.207}$$

There is another possible approach to calculate the terms b_i starting from equation (2.186). In this way, the vectors $\vec{x}$'s in equation (2.205) are not calculated, with $\vec{b}$ directly related to the $\vec{P}_1$, $\vec{P}_2$ and $\vec{Q}$. One new matrix, $\hat{M}_P^{\star}$ must be defined to be used in an equation for $\vec{b}$ analogous to equation (2.201), resulting in a more efficient algorithm, saving the time needed to solve the linear sparse system in equation (2.205). However, the algorithm based on equation (2.205), imposing explicitly the relation of energy conservation in equation (2.191), needs many fewer iterations to converge, and therefore it is globally more efficient.

Numerical tests have been realized in order to compare the performances of the numerical scheme proposed, with the method developed by Rockwood [45]. Four procedures have been verified, different in the algorithm used to calculate the e–e matrices and in performing internal iteration:

RCK0 the matrices $\hat{A}$ and $\hat{B}$ are explicitly calculated following the Rockwood algorithm. No internal iteration is performed.

SPR0 the coefficients a_k and b_k are calculated following equations (2.197)–(2.205). No internal iteration is performed.

RCK1 same as **RCK0** but with internal iterations up to convergence in equation (2.179).

SPR1 same as **SPR0** but with internal iterations following the procedure in equation (2.179).

When only e–e collisions are considered, the calculations have been performed varying the time step dt ($10^{-15} - 10^{-11}$ s), the tolerance $\epsilon_{\max}$ for the energy conservation ($10^{-10} - 10^{-3}$), the number of energy groups N_ε (200–2000), keeping constant the energy domain ($\varepsilon_{\max} = N_\varepsilon \Delta\varepsilon = 20$ eV) and the initial EEDF, considered as a step functions with a variable width ε_Δ as

$$n(\varepsilon, t = 0) = \begin{cases} \dfrac{n_e}{\varepsilon_\Delta} & \text{if} \quad \varepsilon \leqslant \varepsilon_\Delta \\ 0 & \text{if} \quad \varepsilon > \varepsilon_\Delta \end{cases} \tag{2.208}$$

where $K_\Delta = \frac{\varepsilon_\Delta}{N_\varepsilon \Delta\varepsilon}$ assumes one of the values 0.1, 0.25, 0.5 0.75, 0.9. Different initial EEDFs correspond to different mean electron energy related to ε_Δ by

$$\bar{\varepsilon}_e = \frac{1}{n_e}\sum_{i=1}^{N_\varepsilon} n_i \varepsilon_i \Delta\varepsilon = \frac{1}{n_e}\sum_{i=1}^{K_\Delta} \frac{n_e}{\varepsilon_\Delta}\Delta\varepsilon(i - 0.5)\Delta\varepsilon = \frac{\varepsilon_{\max} K_\Delta}{2} \tag{2.209}$$

When only e–e collisions are considered, the stationary solution is a Maxwellian distribution function whose temperature is related to the initial mean energy of the EEDF. Simulations have been performed in order to verify that the numerical scheme proposed is much more efficient than standard numerical techniques and energy conservation in e–e collisions can be fulfilled numerically under the required tolerance. To evaluate the efficiency of the sparse matrix algorithm with respect to Rockwood we report the ratio

$$\tau_{a,\, b} = \frac{\text{CPU time for } \mathbf{RCKa}}{\text{CPU time for } \mathbf{SPRb}} \tag{2.210}$$

Figure 2.4 shows the ratio $\tau_{0,0}$ of **RCK0** and **SPR0** algorithms as a function of the number of energy groups N_ε. The ratio τ has been evaluated when the EEDF reaches the stationary (Maxwellian) distribution The simulations, performed with $dt = 10^{-12}$ s and $K_\Delta = 0.75$, show that for **SPR0** the efficiency grows almost linearly with the mesh size, as expected, because the time needed to calculate the $\hat{T}$ matrix with **RCK0** is proportional to N_ε^2 while with the sparse matrix algorithm it is proportional to N_ε.

In figure 2.5 the relative percentage errors of the electron mean energy for the stationary solutions **RCK0** and **SPR0** algorithms are represented for different time steps with $N_\varepsilon = 200$ and $K_\Delta = 0.1$. The relative errors can reach up to 16% for higher time steps. Even if the stationary solutions are Maxwellian, as expected, the

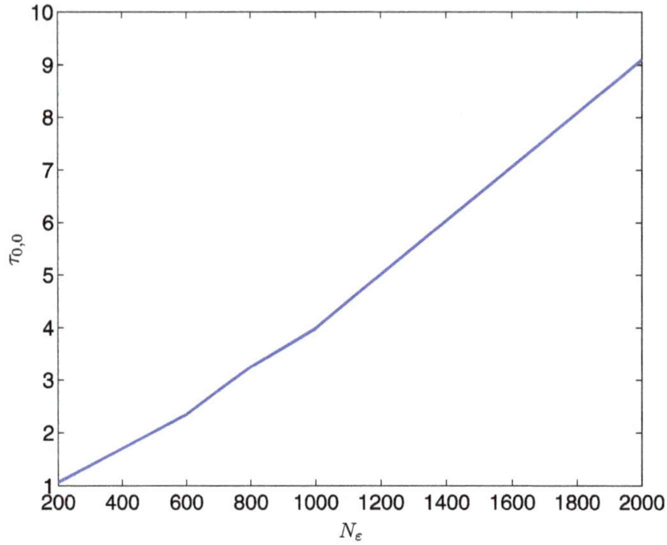

Figure 2.4. Ratio between the CPU times of **RCK0** and **SPR0** algorithms as a function of the number of energy groups N_ε for $dt = 10^{-12}$ s and $K_\Delta = 0.75$. Reproduced from reference [58]. Copyright 2010 with permission from Elsevier.

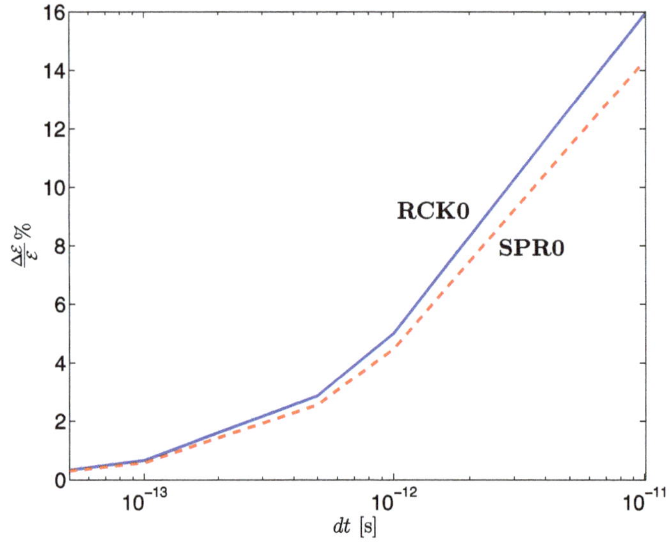

Figure 2.5. Relative percentage errors of the energy density of the stationary distributions calculated by **RCK0** and **SPR0** algorithms as a function of the time step, for $N_\varepsilon = 200$ and $K_\Delta = 0.1$. Reproduced from reference [58]. Copyright 2010 with permission from Elsevier.

calculated electron temperatures do not respect energy conservation. **SPR0** algorithm gives smaller error than **RCK0**, a hint that the former should manifest a little better convergence properties. The difference is so small that we can safely assume that $\tau_{1,1} \approx \tau_{0,0}$.

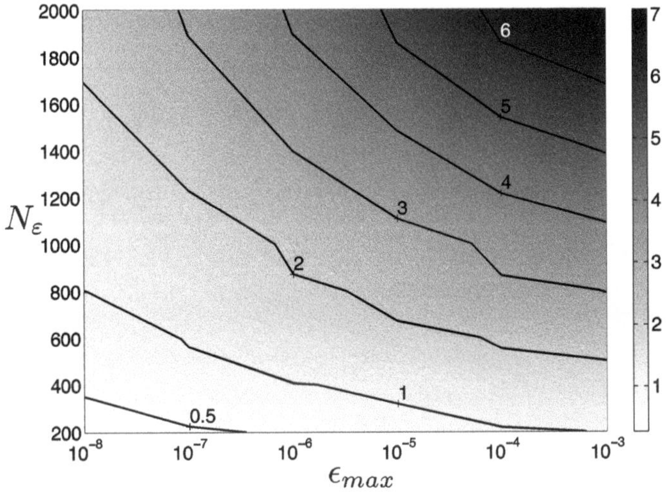

Figure 2.6. Ratio $\tau_{0,1}$ with $dt = 10^{-12}$ s and $K_\Delta = 0.25$. The ratio is evaluated when stationary solutions are reached after $3 \cdot 10^{-10}$ s. Reproduced from reference [58]. Copyright 2010 with permission from Elsevier.

Sparse matrix approach in evaluating the e–e contribution, when energy density conservation in e–e collision is required (**SPR1**) can be even faster than the traditional Rockwood algorithm (**RCK0**), as can be observed in figure 2.6 . The CPU time needed by the **RCK0** algorithm to reach the stationary Maxwellian distribution can be up to six times higher than **SPR1** in the case of large matrices and small tolerance. The sparse matrix procedure has been tested [58] also in a self-consistent code (see chaptes 5 and 9) for a helium plasma. The new algorithm has been proved to be very efficient and accurate limiting the computational time dedicated to the e–e calculation to a small percentage of the total CPU time.

2.9 Appendix: angle integrals

We will demonstrate that functions of the type

$$\Phi(\vec{\varphi}) = \begin{cases} \varphi_i^{\,p} & \text{type (a)} \\ \varphi_i^{\,p}\varphi_{j\neq i} & \text{type (b)} \end{cases} \tag{2.211}$$

can be used in the derivation of the two-term Boltzmann equation, leading to the expressions in equations (2.5) and 2.6. Let us recall that we are considering functions of the type $F(\vec{v}) = G(v)\Phi(\vec{\varphi})$ (see equation (2.4))

The three director cosines are not independent and, to be rigorous, they must be expressed in terms of polar coordinates (ϑ, φ) as

$$\begin{aligned} \varphi_x &= \sin\vartheta\cos\varphi \\ \varphi_y &= \sin\vartheta\sin\varphi \\ \varphi_z &= \cos\vartheta \end{aligned} \tag{2.212}$$

giving for the general expression of the integrals in angle coordinates (see equation (2.5))

$$\iiint_{-\infty}^{\infty} F(\vec{v})d^3v = \frac{\pi}{2}\int_0^{\infty} v^2 dv G(v) \int_{-1}^{1}\int_0^{2\pi} \Phi(\vec{\varphi})d\cos\vartheta d\varphi \qquad (2.213)$$

Logically, the value of integral is obviously independent on which φ_i is considered, because the choice of the reference axis z is arbitrary. Therefore, for sake of simplicity, we will demonstrate only the case for φ_z, leaving to the reader the verification in the other components.

2.9.1 Type (a)

$$\int_{-1}^{1}\int_0^{2\pi} \varphi_z^{a-1} d\cos\vartheta d\varphi \overset{\varphi_z=\cos\vartheta}{\equiv} \int_{-1}^{1} \varphi_z^{a-1} d\varphi_z \int_0^{2\pi} d\varphi = 2\pi\frac{1^a - (-1)^a}{a} = \frac{4\pi}{a}a \Big| 2 \qquad (2.214)$$

2.9.2 Type (b)

$$\int_{-1}^{1}\int_0^{2\pi} \varphi_z^a \varphi_{x/y} d\cos\vartheta d\varphi \overset{\text{eq. 2.212}}{\equiv} \int_0^{\pi} \cos^a\vartheta \sin^2\vartheta d\vartheta \int_0^{2\pi} f_{x/y}(\varphi)d\varphi = 0 \qquad (2.215)$$

because the integral over φ is null being the function $f_{x/y}(\varphi)$ equal to sin or cos (see equation (2.212)).

References

[1] Capriati G, Colonna G, Gorse C and Capitelli M 1992 A parametric study of electron energy distribution functions and rate and transport coefficients in nonequilibrium helium plasmas *Plasma Chem. Plasma Process.* **12** 237–60

[2] Yachi S, Date H, Kitamori K and Tagashira H 1991 A multi-term Boltzmann equation analysis of electron swarms in gases-the time-of-flight parameters *J. Phys. D: Appl. Phys.* **24** 573

[3] Yachi S, Kitamura Y, Kitamori K and Tagashira H 1988 A multi-term Boltzmann equation analysis of electron swarms in gases *J. Phys. D: Appl. Phys* **21** 914

[4] Ness K 1994 Multi-term solution of the Boltzmann equation for electron swarms in crossed electric and magnetic fields *J. Phys. D: Appl. Phys* **27** 1848

[5] Maeda K and Makabe T 1994 Radiofrequency electron swarm transport in reactive gases and plasmas *Phys. Scr.* **1994** 61

[6] Loffhagen D, Winkler R and Braglia G 1996 Two-Term and multi-term approximation of the nonstationary electron velocity distribution in an electric field in a gas *Plasma Chem. Plasma Process.* **16** 287–300

[7] Leyh H, Loffhagen D and Winkler R 1998 A new multi-term solution technique for the electron Boltzmann equation of weakly ionized steady-state plasmas *Comput. Phys. Commun.* **113** 33–48

[8] Braglia G, Wilhelm J and Winkler E 1985 Multi-term solutions of Boltzmann's equation for electrons in the real gases Ar, CH_4 and CO_2 *Lett. Nuovo Cimento* **44** 365–78

[9] Dujko S, White R, Petrović Z L and Robson R 2011 A multi-term solution of the nonconservative Boltzmann equation for the analysis of temporal and spatial non-local effects in charged-particle swarms in electric and magnetic fields *Plasma Sources Sci. Technol.* **20** 024013

[10] Colonna G and Capitelli M 1996 Electron and vibrational kinetics in the boundary layer of hypersonic flow *J. Thermophys. Heat Transfer* **10** 406–12

[11] Colonna G, Capitelli M, DeBenedictis S, Gorse C and Paniccia F 1991 Electron energy distribution functions in CO_2 laser mixture: the effects of second kind collisions from metastable electronic states *Contrib. Plasma Phys.* **31** 575–9

[12] Capitelli M, Colonna G, D'Ammando G, Laporta V and Laricchiuta A 2013 The role of electron scattering with vibrationally excited nitrogen molecules on non-equilibrium plasma kinetics *Phys. Plasmas* **20** 101609

[13] Capitelli M, Colonna G, D'Ammando G, Laporta V and Laricchiuta A 2014 Nonequilibrium dissociation mechanisms in low temperature nitrogen and carbon monoxide plasmas *Chem. Phys.* **438** 31–6

[14] Colonna G, D'Ammando G, Pietanza L and Capitelli M 2014 Excited-state kinetics and radiation transport in low-temperature plasmas *Plasma Phys. Control. Fusion* **57** 014009

[15] Pietanza L, Colonna G, D'Ammando G, Laricchiuta A and Capitelli M 2015 Vibrational excitation and dissociation mechanisms of CO_2 under non-equilibrium discharge and post-discharge conditions *Plasma Sources Sci. Technol.* **24** 042002

[16] Pietanza L D, Colonna G, Laporta V, Celiberto R, D'Ammando G, Laricchiuta A and Capitelli M 2016 Influence of electron molecule resonant vibrational collisions over the symmetric mode and direct excitation-dissociation cross sections of CO_2 on the electron energy distribution function and dissociation mechanisms in cold pure CO_2 plasmas *J. Phys. Chem.* A **120** 2614–28

[17] Colonna G, Pietanza L and D'Ammando G 2012 Self-consistent collisional-radiative model for hydrogen atoms: atom-atom interaction and radiation transport *Chem. Phys.* **398** 37–45

[18] Capitelli M, Celiberto R, Colonna G, Esposito F, Gorse C, Hassouni K, Laricchiuta A and Longo S 2015 *Fundamental Aspects of Plasma Chemical Physics: Kinetics* (Springer Series on Atomic, Optical, and Plasma Physics (SSAOPP, vol 85) (Berlin: Springer) https://doi.org/10.1007/978-1-4419-8185-1

[19] Colonna G, Laporta V, Celiberto R, Capitelli M and Tennyson J 2015 Non-equilibrium vibrational and electron energy distributions functions in atmospheric nitrogen ns pulsed discharges and μs post-discharges: the role of electron molecule vibrational excitation scaling-laws *Plasma Sources Sci. Technol.* **24** 035004

[20] Colonna G, Gorse C, Capitelli M, Winkler R and Wilhelm J 1993 The influence of electron-electron collisions on electron energy distribution functions in N_2 post discharge *Chem. Phys. Lett.* **213** 5–9

[21] Capitelli M, Colonna G, Gicquel A, Gorse C, Hassouni K and Longo S 1996 Maxwell and non-Maxwell behavior of electron energy distribution function under expanding plasma jet conditions: the role of electron-electron, electron-ion, and superelastic electronic collisions under stationary and time-dependent conditions *Phys. Rev.* E **54** 1843

[22] Pietanza L, Colonna G, D'Ammando G, Laricchiuta A and Capitelli M 2016 Electron energy distribution functions and fractional power transfer in "cold" and excited CO_2 discharge and post discharge conditions *Phys. Plasmas* **23** 013515

[23] Kortshagen U, Parker G and Lawler J 1996 Comparison of Monte Carlo simulations and nonlocal calculations of the electron distribution function in a positive column plasma *Phys. Rev.* E **54** 6746

[24] Kortshagen U 1993 A non-local kinetic model applied to microwave produced plasmas in cylindrical geometry *J. Phys. D: Appl. Phys.* **26** 1691

[25] Busch C and Kortshagen U 1995 Numerical solution of the spatially inhomogeneous Boltzmann equation and verification of the nonlocal approach for an argon plasma *Phys. Rev.* E **51** 280

[26] Uhrlandt D and Winkler R 1996 Radially inhomogeneous electron kinetics in the DC column plasma *J. Phys. D: Appl. Phys.* **29** 115

[27] Tsendin L 1995 Electron kinetics in non-uniform glow discharge plasmas *Plasma Sources Sci. Technol.* **4** 200

[28] Colonna G and Capitelli M 2008 Boltzmann and master equations for MHD in weakly ionized gases *J. Thermophys. Heat Transfer* **22** 414–23

[29] Colonna G and Capitelli M 2003 The effects of electric and magnetic fields on high enthalpy plasma flows *34th AIAA Plasmadynamics and Lasers Conference (Orlando, FL, May 2003)* AIAA 2003-4036 https://doi.org/10.2514/6.2003-4036

[30] Frost L and Phelps A 1964 Momentum-transfer cross sections for slow electrons in He, Ar, Kr, and Xe from transport coefficients *Phys. Rev.* **136** A1538

[31] Frost L and Phelps A 1962 Rotational excitation and momentum transfer cross sections for electrons in H_2 and N_2 from transport coefficients *Phys. Rev.* **127** 1621

[32] Engelhardt A and Phelps A 1963 Elastic and inelastic collision cross sections in hydrogen and deuterium from transport coefficients *Phys. Rev.* **131** 2115

[33] Lowke J J, Phelps A V and Irwin B W 1973 Predicted electron transport coefficients and operating characteristics of CO_2-N_2-He laser mixtures *J. Appl. Phys.* **44** 4664–71

[34] Pitchford L and Phelps A 1982 Comparative calculations of electron-swarm properties in N_2 at moderate E/N values *Phys. Rev.* A **25** 540

[35] Phelps A V and Pitchford L C 1985 Anisotropic scattering of electrons by N_2 and its effects on electron transport: tabulation of cross sections and results, JILA report 23, JILA, Univ. of Colorado, Boulder, CO

[36] Phelps A 1991 Cross sections and swarm coefficients for nitrogen ions and neutrals in N_2 and argon ions and neutrals in Ar for energies from 0.1 eV to 10 keV *J. Phys. Chem. Ref. Data* **20** 557–73

[37] Morgan W L 1993 Test of a numerical optimization algorithm for obtaining cross sections for multiple collision processes from electron swarm data *J. Phys. D: Appl. Phys.* **26** 209

[38] Morgan W 1992 A critical evaluation of low-energy electron impact cross sections for plasma processing modeling. I: Cl_2, F_2, and HCl *Plasma Chem. Plasma Process.* **12** 449–76

[39] Morgan W L 1991 Use of numerical optimization algorithms to obtain cross sections from electron swarm data *Phys. Rev.* A **44** 1677

[40] Petrović Z L, Dujko S, Marić D, Malović G, Nikitović Ž, Šaši O, Jovanović J, Stojanović V and Radmilović-Rađenović M 2009 Measurement and interpretation of swarm parameters and their application in plasma modelling *J. Phys. D: Appl. Phys.* **42** 194002

[41] Hagelaar G and Pitchford L 2005 Solving the Boltzmann equation to obtain electron transport coefficients and rate coefficients for fluid models *Plasma Sources Sci. Technol.* **14** 722
[42] LXcat 2016 Plasma Data Exchange Project database (http://fr.lxcat.net/home/)
[43] Mitchner M and Kruger C H J 1973 *Partially Ionized Gases* (New York: Wiley)
[44] Golant V E, Zilinskij A P and Sacharov I E 1980 *Fundamentals of Plasma Physics* (New York: Wiley)
[45] Rockwood S D 1973 Elastic and inelastic cross sections for electron-Hg scattering from Hg transport data *Phys. Rev.* A **8** 2348–60
[46] Loureiro J and Amorim J 2016 *Kinetics and Spectroscopy of Low Temperature Plasmas* (Berlin: Springer) https://doi.org/10.1007/978-3-319-09253-9
[47] Landau L D and Lifschitz E 1981 *Physical Kinetics* (Oxford: Pergamon)
[48] Celiberto R *et al* 2016 Atomic and molecular data for spacecraft re-entry plasmas *Plasma Sources Sci. Technol.* **25** 033004
[49] Gianturco F A and Huo W M (ed) 1995 *Computational Methods for Electron–Molecule Collisions* (Springer Series on Atomic, Optical, and Plasma Physics) (New York: Springer) https://doi.org/10.1007/978-1-4757-9797-8
[50] Starikovskiy A and Aleksandrov N 2016 Plasma assisted combustion mechanism for hydrogen and small hydrocarbons *54th AIAA Aerospace Sciences Meeting* 60 *(San Diego, CA, 4–8 January 2016)* AIAA 2016–0191 https://doi.org/10.2514/6.2016-0191
[51] Elliot C J and Greene A E 1976 Electron energy distribution in e-beam generated Xe and Ar plasmas *J. Appl. Phys.* **47** 2946–53
[52] Laporta V, Celiberto R and Tennyson J 2013 Resonant vibrational-excitation cross sections and rate constants for low-energy electron scattering by molecular oxygen *Plasma Sources Sci. Technol.* **22** 025001
[53] Colonna G 1998 Step adaptive method for vibrational kinetics and other initial value problems *Rend. Circ. Mat. Palermo II Suppl.* **57** 159–63
[54] Kaps P and Rentrop P 1979 Generalized Runge-Kutta methods of order four with stepsize control for stiff ordinary differential equations *Numer. Math.* **33** 55–68
[55] Verwer J G 1994 Gauss-Seidel iteration for stiff ODEs from chemical kinetics *SIAM J. Sci. Comput.* **15** 1243–50
[56] Formaggia L and Scotti A 2011 Positivity and conservation properties of some integration schemes for mass action kinetics *SIAM J. Numer. Anal.* **49** 1267–88
[57] Martiradonna A, Colonna G and Diele F 2020 GeCo: Geometric Conservative nonstandard schemes for biochemical systems *Appl. Numer. Math.* **155** 38–57
[58] D'Angola A, Coppa G, Capitelli M, Gorse C and Colonna G 2010 An efficient energy-conserving numerical model for the electron energy distribution function in the presence of electron-electron collisions *Comput. Phys. Commun.* **181** 1204–11

IOP Publishing

Plasma Modeling (Second Edition)

Methods and applications

Gianpiero Colonna and Antonio D'Angola

Chapter 3

Multiterm and non-local electron Boltzmann equation

Detlef Loffhagen

3.1 Introduction

The electron component plays an essential role for the physical understanding of the temporal and spatial behaviour of weakly ionized plasmas, because the electrons supply energetically the other species of the plasmas. The power supply to the plasma usually takes place by an external electric field which acts on the charged particles and thus maintains the plasma. The power coupled into the plasma is subsequently transferred to the heavy particles by elastic and inelastic electron collision processes. These collision processes lead to considerable changes of the concentrations of the various heavy particle components of the plasma and to the generation of different kinds of excited and charged particles.

Because the mass of the electrons is small compared to that of the heavy particles, the energy exchange between electrons and heavy particles due to elastic collisions is much smaller than the energy exchange between heavy particles with approximately equal masses. This results in a weak energetic contact of the electrons with the heavy particle components of the plasma. Acted upon by the electric field, the electrons are consequently able to attain values of mean kinetic energy which are much larger than the mean energies of the neutral particles and ions of the plasma so that a nonthermal plasma exists.

As a result of the interaction of power gain and power loss by collisions, the electrons are generally not in thermodynamic equilibrium among each other and with the other components of the plasma. This non-equilibrium behaviour of the electron component in nonthermal plasmas cannot be described by the known methods of equilibrium thermodynamics, but it requires an appropriate micro-physical treatment to determine the established non-equilibrium state of the electrons or to describe their temporal and spatial development in nonthermal plasmas [1, 2].

In order to perform microphysical investigations of the behaviour of the electrons, there are basically two quite different approaches with specific advantages and disadvantages available: the solution of the Boltzmann equation of the electrons and particle simulation techniques [3]. The Boltzmann equation method consists of the formulation of an appropriate kinetic equation of the electrons and its solution. Here, the approach for solving the resulting kinetic equation is quite different for investigations of local and non-local plasmas, where the non-locality can take place in time and space. Despite the larger mathematical and numerical effort, the strict kinetic treatment of the electrons using efficient numerical solution techniques became possible for steady-state, time-dependent, spatially one- and two-dimensional as well as time- and space-dependent plasma conditions [4–6].

In contrast to the solution of the electron Boltzmann equation, particle simulation methods are characterized by their simpler applicability to more complex geometries. However, they are generally more time-consuming, because a larger number of electrons has to be treated in any simulation to get sufficient accuracy for the determination of their distribution function in the entire range of kinetic energies and especially at higher energies. The treatment of the kinetic equation of the electrons usually does not have such limitation and is generally characterized by less expenditure of computing time.

The present contribution focuses on methods for solving the electron Boltzmann equation, where the special focus is on techniques to determine the electron velocity distribution function (EVDF) and related macroscopic properties in higher-order accuracy. Such so-called multiterm treatment includes in its lowest order the solution of the electron kinetic equation in the well-known two-term approximation of the EVDF [4–7], which is represented in chapter 2. Notice that the solution of the electron Boltzmann equation has not only been applied to analyse specific problems related to the kinetic behaviour of the electrons. It has also been integrated into a self-consistent description of different nonthermal plasmas by means of hybrid modelling [8–12], where the two-term approximation of the EVDF was used at most.

3.2 Basic relations

The kinetic description of the electron component starts from velocity distribution function $f_e(\vec{r}, \vec{v}, t)$ of the electrons, where $\mathrm{d}n_e(\vec{r}, t) = f_e(\vec{r}, \vec{v}, t)\mathrm{d}\vec{v}$ represents the number of electrons in the velocity space interval $[\vec{v}, \vec{v} + \mathrm{d}\vec{v}]$ with $\mathrm{d}\vec{v} \equiv \mathrm{d}^3v$ at the coordinate space position $\vec{r}$ and at time t. If this velocity distribution $f_e(\vec{r}, \vec{v}, t)$ is known, all relevant macroscopic properties of the electrons, as for instance the electron density $n_e(\vec{r}, t)$, their particle flux $\vec{j}_e(\vec{r}, t)$ and the mean electron energy density $u_e(\vec{r}, t)$, can be determined by appropriate averaging of the EVDF over the velocity space. These three macroscopic properties mentioned are given by the averages [13–15]

$$n_e(\vec{r}, t) = \int f_e(\vec{r}, \vec{v}, t)\mathrm{d}\vec{v}, \tag{3.1}$$

$$\vec{j}_e(\vec{r}, t) = \int \vec{v} f_e(\vec{r}, \vec{v}, t) d\vec{v}, \tag{3.2}$$

$$u_e(\vec{r}, t) = \int \frac{1}{2} m_e v^2 f_e(\vec{r}, \vec{v}, t) d\vec{v}, \tag{3.3}$$

where m_e denotes the electron mass and $v = |\vec{v}|$.

3.2.1 Boltzmann equation of the electrons

The electrons in weakly ionized plasmas are subject to different interactions such as the action of noncollisional forces and different binary collision processes. These interactions cause that the electrons are redistributed in the six-dimensional phase space consisting of the configuration and velocity spaces. In order to describe this complex redistribution of the electrons, an appropriate balance equation for the velocity distribution function $f_e(\vec{r}, \vec{v}, t)$ has to be formulated and solved. According to the concept of binary electron collisions with short-range interaction, this balance equation is the Boltzmann equation [13–15]

$$\frac{\partial f_e}{\partial t} + \vec{v} \cdot \nabla_{\vec{r}} f_e + \vec{a} \cdot \nabla_{\vec{v}} f_e = \sum_i \left(C_i^{\text{el}}(f_e) + \sum_p C_{i,p}^{\text{in}}(f_e) \right). \tag{3.4}$$

Here $\vec{a}$ represents the acceleration due to noncollisional forces like electromagnetic and gravitational forces. Focusing on the action of an electric field $\vec{E}$ in the following, the acceleration is given by

$$\vec{a} = -\frac{e_0}{m_e} \vec{E}, \tag{3.5}$$

where $-e_0$ denotes the charge of the electron. The right-hand side of equation (3.4) describes the change of the EVDF due to binary collisions of electrons with the neutral heavy particle components of the plasma, i.e. atoms and molecules in their ground state as well as in excited states. $C_i^{\text{el}}(f_e)$ and $C_{i,p}^{\text{in}}(f_e)$ are the collision integrals for elastic (el) and various inelastic (in) collisions of electrons with the heavy particle component i. In addition to elastic collisions, the kinetic treatment of the electrons has to include all inelastic electron collision processes, which are important for the plasma under consideration, such as electron impact excitation and de-excitation, dissociation, ionization and attachment.

In the derivation of the kinetic equation (3.4), collisions of electrons with heavy particle species j are assumed to be instantaneous events taking place at a certain point in space and being not affected by any external forces. Thus, the properties of the colliding particles before and after the collision event can be expressed in terms of the momentum and energy conservation laws without considering the specific type of the interaction forces. In the case of binary electron collisions in which the nature of the heavy particle species i with mass M_i remains unchanged, the conservation of momentum and energy in each binary collision is given by the relations [14, 15]

$$m_e\vec{v}_e + M_i\vec{V}_i = m_e\vec{v}_e' + M_i\vec{V}_i', \tag{3.6}$$

$$\frac{m_e v_e^2}{2} + \frac{M_i V_i^2}{2} = \frac{m_e v_e'^2}{2} + \frac{M_i V_i'^2}{2} + \Delta E_{\text{int}}, \tag{3.7}$$

where $\vec{v}_e$ and $\vec{v}_e'$ are the velocity vectors of the electron before and after the collision event, $\vec{V}_i$ and $\vec{V}_i'$ denote the corresponding velocity vectors of the heavy particle, v_e, v_e', V_i and V_i' are the respective absolute values of the velocities and $\Delta E_{\text{int}} = E_{\text{int}}' - E_{\text{int}}$ is the net change in the internal energy of the heavy particle as a result of the collision process. For elastic collisions, ΔE_{int} is obviously equal to zero because they do not influence the internal state of the heavy particle. Among the various inelastic collision processes, collisions of first kind with $\Delta E_{\text{int}} > 0$ such as exciting collisions are distinguished from collisions of second kind with $\Delta E_{\text{int}} < 0$. Collisions causing atoms or molecules to shift from an excited to the ground state are an example of the latter.

The change of momentum and kinetic energy of the electrons in binary collisions with heavy particles can be deduced from the conservation equations (3.6) and (3.7). The momentum change [15]

$$m_e\vec{v}_e' - m_e\vec{v}_e = -\frac{m_e M_i}{m_e + M_i}(1 - \cos\theta)(\vec{v}_e - \vec{V}_i) \approx -m_e\vec{v}_e(1 - \cos\theta) \tag{3.8}$$

is proportional to the relative velocity of the colliding particles, where θ is the scattering angle related to the collision. Because the heavy particle velocity is usually much less than the electron velocity and $m_e \ll M_i$, it can be approximated by the right-hand-side expression of (3.8). The corresponding change of kinetic energy due to elastic collisions is given by [15]

$$\frac{m_e v_e'^2}{2} - \frac{m_e v_e^2}{2} = -\left(\frac{m_e v_e^2}{2} - \frac{M_i V_i^2}{2}\right) 2\frac{m_e}{M_i}(1 - \cos\theta) \tag{3.9}$$

and that resulting from inelastic collisions takes the form

$$\frac{m_e v_e'^2}{2} - \frac{m_e v_e^2}{2} = -\Delta E_{\text{int}}. \tag{3.10}$$

Similar relations can directly be derived for binary electron collisions in which the nature of the heavy particle changes such as electron impact dissociation, ionization or electron attachment.

3.2.2 Expansion of the velocity distribution

The electron Boltzmann equation (3.4) is very complex and comprises a wide range of specific problems of the kinetics of the electrons. Because a general solution of this kinetic equation can hardly be found, various solution methods and numerical techniques have been developed and utilized to analyse different conditions of plasmas such as steady-state conditions, solely time-dependent or space-dependent plasmas and even time- and space-dependent problems [4, 5, 7, 16].

In order to find the solution of the kinetic equation, a decomposition of the velocity distribution function in terms of orthogonal functions in the directions $\vec{v}/v$ of the velocity $\vec{v}$ is generally applied. Spherical harmonics are such functions [14]. Whether such complex expansion with respect to both the angular coordinates of $\vec{v}/v$ is required or a reduced expansion concerning only one angular coordinate of the velocity can be used, that depends on the structure of the electric field, the anticipated inhomogeneity of the plasma under consideration and the boundary conditions imposed on the velocity distribution [17–21]. Here, it is generally assumed that the electric field and additionally a potential inhomogeneity in the plasma are parallel to one preferred direction in space only, taken to be the z direction. For aspects related to the multiterm treatment of the electrons in cylindrical and spherical geometry, the reader is referred to e.g. [19–21].

Because the electric field and the inhomogeneity in the plasma are parallel to the z axis of a given coordinate system, the EVDF $f_e(\vec{r}, \vec{v}, t)$ becomes symmetrical around the field $\vec{E}(z, t) = E(z, t)\vec{e}_z$ with $\vec{e}_z$ being the unit vector in z direction. Thus, it possesses the reduced dependence $f_e(z, v, v_z/v, t)$, where v_z is the z component of $\vec{v}$, and can be expanded with respect to $v_z/v \equiv \cos\theta$ in Legendre polynomials $P_n(v_z/v)$ according to [14, 15]

$$f_e\left(z, v, \frac{v_z}{v}, t\right) = \sum_{n=0}^{\infty} \hat{f}_n(z, v, t) P_n\left(\frac{v_z}{v}\right). \tag{3.11}$$

The usual procedure to get the velocity distribution function f_e is to substitute the expansion (3.11) into the electron Boltzmann equation (3.4) and to equate coefficients of Legendre polynomials. When employing this expansion and using the orthogonality relation of the Legendre polynomials, a chain of equations for the expansion coefficients $\hat{f}_n$ results. In order to get a closed system of equations and its approximate solution, the infinite set of equations is truncated. When replacing the velocity magnitude v by the kinetic energy $U = m_e v^2/2$ of the electrons as usual, the hierarchy of partial differential equations [17, 22]

$$\begin{aligned}
&\left(\frac{m_e}{2}\right)^{1/2} U^{1/2}\frac{\partial}{\partial t} f_0(z, U, t) + \frac{U}{3}\frac{\partial}{\partial z} f_1(z, U, t) - \frac{e_0}{3} E(z, t)\frac{\partial}{\partial U}\big(U f_1(z, U, t)\big) \\
&\quad - \frac{\partial}{\partial U}\left(\sum_i 2\frac{m_e}{M_i} U^2 N_i(z, t)\big(Q_i^{\mathrm{el},0}(U) - Q_i^{\mathrm{el},1}(U)\big) f_0(z, U, t)\right) \\
&\qquad + \sum_i \sum_p U N_i(z, t) Q_{i,p}^{\mathrm{in},0}(U) f_0(z, U, t) \\
&= \sum_i \sum_p \beta_p^2 \tilde{U}_{i,p} N_i(z, t) Q_{i,p}^{\mathrm{in},0}(\tilde{U}_{i,p}) f_0(z, \tilde{U}_{i,p}, t), \qquad \tilde{U}_{i,p} = \beta_p U + \rho_p U_{i,p}^{\mathrm{in}},
\end{aligned} \tag{3.12}$$

$$\begin{aligned}
&\left(\frac{m_e}{2}\right)^{1/2} U^{1/2}\frac{\partial}{\partial t}f_n(z, U, t) + \frac{n}{2n-1}U\frac{\partial}{\partial z}f_{n-1}(z, U, t) \\
&+ \frac{n+1}{2n+3}U\frac{\partial}{\partial z}f_{n+1}(z, U, t) \\
&- e_0E(z, t)\frac{n}{2n-1}\left(U\frac{\partial}{\partial U}f_{n-1}(z, U, t) - \frac{n-1}{2}f_{n-1}(z, U, t)\right) \\
&- e_0E(z, t)\frac{n+1}{2n+3}\left(U\frac{\partial}{\partial U}f_{n+1}(z, U, t) + \frac{n+2}{2}f_{n+1}(z, U, t)\right) \\
&+ \sum_i UN_i(z, t)\left(\left(Q_i^{\mathrm{el},0}(U) - Q_i^{\mathrm{el},n}(U)\right) + \sum_p Q_{i,p}^{\mathrm{in},0}(U)\right)f_n(z, U, t) \\
&= \sum_i \sum_p \beta_p^2 \tilde{U}_{i,p} N_i(z, t) Q_{i,p}^{\mathrm{in},n}(\tilde{U}_{i,p}) f_n(z, \tilde{U}_{i,p}, t), \\
&1 \leqslant n \leqslant l-1, \quad l \geqslant 2, \quad f_l(z, U, t) \equiv 0
\end{aligned} \tag{3.13}$$

is obtained in the framework of the multiterm approximation with l coefficients. This hierarchy describes the time- and space-dependent behaviour of the isotropic ($n = 0$) and the anisotropic ($n = 1, \ldots, l-1$) components

$$f_n(z, U, t) = 2\pi\left(\frac{2}{m_e}\right)^{3/2} \hat{f}_n(z, v(U), t) \tag{3.14}$$

of the velocity distribution function $f_e(z, v, v_z/v, t)$.

In order to derive the set of equations (3.12) and (3.13), further assumptions regarding the collision integrals for elastic and inelastic collisions of electrons with the heavy particle species i are required [23, 24]. As usual and sufficient, an additional expansion with respect to the ratio m_e/M_i of electron mass to heavy particle mass is applied and only the leading terms regarding m_e/M_i are considered in each equation of the hierarchy (3.12) and (3.13). That is, the terms of zeroth order in m_e/M_i are taken into account in the collision integrals relevant to inelastic collisions and the first-order terms are included in the collision integral part relevant to elastic collisions in equation (3.12). Furthermore, the heavy gas particles are assumed to be at rest before the collision events.

The terms with the shifted energy arguments $\tilde{U}_{i,p} = \beta_p U + \rho_p U_{i,p}^{\mathrm{in}}$ occurring on the right-hand side of equations (3.12) and (3.13) describe the inscattering of electrons at the kinetic energy U, which have carried out the inelastic collision process p with the heavy particle component i at the energy $\beta_p U + \rho_p U_{i,p}^{\mathrm{in}}$ and are scattered to the energy U as a result of the change of kinetic energy $\rho_p U_{i,p}^{\mathrm{in}}$ in this process with positive value $U_{i,p}^{\mathrm{in}}$. Here, $\rho_p = 1$ for exciting, dissociating and ionizing electron collisions and $\rho_p = -1$ for de-exciting collision processes. Furthermore, $\beta_p = 1$ for the excitation, de-excitation and the dissociation process. Regarding the ionization process, it has been supposed that the remaining kinetic energy of the colliding electron is equally shared between the two outcoming electrons after each ionizing event, i.e. $\beta_p = 2$ [25]. Aspects regarding different assumptions of the energy sharing in ionizing electron collision processes are discussed, e.g. in [26, 27]. Moreover, electrons that are attached to a molecule in the collision event are completely lost so that there is no inscattering term in that case, i.e. $\beta_p = \rho_p = 0$ for electron attachment processes.

The substitution of the Legendre polynomial expansion (3.11) into the electron Boltzmann equation (3.4) also involves the expansion of the collision integrals on the right-hand side of (3.4). As a result of that expansion, the sequences of generalized total collision cross-sections

$$Q_i^{\mathrm{el},n}(U) = \int \sigma_i^{\mathrm{el}}(U, \cos\theta) P_n(\cos\theta) \sin\theta \, \mathrm{d}\theta \mathrm{d}\varphi, \tag{3.15}$$

$$Q_{i,p}^{\mathrm{in},n}(U) = \int \sigma_{i,p}^{\mathrm{in},n}(U, \cos\theta) P_n(\cos\theta) \sin\theta \, \mathrm{d}\theta \mathrm{d}\varphi \tag{3.16}$$

with $n = 0, \ldots, l - 1$ arising in the hierarchy equations (3.12) and (3.13). These total collision cross-sections result from the integration of the product of the differential collision cross-section $\sigma_i^{\mathrm{el}}(U, \cos\theta)$ and $\sigma_{i,p}^{\mathrm{in},n}(U, \cos\theta)$, respectively, and the Legendre polynomial $P_n(\cos\theta)$ over the entire solid scattering angle $\sin\theta \mathrm{d}\theta \mathrm{d}\varphi$. The series (3.15) and (3.16) describe the impact of anisotropic electron-heavy particle scattering due to elastic and the various inelastic collision processes, respectively. In particular, the difference

$$Q_i^{\mathrm{d}}(U) = Q_i^{\mathrm{el},0}(U) - Q_i^{\mathrm{el},1}(U) \tag{3.17}$$

occurring in (3.12) and in (3.13) for $n = 1$ represents the well-known momentum transfer cross-section for elastic collisions. Because the differential collision cross-section data available is quite limited, isotropic scattering is usually assumed for the inelastic collision processes. In that case, the relations $Q_{i,p}^{\mathrm{in},n}(U) = 0$ for $n \geqslant 1$ are found from (3.16) due to the orthogonality of the Legendre polynomials, and the associated terms on the right-hand side of (3.13) are zero as well. With regard to the elastic collision processes, data for the total (or integral) collision cross-section $Q_i^{\mathrm{el},0}$ for elastic collisions as well as for the elastic momentum transfer cross-section Q_i^{d} (3.17) are known for a number of atoms and molecules. Thus, anisotropic scattering in elastic collisions is taken into consideration in the framework of the two-term approximation ($l = 2$) of hierarchy (3.12) and (3.13) when using the data for Q_i^{d}. In multiterm approximation the corresponding differences $Q_i^{\mathrm{el},0}(U) - Q_i^{\mathrm{el},n}(U)$ with $2 \leqslant n \leqslant l - 1$ occurring in (3.13) are often approximated by the cross-section Q_i^{d} for elastic momentum transfer.

The set (3.12) and (3.13) of partial differential equations forms the basis for the studies and numerical methods discussed below.

3.2.3 Macroscopic balances

The equation system (3.12) and (3.13) provides a microscopic description of the behaviour of the electrons. The consistent macroscopic balance equations of the electrons are obtained by appropriate integration of the hierarchy equations (3.12) and (3.13) with $n = 1$ with respect to the kinetic energy U of the electrons. The resulting particle, momentum and power balance equations have the form

$$\frac{\partial}{\partial t}n_{\rm e}(z,\,t) + \frac{\partial}{\partial z}j_z(z,\,t) = \sum_i\sum_p n_{\rm e}(z,\,t)\nu_{i,p}^{\rm io}(z,\,t) - \sum_i\sum_p n_{\rm e}(z,\,t)\nu_{i,p}^{\rm at}(z,\,t), \quad (3.18)$$

$$\frac{\partial}{\partial t}j_z(z,\,t) + \frac{\partial}{\partial z}\Pi_{zz}(z,\,t) = -\frac{e_0}{m_{\rm e}}n_{\rm e}(z,\,t)E(z,\,t) - \sum_i I_i^{\rm el}(z,\,t) - \sum_{i,p} I_{i,p}^{\rm in}(z,\,t), \quad (3.19)$$

$$\frac{\partial}{\partial t}u_{\rm e}(z,\,t) + \frac{\partial}{\partial z}Q_z(z,\,t) = -e_0 j_z(z,\,t)E(z,\,t) - \sum_i P_i^{\rm el}(z,\,t) - \sum_{i,p} P_{i,p}^{\rm in}(z,\,t). \quad (3.20)$$

The particle density $n_{\rm e}(z,\,t)$, the particle flux $\vec{j}_{\rm e}(z,\,t) = j_z(z,\,t)\vec{e}_z$, the zz component of the momentum flux tensor $\Pi_{zz}(z,\,t)$, the mean energy density $u_{\rm e}(z,\,t)$ and the energy flux $\vec{Q}_{\rm e}(z,\,t) = Q_z(z,\,t)\vec{e}_z$ of the electrons are given by the integral expressions

$$n_{\rm e}(z,\,t) = \int_0^\infty U^{1/2}f_0(z,\,U,\,t)\mathrm{d}U, \quad (3.21)$$

$$j_z(z,\,t) = \frac{1}{3}\left(\frac{2}{m_{\rm e}}\right)^{1/2}\int_0^\infty Uf_1(z,\,U,\,t)\mathrm{d}U, \quad (3.22)$$

$$\Pi_{zz}(z,\,t) = \frac{2}{3m_{\rm e}}\int_0^\infty U^{3/2}\left(f_0(z,\,U,\,t) + \frac{2}{5}f_2(z,\,U,\,t)\right)\mathrm{d}U, \quad (3.23)$$

$$u_{\rm e}(z,\,t) = \int_0^\infty U^{3/2}f_0(z,\,U,\,t)\mathrm{d}U, \quad (3.24)$$

$$Q_z(z,\,t) = \frac{1}{3}\left(\frac{2}{m_{\rm e}}\right)^{1/2}\int_0^\infty U^2f_1(z,\,U,\,t)\mathrm{d}U. \quad (3.25)$$

Notice that the particle and energy fluxes $\vec{j}_{\rm e}$ and $\vec{Q}_{\rm e}$ have only a z component because the velocity distribution function $f_{\rm e}$ is rotationally symmetric around the direction of $\vec{e}_z$ owing to the assumption that the electric field action and the possible inhomogeneity in the plasma take place parallel to this direction only.

The right-hand-side terms of equations (3.18), (3.19) and (3.20) describe the gain and loss of electrons, momentum and power due to collision processes and the action of the electric field. According to the particle balance (3.18), the temporal variation of the electron density $n_{\rm e}(z,\,t)$ at each z position results from the net particle flux into or out of the respective plasma volume, represented by the divergence of the electron particle flux $j_z(z,\,t)$, as well as from the electron gain due to ionization and electron loss due to attachment. The corresponding mean collision frequencies $\nu_{i,p}^{\rm in}(z,\,t)$ for individual inelastic collision processes such as ionization (io) or attachment (at) are given by the representation

$$\nu_{i,p}^{\text{in}}(z,\,t) = \frac{1}{n_\text{e}(z,\,t)}\left(\frac{2}{m_\text{e}}\right)^{1/2}\int_0^\infty UN_i(z,\,t)Q_{i,p}^{\text{in},0}(U)f_0(z,\,U,\,t)\text{d}U. \tag{3.26}$$

The momentum balance (3.19) states that the temporal change of the particle flux $j_z(z, t)$ is determined at each position by the net momentum flux into or out of the plasma volume considered as described by the spatial derivative of $\Pi_{zz}(z, t)$, the momentum gain from the electric field $I^\text{f}(z, t) = -e_0 n_\text{e}(z, t)E(z, t)/m_\text{e}$ and the momentum dissipation due to elastic and inelastic collisions, where

$$I_i^{\text{el}}(z,\,t) = \frac{2}{3m_\text{e}}\int_0^\infty U^{3/2}N_i(z,\,t)Q_i^{\text{d}}(U)f_1(z,\,U,\,t)\text{d}U, \tag{3.27}$$

$$\begin{aligned} I_{i,p}^{\text{in}}(z,\,t) = \frac{2}{3m_\text{e}}\int_0^\infty U\Big[&U^{1/2}N_i(z,\,t)Q_{i,p}^{\text{in},0}(U) \\ &- \beta_p^{1/2}(U-\rho_p U_{i,p}^{\text{in}})^{1/2}N_i(z,\,t)Q_{i,p}^{\text{in},1}(U)\Big]f_1(z,\,U,\,t)\text{d}U. \end{aligned} \tag{3.28}$$

The power balance (3.20) has a structure similar to the momentum balance. It shows that the temporal change of the mean energy density $u_\text{e}(z, t)$ is caused at the position z by the net power transfer in the macroscopic electron transport, as given by the divergence of the electron energy flux $Q_z(z, t)$, the power gain from the electric field $P^\text{f}(z, t) = -e_0 j_z(z, t)E(z, t)$ and the power loss by elastic and inelastic collision processes, where

$$P_i^{\text{el}}(z,\,t) = 2\frac{m_\text{e}}{M_i}\left(\frac{2}{m_\text{e}}\right)^{1/2}\int_0^\infty U^2 N_i(z,\,t)Q_i^{\text{d}}(U)f_0(z,\,U,\,t)\text{d}U, \tag{3.29}$$

$$P_{i,p}^{\text{in}}(z,\,t) = \rho_p U_{i,p}^{\text{in}} n_\text{e}(z,\,t)\nu_{i,p}^{\text{in}}(z,\,t) \quad \text{for in} \neq \text{at}, \tag{3.30}$$

$$P_{i,p}^{\text{at}}(z,\,t) = \left(\frac{2}{m_\text{e}}\right)^{1/2}\int_0^\infty U^2 N_i(z,\,t)Q_{i,p}^{\text{at},0}(z,\,t)f_0(z,\,U,\,t)\text{d}U. \tag{3.31}$$

It should be mentioned that all macroscopic properties of the electrons of interest are obtained by an energy space averaging over the isotropic distribution f_0 as well as of the first (f_1) and second (f_2) contribution to the distribution anisotropy. This characteristic feature of the macroscopic properties is an immediate consequence of the orthogonality properties of the expansion (3.11) applied to the velocity distribution with respect to the angular component v_z/v.

3.3 Numerical treatment

In this section details about the numerical solution of the coupled set of equations (3.12) and (3.13) and results for selected conditions are given. In order to solve this time- and space-dependent hierarchy of equations for f_n with $n = 0, \ldots, (l-1)$ or the corresponding system of equations related to purely time-dependent, solely space-dependent and even time- and space-dependent problems, respectively, quite different solution and numerical techniques have been developed and employed. These techniques are most commonly based on the finite difference method.

When using the finite difference method to solve a partial differential equation being completed by appropriate boundary conditions and, if necessary, initial values, the solution domain of interest spanned by the independent variables is divided into a discrete set of grid points at first. For instance, there are the space coordinate z, the kinetic energy U and the time t as independent variables in the hierarchy (3.12) and (3.13), and the respective grid spacings are Δz, ΔU and Δt. When using the subscripts j, k and m, the grid points (z_j, U_k, t_m) may have the coordinates $j\Delta z$, $k\Delta U$ and $m\Delta t$. Then, the partial differential equations are discretized at the grid points or at points between the grid point such as the centred points, where the derivatives of f_n are approximated by suitable finite difference expressions including function values at neighbouring grid points and the grid spacings. This procedure leads to a system of algebraic equations for the discretized distributions $f_n^{j,k,m}$ at the grid points, whose values are then determined by means of appropriate numerical algorithms. By making the grid spacings sufficiently small, the result for the discretized distributions $f_n^{j,k,m}$ is expected to be a sufficiently close approximation to the distributions f_n at any grid point.

Independent of the problem considered, it is always aimed to have a fast numerical solution. However, aspects regarding, e.g. the stability of the numerical scheme and the convergence and accuracy of the result obtained have to be kept in mind [28–30]. In order to evaluate the accuracy of the solution of system (3.12) and (3.13), the simultaneous fulfilment of the relevant macroscopic balance equations (3.18), (3.19) and (3.20) can be checked [17] similar to the method of manufactured solution. Therefore, all macroscopic properties involved are calculated according to (3.21)–(3.31) using the numerically obtained distributions f_0, f_1 and f_2. By appropriate choice of the respective grid spacings, all macroscopic balances should be simultaneously satisfied up to a certain numerical error.

In the following section 3.3.1 and 3.3.2 details about the numerical method for a multiterm solution of the respective time-dependent or space-dependent electron kinetic equation resulting from the hierarchy equations (3.12) and (3.13) are represented. The applicability of the respective method is demonstrated by results for the temporal and spatial relaxation of plasma electrons acted upon by a constant electric field. In order to illustrate the main aspects of such relaxation behaviour of the electron component and to discuss the impact of higher-order terms of the Legendre polynomial expansion on the results, weakly ionized plasmas in argon and carbon dioxide are considered as representatives of atomic and molecular gas plasmas. The total collision cross-sections used for the investigations are shown in figure 3.1. The individual collision cross-sections for argon and CO_2 have been taken from Hayashi [31, 32]. In addition to the momentum transfer cross-section for elastic collisions Q_i^{d} according to (3.17) being used for all differences $Q_i^{\mathrm{el},0} - Q_i^{\mathrm{el},n}$ with $2 \leqslant n \leqslant l - 1$ occurring in (3.13), several cross-sections for inelastic collisions are contained in the sets of data, where isotropic scattering has been assumed in all inelastic collision processes. With respect to Ar, 25 electronic excitation (ex) cross-sections with the energy losses 11.55, 11.62, 11.72, 11.83, 12.91, 13.08, 13.09, 13.15, 13.17, 13.27, 13.30, 13.33, 13.48, 13.84, 13.90, 13.98, 14.01,

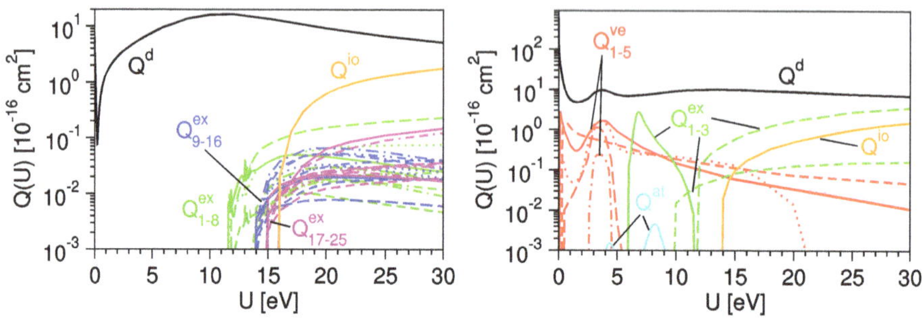

(a) Argon. (b) CO_2.

Figure 3.1. Collision cross-sections for argon and carbon dioxide.

14.06, 14.09, 14.15, 14.21, 14.23, 14.30, 14.71 and 15.20 eV, and the ionization (io) cross-section with the energy loss 15.76 eV have been taken into account. Regarding CO_2, the cross-section set includes five total inelastic collision cross-sections for vibrational excitation (ve), three total cross-sections for electronic excitation (ex) as well as the total cross-sections for electron impact ionization (io) and for dissociative electron attachment (at). The corresponding energy losses are 0.083, 0.172, 0.291, 0.36 and 2.5 eV for the vibrational excitation processes, 5.7, 9.0 and 11.0 eV for the electronic excitation processes and 13.7 eV for the electron impact ionization.

3.3.1 Solution method for time-dependent conditions

The microphysical treatment of the kinetics of the electron component of time-dependent plasmas takes place primarily by means of solving the non-stationary, spatially homogeneous Boltzmann equation of the electrons. Starting from the set of equations (3.12) and (3.13), the behaviour of the EVDF in a purely time-dependent plasma results from the hierarchy of partial differential equations

$$
\begin{aligned}
&\left(\frac{m_\mathrm{e}}{2}\right)^{1/2} U^{1/2}\frac{\partial}{\partial t}f_0(U,t) - \frac{e_0}{3}E(t)\frac{\partial}{\partial U}\big(Uf_1(U,t)\big) \\
&- \frac{\partial}{\partial U}\left(\sum_i 2\frac{m_\mathrm{e}}{M_i}U^2 N_i(t) Q_i^{\mathrm{d}}(U) f_0(U,t)\right) + \sum_i\sum_p U N_i(t) Q_{i,p}^{\mathrm{in},0}(U) f_0(U,t) \\
&= \sum_i\sum_p \beta_p^2 \tilde{U}_{i,p} N_i(t) Q_{i,p}^{\mathrm{in},0}(\tilde{U}_{i,p}) f_0(\tilde{U}_{i,p},t), \qquad \tilde{U}_{i,p} = \beta_p U + \rho_p U_{i,p}^{\mathrm{in}}, \\
&\left(\frac{m_\mathrm{e}}{2}\right)^{1/2} U^{1/2}\frac{\partial}{\partial t}f_n(U,t) - e_0E(t)\frac{n}{2n-1}\left(U\frac{\partial}{\partial U}f_{n-1}(U,t) - \frac{n-1}{2}f_{n-1}(U,t)\right) \\
&- e_0E(t)\frac{n+1}{2n+3}\left(U\frac{\partial}{\partial U}f_{n+1}(U,t) + \frac{n+2}{2}f_{n+1}(U,t)\right) \\
&+ \sum_i UN_i(t)\left(\left(Q_i^{\mathrm{el},0}(U) - Q_i^{\mathrm{el},n}(U)\right) + \sum_p Q_{i,p}^{\mathrm{in},0}(U)\right) f_n(U,t) \\
&= \sum_i\sum_p \beta_p^2 \tilde{U}_{i,p} N_i(t) Q_{i,p}^{\mathrm{in},n}(\tilde{U}_{i,p}) f_n(\tilde{U}_{i,p},t),
\end{aligned}
\tag{3.32}
$$

$$1 \leqslant n \leqslant l-1, \quad l \geqslant 2, \quad f_l(U, t) \equiv 0. \tag{3.33}$$

This hierarchy describes the temporal evolution of the isotropic distribution $f_0(U, t)$ and the anisotropic components $f_n(U, t)$ with $n \geqslant 1$ of the velocity distribution function due to the action of a time-dependent electric field and the binary electron-heavy particle collisions. Thus, a system of l partial differential equations of first order including additional terms $f_n(\beta_p U + \rho_p U_{i,p}^{\text{in}}, t)$ with shifted energy arguments has to be solved within the framework of the multiterm approximation.

The system (3.32) and (3.33) of partial differential equations of first order is solved numerically as an initial-boundary value problem in an appropriate energy region $0 \leqslant U \leqslant U_\infty$ and for times $t \geqslant 0$. Here, the time represents the evolution direction of the kinetic problem and U_∞ denotes a sufficiently large energy, which is used as an upper limit of the kinetic energy in the numerical approach. The initial values for all expansion coefficients $f_n(U, t = 0)$ with $n = 0, \dots, l-1$ are chosen according to the kinetic problem under consideration. Possible boundary conditions of the equation system for all times $t \geqslant 0$ are [17, 33, 34]

$$f_n(U = 0, t) = 0 \quad \text{for } n \geqslant 1, \tag{3.34}$$

$$f_n(U \geqslant U_\infty, t) = 0 \quad \text{for } n \geqslant 0. \tag{3.35}$$

These boundary conditions enforce the suppression of singular contributions to the solution of system (3.32) and (3.33).

In order to get the numerical solution, a finite difference approximation for the system (3.32) and (3.33) is applied using the grid point representation

$$\begin{aligned} &U_k = k\Delta U, \quad \Delta U = U_\infty/n_\infty, \quad 0 \leqslant k \leqslant n_\infty, \\ &t_m = m\Delta t, \quad m \geqslant 0, \\ &f_n^{k,m} = f_n(U_k, t_m), \quad 0 \leqslant n \leqslant l-1 \end{aligned} \tag{3.36}$$

of an equidistant energy grid with the energy boundaries $U_0 = 0$ and $U_{n_\infty} = U_\infty$. In accordance with [17], the partial differential equations are discretized at all centred points $(U_{k+1/2}, t_{m+1/2})$ with $0 \leqslant k \leqslant n_\infty - 1$ and $m \geqslant 0$ using the second-order-correct centred difference expressions

$$\begin{aligned} f_n^{k+1/2,m+1/2} &= \frac{1}{4}\left(f_n^{k+1,m+1} + f_n^{k,m+1} + f_n^{k+1,m} + f_n^{k,m}\right), \\ \left(\frac{\partial}{\partial U} f_n(U, t)\right)^{k+1/2,m+1/2} &= \frac{1}{2\Delta U}\left(f_n^{k+1,m+1} - f_n^{k,m+1} + f_n^{k+1,m} - f_n^{k,m}\right), \\ \left(\frac{\partial}{\partial t} f_n(U, t)\right)^{k+1/2,m+1/2} &= \frac{1}{2\Delta t}\left(f_n^{k+1,m+1} - f_n^{k+1,m} + f_n^{k,m+1} - f_n^{k,m}\right) \end{aligned} \tag{3.37}$$

for the functions f_n and their partial derivatives. Furthermore, the second-order-correct expression with respect to time

$$f_n((\tilde{U}_{i,p})_{k+1/2}, t_{m+1/2}) = \frac{1}{2}\left[f_n((\tilde{U}_{i,p})_{k+1/2}, t_{m+1}) + f_n((\tilde{U}_{i,p})_{k+1/2}, t_m)\right] \tag{3.38}$$

is used for the treatment of the terms with shifted energy argument on the right-hand side of (3.32) and (3.33). The corresponding representation of $f_n((\tilde{U}_{i,p})_{k+1/2}, t_\mu)$ for $\mu = m$ and $m+1$ on the energy grid is obtained by means of a parabolic interpolation using three neighbouring energy grid points. For instance, such an interpolation has the representation [17]

$$\begin{aligned} f_n(U_{k+1/2} + U_{i,p}^{\text{in}}, t_\mu) &= \frac{1}{2}\alpha_{i,p}^{(-3)}\alpha_{i,p}^{(-1)} f_n^{k+k_{i,p},\mu} \\ &\quad - \alpha_{i,p}^{(-3)}\alpha_{i,p}^{(+1)} f_n^{k+k_{i,p}+1,\mu} + \frac{1}{2}\alpha_{i,p}^{(-1)}\alpha_{i,p}^{(+1)} f_n^{k+k_{i,p}+2,\mu}, \\ k_{i,p} = \text{int}\left(\frac{U_{i,p}^{\text{in}}}{\Delta U}\right), &\quad \alpha_{i,p}^{(\tau)} = \frac{U_{i,p}^{\text{in}}}{\Delta U} - k_{i,p} + \frac{\tau}{2}, \quad \tau = -3, -1, +1 \end{aligned} \tag{3.39}$$

for electron impact excitation processes, where $\text{int}(x)$ denotes the integer part of x. Such discretization of (3.32) and (3.33) finally leads to the finite difference equation system

$$\begin{aligned} &v_n^{k,m} f_{n-1}^{k-1,m+1} + w_n^{m} f_{n-1}^{k,m+1} + g_n^{k,m} f_n^{k,m+1} + h_n^{k,m} f_n^{k+1,m+1} \\ &+ r_n^{k,m} f_{n+1}^{k,m+1} + s_n^{k,m} f_{n+1}^{k+1,m+1} = d_n^{k,m} + \tilde{d}_n^{k,m+1} \end{aligned} \tag{3.40}$$

for $0 \leqslant n \leqslant l-1$ with $0 \leqslant k \leqslant n_\infty - 1$ and $m \geqslant 0$, where the coefficients are given by

$$\begin{aligned} g_0^{k,m} &= A^k\frac{\Delta U}{\Delta t} - D^{k,m} + F^{k,m}\frac{\Delta U}{2}, \quad h_0^{k,m} = A^k\frac{\Delta U}{\Delta t} + D^{k,m} + F^{k,m}\frac{\Delta U}{2}, \\ g_n^{k,m} &= A^k\frac{\Delta U}{\Delta t} + H^{k,m}\frac{\Delta U}{2}, \quad h_n^{k,m} = g_n^{k,m}, \quad \text{for } 1 \leqslant n \leqslant l-1, \\ r_n^{k,m} &= \frac{n+1}{2n+3}e_0 E(t_{m+1/2})\left(U_{k+1/2} - \frac{n+2}{2}\frac{\Delta U}{2}\right), \quad \text{for } 0 \leqslant n \leqslant l-2, \\ s_n^{k,m} &= -\frac{n+1}{2n+3}e_0 E(t_{m+1/2})\left(U_{k+1/2} + \frac{n+2}{2}\frac{\Delta U}{2}\right), \quad \text{for } 0 \leqslant n \leqslant l-2, \\ v_n^{k,m} &= \frac{n}{2n-1}e_0 E(t_{m+1/2})\left(U_{k+1/2} + \frac{n-1}{2}\frac{\Delta U}{2}\right), \quad \text{for } 1 \leqslant n \leqslant l-1, \\ w_n^{k,m} &= \frac{n}{2n-1}e_0 E(t_{m+1/2})\left(-U_{k+1/2} + \frac{n-1}{2}\frac{\Delta U}{2}\right), \quad \text{for } 1 \leqslant n \leqslant l-1, \\ v_0^{k,m} &= w_0^{k,m} = r_{l-1}^{k,m} = s_{l-1}^{k,m} = 0, \\ d_n^{k,m} &= -v_n^{k,m} f_{n-1}^{k,m} - w_n^{k,m} f_{n-1}^{k+1,m} + x_n^{k,m} f_n^{k,m} + y_n^{k,m} f_n^{k+1,m} \\ &\quad - r_n^{k,m} f_{n+1}^{k,m} - s_n^{k,m} f_{n+1}^{k+1,m} + \Delta U \sum_i \sum_p I_{i,p,n}^{k,m}\left(\frac{1}{2}\alpha_{i,p}^{(-3)}\alpha_{i,p}^{(-1)} f_n^{k+k_{i,p},m}\right. \\ &\quad \left. - \alpha_{i,p}^{(-3)}\alpha_{i,p}^{(+1)} f_n^{k+k_{i,p}+1,m} + \frac{1}{2}\alpha_{i,p}^{(-1)}\alpha_{i,p}^{(+1)} f_n^{k+k_{i,p}+2,m}\right) \\ x_0^{k,m} &= A^k\frac{\Delta U}{\Delta t} + D^{k,m} - F^{k,m}\frac{\Delta U}{2}, \quad y_0^{k,m} = A^k\frac{\Delta U}{\Delta t} - D^{k,m} - F^{k,m}\frac{\Delta U}{2}, \\ x_n^{k,m} &= A^k\frac{\Delta U}{\Delta t} - H^{k,m}\frac{\Delta U}{2}, \quad y_n^{k,m} = x_n^{k,m}, \quad \text{for } 1 \leqslant n \leqslant l-1, \\ \tilde{d}_n^{k,m+1} &= \Delta U \sum_i \sum_p I_{i,p,n}^{k,m}\left(\frac{1}{2}\alpha_{i,p}^{(-3)}\alpha_{i,p}^{(-1)} f_n^{k+k_{i,p},m+1}\right. \\ &\quad \left. - \alpha_{i,p}^{(-3)}\alpha_{i,p}^{(+1)} f_n^{k+k_{i,p}+1,m+1} + \frac{1}{2}\alpha_{i,p}^{(-1)}\alpha_{i,p}^{(+1)} f_n^{k+k_{i,p}+2,m+1}\right) \end{aligned} \tag{3.41}$$

with the further coefficients

$$A^k = \left(\frac{m_e}{2} U_{k+1/2}\right)^{1/2}, \qquad D^{k,m} = -(U_{k+1/2})^2 \sum_i 2\frac{m_e}{M_i} N_i(t_{m+1/2}) Q_i^{\mathrm{d}}(U_{k+1/2}),$$

$$F^{k,m} = -\, U_{k+1/2} \sum_i 2\frac{m_e}{M_i} N_i(t_{m+1/2}) \left(2Q_i^{\mathrm{d}}(U_{k+1/2}) + U_{k+1/2} \frac{Q_i^{\mathrm{d}}(U_{k+1}) - Q_i^{\mathrm{d}}(U_k)}{\Delta U} \right)$$
$$+\, U_{k+1/2} \sum_i \sum_p N_i(t_{m+1/2}) Q_{i,p}^{\mathrm{in},0}(U_{k+1/2}), \tag{3.42}$$

$$H^{k,m} = U_{k+1/2} \sum_i N_i(t_{m+1/2}) \left(Q_i^{\mathrm{el},0}(U_{k+1/2}) - Q_i^{\mathrm{el},n}(U_{k+1/2}) + \sum_p Q_{i,p}^{\mathrm{in},0}(U_{k+1/2}) \right),$$

$$I_{i,p,n}^{k,m} = \beta_p^2 (\tilde{U}_{i,p})_{k+1/2} N_i(t_{m+1/2}) Q_{i,p}^{\mathrm{in},n}((\tilde{U}_{i,p})_{k+1/2}).$$

In combination with the discrete form [17]

$$f_n^{0,\,m} = 0 \quad \text{for all odd } n \geqslant 1, \qquad f_n^{k \geqslant n_\infty,\, m} = 0 \quad \text{for all even } n \geqslant 0 \tag{3.43}$$

of the selected boundary conditions according to (3.34, 3.35) at $m \geqslant 0$, which has proven to be suitable, system (3.40) constitutes a complete system of linear equations for the determination of the function values $f_n^{k,m+1}$ with $0 \leqslant n \leqslant l - 1$ for $0 \leqslant k \leqslant n_\infty$ at the time step $m + 1$, if all corresponding function values $f_n^{k,m}$ at the preceding time step m or their initial values at $m = 0$ are known. This linear system is tridiagonal in the l dependent variables f_n with $0 \leqslant n \leqslant l - 1$, where the diagonal terms of the coefficient matrix are $g_n^{k,m}$ for even $n \geqslant 0$ and $h_n^{k,m}$ for odd $n \geqslant 1$. It is converted to the general band form and solved using the general band algorithm given in [35].

On closer inspection, system (3.40) also shows an additional dependence on function values $f_n^{k,m+1}$ in the formal inhomogeneities $\tilde{d}_n^{k,m+1}$ given in (3.41). In order to avoid the numerically much more expensive solution of a non-tridiagonal system, an iterative treatment of the function values $f_n^{k,m+1}$ in $\tilde{d}_n^{k,m+1}$ with $0 \leqslant n \leqslant l - 1$ is applied while using the general band algorithm in each iteration cycle. At each time step t_{m+1} the iterative solution process is finished as soon as the absolute deviation of the ratios $f_n^{k,m+1,J+1}/f_n^{k,m+1,J}$ of two successive discrete function values obtained in the iteration cycles $J + 1$ and J from unity is less than 10^{-5} for all expansion coefficients, i.e., $0 \leqslant n \leqslant l - 1$, at all relevant energy points.

The multiterm method for solving equation system (3.32) and (3.33) has been applied to quite different temporal relaxation processes. Here, results for the temporal relaxation of electrons in argon and carbon dioxide plasmas under the action of a time-independent electric field are presented using the collision cross-section data shown in figure 3.1. The solution procedure started at $t = 0$ from the Gaussian distribution $G_t(U) = c\exp(-(U - U_c)^2/U_w^2)$ with the centre energy $U_c = 5$ eV and energy width $U_w = 2$ eV as initial value for the isotropic distribution, i.e. $f_0(U, 0)/n_e(0) = G_t(U)$ with the factor c fixed by the normalization condition (3.21). The initial values $f_n(U, 0)$,

$n \geqslant 1$, of the anisotropic components were chosen to be zero. The following relaxation results belong to the gas density $N = 3.54 \times 10^{16}\,\mathrm{cm}^{-3}$ assuming a gas pressure of 1 Torr at 0 °C. The electric field was set to $E = 10.62\,\mathrm{V\,cm}^{-1}$ corresponding to a reduced electric field strength of $E/N = 30$ Td with 1 Td $= 10^{-17}\,\mathrm{V\,cm}^2$. Figure 3.2 shows the temporal relaxation of the isotropic distribution $f_0(U, t)/n_e(t)$ for argon (a) and carbon dioxide (b) calculated by use of the multiterm method with $l = 8$ expansion coefficients, which was found to yield the converged solution of the electron Boltzmann equation during the entire relaxation course. Starting at $t = 0$ from the same isotropic distribution with a maximum around 5 eV, the temporal evolution finally leads to the establishment of time-independent distributions $f_n(U, t)/n_e(t)$ normalized by the electron density $n_e(t)$. The power input from the field leads to an acceleration of the electrons to larger energies and a widening of $f_0(U, t)/n_e(t)$. Because the electron collision cross-sections are very different and cover quite different ranges of kinetic energies (see figure 3.1), the temporal relaxation takes place in different ways in the two gases and leads to very different final distributions.

The corresponding expansion coefficients $f_n(U, t)/n_e(t)$ with $n = 0, 1, 2$ during the early temporal relaxation course at $t = 3$ ns are displayed in figure 3.3. The first two anisotropic components $f_1(U, t)/n_e(t)$ and $f_2(U, t)/n_e(t)$ of the velocity distribution assume positive and negative function values in the kinetic energy range, which are marked by plus and minus signs in parentheses. The opposite signs in the first contribution to the distribution anisotropy $f_1(U, t)/n_e(t)$ are caused by the structure of the isotropic distribution showing one (figure 3.3(a)) or several (figure 3.3(b)) peaks and those in $f_2(U, t)/n_e(t)$ are mainly induced by the structure of $f_1(U, t)/n_e(t)$. The comparison of the magnitude of these distribution anisotropy coefficients with $f_0(U, t)/n_e(t)$ reveals values especially for CO_2 which are of the same order. This indicates that higher-order terms can have a remarkable impact on the EVDF not only at steady state but also during the temporal relaxation process.

In order to illustrate the impact of higher-order terms of the EVDF expansion on selected macroscopic properties, results for the mean electron energy $u_{\mathrm{m}}(t) = u_e(t)/n_e(t)$ and the electron drift velocity $v_{\mathrm{d}}(t) = j_z(t)/n_e(t) = -\mu_e(t)E(t)$

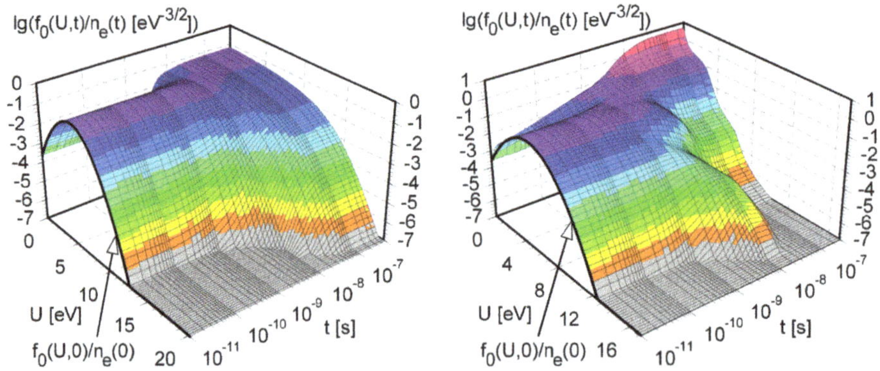

(a) Argon.

(b) CO_2.

Figure 3.2. Temporal relaxation of $f_0(U, t)/n_e(t)$ at $E = 10.62\,\mathrm{V\,cm}^{-1}$.

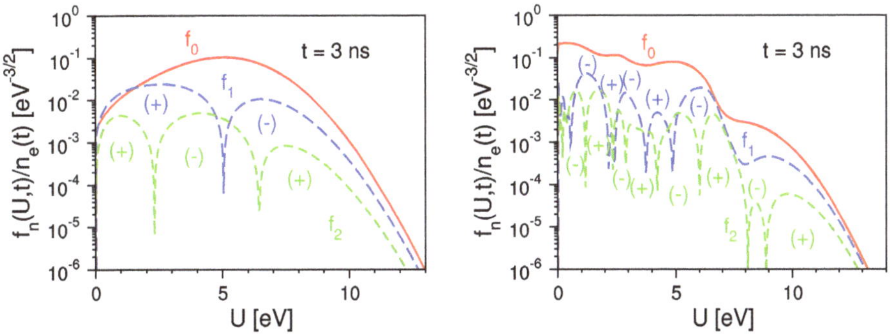

(a) Argon. (b) CO_2.

Figure 3.3. Expansion coefficients $f_n(U, t)/n_e(t)$ with $n = 0, 1, 2$ at $t = 3$ ns.

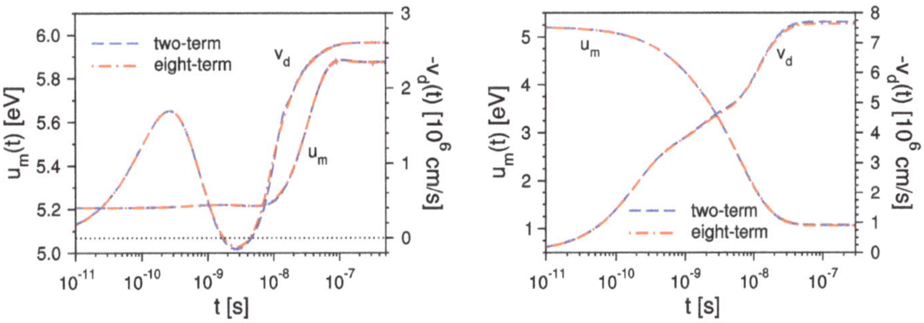

(a) Argon. (b) CO_2.

Figure 3.4. Temporal evolution of the mean energy and drift velocity.

obtained by time-dependent two-term and eight-term approximation are shown in figure 3.4, where μ_e denotes the electron mobility. Certain differences between two-term and converged eight-term results are found for argon during the temporal relaxation course and for carbon dioxide, especially when approaching the time-independent state. Furthermore, the results for argon demonstrate a change of sign of the drift velocity during the temporal evolution around $t = 3$ ns which corresponds to a transient state with negative mobility [36]. This behaviour was found to depend on the initial conditions chosen. In the time-independent state in pure argon the electron mobility assumes a positive value as expected.

Finally, it is to be noted that the adequate method for the estimation of the relaxation time consists in the analysis of the corresponding macroscopic balances resulting from (3.18), (3.19) and (3.20). For time-dependent, spatially homogeneous conditions of plasmas of a single background gas component ($i = 1$), the establishment of time-independent distributions $f_n(U, t)/n_e(t)$, i.e., of a quasi-stationary or hydrodynamic state with constant $[dn_e(t)/dt]/n_e(t)$, can be generally evaluated by considering the gain and loss terms on the right-hand side of the macroscopic balances. Figure 3.5 illustrates the temporal behaviour of the ratios of the particle gain to loss $\nu^{io}(t)/\nu^{at}(t)$, of the momentum gain to loss $I^{f}(t)/[I^{el}(t) + \sum_p I_p^{in}(t)]$ and of

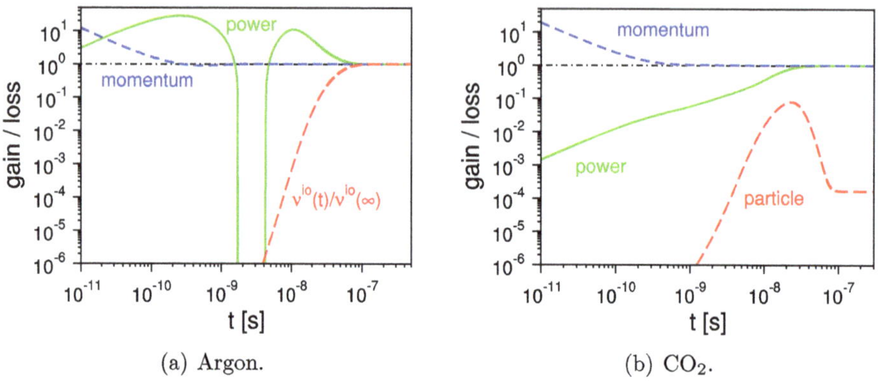

Figure 3.5. Temporal evolution of gain to loss ratios and the ionization frequency.

the power gain to loss $P^{f}(t)/[P^{el}(t) + \sum_p P_p^{in}(t) + P^{at}(t)]$. Because argon does not attach electrons, here the ionization frequency $\nu^{io}(t)/\nu^{io}(\infty)$ normalized by its time-independent value $\nu^{io}(\infty)$ is shown in figure 3.5(a). Quite different temporal evolutions with changes of several orders of magnitude are found for argon and carbon dioxide until finally the time-independent state is established after some 100 ns at $E = 10.62\ \mathrm{V\ cm^{-1}}$ and $N = 3.54 \times 10^{16}\ \mathrm{cm^{-3}}$. This relaxation time is mainly determined by the considerably slower establishment of the particle and power gain and loss compared to the much faster relaxation of the momentum gain to loss ratio towards a constant value within about 1 ns. Notice that for argon (figure 3.5(a)) the ratio of power gain to loss assumes negative values around $t = 3$ ns due to the changes in the electron drift direction (see figure 3.4(a)).

3.3.2 Multiterm solution for space-dependent plasmas

The behaviour of the electron component in space-dependent plasmas is frequently characterized by the fact that the local gain of power and momentum from the electric field cannot be dissipated at the same position by elastic and inelastic collision processes. Starting point for the following kinetic treatment of the electrons in axially inhomogenous plasmas is the stationary, spatially one-dimensional electron Boltzmann equation. Beginning with the hierarchy (3.12) and (3.13), the velocity distribution function of the electrons in such a space-dependent plasma can be determined by solving the system of partial differential equations

$$\begin{aligned} &\frac{1}{3}\left[\frac{\partial}{\partial z}\big(Uf_1(z, U)\big) - e_0E(z)\frac{\partial}{\partial U}\big(Uf_1(z, U)\big)\right] \\ &- \frac{\partial}{\partial U}\left(\sum_i 2\frac{m_e}{M_i}U^2N_i(z)Q_i^{d}(U)f_0(z, U)\right) + \sum_i\sum_p UN_i(z)Q_{i,p}^{in,0}(U)f_0(z, U) \quad (3.44) \\ &= \sum_i\sum_p \beta_p^2\tilde{U}_{i,p}N_i(z)Q_{i,p}^{in,0}(\tilde{U}_{i,p})f_0(z, \tilde{U}_{i,p}), \quad \tilde{U}_{i,p} = \beta_pU + \rho_pU_{i,p}^{in}, \end{aligned}$$

$$
\begin{aligned}
&\frac{n}{2n-1}\left[U\frac{\partial}{\partial z}f_{n-1}(z, U) - e_0E(z)\left(U\frac{\partial}{\partial U}f_{n-1}(z, U) - \frac{n-1}{2}f_{n-1}(z, U)\right)\right]\\
&+\frac{n+1}{2n+3}\left[U\frac{\partial}{\partial z}f_{n+1}(z, U) - e_0E(z)\left(U\frac{\partial}{\partial U}f_{n+1}(z, U) + \frac{n+2}{2}f_{n+1}(z, U)\right)\right]\\
&+\sum_i UN_i(z)\left(\left(Q_i^{\mathrm{el},0}(U) - Q_i^{\mathrm{el},n}(U)\right) + \sum_p Q_{i,p}^{\mathrm{in},0}(U)\right)f_n(z, U)\\
&=\sum_i\sum_p \beta_p^2\tilde{U}_{i,p}N_i(z)Q_{i,p}^{\mathrm{in},n}(\tilde{U}_{i,p})f_n(z, \tilde{U}_{i,p}),\\
&1 \leqslant n \leqslant l-1, \quad l \geqslant 2, \quad f_l(z, U) \equiv 0.
\end{aligned}
\tag{3.45}
$$

This equation system describes the development of the isotropic distribution $f_0(z, U)$ and the contributions $f_n(z, U)$ with $n = 1, \ldots, l-1$ to the anisotropy of the EVDF in the space of the coordinate z and the kinetic energy U of the electrons under the impact of a space-dependent electric field and binary collisions of electrons with heavy particles. Consequently, a system of l first-order partial differential equations comprising additional terms $f_n(z, \beta_p U + \rho_p U_{i,p}^{\mathrm{in}})$ with shifted energy argument $(\beta_p U + \rho_p U_{i,p}^{\mathrm{in}})$ has to be solved within the framework of the multiterm approximation.

The direct numerical treatment of the system (3.44) and (3.45) has proven to be hardly practicable so far. This can already be understood on the basis of an analysis of the two-term ($l = 2$) approximation with isotropic scattering in inelastic collision processes. Then, a partial differential equation of second order for the isotropic distribution $f_0(z, U)$ results from (3.44) and (3.45). This second-order partial differential equation includes cross derivative terms that are known to cause difficulties during the numerical solution.

In order to overcome this problem and to simplify the partial differential equation system (3.44) and (3.45), the kinetic energy U is replaced by the total energy ϵ according to

$$
\epsilon = U + W(z), \tag{3.46}
$$

i.e., by the sum of kinetic energy U and potential energy $W(z)$ of the electrons in the electric field given by

$$
W(z) = e_0\int_0^z E(\tilde{z})\mathrm{d}\tilde{z}. \tag{3.47}
$$

This transformation leads to the system of first-order partial differential equations

$$
\begin{aligned}
&\frac{1}{3}\left(U_{z\epsilon}\frac{\partial}{\partial z}\tilde{f}_1(z, \epsilon) - e_0E(z)\tilde{f}_1(z, \epsilon)\right) - \frac{\partial}{\partial \epsilon}\left(\sum_i 2\frac{m_\mathrm{e}}{M_i}U_{z\epsilon}^2N_i(z)Q_i^{\mathrm{d}}(U_{z\epsilon})\tilde{f}_0(z, \epsilon)\right)\\
&+\sum_i\sum_p U_{z\epsilon}N_i(z)Q_{i,p}^{\mathrm{in},0}(U_{z\epsilon})\tilde{f}_0(z, \epsilon) = \sum_i\sum_p \beta_p^2\hat{U}_{i,p}^{z\epsilon}N_i(z)Q_{i,p}^{\mathrm{in},0}(\hat{U}_{i,p}^{z\epsilon})\tilde{f}_0(z, \hat{\epsilon}_{i,p})\\
&\hat{U}_{i,p}^{z\epsilon} = \beta_p U_{z\epsilon} + \rho_p U_{i,p}^{\mathrm{in}}, \quad \hat{\epsilon}_{i,p} = \beta_p\epsilon - (\beta_p - 1)W(z) + \rho_p U_{i,p}^{\mathrm{in}},
\end{aligned}
\tag{3.48}
$$

$$
\begin{aligned}
&\frac{n}{2n-1}\left(U_{z\epsilon}\frac{\partial}{\partial z}\tilde{f}_{n-1}(z,\epsilon)+\frac{n-1}{2}e_0E(z)\tilde{f}_{n-1}(z,\epsilon)\right)\\
&+\frac{n+1}{2n+3}\left(U_{z\epsilon}\frac{\partial}{\partial z}\tilde{f}_{n+1}(z,\epsilon)-\frac{n+2}{2}e_0E(z)\tilde{f}_{n+1}(z,\epsilon)\right)\\
&+\sum_i U_{z\epsilon}N_i(z)\left(\left(Q_i^{\mathrm{el},0}(U_{z\epsilon})-Q_i^{\mathrm{el},n}(U_{z\epsilon})\right)+\sum_p Q_{i,p}^{\mathrm{in},0}(U_{z\epsilon})\right)\tilde{f}_n(z,\epsilon)\\
&=\sum_i\sum_p \beta_p^2\hat{U}_{i,p}^{z\epsilon}N_i(z)Q_{i,p}^{\mathrm{in},n}(\hat{U}_{i,p}^{z\epsilon})\tilde{f}_n(z,\hat{\epsilon}_{i,p}),\\
&1\leqslant n\leqslant l-1,\quad l\geqslant 2,\quad \tilde{f}_l(z,\epsilon)\equiv 0
\end{aligned}
\tag{3.49}
$$

for the transformed expansion coefficients $\tilde{f}_n(z,\epsilon)=f_n(z,U_{z\epsilon})$ with $U_{z\epsilon}\equiv\epsilon-W(z)$.

This system of l partial differential equations including additional terms $\tilde{f}_n(z,\beta_p\epsilon-(\beta_p-1)W(z)+\rho_pU_{i,p}^{\mathrm{in}})$ with shifted energy arguments is solved numerically as an initial-boundary value problem for given spatial dependences of the electric field and heavy particle densities. The originally rectangular solution domain $(0\leqslant z\leqslant z_{\max},\ 0\leqslant U\leqslant U_\infty)$ with the spatial margins $z=0$ and $z=z_{\max}$ changes its shape due to the transformation to the total energy. The resulting transformed solution area $(0\leqslant z\leqslant z_{\max},\ W(z)\leqslant\epsilon\leqslant W(z)+U_\infty)$ is nonregtangular and its boundaries are determined in part by the spatial course of the electric potential.

The total energy ϵ represents the evolution direction of the kinetic problem. The solution of system (3.48) and (3.49) as an initial-boundary value problem takes place from higher to lower total energies. It starts at a sufficiently large total energy $\epsilon=\epsilon_\infty$, where all expansion coefficients can be assumed to be negligibly small in accordance with equation (3.35). Thus, the relation $\tilde{f}_n(z,\epsilon\geqslant\epsilon_\infty)=0$ for all $n=0,\ldots,l-1$ provides appropriate initial values. By analogy with equation (3.34), possible conditions at the total energy boundary $\epsilon=W(z)$ for all z are

$$\tilde{f}_n(z,\epsilon=W(z))=0\quad\text{for } n\geqslant 1. \tag{3.50}$$

Finally, appropriate conditions at the spatial margins of the solution domain, i.e., at $z=0$ and $z=z_{\max}$ have to be chosen according to the specific problem under consideration. Examples of such conditions can be found e.g. in [37–39].

In order to numerically solve the equation system (3.48) and (3.49) using the finite difference method, it can be discretized on an equidistant grid in the coordinate z with

$$z_j=j\Delta z,\qquad \Delta z=z_{\max}/j_{\max},\qquad 0\leqslant j\leqslant j_{\max}. \tag{3.51}$$

For example, when studying the spatial relaxation of electrons in an idealized steady-state Townsend (SST) experiment with plane-parallel geometry similar to e.g. [4, 40], where the electrons start at $z=0$ and are accelerated in the positive z direction with $z_{\max}>0$, the electric field $E(z)$ is assumed to be negative everywhere. Thus, the potential energy $W(z)$ given by (3.47) is negative for all $z>0$ as well.

The corresponding total energy grid consists of a part for $\epsilon \geqslant 0$ and a part belonging to negative total energies [37], and it is given by

$$\begin{aligned} \epsilon_k &= k\Delta\epsilon, \quad \Delta\epsilon = U_\infty/n_\infty, \quad 0 \leqslant k \leqslant n_\infty, \quad n_\infty < j_{\max}, \\ \epsilon_k &= W(z_{-k}), \quad k = -1, -2, \dots, -j_{\max} + n_\infty. \end{aligned} \tag{3.52}$$

The corresponding discretized distributions at the grid points have the representation

$$\tilde{f}_n^{j,k} = \tilde{f}_n(z_j, \epsilon_k), \quad 0 \leqslant n \leqslant l-1. \tag{3.53}$$

The discretization of the system of partial differential equations (3.48) and (3.49) can be performed at all centred points $(z_{j+1/2}, \epsilon_{k+1/2})$ [22, 37] using second-order-correct centred difference expressions for all functions $\tilde{f}_n$ with $0 \leqslant n \leqslant l-1$ and their first derivatives analogous to (3.37), where $j = j_0, \dots, j_0 + n_\infty$ with $j_0 = -\min(0, k)$ and $k = n_\infty - 1, \dots, -j_{\max} + n_\infty$. In addition, the second-order-correct expression

$$\tilde{f}_n(z_{j+1/2}, (\hat{\epsilon}_{i,p})_{k+1/2}) = \frac{1}{2}\left[\tilde{f}_n(z_j, (\hat{\epsilon}_{i,p})_{k+1/2}) + \tilde{f}_n(z_{j+1}, (\hat{\epsilon}_{i,p})_{k+1/2})\right] \tag{3.54}$$

with respect to the space coordinate is applied to treat the terms with shifted total energy argument on the right-hand side of (3.48) and (3.49). The transformed expansion coefficients $\tilde{f}_n(z_\xi, (\hat{\epsilon}_{i,p})_{k+1/2})$ with $\xi = j$ and $j+1$ are represented on the same energy grid by means of a parabolic interpolation using the three neighbouring energy grid points. The discrete form of the equation system (3.48) and (3.49) finally reads

$$\begin{aligned} &v_n^{j,k}\tilde{f}_{n-1}^{j,k} + w_n^{j,k}\tilde{f}_{n-1}^{j+1,k} + g_n^{j,k}\tilde{f}_n^{j,k} + g_n^{j,k}\tilde{f}_n^{j+1,k} \\ &\quad + r_n^{j,k}\tilde{f}_{n+1}^{j,k} + s_n^{j,k}\tilde{f}_{n+1}^{j+1,k} = d_n^{j,k+1} + \tilde{d}_n^{j,k} \end{aligned} \tag{3.55}$$

for $0 \leqslant n \leqslant l-1$. It has the coefficients

$$\begin{aligned} g_0^{j,k} &= -D_0^{j,k}\frac{\Delta z}{\epsilon_{k+1} - \epsilon_k} + F_0^{j,k}\frac{\Delta z}{2}, \quad h_0^{j,k} = -D_0^{j,k}\frac{\Delta z}{\epsilon_{k+1} - \epsilon_k} - F_0^{j,k}\frac{\Delta z}{2}, \\ g_n^{j,k} &= \frac{\Delta z}{2}\sum_i H_{i,n}^{j,k}, \quad h_n^{j,k} = -g_n^{j,k} \quad \text{for } 1 \leqslant n \leqslant l-1, \\ r_n^{j,k} &= -A_n^{j,k} + B_n^j\frac{\Delta z}{2}, \quad s_n^{j,k} = A_n^{j,k} + B_n^j\frac{\Delta z}{2} \quad \text{for } 0 \leqslant n \leqslant l-2, \\ v_n^{j,k} &= -C_n^{j,k} + G_n^j\frac{\Delta z}{2}, \quad w_n^{j,k} = C_n^{j,k} + G_n^j\frac{\Delta z}{2} \quad \text{for } 1 \leqslant n \leqslant l-1, \\ v_0^{j,k} &= w_0^{j,k} = r_{l-1}^{j,k} = s_{l-1}^{j,k} = 0, \\ d_n^{j,k+1} &= -v_n^{j,k}\tilde{f}_{n-1}^{j,k+1} - w_n^{j,k}\tilde{f}_{n-1}^{j+1,k+1} + h_n^{j,k}\tilde{f}_n^{j,k+1} + h_n^{j,k}\tilde{f}_n^{j+1,k+1} \\ &\quad - r_n^{j,k}\tilde{f}_{n+1}^{j,k+1} - s_n^{j,k}\tilde{f}_{n+1}^{j+1,k+1}, \\ \tilde{d}_n^{j,k} &= \Delta z\sum_i\sum_p \beta_p^2 I_{i,p,n}^{j,k}\left(\tilde{f}_n(z_j, (\hat{\epsilon}_{i,p})_{k+1/2}) + \tilde{f}_n(z_{j+1}, (\hat{\epsilon}_{i,p})_{k+1/2})\right) \end{aligned} \tag{3.56}$$

with the further terms

$$
\begin{aligned}
D_0^{j,k} &= -\sum_i 2\frac{m_e}{M_i}\left[U_{z\epsilon}^2 N_i(z) Q_i^{d}(U_{z\epsilon})\right]_{j+1/2,k+1/2},\\
F_0^{j,k} &= -\sum_i 2\frac{m_e}{M_i}\left(\left[2U_{z\epsilon}N_i(z)Q_i^{d}(U_{z\epsilon})\right]_{j+1/2,k+1/2}\right.\\
&\quad \left. + (U_{z\epsilon})^2_{j+1/2,\,k+1/2}\frac{\left[N(z)Q_i^{d}(U_{z\epsilon})\right]_{j+1,k+1} - \left[N(z)Q_i^{d}(U_{z\epsilon})\right]_{j+1,k}}{\epsilon_{k+1}-\epsilon_k}\right)\\
&\quad + \sum_i\sum_p\left[U_{z\epsilon}N_i(z)Q_{i,p}^{\mathrm{in},0}(U_{z\epsilon})\right]_{j+1/2,k+1/2},\\
H_{i,n}^{j,k} &= \left[U_{z\epsilon}N_i(z)\left(\left(Q_i^{\mathrm{el},0}(U_{z\epsilon}) - Q_i^{\mathrm{el},n}(U_{z\epsilon})\right) + \sum_p Q_{i,p}^{\mathrm{in},0}(U_{z\epsilon})\right)\right]_{j+1/2,k+1/2},\\
A_n^{j,k} &= \frac{n+1}{2n+3}(U_{z\epsilon})_{j+1/2,k+1/2}, \quad B_n^j = -\frac{n+1}{2n+3}\frac{n+2}{2}e_0E(z_{j+1/2}),\\
C_n^{j,k} &= \frac{n}{2n-1}(U_{z\epsilon})_{j+1/2,k+1/2}, \quad G_n^j = \frac{n}{2n-1}\frac{n-1}{2}e_0E(z_{j+1/2}),\\
I_{i,p,n}^{j,k} &= \left[\hat{U}_{i,p}^{z\epsilon}N_i(z)Q_{i,p}^{\mathrm{in},n}(\hat{U}_{i,p}^{z\epsilon})\right]_{j+1/2,k+1/2}.
\end{aligned}
\tag{3.57}
$$

In order to complete this finite difference equation system, one boundary condition for each of the l expansion coefficients $\tilde{f}_n$ with $0 \leqslant n \leqslant l-1$ is required. It has been found that the conditions [37]

$$\tilde{f}_n^{j_0=0,\,k} = \hat{f}_n^{k}, \quad \tilde{f}_n^{j_0>0,\,k} = 0 \quad \text{for } n = 1, 3, 5, \dots, \tag{3.58}$$

$$\tilde{f}_n^{j_0+n_\infty,\,k} = 0 \quad \text{for } n = 0, 2, 4, \dots \tag{3.59}$$

represent a proper choice for the spatial relaxation studies. The condition (3.58) for $j_0 = 0$ fixes the inflow of electrons at the spatial boundary $z = 0$, where the $\hat{f}_n^k$ with odd n are the discrete forms of appropriate functions $\hat{f}_n(\epsilon)$ of the total energy to be specified according to the problem considered. The condition (3.58) for $j_0 > 0$ is the discrete form of the boundary condition (3.50) for odd n. At $z_{j_0+n_\infty}$ the kinetic energy values are always equal to or larger than U_∞ so that the condition (3.59) is in accordance with equation (3.35) for even n. This choice of boundary conditions allows one to suppress singular contributions to the solution of physical relevance at large kinetic energies.

The combination of the equation system (3.55) and the boundary conditions (3.58) and (3.59) represent a complete set of linear equations to determine the function values $\tilde{f}_n^{j,k}$ with $0 \leqslant n \leqslant l-1$ for $j = j_0, \dots, j_0 + n_\infty$ at the fixed total energy with index k, if all corresponding function values $\tilde{f}_n^{j,k+1}$ at the larger total energy with index $k+1$ or at their initial values at $k = n_\infty$ are known. This complete linear equation system is again tridiagonal in the l independent variables $\tilde{f}_n$ with $0 \leqslant n \leqslant l-1$. After its conversion to the general band form, it is resolved by using the general band algorithm given in [35]. Finally, the transformation back from the total energy ϵ to the kinetic energy U yields the expansion coefficients $f_n(z, U)$ with $0 \leqslant n \leqslant l-1$ of the EVDF and the related macroscopic properties.

A more detailed examination of the formal inhomogeneities $\tilde{d}_n^{j,k}$ in (3.55) also makes clear that the progression from larger to lower total energies induces that the discrete function values $\tilde{f}_n(z_j, (\hat{\epsilon}_{i,p})_{k+1/2})$ with shifted energy argument $\hat{\epsilon}_{i,p}$ are already known for all j in the case of exciting, dissociating and ionizing collision processes and can thus be directly determined. Only de-exciting electron collision processes comprising function values at total energies smaller than the actual one have to be treated iteratively in a similar way as discussed in section 3.3.1.

Another issue concerns the discretization of the set of equations (3.48) and (3.49). Here, recent studies indicate that a first-order accurate, fully implicit differencing scheme with respect to the total energy ϵ at the points $(z_{j+1/2}, \epsilon_k)$ appears to be preferable to obtain converged results especially for the analysis of molecular gases like CO_2.

In the following, results for the spatial relaxation of electrons in argon and CO_2 plasmas with a gas density of $N = 3.54 \times 10^{16}\ \text{cm}^{-3}$ acted upon by a space-independent electric field of $E = -10.62\ \text{V cm}^{-1}$ are represented. The results of these studies resembling an idealized SST experiment with plane-parallel geometry have been obtained by solving the equation system (3.48, 3.49) by means of the multiterm method using typically $l = 8$ expansion coefficients. In order to initiate the spatial electron relaxation at $z = 0$, a Gaussian-like distribution $G_z(U) = UG_t(U)$ with the centre $U_c = 5\ \text{eV}$ and width $U_w = 2\ \text{eV}$ was used to prescribe the first anisotropic component $f_1(z = 0, U)$ of the EVDF and the further anisotropic components $f_n(z = 0, U)$ with odd $n \geqslant 3$ were chosen to be zero. The corresponding boundary values in the transformed space read $\hat{f}_1(\epsilon) = G_z(\epsilon)$ and $\hat{f}_{n=3,5,\ldots}(\epsilon) = 0$. The factor c in G_z is used to normalize all expansion coefficients on the electron density $n_e(0)$ at $z = 0$. Because a negative value is chosen for the constant electric field E, the electrons are accelerated in the positive z direction and therefore the spatial relaxation of the properties of the electrons happens in that direction as well.

Figure 3.6 illustrates the spatial relaxation of the isotropic distribution $f_0(z, U)/n_e(z)$ normalized by the electron density $n_e(z)$ for argon (a) and carbon dioxide (b). It clearly demonstrates that the spatial relaxation behaviour of $f_0(z, U)/n_e(z)$ initiated by the boundary value of the anisotropic distribution $f_1(0, U)/n_e(0)$ is very different for the two gases considered. A strongly damped spatial relaxation with a comparatively small relaxation length takes place in the molecular gas CO_2 (figure 3.6(b)), where the normalized space-independent distributions $f_n(z, U)/n_e(z)$ belonging to the hydrodynamic state with constant $[\mathrm{d}n_e(z)/\mathrm{d}z]/n_e(z)$ appear already after about 1.5 cm. In contrast, a weakly damped, pronouncedly periodic evolution is found for the rare gas Ar. Therefore, only the initial part of the spatial relaxation of $f_0(z, U)/n_e(z)$ is presented in figure 3.6(a). This periodic behaviour results from the interplay of the electron acceleration caused by the electric field action and the energetic backscattering of electrons due to

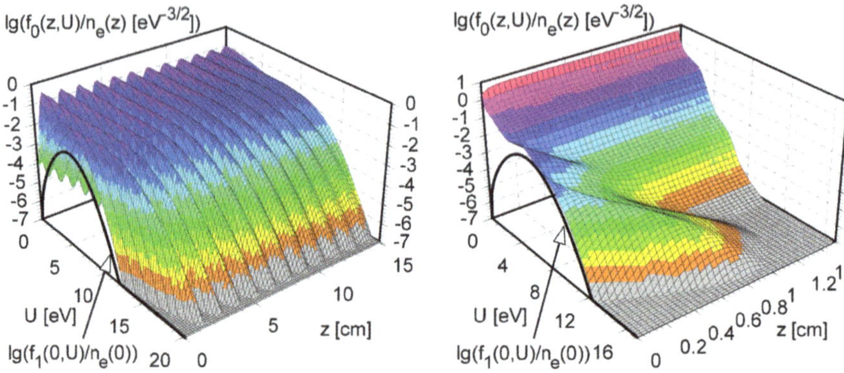

(a) Argon. (b) CO_2.

Figure 3.6. Spatial relaxation of $f_0(z,\ U)/n_e(z)$ in argon and CO_2 at $E = -10.62$ V cm^{-1}.

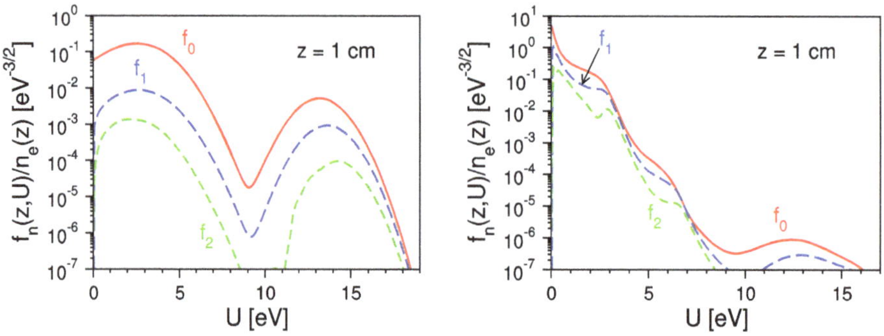

(a) Argon. (b) CO_2.

Figure 3.7. Expansion coefficients $f_n(z,\ U)/n_e(z)$ with $n = 0,\ 1,\ 2$ at $z = 1$ cm.

inelastic collision processes. It is more pronounced in the case of argon due to the larger value of the lowest threshold energy of 11.55 eV compared to CO_2.

Figure 3.7 presents the corresponding expansion coefficients $f_n(z,\ U)/n_e(z)$ with $n = 0,\ 1,\ 2$ during the initial part of the spatial relaxation process at $z = 1$ cm. Two peaks are obvious for the expansion coefficients in argon (figure 3.7(a)), which are separated by about 11 eV and are caused by the interaction of the electric field action and the inelastic collision processes mentioned. The impact of expansion coefficients higher than $l = 2$ is found to be generally small during the entire spatial relaxation in argon. In contrast, the anisotropy coefficients $f_n(z,\ U)/n_e(z)$ with $n = 1$ and 2 reach values of the order of the isotropic distribution $f_0(z,\ U)/n_e(z)$ for CO_2 (figure 3.7(b)) indicating the impact of higher-order terms on the EVDF already during the course of the spatial relaxation.

The spatial relaxation of the mean energy $u_m(z) = u_e(z)/n_e(z)$ and mean velocity $v_m(z) = j_z(z)/n_e(z)$ of the electrons in argon and carbon dioxide at $E = -10.62$ V cm^{-1} and $N = 3.54 \times 10^{16}$ cm^{-3} is shown in figure 3.8. The mean

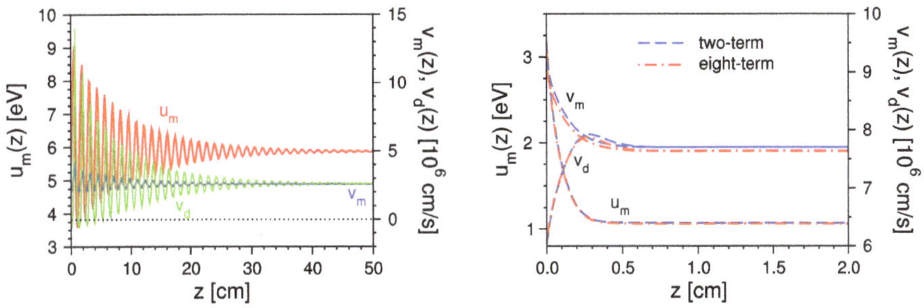

(a) Argon.

(b) CO_2.

Figure 3.8. Spatial relaxation of mean energy, mean velocity and drift velocity.

velocity is composed of the electron drift velocity $v_d(z) = -\mu_e(z)E(z)$ and the diffusion velocity part, where the electron mobility μ_e is given by [41]

$$\mu_e(z) = -\frac{e_0}{3n_e(z)}\left(\frac{2}{m_e}\right)^{1/2}\int_0^\infty \frac{1}{\sum_i N_i(z)\left(Q_i^d(U) + \sum_p Q_{i,p}^{in,0}(U)\right)} \left[U\left(\frac{\partial}{\partial U}f_0(z, U) + \frac{2}{5}\frac{\partial}{\partial U}f_2(z, U)\right) + \frac{3}{5}f_2(z, U)\right] \tag{3.60}$$

in multiterm approximation. Thus, the figure displays the electron drift velocity v_d in addition. It is found for argon (figure 3.8(a)) that the macroscopic properties are subject to larger periodic alterations during the weakly damped relaxation process and run into the hydrodynamic state after more than 50 cm. There the mean velocity is about 0.5% smaller than the drift velocity because of a small electron density increase due to ionization in accordance with [40]. Furthermore, the electron drift velocity v_d changes its signs several times and becomes negative during the early relaxation course up to about $z = 5$ cm so that states with negative mobility appear again. Compared with this, the mean electron velocity v_m always remains positive due to the impact of the diffusive flux.

The strongly damped spatial relaxation process in CO_2 (figure 3.8(b)) does not exhibit any negative mobility. There, the mean velocity v_m is greater than the drift velocity v_d during the early relaxation course. When approaching the hydrodynamic state, v_m and v_d are almost equal because the impact of the dissociative electron attachment is very small. At the same time, certain differences between two-term and converged eight-term results, especially for the velocities, are obvious during the entire spatial relaxation.

In order to assess the relaxation length, the analysis of the corresponding macroscopic balances resulting from (3.18), (3.19) and (3.20) similar to that discussed at the end of section 3.3.1 is the most appropriate method. Figure 3.9 represents the spatial relaxation of the corresponding particle, momentum and power gain to loss ratios $\nu^{io}(z)/\nu^{at}(z)$, $I^f(z)/[I^{el}(z) + \sum_p I_p^{in}(z)]$ and

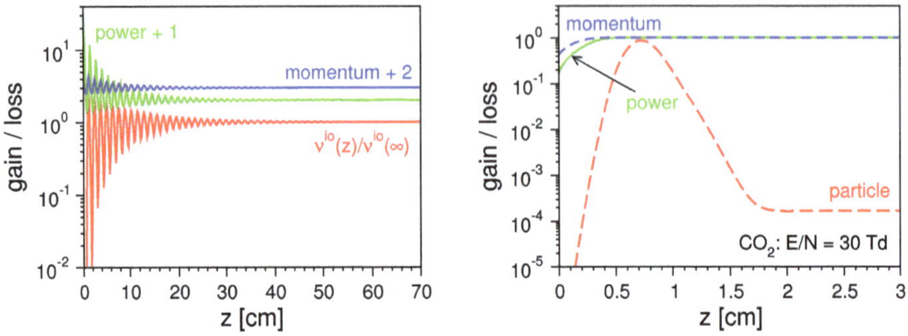

(a) Argon. (b) CO_2.

Figure 3.9. Spatial relaxation of gain to loss ratios and the ionization frequency.

$P^{f}(z)/[P^{el}(z) + \sum_p P_p^{in}(z) + P^{at}(z)]$, respectively. For argon (figure 3.9(a)), the ionization frequency $\nu^{io}(z)/\nu^{io}(\infty)$ normalized by its space-independent value $\nu^{io}(\infty)$ is shown and a shift by 1 and 2, respectively, was added to the power and momentum gain to loss ratio for a clearer presentation. Contrary to the temporal relaxation, the spatial relaxation process of the power and momentum balances takes place on the same spatial scale in accordance with [4]. However, the relaxation length required to reach hydrodynamic conditions is finally determined by the establishment of a constant particle gain to loss ratio, as especially figure 3.9(b) clearly shows.

3.4 Concluding remarks

An introduction into the kinetic treatment of the electrons by solving their Boltzmann equation has been given. The focus is on the determination of the velocity distribution function and related macroscopic properties of the electrons in the framework of a generally valid multiterm description. Basic relations of this multiterm approach have been represented and details about the numerical methods for solving both the time-dependent and the spatially one-dimensional electron Boltzmann equation in higher-order accuracy have been discussed.

The corresponding numerical multiterm methods have been applied to study the electron relaxation in a constant electric field in argon and carbon dioxide. The results of the investigations have illustrated the pronounced non-equilibrium behaviour of the electrons including similarities and differences for the two gases considered. They have also shown the impact of terms higher than two of the Legendre polynomial expansion of the EVDF. Furthermore, the analysis has made it possible to get a clear assessment of the non-local response of the electrons during the temporal and spatial course of the relaxation processes. The resulting relaxation times and relaxation lengths to approach steady states are to be estimated on the basis of the macroscopic balances using the method presented.

The spectrum of collision processes considered in the kinetic treatment covers elastic electron collisions as well as several binary inelastic collision processes such as excitation, de-excitation, dissociation, ionization and attachment. In order to allow for an analysis of various other aspects which can affect the non-local behaviour of the electrons, an extension of the kinetic description and the resulting numerical methods is required. These may include, for example, the effect of the neutral gas temperature in elastic collision processes [14, 22, 42], the sharing of the remaining energy between the two electrons after an ionization event [26, 27], the additional impact of the electron–electron interaction [14, 42] and the influence of further non-conservative electron collision processes such as the electron generation due to Penning ionization, chemo-ionization, collisional and associative detachment or the electron loss due to electron–ion recombination, three-body electron attachment and recombination [12, 26]. The integration of these aspects into the multiterm treatment can immediately be done and has partly already been realized.

Another important point concerns the integration of both the time-dependent and space-dependent multiterm method into a self-consistent description of nonthermal plasmas by means of hybrid modelling [9]. The realization of this numerical approach is still an open task. While the inclusion of the time-dependent multiterm method in such a more complex plasma description can be realized relatively easily, the corresponding integration of the space-dependent multiterm approach is more challenging. This results, for instance, from the transformation to total energy needed for the solution, which requires special care in particular at conditions when electric field reversal occurs [43]. Here, the use of an equidistant grid in coordinate and total energy appears to be necessary [44] to allow for an effective iterative solution approach.

References

[1] Robson R E, White R D and Petrović Z Lj 2005 *Colloquium*: Physically based fluid modeling of collisionally dominated low-temperature plasmas *Rev. Mod. Phys.* **77** 1303–20

[2] Tsendin L D 2009 Electron kinetics in glows–from Langmuir to the present *Plasma Sources Sci. Technol.* **18** 014020

[3] Donkó Z 2011 Particle simulation methods for studies of low-pressure plasma sources *Plasma Sources Sci. Technol.* **20** 024001

[4] Winkler R, Loffhagen D and Sigeneger F 2002 Temporal and spatial relaxation of electrons in low temperature plasmas *Appl. Surf. Sci.* **192** 50–71

[5] Winkler R, Arndt S, Loffhagen D, Sigeneger F and Uhrlandt D 2004 Progress of the electron kinetics in spatial and spatiotemporal plasma structures *Contrib. Plasma Phys.* **44** 437–49

[6] Loffhagen D, Sigeneger F and Winkler R 2008 Electron kinetics in weakly ionized plasmas *Low Temperature Plasmas. Fundamentals, Technologies and Techniques* ed R Hippler, H Kersten, M Schmidt and K H Schoenbach 2nd edn (New York: Wiley) ch 2

[7] Hagelaar G J M and Pitchford L C 2005 Solving the Boltzmann equation to obtain electron transport coefficients and rate coefficients for fluid models *Plasma Sources Sci. Technol.* **14** 722–33

[8] Porokhova I A, Golubovskii Yu B, Bretagne J, Tichy M and Behnke J F 2001 Kinetic simulation model of magnetron discharges *Phys. Rev.* E **63** 056408

[9] Loffhagen D and Sigeneger F 2009 Advances in Boltzmann equation based modelling of discharge plasmas *Plasma Sources Sci. Technol.* **18** 034006

[10] Dyatko N A, Ionikh Yu Z, Meshchanov A V, Napartovich A P and Barzilovich K A 2010 Specific features of the current-voltage characteristics of diffuse glow discharges in Ar:N_2 mixtures *Plasma Phys. Rep.* **36** 1040–64

[11] Pintassilgo C D, Guerra V, Guaitella O and Rousseau A 2010 Modelling of an afterglow plasma in air produced by a pulsed discharge *Plasma Sources Sci. Technol.* **19** 055001

[12] Hübner M, Gortschakow S, Guaitella O, Marinov D, Rousseau A, Röpcke J and Loffhagen D 2016 Kinetic studies of NO formation in pulsed air-like low-pressure dc plasmas *Plasma Sources Sci. Technol.* **25** 035005

[13] Desloge E A 1966 *Statistical Physics* (New York: Holt, Rinehart and Winston)

[14] Shkarofsky I P, Johnston T W and Bachynski M P 1966 *The Particle Kinetics of Plasmas* (London: Addison-Wesley)

[15] Golant V E, Zhilinsky A P and Sakharov I E 1980 *Fundamental of Plasma Physics* (New York: Wiley)

[16] White R D, Robson R E, Dujko S, Nicoletopoulos P and Li B 2009 Recent advances in the application of Boltzmann equation and fluid equation methods to charged particle transport in non-equilibrium plasmas *J. Phys. D: Appl. Phys.* **42** 194001

[17] Loffhagen D and Winkler R 1996 Time-dependent multi-term approximation of the velocity distribution in the temporal relaxation of plasma electrons *J. Phys. D: Appl. Phys.* **29** 618–27

[18] Winkler R, Petrov G, Sigeneger F and Uhrlandt D 1997 Strict calculation of electron energy distribution functions in inhomogeneous plasmas *Plasma Sources Sci. Technol.* **6** 118–32

[19] Robson R E, Winkler R and Sigeneger F 2002 Multiterm spherical tensor representation of Boltzmann's equation for a nonhydrodynamic weakly ionized plasma *Phys. Rev.* E **65** 056410

[20] Sigeneger F and Winkler R 2002 Study of the electron kinetics in cylindrical hollow cathodes by a multi-term approach *Eur. Phys. J. Appl. Phys.* **19** 211–23

[21] Porokhova I A, Golubovskii Yu B and Behnke J F 2005 Anisotropy of the electron component in a cylindrical magnetron discharge. I. Theory of the multiterm analysis *Phys. Rev.* E **71** 066406

[22] Hannemann M, Hardt P, Loffhagen D, Schmidt M and Winkler R 2000 The electron kinetics in the cathode region of H_2/Ar/N_2 discharges *Plasma Sources Sci. Technol.* **9** 387–99

[23] Wilhelm J and Winkler R 1968 Tensorielle Kugelflächenfunktionsentwicklung in der Boltzmann–Gleichung für das Lorentz-Plasma bzw. -Gas *Beitr. Plasmaphys.* **8** 167–88

[24] Winkler R, Braglia G L and Wilhelm J 1988 Impact of nonisotropic elastic and inelastic scattering on the electron velocity distribution in weakly ionized plasmas *Il Nuovo Cimento* D **10** 1209–34

[25] Margenau H 1948 Theory of high frequency gas discharges. I. Methods for calculating electron distribution functions *Phys. Rev.* **73** 297–308

[26] Loffhagen D and Winkler R 1994 A new nonstationary Boltzmann solver in self-consistent modelling of discharge pumped plasmas for excimer lasers *J. Comput. Phys.* **112** 91–101

[27] Grubert G K and Loffhagen D 2014 Boltzmann equation and Monte Carlo analysis of the electrons in spatially inhomogeneous, bounded plasmas *J. Phys. D: Appl. Phys.* **47** 025204

[28] Richtmyer R D and Morton K W 1967 *Difference Methods for Initial Value Problems* 2nd edn (New York: Wiley)

[29] Thomée V 1969 Stability theory for partial difference operators *SIAM Rev.* **11** 152–95

[30] Thomas J W 1995 *Numerical Partial Differential Equations: Finite Difference Methods* (Texts in Applied Mathematics vol 22) (New York: Springer)

[31] Hayashi M 1992 Electron collision cross sections *Plasma Material Science Handbook* ed Japan Society for the Promotion of Science (Tokyo: Ohmsha Ltd) pp 748–66

[32] Hayashi M 2003 Bibliography of electron and photon cross sections with atoms and molecules published in the 20th century—argon *Technical Report* NIFS-DATA-72, National Institute for Fusion Science

[33] Winkler R, Braglia G L, Hess A and Wilhelm J 1984 Fundamentals of a technique for determining electron distribution functions by multi-term even-order expansion in Legendre polynomials. 1. Theory *Beitr. Plasmaphys.* **24** 657–74

[34] Winkler R, Braglia G L and Wilhelm J 1988 Modification of the electron velocity distribution function in weakly ionized plasmas by nonisotropic elastic-scattering processes *Il Nuovo Cimento* D **10** 1031–60

[35] von Rosenberg D U 1969 *Methods for the Numerical Solution of Partial Differential Equations* (Modern Analytic and Computational Methods in Science and Mathematics vol 16) (New York: Elsevier)

[36] Dyatko N A, Kochetov I V and Napartovich A P 2014 Non-thermal plasma instabilities induced by deformation of the electron energy distribution function *Plasma Sources Sci. Technol.* **23** 043001

[37] Petrov G and Winkler R 1997 Multi-term treatment of electron kinetics in inhomogeneous nonequilibrium plasmas *J. Phys. D: Appl. Phys.* **30** 53–66

[38] Loffhagen D, Sigeneger F and Winkler R 2002 Study of the electron kinetics in the anode region of a glow discharge by a multiterm approach and Monte Carlo simulations *J. Phys. D: Appl. Phys.* **35** 1768–76

[39] Becker M M, Grubert G K and Loffhagen D 2010 Boundary conditions for the electron kinetic equation using expansion techniques *Eur. Phys. J. Appl. Phys.* **51** 11001

[40] Dujko S, White R D and Petrović Z L 2008 Monte Carlo studies of non-conservative electron transport in the steady-state Townsend experiment *J. Phys. D: Appl. Phys.* **41** 245205

[41] Grubert G K, Becker M M and Loffhagen D 2009 Why the local-mean-energy approximation should be used in hydrodynamic plasma descriptions instead of the local-field approximation *Phys. Rev.* E **80** 036405

[42] Loffhagen D 2005 Impact of electron-electron collisions on the spatial electron relaxation in non-isothermal plasmas *Plasma Chem. Plasma Process.* **25** 519–38

[43] Loffhagen D, Sigeneger F and Winkler R 2004 The effect of a field reversal on the spatial transition of the electrons from an active plasma to a field-free remote plasma *Eur. Phys. J. Appl. Phys.* **25** 45–56

[44] Grubert G K and Loffhagen D 2007 Nonequilibrium properties of electrons in oxygen plasmas *IEEE Trans. Plasma Sci.* **35** 1215–22

Chapter 4

Particle-based simulation of plasmas

Giovanni Lapenta

4.1 Types of interacting systems

The first step to decide how to model a system of interacting particles is to distinguish between weakly and strongly interacting systems. The distinction has a profound meaning and a direct impact on how to model the system on a computer. Figure 4.1 summarizes visually the situation. Let us consider a system made of a collection of charged particles in a box with the side of the Debye length, λ_D (the box is 3D but is depicted as 2D for convenience). We choose the Debye length because a basic property of plasmas is to shield the effects of localized charges over distances exceeding the Debye length. Of course the shielding is exponential and the effect is not totally cancelled over one Debye length, but such a length provides a conventional reasonable choice for the interaction range. The electric field in each point of the box is computed by the superposition of the contributions of each particle.

Let us conduct an ideal thought experiment based on using a device able to detect the local electric field in one spatial position. We identify in the figure such an ideal measurement device with an icon connected to a point in the system. We try to conduct a thought experiment where in no step any law of physics is violated but where the difficulties of experimental work are eliminated.

If we consider the configuration in figure 4.1(a), we note that within the domain there are few particles and the measurement obtained by our fantastic electric field meter would be very jumpy. The particles in the box move constantly, interacting with each other and agitated by their thermal motion. As a particle passes by the detector, the measurement detects a jump up and when a particle moves away it detects a jump down. On average at any given time very few particles are near the detector and their specific positions are key in determining the value measured. The effect of a given particle on the electric field at the location of measurement decays very rapidly with the distance and only when the particle is nearby is the effect strong.

The same effect is detected by each of the particles in the system. The electric field each particle feels is a sum of the contributions of all others but only when another

doi:10.1088/978-0-7503-3559-1ch4

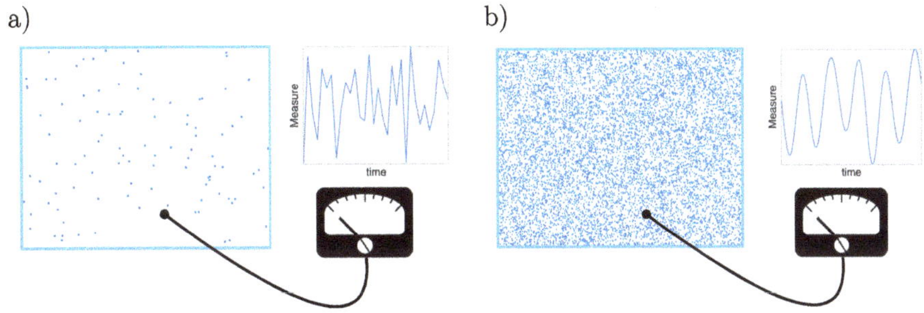

Figure 4.1. Thought experiment for strongly (a, left) versus weakly (b, right) coupled systems.

particle passes by the electric field would it register a jump: in a common term this event is called a *collision*. The particle trajectories would then be affected by a series of close encounters registered as jumps (collisions) in the trajectory.

The system described goes in the language of kinetic theory by the name of *strongly coupled system*, a system where the evolution is determined by the close encounters and by the relative configuration of any two pairs of particles. The condition just described is characterized by the presence of few particles in the Debye box.

The opposite situation is that of a weakly coupled system. The corresponding configuration is described in figure 4.1(b).

Now the system is characterized by being composed by an extremely large number of particles in the Debye box. At any given point, the number of particles contributing to the electric field is very large. Regardless of the particle motion, the field is given by the superposition of many contributions. As a consequence, by simple averaging of the effects of all the particles contributing to the measurement, the measurement is smooth and does not jump in time. Similarly the trajectory of a particle is at any time affected by a large number of other particles. The trajectory is smooth and without jumps. These systems are called *weakly coupled.* If in the strongly coupled system, the characteristic feature was the presence of a succession of collisions, in the weakly coupled system, the characteristic feature is the *mean field* produced by the superposition of contributions from a large number of particles.

4.1.1 Strength of interaction

The discriminant factor in the previous discussion was the number of particles present inside the box under consideration. If we choose the conventional box with side equal to the Debye length, the number of particles present is

$$N_D = n\lambda_D^3 \tag{4.1}$$

where n is the plasma density.

A system is considered weakly coupled when N_D is large and strongly coupled when N_D is small.

This concept can be further elaborated by considering the energies of the particles in the system. The particles in the box are distributed in a non-uniform, random way, but on average, the volume associated with each particle is simply the volume of the box,

λ_D^3, divided by the number of particles in the box, N_D. This volume, $V_p = n^{-1}$, can be used to determine the average interparticle distance, $a = V_p^{1/3} \equiv n^{-1/3}$. This relation provides an average statistical distance. The particles are distributed randomly and their distances are also random, but on average the interparticle distance is a.

The electrostatic potential energy between two particles with separation a is

$$E_{\text{pot}} = \frac{q^2}{4\pi\epsilon_0 a} \tag{4.2}$$

where we have assumed equal charge q for the two particles. Conversely, from statistical physics, the kinetic energy of the particles is of the order of the thermal energy:

$$E_{\text{th}} = kT \tag{4.3}$$

where k is the Boltzmann constant.

A useful measure of the plasma coupling is given by the so-called *plasma coupling parameter*, Γ, defined as:

$$\Gamma = \frac{E_{\text{pot}}}{E_{\text{th}}} = \frac{q^2}{4\pi\epsilon_0 akT} \tag{4.4}$$

Recalling the definition of Debye length ($\lambda_D = (\epsilon_0 kT/nq^2)^{1/2}$) and the value of a obtained above, it follows that:

$$\Gamma = \frac{q^2 n^{1/3}}{4\pi\epsilon_0 kT} \equiv \frac{1}{4\pi N_D^{2/3}} \tag{4.5}$$

The plasma coupling parameter gives a new physical meaning to the number of particles per Debye cube. When many particles are present in the Debye cube, the coupling parameter is small. In this case, the thermal energy far exceeds the potential energy, making the trajectory of each particle little influenced by the interactions with the other particles: this is the condition outlined above for the weakly coupled systems. Conversely, when the number of particles per Debye cube is small, the coupling parameter is large, the potential energy dominates and the trajectories are strongly affected by the near neighbour interactions: this is the condition typical of strongly coupled systems.

4.2 Computer simulation of interacting systems

A computer simulation of a system of interacting particles can be done in principle by simply following each particle in the system. The so-called *particle–particle* (PP) approach describes the motion of N particles by evolving the equations of Newton for each of the N particles taking as a force acting on the particle the combined effect of all the other particles in the system.

The evolution is discretized in many temporal steps Δt, each chosen so that the particles move only a small distance. After each move, the force is recomputed and a new move is made for all the particles. The main cost of the effort is the computation of the force which requires to sum over all the particles in the system. Once the force is computed the new velocities can be computed. Then the new positions can be computed and the cycle can be repeated indefinitely.

For each particle, the number of terms to sum to compute the force is $N - 1$, and considering that there are N particles, but that each pair needs to be computed only once, the total number of force computations is $N(N - 1)/2$.

For strongly coupled systems, where the number of particles per Debye cube is small, the PP approach is feasible and forms the basis of the very successful *molecular dynamics method* used in condensed matter and in biomolecular studies. We refer the reader to a specific textbook on molecular dynamics to investigate the approach more in depth [1].

The PP approach is also used in the study of gravitational interactions, for example in the cosmological studies of the formation and distribution of galaxies. In that case, specifically the dark matter is studied with a PP approach. The PP approach can be made more efficient by using the *Barnes–Hut* or *tree algorithm* [2] that can reduce the cost (but not without loss of information [3]) to $O(N \log N)$.

Even with the reduced cost of the tree algorithm, PP methods cannot be practical for weakly coupled systems where the number of particles is very large. As the number of particles increases, the cost scales quadratically (or as $N \log N$) and makes the computational effort unmanageable. In that case, one cannot simply describe every particle in the system and a method must be devised to reduce the description to just a statistical sample of the particles. This is the approach of the *particle in cell* (PIC) method (sometimes referred to as particle-mesh, PM).

The key idea behind the simulation of weakly coupled systems is to use as the building block of the model not single particles but rather collective clouds of them: each *computational particle* (referred to sometimes as a superparticle) represents a group of particles and can be visualized as a small piece of phase space. The concept is visualized in figure 4.2.

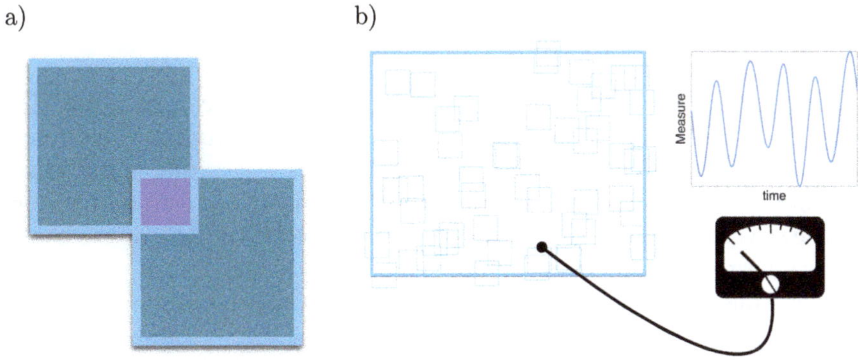

Figure 4.2. Finite size computational particles. Panel (a) shows the overlap of two finite size particles: the part of the particle clouds that overlap does not contribute to the force between the two clouds. As two clouds get closer, the force increases only till the overlap starts. After the overlap starts increasing, the force starts to decrease to become zero for complete overlap. Panel (b) shows a Debye cube filled by a set of computational particles to illustrate the idea of the PIC method.

The fundamental advantage of the finite-size particle approach is that the computational particles, being of finite size, interact more weakly than point particles. When two point particles interact, for example via the Coulomb force, the repulsive or attractive force grows as the particles approach, reaching a singularity at zero separation. Finite size particles, instead, behave as point particles until their respective domains start to overlap. Once overlap occurs the overlap zone is neutralized, not contributing to the force between the particles. At zero distances when the particles fully overlap (assuming here that all particles have the same size) the force becomes zero. The figure presented is 2D for convenience but the reader is invited to try to visualize this also in 3D.

The use of finite-size computational particles allows us to reduce the interaction among particles. A system of finite size particles behaves as a weakly coupled system. Let us return to our fantastic electric field instrument. The passage of a computational particle is the passage of a cloud of finite size that is affecting the instrument more weakly at a point particle. At any given time multiple clouds are in the vicinity rendering the measurement smooth in time. In the same way, the motion of every cloud is affected by the other clouds in the vicinity. When two clouds begin to overlap their relative force weakens. As a result the motion of each cloud is smooth. Even though we have few computational particles in the Debye cube, the behaviour resembles that of weakly coupled systems.

Recalling the definition of plasma parameter, the use of finite-size particles in practice results in a reduction of the potential energy for the same kinetic energy. The beneficial consequence is that the correct plasma parameter can be achieved by using fewer particles than in the physical system. The conclusion is that the correct coupling parameter is achieved by fewer particles interacting more weakly. The realistic condition is recovered in essence making two mistakes that cancel each other: using fewer particles and using weaker forces. Obviously great mathematical insight has gone into making this rigorous [4, 5]. Let us now see how that can be achieved.

4.3 Particle-in-cell method

The kinetic description of a plasma aims at describing the system via electromagnetic fields and distribution functions. A distribution function expresses the probability of finding a particle with certain properties (e.g. position and velocity). In its nature it is a probabilistic description of Nature. It gives up on the hope of following every particle in the system and contents itself by describing the ensemble in statistical terms. A preliminary knowledge of the basic tenants of plasma kinetic theory is assumed in the remainder of this chapter. Classic textbooks provide the necessary preliminary knowledge (see e.g. [6]).

In extreme summary, the starting point of a kinetic simulation model of plasmas is the Boltzmann equation. The phase space distribution function $f_s(\mathbf{x}, \mathbf{v}, t)$ for a given species s (electrons or ions) is defined as the number of particles at time t in the infinitesimal element of phase space $d\mathbf{x}$ and $d\mathbf{v}$ around a certain phase space point $(\mathbf{x}, \mathbf{v})$. The distribution of each species is governed by the Boltzmann equation:

$$\frac{\partial f_s}{\partial t} + \mathbf{v} \cdot \frac{\partial f_s}{\partial \mathbf{x}} + \frac{q_s}{m_s}(\mathbf{E} + \mathbf{v} \times \mathbf{B}) \cdot \frac{\partial f_s}{\partial \mathbf{v}} = St(f_s) \tag{4.6}$$

where q_s and m_s are the charge and mass of the species, respectively. The last term is the collision operator that summarizes collisions with a neutral medium (for partially ionized plasmas or for plasmas with immersed dust) or among the charged particles of the plasma. By its definition it is obvious that the integral of f_s over all the phase space is the total number of particles in the system.

The derivatives are written in the standard vector notation in the three-dimensional configuration space $\mathbf{x}$ and three-dimensional velocity space $\mathbf{v}$, forming together the classical phase space.

The electric and magnetic fields are given by the two curl Maxwell's equations,

$$\begin{aligned} \nabla \times \mathbf{E} &= -\frac{\partial \mathbf{B}}{\partial t} \\ \nabla \times \mathbf{B} &= \mu_0 \epsilon_0 \frac{\partial \mathbf{E}}{\partial t} + \mu_0 \mathbf{j} \end{aligned} \tag{4.7}$$

supplemented by the two divergence equations:

$$\begin{aligned} \epsilon_0 \nabla \cdot \mathbf{E} &= \rho \\ \nabla \cdot \mathbf{B} &= 0 \end{aligned} \tag{4.8}$$

where the net charge density is computed from the distribution functions as:

$$\rho(\mathbf{x}, t) = \sum_s q_s \int_{\mathbb{V}} f_s(\mathbf{x}, \mathbf{v}, t) \mathrm{d}\mathbf{v} \tag{4.9}$$

and the current density as:

$$\mathbf{j}(\mathbf{x}, t) = \sum_s q_s \int_{\mathbb{V}} \mathbf{v} f_s(\mathbf{x}, \mathbf{v}, t) \mathrm{d}\mathbf{v} \tag{4.10}$$

where $\mathbb{V}$ is the velocity space.

4.3.1 Mathematical formulation of PIC

The PIC method can be regarded as a representation of the distribution function of each species by a superposition of moving elements, each representing a cloud of physical particles. We follow here the specific approach described in reference [7].

The mathematical formulation of the PIC method is obtained by assuming that the distribution function of each species is given by the superposition of several elements (called computational particles or superparticles):

$$f_s(\mathbf{x}, \mathbf{v}, t) = \sum_p f_p(\mathbf{x}, \mathbf{v}, t) \tag{4.11}$$

Each element represents a large number of physical particles that are near each other in phase space. For this reason, the choice of the elements is made in order to be at the same time physically meaningful (i.e. to represent a group of particles near each other) and mathematically convenient (i.e. to allow the derivation of a manageable set of equations) [70].

The PIC method is based upon assigning to each computational particle a specific functional form for its distribution, a functional form with a number of free parameters whose time evolution will determine the numerical solution of the Vlasov equation [8]. In the standard PIC methods [4, 5], the choice is made to have two free parameters in the functional shape for each spatial dimension. The free parameters will acquire the physical meaning of position and velocity of the computational particle. A more advanced method takes into account also the evolving shape of the physical cloud the superparticle is supposed to represent [8].

In the standard PIC model, the functional dependence is assumed to be a tensor product of the shape in each direction of phase space [5]:

$$f_p(\mathbf{x}, \mathbf{v}, t) = N_p S_\mathbf{x}(\mathbf{x} - \mathbf{x}_p(t)) S_\mathbf{v}(\mathbf{v} - \mathbf{v}_p(t)) \tag{4.12}$$

where $S_\mathbf{x}$ and $S_\mathbf{v}$ are the *shape functions* for the computational particles and N_p is the number of physical particles that are present in the element of phase space represented by the computational particle.

A number of properties of the shape functions come from their definition:

1. The support of the shape functions is compact, to describe a small portion of phase space, (i.e. it is zero outside a small range).
2. Their integral over the domain is unitary:

$$\int_{V_\xi} S_\xi(\xi - \xi_p) \mathrm{d}\xi = 1 \tag{4.13}$$

   where ξ stands for any coordinate or any velocity direction.
3. While not strictly necessary, Occam's razor suggests to choose symmetric shapes:

$$S_\xi(\xi - \xi_p) = S_\xi(\xi_p - \xi) \tag{4.14}$$

While these definitions still leave very broad freedom in choosing the shape functions, traditionally the choices actually used in practice are very few.

4.3.2 Selection of the particle shapes

A critical choice in the definition of a PIC algorithm is the choice of the shape functions.

For the velocity, $S_\mathbf{v}$, virtually all PIC methods assume a Dirac's delta in each direction:

$$S_\mathbf{v}(\mathbf{v} - \mathbf{v}_p) = \delta(v_x - v_{xp})\delta(v_y - v_{yp})\delta(v_z - v_{zp}) \tag{4.15}$$

This choice has the fundamental advantage that if all particles within the element of phase space described by one computational particle have the same speed, they remain closer in phase space during the subsequent evolution.

The original PIC methods developed in the 1950s were based on using a Dirac's delta also as the shape function in space. But now for the spatial shape functions, all commonly used PIC methods are based on the use of the so-called b-splines [9]. The b-spline functions are a series of consecutively higher order functions obtained from each other by integration. The first b-spline is the flat-top function $b_0(\xi)$ defined as:

$$b_0(\xi) = \begin{cases} 1 & \text{if } |\xi| < 1/2 \\ 0 & \text{otherwise} \end{cases} \tag{4.16}$$

The subsequent b-splines, b_ℓ, are obtained by successive integration via the following generating formula:

$$b_\ell(\xi) = \int_{-\infty}^{\infty} d\xi' b_0(\xi - \xi') b_{\ell-1}(\xi') \tag{4.17}$$

figure 4.3 shows the first three b-splines.

Two crucial properties of splines are :

(a) when a b-spline of any order is evaluated on a uniform grid of step 1, the sum over all points of evaluation is unitary, regardless of the central point ξ of the b-spline:

$$\sum_i b_\ell(\xi + i) = 1, \tag{4.18}$$

a property of great convenience in particle interpolation;

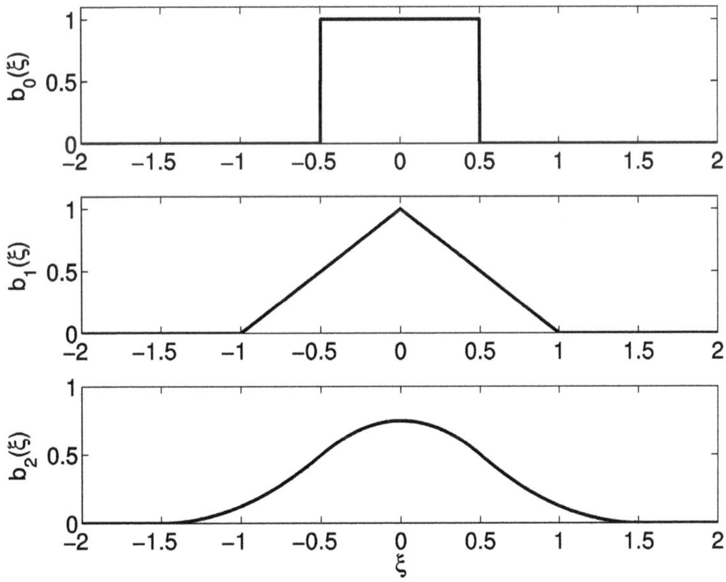

Figure 4.3. First three b-spline functions, $b_\ell(\xi)$.

(b) The integral of b-splines of any order is unitary:

$$\int_{-\infty}^{\infty} b_\ell(\xi)\mathrm{d}\xi = 1. \tag{4.19}$$

(c) The Dirac's delta $\delta(\xi)$ can be regarded as $b_{-1}(\xi)$.

Based on the b-splines, the spatial shape function of PIC methods is chosen as:

$$S_{\mathbf{x}}(\mathbf{x} - \mathbf{x}_p) = \frac{1}{\Delta x_p \Delta y_p \Delta z_p} b_\ell\left(\frac{x - x_p}{\Delta x_p}\right) b_\ell\left(\frac{y - y_p}{\Delta y_p}\right) b_\ell\left(\frac{z - z_p}{\Delta z_p}\right) \tag{4.20}$$

where Δx_p, Δy_p and Δz_p are the lengths of the support of the computational particles (i.e. its size) in each spatial dimension.

A common choice in PIC is to use the shape functions equal to the b-splines of order 0, a choice referred to as *cloud-in-cell* because the particle is a uniform square cloud in space with infinitesimal span in the velocity directions.

4.3.3 Derivation of the equations of motion

Following the procedure in reference [7], to derive the evolution equations for the free parameters $\mathbf{x}_p$ and $\mathbf{v}_p$, we require the first moments of the Vlasov equation to be exactly satisfied by the functional forms chosen for the elements. This procedure requires some explanation.

First, the Vlasov equation is formally linear in f_s and the equation satisfied by each element is still the same Vlasov equation. The linear superposition of the elements gives the total distribution function and if each element satisfies the Vlasov equation, the superposition does too. A caveat is that the fields really depend on f_s making the Vlasov equation non-linear. As a consequence, the fields used in each Vlasov equation for each element must be the total fields due to all elements, the same entering the complete Vlasov equation for f_s. We can then write the Vlasov equation for each element f_p:

$$\frac{\partial f_p}{\partial t} + \mathbf{v} \cdot \frac{\partial f_p}{\partial \mathbf{x}} + \frac{q_s}{m_s}(\mathbf{E} + \mathbf{v} \times \mathbf{B}) \cdot \frac{\partial f_p}{\partial \mathbf{v}} = 0 \tag{4.21}$$

and observe that their sum reproduces correctly the Vlasov equation for the complete distribution function f_s.

Second, the arbitrary functional form chosen for the elements does not satisfy exactly the Vlasov equation. If we substitute the definition (4.12) in equation (4.21), the equation cannot be satisfied exactly at every point. This circumstance is typical of finite element methods [10] and is addressed by requiring that the equations be satisfied only in a weak sense, meaning that some averages are satisfied even though locally point by point the equation in not exactly valid.

In analogy with the usual procedure for finite element methods, we require that the moments of the equation are satisfied. The moments of the Vlasov equation are obtained multiplying by powers of $\mathbf{x}$ and $\mathbf{v}$ and integrating. Using the zeroth order moment and the six first order moments (one for each variable in phase space), the following complete set of evolution equations is obtained for the parameters defining the functional dependence of the distribution within each computational particle [7]:

$$\begin{aligned} \frac{\mathrm{d}N_p}{\mathrm{d}t} &= 0 \\ \frac{\mathrm{d}x_p}{\mathrm{d}t} &= \mathbf{v}_p \\ \frac{\mathrm{d}v_p}{\mathrm{d}t} &= \frac{q_s}{m_s}(\mathbf{E}_p + \mathbf{v}_p \times \mathbf{B}_p) \end{aligned} \tag{4.22}$$

A great advantage of the PIC method is that its evolution equations resemble the same Newton equations as followed by the regular physical particles. The key difference is that the field is computed as average over the shape function based on the definition of $\mathbf{E}_p$ and $\mathbf{B}_p$ as:

$$\mathbf{E}_p = \int S_{\mathbf{x}}(\mathbf{x} - \mathbf{x}_p)\mathbf{E}(\mathbf{x})\mathrm{d}\mathbf{x} \tag{4.23}$$

$$\mathbf{B}_p = \int S_{\mathbf{x}}(\mathbf{x} - \mathbf{x}_p)\mathbf{B}(\mathbf{x})\mathrm{d}\mathbf{x} \tag{4.24}$$

obtained in the course of the procedure for imposing the validity of the moments of the Vlasov equation.

The set of equations above provide a closed description for the Vlasov equation. Once accompanied by an algorithm to solve Maxwell's equations the full Vlasov–Maxwell system can be solved.

In summary, the particle method, as derived above, is a discretization of phase space based on a finite set of computational particles. The continuum distribution function is replaced by a discrete mathematical representation provided by the superposition of moving fixed-shape computational particles of finite size.

The pitfalls of this representation have been studied in depth in the past. A full discussion of this aspect is beyond the scope of the present review. We refer the reader to references [4, 5, 8] for additional comments. In extreme summary, an element of phase space initially described correctly by the shape functions chosen for the discretization would be distorted by non-uniform electric and magnetic fields. Instead, the computational particles have fixed shape and neglect this effect. The Liouville theorem requires the conservation of the phase-space volume of each element, a process that is still correctly described by the discretized system.

Alternatives to PIC have been developed to remove this error by introducing additional degrees of freedom for each computational element (superparticles) [8, 11]. The additional degrees of freedom can describe the particle shape and represent its distortion in phase space. However, the additional complexity of such methods has so far prevented their widespread use.

4.4 Coupling with the field equations: spatial discretization on a grid

So far the attention has focused on the particle discretization of the Vlasov part of the Vlasov–Maxwell system. To complete the simulation method, Maxwell's equations need to be discretized as well.

A wide variety of methods have been developed to do so. Essentially all electromagnetic solvers can be coupled with particle methods [4, 5, 12]. We use a general notation where the continuum operator ∇ is replaced by a discrete operator that approximates it, using the notation ∇_g. The discretized Maxwell's equations become:

$$\begin{aligned} \nabla_g \times \mathbf{E} &= -\frac{\partial \mathbf{B}}{\partial t} \\ \nabla_g \times \mathbf{B} &= \mu_0 \mathbf{J}_g + \mu_0 \epsilon_0 \frac{\partial \mathbf{E}}{\partial t} \end{aligned} \tag{4.25}$$

supplemented by the two divergence conditions:

$$\begin{aligned} \nabla_g \cdot \mathbf{E} &= \frac{\rho_g}{\epsilon_0} \\ \nabla_g \cdot \mathbf{B} &= 0 \end{aligned} \tag{4.26}$$

Maxwell's equations require two inputs, the sources ρ_g and $\mathbf{j}_g$ that come from the information carried by the particles. In turn, the outputs are the electric and magnetic fields (or, equivalently the vector and scalar potential). The main concern here is the coupling of Maxwell's equations with the particles and its consequences.

We assume that the fields are known at grid points, not necessarily the same for both fields, and are defined as point values or averages over control volumes: $\mathbf{E}_g$, $\mathbf{B}_g$ where the index g labels the discrete points. Similarly, we assume that the chosen Maxwell scheme needs the sources at discrete grid points defined as averages over control volumes V_g defining each discrete point:

$$\begin{aligned} \rho_g &= \sum_s \frac{q_s}{V_g} \int_{\mathbb{V}} \int_{V_g} f_s \mathbf{dvdx} \\ \mathbf{j}_g &= \sum_s \frac{q_s}{V_g} \int_{\mathbb{V}} \int_{V_g} \mathbf{v} f_s \mathbf{dvdx} \end{aligned} \tag{4.27}$$

In all the formulas above we indicated the grid index as g generically. The cell positions can be chosen differently for different fields and different sources. Staggered schemes and Yee lattices are commonly used [13, 14]. The considerations below are independent of the details on how space is discretized.

Recalling equations (4.11) and (4.12), the definition of the sources for Maxwell's equations requires a three-dimensional integral of the particle shape over the control volume of each grid point:

$$\rho_g = \sum_p \frac{1}{V_g} \int_{V_g} S_{\mathbf{x}}(\mathbf{x} - \mathbf{x}_p)\mathrm{d}\mathbf{x}$$
$$\mathbf{j}_g = \sum_p \mathbf{v}_p \frac{1}{V_g} \int_{V_g} S_{\mathbf{x}}(\mathbf{x} - \mathbf{x}_p)\mathrm{d}\mathbf{x} \tag{4.28}$$

where the integrations in velocity space were done easily recalling the definition of $S_{\mathbf{v}}$. The summation is over all particles of all species (for this reason the summation over the species is not indicated explicitly).

The integral in space is straightforward but can involve complicated geometric operations if one allows general shapes for the control volumes. Instead, the use of b-splines introduced above simplifies this operation tremendously. If the particle shapes are chosen as in equation (4.20) and the grid is chosen uniform, Cartesian and with cell sizes in each dimension equal to the particle sizes, a simple and elegant reformulation of the integration step can be obtained recalling property (4.17) of b-splines. Each dimension can be done separately thanks to the choice of a Cartesian uniform grid and of the shape of particles as products of shapes in each direction. If we use the properties of the top hat function (i.e. the b-spline of order 0), the integral needed in each direction can be reformulated as

$$\int_{\Delta x_g} S_x(x - x_p) = \int_{-\infty}^{\infty} S_x(x - x_p) b_0\left(\frac{x - x_g}{\Delta x_g}\right) \mathrm{d}x \tag{4.29}$$

where Δx_g is the interval in x relative to control volume V_g. Choosing, $\Delta x_g = \Delta x_p$ (called simply Δx), recalling the definition of S_x, equation (4.20) and using equation (4.17):

$$\int_{-\infty}^{\infty} S_x(x - x_p) b_0\left(\frac{x - x_g}{\Delta x}\right) \mathrm{d}x = b_{l+1}\left(\frac{x_g - x_p}{\Delta x}\right) \tag{4.30}$$

Collecting the integral in each direction, the so-called interpolation function can be defined:

$$W(\mathbf{x}_g - \mathbf{x}_p) = b_{l+1}\left(\frac{x_g - x_p}{\Delta x}\right) b_{l+1}\left(\frac{y_g - y_p}{\Delta y}\right) b_{l+1}\left(\frac{z_g - z_p}{\Delta z}\right) \tag{4.31}$$

The summation property of the b-splines, equation (4.18), ensures that:

$$\sum_g W(\mathbf{x}_g - \mathbf{x}_p) = 1 \tag{4.32}$$

and the sum of the fractional contributions of a particle to the grid is unitary regardless of the position of the particle.

The interpolation function is a direct consequence of the choice made for the shape function. Using b-splines proves a powerful choice as it allows us to write the interpolation function just as the b-spline one order higher than that used for the shape. The calculation of the sources for Maxwell's equations becomes then just the sum of a number of function evaluations without requiring geometrically complex integrals:

$$\rho_g = \frac{1}{V_g}\sum_p q_p W(\mathbf{x}_g - \mathbf{x}_p)$$
$$\mathbf{j}_g = \frac{1}{V_g}\sum_p q_p \mathbf{v}_p W(\mathbf{x}_g - \mathbf{x}_p) \tag{4.33}$$

An analogous set of operations can be carried out for connecting the output of Maxwell's equations to the particle equations of motion. The needed quantities are the electric and magnetic fields derived above in equations (4.23) and (4.24). The definitions for the particle fields require continuum fields. These can be obtained once again using interpolation:

$$\mathbf{E}(\mathbf{x}) = \sum_g \mathbf{E}_g S_\mathbf{E}(\mathbf{x} - \mathbf{x}_g)$$
$$\mathbf{B}(\mathbf{x}) = \sum_g \mathbf{B}_g S_\mathbf{B}(\mathbf{x} - \mathbf{x}_g) \tag{4.34}$$

where $S_\mathbf{E}(\mathbf{x} - \mathbf{x}_g)$ and $S_\mathbf{B}(\mathbf{x} - \mathbf{x}_g)$ are the fields interpolation functions. Upon substitution in the definition of the particle fields,

$$\mathbf{E}_p = \sum_g \int S_\mathbf{x}(\mathbf{x} - \mathbf{x}_p)\mathbf{E}_g S_\mathbf{E}(\mathbf{x} - \mathbf{x}_g)\mathrm{d}\mathbf{x}$$
$$\mathbf{B}_p = \sum_g \int S_\mathbf{x}(\mathbf{x} - \mathbf{x}_p)\mathbf{B}_g S_\mathbf{B}(\mathbf{x} - \mathbf{x}_g)\mathrm{d}\mathbf{x} \tag{4.35}$$

The geometrical complexity of the integrals can be avoided choosing the interpolation functions for the electric and magnetic fields as b-splines of order 0:

$$S_{\mathbf{E},\mathbf{B}}(\mathbf{x} - \mathbf{x}_g) = b_0\left(\frac{x_g - x_p}{\Delta_x}\right)b_0\left(\frac{y_g - y_p}{\Delta_y}\right)b_0\left(\frac{z_g - z_p}{\Delta_z}\right) \tag{4.36}$$

leading to

$$\mathbf{E}_p = \sum_g \mathbf{E}_g W(\mathbf{x}_g - \mathbf{x}_p)\mathrm{d}\mathbf{x}$$
$$\mathbf{B}_p = \sum_g \mathbf{B}_g W(\mathbf{x}_g - \mathbf{x}_p)\mathrm{d}\mathbf{x} \tag{4.37}$$

with the same interpolation function used for the sources.

The majority of PIC codes use b-spline interpolations and can take advantage of the convenient interpolation function defined above. For staggered grid, the interpolation will happen to different grid points for different quantities, but using the same interpolation function. In some implementations, different orders of interpolations can be used for different quantities [15]. A more radical solution is to consider non-uniform [16, 17] or even unstructured grids [18]. In the most complex cases, the exchange of information between particles and cells cannot use the interpolation function derived using b-splines and the integral definitions need to be used directly.

4.5 Temporal discretization of the particle methods

The coupled Vlasov–Maxwell system requires three discretizations. We have first described how to discretize phase space in the Vlasov equation using finite-size computational particles that resemble flat strings of finite size in space but of infinitesimal size in velocity. Next we have given some indications on how to discretize space. Space needs to be discretized on Maxwell's equations and a vast array of possibilities exist in the computational electromagnetism. We focused on how to couple these discretizations with the computational particle information.

The last and most important discretization remaining is time. Again for Maxwell's equations the temporal discretization can be done in any of the various choices used in computational electromagnetism. What we need to focus on is how to combine the time discretization of the Maxwell equations with that of the particles. Figure 4.4 illustrates the concept. The field equations are solved on a grid where the fields are advanced based on the sources provided by the particles. The

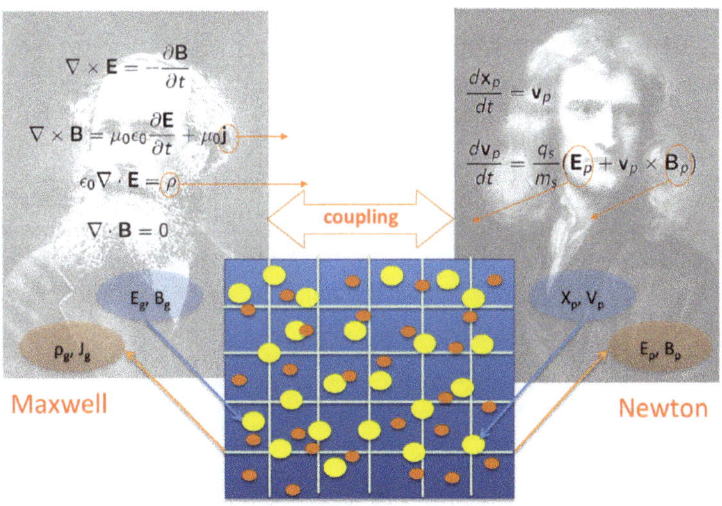

Figure 4.4. Illustration of the coupling between Maxwell's and Newton's equations. Newton's equations need the electric and magnetic fields and Maxwell's equations need the particle positions to compute the sources: current and density. The computational particles are indicated as coloured dots, bigger yellow ions and smaller orange electrons. The dot size is indicative of the mass of the particles, not their sizes. All computational particles have the same size: their size is identical to the size of the cells but with their centre on the centre of the particles.

particle equations are solved using the fields computed on the grid. The interpolation formulas derived above provide the rule to exchange information between grid and particles.

The question is how to deal with the coupling of Maxwell's equations for the fields and Newton's equations for the computational particles. In principle as the particles move, the fields evolve as well, any change in one side of figure 4.4 reflects on the other. The first point to keep in mind is the vastness of the information being exchanged. Modern PIC methods on supercomputers (i.e. parallel computers made of hundreds of thousands of processors) use millions of cells and billions of particles. In fact, the state of the art is starting to reach the trillion particle level. The desire to use more and more particles to cover phase space more accurately suggests to keep the operations per particle as simple as possible.

This guiding principle has long suggested using explicit methods, a decision whose wisdom will be questioned below. In explicit methods, the two sets of equations, Maxwell's and Newton's are solved in a marching order. Using the visualization scheme in figure 4.4, we can assume that each side can be advanced for a small time step, while the other is assumed temporarily frozen. We can advance the particles for a small time step in given fields. We use then the new particle positions and speeds to compute the current and density to advance the fields for the same small time step. If the time step is small enough, this procedure has a small error. The method just described is called explicit. The advantage of the explicit approach is to be extremely simple but the disadvantage is that the time step has to be very small.

4.5.1 Explicit temporal discretization of the particle equations

Let us start with discretizing in time Newton's equations for the particles. The simplest and most widely used explicit algorithm is the so-called *leap-frog algorithm* based on staggering the time levels of the velocity and position by half time step: $\mathbf{x}_p(t = n\Delta t) \equiv \mathbf{x}_p^n$ and $\mathbf{v}_p(t = (n + 1/2)\Delta t) \equiv \mathbf{v}_p^{n+1/2}$. The update of position from time level n to time level $n + 1$ uses the velocity at mid-point $\mathbf{v}_p^{n+1/2}$, and similarly the update of the velocity from time level $n - 1/2$ to $n + 1/2$ uses the mid-point position x_p^n. This stepping of velocity over position and of position over velocity gives the method its name for its resemblance to the children's game bearing the same name (see figure 4.5).

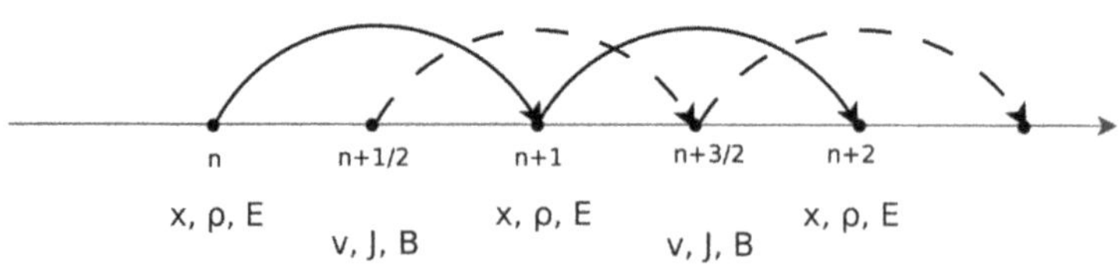

Figure 4.5. Visual representation of the leap-frog algorithm. The time discretization is staggered, with the electric field, charge density and particle positions at integer times and the magnetic field, current and particle velocity at half time steps.

The scheme is summarized by:

$$\begin{aligned} \frac{\mathbf{x}_p^{n+1} - \mathbf{x}_p^n}{\Delta t} &= \mathbf{v}_p^{n+1/2} \\ \frac{\mathbf{v}_p^{n+1/2} - \mathbf{v}_p^{n-1/2}}{\Delta t} &= \frac{q_s}{m_s}\mathbf{E}^n(x_p^n) + \frac{q_s}{m_s}\left(\frac{\mathbf{v}_p^{n+1/2} + \mathbf{v}_p^{n-1/2}}{2}\right) \times \mathbf{B}^n(x_p^n) \end{aligned} \tag{4.38}$$

where the use of $\mathbf{E}_p^n$ and $\mathbf{B}_p^n$ implies knowing the solution of Maxwell's equations given the particle information. The first equation is clearly explicit, the second can be formulated explicitly as a roto-translation of the vector $\mathbf{v}_p^{n-1/2}$. The velocity equation can be rewritten in the equivalent explicit form:

$$\mathbf{v}_p^{n+1/2} = 2\left(\hat{\mathbf{v}}_p + \beta_s \widehat{\mathbf{E}}_p\right) - \mathbf{v}_p^{n-1/2} \tag{4.39}$$

where hatted quantities have been rotated by the magnetic field:

$$\begin{aligned} \hat{\mathbf{v}}_p &= \alpha_p^n \mathbf{v}_p^{n-1/2} \\ \widehat{\mathbf{E}}_p &= \alpha_p^n \mathbf{E}_p^n \end{aligned} \tag{4.40}$$

via a rotation matrix α_p^n defined as:

$$\alpha_p^n = \frac{1}{1 + (\beta_s B_p^n)^2}\left(\mathbb{I} - \beta_s \mathbb{I} \times \mathbf{B}_p^n + \beta_s^2 \mathbf{B}_p^n \mathbf{B}_p^n\right) \tag{4.41}$$

where $\mathbb{I}$ is the dyadic tensor (matrix with diagonal of 1) and $\beta_s = q_p \Delta t / 2m_p$ (independent of the particle weight and unique to a given species). The shorter notation $\mathbf{E}^n(\mathbf{x}_p^n) = \mathbf{E}_p^n$ has been introduced for all fields. An alternative implementation of the rotation procedure above is called the Boris mover [4].

Note that technically the leap-frog algorithm is second-order accurate, when instead the regular explicit Euler-scheme is only first order. Nevertheless, the two differ in practice only for the fact that the velocity is staggered by half time step. This staggering is achieved by moving the initial velocity of the first time cycle by half a time step using an explicit method:

$$\frac{\mathbf{v}_p^{1/2} - \mathbf{v}_p^0}{\Delta t/2} = \frac{q_s}{m_s}\mathbf{E}_p^0(x_p^0) + \frac{q_s}{m_s}\left(\frac{\mathbf{v}_p^{1/2} + \mathbf{v}_p^0}{2}\right) \times \mathbf{B}_p^0(x_p^0) \tag{4.42}$$

after the first step all subsequent steps would be identical to the Euler explicit scheme.

4.5.2 Explicit PIC cycle

After discretizing the particles equation of motion, Maxwell's equations need also to be discretized in time. As for space, also in time different methods are available and the interested reader is referred to textbooks on that subject [14]. The simplest approach is to use again the leap-frog algorithm also for the fields. The electric field is computed at the integer levels while the magnetic field is computed at the intermediate times (see figure 4.5). Starting from the partially discretized equations (4.25) introduced above, time can be discretized as:

$$\begin{aligned} \nabla_g \times \mathbf{E}^n &= \frac{\mathbf{B}^{n+1/2} - \mathbf{B}^{n-1/2}}{\Delta t} \\ \nabla_g \times \mathbf{B}^{n+1/2} &= \mu_0 \mathbf{J}_g^{n+1/2} + \mu_0 \epsilon_0 \frac{\mathbf{E}^{n+1} - \mathbf{E}^n}{\Delta t} \end{aligned} \tag{4.43}$$

where all differences are time-centered, resulting in a second-order accurate scheme.

The current needs to be accumulated from the particle information at intermediate times $n + 1/2$. Smartly, the leap-frog algorithm for the particles has the particle velocities at that same time level, but unfortunately the position is staggered by half time step. A simple choice is to use an averaging method:

$$\mathbf{j}_g^{n+1/2} = \frac{1}{V_g} \sum_p q_p \mathbf{v}_p^{n+1/2} \frac{W(\mathbf{x}_g - \mathbf{x}_p^n) + W(\mathbf{x}_g - \mathbf{x}_p^n)}{2} \tag{4.44}$$

In a similar manner the magnetic field at integer times needed in the mover is obtained by averaging from the neighbouring intermediate times:

$$\mathbf{B}^n = \frac{1}{2}(\mathbf{B}^{n+1/2} + \mathbf{B}^{n-1/2}) \tag{4.45}$$

In the continuum, the divergence conditions are valid at all times if they are valid at the initial time and if the charge continuity equation is satisfied. In PIC the latter condition is not necessarily valid. The divergence of the magnetic field is always zero, if the discretized operator retains the property that the divergence of the curl is zero. This condition is often easily satisfied. But the other divergence condition is a major issue in PIC.

Historically two solutions have been adopted. First, the charge and current interpolation from the particle to the grid can be modified in a manner that ensures the validity of the charge continuity equation: these schemes are called charge conserving [19]. Second, the divergence condition can be applied as a correction to the electric field. At each time step, or at regular intervals, the electric field can be *cleaned* for any component not satisfying the Gauss law. The operation is called divergence cleaning and is numerically rather expensive because it requires solving an elliptic problem with an iterative method. Simplified approaches have been developed [20].

In practice, the two methods can be used in combination because even though a charge conserving scheme might do a good job at keeping the divergence correct, numerical errors will start to accumulate and a divergence cleaning operation can be judiciously applied regularly after a number of time steps.

The end result is a simple marching algorithm that is illustrated in figure 4.6. Each step in the figure is self-contained and requires only information that is available

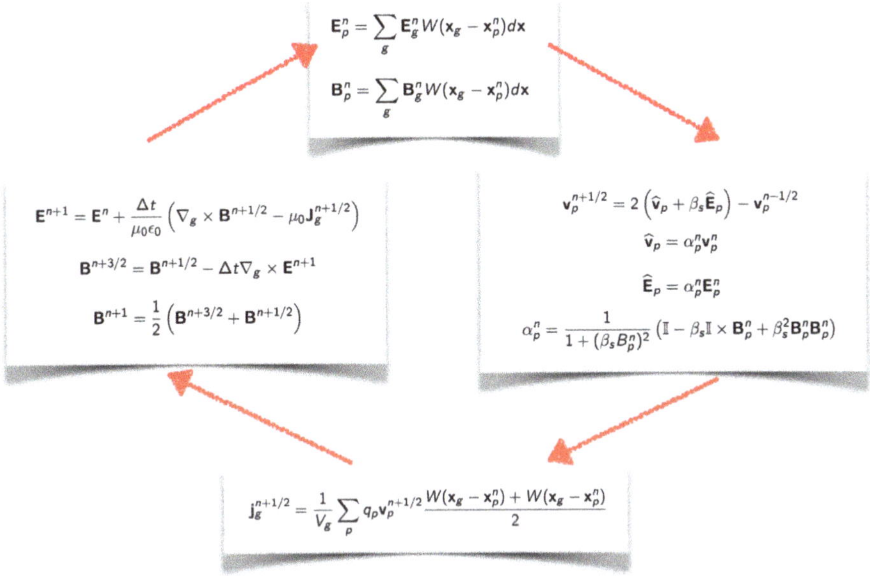

Figure 4.6. Explicit PIC cycle. Four steps are followed in sequence over a time step Δt. Starting for example from the top, the electric and magnetic fields from the previous time step are interpolated to the particles. Next the particles are advanced over Δt to find the new position and velocity. The particle information is then interpolated to the grid to obtain the current density in each cell. The fields are then advanced for the same Δt on the grid. A new time step is ready to start.

from the previous steps in the cycle. There is no iteration needed. This is the critical advantage of explicit PIC: the algorithm advances forward without any iterations. But this comes at a price. We have assumed in each step that the others remain frozen. During the time step Δt of the cycle, the fields are frozen while the particles move and the particles are frozen while the fields evolve. This is only acceptable if Δt is small. Very small.

4.5.3 Electrostatic explicit methods

In many applications, the currents are sufficiently small to justify an electrostatic approximation. In this case the two curl equations can be ignored and the only remaining equation in the Maxwell system is the Gauss law that can be rewritten in terms of the electrostatic potential:

$$\Delta_g \phi = -\frac{\rho_g}{\epsilon_0} \tag{4.46}$$

where the Laplacian operator is discretized on the grid.

In this case there is no escape form the need of solving the elliptic problem with the curious consequence that explicit electrostatic PIC is in fact more complex and expensive than explicit electromagnetic PIC. But note that using an electromagnetic PIC for an electrostatic case requires one to make sure the divergence cleaning is very accurate. In essence this requires again solving the elliptic Poisson equation and the cost is the same. There is no forgiveness in electromagnetism.

4.5.4 Stability of the explicit PIC method

Explicit PIC is subject to three stability conditions. Exceed them and energy explodes rendering the simulation meaningless. Users must never exceed these limits.

4.5.4.1 Stability of the particle mover

The discretization of the equations of motion using the leap-frog algorithm (or other similar explicit time differencing) introduces a stability constraint. As usual, stability can be studied with the Von Neumann analysis [4, 5, 21]: the equations are linearized and their stability is studied by Fourier analysis. Note that the leap-frog algorithm is implicit in the Lorentz force term, but explicit on the electric force. As a consequence only the latter is of primary concern and stability can be studied in the limit of zero magnetic field.

Stability needs to address all temporal scales in the system. Obviously the fastest are the most challenging for an explicit scheme. Among the plasma waves, the fastest electrostatic wave is the Langmuir (i.e. electron plasma) wave.

We proceed using the Von Neumann analysis [21], meaning we assume the temporal evolution to be harmonic so that all quantities depend on time as $e^{i\omega t}$:

$$\begin{aligned}\mathbf{x}_p^n &= \tilde{\mathbf{x}}_p e^{i\omega n\Delta t}\\ \mathbf{v}_p^{n+1/2} &= \tilde{\mathbf{v}}_p e^{i\omega(n+1/2)\Delta t}\\ \mathbf{E}_p^n &= \tilde{\mathbf{E}}_p e^{i\omega n\Delta t}\end{aligned} \tag{4.47}$$

where ω is the temporal frequency of the numerical response.

We consider the mover introduced in equation (4.38) repeated here for the case of a Langmuir wave (i.e. without the magnetic field) in a single spatial and single velocity dimension:

$$\begin{aligned}x_p^{n+1} &= x_p^n + \Delta t v_p^{n+1/2}\\ v_p^{n+1/2} &= v_p^{n-1/2} + \frac{q_p\Delta t}{m_p}E_p^n\end{aligned} \tag{4.48}$$

Substituting in each term the harmonic dependence, we obtain in the Fourier transformed space:

$$\begin{aligned}\tilde{x}_p(e^{i\omega\Delta t} - 1) &= \tilde{v}_p\Delta t e^{i\omega\Delta t/2}\\ \tilde{v}_p(e^{i\omega\Delta t/2} - e^{-i\omega\Delta t/2}) &= \frac{q_s\Delta t}{m_s}\tilde{E}_p\end{aligned} \tag{4.49}$$

Recalling now the basic property of cold plasma Langmuir waves that relate the variation of the particle displacement with the electric field in Fourier space as [22]:

$$\frac{q_s\tilde{E}}{m_s} = -\omega_{pe}^2\tilde{x} \tag{4.50}$$

the final solution is given by the algebraic system:

$$\begin{pmatrix} e^{i\omega\Delta t} - 1 & -\Delta t e^{i\omega\Delta t/2} \\ \omega_{pe}^2\Delta t & e^{i\omega\Delta t/2} - e^{-i\omega\Delta t/2} \end{pmatrix}\begin{pmatrix} \tilde{x}_p \\ \tilde{v}_p \end{pmatrix} = 0 \tag{4.51}$$

The dispersion relation can be computed as the determinant of the system matrix [4, 5]:

$$\left(\frac{\omega_{\mathrm{pe}}\Delta t}{2}\right)^2 = \sin^2\left(\frac{\omega\Delta t}{2}\right) \tag{4.52}$$

where the Euler formula for the sin function was used. The dispersion relation (4.52) expresses the frequency of the numerical solution for particle motion in Langmuir waves with physical frequency ω_{pe}. The numerical frequency is not the same as the physically correct frequency.

Notoriously the sin function of real arguments has values only in the $[-1, 1]$ interval. As a consequence, for any values of $\omega_{pe}\Delta t > 2$, the Von Neumann analysis leads to a complex numerical oscillation frequency, ω. Indeed with the problem being real to begin with, two complex conjugate solutions ω arise, one of the two being damped and the other a growing exponential. The numerical solution becomes unstable, with a numerical growth rate that has no physical equivalence. In practice, the particles heat unboundedly and quickly and the simulation fails within a few time steps.

The first cardinal rule of explicit plasma simulation is that the condition

$$\omega_{pe}\Delta t < 2 \tag{4.53}$$

should never be violated. Indeed, practice advises the use of a considerably smaller time step of order $\omega_{pe}\Delta t = .1$, to avoid numerical heating [4].

The temporal stability constraint produces a first great challenge to the application of PIC methods in practice. The range of time scales typical in plasmas covers several orders of magnitude. The electrons are much faster than the ions (due to the mass difference). What the stability constraint imposes is to choose a time step that is able to resolve the fastest frequency in the system: the electron plasma frequency. If we are interested in following longer time scales, we still need to use a time step that follows the Langmuir waves. Even if our insight tells us there is nothing of interest in those small scales.

4.5.4.2 Stability due to the explicit time differencing of the field equations

Discretizing the Maxwell's equations explicitly leads to the imposition of the Courant–Friedrich–Levy (CFL) condition. This condition arises whenever a hyperbolic equation is discretized explicitly [23]. The CFL condition states that the time step must not exceed the time taken by the characteristic signals of the hyperbolic equation to travel one cell. In the case of the Maxwell's equations the characteristic speed of the signal is the speed of light, c. The CFL condition then requires:

$$\Delta t < \Delta x/c \tag{4.54}$$

This condition is in a sense less damaging than the previous one because it can be satisfied by a large Δt if Δx is also large. Often in practice, one needs to resolve long term processes developing over macroscopic scales. In this case both Δt and Δx are large. Additionally it is relatively easy to discretize just the Maxwell's equations implicitly removing just this constraint [24]. As we will see below, what is difficult is to deal with the coupling of particles and fields implicitly. But just discretizing the Maxwell's equations implicitly is a simple task by comparison.

4.5.4.3 Space resolution and finite grid instability

The introduction of a grid to compute the field characterizes the particle methods for plasma physics as particle-in-cell (PIC) methods. The particles move in a continuum space, but their information is projected to grid points using integration over control volumes. This operation reduces the information, collapsing the continuum particle shapes to discrete contributions in a set of points (the grid values). The loss of information allows one numerical instability to creep in: the finite grid instability.

The mathematical study of the finite grid instability is complex, requiring a careful and complete analysis of all computational steps with the Laplace transformation of time and the Fourier transformation of space. The reader interested in the details is referred to the textbooks on the subject [4, 5] and to the original work cited therein. The summary of the analysis is that to avoid the finite grid instability in explicit particles methods, the grid spacing must be chosen to satisfy the constraint

$$\Delta x/\lambda_{\mathrm{De}} < \varsigma \tag{4.55}$$

where λ_{De} is the Debye length [4, 5] and the parameter ς is of order one and depends on the details of the implementation. For the widely used choice of shape functions as b-spline of order 0 and consequently interpolation functions of order 1, the scheme referred to as cloud-in-cell (CIC), the literature reports, $\varsigma \approx \pi$. Using filtering methods and higher order interpolation, this number can be increased significantly allowing the use of larger cells.

The practical consequence of the finite grid instability is a tremendous numerical heating of the plasma characterized by an alternately positive–negative variation of the electric field accompanied by a correlated zig-zag perturbation of the phase-space. In a very short time, the energy reaches the bounds of overflow in the machine representation of numbers. The finite grid instability must be avoided, requiring its stability constraint to be respected everywhere in the domain for all directions.

The limit imposed by this stability constraint is devastating. The spatial scale to resolve is the Debye length in the highest density region of the domain where the Debye length is smallest. If using uniform grids, this constraint requires us to use cell sizes that can be several orders of magnitude smaller than the scales of interest. Note that this consideration has two sides.

First, often the processes of interest do not reach down to the Debye scale. For example most electromagnetic processes are at their smallest on the electron inertial length, $d_e = c/\omega_{\rm pe}$. This scale can be orders of magnitude larger than the Debye length. We recall that the Debye length is the electron thermal speed divided by the plasma frequency while the electron inertial length is the speed of light divided by the plasma frequency: the inertial length is obviously larger than the Debye length by the ratio $v_{\rm the}/c$. Having to resolve the Debye length purely for numerical stability reasons implies a great waste in each direction. When done in 3D, this waste compounds. If the time scale constraint encountered before hits the explicit approach once, the spatial stability constraint hits it three times, once per spatial dimension. The combined effect of these two constraints makes it imperative to look beyond explicit particle methods for many practical applications. The ability to relax these two constraints by one order of magnitude, for example, would result in a gain of four orders of magnitude: one for time and three for the three spatial dimensions.

Second, often plasmas present regions of high density in contact with regions of lower density, resulting in a large range of Debye lengths. To avoid the finite grid instability, uniform grids need to resolve the smallest Debye length in the system. Adaptive schemes can be devised to resolve in each region only the local Debye length, resulting in considerable savings [25–27]. However, even in this case, often the local scales of interest are much larger than the Debye length and the local grid resolution would be more desirably chosen according to other considerations than the need to resolve the Debye length.

4.6 Implicit particle methods

An approach to avoid the prohibitively small time steps and grid spacing required by explicit methods is to reduce the physics model used.

For example, the speed of light limitation can be eliminated by retaining only some of the full electromagnetic physics while removing from the Maxwell's equations the terms allowing the propagation of light. The approximation is called Darwin [28] and is used when slow-varying magnetic fields are present without any important role of electromagnetic waves.

Other approaches are based on eliminating some of the electron physics to eliminate the need for resolving the Debye length and the plasma frequency. If electrons are treated as fluids and neutrality is enforced, both conditions can be eliminated. The approach is called hybrid and can be implemented in varying degrees of complexity [29]. Another very important approximation is that of averaging over the gyration of particles in large magnetic fields. This approximation is called gyrokinetic and is widely used in fusion research and in many systems where large magnetic fields are present [30].

The other approach to eliminate the stability constraints is to retain the full first principle description of the electron physics and of the electromagnetic fields but to rely on more advanced numerical methods that can handle the presence of multiple scales in space and time. To avoid the prohibitively large number of time steps and the exceedingly high resolution needed to run realistic problems with explicit particle methods, implicit methods have been considered for several decades [31–34].

The constraints of the explicit methods come from the artificial freezing of the coupling between fields and particles during the time step. The obvious solution is to reintroduce this coupling and solve the field equations on the grid and the particle equations maintaining their coupling. This coupling of course is non-linear as the particles are the sources of the current and density in the Maxwell's equations and the fields in the particle equations are produced by the Maxwell's equations.

In the 1970s, 1980s and 1990s, the consensus was that solving the full non-linear coupled system of field equations and particle equations was not feasible. But the situation has changed in recent years. First, computers are of course vastly more powerful. Second and more importantly, the progress made in the domain of Newton–Krylov solvers made this task possible [35, 36].

Two independent teams proposed fully implicit PIC codes [37, 38] capable of solving electrostatic, full electromagnetic [37], fully relativistic [39] or reduced Darwin descriptions of plasmas [40]. The key aspect of these methods is that they completely remove the need to satisfy any of the constraints above. Additionally and equally importantly, these methods are exactly energy conserving.

All previous PIC methods did not conserve energy. Explicit PIC methods can be readily formulated to conserve momentum exactly [41]. With respect to energy conservation, unfortunately the situation is far less rosy. The random aspect linked with the presence of discrete particles induces a tendency for the explicit PIC method to generate numerical heating: random fluctuations in the particle information lead to the generation of electromagnetic waves at small scales (noise). Energy is not only not conserved but it tends to increase in time [4, 5]. The increase tends to be secular at small time steps but it becomes exponential in the presence of finite-grid instability [42]. The aforementioned stability constraints are then supplemented by the additional requirement of maintaining good energy conservation, typically limiting the time step by an additional order of magnitude to $\omega_{pe}\Delta t \sim 0.1$, but sometimes requiring even smaller time steps to retain an accurate description of particle energization [39].

A class of so-called explicit 'energy conserving' PIC [43] (the quotation marks are used in the literature [44]) has been developed where the energy was conserved only in the limit of zero time step. Their peculiarity was the use of different interpolation functions for charge and current and for the magnetic and electric fields, resulting in improved energy conservation. In practice energy conservation was marginally better than in other explicit PIC, but still energy was by no means exactly conserved. For example, the code Celeste belonged to the class of 'energy conserving' PIC of the old definition [15].

The new fully implicit PIC are exactly energy conserving, the acronym ECPIC (energy conserving PIC) is then properly deserved. One can prove that if the discretized equations defining the ECPIC are solved exactly, energy is conserved exactly. The ECPIC relies on a Newton iteration to solve the model equations. Of course, the Newton iteration is stopped at a given tolerance and the energy will then also have the same tolerance in the error. But the method is in principle exactly energy conserving. In practice one can easily achieve relative energy errors below 10^{-9} as in the examples below or any desired accuracy within machine precision. This accuracy is many orders of magnitude better than anything published before.

The basic common idea of the ECPIC methods is to replace the leap-frog algorithm with an implicit particle mover where the fields appear at the advanced time level, requiring an iteration with the field equations. For example, reference [37] proposes the θ-scheme:

$$\begin{aligned} \mathbf{x}_p^{n+1} &= \mathbf{x}_p^n + \Delta t \mathbf{v}_p^{n+\theta} \\ \mathbf{v}_p^{n+1} &= \mathbf{v}_p^n + \frac{q_p \Delta t}{m_p}\left(\mathbf{E}_p^{n+\theta}(\bar{\mathbf{x}}_p) + \bar{\mathbf{v}}_p \times \mathbf{B}_p^{n+\theta}(\bar{\mathbf{x}}_p)\right) \end{aligned} \tag{4.56}$$

where barred variables are average between time level n and $n+1$, $\bar{\mathbf{x}} = (\mathbf{x}^{n+1} + \mathbf{x}^n)/2$ and variables of time index $n+\theta$ are defined as $\mathbf{x}^{n+\theta} = \theta\mathbf{x}^{n+1} + (1-\theta)\mathbf{x}^n$. The same notation applies to all variables, particles and fields.

The implicit scheme is unconditionally stable. If the same Von Neumann analysis done for the leap-frog scheme is repeated for equation (4.56), the following relation is obtained:

$$\left(\frac{\omega_{\rm pe}\Delta t}{2}\right)^2 = \tan^2\left(\frac{\omega\Delta t}{2}\right) \tag{4.57}$$

Now the tan function is unbounded and any values of Δt lead to real values for ω without any numerical instability. Figure 4.7 reports the stability analysis for explicit and implicit methods. The frequency of the numerical solution is compared with the

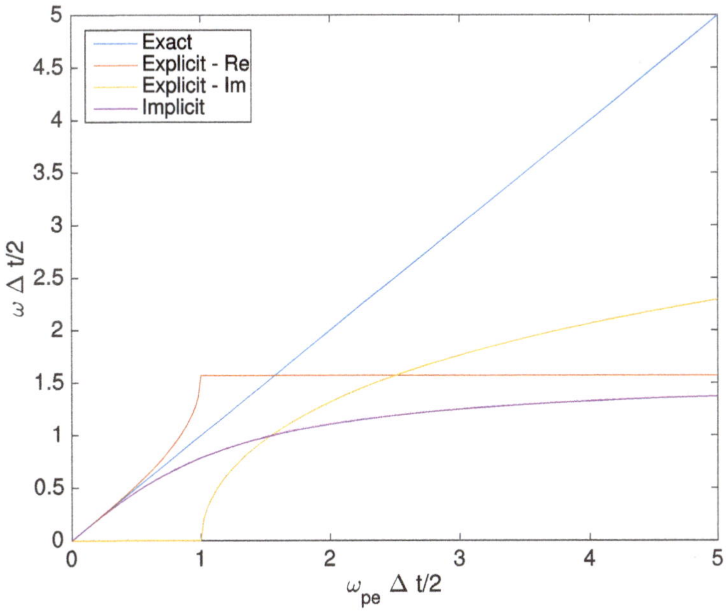

Figure 4.7. Results of the stability analysis of PIC particle movers. The explicit leap-frog mover is compared with the implicit mover in [37]. The explicit method is stable only when $\omega_{pe}\Delta t < 2$. For larger Δt, the numerical solution has an imaginary component of the frequency that leads to numerical growth. The implicit method is always stable.

actual correct analytical solution in the presence of Langmuir waves. As Δt is increased, the numerical solution of the explicit method increases compared with the correct solution, until the stability limit is reached. When the limit is exceeded, an imaginary part to the numerical frequency appears, leading to growing, numerically unstable solutions. The implicit method instead is always stable.

When $\theta = 1/2$, the method is exactly energy conserving [37, 38], provided the field and particle equations are solved to non-linear consistency within a Newton iteration.

Of course, accuracy still imposes bounds on the time step. Since the mover uses only one punctual value of the fields, if a particle travels more than one cell per time step the formula becomes inaccurate, leading to severe non-momentum conservation. Even this limitation is removed if sub-cycling is used [38].

As mentioned above, this modern solution was not available until recent times. Even today it remains computationally challenging and it requires the deployment of Newton–Krylov non-linear solvers, complicating significantly the optimization of massively parallel PIC codes. The use of non-linear solvers requires global communication and the storage of multiple states of the unknown to form the Krylov space orthogonalization. When applied to a full particle-fields coupled system, this leads to extreme memory requirements. This problem can be circumvented with particle-enslavement and other preconditions [37, 45] but at the cost of further complexity in code development.

The solution adopted in the earlier days was that of linearizing the particle-field coupling and solving the resulting linear system. This method is semi-implicit in the sense that the formulating equations are still fully coupled and require an iteration but the iteration is linear and not solved to non-linear consistency. Two approaches have been presented.

The direct implicit method, developed at Lawrence Livermore National Laboratory and implemented in codes such as AVANTI and the LSP [34, 47, 46, 48, 49]. This method is based on replacing the leap-frog algorithm with more advanced implicit movers. For example, the so-called D_1 algorithm (formulated with recursive filter on the fields [50]):

$$\begin{aligned} \mathbf{x}_p^{n+1} &= \mathbf{x}_p^n + \Delta t \mathbf{v}_p^{n+1/2} \\ \mathbf{v}_p^{n+1/2} &= \mathbf{v}_p^{n-1/2} + \frac{q_p}{m_p}\left(\bar{\bar{\mathbf{E}}}_n + \frac{\mathbf{v}_p^{n+1/2} + \mathbf{v}_p^{n-1/2}}{2} \times \mathbf{B}^n(\mathbf{x}_p^n)\right) \end{aligned} \tag{4.58}$$

where the only difference with the standard leap-frog method of equation (4.5) is the calculation of the electric field acceleration as:

$$\bar{\bar{\mathbf{E}}}_n = (\mathbf{E}_{n-1}(\mathbf{x}_n) + \mathbf{E}_{n+1}(\mathbf{x}_n))/2 \tag{4.59}$$

A small formal difference but a huge numerical difference. The scheme now requires the new advanced field that can be calculated only after moving the particles. This non-linear iteration is avoided by the direct implicit method by computing the linear response of the plasma.

In the direct implicit method, the plasma response (i.e. the changes in the electric field in response to the particle motion) is obtained by linearizing the particle equations of motion about an estimate (extrapolation) for their unknowns at the new time level $n + 1$. The mathematical procedure used can vary from author to author, but it leads to a form of sensitivity matrix of the linearization that assumes the form of a plasma susceptibility [51]. The field equations can then be formulated using this matrix and solved alone without iterating with the particles. After the new field is computed, the particles can be moved.

An alternative approach, also going back to the late 1970s, is the implicit moment method (IMM), developed at Los Alamos National Laboratory [31, 33, 51]. In this approach, the linearization is obtained by linearizing the moments rather than the particle equations of motion. The method has been used in a family of codes originating with Venus [33], followed by Celeste [15, 52], Parsek2D [53] and now used in iPic3D [54].

The implicit moment method is based on a scheme similar to the θ-scheme described above in the fully implicit ECPIC method:

$$\begin{aligned} \mathbf{x}_p^{n+1} &= \mathbf{x}_p^n + \Delta t \mathbf{v}_p^{n+\theta} \\ \mathbf{v}_p^{n+1} &= \mathbf{v}_p^n + \frac{q_p \Delta t}{m_p}\left(\mathbf{E}^{n+\theta}(\bar{\mathbf{x}}_p) + \bar{\mathbf{v}}_p \times \mathbf{B}^n(\bar{\mathbf{x}}_p)\right) \end{aligned} \tag{4.60}$$

differing only in the time level of the magnetic field. As in the case of the leap-frog algorithm, equation (4.39), the velocity equation can be rewritten in the equivalent form:

$$\bar{\mathbf{v}}_p = \hat{\mathbf{v}}_p + \beta_s \widehat{\mathbf{E}}_p \tag{4.61}$$

where the same notation is used as in equations (4.40)–(4.41) with the critical difference that now $\widehat{\mathbf{E}}_p = \alpha_p^n \mathbf{E}_p^{n+\theta}$ is based on the new advanced electric field that can only be known once the field equations are advanced.

The key defining property of the IMM is that the θ-scheme is now coupled to the Maxwell equations via a linearization procedure. The two curl Maxwell equations are discretized in time with another θ-scheme:

$$\begin{aligned} \nabla_g \times \mathbf{E}^{n+\theta} + \frac{1}{c}\frac{\mathbf{B}_g^{n+1} - \mathbf{B}_g^n}{\Delta t} &= 0 \\ \nabla_g \times \mathbf{B}^{n+\theta} - \frac{1}{c}\frac{\mathbf{E}_g^{n+1} - \mathbf{E}_g^n}{\Delta t} &= \frac{4\pi}{c}\bar{\mathbf{J}}_g \end{aligned} \tag{4.62}$$

The spatial operators in equation (4.62) are discretized on a grid. Consistent with the notation introduced earlier, we index the grid elements generically with the index g. ∇_g is a shorthand for the spatial discretization used. For example in the IMM family of codes the discretization used is different from the Yee scheme [55] and locates on the cell centres the magnetic field and on the cell vertices for the electric field [7, 13, 54]. But all the derivations below are not critically dependent on which spatial discretization is used.

The current is computed as average of the old and new time level, requiring again to solve the particle equations of motion before the fields can be computed and vice versa. This iteration is avoided by expanding the current in Taylor series truncated at the first order [15]. Computing the current separately for each species s, one obtains:

$$\bar{\mathbf{J}}_{sg} = \widehat{\mathbf{J}}_{sg} - \frac{\Delta t}{2}\mu_{sg} \cdot E_\theta - \frac{\Delta t}{2}\nabla \cdot \widehat{\Pi}_{sg} \tag{4.63}$$

where the following expressions were defined:

$$\begin{aligned} \widehat{\mathbf{J}}_{sg} &= \sum_p q_p \widehat{\mathbf{v}}_p W(\mathbf{x}_g - x_p^n) \\ \widehat{\Pi}_{sg} &= \sum_p q_p \widehat{\mathbf{v}}_p \widehat{\mathbf{v}}_p W(\mathbf{x}_g - \mathbf{x}_p^n) \end{aligned} \tag{4.64}$$

with the obvious meaning, respectively, of current and pressure tensor based on the hatted velocities.

An effective dielectric tensor is defined to express the feedback of the electric field on the plasma current and density:

$$\mu_{sg}^n = -\frac{q_s \rho_s^n}{m_s} \alpha_{sg}^n \tag{4.65}$$

where the rotation matrix α_s^n is defined in the same manner as that of the particles but based on the local field on the grid rather than the particle field:

$$\alpha_{sg}^n = \frac{1}{1 + (\beta_s B_g^n)^2}\left(\mathbb{I} - \beta_s \mathbb{I} \times \mathbf{B}_g^n + \beta_s^2 \mathbf{B}_g^n \mathbf{B}_g^n\right) \tag{4.66}$$

With this approach, the current needed for the Maxwell equations can be computed using exclusively particle data from the time level n and no iteration is needed between particles and fields.

Regardless of the use of the direct implicit or IMM, two advantages arise in using semi-implicit methods.

First, these methods avoid any non-linear iteration. The resulting field equations obtained with the procedure described above are linear and can be solved with any linear solver. The three latest implementations of the IMM (Celeste, Parsek and iPic3D) use the GMRES approach [56, 57]. These methods are still requiring global communication when implemented in parallel but converge very quickly due to the diagonal dominance of the two curl Maxwell equations. Furthermore these solvers deal only with the grid and the memory requirement is vastly smaller than that of an iteration scheme dealing also with particles [58, 59]. There are typically at least hundreds of particles per cell, the memory requirement of the iteration method is then reduced by that same factor.

Second, the semi-implicit method retains the same structure of the time step loop of the usual explicit PIC (see figure 4.6): the field equations are solved on the grid, the fields are interpolated to the particles, the particles are moved and then the moments on the grid are computed and the next cycle can continue. The field solver is somewhat more complex and the particle mover is also more complex but the overall scheme is the same as in explicit PIC.

It is important to realize what the added complexity is. The field solver, at first sight, is more complex than the explicit field solver of explicit PIC. In the simplest explicit PIC methods, the fields are marched in a leap-frog manner and require no linear solver. However, even for the so-called charge conserving schemes the two divergence equations are not satisfied exactly, requiring periodic elliptic solvers to apply the so-called divergence cleaning [60]. The need for elliptic solvers is completely eliminated in IMM because of the natural tendency of the IMM to damp any divergence error [57]. As noted above, the application of GMRES to the two curl equations is much simpler and converges faster than for elliptic divergence cleaners [59].

The particle mover is also more complex. The mover in equation (4.60) is not explicit in itself. The equation for the velocity requires one to compute the fields at the position $\bar{\mathbf{x}}_p$ that is not known until the particle is moved, but to move the particle the velocity needs to be known. This requires a separate iteration for each particle. The iteration is just among the six equations of motions for each particle. All codes in the IMM family use a predictor correct approach that is a Picard-type iteration between the two Newton equations for each particle. The iteration can be truncated at a specified number (typically three is used) or until a convergence criterion is satisfied [61, 62]. As we will see below, this iteration, even though it is local, is still an expensive operation because it is applied to each particle.

As a rule of thumb, compared with an explicit PIC the IMM is about three times more expensive per time step. The mover requires three steps where the leap-frog algorithm requires only one. The field solver is a small fraction of the cost. The IMM is not bound by the stability constraint and can use larger grid spacing and time steps, winning a large victory in terms of the time needed to complete the simulation of a given box size for a given time interval.

The IMM does not have to satisfy any stability constraints because its model equations are essentially the same as those of the ECPIC. A complete stability analysis of the IMM shows that waves and scales not resolved by a certain simulation are averaged and damped at the sub-grid level [33]. There is however a catch. The two great advantages of the semi-implicit methods compared with the ECPIC come at a great loss: energy is no longer conserved. The lack of full non-linear consistency breaks energy conservation and introduces an accuracy constraint. The average particle motion cannot exceed one cell per time step. This limitation comes mathematically from the need for the Taylor series expansion truncated at the first order used above to be a valid approximation. Physically, the requirement is that the moment description introduced by the series expansion can be a good approximation of the particle evolution. The procedure of the IMM can be regarded as a first order Chapman–Enskog expansion [63]. The IMM in this regard uses a fluid moment description to guess the plasma response to the changes happening to the particles during one time step. This description is not completely consistent with the actual particle evolution. To keep the error under control the time step becomes limited. The limit is usually expressed as $v_{the}\Delta t/\Delta x < 1$ [33]. If this condition is satisfied the method is accurate and the energy error is under control.

A new variant of the IMM has been recently developed to exactly conserve energy [64]. The new method is a variant of the IMM just described but introduces a new approach to compute the current to enforce exact energy conservation. The advantage is that the finite grid instability is completely eliminated. The method is very new at the moment of writing but it holds promise of being a quantum step forward. More research is needed to verify if this promise is realized in practice.

4.7 Annotated python code

To make it all more concrete, we present now a complete functioning PIC code. We choose the simplest case: 1D and electrostatic with electrons being followed as particles but ions forming a fixed background of uniform density. This last assumption is justified by the heavy mass of the ions. Once this code is fully understood going to full 3D electromagnetic is truly a simple step.

We choose the python language, but on the website of the author other versions (classical, relativistic, electrostatic, electromagnetic, implicit, explicit) are available: https://pershttp://www.kuleuven.be/ u0052182/. The graphical parts to output the data are also available on the website but are not discussed here.

4.7.1 Initialization

As a first step, some python initializations are needed.

```
import numpy as np
import pylab as plt
from scipy import sparse
from scipy.sparse import linalg
```

We then need to initialize the definition of the simulation, in terms of domain size and of the grid used to discretize it. Next, the time step is set with the number of cycles to be run. The next step is to define the plasma density. We use normalized units where the plasma frequency is set to be unitary. But physical units can be used.

The ion uniform background is set to exactly balance the charge of the particles (that are all electrons in this simple model).

```
# Simulation parameters
L = 20*np.pi #20*np.pi # Domain size

NG = 80 # Number of grid cells
N = NG * 200 # Number of particles (200 per cell for example)
dx = L / NG # Cell size

DT = 0.005 # Time step
NT = 50000 # Number of time steps
```

```
WP = 1. # Plasma frequency
QM = -1. # Charge/mass ratio
Q = WP**2 / (QM*N/L) # rho0*L/N: charge carried by a single
particle rho_back = -Q*N/L # Background charge density
```

4.7.2 Particle initialization

We then initialize the electrons. We choose a classic problem of two-stream instability [65] where the initial electrons are subdivided into two equal beams of opposite mean velocity but equal density and thermal speed.

```
#Particle initial properties
V0 = 0.9 # Stream velocity
VT = 0.000 000 1 # Thermal speed

# perturbation
XP1 = 1.0
mode = 1

# particles (electrons)
xp = np.linspace(0, L-L/N, N).T # Particle positions
vp = VT * np.random.randn(N) # Particle momentum, initially
Maxwellian
pm = np.arange(N)
pm = 1-2 * np.mod(pm+1, 2) # Even and odd particles have
opposite speed
vp += pm * V0 # Momentum + stream velocity
np.random.shuffle(v) # We reshuffle the indices to avoid any bias

# Add electron perturbation to excite the desired mode
xp += XP1 * (L/N) * np.sin(2 * np.pi * xp / L * mode)
xp[np.where(xp <0)] += L
xp[np.where(xp >= L)] -= L
```

4.7.3 Grid initialization

In the 1D electrostatic limit, the Maxwell equations reduce to just the Poisson equation:

$$-\epsilon_0 \nabla^2 \phi = en_i - en_e \tag{4.67}$$

where n_i is the uniform background ion density and n_e is the electron density projected to the grid from the particles. The ∇^2 operator is discretized as in the simplest textbook finite difference method [23]:

$$-\epsilon_0(\phi_{i+1} + \phi_{i-1} - 2\phi_i) = (en_i - en_e)\Delta x^2 \tag{4.68}$$

The potential is computed in the cell centres. The continuum is subdivided in cells indexed by i with $i \in [0, \text{NG} - 1]$ (note than python as C and C++ counts the indices in vectors from 0, in MATLAB the same index range would be 1 to NG).

Of course, the first cell and the last cell have a problem: we do not have the neighbouring cell outside. This challenge is called *boundary condition* and it is the worst nightmare any computational scientist can experience. Boundary conditions are pure distilled evil—an evil we will avoid with a double trick. First, we use periodic boundary conditions: the left neighbour of the first cell is the last cell and the right neighbour of the last cell is the first cell. Second, the potential is defined minus a constant and we use this freedom to set the potential in the last cell to 0. The number of unknown potentials then is just NG-1. This tricks works in 1D for periodic boundary conditions, more general boundary conditions in 3D quickly become the aforementioned nightmare.

With these assumptions, the matrix of the discretized Poisson equation is then very simple, it has –2 on the main diagonal and 1 on the two neighbouring ones.

```
# Auxiliary vectors
p = np.concatenate([np.arange(N), np.arange(N)]) # Particle
indices up to N
Poisson = sparse.spdiags(([1, -2, 1] * np.ones((1, NG-1),
dtype=int).T).T, [-1, 0, 1], NG-1, NG-1)
Poisson = Poisson.tocsc()
```

4.7.4 Main cycle

At this point we have prepared the simulation for the main cycle where the four steps of PIC are followed. First, the position is advanced, then the particles are projected to the grid to compute the density. With the density, the Poisson equation is solved. The electric field is computed to obtain the force on the particles and advance the particle velocity.

```
# Main cycle
for it in xrange(NT+1):
  # update particle position xp
  xp += vp * DT
  # Periodic boundary condition
  xp[np.where(xp <0)] += L
  xp[np.where(xp >= L)] -= L

  # Project particles->grid
  csi = xp/dx
  g1 = np.floor(csi–0.5) # Distance from the centre of the cell
  g = np.concatenate((g1, g1+1))
  fraz1 = 1–np.abs(xp/dx–g1-0.5)
  fraz = np.concatenate((fraz1, 1-fraz1))
  g[np.where(g <0)] += NG
```

```
g[np.where(g >NG-1)] -= NG
mat = sparse.csc_matrix((fraz, (p, g)), shape=(N, NG))
rho = Q / dx * mat.toarray().sum(axis=0) + rho_back

# Compute electric field potential
Phi = linalg.spsolve(Poisson, -dx**2 * rho[0:NG-1])
Phi = np.concatenate((Phi,[0]))

# Electric field on the grid
Eg = (np.roll(Phi, 1) - np.roll(Phi, -1)) / (2*dx)

# interpolation grid->particle and velocity update
vp += mat * QM * Eg * DT
```

Let us comment the particle steps in each of these phases:

1. *Position advancement*: When particles are advanced, the possibility exists for the particles to exit the system. Consistent with the periodicity used above, also particles exiting form the left re-enter on the right and vice versa.
2. *Particle projection*: We introduce a *logical coordinate* defined as $\xi = x/\Delta x$. The particles and the cells have logical coordinates, where the interpolation function is more easily defined as $W_{ip} = b_\ell(\xi_p - \xi_i)$. Cell centres have coordinates $\xi_i = (i + 1/2)$, $i \in [0, \mathrm{NG} - 1]$ and particles have $\xi_p = x/\Delta x$. We consider the case of particle shapes given by b-splines of order 0 and interpolation functions, consequently, of order 1:

$$W_{ip} = b_1(\xi_p - \xi_i) \equiv \begin{cases} 1 - |\xi_p - \xi_i|, & \text{if: } |\xi_p - \xi_i| < 1 \\ 0, & \text{otherwise} \end{cases} \tag{4.69}$$

   With this choice of interpolation, each particle can only contribute to a maximum of two cells. The choice of making the particle size equal to the cell size means that it is impossible for a particle to overlap more than two cells. Figure 4.8 shows the concept. The contribution of a particle of finite length to

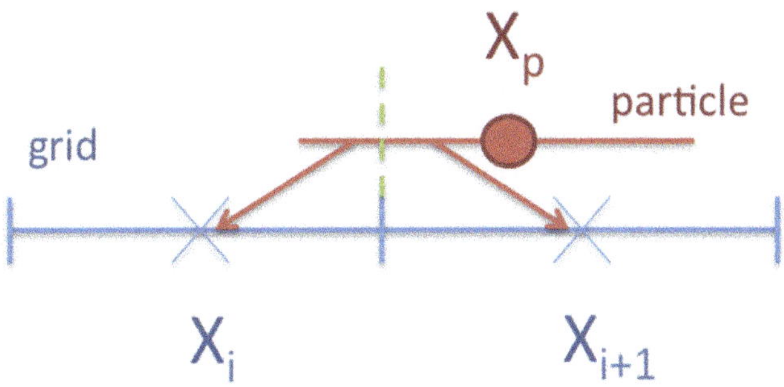

Figure 4.8. Particle interpolation in 1D: the contribution of a particle to a cell is proportional to the length of the interval of overlap with that cell. When the particle shape is a b-spline of order zero (a case called cloud-in-cell), the particle is subdivided to the cells according to its overlap.

a cell is equal to the fraction of the length of the particle that overlaps that cell. This is computed by first identifying the leftmost cell using the floor command and then computing the two fractions (one being 1 minus the other since the total contribution is unity). Once the two fractions are computed the contribution to the charge of a cell is the fraction times the charge of the particle. Periodic boundary conditions are applied also to charge projection. The code in the example uses a matrix notation to project the particle avoiding loops but a simple loop over the particles would work as well. The matrix 'mat' is equal to the interpolation function $W_{ip} = W(x_p - x_i)$ that indeed has two indexes. The charge is than the matrix W_{ip} applied to the vector q_p:

$$q_i = W\mathbf{q} \equiv \sum_p W_{ip} q_p \tag{4.70}$$

The matrix notation can in some computer languages produce a faster execution, but this issue is largely eliminated in the latest version of most languages. The computational performance of the interpolation steps depends dramatically on the precise style of code writing, on the language and on the hardware [66–69]. Alternative to a matrix formulation, a loop over the particles can be used:

```
qodx = Q / dx
rho = np.zeros(NG)
for i in range(N):
  csi = xp[i]/dx
  g1 = np.floor(csi-0.5)
  g2 = g1 + 1
  fraz1 = 1-np.abs(csi-g1-0.5)
  fraz2 = 1.0-fraz1
  rho[j1] = rhoe[j1] + qodx*fraz1
  rho[j2] = rhoe[j2] + qodx*fraz2
rho[0] += rho[NG-1]
```

In both cases the charge in the cell must be divided by Δx to obtain the density.

3. *Field solver*: In 1D electrostatic there are many alternatives. Some are very simple [5], but we use here the most general approach that would work also with more complex boundary conditions and in more dimensions. We use a linear matrix solver provided by Python. The boundary conditions are then imposed (setting the last cell to 0) and the electric field is computed from the potential:

$$E_i = -\frac{\phi_{i+1} - \phi_{i-1}}{2\Delta x} \tag{4.71}$$

4. *Velocity update*: The electric field is used to advance the particle position. The same interpolation method used for the charge is used for the electric field.

Note that we have chosen the position update as our initial step in the cycle but the starting point is completely arbitrary, the cycle can start at any point. This means also that logically position and velocity advancement happen in sequence, the last operation of one cycle being followed by the first of the next.

References

[1] Frenkel D and Smit B 2002 *Understanding Molecular Simulation* (San Diego, CA: Academic)
[2] Barnes J and Hut P 1986 A hierarchical O(N logN) force-calculation algorithm *Nature* **324** 4446–9
[3] Salmon J K and Warren M S 1994 Skeletons from the treecode closet *J. Comput. Phys.* **111** 136–55
[4] Birdsall C and Langdon A 2004 *Plasma Physics Via Computer Simulation* (London: Taylor and Francis) https://doi.org/10.1201/9781315275048
[5] Hockney R and Eastwood J 1988 *Computer Simulation Using Particles* (London: Taylor and Francis) https://doi.org/10.1201/9780367806934
[6] Nicholson D 1983 *Introduction to Plasma Theory* (New York: Wiley) https://doi.org/10.1017/S0263034600000677
[7] Lapenta G 2012 Particle simulations of space weather *J. Comput. Phys.* **231** 795–821
[8] Coppa G G M, Lapenta G, Dellapiana G, Donato F and Riccardo V 1996 Blob method for kinetic plasma simulation with variable-size particles *J. Comput. Phys.* **127** 268–84
[9] Boor C D 1978 *A Practical Guide to Splines* (Berlin: Springer)
[10] Thomée V 1984 *Galerkin Finite Element Methods for Parabolic Problems* (Springer Series in Computational Mathematics vol 25) (Berlin: Springer) https://doi.org/10.1007/978-3-662-03359-3
[11] Bateson W B and Hewett D W 1998 Grid and particle hydrodynamics: beyond hydrodynamics via fluid element particle-in-cell *J. Comput. Phys.* **144** 358–78
[12] Grigoryev Y N, Vshivkov V A and Fedoruk M P 2005 *Numerical Particle-in-Cell Methods: Theory and Applications* (AH Zeist: VSP BV)
[13] Sulsky D and Brackbill J 1991 A numerical method for suspension flow *J. Comput. Phys.* **96** 339–68
[14] Taflove A and Hagness S C 2005 *Computational Electrodynamics: The Finite-Difference Time-Domain Method* (Norwood: Artech House Publishers)
[15] Vu H X and Brackbill J U 1992 CELEST1D: An implicit, fully-kinetic model for low-frequency, electromagnetic plasma simulation *Comput. Phys. Commun.* **69** 253
[16] Lapenta G 2011 DEMOCRITUS: An adaptive particle in cell (PIC) code for object-plasma interactions *J. Comput. Phys.* **230** 4679–95
[17] Delzanno G L, Camporeale E, Moulton J D, Borovsky J E, MacDonald E A and Thomsen M F 2013 CPIC: A curvilinear particle-in-cell code for plasma-material interaction studies *IEEE Trans. Plasma Sci.* **41** 3577–87
[18] Jacobs G B and Hesthaven J S 2006 High-order nodal discontinuous Galerkin particle-in-cell method on unstructured grids *J. Comput. Phys.* **214** 96–121
[19] Esirkepov T 2001 Exact charge conservation scheme for particle-in-cell simulation with an arbitrary form-factor *Comput. Phys. Commun.* **135** 144–53
[20] Langdon A B 1992 On enforcing Gauss' law in electromagnetic particle-in-cell codes *Comput. Phys. Commun.* **70** 447–50

[21] Isaacson E and Keller H B 1994 *Analysis of Numerical Methods* (Chelmsford, MA: Courier Corporation)

[22] Dendy R O 1990 *Plasma Dynamics* (Oxford: Oxford University Press)

[23] Morton K and Mayers D 1995 *Iterative Methods for Linear and Nonlinear Equations* (Philadelphia, PA: SIAM)

[24] Morse R L and Nielson C W 1971 Numerical simulation of the Weibel instability in one and two dimensions *Phys. Fluids* **14** 830–40

[25] Fujimoto K and Sydora R D 2008 Electromagnetic particle-in-cell simulations on magnetic reconnection with adaptive mesh refinement *Comput. Phys. Commun.* **178** 915–23

[26] Innocenti M E, Lapenta G, Markidis S, Beck A and Vapirev A 2013 A multi level multi domain method for particle in cell plasma simulations *J. Comput. Phys.* **238** 115–40

[27] Vay J-L, Colella P, Friedman A, Grote D P, McCorquodale P and Serafini D 2004 Implementations of mesh refinement schemes for particle-in-cell plasma simulations *Comput. Phys. Commun.* **164** 297–305

[28] Degond P and Raviart P-A 1992 An analysis of the Darwin model of approximation to Maxwell's equations *Forum Mathematicum* **4** 13–44

[29] Lipatov A S S 2002 *The Hybrid Multiscale Simulation Technology* (Berlin: Springer)

[30] Lee W 1987 Gyrokinetic particle simulation model *J. Comput. Phys.* **72** 243–69

[31] Mason R 1981 Implicit moment particle simulation of plasmas *J. Comput. Phys.* **41** 233

[32] Denavit J 1981 Time-filtering particle stimulations with $\omega_{pe}\Delta t \gg 1$ *J. Comput. Phys.* **42** 337

[33] Brackbill J and Forslund D 1982 Simulation of low frequency, electromagnetic phenomena in plasmas *J. Comput. Phys.* **46** 271

[34] Langdon A, Cohen B and Friedman A 1983 Direct implicit large time-step particle simulation of plasmas *J. Comput. Phys.* **51** 107–38

[35] Kelley C T 2003 *Solving Nonlinear Equations with Newton's Method* vol 1 (Philadelphia, PA: SIAM)

[36] Knoll D A and Keyes D E 2004 Jacobian-free Newton-Krylov methods: a survey of approaches and applications *J. Comput. Phys.* **193** 357–97

[37] Markidis S and Lapenta G 2011 The energy conserving particle-in-cell method *J. Comput. Phys.* **230** 7037–52

[38] Chen G, Chacón L and Barnes D C 2011 An energy- and charge-conserving, implicit, electrostatic particle-in-cell algorithm *J. Comput. Phys.* **230** 7018–36

[39] Lapenta G and Markidis S 2011 Particle acceleration and energy conservation in particle in cell simulations *Phys. Plasmas* **18** 072101

[40] Chen G and Chacon L 2015 A multi-dimensional, energy-and charge-conserving, non-linearly implicit, electromagnetic Vlasov-Darwin particle-in-cell algorithm *Comput. Phys. Commun.* **197** 73–87

[41] Grigoryev Y N, Vshivkov V A and Fedoruk M P 2002 *Numerical Particle-in-Cell Methods: Theory and Applications* (Berlin: De Gruyter)

[42] Ueda H, Omura Y, Matsumoto H and Okuzawa T 1994 A study of the numerical heating in electrostatic particle simulations *Comput. Phys. Commun.* **79** 249–59

[43] Lewis H R 1970 Energy-conserving numerical approximations for Vlasov plasmas *J. Comput. Phys.* **6** 136–41

[44] Langdon A B 1973 'Energy-conserving' plasma simulation algorithms *J. Comput. Phys.* **12** 247–68

[45] Chen G, Chacón L, Leibs C A, Knoll D A and Taitano W 2014 Fluid preconditioning for Newton-Krylov-based, fully implicit, electrostatic particle-in-cell simulations *J. Comput. Phys.* **258** 555–67

[46] Hewett D W and Langdon A B 1987 Electromagnetic direct implicit plasma simulation *J. Comput. Phys.* **72** 121–55

[47] Cohen B I, Langdon A B, Hewett D W and Procassini R J 1989 Performance and optimization of direct implicit particle simulation *J. Comput. Phys.* **81** 151

[48] Drouin M, Gremillet L, Adam J-C and Héron A 2010 Particle-in-cell modeling of relativistic laser-plasma interaction with the adjustable-damping, direct implicit method *J. Comput. Phys.* **229** 4781–12

[49] Welch D R, Rose D V, Clark R E, Genoni T C and Hughes T 2004 Implementation of an non-iterative implicit electromagnetic field solver for dense plasma simulation *Comput. Phys. Commun.* **164** 183–8

[50] Barnes D, Kamimura T, Leboeuf J-N and Tajima T 1983 Implicit particle simulation of magnetized plasmas *J. Comput. Phys.* **52** 480–502

[51] Brackbill J U and Cohen B I 1985 *Multiple Time Scales* vol 3 (New York: Academic)

[52] Lapenta G, Brackbill J U and Ricci P 2006 Kinetic approach to microscopic-macroscopic coupling in space and laboratory plasmas *Phys. Plasmas* **13** 055904

[53] Markidis S, Camporeale E, Burgess D, Rizwan-Uddin and Lapenta G 2009 Parsek2D: an implicit parallel particle-in-cell code *Numerical Modeling of Space Plasma Flows: ASTRONUM-2008 ASP Conference Series*vol 406 *(US Virgin Islands, 8–13 June)* N V Pogorelov, E Audit, P Colella and G P Zank (San Francisco, CA: Astronomical Society of the Pacific) 237–42

[54] Markidis S, Lapenta G and Rizwan-Uddin 2010 *Math. Comput. Simul.* **80** 1509–19

[55] Yee K S and Chen J S 1997 The finite-difference time-domain (FDTD) and the finite-volume time-domain (FVTD) methods in solving Maxwell's equations *IEEE Trans. Antennas Propag.* **45** 354–63

[56] Saad Y and Schultz M H 1986 GMRES: A generalized minimal residual algorithm for solving nonsymmetric linear systems *SIAM J. Sci. Stat. Comput.* **7** 856–69

[57] Ricci P, Lapenta G and Brackbill J 2002 A simplified implicit Maxwell solver *J. Comput. Phys.* **183** 117–41

[58] Mallon D A, Eicker N, Innocenti M E, Lapenta G, Lippert T and Suarez E 2012 On the scalability of the clusters-booster concept: a critical assessment of the deep architecture *Proceedings of the Future HPC Systems: The Challenges of Power-Constrained Performance* (New York: ACM)

[59] Kumar P, Markidis S, Lapenta G, Meerbergen K and Roose D 2013 High performance solvers for implicit particle in cell simulation *Procedia Comput. Sci.* **18** 2251–8

[60] Munz C, Omnes P, Schneider R, Sonnendrücker E and Voß U 2000 *J. Comput. Phys.* **161** 484–511

[61] Peng I B, Markidis S, Vaivads A, Vencels J, Amaya J, Divin A, Laure E and Lapenta G 2015 The formation of a magnetosphere with implicit particle-in-cell simulations *Procedia Comput. Sci.* **51** 1178–87

[62] Noguchi K, Tronci C, Zuccaro G and Lapenta G 2007 Formulation of the relativistic moment implicit particle-in-cell method *Phys. Plasmas* **14** 042308

[63] Chapman S and Cowling T G 1970 The mathematical theory of non-uniform gases: an account of the kinetic theory of viscosity *Thermal Conduction and Diffusion in Gases* (Cambridge: Cambridge University Press)

[64] Lapenta G 2017 Exactly energy conserving implicit moment particle in cell formulation *J. Comput. Phys.* **334** 349–66

[65] Goldston R and Rutherford P 1995 *Introduction to Plasma Physics* (London: Taylor and Francis)

[66] Markidis S, Lapenta G, VanderHeyden W and Budimlić Z 2005 Implementation and performance of a particle-in-cell code written in Java *Concurr. Comput. Pract. Exp.* **17** 821–37

[67] Bowers K J, Albright B J, Yin L, Daughton W, Roytershteyn V, Bergen B and Kwan T J T 2009 Advances in petascale kinetic plasma simulation with VPIC and Roadrunner *J. Phys. Conf. Ser.* **180** 012055

[68] Vapirev A, Deca J, Lapenta G, Markidis S, Hur I and Cambier J-L 2015 Initial results on computational performance of Intel many integrated core, sandy bridge, and graphical processing unit architectures: implementation of a 1D c++/OpenMP electrostatic particle-in-cell code *Concurr. Comput. Pract. Exp.* **27** 581–93

[69] Decyk V K 2015 Skeleton particle-in-cell codes on emerging computer architectures *Comput. Sci. Eng.* **17** 47–52

[70] Dawson J M 1983 Particle simulation of plasmas *Rev. Mod. Phys.* **55** 403

Chapter 5

The ergodic method: plasma dynamics through a sequence of equilibrium states

Gianni Coppa, Luis O Silva, Antonio D'Angola, Fabio Peano and Federico Peinetti

5.1 Introduction to the ergodic method

In problems of plasma physics, characterized by the simultaneous presence of phenomena spanning over well separated time scales, theoretical methods can be devised in which time averages over the more rapid phenomenon are performed in order to calculate the parameters for the study of the evolution of the slower one. In some respects, the idea behind these methods is similar to the concept of ergodic theory in kinetic theory [1].

In this chapter, investigations on two different problems are described. The first (section 5.2) concerns the expansion of a spherical plasma, in which the dynamics of ions and electrons have characteristic times whose ratio is of the order of $10^{-3} \div 10^{-4}$. In practice, at any time in which ion dynamics is followed, an appropriate time average is replaced to the real position of electrons.

The second problem (section 5.3) regards the electron dynamics in a Penning trap in which a rarefied gas is present. In this case, the slow phenomenon is represented by the collisions of the electrons with atoms and with the walls of the trap, which modify the trajectories of the electrons and their number, through the ionizations and the phenomena of absorption and secondary emission from the walls. The separation of multiple time scales leads to a significant computational advantage, as one can use time-steps suitable to follow only the slow phenomena of interest. The two cases presented in the chapter show the potential of the ergodic method.

5.2 Expansion of spherical nanoplasmas

The irradiation of solid targets with ultraintense lasers can induce the prompt formation of hot, dense plasmas, which rapidly expand into a vacuum [2]. While these phenomena have been extensively studied, the expansion of spherical nanoplasmas, which are generated upon interaction of ultraintense lasers with atomic or

doi:10.1088/978-0-7503-3559-1ch5

molecular clusters [3–7], has not been analyzed as thoroughly. An accurate knowledge of the expansion, accounting for the self-consistent dynamics of ions and electrons, is relevant in particular situations where control over the expansion is necessary [8]. Actually, exact solutions for spherical expansions exist only for ideal cases, such as the Coulomb explosion [9] of a pure ion plasma, which occurs when all the electrons are suddenly swept away from the cluster by the laser field. In opposite conditions, when most of the electrons are not pulled away from the cluster, hydrodynamic models can be employed to estimate the basic features of the expansion [10]; in the quasineutral limit, a kinetic solution for the adiabatic expansion of plasma bunches into a vacuum has also been derived [11]. However, in more general situations, in which a fraction of the electrons remains bound to the cluster, the expansion process is strongly dependent on the self-consistent dynamics of ions and of trapped electrons and it can be described accurately only by the Vlasov–Poisson theory.

In this section, a kinetic analysis is presented for the collisionless expansion of spherical plasmas, based on a peculiar ergodic model, which accounts for the radial motion of the ions and for the three-dimensional motion of the electrons [12, 13]. Furthermore, a procedure to study the initial charging transient, during which the faster electrons leave the cluster is presented within the framework of the ergodic model. The precise knowledge of the electron distribution after the transient is important to set the correct initial conditions for the long-term plasma expansion.

Kinetic models for the expansion

In the electrostatic, nonrelativistic limit, the dynamics of a collisionless electron–ion plasma is described rigorously by the Vlasov–Poisson set of equations:

$$\begin{cases} \dfrac{\partial f_e}{\partial t} = -\dfrac{\mathbf{p}}{m}\cdot\dfrac{\partial f_e}{\partial \mathbf{x}} - e\dfrac{\partial \Phi}{\partial \mathbf{x}}\cdot\dfrac{\partial f_e}{\partial \mathbf{p}}, \\ \dfrac{\partial f_i}{\partial t} = -\dfrac{\mathbf{p}}{M}\cdot\dfrac{\partial f_i}{\partial \mathbf{x}} + e\dfrac{\partial \Phi}{\partial \mathbf{x}}\cdot\dfrac{\partial f_i}{\partial \mathbf{p}}, \\ \nabla^2\Phi = 4\pi e\left(\int f_e \mathrm{d}^3p - \int f_i \mathrm{d}^3p\right), \end{cases} \tag{5.1}$$

where $f_e(\mathbf{x}, \mathbf{p}, t)$ and $f_i(\mathbf{x}, \mathbf{p}, t)$ are the distribution functions in phase space for electrons (having mass m and charge $-e$) and ions (having mass M and charge $+e$; for simplicity, only $Z = 1$ ions are considered), respectively, and $\Phi(\mathbf{x}, t)$ is the electrostatic potential; Φ is set to zero at infinity, so that the energy of a single electron, $\mathscr{E} = \frac{1}{2}mv^2 - e\Phi$, is negative if it is trapped. In the following, the attention is focused on the expansion of a plasma sphere with initial radius R_0 composed of cold ions with initial uniform density n_{i0} and hot electrons with initial uniform density $n_{e0} = n_{i0}$ and temperature T_0. The general initial conditions for equations (5.1) can be cast in the form:

$$f_{e0}(\mathbf{x}, \mathbf{p}) = n_{e0}g(v)\Theta\left(1 - \frac{r}{R_0}\right), \; f_{i0}(\mathbf{x}, \mathbf{p}) = n_{i0}\delta(\mathbf{p})\Theta\left(1 - \frac{r}{R_0}\right), \tag{5.2}$$

where Θ is the step function and g is the Maxwell–Boltzmann distribution:

$$g(v) = \left(\frac{m}{2\pi k_B T_0}\right)^{3/2} \exp\left(-\frac{mv^2}{2k_B T_0}\right). \tag{5.3}$$

Actually, the dynamics of the system is fully determined by the mass ratio m/M and by $\widehat{T_0} = 3\lambda_{D,0}^2/R_0^2$, being $\lambda_{D,0}$ the electron Debye length at $t = 0$ [12]. In practice, the electrons are supposed to be instantaneously heated by an infinitely short laser pulse, without expanding. The expansion process can be split in two stages: the first is a rapid expansion of the electrons, which leads to an equilibrium before the ions move appreciably; the second is a slow expansion of the plasma bulk, driven by the positive charge build-up formed in the first stage. Due to the large mass disparity between ions and electrons, a simplified model can be derived, in which the ions are assumed fixed during the former process, whereas the electrons can be considered instantaneously at equilibrium with the electrostatic potential during the latter stage. A self-consistent theoretical framework can be developed, which allows one to determine accurately both the initial equilibrium and the bulk expansion by treating the electron dynamics as a sequence of equilibrium configurations with frozen ions. The model is obtained by exploiting the functional relation existing between n_e and Φ at equilibrium, and by calculating the energy variation of the electrons under the hypothesis of slow variations of Φ.

Equilibrium solutions of the Vlasov equation for the electrons must depend on $\mathbf{x}$ and $\mathbf{p}$ only through the invariants of motion. Under the hypothesis of spherical symmetry, the only invariants of motion to be considered are the Hamiltonian $\mathcal{H}(\mathbf{x}, \mathbf{p}) = p^2/2m - e\Phi(r)$ and the angular momentum, $\mathbf{L} = \mathbf{x} \times \mathbf{p}$. Consequently, the equilibrium distribution function can be written as $f_e(\mathbf{x}, \mathbf{p}) = F(\mathcal{H}(\mathbf{x}, \mathbf{p}), \mathbf{x} \times \mathbf{p})$. If a generic space point $\mathbf{x} = r\hat{\mathbf{e}}_r$ and a generic momentum $\mathbf{p} = p_r\hat{\mathbf{e}}_r + p_\perp\hat{\mathbf{e}}_\perp$ are considered, the phase-space density is $F\left(\frac{1}{2m}\left(p_r^2 + p_\perp^2\right) - e\Phi(r), L, \hat{\mathbf{e}}_r \times \hat{\mathbf{e}}_\perp\right)$ being $L = |\mathbf{L}| = mrv_\perp$. Due to the symmetry with respect to rotations around $\hat{\mathbf{e}}_r$, the phase-space density cannot depend upon $\hat{\mathbf{e}}_\perp$, and, consequently, F depends only on $\mathcal{H}$ and L.

The energy-angular momentum distribution for the electrons, $\sigma_e(\mathcal{E}, L)$, is defined as

$$\begin{aligned}\sigma_e(\mathcal{E}, L) &= \iint F(\mathcal{H}(\mathbf{x}, \mathbf{p}), L)\delta(\mathcal{H}(\mathbf{x}, \mathbf{p}) - \mathcal{E})\delta(|\mathbf{x} \times \mathbf{p}| - L)\mathrm{d}^3x\mathrm{d}^3p \\ &= \frac{8\pi^2\sqrt{2}}{m^{3/2}}F(\mathcal{E}, L)\int_{R_1(\mathcal{E}, L)}^{R_2(\mathcal{E}, L)}\left[\mathcal{E} - \frac{L^2}{2mr^2} + e\Phi(r)\right]^{-\frac{1}{2}}\mathrm{d}r,\end{aligned} \tag{5.4}$$

where $R_1(\mathcal{E}, L)$ and $R_2(\mathcal{E}, L)$ are the radial turning points such that $\mathcal{E} - \frac{L^2}{2mr^2} + e\Phi(r) = 0$. The quantity $\sigma_e(\mathcal{E}, L)d\mathcal{E}dL$ represents the number of electrons having energy in $[\mathcal{E}, \mathcal{E} + d\mathcal{E}]$ and angular momentum in $[L, L + dL]$. The electron space density, n_e, can be written as

$$n_e(r) = \frac{1}{4\pi r^2} \iint \sigma_e(\mathscr{E}, L) \mathcal{P}(r, \mathscr{E}, L; \{\Phi\}) \mathrm{d}\mathscr{E} \mathrm{d}L, \tag{5.5}$$

where

$$\mathcal{P}(r, \mathscr{E}, L) = \frac{\left[\mathscr{E} - \frac{L^2}{2mr^2} + e\Phi(r)\right]^{-\frac{1}{2}}}{\int_{R_1(\mathscr{E},\, L)}^{R_2(\mathscr{E},\, L)} \left[\mathscr{E} - \frac{L^2}{2mr'^2} + e\Phi(r')\right]^{-\frac{1}{2}} \mathrm{d}r'}; \tag{5.6}$$

$\mathcal{P}(r; \mathscr{E}, L)dr$ gives the probability for an electron with energy $\mathscr{E}$ and angular momentum L, to be found in $[r, r + dr]$. If variations of Φ, due to the ion motion, are slow with respect to the period of the radial oscillation of the electrons, the mean value of $\mathrm{d}\mathscr{E}/\mathrm{d}t$ can be evaluated by using the ensemble average of $\frac{\partial \Phi}{\partial t}$, as

$$\left\langle \frac{\mathrm{d}\mathscr{E}}{\mathrm{d}t} \right\rangle = \left\langle -e\frac{\partial \Phi}{\partial t}(r(t), t) \right\rangle = -e \int_{R_1(\mathscr{E},\, L)}^{R_2(\mathscr{E},\, L)} \frac{\partial \Phi}{\partial t}(r, t) \mathcal{P}(r, \mathscr{E}, L) \mathrm{d}r; \tag{5.7}$$

this is equivalent [14] to preserving the adiabatic invariant:

$$\mathcal{I}(\mathscr{E}(t), L, t) = \text{Const} \cdot \int_{R_1(\mathscr{E},\, L)}^{R_2(\mathscr{E},\, L)} \left[\mathscr{E} - \frac{L^2}{2mr^2} + e\Phi(r, t)\right]^{\frac{1}{2}} \mathrm{d}r. \tag{5.8}$$

equations (5.4)–(5.7), coupled with Poisson's equation and the dynamic equation for the radial motion of the ions, provide a self-consistent model for the expansion of a spherical plasma.

In general, there is a precise relationship between time scales and the proper number of parameters to be used to describe a phenomenon: in the present case, the system (5.1) allows one to follow the expansion dynamics on the time scale of the fastest particles. To study the ion expansion, a quasi-equilibrium model can be used, in which the stationary solution of the Vlasov equation for the electrons is employed (equations (5.4)–(5.7)). In fact, as the kinetic model is collisionless, the equations are time-reversible and a physical mechanism leading towards the equilibrium does not exist. To justify the use of the equilibrium distribution, $f(\mathbf{x}, \mathbf{p}) = f(\mathscr{H}(\mathbf{x}, \mathbf{p}), |\mathbf{x} \times \mathbf{p}|)$, one must suppose that the stationary solution of the Vlasov equation is a good representation of the real electron distribution, once high-frequency fluctuations are eliminated; formally, this can be performed by introducing a dissipation mechanism, i.e., a suitable collision term into the Vlasov equation. In general, an approximate kinetic model can be regarded as the result of introducing a particular collision term. For example, by using a binary collision term with sufficiently high collision frequency, f_e tends towards the Maxwell–Boltzmann distribution; in this case, a hydrodynamic description is obtained, whose domain of validity is confined to situations where $\widehat{T_0} \ll 1$. For larger values of $\widehat{T_0}$, the use of a proper energy spectrum is fundamental; in fact, the energy distribution presents a cutoff for $\mathscr{E} = 0$, because for $\mathscr{E} > 0$ the electrons are not confined and their stationary density must vanish; instead, a Maxwellian distribution has a non-negligible fraction of electrons with $\mathcal{E} > 0$. Within this framework, the approach of reference [12] can be introduced by considering a collision term of the form

$$J(f_e) = -\nu(f_e - \bar{f}_e), \qquad \bar{f}_e = \frac{1}{4\pi}\int f_e(\mathbf{x}, p\mathbf{\Omega}, t)\mathrm{d}^2\Omega \tag{5.9}$$

where $\mathbf{\Omega}$ is the unit vector indicating the electron direction and ν is the collision frequency; $1/\nu$ must be smaller than the characteristic time of the ion expansion. In this case, the kinetic Vlasov equation for the electrons is replaced by the collisional equation

$$\frac{\partial f_e}{\partial t} = -\mathbf{p}\cdot\frac{\partial f_e}{\partial \mathbf{x}} - \frac{e}{m}\frac{\partial \Phi}{\partial \mathbf{x}}\cdot\frac{\partial f_e}{\partial \mathbf{p}} - \nu f + \frac{\nu}{4\pi}\int f(\mathbf{x}, v\mathbf{\Omega}, t)\mathrm{d}^2\Omega, \tag{5.10}$$

The collisions do not alter the electron energy, but change randomly their direction $\mathbf{\Omega}$, driving f_e towards an equilibrium distribution having the form $f_e(\mathbf{x}, \mathbf{p}) = f[\mathcal{H}(\mathbf{x}, \mathbf{p})]$, a sort of ergodic density such that each electron has an equal probability to be found in every point of the surface of phase space having equation $\mathcal{H}(\mathbf{x}, \mathbf{p}) = \mathcal{E}$. This fact can be proved rigorously by defining the electron entropy, $S_e = -\iint f_e \log(f_e)\mathrm{d}^3x\mathrm{d}^3p$, and noticing that its derivative:

$$\frac{\mathrm{d}S_e}{\mathrm{d}t} = \nu\iint \log\left(\frac{f_e}{\bar{f}_e}\right)\cdot(f_e - \bar{f}_e)\mathrm{d}^3x\mathrm{d}^3p, \tag{5.11}$$

is always nonnegative unless $f_e = \bar{f}_e$. Therefore, a necessary condition for the distribution to be stationary is that f_e must not depend on $\mathbf{\Omega}$. Finally, the equilibrium distribution must be a function of $\mathcal{H}$ and L; as it does not depend on $\mathbf{\Omega}$, it is a function of $\mathcal{H}$ only. The approach can be called single-particle ergodic method[1]. According to this approach, the equilibrium distribution function can be written simply as $f_e(\mathbf{x}, \mathbf{p}) = f[\mathcal{H}(\mathbf{x}, \mathbf{p})]$; the dependence on L is lost, and equations (5.4)–(5.6) are replaced by

$$\begin{aligned}\rho_e(\mathcal{E}) &= \iint f(\mathcal{H}(\mathbf{x}, \mathbf{p}))\delta(\mathcal{H}(\mathbf{x}, \mathbf{p}) - \mathcal{E})\mathrm{d}^3x\mathrm{d}^3p \\ &= 16\pi^2\sqrt{2}m^{3/2}f(\mathcal{E})\int_{\mathcal{D}(\mathcal{E})}[\mathcal{E} + e\Phi(r)]^{\frac{1}{2}}r^2\mathrm{d}r,\end{aligned} \tag{5.12}$$

$$n_e(r) = \frac{1}{4\pi r^2}\int \rho_e(\mathcal{E})\mathcal{Q}(r, \mathcal{E})\mathrm{d}\mathcal{E}, \tag{5.13}$$

$$\mathcal{Q}(r, \mathcal{E}) = \frac{r^2[\mathcal{E} + e\Phi(r)]^{\frac{1}{2}}}{\int_{\mathcal{D}(\mathcal{E})} r'^2[\mathcal{E} + e\Phi(r')]^{\frac{1}{2}}\mathrm{d}r'}, \tag{5.14}$$

[1] This is different from the usual ergodic theory of the statistical mechanics, in which the state of a system of N particles can be found with equal probability on the phase-space surface $\mathcal{H}(\mathbf{x}_1, \mathbf{x}_2, \ldots, \mathbf{x}_N; \mathbf{p}_1, \mathbf{p}_2, \ldots, \mathbf{p}_N) =$ Const.

where $\mathscr{D}(\mathscr{E})$ is the integration domain, such that $r \in \mathscr{D}(\mathscr{E}) \Rightarrow \mathscr{E} + e\Phi(r) \geqslant 0$ (for monotonic potentials, $\mathscr{D}(\mathscr{E}) = [0, R(\mathscr{E})]$, where $\mathscr{E} + e\Phi(R(\mathscr{E})) = 0$), and the adiabatic invariant, equation (5.8), is replaced by the ergodic invariant [15]

$$\mathcal{J}[\mathscr{E}(t), t] = \text{Const} \cdot \int_{\mathscr{D}(\mathscr{E})} [\mathscr{E} + e\Phi(r, t)]^{\frac{3}{2}} r^2 \mathrm{d}r, \tag{5.15}$$

which represents the volume of the region of $\mathbb{R}^6$ enclosed by the surface of equation $\frac{1}{2m}p^2 - e\Phi(r) = \mathscr{E}$. Moreover, $\mathcal{Q}$ represents the density distribution associated with an electron of energy $\mathscr{E}$.

As shown in references [12, 13], this approach provides excellent results for the expansion of a spherical plasma in a wide range of the parameter $\widehat{T}_0$. There are a number of reasons to explain its success. First, even though hydrodynamic models can provide a qualitative agreement with the real expansion dynamics, the ergodic model is more flexible in describing the energy distribution of the electrons. In addition, the initial phase-space distribution, equation (5.2), is assumed to be an ergodic function. Finally, it must be noticed that the angular momentum is invariant only in the case of perfect spherical symmetry. In practical situations, perturbations breaking that symmetry, such as an initial shape which is not perfectly spherical, or collisions with heavy particles, would cause a mixing in L distribution, and their effect can be taken into account by introducing a collision term such as the one in equation (5.9).

Single-particle ergodic model

Using the hypothesis of single-particle ergodic distribution, a self-consistent model for the expansion of a spherical plasma can be formulated, starting from equations (5.12)–(5.15), as follows [12]. A Lagrangian approach can be used both for the ions and the electrons. Ions move in the radial direction, starting from the initial position r_0 with zero velocity, while the electron energy $\mathscr{E}$ evolves in time starting from the initial value $\mathscr{E}_0$. In this way, the ion trajectories $r_\mathrm{i}(r_0, t)$, the electron energies $\mathscr{E}(\mathscr{E}_0, t)$, the ion density $n_\mathrm{i}(r, t)$, the electron density, $n_\mathrm{e}(r, t)$, the electron energy distributions, $\rho_\mathrm{e}(\mathscr{E}, t)$, and the potential $\Phi(r, t)$ are determined by solving the set of equations:

$$\begin{cases} \dfrac{\partial^2 r_\mathrm{i}}{\partial t^2} = -\dfrac{e}{M}\dfrac{\partial \Phi}{\partial r}(r_\mathrm{i}), \\ \dfrac{1}{r^2}\dfrac{\partial}{\partial r}\left(r^2 \dfrac{\partial \Phi}{\partial r}\right) = 4\pi e(n_\mathrm{e} - n_\mathrm{i}), \\ n_\mathrm{i}(r_\mathrm{i}) = n_{\mathrm{i},0}(r_0)\dfrac{(r_0/r_\mathrm{i})^2}{\partial r_\mathrm{i}/\partial r_0}, \\ n_\mathrm{e} = \displaystyle\int \rho_\mathrm{e}(\mathscr{E})\mathcal{Q}(r, \mathscr{E}, \{\Phi\})\mathrm{d}\mathscr{E}, \\ \rho_\mathrm{e}(\mathscr{E}, t) = \dfrac{\rho_{\mathrm{e},0}(\mathscr{E}_0)}{\partial \mathscr{E}/\partial \mathscr{E}_0}, \\ \dfrac{\mathrm{d}}{\mathrm{d}t}\mathcal{J}(\mathscr{E}(\mathscr{E}_0, t), t) = 0, \end{cases} \tag{5.16}$$

In equations (5.16), the evolution equations for the radial coordinates of the ions and the electron energies are coupled via the Poisson's equation. The expansion dynamics is determined once the initial ion density $n_{i,0}$ and the electron energy distribution $\rho_{e,0}$ are given. In equations (5.16), the electron density n_e is calculated by summing the number of electrons having energy in $[\mathscr{E}, \mathscr{E} + d\mathscr{E}]$, $\rho_e(\mathscr{E})d\mathscr{E}$, multiplied by the probability for an electron to be found at the radius r. For simplicity, the ion density n_i is written under the hypothesis of no ion overtaking ($\partial r_i/\partial r_0 \neq 0$).

The set of equations (5.16) describes the expansion dynamics on the ion time scale; therefore, its numerical solution is much faster than solving the full Vlasov–Poisson model, where the electron time scale must be resolved. The model is solved by calculating the radial trajectories of a set of representative ions and the energy variations of a set of computational particles. Each computational particle represents a given number of electrons, whose radial distribution is given by equation (5.14). This description of the energy dependence corresponds to a suitable discretization of the integral in equation (5.13), which is similar to the description of the spatial dependence commonly adopted in particle-in-cell approach [16].

Drift-diffusion approximation

As an alternative means of solving equation (5.10), one can consider the drift-diffusion equation

$$\frac{\partial \Psi}{\partial t} = e\frac{\partial \Phi}{\partial t}\frac{\partial \Psi}{\partial \mathscr{E}} + \frac{2}{3m\nu}\frac{1}{r^2}\frac{\partial}{\partial r}\left[r^2\left((\mathscr{E} + e\Phi)\frac{\partial \Psi}{\partial r} - \frac{e}{2}\frac{\partial \Phi}{\partial r}\Psi\right)\right], \tag{5.17}$$

obtained from equation (5.10) by approximating f_e as $f_0(r, p) + \mathbf{p} \cdot \mathbf{f}_1(r, p)$ [17]. In equation (5.17), the quantity $\Psi(r, \mathscr{E}, t)$ represents the space-energy distribution of the electrons. The self-consistent potential Φ is determined by solving Poisson's equation

$$\frac{1}{r^2}\frac{\partial}{\partial r}\left(r^2\frac{\partial \Phi}{\partial r}\right) = 4\pi e\left(\int \Psi d\mathscr{E} - n_{i0}\right). \tag{5.18}$$

Asymptotically, for $t \to \infty$, the solution approaches a stationary solution of equation (5.17), Ψ_∞, such that

$$\frac{1}{2}e\frac{d\Phi_\infty}{dr}\Psi_\infty - (\mathscr{E} + e\Phi_\infty)\frac{\partial \Psi_\infty}{\partial r} = 0. \tag{5.19}$$

By solving equation (5.19) with respect to Ψ_∞, one finds

$$\Psi_\infty(r, \mathscr{E}) = \frac{\rho_\infty(\mathscr{E})[\mathscr{E} + e\Phi_\infty(r)]^{\frac{1}{2}}}{4\pi \int_{\mathscr{D}(\mathscr{E})} r'^2[\mathscr{E} + e\Phi_\infty(r')]^{\frac{1}{2}}dr'}, \tag{5.20}$$

where $\rho_\infty(\mathscr{E}) = 4\pi \int \Psi_\infty(r, \mathscr{E})r^2 dr$ is the energy distribution, which corresponds to the ergodic distribution expressed by equations (5.12)–(5.14).

Charging transient

Since the initial configuration considered here, equation (5.2), is far from equilibrium, the proper energy spectrum $\rho_{e,0}$ to be used in equations (5.16) must be determined as the equilibrium configuration after the initial charging transient. Apparently, the ergodic method cannot be of help to this purpose, since it is valid only for sufficiently smooth variations of Φ in time, a condition which is not met in the early stage, when the hot electrons are suddenly allowed to expand (as if a rigid wall, initially confining them, were instantaneously brought to infinity). However, a procedure can be envisaged, which makes these equations suitable also for the analysis of the initial electron equilibrium.

To this purpose, the real charging transient is replaced by a virtual one, in which an external potential barrier, initially confining the electrons, is gradually moved from $R_b = R_0$ to infinity with a series of small radial displacements. Each time the barrier is moved farther by δR_b the new self-consistent potential Φ is calculated and the energy of the electrons is updated. In order to actually simulate an expansion into vacuum (which the real transient is), the electrons and the expanding barrier must not exchange energy, i.e., the electron energy must vary only because of Φ variations. This implies that the ergodic invariant (5.15) is not conserved during the initial stage. Should one conserve $\mathcal{J}$ when displacing the barrier from a given radius R_b to $R_b + \delta R_b$, the corresponding electron energy variation, $\delta\mathcal{E}$, would be

$$\delta\mathcal{E} = -e\int_0^{R_b} \delta\Phi Q(r,\,\mathcal{E})\mathrm{d}r - \delta W, \tag{5.21}$$

where δW, defined as

$$\delta W = \frac{2}{3}[\mathcal{E} + e\Phi(R_b)]Q(R_b;\,\mathcal{E})\delta R_b, \tag{5.22}$$

represents the expansion work, done by an electron having energy $\mathcal{E}$, against the expanding barrier. Thus, conserving $\mathcal{J}$ would cause the overestimation of the electron cooling as the system would lose an extra amount of energy corresponding to the expansion work. In order to obtain an energy balance equivalent to that of a vacuum expansion, the energy loss associated with the expansion work is set to zero in equation (5.21).

The physical process simulated with the barrier method can be regarded as an infinitely slow expansion during which some external energy source compensates exactly for the expansion work δW against the barrier; alternatively, it can be considered as a series of instantaneous, small displacements of the barrier, where, after each displacement, one waits for a new equilibrium configuration to establish.

Reference results for the transient leading to the initial space-charge distribution of the plasma have been determined by solving equations (5.1) numerically, in the hypothesis of immobile ions, using the so called 'shell method'[2]. The same framework has been used also to investigate numerically the effect of the presence of the

---

[2] According to this scheme [9, 18], computational particles representative of a given number of electrons are moved in space, under the action of the sum of the self-consistent electric field and the electric field due to the ion distribution. By resorting the spherical symmetry of the system, the field generated by the electrons is evaluated using Gauss' law, as if each particle were actually a spherical shell, thus avoiding the use of a computational grid for solving Poisson's equation, and allowing for an infinite radial domain.

collision term, equation (5.10), which forces the system towards an ergodic distribution. Such perturbations are introduced in the model by scattering randomly the computational particles, without changing their energy, according to the collision frequency ν. Figure 5.1 shows the evolution of the electronic charge contained within the ion sphere, for low-temperature ($\widehat{T_0} = 7.2 \times 10^{-3}$) and high-temperature ($\widehat{T_0} = 7.2 \times 10^{-2}$) cases, as obtained with the Vlasov–Poisson equations, i.e., with $\nu = 0$, and with the collisional model, equation (5.10), using $\nu = \omega_{pe}$. In the collisionless case, the charge transient exhibits small-amplitude oscillations. In the collisional model, for $\nu \gtrsim \omega_{pe}$, the oscillations are strongly damped and the system rapidly reaches an equilibrium configuration, as predicted theoretically. In figure 5.2, the electron density is plotted, along with the corresponding electric field. Figure 5.3 shows the equilibrium energy distribution $\rho_{e,0}$, to be used as initial condition for the bulk expansion. The excellent agreement between different models confirms the validity of

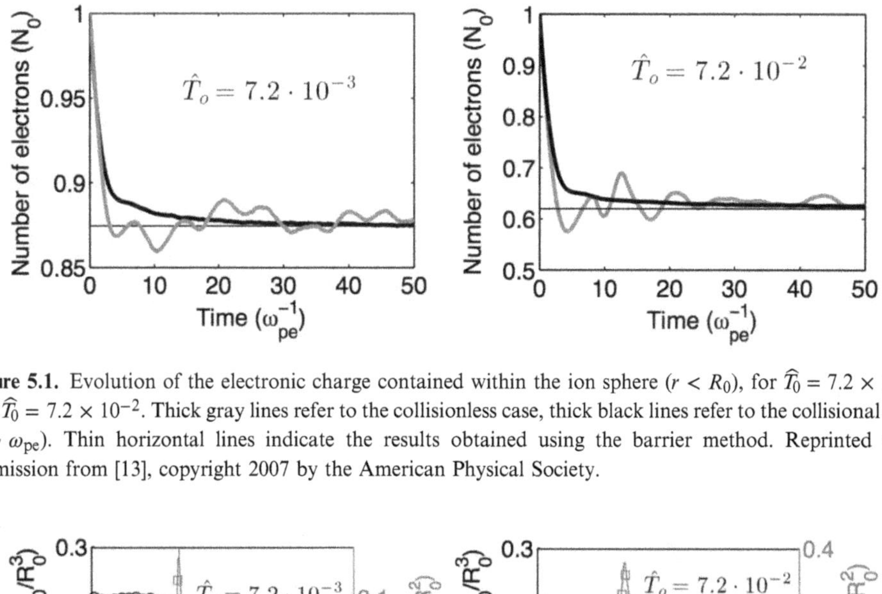

Figure 5.1. Evolution of the electronic charge contained within the ion sphere ($r < R_0$), for $\widehat{T_0} = 7.2 \times 10^{-3}$ and $\widehat{T_0} = 7.2 \times 10^{-2}$. Thick gray lines refer to the collisionless case, thick black lines refer to the collisional case ($\nu \sim \omega_{pe}$). Thin horizontal lines indicate the results obtained using the barrier method. Reprinted with permission from [13], copyright 2007 by the American Physical Society.

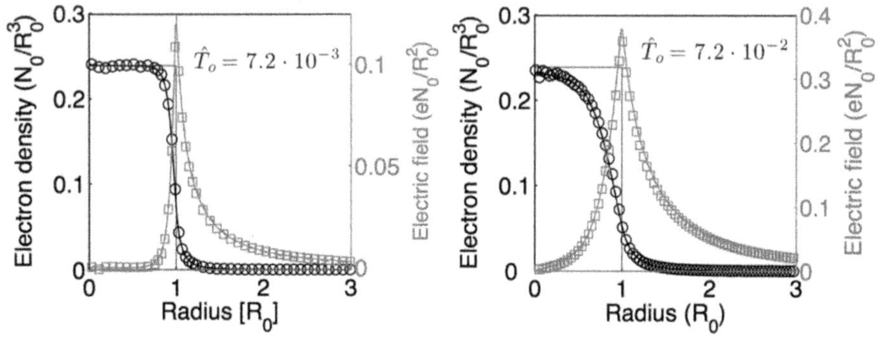

Figure 5.2. Equilibrium electron density (black) and radial electric field (gray), for $\widehat{T_0} = 7.2 \times 10^{-3}$ and $\widehat{T_0} = 7.2 \times 10^{-2}$. Solid lines refer to results from the ergodic model, markers to results from the Vlasov–Poisson model, and dotted lines to results from the drift-diffusion model of equations (5.17) and (5.18) (the curves obtained with the ergodic model and those obtained with the drift-diffusion model are practically undistinguishable). Reprinted with permission from [13], copyright 2007 by the American Physical Society.

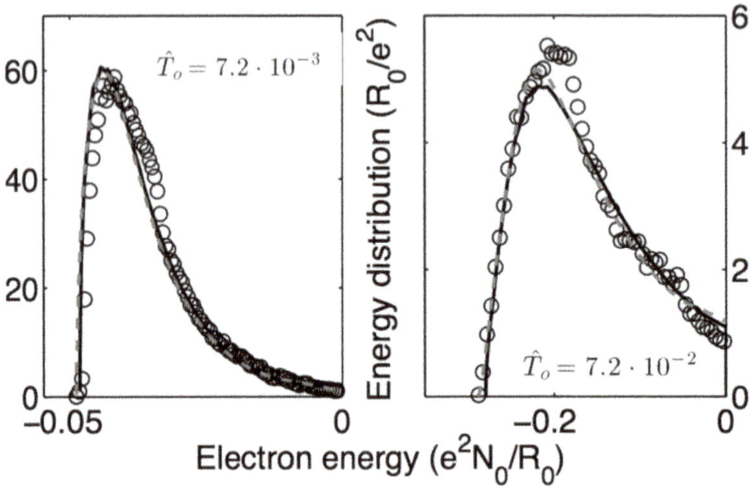

Figure 5.3. Equilibrium energy spectrum of trapped electrons, (i.e., with $\mathscr{E} = m\mathbf{p}^2/2 - e\Phi < 0$) for $\widehat{T}_0 = 7.2 \times 10^{-3}$ and $\widehat{T}_0 = 7.2 \times 10^{-2}$. Solid lines refer to results from the ergodic model, dots to results from the Vlasov–Poisson model, and dotted lines to results from the drift-diffusion model. Reprinted with permission from [13], copyright 2007 by the American Physical Society.

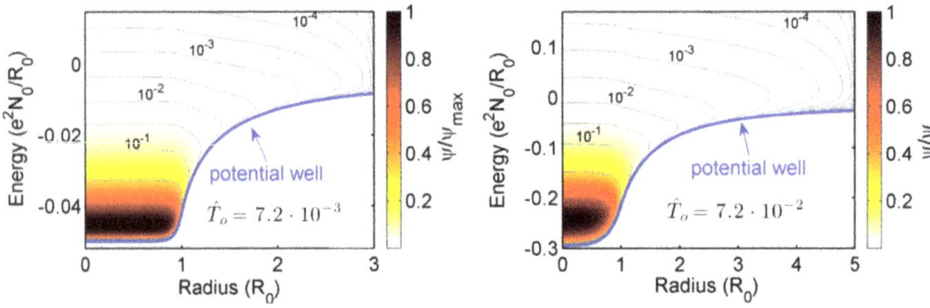

Figure 5.4. Equilibrium distribution $\Psi(r, \mathscr{E})$ (normalized to its maximum value) as obtained from the drift-diffusion model of equations (5.17) and (5.18), for $\widehat{T}_0 = 7.2 \times 10^{-3}$ and $\widehat{T}_0 = 7.2 \times 10^{-2}$. Reprinted with permission from [13], copyright 2007 by the American Physical Society.

the barrier method. Figure 5.4 shows the asymptotic solution of equations (5.17) and (5.18), $\Psi_\infty(r, \mathscr{E})$: the corresponding electron density and energy distribution (figures 5.2 and 5.3) have been calculated as $\int \Psi_\infty(r, \mathscr{E})\mathrm{d}\mathscr{E}$ and $4\pi \int \Psi_\infty(r, \mathscr{E})r^2\mathrm{d}r$, respectively.

Bulk expansion

Using the tools presented above, the expansion of ions and electrons can be investigated for a wide range of the parameter $\widehat{T}_0$. The results of the study show that the expansion dynamics changes smoothly from a hydrodynamic-like regime ($\widehat{T}_0 \ll 1$) to a Coulomb expansion regime ($\widehat{T}_0 \sim 1$). In the former case the outer ions expand first and a rarefaction front propagates inward, in the latter all ions start expanding at the same time. The evolution of the ion phase-space profile and of the electron and ion densities, starting from the initial equilibrium of figure 5.2, are shown in figures 5.5 and 5.6, respectively. In case (a), the ion expansion starts from

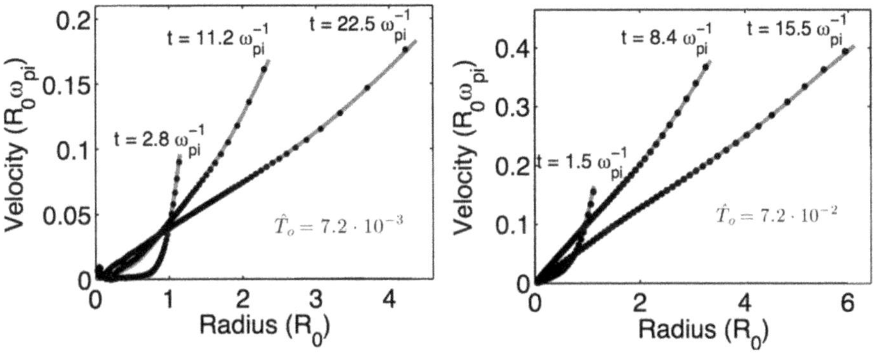

Figure 5.5. Evolution of the ion phase-space profile, for $\widehat{T_0} = 7.2 \times 10^{-3}$ and $\widehat{T_0} = 7.2 \times 10^{-2}$. Lines refer to results from the ergodic model, dots to results from the full Vlasov–Poisson model. Reprinted with permission from [13], copyright 2007 by the American Physical Society.

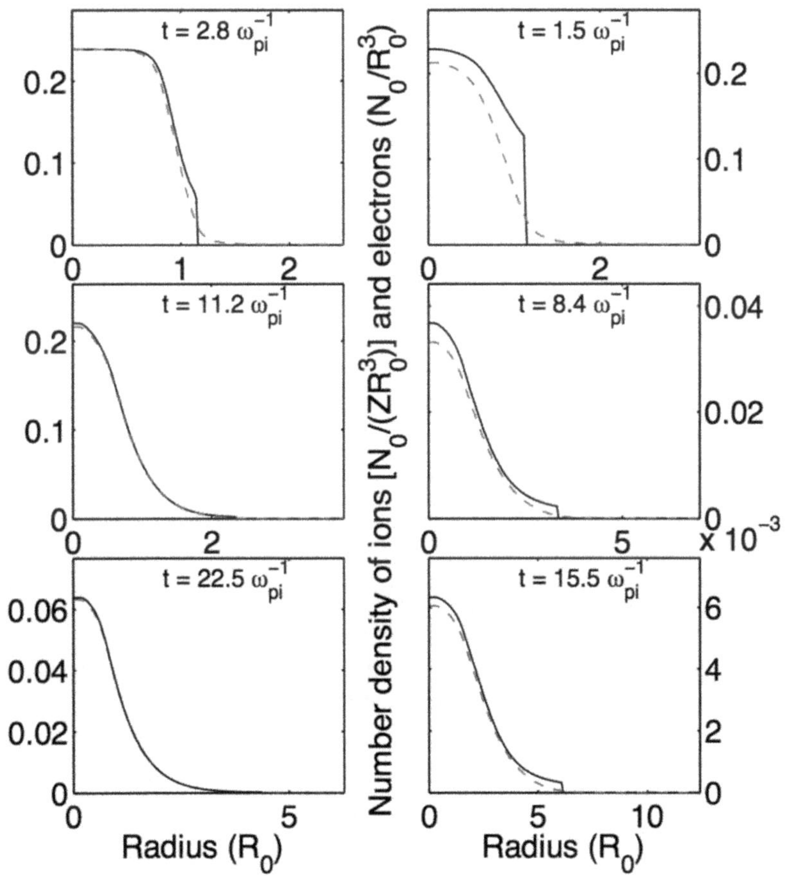

Figure 5.6. Evolution of ion (solid line) and electron (dashed curve) densities, for $\widehat{T_0} = 7.2 \times 10^{-3}$ (left) and $\widehat{T_0} = 7.2 \times 10^{-2}$ (right). Reprinted with permission from [13], copyright 2007 by the American Physical Society.

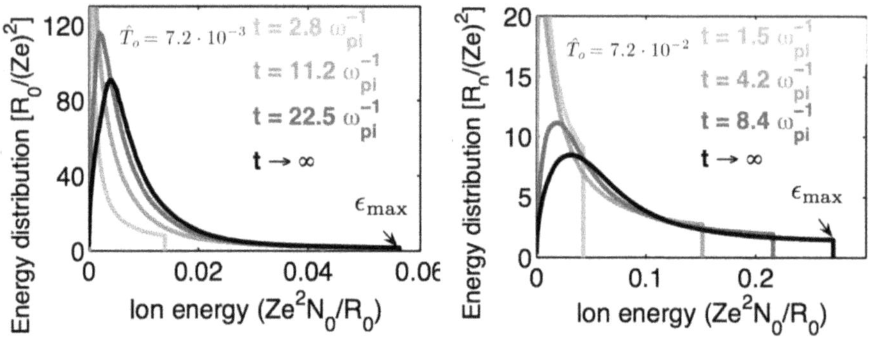

Figure 5.7. Evolution of the ion energy spectrum (from gray to black), for $\widehat{T_0} = 7.2 \times 10^{-3}$ and $\widehat{T_0} = 7.2 \times 10^{-2}$. Reprinted with permission from [13], copyright 2007 by the American Physical Society.

the periphery and a rarefaction front is clearly observed to propagate inward until it reaches the center of the distribution; during the expansion, the plasma remains approximately neutral, apart from the thin double-layer at the ion front. These features, typical of quasineutral, hydrodynamic expansions, are lost in case (b), in which all the ions are promptly involved in the expansion (figure 5.5(right)) and the distribution remains non-neutral during the whole process (figure 5.6(right)). In both cases, as the ions expand and gain kinetic energy, the electrons cool down and the charge build-up within the ion sphere decreases. Asymptotically, the sphere enveloped by the expanding ion front encloses all trapped electrons, and a ballistic regime is reached for both species [19]. The self-consistent behavior of the electrons affects strongly the ion dynamics and their resulting energy spectrum. Figure 5.7 illustrates the evolution of the ion energy spectrum towards its asymptotic form: in both cases, the spectrum develops a local maximum, far from the maximum value of the ion energy. This feature is absent in Coulomb explosion, where the asymptotic spectrum is always monotonic, so the maximum in the spectrum is expected to disappear when increasing $\widehat{T_0}$ further. The transition from non-monotonic to monotonic ion spectra occurs about $\widehat{T_0} = 0.5$, as shown in figure 5.8.

5.3 Electron dynamics in a Penning trap for technology applications

A Penning trap is a device used to confine non-neutral plasmas, in which an axial magnetic field and a voltage applied to cylindric electrodes provide radial and axial confinement. Penning traps are used in plasma physics investigations [20–23] and in many advanced technological applications such as in ultra-high vacuum systems [24, 25], to determine fundamental physical constants [26, 27], in high precision mass spectrometry [28–32], in Zeeman spectroscopy [33], in antihydrogen production and trapping [34, 35]

The electron motion inside a Penning trap is the superposition of Larmor and diocotron orbits with frequencies ω_c and ω_D, respectively, and axial bounce between the cathodes, with frequency ω_b. In ideal traps [20–23, 34, 36, 37], the three characteristic frequencies of these orbits differ by orders of magnitude [38], in

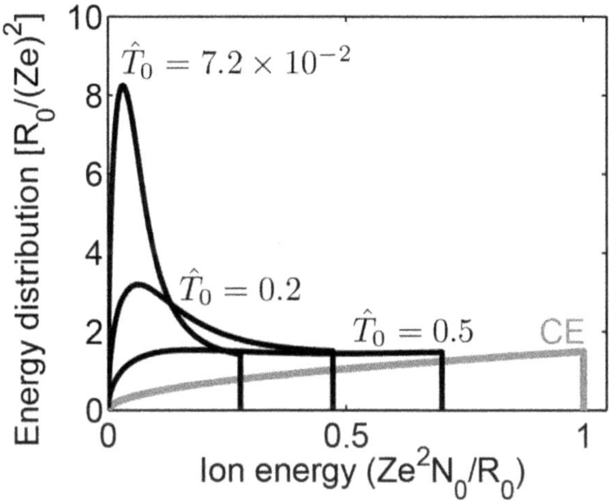

Figure 5.8. Asymptotic energy spectrum of the ions for different values of $\widehat{T_0}$, compared with the asymptotic spectrum for a Coulomb explosion (gray). Reprinted with permission from [13], copyright 2007 by the American Physical Society.

particular $\omega_c \gg \omega_b \gg \omega_D$. These differences arise from the high intensity of the magnetic field and from the large value of the aspect ratio L_z: R ($2L_z$ and R being the length and the radius of the trap). Moreover, collisions between electrons and atoms are negligible due to the ultra-high vacuum condition. In contrast, in Penning traps used in some technological applications, such as for ion pumps, L_z and R are comparable; moreover, the magnetic field, which is usually generated by permanent magnets, is significantly lower and electron–neutral collisions are important. Actually, the working principle of ion pumps is the electron-induced ionization of gas atoms which are then removed by the electric field [24]. For these devices, Larmor, bouncing and diocotron frequencies are of the same order of magnitude, while they differ by several orders of magnitude with respect to the electron–gas collision frequency, ν_{el} (for a gas with a pressure of 10^{-6} Torr at room temperature, the collision frequency for an electron is $\nu_{\mathrm{el}} \sim 10^4$ Hz [39]). The large difference between the two characteristic time scales of the phenomena under examination, along with the need of analyzing the system on the slow time scale, suggests the use of a suitable average of the electron dynamics on the fast time scale.

In spite of its apparent simplicity, the physics involved in Penning traps used in ion pumps is quite rich. If electrons are introduced into the device and a discharge is set up, gas atoms or molecules can be ionized inside the cylindrical cavity. The biasing of the electrodes creates a potential well for the electrons, and hence the positive ions produced in the electron–gas collisions are accelerated toward the negatively-biased planar cathodes. Consequently, ions impinge on these electrodes with a relevant kinetic energy and are removed from the system [40–42]. In contrast, electrons are confined by the electric and magnetic fields, and therefore they can induce further ionizations. In this way, an increasing negative charge is set up, which induces large modifications to the electrostatic potential. Eventually, stationary configurations are achieved due to the equilibrium between the electrons generated by ionization and secondary emission (induced by the ions impinging on the cathodes) and electron radial transport to the anode. Between two successive collisions, the electron trajectory is characterized by two constants of the motion: the total energy and the canonical angular momentum. The knowledge of these constants provides useful information: in particular, for each electron a space region can be determined in which the trajectory is constrained. Therefore, each electron can be regarded as a space-charge distribution, which represents the time average of the positions spanned during the motion. The effect of collisions of the electrons with neutral atoms or molecules can be calculated in a simple way, as long as the gas density is sufficiently low: in practice, the mean time between two collisions must be higher with respect to the time required for an electron trajectory to fill the region in which it is constrained. In this case, the effect of an elastic collision is simply a variation of the constants of the motions of the electron, and, consequently, of the shape of the region that is associated with them. Ionization collisions and secondary emissions can be taken into account in a similar way.

A typical Penning trap for ion pumps is made of a cylindrical anode of radius R which is inserted between two planar cathodes whose distance is L_z (R and L_z are of the order of few centimeters), biased at a negative potential $-V_0$, typically of the order of a few kV.

The electric field confines the electrons in z direction, causing a bounce motion between the two planar cathodes; in addition, the cell is immersed in a uniform magnetic field B_0 of the order of 10^{-1} T, parallel to the axis of the cylinder, whose purpose is confining the electrons in radial direction (figure (5.9)) [21].

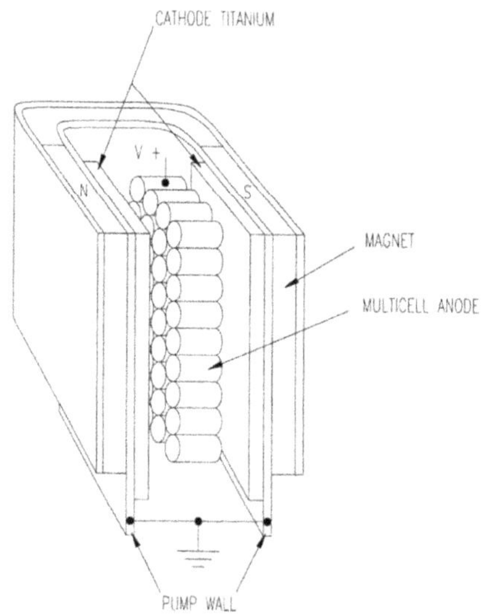

Figure 5.9. Scheme of a Penning trap for ion pump technology (from [43], reprinted with permision of Agilent Technologies).

Hyperbolic trap

In order to study the general properties of the electron trajectories, one can start from a simplified two-dimensional case, in which a constant magnetic field in z direction is present and an electrostatic potential $\Phi(x, z)$ is generated by the electrodes [44].

Between two successive collisions, the motion of a particle of charge q is governed by the equations:

$$\begin{cases} \dfrac{\mathrm{d}v_x}{\mathrm{d}t} = -\dfrac{q}{m}\dfrac{\partial\Phi}{\partial x} - \omega_c v_z, \\ \dfrac{\mathrm{d}v_z}{\mathrm{d}t} = -\dfrac{q}{m}\dfrac{\partial\Phi}{\partial z}, \\ \dfrac{\mathrm{d}v_z}{\mathrm{d}t} = \omega_c v_x \end{cases} \tag{5.23}$$

where $\omega_c = qB/(mc)$ is the Larmor frequency. From the third (5.23), one obtains immediately $v_z = \omega_c x + \text{Const}$, which states that the canonical momentum

$$p_z = m(v_z - \omega_c x) \tag{5.24}$$

is a constant of the motion. Therefore v_z can be eliminated from the first (5.23) and the equations of motion in the (x, z) plane can be written as:

$$\frac{\mathrm{d}p_x}{\mathrm{d}t} = -q\frac{\partial\Phi}{\partial x} - m\omega_c\left(\omega_c x + p_z/m\right), \quad \frac{\mathrm{d}p_z}{\mathrm{d}t} = -q\frac{\partial\Phi}{\partial z}. \tag{5.25}$$

By defining the effective potential, $\mathscr{U}$, as:

$$\mathscr{U}(x, z) = q\Phi(x, z) + \frac{1}{2m}(m\omega_c x + p_z)^2, \tag{5.26}$$

the equations of motion can be deduced from the Hamiltonian:

$$\begin{aligned} \mathscr{H}(x, z, p_x, p_z) &= \frac{1}{2m}\left(p_x^2 + p_z^2\right) + \mathscr{U}(x, z) \\ &= \frac{1}{2m}\left[p_x^2 + p_z^2 + (m\omega_c x + p_z)^2\right] + q\Phi(x, z). \end{aligned} \tag{5.27}$$

As $\mathscr{H}$ does not depend explicitly on time, its value is a constant of the motion:

$$\mathscr{H}\big(x(t), z(t), p_x(t), p_z(t)\big) = \text{Const} = \mathscr{E}, \tag{5.28}$$

and it coincides with the total energy of an electron in the potential Φ. The level curves for $\mathscr{U}(x, z)$ and $\Phi(x, z)$ have the following property: given a generic point (x_0, z_0), if $p_z = -m\omega_c x_0$ then the two curves $\Phi(x, z) = \Phi(x_0, z_0)$ and $\mathscr{U}(x, z) = q\Phi(x_0, z_0)$ are tangent in (x_0, z_0). This can be proved by considering that the tangent line to Φ= Const is such that $\mathrm{d}\Phi = 0$ and so $\frac{\partial\Phi}{\partial x}\mathrm{d}x + \frac{\partial\Phi}{\partial z}\mathrm{d}z = 0$, while the tangent line to $\mathscr{U}$= Const is such that $0 = \mathrm{d}\mathscr{U} = \mathrm{d}\Phi + m\omega_c(\omega_c x - \omega_c x_0)\mathrm{d}x$, so, if $x = x_0$, then $\mathrm{d}\mathcal{U} = \mathrm{d}\Phi$. Therefore, the two lines have the same equation. The property can be interpreted as follows: if (x_0, z_0) is a point of the cathode and in this point an electron with zero velocity is emitted ($p_x = 0$, $p_z = -m\omega_c x_0$) then the electron moves in a part of the phase-space such that $\mathcal{H} = q\Phi(x_0, z_0) = \mathcal{U}(x_0, z_0)$. Being $\mathcal{H}(x, z, p_x, p_z) \geqslant \mathcal{U}(x, z)$, the region of $\mathbb{R}^2$ in which the electron can be found must satisfy the condition

$$\mathcal{U}(x, z) < \mathcal{E} = \mathcal{U}(x_0, z_0), \tag{5.29}$$

and it is delimited by the curve $\mathcal{U}(x, z) = \mathcal{U}(x_0, z_0)$. In addition, if the curves $\Phi(x, z) = \Phi(x_0, z_0)$ and $\mathcal{U}(x, z) = \mathcal{U}(x_0, z_0)$ were not tangent in (x_0, z_0), the region of the motion, equation (5.29), should have points in the region $q\Phi(x, z) > q\Phi(x_0, z_0)$, but this is impossible, as

$$\mathcal{E} = q\Phi(x, z) + \frac{1}{2}mv^2 = q\Phi(x_0, z_0) \Rightarrow q\Phi(x, z) < q\Phi(x_0, z_0). \tag{5.30}$$

In the case of potential $\Phi(x, z)$ produced by using a pair of cathodes and anodes having a hyperbolic shape, i.e.,

$$q\Phi(x, z) = a^2(z^2 - x^2), \tag{5.31}$$

the effective potential $\mathcal{U}$ assumes the form

$$\mathcal{U}(x, z) = \mathcal{U}_x(x) + \mathcal{U}_z(z) + \mathcal{E}_m(p_z), \tag{5.32}$$

with

$$\mathcal{U}_x(x) = b^2(x - x_m)^2, \quad \mathcal{U}_z(z) = a^2 z^2, \quad \mathcal{E}_m(p_z) = -\frac{a^2 p_z^2}{2mb^2}, \tag{5.33}$$

being $b^2 = m\omega_c^2/2 - a^2$ and $x_m(p_z) = -\omega_c p_z/(2b^2)$. Although the potential $\Phi(x, z)$ confines electron only in z direction, a sufficiently strong magnetic field can provide the confinement also in the x direction if $b^2 > 0$, i.e., if $|\omega_c| > a(2/m)^{\frac{1}{2}}$.

The electron motion is the composition of two harmonic oscillations in x and in z directions, according to the following equations:

$$\begin{aligned} \frac{\mathrm{d}^2 x}{\mathrm{d}t^2} &= -\,\omega_x^2(x - x_m), \quad & \omega_x &= b\left(\frac{2}{m}\right)^{\frac{1}{2}}, \\ \frac{\mathrm{d}^2 z}{\mathrm{d}t^2} &= -\,\omega_z^2 z, \quad & \omega_z &= a\left(\frac{2}{m}\right)^{\frac{1}{2}}, \end{aligned} \tag{5.34}$$

which are obtained starting from the potential equation (5.32). Therefore, the motion is confined in the rectangular region $\mathcal{R} \subset \mathbb{R}^2$ defined by

$$\mathcal{R} = \left[x_m - \frac{\mathcal{E}_x^{1/2}}{b}, x_m + \frac{\mathcal{E}_x^{1/2}}{b}\right] \times \left[-\frac{\mathcal{E}_z^{1/2}}{a}, \frac{\mathcal{E}_z^{1/2}}{a}\right], \tag{5.35}$$

where $\mathscr{E}_x$, $\mathscr{E}_z$ are the energies associated with the oscillatory motion in x and z directions, respectively; the total energy, $\mathscr{E}$, is simply given by $\mathscr{E} = \mathscr{E}_x + \mathscr{E}_z + \mathscr{E}_m(p_z)$. Four constants of the motion exist, i.e., the energies along x and z and the two phases, φ_x and φ_z, for the oscillations in x and z directions. While the value of two energies determines the rectangular region $\mathscr{R}$ in which the motion is allowed, the difference $\varphi_x - \varphi_z$ determines the particular kind of trajectory. Only in the very particular case in which ω_x/ω_z is a rational number is the trajectory a closed Lissajous curve, otherwise, for $t \to +\infty$, the curve tends to 'cover' the entire rectangular region.

Figure 5.10 shows the elliptic region defined by equation (5.29) and the corresponding region $\mathscr{R}$ for an electron emitted by the cathode. According to the theory, the ellipse is tangent to the cathode. In summary, each electron can be associated with a phase-space distribution depending on three parameters $\{p_z, \mathscr{E}_x, \mathscr{E}_z\}$. Alternatively, a simplified 'ergodic' model can be used, in which each particle is associated with the distribution $\delta\left(\frac{m}{2}(v_x^2 + v_z^2) + \mathscr{U}(x, z) - \mathscr{E}\right)$, with $\mathscr{E}$ being the total energy. The last assumption is rigorously correct in the presence of perturbations of the potential, otherwise it represents a simplified description of the real physical system.

Real Penning trap

In a real Penning trap, the motion of an electron can be described by the Lagrangian $\mathscr{L}$ of a particle of mass m and electric charge $-e$ moving in a homogeneous magnetic field $\boldsymbol{B} = B_0\boldsymbol{e}_z$ and in an electric field $\boldsymbol{E}(r, z) = -\nabla\Phi(r, z)$, as [21]:

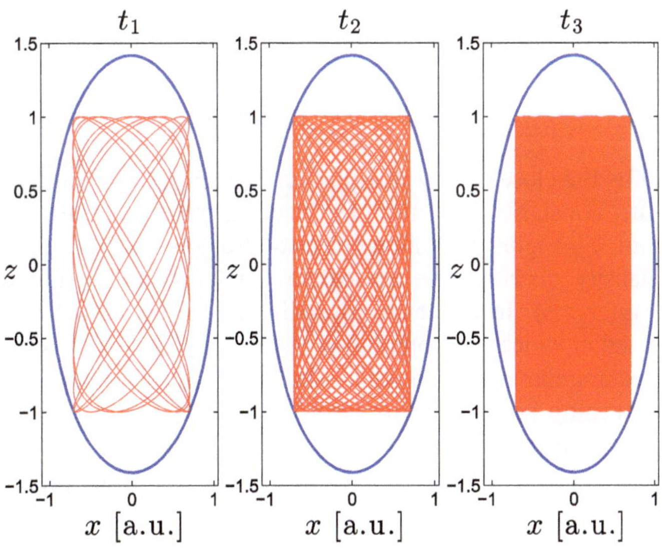

Figure 5.10. Electron trajectories for the hyperbolic potential, Equation (5.31), for $t \in [0, t_i]$ with $t_1 < t_2 \ll t_3$. The ellipse $\mathscr{U}(x, z) = \mathscr{E}$ is also shown.

$$\begin{aligned}\mathscr{L}(r, \vartheta, z, \dot{r}, \dot{\vartheta}, \dot{z}) &= \frac{m}{2}v^2 + e\left[\Phi(r, z) - \frac{\mathbf{v}}{c} \cdot \boldsymbol{A}\right] \\ &= \frac{m}{2}\left(\dot{r}^2 + r^2\dot{\vartheta}^2 + \dot{z}^2\right) + e\left[\Phi(r, z) - \frac{1}{2}\frac{B_0}{c}r^2\dot{\vartheta}\right],\end{aligned} \tag{5.36}$$

where $\boldsymbol{A} = \frac{1}{2}B_0 r\boldsymbol{e}_\vartheta$ is the vector potential. As the Lagrangian defined by equation (5.36) does not depend on ϑ, the corresponding canonical momentum $p_\vartheta = mr^2\left(\dot{\vartheta} - \frac{1}{2}\omega_c\right)$ is a constant of the motion and therefore v_ϑ can be expressed as function of r and p_ϑ, and an effective potential, $\mathscr{U}$, can be defined as

$$\begin{aligned}\mathscr{U}(r, z; p_\vartheta) &= -e\Phi(r, z) + \frac{m}{2}v_\vartheta^2(r, p_\vartheta) \\ &= -e\Phi(r, z) + \frac{m}{2}\left(\frac{p_\vartheta}{mr} + \frac{\omega_c r}{2}\right)^2.\end{aligned} \tag{5.37}$$

Introducing the Hamiltonian, $\mathscr{H}$:

$$\begin{aligned}\mathscr{H}(r, z, p_r, p_z; p_\vartheta) &= \frac{m}{2}(v_r^2 + v_z^2) + \frac{m}{2}v_\vartheta^2(r, p_\vartheta) - e\Phi(r, z) = \\ &= \frac{1}{2m}\left[p_r^2 + p_z^2\right] + \mathscr{U}(r, z; p_\vartheta)\end{aligned} \tag{5.38}$$

the dynamics of the electrons can be reduced to the (r, z) domain. The Hamiltonian is a constant of the motion as it does not depend explicitly on time; its value, $\mathscr{E}$, coincides with the total energy of an electron in the potential field Φ: $\mathscr{E} = \frac{1}{2}mv^2 - e\Phi$.

Between two successive collisions, electrons conserve the total energy $\mathscr{E}$ and the canonical angular momentum p_ϑ and, being $\mathscr{H}(r, z, p_r, p_z; p_\vartheta) \geqslant \mathscr{U}(r, z; p_\vartheta)$, the particle can be found in the region $\mathscr{R}_E \subset \mathbb{R}^2$ defined by

$$\mathscr{U}(r, z; p_\vartheta) \leqslant \mathscr{E}. \tag{5.39}$$

$\mathscr{R}_E$ is delimited by the closed curve of equation $\mathscr{U}(r, z; p_\vartheta) = \mathscr{E}$. Actually, due to the existence of other constants of motion, the electron trajectories are constrained in a subdomain $\mathscr{R}_0$ of $\mathscr{R}_E$. Figure 5.11 shows electron trajectory and the associate density distribution (density distributions are calculated by sampling the trajectories at constant time intervals). In order to approximate the accessible region, a simple model can be used in which the potential $\mathscr{U}$ is decoupled along directions r and z. In this way, the rectangular region in which the electron is confined is determined by calculating the turning points. For an electron located in (r_0, z_0) and with velocity (v_{r0}, v_{z0}), the energy associated to the motion along z direction has been defined as $\frac{1}{2}mv_{z0}^2 + \mathscr{U}(r_0, z_0; p_\vartheta)$; then, the axial turning points, $\pm z_b$, have been calculated by solving the equation

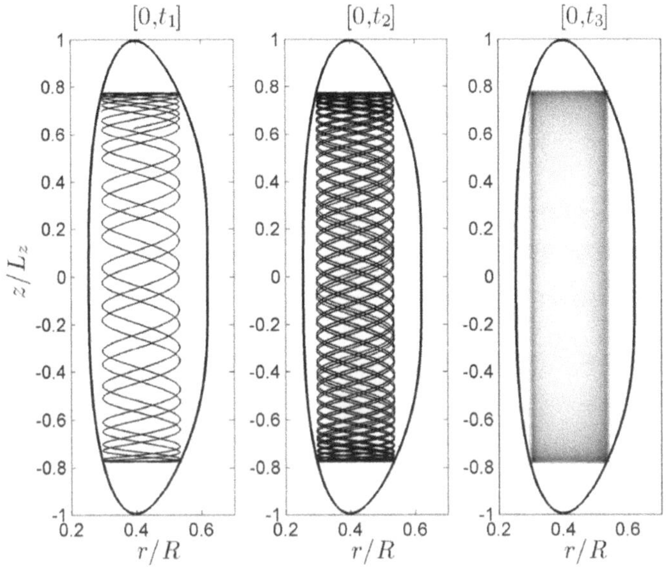

Figure 5.11. Electron trajectories for $t = [0,\ t_1]$, $[0,\ t_2]$, with $t_2 \gg t_1$, and density distribution calculated by sampling the trajectory at constant time intervals for $t = [0,\ t_3]$ with $t_3 \gg t_2$ (B_0= 0.1 T). The curve $\mathcal{U}(r,\ z;\ p_\vartheta) = \mathcal{E}$ is also shown. Reproduced with permission from [45]. Copyright 2012 AIP Publishing.

$$\mathcal{U}(r_0,\ z_b;\ p_\vartheta) = \frac{1}{2}mv_{z0}^2 + \mathcal{U}(r_0,\ z_0;\ p_\vartheta). \tag{5.40}$$

Then, the turning points r_1 and r_2 along r direction have been obtained as the solutions of the equation

$$\mathcal{U}(r_{1,2},\ z_b;\ p_\vartheta) = \frac{1}{2}mv_{r0}^2 + \mathcal{U}(r_0,\ z_b;\ p_\vartheta) \tag{5.41}$$

The turning points belong to the boundary of the region $\mathcal{R}_E$; making use of equation (5.40), equation (5.41) can be rewritten as $\mathcal{U}(\pm r_{1,2},\ z_b;\ p_\vartheta) = \mathcal{E}$. In general, the effective potentials $\mathcal{U}$ may present a non-monotonic behavior and, depending on the value of total energy, from equations (5.40 and 5.41) multiple turning points can be found along r and z directions. In this case, the domain $\hat{\mathcal{R}}$ can be univocally calculated taking into account the place where collisions happen.

The effectiveness of the approximation can be proved by comparing the rectangular domain $\hat{\mathcal{R}} = [r_1,\ r_2] \times [-z_b,\ z_b]$ with the real trajectory; in fact, a more significant test is provided by the probability distribution of finding an electron at a given spatial position. This quantity is very useful in the study of the dynamics of the electrons: as collision frequencies are much smaller with respect to all the characteristic frequencies of the motion, one can consider the time average of the trajectory instead of calculating the position of the electron at each time [44–46]. The calculation is simple in the case of the separable potential (i.e., $\mathcal{U}_r(r) + \mathcal{U}_z(z)$), for which one can express the time-averaged phase-space distribution associated with the electron motion as:

$$f(r, z, p_r, p_z) = C \cdot \delta\left(\frac{p_r^2}{2m} + \mathcal{U}_r(r) - \mathcal{E}_r\right)\delta\left(\frac{p_z^2}{2m} + \mathcal{U}_z(z) - \mathcal{E}_z\right), \tag{5.42}$$

with $\mathcal{E}_r$, $\mathcal{E}_z$ being the energies associated with the motion along r and z, respectively, (i.e., the values of $\mathcal{U}_r$ and $\mathcal{U}_z$ at turning points), and C is such that $\iint f\,\mathrm{d}^2x\mathrm{d}^2p = 1$. In general, $\mathcal{U}$ is not separable, but, according to the above-introduced approximation, the following expressions can be used:

$$\begin{aligned} \mathcal{U}_r(r; p_\vartheta) &= \mathcal{U}(r, z_b; p_\vartheta) \\ \mathcal{U}_z(z; p_\vartheta) &= \frac{1}{2}\left[\mathcal{U}(r_1, z; p_\vartheta) + \mathcal{U}(r_2, z; p_\vartheta)\right] \\ \mathcal{E}_r &= \mathcal{E}_z = \mathcal{E} \end{aligned} \tag{5.43}$$

From the density f, the probability for a particle to be in $[r - \Delta r/2, r + \Delta r/2] \times [z - \Delta z/2, z + \Delta z/2]$ is evaluated as $\mathcal{P}(r, z)\Delta r\Delta z$ where

$$\mathcal{P}(r, z) = \iint f(r, z, p_r, p_z)\mathrm{d}p_r\mathrm{d}p_z = \mathcal{P}_r(r)\mathcal{P}_z(z) \tag{5.44}$$

where

$$\mathcal{P}_r(r) = \begin{cases} C_r\left[\mathcal{E} - \mathcal{U}_r(r; p_\vartheta)\right]^{-1/2}, & r_1 < r < r_2 \\ 0, & \text{otherwise} \end{cases} \tag{5.45}$$

$$\mathcal{P}_z(z) = \begin{cases} C_z\left[\mathcal{E} - \mathcal{U}_z(z; p_\vartheta)\right]^{-1/2}, & |z| < z_b \\ 0, & \text{otherwise} \end{cases} \tag{5.46}$$

The constants C_r, C_z are determined by requiring that $\int \mathcal{P}_r(r)\mathrm{d}r = \int \mathcal{P}_z(z)\mathrm{d}z = 1$. Figures 5.12 and 5.13 show the excellent agreement between the normalized probability $\mathcal{P}(r, z)/\bar{\mathcal{P}}$ (where $\bar{\mathcal{P}} = 1/(2RL_z)$) as given by equations (5.45) and (5.46) and the probability obtained by solving numerically equations of motion (as before, the 'exact' probability has been calculated by considering a grid in (r, z) domain and sampling the solution at constant time intervals).

In addition to the method presented above, a simpler description of electron dynamics can be considered, in which the time average of the trajectories is calculated using an ergodic distribution for each particle. The use of phase-space ergodic distribution is rigorously justified if a physical system is perturbed in a way that only 'strong' constants of the motion, such as the energy, are conserved. As an example, in figure 5.14 the trajectory of an electron is shown for a case, in which a suitable perturbation $\delta\Phi(r, z)$ is added to the electrostatic potential [45]. In this case, the trajectory tends to cover uniformly the region $\mathcal{R}_E$ and not only $\hat{\mathcal{R}}$. This fact can be explained by considering that a weak perturbation does not change appreciably the total energy $\mathcal{E}$; instead, it produces a different distribution between the axial and the radial energies. In real situations, as the perturbation may involve also the ϑ coordinate, one should control the sensitivity of p_ϑ to the perturbation. In other

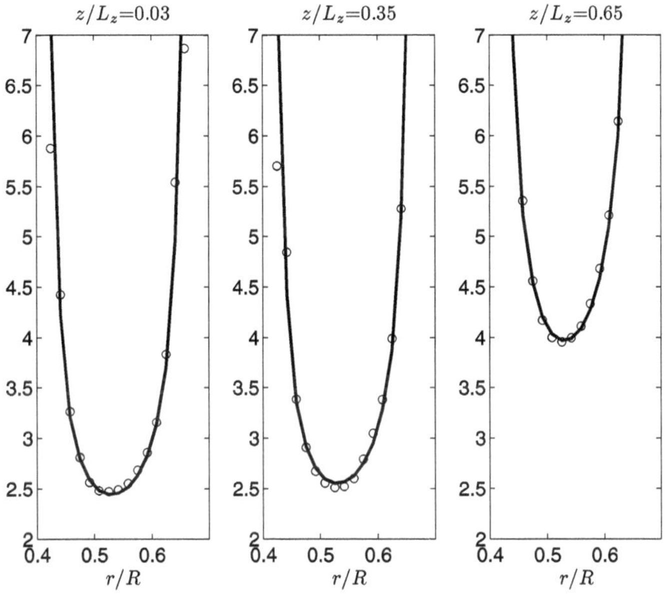

Figure 5.12. Example of comparisons between the probability density $\mathscr{P}(r, z)/\bar{\mathscr{P}}$ at different z given by equation (5.45) (full line) with the one obtained by solving numerically the equations of motion (∘). Reproduced with permission from [45]. Copyright 2012 AIP Publishing.

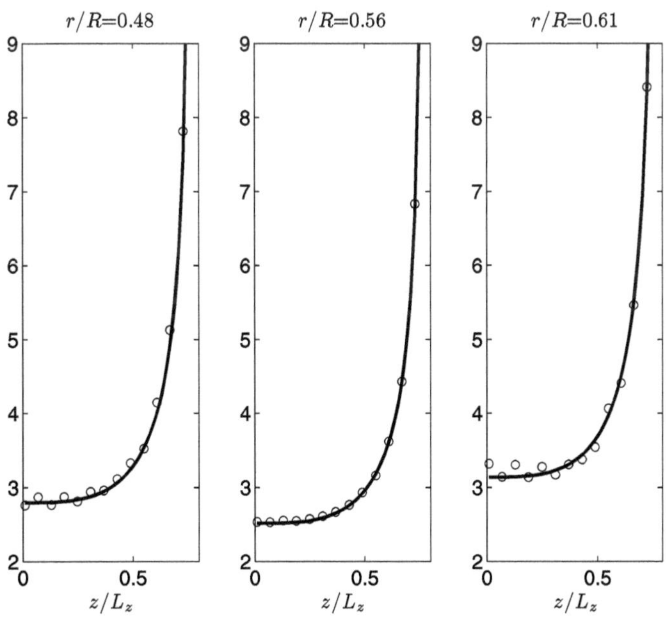

Figure 5.13. Comparison between the probability density $\mathscr{P}(r, z)/\bar{\mathscr{P}}$ at different r given by equation (5.46) (full line) with the one obtained by solving numerically the equations of motion (∘) for the same case reported in figure 5.12. Reproduced with permission from [45]. Copyright 2012 AIP Publishjing.

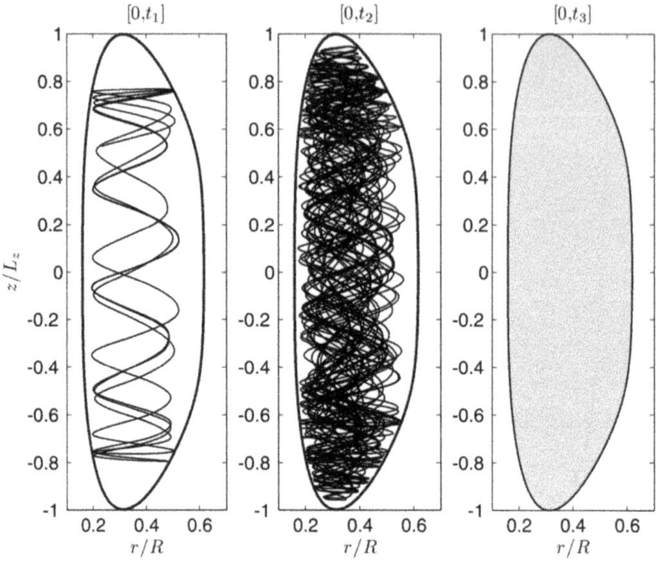

Figure 5.14. Electron trajectories when a perturbation to the potential is considered for $t = [0, t_1], [0, t_2]$, with $t_2 \gg t_1$, and density distribution calculated by sampling the trajectory at constant time intervals for $t = [0, t_3]$ with $t_3 \gg t_2$ (B_0= 0.08 T). The curve $\mathcal{U}(r, z; p_\vartheta) = \mathcal{E}$ is also shown. Reproduced with permission from [45]. Copyright 2012 AIP Publishing.

terms, p_ϑ may not be a 'strong' constant. However, independently from the presence and the possible effects of a perturbation to the potential, the use of ergodic distribution provides a useful tool for a simplified description of the electron density [12, 13]. In this case the time-averaged phase-space density has the form

$$f_E(\boldsymbol{x}, \boldsymbol{p}; p_\vartheta) = C \cdot \delta\left(\frac{\boldsymbol{p}^2}{2m} + \mathcal{U}(\boldsymbol{x}; p_\vartheta) - \mathcal{E}\right), \tag{5.47}$$

which depends only on two parameters, $\mathcal{E}$ and p_ϑ, and the probability $\mathcal{P}_E$ is given by

$$\mathcal{P}_E(\boldsymbol{x}; p_\vartheta) = \begin{cases} \dfrac{1}{\mathcal{A}(\mathcal{E}, p_\vartheta)}, & \mathcal{U}(\boldsymbol{x}; p_\vartheta) \leqslant \mathcal{E}, \\ 0, & \mathcal{U}(\boldsymbol{x}; p_\vartheta) > \mathcal{E}, \end{cases} \tag{5.48}$$

where $\mathcal{A}(\mathcal{E}, p_\vartheta) = \int_{\mathcal{U} \leqslant \mathcal{E}} d^2x$ is the area of the permitted region of motion.

In figure 5.15, the probability densities of time-average and ergodic models, given, respectively, by equations (5.44) and (5.48), are shown for trajectories with same initial conditions.

Collisions with atoms

When phenomena having time scale much larger than the ones characterizing the electron motion are considered, each electron can be usefully regarded not as a point but as an 'object' having a spatial charge distribution, $\rho(r, z)$, given by

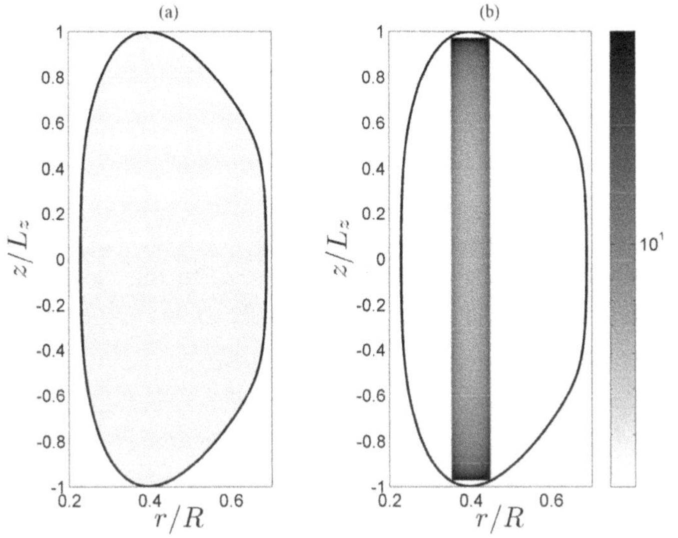

Figure 5.15. Normalized densities $\mathscr{P}_E(r, z)/\bar{\mathscr{P}}$ (a) and $\mathscr{P}(r, z)/\bar{\mathscr{P}}$ (b) for an electron with the same initial conditions. Reproduced with permission from [45]. Copyright 2012 AIP Publishing.

$$\rho(r, z) = -\frac{e\mathscr{P}(r, z)}{2\pi r} \tag{5.49}$$

In this way, each electron is described by its constants of the motion, $\mathscr{E}$, p_ϑ and the turning point z_b. In practice, probabilities given by equations (5.45) and (5.46) can be calculated if $\mathscr{E}$, p_ϑ and z_b are known, considering that radial turning points $r_{1,2}$ are solutions of equation (5.41). The density distribution associated with the electron can be described by a set of parameters $\{\mathscr{E}, p_\vartheta, z_b\}$, whose values change after each collision. The instantaneous collision frequency, $\nu(t)$, is given by $v(t)\sigma(v(t))\mathscr{N}_a$, with $\mathscr{N}_a$ the atom density and $\sigma(v)$ the cross-section of the considered process. Instead, the time-averaged collision frequency, $\bar{\nu}$, can be calculated as

$$\bar{\nu} = \iint v\sigma(v)\mathscr{N}_a f(\boldsymbol{x}, \boldsymbol{p}; p_\vartheta)\mathrm{d}^2x\mathrm{d}^2p, \tag{5.50}$$

where $\boldsymbol{x} = (r, z)$, $\boldsymbol{p} = (p_r, p_z)$ and $v(\mathscr{E}, \boldsymbol{x}) = [2(\mathscr{E} + e\Phi(\boldsymbol{x}))/m]^{\frac{1}{2}}$. Using equations (5.42) and (5.44), one obtains:

$$\bar{\nu}(\mathscr{E}, p_\vartheta, z_b) = \int v(\mathscr{E}, \boldsymbol{x})\sigma(v(\mathscr{E}, \boldsymbol{x}))\mathscr{N}_a\mathscr{P}(\boldsymbol{x}; \mathscr{E}, p_\vartheta, z_b)\mathrm{d}^2x. \tag{5.51}$$

The frequency $\bar{\nu}$ is the sum of the frequency of elastic collisions, $\bar{\nu}_{\mathrm{el}}$ and the frequency of ionizations, $\bar{\nu}_i$. The two frequencies are calculated using equation (5.51) by putting $\sigma = \sigma_{\mathrm{el}}$ or $\sigma = \sigma_i$ in it. If an elastic collision has place in $\boldsymbol{x}$ (with density probability $v(\mathscr{E}, \boldsymbol{x})\sigma_{\mathrm{el}}(v(\mathscr{E}, \boldsymbol{x}))\mathscr{N}_a/\bar{\nu}_{\mathrm{el}}$), by considering that the mass of the atoms are much larger than the electron mass, after the collision the electron has velocity $v' = v(\mathscr{E}, \boldsymbol{x})\boldsymbol{n}'$, where $\boldsymbol{n}'$ is a randomly directed unit vector. After the collision, the electron trajectory is described by a set of new constants of the motion, $\{\mathscr{E}', p'_\vartheta, z'_b\}$, as:

$$p'_\vartheta = mr\left(v'_\vartheta - \frac{1}{2}\omega_c r\right), \ \mathscr{E}' = \mathscr{E}, \ z'_b \text{ from equation (5.40).} \tag{5.52}$$

Ionizations happen in $\boldsymbol{x}$ with density probability $v(\mathscr{E}, \boldsymbol{x})\sigma_i(v(\mathscr{E}, \boldsymbol{x}))/\bar{v}_i$ only if $\frac{1}{2}mv^2(\mathscr{E}, \boldsymbol{x})$ is greater than the ionization energy, ϵ_i. In this case, one must specify the constants of motion of the primary (I_1) and secondary (I_2) electrons, as:

$$\begin{cases} v'_{I_1} = (v^2 - 2\epsilon_i/m)^{1/2}, \ \mathbf{v}'_{I_1} = v'_{I_1}\boldsymbol{n}', \ v'_{I_2} = 0, \\ \mathscr{E}'_{I_1} = \mathscr{E}'_{r, I_1} = \mathscr{E} - \epsilon_i, \ p'_{\vartheta, I_1} = mr\left(v'_{I_1, \vartheta} - \frac{1}{2}\omega_c r\right), \ z'_{b, I_1} \text{ from equation (5.40)}, \\ \mathscr{E}'_{I_2} = \mathscr{E}'_{r, I_2} = q\Phi(\boldsymbol{x}), \ p'_{\vartheta, I_2} = -\frac{1}{2}m\omega_c r^2, \ z'_{b, I_2} \text{ from equation (5.40).} \end{cases} \tag{5.53}$$

The positive ions produced from ionizations impact on the cathodes and, according to the mechanism of secondary emission, they emit electrons. If an ion impacts on the cathode in $\boldsymbol{x}_s = (r_s, \pm L_z)$, the constants of motion of the secondary emitted electron (S) are:

$$\begin{cases} v'_S = 0 \\ \mathscr{E}'_S = \mathscr{E}'_{r, S} = q\Phi(\boldsymbol{x}_s), \ p'_{\vartheta, S} = -\frac{1}{2}m\omega_c r_s^2, \ z'_{b, S} \text{ from equation (5.40).} \end{cases} \tag{5.54}$$

If the region occupied by an electron intersects the anode, the particle is removed as it is absorbed by the electrode. The time-averaged frequency, $\bar{v}_E(\mathscr{E}, p_\vartheta)$, is obtained from equation (5.51) by replacing $\mathscr{P}$ with $\mathscr{P}_E$. Using the ergodic approximation each electron is described by indicating only the two constants of the motion, $\{\mathscr{E}, p_\vartheta\}$ and after elastic collision, ionization or secondary emission, the electrons change their parameters as follows:

- for elastic collision:

$$\mathscr{E}' = \mathscr{E}, \quad p'_\vartheta = mr\left(v'_\vartheta - \frac{1}{2}\omega_c r\right); \tag{5.55}$$

- for ionization:

$$\begin{cases} v'_{I_1} = (v^2 - 2\epsilon_i/m)^{1/2}, \ \mathbf{v}'_{I_1} = v'_{I_1}\boldsymbol{n}', \ \mathscr{E}'_{I_1} = \mathscr{E} - \epsilon_i, \ p'_{\vartheta,} \\ I_1 = mr\left(v'_{I_1, \vartheta} - \frac{1}{2}\omega_c r\right), \\ v'_{I_2} = 0, \ \mathscr{E}'_{I_2} = q\Phi(\boldsymbol{x}), \ p'_{\vartheta, I_2} = -\frac{1}{2}m\omega_c r^2; \end{cases} \tag{5.56}$$

- for secondary emission:

$$v'_S = 0, \ \mathcal{E}'_S = q\Phi(\boldsymbol{x}_s), \ p'_{\vartheta, S} = -\frac{1}{2}m\omega_c r_S^2. \tag{5.57}$$

Monte Carlo simulation of the charge loading and saturation

Making use of the time-averaged trajectories, the initial stage of the charge loading can be investigated numerically on the time scale of electron–neutral collisions. A particle technique can be adopted, in which electron–neutral elastic collisions, ionizations and secondary emissions are taken into account by using a Monte Carlo approach. During this initial stage, the electron density is sufficiently low and the self-consistent field can be neglected. Therefore, only the potential Φ due to the electrodes has been considered by solving the Laplace equation. When the charge density inside the trap becomes higher, the self-consistent potential must be considered by solving a self-consistent Poisson equation for Φ, which is nonlinear, as the charge density for each electron depends on the accessible region $\hat{\mathcal{R}}$, which is a function of Φ itself.

Initially, a set of N computational particles is generated having zero velocity at different positions of the cathode and the relative set of $\mathcal{E}$, p_ϑ and z_b parameters is calculated. Then, at each time step a subset of colliding particles is selected. For each electron the type of collision (elastic or ionization), the spatial position of the event and the new direction $\boldsymbol{n}'$ are chosen by using Monte Carlo techniques. In this way, the set of characteristic parameters is updated at each time step.

The calculations are simpler, as long as the effect of the self-consistent electric field can be neglected. When the charge density increases, the electric field generated by the electron charge must be included in the calculation by solving the nonlinear Poisson equation:

$$\nabla^2\Phi(r, z) = 4\pi e \sum_{p=1}^{N} \frac{\mathcal{P}(r, z; \mathcal{E}_p, p_{\vartheta,p}, z_{b,p}; \{\Phi\})}{2\pi r}, \tag{5.58}$$

where probabilities $\mathcal{P}$ are given by equations (5.44)–(5.46). Moreover, a change $\delta\Phi$ on the potential induces a variation of the electron energy, $\delta\mathcal{E}$, given by:

$$\delta\mathcal{E} = -e\int_{\hat{\mathcal{R}}} \delta\Phi(\mathbf{x})\mathcal{P}(\mathbf{x})\mathrm{d}^2x. \tag{5.59}$$

Finally, from $\delta\Phi$ and $\delta\mathcal{E}$ the new turning points are calculated. As $\mathcal{E}_p$ depends on Φ and Φ depends on $\mathcal{E}_p$, equations (5.58, 5.59) can be solved numerically by using simple iterative procedures.

Typical results obtained with the proposed methods are shown in figure 5.16, in which the time evolution of the charge inside a trap has been reported for different values of the magnetic field ($B_0 = 0.05$, 0.08 and 0.3 T). For higher values of the magnetic field, for which $\hat{\mathcal{R}} \sim \mathcal{R}_E$, the results obtained with the approximate ergodic technique are close to the ones of the three-parameter model. In the simulation, 10^3 computational particles have been initially emitted by the cathodes in the region $[0, 0.4R]$.

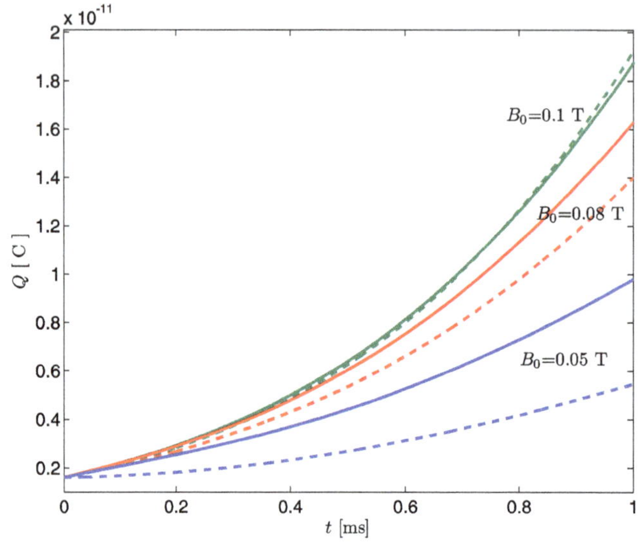

Figure 5.16. Time evolution of the total charge for different values of the magnetic field ($B_0 = 0.05, 0.08, 0.3$ T). Dashed and full lines refer, respectively, to the ergodic and to the three-parameter model. Reproduced with permission from [45]. Copyright 2012 AIP Publishing.

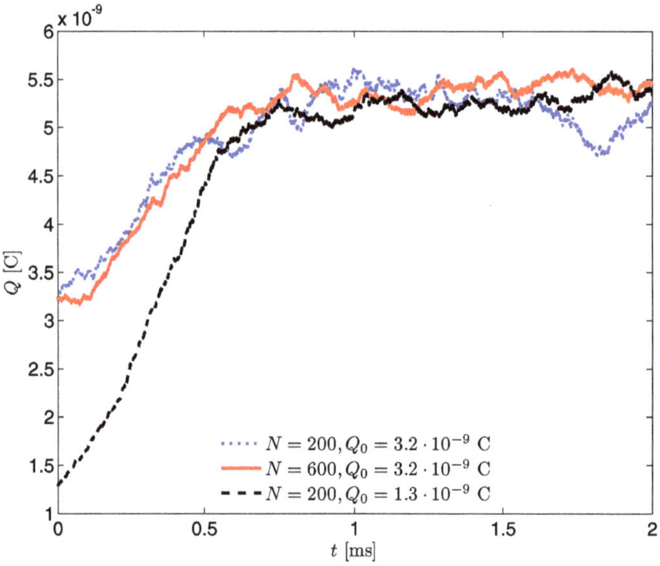

Figure 5.17. Time evolution of the total charge Q by using the three-parameter description for $B_0 = 0.06$ T, using a different number of particles and different initial charge. Reproduced with permission from [45]. Copyright 2012 AIP Publishing.

Figure 5.17 shows the phenomenon of saturation of the total charge, for $B_0 = 0.06$ T. Two calculations have been performed using the three-parameter description with the same initial charge and a different number of computational particles ($N = 200$ and $N = 600$). From the two simulations, the behavior of the transient is the same (apart

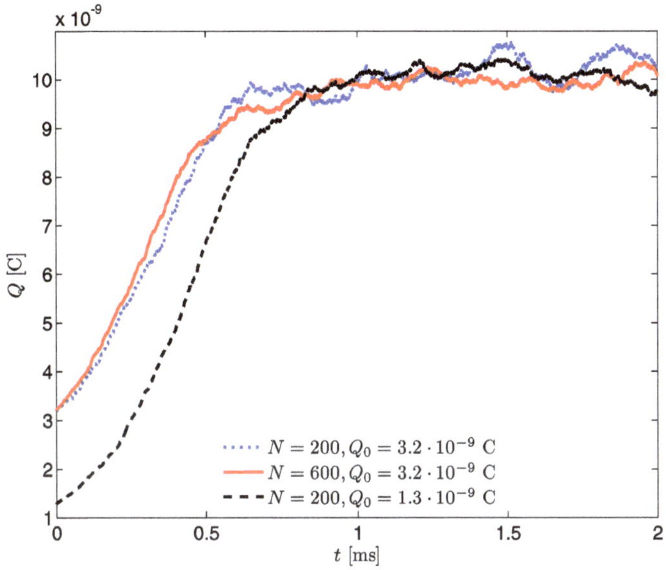

Figure 5.18. Same as figure 5.17, for $B_0 = 0.1$ T. Reproduced with permission from [45]. Copyright 2012 AIP Publishing.

from the numerical noise, which is greater for the case of $N = 200$) and, in particular, the same value of the saturation charge is obtained. In the third calculation, the initial charge is a fraction (2/5) of the one of the previous simulations, and, although the transient is different, the asymptotic value of the total charge is the same. Similar calculations have been repeated for $B_0 = 0.1$ T; the results are presented in figure 5.18. As shown in these examples, for its simplicity the method provides a useful tool to study the electron dynamics in Penning traps on long time scales.

References

[1] Huang K 1987 *Statistical Mechanics* (New York: Wiley)

[2] Dawson J M 1964 *Phys. Fluids* **7** 981–7

[3] Ditmire T *et al* 1997 *Nature* **398** 54
Ditmire T *et al* 1999 *Nature* **398** 489

[4] Zweiback J *et al* 2000 *Phys. Rev. Lett.* **84** 2634
Zweiback J *et al* 2000 *Phys. Rev. Lett.* **85** 3640
Madison K W *et al* 2004 *Phys. Plasmas* **11** 270
Grillon G *et al* 2002 *Phys. Rev. Lett.* **89** 065005
Madison K W *et al* 2004 *Phys. Rev. A* **70** 053201
Sakabe S *et al* 2004 *Phys. Rev. A* **69** 023203
Hirokane M *et al* 2004 *Phys. Rev. A* **69** 063201

[5] Sakabe S *et al* 2006 *Phys. Rev.* A **74** 043205

[6] Last I and Jortner J 2004 *J. Chem. Phys.* **120** 1336
Last I and Jortner J 2004 *J. Chem. Phys.* **120** 1348

[7] Last I and Jortner J 2006 *Phys. Rev. Lett.* **97** 173401
Heidenreich A, Jortner J and Last I 2006 *Proc. Natl Acad. Sci. U.S.A.* **103** 10589

[8] Peano F, Fonseca R A and Silva L O 2005 *Phys. Rev. Lett.* **94** 033401
[9] Boella E *et al* 2016 *J. Plasma Phys.* **82** 905820110
[10] Ditmire T *et al* 1996 *Phys. Rev.* A **53** 3379
Ditmire T *et al* 1998 *Phys. Rev. A* **57** 369
Krainov V P and Smirnov M B 2002 *Phys. Rep.* **370** 237
Milchberg H M, McNaught S J and Parra E 2001 *Phys. Rev. E* **64** 056402
Lezius M *et al* 1998 *Phys. Rev. Lett.* **80** 261
Zweiback J *et al* 2002 *Phys. Plasmas* **9** 3108
Liu C S and Tripathi V K 2003 *Phys. Plasmas* **10** 4085
Murakami M and Basko M M 2006 *Phys. Plasmas* **13** 012105
[11] Kovalev V F and Yu Bychenkov V 2003 *Phys. Rev. Lett.* **90** 185004
[12] Peano F *et al* 2006 *Phys. Rev. Lett.* **96** 175002
[13] Peano F *et al* 2007 *Phys. Rev.* E **75** 066403
[14] Goldstein H 1980 *Classical Mechanics* 2nd edn (London: Addison-Wesley)
[15] Ott E 1979 *Phys. Rev. Lett.* **42** 1628
[16] Birdsall C K and Langdon A B 1985 *Plasma Physics via Computer Simulations* (New York: McGraw-Hill) https://doi.org/10.1201/9781315275048
[17] Raizer Yu P 1991 *Gas Discharge Physics* (Berlin: Springer)
[18] D'Angola A, Boella E and Coppa G 2014 *Phys. Plasmas* **21** 082116
[19] Manfredi G, Mola S and Feix M R 1993 *Phys. Fluids* B **90** 388
[20] Coppa G, D'Angola A, Delzanno G L and Lapenta G 2001 *Phys. Plasmas* **8** 1133–40
[21] Davidson R C 2001 *Physics of Nonneutral Plasmas* (Singapore: World Scientific)
[22] Huang X-P *et al* 1997 *Phys. Rev. Lett.* **78** 875–8
[23] Kriesel J M and Driscoll C F 2001 *Phys. Rev. Lett.* **87** 135003
[24] Audi M and de Simon D 1987 Ion pumps *Vacuum* **37**
[25] Jepsen R L 1968 On the physics of sputter-ion pumps *Fourth Int. Vacuum Congress (Manchester (UK))*
[26] Beier T *et al* 2001 *Phys. Rev. Lett.* **88** 011603
[27] Van Dyck R S, Schwinberg P B and Dehmelt H G 1987 *Phys. Rev. Lett.* **59** 26–9
[28] Bollen G, Moore R B, Savard G and Stolzenberg H 1990 *J. Appl. Phys.* **68** 4355–74
[29] Cornell E A *et al* 1989 *Phys. Rev. Lett.* **63** 1674–7
[30] Gerz C, Wilsdorf D and Werth G 1990 *Nucl. Instrum. Methods Phys. Res.* B **47** 453–61
[31] Gerz C, Wilsdorf D and Werth G 1990 *Z. Phys.* D **17** 119–21
[32] Moore F L, Farnham D L, Schwinberg P B and van Dyck R S Jr 1989 *Nucl. Instrum. Methods Phys. Res.* B **43** 425–30
[33] Wineland D J, Bollinger J J and Itano W M 1983 *Phys. Rev. Lett.* **50** 628–31
[34] O'Neil T M 1995 *Phys. Scr.* **T59** 341
[35] Trapp S, Tommaseo G, Revalde G, Stachowska E, Szawiola G and Werth G 2003 *Eur. Phys. J.* D **26** 237–44
[36] Andresen G *et al* 2007 *Phys. Rev. Lett.* **98** 023402
[37] Fajans J *et al* 2005 *Phys. Rev. Lett.* **95** 155001
[38] Werth G 1985 *Contemp. Phys.* **26** 241–56
[39] Lieberman M A and Lichtenberg A l J 2005 *Principles of Plasma Discharges and Materials Processing* 2nd edn (New York: Wiley)
[40] Helmer J C and Jespen R L 1961 Electrical characteristic of a penning discharge *Proc. of the IRE*

[41] Jepsen R L 1961 *J. Appl. Phys.* **32** 2619–26
[42] Redhead P A 1988 *Vacuum* **38** 901–6
[43] Agilent Ion Pumps *Catalog* (Agilent Technologies)
[44] Coppa G, D'Angola A and Mulas R 2010 *Nuovo Cimento* C **33** 87–94
[45] Coppa G, D'Angola A and Mulas R 2012 *Phys. Plasmas* **19** 062507
[46] Coppa G and Ricci P 2002 *Phys. Rev.* E **66** 046409

Part II

Fluid and hybrid models

IOP Publishing

Plasma Modeling (Second Edition)
Methods and applications
Gianpiero Colonna and Antonio D'Angola

Chapter 6

Fluid models for collisionless magnetic reconnection

Daniela Grasso and Dario Borgogno

Magnetic reconnection is a basic fundamental process that can be defined as a change in the magnetic field topology. Solar flares in the solar corona and sawtooth oscillations in tokamak devices are typical phenomena related to reconnection events. In reference [1] we introduced this phenomenon and gave a first description in terms of the resistive reduced magnetohydrodynamic model. But in most relevant plasmas resistivity can be very low and such a single fluid model may not be suitable to describe the reconnection process. In this case it is more appropriate to consider the two-fluid description of the plasma, that allows the explanation of reconnection process in collisionless plasmas. In this context the so called *generalized Ohm's law* [1], which relates the electric field to the current density, includes other terms than resistivity that drive the reconnection process. Here we will consider the electron inertia as driving mechanism for such a process to occur.

In this chapter we will derive, starting from the two-fluid equations, a reduced model for describing magnetic reconnection in collisionless plasmas. Then we will show how this model can be recast in non-canonical Hamiltonian formalism, that allows insights on the evolution of the reconnection process itself. Finally, as an application of the model, we will show numerical simulations of the magnetic reconnection in this collisionless regime, with a special focus on the effects of the presence of shear flows on the reconnecting instability.

6.1 Two-fluid model

We start the derivation from the continuity and momentum equation for the electrons and the total momentum equation. The equations are, respectively:

$$\frac{\partial n}{\partial t} + \nabla \cdot (n\mathbf{v}_e) = 0 \tag{6.1}$$

doi:10.1088/978-0-7503-3559-1ch6

$$m_e n\left(\frac{\partial \mathbf{v}_e}{\partial t} + \mathbf{v}_e \cdot \nabla \mathbf{v}_e\right) = -en\left(\mathbf{E} + \frac{1}{c}\mathbf{v}_e \times \mathbf{B}\right) - \nabla p_e - \nabla \cdot \mathbf{\Pi}_e - m_e n \nu_e \mathbf{v}_e \tag{6.2}$$

$$m_i n\left(\frac{\partial \mathbf{v}}{\partial t} + \mathbf{v} \cdot \nabla \mathbf{v}\right) = \frac{1}{c}(\mathbf{J} \times \mathbf{B}) - \nabla p \tag{6.3}$$

We adopt the reduction scheme typical of reduced magnetohydrodynamics [2] and we assume the following orderings:

$$\frac{\partial}{\partial x} \sim 1, \quad \frac{\partial}{\partial y} \sim 1, \quad \frac{\partial}{\partial z} \sim \epsilon \tag{6.4}$$

$$B_x \sim \epsilon, \quad B_y \sim \epsilon, \quad B_z \sim 1 + \epsilon^2 \tag{6.5}$$

Fields are assumed to be composed by an equilibrium part plus a perturbation that is of order $\epsilon \ll 1$.

In the following we assume a magnetic field such that:

$$\mathbf{B}(x, y, t) = B_0 \mathbf{e}_z + \nabla \psi(x, y, t) \times \mathbf{e}_z \tag{6.6}$$

where $\mathbf{e}_z$ is the unit vector along the guide field. From Ampère's law we get the expression for the current density as:

$$J_z \equiv J = -\frac{c}{4\pi}\nabla^2 \psi \approx -env_z \tag{6.7}$$

where $v_z \approx v_{ez}$ since $v_{ez} \gg v_{iz}$.

From Faraday's equation we derive the expression for the electric field as:

$$\mathbf{E} = -\nabla \Phi - \frac{1}{c}\frac{\partial \psi}{\partial t}\mathbf{e}_z \tag{6.8}$$

From equation (6.2), neglecting electron inertia and collisionality, by taking the external product with $\mathbf{e}_z$ we get an expression for the perpendicular velocity:

$$v_{e\perp} = \frac{c}{B_0}\mathbf{e}_z \times \nabla \Phi - \frac{c}{enB_0}\mathbf{e}_z \times \nabla p_e + \frac{1}{B_0}v_z \nabla \psi \times \mathbf{e}_z \tag{6.9}$$

where the last term is supposed to be small compared to the $\mathbf{E} \times \mathbf{B}$ drift.

From the electron momentum equation to the leading order, we get:

$$\frac{\partial n}{\partial t} + \mathbf{v}_{e\perp} \cdot \nabla n + n_0 \nabla_\perp \cdot \mathbf{v}_{e\perp} + n_0 \frac{\partial v_z}{\partial z} = 0 \tag{6.10}$$

and after substituting the perpendicular and parallel velocity:

$$\frac{\partial n}{\partial t} + \frac{c}{B_0}\mathbf{e}_z \times \nabla \Phi \cdot \nabla n - \frac{1}{eB_0}\nabla \psi \times e_z \cdot \nabla J - \frac{1}{e}\frac{\partial J}{\partial z} = 0 \tag{6.11}$$

Let us take the projection along $\mathbf{e}_z$ of the electron momentum equation. Neglecting terms of order $O(\epsilon^2)$ one obtains:

$$m_e n\left(\frac{\partial}{\partial t} + \mathbf{v}_{e\perp}\cdot\nabla_\perp\right)v_z = -en\left(-\nabla\Phi\cdot\mathbf{e}_z - \frac{1}{c}\frac{\partial\psi}{\partial t} - \frac{1}{c}\mathbf{v}_{e\perp}\cdot\nabla_\perp\psi\right) - \nabla p_e\cdot\mathbf{e}_z - (\nabla\cdot\mathbf{\Pi}_e)\cdot\mathbf{e}_z - m_e n\nu_e v_z, \tag{6.12}$$

which gives:

$$\begin{aligned} m_e n\left(\frac{\partial v_z}{\partial t} + \frac{c}{B_0}\nabla\Phi\times\nabla v_z\cdot\mathbf{e}_z - \frac{c}{enB_0}\nabla p_e\times\nabla v_z\cdot\mathbf{e}_z\right) \\ = en\frac{\partial\Phi}{\partial z} + \frac{en}{c}\left(\frac{\partial\psi}{\partial t} + \frac{c}{B_0}\nabla\Phi\times\nabla\psi\cdot\mathbf{e}_z\right) - \frac{1}{B_0}\nabla p_e \\ \times\nabla\psi\cdot\mathbf{e}_z - \frac{\partial p_e}{\partial z} + \frac{cm_e}{eB_0}\nabla p_e\times\mathbf{e}_z\cdot\nabla v_z - m_e n\nu_e v_z \end{aligned} \tag{6.13}$$

We introduce the Poisson bracket $[f, g] = \nabla f\times\nabla g\cdot\mathbf{e}_z$ and rewrite equation (6.11) and equation (6.13), where it has been exploited the gyroviscous cancellation, as:

$$\frac{\partial n}{\partial t} + \frac{c}{B_0}[\Phi, n] + \frac{1}{eB_0}[\psi, J] = \frac{1}{e}\frac{\partial J}{\partial z} \tag{6.14}$$

$$\begin{aligned} m_e n\left(\frac{\partial v_z}{\partial t} + \frac{c}{B_0}[\Phi, v_z]\right) = \frac{en}{c}\left(\frac{\partial\psi}{\partial t} + \frac{c}{B_0}[\Phi, \psi]\right) + en\frac{\partial\Phi}{\partial z} - \frac{1}{B_0}[p_e, \psi] \\ - \frac{\partial p_e}{\partial z} - m_e n\nu_e v_z. \end{aligned} \tag{6.15}$$

With some straightforward algebra one gets:

$$\left(\frac{\partial}{\partial t} + \frac{c}{B_0}[\Phi]\right)(\psi - d_e^2\nabla^2\psi) = -c\frac{\partial\Phi}{\partial z} + \frac{c}{enB_0}[p_e, \psi] + \frac{c}{en}\frac{\partial p_e}{\partial z} + \frac{\eta c^2}{4\pi}\nabla^2\psi. \tag{6.16}$$

Finally, in agreement with the reduction ordering and the constant electron temperature assumption, we write

$$p = n_0 T_{e0} + nT_{e0}. \tag{6.17}$$

In analogy with what we presented in reference [1], we focus on a 2D configuration and neglecting derivative respect to z we obtain:

$$\left(\frac{\partial}{\partial t} + \frac{c}{B_0}[\Phi]\right)(\psi - d_e^2\nabla^2\psi) = \frac{cT_{e0}}{en_0B_0}[n, \psi] + \frac{\eta c^2}{4\pi}\nabla^2\psi. \tag{6.18}$$

By applying the $\mathbf{e}_z\cdot\nabla\times$ operator to the total momentum equation we get the third equation of our model. We consider first that $\mathbf{v}\approx\mathbf{v}_\perp\approx\frac{c}{B_0}\mathbf{e}_z\times\nabla\Phi$ to obtain that $\mathbf{e}_z\cdot\nabla\times\mathbf{v} = \frac{c}{B_0}\nabla^2\Phi$. Secondly, using the vectorial identities we show that

$$\mathbf{e}_z \cdot \nabla \times (\mathbf{J} \times \mathbf{B}) = (\nabla J \times \nabla \psi) \cdot \mathbf{e}_z = [J, \psi].$$

In the end we obtain the following form for equation (6.18):

$$\left(\frac{\partial}{\partial t} + \frac{c}{B_0}[\Phi]\right)\frac{c}{B_0}\nabla^2\Phi = \frac{1}{m_i n_0 c}[J, \psi]. \tag{6.19}$$

Finally, we observe that in the 2D-limit $\nabla_{\|} = \frac{\mathbf{B}\cdot\nabla}{B_0} = -[\psi, \cdot]/B_0$ and, defining $F \equiv \psi - d_e^2\nabla^2\psi$, introducing the stream function $\varphi = \frac{c}{B_0}\Phi$, the vorticity $U \equiv \nabla^2\varphi$ and exploiting the properties of the Poisson brackets, it is possible to rewrite the model equations as:

$$\frac{\partial n}{\partial t} + [\varphi, n] + \frac{1}{eB_0}[\psi, J] = 0 \tag{6.20}$$

$$\frac{\partial F}{\partial t} + [\varphi, F] = \frac{cT_{e0}}{en_0B_0}[n, \psi] - \eta c J \tag{6.21}$$

$$\frac{\partial U}{\partial t} + [\varphi, U] = \frac{1}{n_0 m_i c}[J, \psi]. \tag{6.22}$$

These equations must be solved together with the relations:

$$F = \psi + d_e^2 J \equiv \psi - d_e^2\nabla^2\psi \quad U = \nabla^2\varphi. \tag{6.23}$$

6.1.1 Normalization

We proceed now with the normalization of the system of equations (6.20)–(6.22), choosing as normalization factors the macroscopic equilibrium magnetic field scale length, L, and the Alfvèn time, $\tau_A = L/v_A$, where $v_A = B_0/(4\pi m_i n)^{1/2}$. Starting from equation (6.22) there follows straightforwardly the normalization of φ, U, ψ and J as:

$$\varphi \to \frac{\varphi}{Lv_A}, \quad U \to \tau_A U, \quad \psi \to \frac{\psi}{LB_0}, \quad J \to \frac{4\pi JL}{cB_0}, \tag{6.24}$$

while from equation (6.21) we get the normalization for the other fields and parameters as:

$$F \to \frac{F}{B_0 L}, \quad d_e \to \frac{d_e}{L}, \quad n \to \frac{n}{\hat{d}_i n_0}, \quad \eta \to \epsilon_\eta = \frac{\eta c^2 \tau_A}{4\pi L^2}, \tag{6.25}$$

where the hat over the ion skin depth means that it has been normalized to L. Hence, the equations can be recast as:

$$\frac{\partial n}{\partial t} + [\varphi, n] + [\psi, J] = 0 \tag{6.26}$$

$$\frac{\partial F}{\partial t} + [\varphi, F] = \rho_s^2[n, \psi] - \epsilon_\eta J \tag{6.27}$$

$$\frac{\partial U}{\partial t} + [\varphi, U] = [J, \psi]. \tag{6.28}$$

Note that by comparing equation (6.26) and equation (6.28) it follows that $n = U$. The first equation becomes redundant and the system reduces only to the last two equations.

6.2 Collisionless plasmas

High temperature and low density plasmas are most of the time collisionless implying a crucial role of the electron skin depth as driving mechanism for reconnection to occur. Let's rewrite equation (6.21), which has been derived from the electron momentum equation, also called *generalized Ohm's law* [1], highlighting the role of the electron skin depth, as:

$$\frac{\partial(\psi + d_e^2 J)}{\partial t} + [\varphi, \psi + d_e^2 J] = \rho_s^2[n, \psi] - \epsilon_\eta J. \tag{6.29}$$

In equation (6.29) we have assumed the normalized Spitzer resistivity, $\eta = c^2 m_e \nu_e/(4\pi 2ne^2)$, where ν_e is the electron–ion collision frequency. If we compare this term with the electron inertial term, where $d_e = c/\omega_{\rm pe} = c/\sqrt{4\pi ne^2/me}$, we can easily infer that the latter dominates over the collisional term when $d/dt > \nu_{ei}$. Indeed, the conditions such that the process can occur on a time scale comparable with, or faster than the collision time, are common in many laboratory and astrophysical plasmas. In the past three decades, much attention has been paid to the problem of fast reconnection. Starting in the 1990s, many studies [3–6] focused on electron inertia as the physical mechanism that drives fast reconnection in high temperature and low collisional plasmas. If we consider equation (6.29) and let the resistivity and the ion sound Larmor radius go to zero, we can recast it as a Lagrangian advection equation for the field $F = \psi + d_e^2 J$, as

$$\frac{\partial F}{\partial t} + [\varphi, F] = 0, \tag{6.30}$$

which transfers the topological constraint on the magnetic field, expressed by *frozen-in law* [1], to F, while the topological transition in the magnetic field leads to a continuous cascade of energy toward small scales, thus causing the narrowing of the current and vorticity layers that typically accompany a reconnection event.

We rewrite here the equations considering that the equation for the density is redundant and replacing the density by the vorticity in the equation for the magnetic flux, as:

$$\frac{\partial F}{\partial t} + [\varphi, F] = \rho_s^2[U, \psi] \tag{6.31}$$

$$\frac{\partial U}{\partial t} + [\varphi, U] = [J, \psi] \tag{6.32}$$

This system of equations can be cast in conservative form for the quantities [7]

$$G_{\pm} \equiv F \pm d_e \rho_s U. \tag{6.33}$$

Let us define the generalized stream functions:

$$\hat{\varphi}_{\pm} = \varphi - \rho_s^2 U \mp \rho_s d_e J. \tag{6.34}$$

We observe that the r.h.s. of equation (6.32) can be written as $d_e^{-2}[F, \psi]$. Thus, multiplying equation (6.32) by $d_e \rho_s$ and adding and subtracting the resulting equation to equation (6.31) we obtain the alternative form,

$$\frac{\partial G_{\pm}}{\partial t} + [\hat{\varphi}_{\pm}, G_{\pm}] = 0. \tag{6.35}$$

We note that $\hat{\varphi}_{\pm}$ are not uniquely defined, since equation (6.35) is unaffected by the change $\hat{\varphi}_{\pm} \rightarrow \hat{\varphi}_{\pm} + h(G_{\pm})$, with h an arbitrary function of $G_{\pm}$, since $[h(G_{\pm}), G_{\pm}] = 0$. So, it is possible to define [8],

$$\varphi_{\pm} \equiv \hat{\varphi}_{\pm} \pm (\rho_s/d_e) G_{\pm} = \varphi \pm (\rho_s/d_e)\psi. \tag{6.36}$$

In the following we will adopt this last definition. Equation (6.35) expresses the Lagrangian conservation of the fields $G_{\pm}$ on streamlines associated with the effective velocities $v_{\pm} = \mathbf{e}_z \times \nabla \varphi_{\pm}$. Since $\varphi_{\pm}(x, y, t)$ are smooth functions, equation (6.35) implies that the topology of the fields $G_{\pm}$ is preserved in the course of the dynamical evolution. This sets an important constraint on the collisionless reconnection process, and it is the key to understanding the spatial structures of the reconnection region that develop nonlinearly.

6.3 Linear dispersion relation

Similarly to what has been done in reference [1] we analyze here the linear version of equations (6.31) and (6.32) adopting standard boundary layers and asymptotic matching techniques. We note that in these equations there are two parameters d_e and ρ_s which are much smaller than the equilibrium magnetic field scale length, L. The electron skin depth is the parameter that allows the reconnection process, and it is associated with higher order derivatives compared to the other terms in the equation. While the ion sound Larmor radius does not cause the reconnection and becomes relevant only if it is greater than or equal to d_e. So, except for the regions close to the *rational surfaces*, where the magnetic field vanishes and the terms proportional to d_e become important, we can neglect them, reducing equation (6.31) to the ideal magnetic flux conservation equation. We give here the growth rate of the collisionless tearing mode instability treated in reference [9] in the context of resistive plasmas. The authors consider there a plane current sheet located in an infinite domain that undergoes spontaneous reconnection. The classical reference configuration is the Harris sheet [10], where the equilibrium magnetic field, **B**, the flux function, ψ, and the stream function, φ, are given by:

$$\mathbf{B}_{eq} = -\tanh(x)\mathbf{e}_y, \quad \psi_{eq} = \log(\cosh(x)), \quad \varphi_{eq} = 0. \tag{6.37}$$

The *rational surface* is located at $x = 0$. In the two-dimensional setting defined by the two-fluid equations the change of topology caused by the reconnection process appears via formation of a magnetic island around the rational surface. Linearizing equations (6.31) and (6.32) and solving them for the perturbed quantities

$$\tilde{\psi} = \delta\psi(x)\exp(iky + \gamma t), \quad \tilde{\varphi} = \delta\varphi(x)\exp(iky + \gamma t),$$

where $k = 2\pi m/L_y$ and m is the mode number, gives two ordinary differential equations for $\delta\psi$ and $\delta\varphi$. The linearized equations are:

$$\begin{aligned}\gamma\delta\psi - ik\delta\varphi\partial_x\psi_{eq} &= d_e^2[\gamma(\partial_x^2 - \gamma k^2)\delta\psi - ik\delta\varphi\partial_x^3\psi_{\text{eq}}] \\ &\quad - ik\rho_s^2\partial_x\psi_{\text{eq}}(\partial_x^2 - k^2)\delta\varphi \\ \gamma(\partial_x^2 - k^2)\delta\varphi &= ik[\partial\psi_{eq}(\partial_x^2 - \gamma k^2)\delta\psi - \partial_x^3\psi_{eq}\delta\psi]\end{aligned} \tag{6.38}$$

This set of equations is then solved by distinguishing between an inner layer, close to the magnetic flux surface, where the magnetic field vanishes and the reconnection takes place and an outer region, far from the rational surface, where it is possible to solve the system of equations neglecting d_e and ρ_s. It is important to note that the presence of a second scale length implies that the inner layer is divided in two layers itself and a double asymptotic matching is applied. We do not enter into the details of all the calculations, that can be found in reference [11]. We simply return the dispersion relation valid in the two limits $\rho_s \to 0$ and $\rho_s \gtrsim d_e$.

6.3.1 The $\rho_s \to 0$ case

In this limit the inner solution will depend only on the electron skin depth parameter and it is matched directly with the outer ideal solution. The dispersion relation is:

$$\gamma = -\frac{k\pi}{\Delta'}\left(\frac{Q}{4}\right)^{3/2}\frac{\Gamma\left(\dfrac{Q-1}{4}\right)}{\Gamma\left(\dfrac{Q+5}{4}\right)}, \tag{6.39}$$

where Δ' is a parameter, which depends on the solution for the magnetic flux in the outer region and expresses its jump across the rational surface, located at $x = 0$, and

$$Q = \frac{\gamma}{kd_e}$$

The Δ' parameter, first introduced in reference [9], is a measure of the instability, which depends only on the wave vector k along the y-direction. The larger the parameter the more unstable the mode we are considering. The solution of the dispersion relation gives the growth rate γ of the reconnection instability in terms of the wave vector k. In particular we highlight two limits of the relation (6.39).

• **Small Δ' regime** This regime, known as *collisionless tearing regime*, is characterized by $\Delta' d_e \ll 1$ and the corresponding growth rate is:

$$\gamma = 0.45\frac{d_e^3 \Delta'^2}{\pi^2} \tag{6.40}$$

The flow cells are localized around the rational surface and the magnetic island grows very slowly, considering realistic values of d_e.

• **Large Δ' regime** This regime is characterized by $\Delta' d_e \gg 1$ and the corresponding growth is:

$$\gamma = d_e. \tag{6.41}$$

The flow cells here have a macroscopic size and the magnetic island grows up to a width of the order of the equilibrium scale length.

6.3.2 The $\rho_s \geqslant d_e$ case

This regime is characterized by having two distinct layers in the inner region. One layer is close to the rational surface where the electron skin depth affects the magnetic field line motion. This layer must be matched with a second layer, where the effects of the ion sound Larmor radius are felt. Finally the solution has to be matched with the ideal region. This procedure gives the following dispersion relation [11]:

$$\frac{\pi\gamma^2}{2k^2} = \rho_s\frac{-\pi}{\Delta'} + \frac{k\rho_s^2 d_e}{\gamma}. \tag{6.42}$$

Again we highlight the small and large Δ' regimes of this relation.

• **Small Δ' regime** In this limit the growth rate is:

$$\gamma = \frac{\Delta'}{\pi}\rho_s d_e \tag{6.43}$$

• **Large Δ' regime** The growth is:

$$\gamma = \left(\frac{2}{\pi}d_e\rho_s^2\right)^{1/3}. \tag{6.44}$$

6.4 Hamiltonian formulation

In this section we will examine the Hamiltonian properties of the set of equations (6.31) and (6.32) derived in section 6.2. This set of equations admits a Hamiltonian formulation [12, 13]. If we introduce the new variables,

$$\xi_1 = F \quad \xi_2 = d_e\rho_s U \tag{6.45}$$

we can cast our system of equations in non-canonical Hamiltonian form

$$\frac{\partial \xi_i}{\partial t} = \xi_i, H \quad i = 1, 2 \tag{6.46}$$

where the generalized Poisson brackets [13] are defined as

$$P, Q = \int d^2x \mathcal{J}^{ij} \left[\frac{\delta P}{\delta \xi_i}, \frac{\delta Q}{\delta \xi_j} \right] \tag{6.47}$$

and

$$\mathcal{J}^{ij} = -\begin{pmatrix} \xi_2 & \xi_1 \\ \xi_1 & \xi_2 \end{pmatrix} \tag{6.48}$$

The Hamiltonian is

$$H = \frac{1}{2} \int d^2x (|\nabla \psi|^2 + d_e^2 J^2 + \rho_s^2 U^2 + |\nabla \varphi|^2). \tag{6.49}$$

In equation (6.49) we recognize the magnetic energy, the kinetic energy due to the motion of the electrons along field lines, the electron potential energy and the ion kinetic energy.

The Poisson brackets defined in equation (6.47) admit two infinite sets of Casimirs, depending only on the electron dynamics

$$C_\pm = \int d^2x h_\pm (\xi_1 \pm \xi_2) = \int d^2x h_\pm (G_\pm) \tag{6.50}$$

with $h_\pm$ arbitrary functions. The Casimirs are constants of the motion which commute with all functionals, i.e. $\{C, F\} = 0$. Expanding $h_\pm$ to first order in ρ_s we find from equation (6.50) the Casimirs in the cold-electron limit, when $\rho_s \to 0$,

$$\begin{aligned} C_1 &= \int d^2x h_1(F) \\ C_2 &= \int d^2x U h_2(F). \end{aligned} \tag{6.51}$$

We see from equation (6.51) that if only the electron skin depth effects are retained in the equations, the generalized magnetic flux F is conserved and its topological structure does not change in time, while if we take into account the effects due to the scale length ρ_s, F can undergo reconnection. We shall see that the presence of conserved quantities causes the development of small scale lengths shrinking with time during the nonlinear evolution [3]. This behavior is typical of dissipationless models, as discussed in [19] in the case of the electron magneto-hydrodynamics. The presence of invariants will also determine the topological structure of the physically relevant quantities.

6.5 Numerical simulations of collisionless reconnection

In this section we present the numerical solution of the system equations (6.31) and (6.32) together with equation (6.23) highlighting the role of the invariants of the model on the nonlinear evolution of the reconnection process and reviewing the original work in reference [14]. We adopt in the (x, y) plane the equilibrium equation (6.37) defined in section 6.3 to study the linear solution. The boundary conditions have been chosen in such a way that the solution goes to zero at the edge of the integration domain along the x-direction and periodic in the y-direction. The code we use [15] is based on a mixing of numerical techniques: a compact finite difference discretization scheme [16] has been implemented on a variable size grid along the x-direction, while a pseudo-spectral discretization has been chosen for the periodic y-direction. The time discretization is explicit. We start the simulation by perturbing the current density, J, according to the following expression for δJ:

$$\delta J(x, y) = \delta \hat{J}(x)\ e^{iky} \tag{6.52}$$

where $k = 2\pi m/L_y$, m is the mode wave number along y and $\delta \hat{J}(x)$ is a function localized within a width of order d_e, around the rational surface located at $x = 0$. The integration box is defined by $-L_x < x < L_x$ and $-L_y < y < L_y$ with $L_x = -11.3$ and $L_y = 8\pi$, which has been resolved with 1024×4096 grid points. For the physical parameters we chose $d_e = 0.2$ and $\rho_s = 0.4$. The growth of the reconnection instability is shown in figure 6.1. The linear growth rate is $\gamma = 0.025$ and the linear phase lasts until $t \approx 190$. The difference with the theoretical value $\gamma_{\text{theor}} = 0.034$ can be ascribed to the fact that for the chosen parameters we are far from the asymptotic large Δ' regime. After $t \approx 190$ a super-exponential phase follows, in which the island grows up to a macroscopic size. As shown in figure 6.2 this process is accompanied by the formation of increasingly small spatial scales corresponding to strong, localized, current and vorticity layers that effectively decouple ψ from the invariants $G_\pm$. These layers tend to be aligned with the separatrix of the magnetic island, forming the typical cross structure around the X-point [17]. The G_+ field is shown in the left panel of figure 6.3, shifted in y-direction in such a way to stress the filamentation occurring inside the magnetic island. The G_- field, not shown here, is

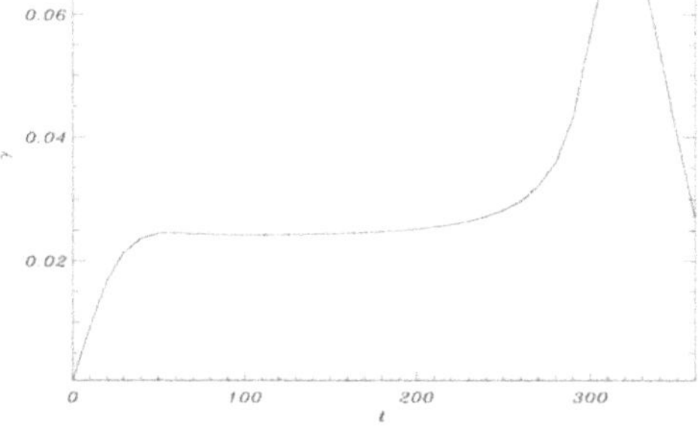

Figure 6.1. Growth rate of the magnetic island.

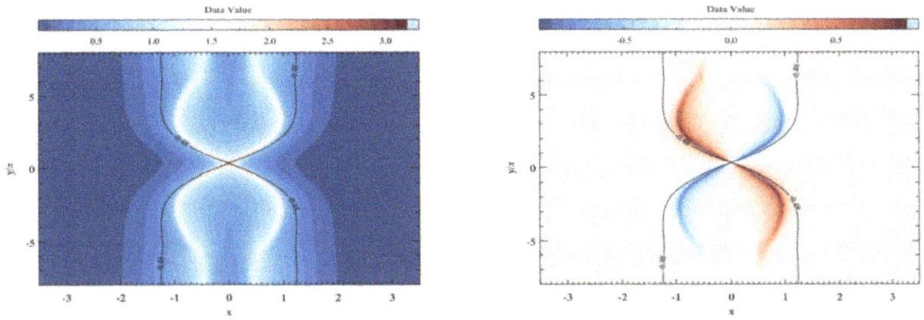

Figure 6.2. Contour plots of the current density (left) and of the vorticity in the nonlinear phase at $t = 330\tau_A$. Superimposed to the contours is the magnetic island.

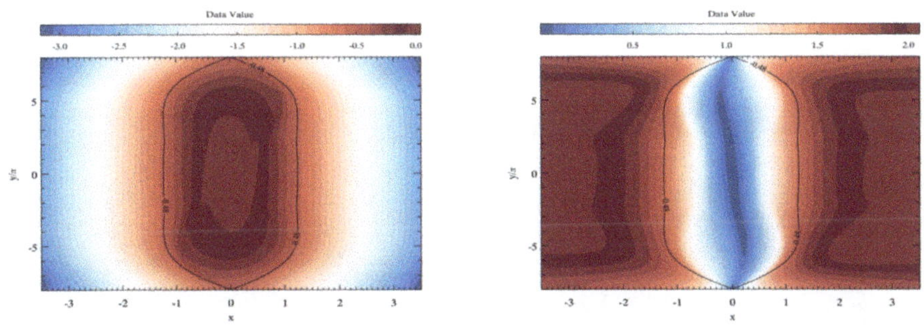

Figure 6.3. Contour plots of the invariant field $G =$ (left panel) and of its corresponding advecting velocity, v_+ (right panel) at $t = 330\tau_A$. Superimposed to the contours is the magnetic island.

the mirror image of G_+. The small scales occur because of the *phase-mixing* [14] of the Lagrangian invariants, which are advected by the generalized velocity fields $v_\pm$ along two mirror-symmetric rotation patterns inside the magnetic island. In the right panel of figure 6.3 is shown the contour plot of the v_+ velocity.

6.5.1 The $\rho_s \to 0$ limit

Here we consider the limiting case of cold electrons. This case is interesting as it also allows one to observe on the top of the reconnecting instability a fluid type instability like the Kelvin–Helmholtz one. This case was first studied in references [18–20] in the context of symmetric magnetic configurations like the ones considered here, and more recently also in the framework of asymmetric configurations [21]. The simulations shown here refer to a case with $d_e = 0.3$, $\rho_s = 0$, which has been integrated on a domain with $L_x = -11.3$ and $L_y = 4\pi$, which has been resolved with 1024×2048 grid points. In figure 6.4 the contour plots of the current density and vorticity are shown at time $t = 220\tau_A$, with superimposed the contour line corresponding to the magnetic island. At this time these fields are still characterized by the typical laminar structure, whose width is shrinking in time. After this time the vorticity becomes unstable to a secondary instability of fluid type and develops a turbulent structure. The occurrence of the Kelvin–Helmholtz instability is detailed in

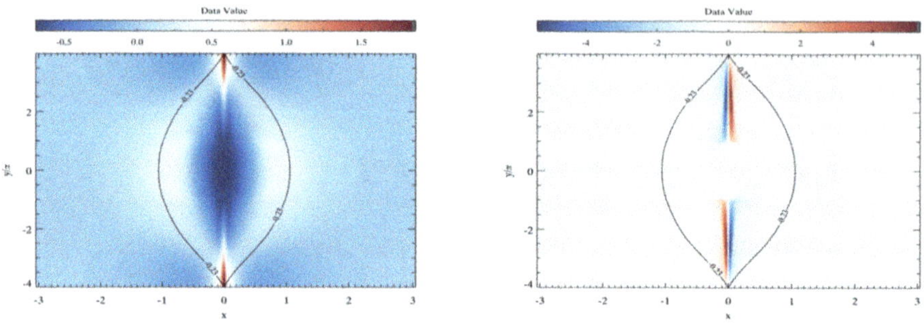

Figure 6.4. Contour plot of the current density (left frame) and vorticity (right frame) at $t = 220\tau_A$ for the case $de = 0.3$, $\rho_s = 0$. At this time the two layers still present a laminar structure, whose width is shrinking in time.

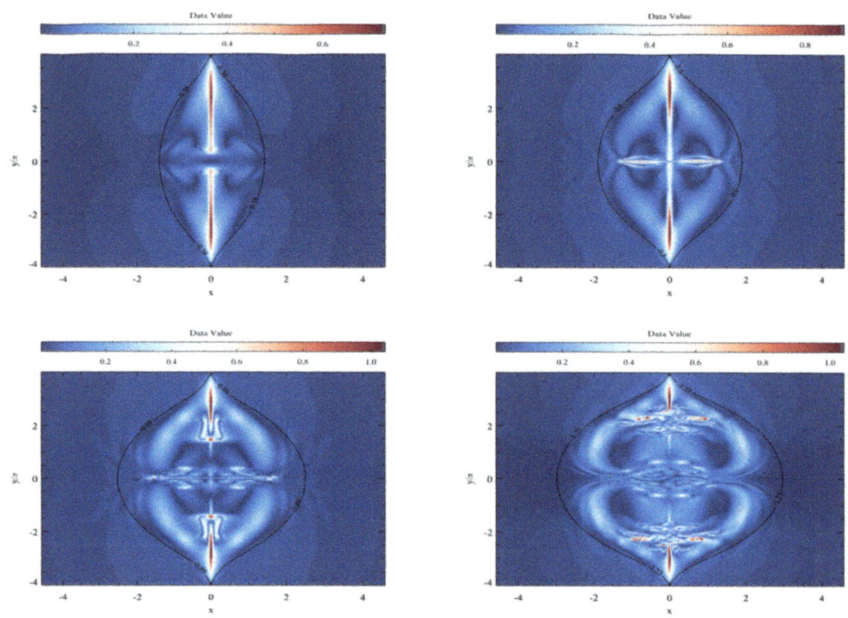

Figure 6.5. Sequence in time in the range $t = 230\tau_A - 260\tau_A$ of the velocity field for the case $de = 0.3$, $\rho_s = 0$, showing the evolution of the jets starting from the *X*-point, colliding at the *O*-point and moving toward the edge of the magnetic island, superimposed in black, and giving rise to turbulence.

figure 6.5, that shows a sequence over time of the distribution of the velocity field in the (x, y) plane. At the beginning there are two jets that start from the *X*-point and move along the y-direction towards each other until they collide at the *O*-point where they are redirected in the x-direction. The shear velocity increases more and more until these layers become unstable to a hydrodynamic-like instability. We remark that in this symmetric configuration the turbulence remains confined to the area enclosed by the magnetic island. Incidentally we report that recently it has been discovered that turbulence can spread out of the magnetic island when asymmetric equilibrium configurations are taken into account [21].

6.6 Shear flow effects on the reconnecting instability

The presence of strong sheared plasma flows is a characteristic feature in magnetized plasmas. They are the results of the topological constraints imposed by the confining magnetic field on the plasma dynamics. The magnetic field can also strongly affect the stability of the plasma flow, depending on the spatial orientation, on the shape and on the magnitude of the magnetic and the velocity field. In particular the magnetic field effect is maximum when it is parallel to the velocity field, while it is almost negligible in the case of perpendicular directions. In an ideal plasma a vortex sheet, i.e. a plasma flow with a step function profile, for instance, is stabilized when the Alfvèn velocity associated with a parallel uniform magnetic field exceeds the flow velocity [22].

Since we want to discuss the effect of a sheared flow on the magnetic reconnection, we focus on the evolution of a magnetized jet, a plasma flow of the form

$$\mathbf{V}_{eq}(x) = \frac{V_{eq_0}}{\cosh^2(x)}\mathbf{e}_y, \tag{6.53}$$

the so called Bickley jet, embedded in the following sheared magnetic field

$$\mathbf{B}_{eq}(x) = B_{eq_0}\tanh(x)\mathbf{e}_y. \tag{6.54}$$

The linear solution of equation (6.29) and equation (6.32), starting from the equilibrium configuration defined by the two previous equations, leads to two classes of eigenmodes:

1. φ even (Kelvin–Helmholtz kink mode) and ψ odd;
2. φ odd (pinch mode) and ψ even.

In the case of an ideal plasma with $d_e = \eta = \rho_s = 0$ the numerical solution of the linearized equations shows the stabilizing effect of the magnetic field on both the classes of solutions. The range of the linearly unstable modes and their growth rates progressively shrink with increasing B_{eq_0}, even though the stabilization proceeds faster for the Kelvin–Helmholtz modes than for the pinch modes [22].

In the presence of non-ideal contributions in the plasma Ohm's law, such as the terms proportional to d_e and η, the pinch mode solutions are strongly affected by the presence of the equilibrium velocity field. Figure 6.6 shows the linear growth rates of the pinch modes in the case $V_{eq_0} = 0$ (dashed lines) and $V_{eq_0} \neq 0$ (solid lines) for the set of parameters $\eta = 0.003$, $d_e = \rho_s = 0$ (blue lines), $d_e = 0.2$, $\eta = \rho_s = 0$ (yellow lines) and $d_e = 0.2$, $\rho_s = 0.4$, $\eta = 0$ (black lines). In all the cases we set $B_{eq_0} = 0.8$ and

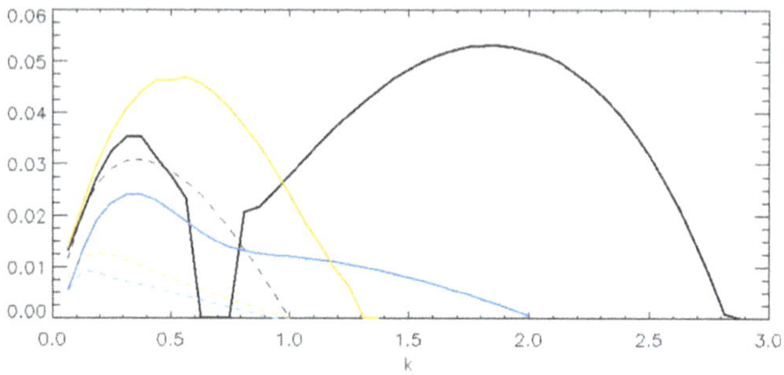

Figure 6.6. Growth rates for of the pinch-tearing modes in the regimes $\eta = 0.003$, $d_e = \rho_s = 0$ (blue curves), $d_e = 0.2$, $\eta = \rho_s = 0$ (yellow curves), $d_e = 0.2$, $\rho_s = 0.4$, $\eta = 0$ (black curves) in the presence of a sheared velocity and magnetic equilibrium, i.e. $V_{eq_0} = 1$, $B_{eq_0} = 0.8$, (solid curves) and of a static sheet pinch, $V_{eq_0} = 0$, $B_{eq_0} = 0.8$ (dashed curves).

$V_{eq_0} = 1$. By comparing the cold-electron regimes, where $\rho_s = 0$, we see that the presence of the plasma jet results in a significant increasing of the pinch-tearing mode growth rate compared with the flowless configuration, especially in the collisionless case. Moreover, the peaks of the reconnection rates are shifted towards higher mode numbers when $V_{eq_0} \neq 0$. It is noteworthy that the presence of the equilibrium shear flow broadens the range of the linearly unstable tearing modes, in particular for the resistive case. The threshold $k = 1$ derived from the stability condition $\Delta' > 0$ for the specific sheared magnetic field we assumed becomes $k = 1.3$ for the $d_e = 0.2$ case and $k = 2$ when $\eta = 0.0003$.

In the finite electron temperature regimes the plasma response to the combined action of the sheared flow and the sheared magnetic field is completely changed. The linear growth rates of the tearing instability for the small k modes, i.e. $k \lesssim 0.5$, are almost the same as the growth rates in the static sheet pinch, where $V_{eq_0} = 0$. By contrast, the plasma flow stabilizes the pinch modes with moderate value of k, i.e. $0.6 \lesssim k \lesssim 0.9$. For higher modes the dispersion relation in the magnetized jet exhibits a second hump, extending up to $k = 2.8$. This is similar to the resistive plasma feature reported in reference [22], but the range of the unstable modes for the hot electron collisionless case is much larger as well as the magnitude of the growth rates.

Note that the non-ideal effects for the magnetized plasma jet do not affect the Kelvin–Helmholtz kink modes that are stable, as observed in the ideal plasmas.

Figure 6.7 shows the spatial distributions of the vorticity field at a given nonlinear stage of the magnetized jet evolution in the three non-ideal regimes previously described. The contour plot of the initial vorticity is also plotted as a reference (top-left frame). According to the Kelvin–Helmholtz instability suppression, the shape of the jet maintains its original structure, just distorted by the magnetic field.

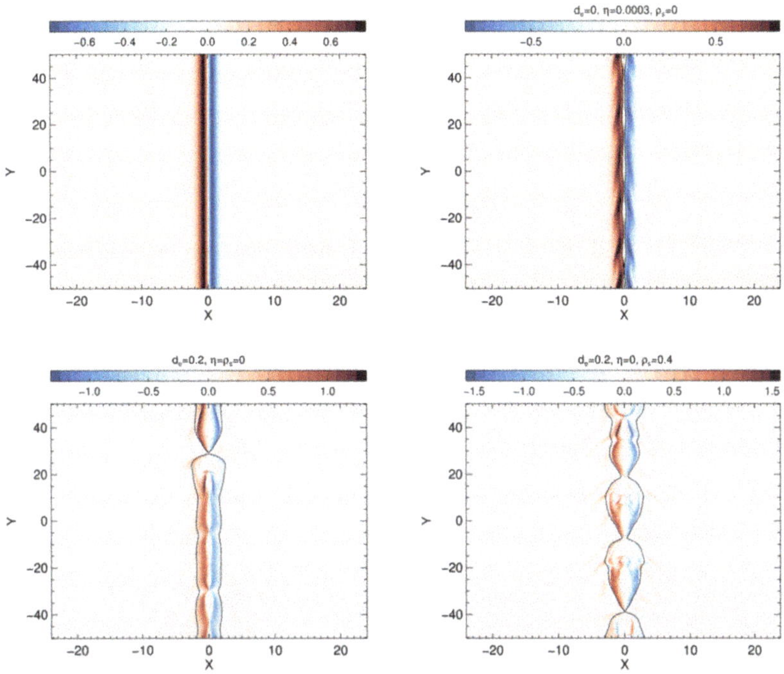

Figure 6.7. Contour plots of the vorticity for the unperturbed magnetized jet (top-left frame) and for the regimes $\eta = 0.0003$, $d_e = \rho_s = 0$ (top-right frame), $d_e = 0.2$, $\rho_s = \eta = 0$ (low-left frame) and $d_e = 0.2$, $\rho_s = 0.4$, $\eta = 0$ (low-right frame) at a given nonlinear stage of the evolution. The superposed black lines show the separatrices of the corresponding magnetic islands.

References

[1] Grasso D 2016 Fluid equations: derivation and simulation *Plasma Modeling: Methods and Applications* (Bristol: IOP Publishing) ch 6

[2] Strauss H R 1976 Nonlinear, three-dimensionalmagnetohydrodynamics of noncircular tokamaks *Phys. Fluids* **19** 134

[3] Birn J *et al* 2001 Geospace Environmental Modeling (GEM) magnetic reconnection challenge *J. Geophys. Res.* **106** 3715

[4] Biskamp D 1997 Collisional and collisionless magnetic reconnection *Phys. Plasmas* **4** 1964–8

[5] Ottaviani M and Porcelli F 1993 Nonlinear collisionless magnetic reconnection *Phys. Rev. Lett.* **71** 3802–5

[6] Zakharov L and Rogers B 1992 Two-fluid magnetohydrodynamic description of the internal kink mode in tokamaks *Phys. Fluids* B **4** 3285–301

[7] Schep T, Pegoraro F and Kuvshinov B 1994 Generalized two-fluid theory of nonlinear magnetic structures *Phys. Plasmas* **1** 2834–52

[8] Kuvshinov B, Lakhin V, Pegoraro F and Schep T 1998 Hamiltonian vortices and reconnection in a magnetized plasma *J. Plasma Phys.* **59** 727–36

[9] Furth H, Killeen J and Rosenbluth M 1963 Finite-resistivity instabilities of a sheet pinch *Phys. Fluids* **6** 459–84

[10] Harris E 1962 On a plasma sheath separating regions of oppositely directed magnetic field *Il Nuovo Cimento* **XXIII** 115–21
[11] Porcelli F 1991 Collisionless m = 1 tearing mode *Phys. Rev. Lett.* **66** 425
[12] Kuvshinov B, Pegoraro F and Schep T 1994 Hamiltonian formulation of low-frequency, nonlinear plasma dynamics *Phys. Lett.* A **191** 296–300
[13] Morrison P J and Hazeltine R D 1984 Hamiltonian formulation of reduced magnetohydrodynamics *Phys. Fluids* **27** 886–97
[14] Grasso D, Califano F, Pegoraro F and Porcelli F 2001 Phase mixing and island saturation in Hamiltonian reconnection *Phys. Rev. Lett.* **86** 5051–4
[15] Militello F, Grasso D and Borgogno D 2014 The deceiving Δ': On the equilibrium dependent dynamics of nonlinear magnetic islands *Phys. Plasmas* **21** 102514
[16] Lele S 1992 Compact finite difference schemes with spectral-like resolution *J. Comput. Phys.* **103** 16–42
[17] Cafaro E, Grasso D, Pegoraro F, Porcelli F and Saluzzi A 1998 Invariants and geometric structures in nonlinear Hamiltonian magnetic reconnection *Phys. Rev. Lett.* **80** 4430–3
[18] Del Sarto D, Califano F and Pegoraro F 2003 Secondary instabilities and vortex formation in collisionless-fluid magnetic reconnection *Phys. Rev. Lett.* **91** 235001
[19] Del Sarto D, Califano F and Pegoraro F 2005 Current layer cascade in collisionless electron-magnetohydrodynamic reconnection and electron compressibility effects *Phys. Plasmas* **12** 012317
[20] Grasso D, Borgogno D and Pegoraro F 2007 Secondary instabilities in two- and three-dimensional magnetic reconnection in fusion relevant plasmas *Phys. Plasmas* **14** 055703
[21] Grasso D, Borgogno D, Tassi E and Perona A 2020 Asymmetry effects driving secondary instabilities in two-dimensional collisionless magnetic reconnection *Phys. Plasmas* **27** 012302
[22] Biskamp D 2000 *Magnetic Reconnection in Plasmas* (Cambridge: Cambridge University Press)

IOP Publishing

Plasma Modeling (Second Edition)
Methods and applications
Gianpiero Colonna and Antonio D'Angola

Chapter 7

Magnetohydrodynamics equations

Andrea Cristofolini

Magnetohydrodynamics is a branch of physics and engineering that investigates the effects of magnetic field on the motion of an electrically conducting fluid. Although the term 'magnetohydrodynamics' and its acronym MHD can be somewhat misleading, it is often used to designate physical phenomena in which the conducting fluid is not a liquid, but rather a compressible ionized gas or a plasma. Some authors prefer more specific terms such as magneto-fluid-dynamics or magneto-gas-dynamics.

Since the pioneering experiment of Michael Faraday in 1832, this appeared as an appealing way to manipulate a fluid without the intervention of mechanical means. Faraday failed in his attempt to measure the electric current produced by the MHD interaction taking place in the water of the river Thames flowing through the Earth's magnetic field. However, his experimental activities opened the way to the study of a mechanism which has fundamental implications in plasma physics, astrophysics and engineering.

The MHD interaction principles have been proposed for a number of technological applications. A first example can be found in MHD generators, in which the MHD interaction is used for the direct conversion of the working fluid energy into electrical energy [1, 2]. MHD generators have been intensively investigated as a means to increase the efficiency of electrical power plants and as a viable alternative to the gas turbine. MHD electric generators have also been proposed in nuclear-powered spacecraft for deep space missions.

Reversing the power flow, the MHD interaction can be exploited to produce a body force on a conducting fluid. Electromagnetic pumps based on the MHD principle have been implemented in fast nuclear reactors, in which a liquid like sodium is used to remove from the core the heat produced by fission. The application of electromagnetic forces has been also successfully implemented in material processing, especially in the fields of steelmaking and advanced metallurgy [3, 4]. In this field, many processes involve the handling of liquid metal in a technologically difficult environment. Electromagnetic stirring and flow control in continuous

doi:10.1088/978-0-7503-3559-1ch7

casting are examples of the application of the MHD interaction in the metallurgical processing.

There are also examples of MHD propulsion in naval and aerospace applications. In 1992, Mitsubishi Industries presented the Yamato 1 [5], the first prototype of an MHD propelled ship. Yamato 1 included a large number of innovative technological solutions, including the adoption of superconducting magnets to generate the magnetic field necessary to produce the MHD interaction, and it took its maiden first dip in the water in Kobe Harbour. During the tests, the vessel reached a velocity of 7.5 knots, demonstrating the feasibility of MHD ship propulsion [6]. In space propulsion, the magnetoplasmadynamic (MPD) thruster has been actively investigated since the late 1960s in ground-based test facilities [7]. The MPD thruster is a plasma accelerator basically constituted by two coaxial electrodes between which a high current discharge ionizes a gas and accelerates it to high exhaust velocities by means of the MHD interaction. The magnetic field responsible for the interaction can be self-induced by the current. Though the magnetic field induced by the electric current is sufficient to support an MHD interaction, many thrusters have been proposed and studied in which the field is reinforced by external coils, primarily for stabilization purpose.

In addition to the electric propulsion, during the last decades the MHD mechanism has received significant interest in the field of aerothermodynamics and aerospace in applications concerning the control of the hypersonic flow of ionized gases [8, 9]. The utilization of the MHD interaction in the hypersonic flight has been investigated since the late '50s [8], and a wide range of possible applications has been proposed and investigated. Many of the applications are directed to the solution of some specific aspects of a spacecraft during the re-entry phase in the planetary atmosphere such as thermal flux mitigation [10, 11], trajectory control through the parachute effect [12, 13] and the blackout time reduction [14]. Other investigations have been focused toward the shock control at the inlet of an air breathing thruster. In the AJAX project, MHD techniques are utilized to by-pass kinetic energy of the working fluid from the supersonic diffuser to the nozzle. By doing this, the flow velocity in the combustion chamber is reduced to acceptable values, even for high Mach numbers.

In this chapter, a general formulation of the MHD model will be derived and discussed. The approximations underlying the model will be examined, and two formulations of the model will be obtained from the general one, in order to adequately describe the limit cases of convective and diffusive magnetic regimes. Finally, an MHD model based on a low magnetic Reynolds number approximation will be described, and some numerical results will be presented and discussed.

7.1 MHD models

7.1.1 Model foundation

A magneto-fluid-dynamic model is generally based on the assumption that the plasma can be regarded as a continuum, and thus may be characterized by relatively few macroscopic quantities. The requirement for the applicability of this approach is that the plasma is collision dominated. Under this assumption, the basis for an MHD model is constituted by the continuity equations for mass, momentum and energy, and by the Maxwell equations. In order to evaluate the plasma properties in thermochemical non-equilibrium, microscopic scale phenomena are taken into account by means of continuity equations for relevant species. For each continuity equation, the proper production rates has to be known.

In order to consider non-equilibrium plasmas, particularly plasmas where electron temperature can deviate from bulk temperature, a supplementary conservation equation for energy has to be added. A variety of approaches, based on different assumptions and addressed to different scientific and technological fields, are available in literature. The assumption of purely convective magnetic regime (ideal MHD with infinite electrical conductivity) is widely utilized in astrophysics and plasma physics. In this case, the equations of the fluid dynamics and electrodynamics are integrated into a single set which yields a propagation system of seven waves. The magnetic field and the flow field are strongly coupled and jointly react to perturbations. The complementary limit case is the low magnetic Reynolds number regime. This approach, very popular in the MHD power generation field, yields a separate set of equations for fluid dynamics and electrodynamics. The intermediate case is the diffusive-convective magnetic regime which is characterized by a dispersed propagation system.

Regardless of the specific assumptions, an MHD model is usually obtained treating the plasma as a continuum medium, and then coupling fluid dynamics with electrodynamics. In the frame of the continuum assumption, fluid dynamics is described by the Navier–Stokes equations. Treating the plasma as a viscous fluid subject to a force per unit volume $\vec{f}$ and supplied with a rate of heat per unit volume $\dot{Q}$, the balance equations for mass, momentum and internal energy can be written as:

$$\frac{\partial \rho}{\partial t} + \nabla \cdot (\rho \, \vec{v}) = 0 \tag{7.1}$$

$$\frac{\partial \rho \, \vec{v}}{\partial t} + \nabla \cdot (\rho \, \vec{v}\vec{v} + \vec{\vec{I}} \, p - \vec{\vec{\tau}}) = \vec{f} \tag{7.2}$$

$$\frac{\partial \rho \, e_i}{\partial t} + \nabla \cdot (\rho \, e_i \vec{v}) + p \, \nabla \cdot \vec{v} - \vec{\vec{\tau}} : \vec{v} + \nabla \cdot \vec{\Gamma}_T = \dot{Q} \tag{7.3}$$

An equation for the energy per unit volume $\rho(e_i + \rho v^2/2)$ can be derived from (7.1), (7.2), and (7.3) after some manipulation:

$$\frac{\partial}{\partial t}\left[\rho\left(e_i + \frac{v^2}{2}\right)\right] + \nabla \cdot \left[\rho\left(e_i + \frac{v^2}{2}\right)\vec{v} + (\vec{\vec{I}}p - \vec{\vec{\tau}}) \cdot \vec{v} + \vec{\Gamma}_T\right] = \dot{Q} + \vec{f} \cdot \vec{v} \tag{7.4}$$

The viscosity tensor $\vec{\vec{\tau}}$ describes the momentum fluxes due to diffusion, and depends on the mobility of the species constituting the fluid. Assuming that the fluid behaves as a Newtonian fluid, the viscous stress tensor can be expressed as:

$$\vec{\vec{\tau}} = \mu\,[\nabla\vec{v} + (\nabla\vec{v})^T] + (\lambda\nabla\cdot\vec{v})\,\vec{\vec{I}} \tag{7.5}$$

where μ is the dynamic viscosity and λ the bulk viscosity. According to the Stokes hypothesis, $\lambda = -2\,\mu/3$.

A description is also needed to model the thermal diffusion mechanisms. The diffusive thermal flux $\vec{\Gamma}_T$ can be expressed by the Fourier's law:

$$\vec{\Gamma}_T = -\,k\,\nabla T \tag{7.6}$$

The fluid dynamic model is closed by adding the equation of state, which specifies the thermodynamic behaviour of the fluid. Assuming a calorically perfect gas, we can relate the pressure p and the mass density ρ to the temperature T:

$$p = \rho\,R\,T \tag{7.7}$$

and we can express the internal energy as:

$$e_i = c_v\,T \tag{7.8}$$

where the ideal gas constant can be expressed as $R = c_p - c_v$ and $c_p/c_v = \gamma$. In the frame of classical theory, the formulation of electromagnetics in a continuum medium is described by the Maxwell equations:

$$\nabla\times\vec{E} = -\frac{\partial\,\vec{B}}{\partial\,t} \tag{7.9}$$

$$\nabla\times\vec{H} = \vec{J} + \frac{\partial\,\vec{D}}{\partial\,t} \tag{7.10}$$

$$\nabla\cdot\vec{J} = -\frac{\partial\,\rho_c}{\partial\,t} \tag{7.11}$$

From equations (7.9)–(7.11), three further equations may then be derived:

$$\nabla\cdot\vec{D} = \rho_c \tag{7.12}$$

$$\nabla\cdot\vec{B} = 0 \tag{7.13}$$

$$\nabla\cdot\left(\vec{J} + \frac{\partial\vec{D}}{\partial\,t}\right) = 0 \tag{7.14}$$

The electromagnetic model is completed by three constitutive relations. Neglecting magnetization and electric polarization, one may write:

$$\vec{D} = \epsilon_0\vec{E} \tag{7.15}$$

$$\vec{B} = \mu_0\vec{H} \tag{7.16}$$

The third relevant constitutive relation is obtained considering the momentum balance for the electric charge carriers, i.e. electrons and ions. In a weakly ionized gas, assuming that the characteristic time of the macroscopic phenomenon is much greater than the average electron collision time, one can obtain the generalized Ohm's law [15]:

$$\vec{J} = \sigma\left(\vec{E} + \vec{v} \times \vec{B} + \frac{\nabla T_e}{e\, n_e}\right) - \beta_e \frac{\vec{J} \times \vec{B}}{B} - s\, \frac{\vec{B} \times (\vec{J} \times \vec{B})}{B^2} \tag{7.17}$$

The electrical conductivity σ can be expressed as:

$$\sigma = \frac{n_e^2\, e}{m_e \nu_{\mathrm{eH}}}. \tag{7.18}$$

Since the electron mobility is $\mu_e = e/(m_e\, \nu_{\mathrm{eH}})$, we can also write:

$$\sigma = n_e\, e\, \mu_e. \tag{7.19}$$

The electron Hall parameter β_e is defined as the radians around the gyration center an electron travels between two collisions in its cyclotron motion induced by the presence of the magnetic field. The electron Hall parameter can formally be expressed as the ratio between the cyclotron angular frequency ω_c and the electron-heavy particles collision frequency ν_{eH}:

$$\beta_e = \frac{\omega_c}{\nu_{\mathrm{eH}}} = \frac{e\, B}{m_e\, \nu_{\mathrm{eH}}}. \tag{7.20}$$

and, recalling the definition of μ_e:

$$\beta_e = \mu_e B. \tag{7.21}$$

We will assume then that the contribution of the gradient of electron pressure $p_e = n_e\, k_B\, T_e$ to the conduction electric current density is negligible. We will neglect as well the ion slip factor term ($s = \beta_e\, \beta_i < <1$), which is equivalent to considering the electrons as the only ones responsible for the electric charge transport. Under these assumptions, (7.17) can be rewritten in the form:

$$\vec{J} = \sigma\left(\vec{E} + \vec{v} \times \vec{B}\right) - \beta_e \frac{\vec{J} \times \vec{B}}{B}. \tag{7.22}$$

or, introducing the reduced electric field $\vec{E}' = \vec{E} + \vec{v} \times \vec{B}$, that is the electric field as seen in a reference system moving with the fluid:

$$\vec{E}' = \frac{1}{\sigma}\left(\vec{J} - \beta_e \frac{\vec{J} \times \vec{B}}{B}\right). \tag{7.23}$$

In a cartesian coordinate system, defining the resistivity tensor $\vec{\vec{\eta}}$ as:

$$\vec{\vec{\eta}} = \frac{1}{\sigma}\begin{bmatrix} 1 & \beta_{e,z} & -\beta_{e,y} \\ -\beta_{e,z} & 1 & -\beta_{e,x} \\ \beta_{e,y} & -\beta_{e,x} & 1 \end{bmatrix}, \tag{7.24}$$

where:

$$\beta_{e,x} = \frac{\beta_e}{B}B_x = \mu_e\, B_x, \qquad \beta_{e,y} = \frac{\beta_e}{B}B_y = \mu_e\, B_y, \qquad \beta_{e,x} = \frac{\beta_e}{B}B_z = \mu_e\, B_z,$$

we can rewrite (7.23) in a more compact form:

$$\vec{E}' = \vec{\vec{\eta}}\vec{J}, \tag{7.25}$$

or also

$$\vec{J} = \vec{\vec{\Sigma}}\vec{E}'. \tag{7.26}$$

where $\vec{\vec{\Sigma}} = \vec{\vec{\eta}}^{\,-1}$ is the conductivity tensor:

$$\vec{\vec{\Sigma}} = \frac{\sigma}{(1+\beta_e^2)}\begin{bmatrix} 1+\beta_{e,\,x}^2 & \beta_{e,x}\beta_{e,y} - \beta_{e,z} & \beta_{e,x}\beta_{e,z} + \beta_{e,y} \\ \beta_{e,y}\beta_{e,x} + \beta_{e,z} & 1+\beta_{e,\,y}^2 & \beta_{e,y}\beta_{e,z} - \beta_{e,x} \\ \beta_{e,z}\beta_{e,x} - \beta_{e,y} & \beta_{e,z}\beta_{e,y} + \beta_{e,x} & 1+\beta_{e,\,z}^2 \end{bmatrix}. \tag{7.27}$$

In order to define the terms appearing at the right hand side of (7.2), (7.3) and (7.4), a convenient expression for $\vec{f}$ and $\dot{Q}$ due to the action of the electromagnetic field has to be developed. Indeed, the body force per unit volume in a fluid subject to an electric field $\vec{E}$ and moving with a velocity $\vec{v}$ across a magnetic field $\vec{B}$ can be expressed as [15]:

$$\vec{f} = \rho_c\vec{E}' + \vec{J}\times\vec{B}. \tag{7.28}$$

The total work per time and volume unit p_{EM} done by the electromagnetic field can then be expressed as:

$$p_{\mathrm{EM}} = \rho_c\vec{v}\cdot\vec{E} + \vec{J}\cdot\vec{E}, \tag{7.29}$$

while the work per time and volume unit delivered to the fluid by the force $\vec{f}$ is:

$$\vec{v}\cdot\vec{f} = \rho_c\vec{v}\cdot\vec{E}' + \vec{v}\cdot(\vec{J}\times\vec{B}). \tag{7.30}$$

The difference between the total power density delivered to the fluid and the mechanical work per time and volume unit on the fluid represents the thermal power per unit volume $\dot{Q}$ that the charge carriers transmit to the fluid through collisions:

$$\dot{Q} = p_{\mathrm{EM}} - \vec{v}\cdot\vec{f} = \rho_c\vec{v}\cdot\vec{E} + \vec{J}\cdot\vec{E} - \left[\rho_c\vec{v}\cdot\vec{E}' + \vec{v}\cdot(\vec{J}\times\vec{B})\right].$$

Since

$$\vec{v} \cdot \vec{E}' = \vec{v} \cdot (\vec{E} + \vec{v} \times \vec{B}) = \vec{v} \cdot \vec{E}$$

and

$$\vec{v} \cdot (\vec{J} \times \vec{B}) = -\vec{J} \cdot (\vec{v} \times \vec{B}),$$

we obtain that:

$$\dot{Q} = \vec{J} \cdot (\vec{E} + \vec{v} \times \vec{B}) = \vec{J} \cdot \vec{E}'. \tag{7.31}$$

From (7.23), noting that $\vec{J} \times \vec{B} \perp \vec{J}$, we obtain:

$$\dot{Q} = \frac{J^2}{\sigma}. \tag{7.32}$$

Thus, the equations (7.31) and (7.32) can be used to express the right hand side term in (7.3). The right hand side term of (7.4) $\dot{Q} + \vec{f} \cdot \vec{v}$ is given by the total electromagnetic work per time and volume unit p_{EM} expressed by (7.29).

7.1.2 MHD approximation

We briefly report here on the MHD approximation [15] . Since (7.22) and (7.23) has been derived assuming that electrons are the only charge carriers, the relevant collision time is given by the inverse of the electrons collision frequency ν_{eH}, and the condition that has to be verified is then:

$$t_c > > \nu_{\mathrm{eH}} \tag{7.33}$$

By comparing the ratio between the displacement current density, given by the time derivative of the displacement field $\vec{D}$, and the conduction current density, one obtain:

$$\frac{\dfrac{\partial D}{\partial t}}{J} = \frac{\dfrac{\partial D}{\partial t}}{\sigma E} \approx \frac{\epsilon_0}{t_c\, \sigma}$$

Thus, displacement current is neglected assuming that:

$$t_c > > \frac{\epsilon_0}{\sigma} \tag{7.34}$$

From (7.18), (7.33) and (7.34), remembering also that the plasma angular frequency as $\omega_p = \sqrt{(n_e e^2)/(\epsilon_0 m_e)}$, the following condition can be derived:

$$t_c >> \frac{1}{\omega_p} \tag{7.35}$$

When conditions (7.34) hold, the ionized gas may be assumed as globally neutral, and the effects of the electric charge density on the charge transport and on the electromagnetic body force can be neglected. Indeed:

$$\frac{\rho_c\, v}{J} \approx \frac{\epsilon_0}{t_c\, \sigma}$$

and

$$\frac{\rho_c\, E}{|\vec{J} \times \vec{B}|} \approx \frac{\epsilon_0}{t_c\, \sigma}$$

Consequently, the force per unit volume becomes:

$$\vec{f} = \vec{J} \times \vec{B}. \tag{7.36}$$

and the total work per unit time and volume delivered to the fluid becomes:

$$p_{\mathrm{EM}} = \vec{v} \cdot \vec{f} + \dot{Q} = \vec{J} \cdot \vec{E}. \tag{7.37}$$

As a result, the equations describing the behaviour of the system can be modified in accordance with the assumptions of MHD approximation so far discussed. The general formulation of the governing equations under the MHD approximation is here reported:

$$\frac{\partial \rho}{\partial t} + \nabla \cdot (\rho\, \vec{v}) = 0 \tag{7.38}$$

$$\frac{\partial \rho\, \vec{v}}{\partial t} + \nabla \cdot (\rho\, \vec{v}\vec{v} + \vec{I}\, p - \vec{\tau}) = \vec{J} \times \vec{B} \tag{7.39}$$

$$\frac{\partial}{\partial t}\left[\rho\left(e_i + \frac{v^2}{2}\right)\right] + \nabla \cdot \left[\rho\left(e_i + \frac{v^2}{2}\right)\vec{v} + (\vec{I}p - \vec{\tau}) \cdot \vec{v} + \vec{\Gamma}_T\right] = \vec{E} \cdot \vec{J} \tag{7.40}$$

$$\nabla \times \vec{E} = -\frac{\partial \vec{B}}{\partial t} \tag{7.41}$$

$$\nabla \times \vec{H} = \vec{J} \tag{7.42}$$

$$\nabla \cdot \vec{J} = 0 \tag{7.43}$$

$$\nabla \cdot \vec{B} = 0 \tag{7.44}$$

7.1.3 Non-equilibrium conditions

As previously stated, a set of closure equations is necessary to close the system of the governing equations of fluid dynamics. The equations of state for ideal gases (7.7) and (7.8) are well suited for describing a wide range of physical situations. However, in ionized reactive gas flows the thermodynamic description of the system increases its complexity, since the equations relating the thermodynamic quantities depend on the local composition of the fluid, which is not uniform and varies in time following the evolution of the system. Diffusive transport parameters (i.e. viscosity, thermal and electrical conductivity, Hall parameter …) also depend on the temperature and on the number density of the species in the mixture. When a condition of thermochemical equilibrium can be assumed, the composition of the gas can be derived as a function of the thermodynamic macroscopic quantities. Nevertheless, the characteristic times of the reactions involved in the investigated phenomenon are very often comparable with the diffusion and convection times. In this case, the relevant species have to be tracked down writing a balance equation for their number densities. For the generic species s with number density n_s and a the rate of production per unit volume $\dot{\Omega}_s$:

$$\frac{\partial n_s}{\partial t} + \nabla \cdot (n_s \vec{v} + \vec{\Gamma}_s) = \dot{\Omega}_s. \tag{7.45}$$

The diffusive fluxes can be expressed as $\Gamma_s = -D_s \nabla n_s$, where D_s is the diffusivity of the species.

In ionized flows where the elastic collision rate of the electrons with the heavy particles is not sufficient a condition of partial thermal equilibrium occurs. When this happens, the electrons temperature T_e can significantly deviate from bulk temperature T. Since the electronic transport parameters and the pressure gradient in (7.17) depend on T_e, an energy equation for electrons has to be added to the model:

$$\begin{aligned} &\frac{\partial}{\partial t}\left(\frac{3}{2}kT_e\right) + \nabla \cdot \left(\frac{5}{2}kT_e\vec{v}\right) + \nabla \cdot \vec{\Gamma}_e \\ &\qquad = \vec{E}' \cdot \vec{J} - \sum_s \nu_{es}\frac{2\, m_e}{m_s} n_e \frac{3}{2}k\,(T_e - T) - \dot{N}. \end{aligned} \tag{7.46}$$

The sum at the right hand side of (7.46) represents the rate of energy per unit volume which electrons transfer to heavy species in the plasma due to elastic collisions. The rate of energy loss per unit volume due to elastic collisions is taken into account by the term $\dot{N}$.

7.1.4 Magnetoquasistatics

The mathematical description of electromagnetics constituted by (7.9), (7.10), (7.12) and (7.13) leads to the well known wave equations for the scalar and vector potentials. The retarded potential solution of wave equations show how the electromagnetic field propagates through space with a velocity $c = (\mu_0 \epsilon_0)^{-1/2}$. Thus, a travelling time L/c exists between the instant in which the field is produced and the instant in which the same field is observed at a distance L from its source. The electromagnetic wave mechanism is related to the interaction between the Faraday–Neumann–Lenz law (7.9) and the Ampère–Maxwell law (7.10): in a plane wave, a wave of electric field $\vec{E}$ travelling at a velocity c is seen by a fixed observer as a time varying field, thus associated with a displacement current density $\frac{\partial \vec{D}}{\partial t}$.

The displacement current density, due to (7.10), produces a space varying magnetic field $\vec{H}$, which travels with $\vec{E}$ at the same velocity c. The travelling wave of magnetic field is seen by a fixed observer as a time varying magnetic field, which closes the loop producing an electric field due to (7.9). In many physical and technological applications, however, the propagative behaviour of the electromagnetic field can be neglected. This is the case when the time L_c/c it takes an electromagnetic wave to travel a distance L_c equal to the characteristic length of the problem considered is much less than the characteristic time t_c. In these cases, it can be safely assumed that the electromagnetic waves instantaneously propagate through the entire space constituting the domain of the considered problem, and that the time behaviour is governed by other mechanisms. When the condition $L_c/c << t_c$ is verified, it is convenient to refer to a quasi static formulation of the electromagnetism, which is obtained when one of the two time derivatives appearing in (7.9) and (7.10) is neglected. In doing so, the previously described mechanism which sustains the electromagnetic wave is interrupted, and the propagation time no longer exists. The quasi static formulation of electromagnetism obtained neglecting the displacement current density is called magnetoquasistatics, as opposed to electroquasistatics, which is obtained neglecting the time derivative of the magnetic induction field $\vec{B}$ in (7.9).

According to the assumptions constituting the MHD approximation, electromagnetism is described by a magnetoquasistatic formulation. We are interested in finding an equation for the magnetic induction field $\vec{B}$, which will uniquely describe the electromagnetic behaviour of the problem. Introducing (7.25) in (7.42), we obtain:

$$\vec{E}' = (\vec{E} + \vec{v} \times \vec{B}) = \frac{1}{\mu}\vec{\vec{\eta}} \cdot \nabla \times \vec{B}.$$

Applying the curl operator to both sides, expressing $\nabla \times \vec{E}$ by means of (7.9) and remembering that $\vec{\vec{\Sigma}}^{-1} = \vec{\vec{\eta}}$, we derive then:

$$-\frac{\partial \vec{B}}{\partial t} + \nabla \times (\vec{v} \times \vec{B}) = \nabla \times \left[\frac{1}{\mu}\vec{\vec{\eta}} \cdot (\nabla \times \vec{B})\right].$$

Reordering and applying the vector identity:

$$\nabla \times (\vec{a} \times \vec{b}) = \nabla \cdot (\vec{b}\vec{a} - \vec{a}\vec{b}),$$

we can write the convection–diffusion equation for the $\vec{B}$ field:

$$\frac{\partial \vec{B}}{\partial t} + \nabla \times \left[\frac{1}{\mu}\vec{\vec{\eta}} \cdot (\nabla \times \vec{B})\right] - \nabla \cdot (\vec{B}\vec{v} - \vec{v}\vec{B}) = 0. \tag{7.47}$$

It is also useful to develop a balance equation for the magnetic energy density $B^2/(2\mu_0)$, which will be conveniently utilized to derive an equation for the total energy. Taking the dot product of (7.41) with $\vec{H}$ and subtracting from the equation obtained dot multiplying (7.42) by $\vec{E}$ we get:

$$\vec{E} \cdot (\nabla \times \vec{H}) - \vec{H} \cdot (\nabla \times \vec{E}) = \vec{E} \cdot \vec{J} + \vec{H} \cdot \frac{\partial \vec{B}}{\partial t}$$

Introducing the vector identity:

$$\nabla \cdot (\vec{a} \times \vec{b}) = \vec{b} \cdot (\nabla \times \vec{a}) - \vec{a} \cdot (\nabla \times \vec{b})$$

and reordering, one obtains:

$$-\frac{\partial}{\partial t}\left(\frac{B^2}{2\,\mu_0}\right) - \nabla \cdot (\vec{E} \times \vec{H}) = \vec{E} \cdot \vec{J}. \tag{7.48}$$

Equation (7.48) is the formulation of the Poynting theorem in the magnetoquasi-static assumption and constitutes a balance of electromagnetic energy for an infinitesimal volume dV. The Poynting vector $\vec{S} = \vec{E} \times \vec{H}$ appearing in (7.48) represents a rate of electromagnetic energy per unit area and its divergence takes into account the net balance energy per unit time leaving the volume. The theorem states that the total work per unit time $(\vec{E} \cdot \vec{J})dV$ delivered to the volume by the electromagnetic field is equal to the sum of the rates of decreasing of the magnetic energy $B^2/(2\mu)dV$ and the electromagnetic energy per unit time $-\nabla \cdot (\vec{E} \times \vec{H})dV$ entering in the infinitesimal volume through its bounding surface. Since, from (7.23),

$$\vec{E} = \vec{\vec{\eta}} \cdot \vec{J} - \vec{v} \times \vec{B}$$

the Poynting vector $\vec{S}$ can be expressed as:

$$\vec{S} = \vec{\vec{\eta}} \cdot \vec{J} \times \vec{H} - (\vec{v} \times \vec{B}) \times \vec{H}$$

and, using (7.16) and (7.10):

$$\vec{S} = \frac{1}{\mu_0}\left[\frac{1}{\mu_0}\vec{\vec{\eta}} \cdot (\nabla \times \vec{B}) \times \vec{B} - (\vec{v} \times \vec{B}) \times \vec{B}\right]. \tag{7.49}$$

The term $(\nabla \times \vec{B}) \times \vec{B}$ appearing in (7.49) can be treated using the vector identity:

$$\nabla(\vec{a} \cdot \vec{b}) = \vec{a} \cdot (\nabla\vec{b}) + \vec{b} \cdot (\nabla\vec{a}) + \vec{a} \times (\nabla \times \vec{b}) + \vec{b} \times (\nabla \times \vec{a}).$$

We can then write:

$$(\nabla \times \vec{B}) \times \vec{B} = -\vec{B} \times (\nabla \times \vec{B}) = \vec{B} \cdot (\nabla\vec{B}) - \frac{1}{2}\nabla(B^2),$$

Noting that for (7.13) we have:

$$\nabla \cdot (\vec{B}\vec{B}) = \vec{B} \cdot (\nabla\vec{B}) + \vec{B}(\nabla \cdot \vec{B}) = \vec{B} \cdot (\nabla\vec{B}),$$

we can finally write:

$$(\nabla \times \vec{B}) \times \vec{B} = \nabla \cdot \left[\vec{B}\vec{B} - \frac{1}{2}\nabla(B^2)\right],$$

Introducing now the Maxwell stress tensor $\vec{\vec{T}}$ that, under the current approximation, is constituted only by its magnetic component:

$$\vec{\vec{T}} = \frac{1}{\mu_0}\left(\vec{B}\vec{B} - \frac{1}{2}\vec{\vec{I}}B^2\right) \tag{7.50}$$

we can conclude that:

$$\frac{1}{\mu}(\nabla \times \vec{B}) \times \vec{B} = \nabla \cdot \vec{\vec{T}}. \tag{7.51}$$

Using the vector identity

$$(\vec{a} \times \vec{b}) \times \vec{c} = \vec{a} \cdot (\vec{c}\vec{b}) - \vec{b} \cdot (\vec{c}\vec{a})$$

and recalling that $\vec{b} \cdot (\vec{c}\vec{a}) = (\vec{b} \cdot \vec{c})\vec{a}$, we can manipulate the term $(\vec{v} \times \vec{B}) \times \vec{B}$ in (7.49) :

$$(\vec{v} \times \vec{B}) \times \vec{B} = \vec{v} \cdot (\vec{B}\vec{B}) - \vec{B} \cdot (\vec{B}\vec{v}) = \vec{v} \cdot (\vec{B}\vec{B}) - B^2\vec{v},$$

and, since $\vec{v} \cdot (\vec{B}\vec{B}) = (\vec{B}\vec{B}) \cdot \vec{v}$:

$$(\vec{v} \times \vec{B}) \times \vec{B} = \left(\vec{B}\vec{B} - \vec{\vec{I}}B^2\right) \cdot \vec{v}.$$

We can now introduce the definition of the Maxwell stress tensor given in (7.50) to conclude that:

$$\frac{1}{\mu_0}(\vec{v} \times \vec{B}) \times \vec{B} = \left(\vec{\vec{T}} - \frac{1}{2\,\mu_0}\vec{\vec{I}}B^2\right) \cdot \vec{v}. \tag{7.52}$$

Using (7.51) and (7.52), the expression for the Poynting vector given in (7.49) can be rewritten as:

$$\vec{S} = \left[\frac{1}{\mu_0}\vec{\eta} \cdot (\nabla \cdot \vec{\vec{T}}) - \left(\vec{\vec{T}} - \frac{1}{2\,\mu_0}\vec{\vec{I}}B^2\right) \cdot \vec{v}\right], \tag{7.53}$$

while (7.48) becomes:

$$-\frac{\partial}{\partial t}\left(\frac{B^2}{2\,\mu_0}\right) - \nabla \cdot \left[\frac{1}{\mu_0}\vec{\eta} \cdot (\nabla \cdot \vec{\vec{T}}) - \left(\vec{\vec{T}} - \frac{1}{2\,\mu_0}\vec{\vec{I}}B^2\right) \cdot \vec{v}\right] = \vec{E} \cdot \vec{J}. \tag{7.54}$$

7.1.5 General model

The general model a resistive MHD flow is obtained coupling the fluid dynamic equations (7.38), (7.39) and (7.40) with the magnetic convection–diffusion equation

(7.47) and the magnetic energy balance equation (7.54). The right hand side term in equation (7.39) is expressed according to equation (7.51):

$$\vec{J} \times \vec{B} = \nabla \cdot \vec{\vec{T}}$$

and an equation for the total energy per unit volume E_T defined as:

$$E_T = \rho\left(e_i + \frac{v^2}{2}\right) + \frac{B^2}{2\,\mu_0} \tag{7.55}$$

is obtained introducing equation (7.54) in (7.40). The complete set of equations reads as follows:

$$\frac{\partial \rho}{\partial t} + \nabla \cdot (\rho\,\vec{v}) = 0 \tag{7.56}$$

$$\frac{\partial \rho\,\vec{v}}{\partial t} + \nabla \cdot (\rho\,\vec{v}\vec{v} + \vec{\vec{I}}\,p - \vec{\vec{\tau}} - \vec{\vec{T}}) = 0 \tag{7.57}$$

$$\frac{\partial E_T}{\partial t} + \nabla \cdot \left[(E_T + p)\,\vec{v} - (\vec{\vec{\tau}} + \vec{\vec{T}}) \cdot \vec{v} + \frac{1}{\mu_0}\vec{\vec{\eta}} \cdot (\nabla \cdot \vec{\vec{T}}) + \vec{\Gamma}_T\right] = 0 \tag{7.58}$$

$$\frac{\partial \vec{B}}{\partial t} + \nabla \times \left[\frac{1}{\mu_0}\vec{\vec{\eta}} \cdot (\nabla \times \vec{B})\right] - \nabla \cdot (\vec{B}\vec{v} - \vec{v}\vec{B}) = 0. \tag{7.59}$$

One can recognize that the only electromagnetic quantity in the above set of equations is the magnetic induction fields $\vec{B}$. The current density and the electric field can be derived from equations (7.42) and (7.25), respectively.

The transport of the magnetic induction and of the magnetic energy density occurs due to two mechanisms, convection and diffusion, that can be observed in equations (7.58) and (7.59). The physical mechanism of magnetic diffusion is more easily understood as a diffusion of the electric current density. Indeed, a non uniform distribution of current density in a conductive medium tends to spread over all the available paths. However, a redistribution in the current density lines causes a time variation in the magnetic induction field produced by the current, which in turn generates through an induced electric field. The induced electric field generates a current density countering the initial variation, hindering the redistribution of the current density. Assuming a scalar conductivity, one can recognize that the parameter governing the magnetic diffusion is the magnetic diffusivity $1/(\sigma\,\mu_0)$.

An important parameter characterizing the MHD regime of the considered problem is the magnetic Reynolds number Re_m that, analogously to its fluid dynamic counterpart, is defined as the ratio of the magnetic convection and diffusion effects. Considering the equation for the magnetic induction, the order of magnitude of the convective term is:

$$\nabla \cdot (\vec{B}\vec{v} - \vec{v}\vec{B}) \sim \frac{B\,v}{L_c}$$

while, for the convective term:

$$\nabla \times \left[\frac{1}{\mu_0} \vec{\vec{\eta}} \cdot (\nabla \times \vec{B}) \right] \sim \frac{B}{\mu_0 \, \sigma \, L_c^2}.$$

We have then:

$$Re_m = \frac{(B \, v)/L_c}{B/(\mu_0 \, \sigma \, L_c^2)} = \mu_0 \, \sigma \, v \, L_c. \tag{7.60}$$

In high Re_m MHD regimes, the magnetic convection is the dominant transport mechanism, and the magnetic behaviour of the system is strongly coupled with the flow field. Conversely, for low Re_m, the magnetic induction field equation is mainly governed by diffusion, and is rather insensitive to the velocity distribution. Two representative cases will now be briefly presented and discussed.

7.1.6 Ideal MHD

The ideal MHD model is obtained by the viscosity and the conductive thermal fluxes in the fluid, and assuming an infinite electrical conductivity. The only transport mechanisms represented in the model are then related to convection. The governing equations are derived by dropping all the diffusive terms in the MHD governing equations:

$$\frac{\partial \rho}{\partial t} + \nabla \cdot (\rho \, \vec{v}) = 0 \tag{7.61}$$

$$\frac{\partial \rho \, \vec{v}}{\partial t} + \nabla \cdot (\rho \, \vec{v}\vec{v} + \vec{\vec{I}} \, p - \vec{\vec{T}}) = 0 \tag{7.62}$$

$$\frac{\partial E_T}{\partial t} + \nabla \cdot \left[(E_T + p) \, \vec{v} - \vec{\vec{T}} \cdot \vec{v} \right] = 0 \tag{7.63}$$

$$\frac{\partial \vec{B}}{\partial t} - \nabla \cdot (\vec{B}\vec{v} - \vec{v}\vec{B}) = 0. \tag{7.64}$$

The basic assumptions of the ideal MHD model do not allow one to adequately treat the presence of viscous and thermal boundary layers. Additionally, ideal MHD equations are applicable to describe physical problems in which the magnetic Reynolds number can be considered infinite. High density astrophysical plasmas such as those found in a star are good examples of physical systems that can be treated by ideal MHD. Actually, the model was firstly formulated by Alfvén with purposes related to astrophysics [16]. It is worth noting that the infinite electrical conductivity assumption involves the cancellation of the $\dot{Q} = \vec{J} \cdot \vec{E}'$ term at the right hand side in the equation (7.4) of internal energy. We can then manipulate equation (7.4) assuming a calorically perfect gas to obtain an equation for pressure p:

$$\frac{\partial p}{\partial t} + \vec{v} \cdot \nabla p + \gamma p \nabla \cdot \vec{v} = 0 \tag{7.65}$$

which can substitute equation (7.63). Introducing the convective derivative $\frac{d\{\cdot\}}{dt} = \frac{\partial\{\cdot\}}{\partial t} + \vec{v} \cdot \nabla\{\cdot\}$, equation (7.65) can be rewritten in the form:

$$\frac{d}{dt}\left(\frac{p}{\rho^{\gamma}}\right) = 0 \tag{7.66}$$

equation (7.66) expresses the fact that an element dV moving with the fluid does not receive any energy contributions and therefore undergoes only adiabatic processes.

A distinctive feature of MHD regimes described by the ideal model is that the induction flux linked to a curve that moves with the fluid cannot undergo variations in time. This condition, known as 'frozen flow' condition, is a direct consequence of the electromagnetic induction law (7.9), and gives an insight on the mechanism underlying the magnetic convection phenomenon: a closed curve Γ moving within an electrically conducting fluid through a region where a non uniform magnetic field is present will experience a time variation in the linked magnetic flux. The time variation of the magnetic flux will produce, due to equation (7.9), an electric field and a system of electric current densities that will in turn produce a reaction magnetic induction field in order to counterbalance the initial variation. In a perfectly conductive medium, the reaction field perfectly matches the initial time variation, and the linked magnetic flux is not allowed to vary in time, and is convected along with the Γ curve.

The ideal MHD model also constitutes the foundation for developing the theory of wave propagation in a plasma. While the conservative formulation given by equations (7.61), (7.62), (7.63) and (7.64) can be attractive by the numerical point of view, a more convenient form of the governing equation can be obtained reformulating the problem in the primitive variables $\rho, \vec{v}, p$ and $\vec{B}$. The equations for ρ and $\vec{B}$ are given by equations (7.61) and (7.64), respectively, and we already derived equation (7.65) for pressure. The equation for velocity can be derived after some manipulation from (7.62):

$$\frac{\partial \vec{v}}{\partial t} + \vec{v} \cdot \nabla \vec{v} + \frac{1}{\rho} \nabla \cdot (\vec{\vec{I}}\, p - \vec{\vec{T}}) = 0 \tag{7.67}$$

We now assume a one-dimensional problem, where the primitive variables depend only on the time t and on the space variable x. Noting that, since $\vec{B}$ is a solenoidal field, in the chosen geometry we must have:

$$\frac{\partial B_x}{\partial x} = 0,$$

we can rewrite equations (7.61), (7.67), (7.65) and (7.64) in the quasilinear form:

$$\frac{\partial \boldsymbol{u}}{\partial t} + A\frac{\partial \boldsymbol{u}}{\partial x} = 0, \tag{7.68}$$

where:

$$u = \begin{Bmatrix} \rho \\ v_x \\ v_y \\ v_z \\ p \\ B_y \\ B_z \end{Bmatrix} \quad A = \begin{bmatrix} v_x & \rho & 0 & 0 & 0 & 0 & 0 \\ 0 & v_x & 0 & 0 & 1/\rho & B_y/(\rho\mu_0) & B_z/(\rho\mu_0) \\ 0 & 0 & v_x & 0 & 0 & -B_x/(\rho\mu_0) & 0 \\ 0 & 0 & 0 & v_x & 0 & 0 & -B_x/(\rho\mu_0) \\ 0 & \gamma p & 0 & 0 & v_x & 0 & 0 \\ 0 & B_y & -B_x & 0 & 0 & v_x & 0 \\ 0 & B_z & 0 & -B_x & 0 & 0 & v_x \end{bmatrix}.$$

The eigenvalues of matrix A can be expressed defining the speeds c_a, c_{ax}, and c_s:

$$c_a^2 = \frac{B^2}{\rho\,\mu_0}, \quad c_{ax}^2 = \frac{B_x^2}{\rho\,\mu_0}, \quad c_s^2 = \frac{\gamma p}{\rho}.$$

The seven eigenvalues are then

$$\lambda_{1,7} = v_x \pm c_{mf}; \quad \lambda_{2,6} = v_x \pm c_{ax}; \quad \lambda_{3,5} = v_x \pm c_{fs}; \quad \lambda_4 = v_x.$$

The speed c_{ax} is called Alfvén speed, while c_{mf} and c_{ms} are the speeds of the fast and slow magnetoacoustic (or magnetosonic) waves, respectively:

$$c_{mf}^2 = \frac{1}{2}\left[(c_a^2 + c_s) + \sqrt{(c_a^2 + c_s)^2 - 4\,c_a^2 c_s^2}\right]$$

$$c_{ms}^2 = \frac{1}{2}\left[(c_a^2 + c_s) - \sqrt{(c_a^2 + c_s)^2 - 4\,c_a^2 c_s^2}\right]$$

In the given geometry, Alfvén waves [17] propagate along the x direction with speed c_{ax}, introducing a transverse perturbation to the B_x field. Alfvén waves do not produce pressure or density variations in the fluid. On the other hand, magnetoacoustic waves travelling in the x direction produce compressions and rarefactions in pressure and in the magnetic field components perpendicular to x. A more detailed discussion of MHD waves can be found in [18] and [19].

The numerical solution of ideal MHD has been intensively investigated, and various schemes are available in literature. Many of them have been obtained as an extension of well known fluid dynamic solvers (Roe [20–22], HLL [23], HLLC [24], HLLD [25]).

7.1.7 Low magnetic Reynolds number model

Under the assumption of $Re_m \ll 1$, some useful approximations can be applied. Most notably, when an externally generated magnetic flux density $\vec{B}_0$ is applied to the plasma flow (i.e. a field produced by a magnet), the magnetic flux density $\vec{B}_0$ produced by the current density flowing in the plasma can be neglected. Indeed, with

$J \sim \sigma\, v\, B$ being the order of magnitude of the current density, then from (7.42), we obtain that B_i is in the order of:

$$B_i \sim \mu_0\, \sigma\, v\, B\, L_c = Re_m\, B$$

Thus, when $Re_m \ll 1$ the total magnetic induction $\vec{B}$ can be considered equal to the externally applied field $\vec{B}_0$. The electromagnetic behaviour of the system is described by (7.59), dropping the convecting term $\nabla \cdot (\vec{B}\vec{v} - \vec{v}\vec{B})$. Comparing the order of magnitude of the time derivative with the diffusive term at the right hand side of (7.59) and taking $t_c = v/L_c$ as the characteristic time, we get:

$$\frac{\dfrac{\partial \vec{B}}{\partial t}}{\nabla \times \left[\dfrac{1}{\mu_0}\vec{\vec{\eta}} \cdot (\nabla \times \vec{B})\right]} \sim \frac{B/t_c}{B/(\mu_0\, \sigma\, L_c^2)} \sim \mu_0\, \sigma\, v\, L_c = Re_m$$

which shows that in a low Re_m regime the rate of change of the magnetic field is negligible, and that electromagnetics is basically described by a steady state formulation. In other words, the electromagnetic characteristic time is negligible when compared with the fluid dynamics, and electromagnetic quantities are assumed to adapt themselves instantaneously to the time variations of the flow field. As a result, equation (7.59) is no longer representative of the problem, as it relates to the $\vec{B}$ field which is essentially a known quantity, and the description of the electromagnetic behaviour has to be rewritten in an electrodynamic formulation. Under the assumption of $Re_m \ll 1$, the electrodynamic formulation is derived assuming that the applied magnetic induction does not vary in time. The electric field $\vec{E}$ becomes then a curl-free field due to equation (7.41) and can be expressed as the gradient of an electric scalar potential φ:

$$\vec{E} = -\nabla\, \varphi \tag{7.69}$$

Introducing (7.69) in (7.26):

$$\vec{J} = -\vec{\vec{\Sigma}} \cdot (\nabla\, \varphi + \vec{v} \times \vec{B}). \tag{7.70}$$

and taking into account of (7.43), a steady state electrodynamics, described by means of an elliptical partial differential equation is derived:

$$\nabla \cdot \left[\vec{\vec{\Sigma}} \cdot (-\nabla\, \varphi + \vec{v} \times \vec{B})\right] = 0. \tag{7.71}$$

Once the velocity, the applied magnetic induction field and the conductivity tensor are known at a generic instant t, (7.71) one can evaluate the distribution of the electric potential and, consequently, the electric field and the current density through equations (7.69) and (7.70). The MHD model in the low Re_m approximation is then constituted by the fluid dynamic equations (7.38), (7.39) and (7.40) leading the evolution in time and driving a steady state electrodynamics described by (7.71).

These quantities can then be utilized to evaluate the right hand side terms $\vec{J} \times \vec{B}$ and $\vec{J} \cdot \vec{E}$ appearing in the momentum and energy equations.

7.2 Numerical model

In this section a brief description of a numerical model of the MHD interaction developed around a body immersed in a hypersonic flow. The model is oriented toward the design and the analysis of experiments aimed to investigate the MHD phenomena concerning the re-entry phase of a space vehicle.

Since the expected value of electrical conductivity ranges between some Siemens per meter and some hundreds and the typical flow velocity is in the range of some thousands of meters per second, a low Re_m approach has been chosen.

As discussed before, the physical coupling between the fluid dynamic and the electrodynamic regimes in a low Re_m condition is rather weak. As a result, the two problems can be solved by means of two distinct solvers which reciprocally interact by exchanging data according to the scheme shown in figure 7.1. Fluid dynamics can be solved by means of an appropriate time dependent algorithm which, at each time step, receives from the electrodynamic solver the distribution of the electric field and of the current density. These quantities are utilized to compute the right hand side terms in the momentum and energy equations. On the other side, the electrodynamic solver obtains from the fluid dynamic solver the velocity field needed to compute the forcing term in equation (7.71) and the data that allow the evaluation of the conductivity tensor.

The fluid dynamic solver utilized to obtain the results presented in the following section is based on a cell-centered finite volume formulation. The control cells are defined as a dual tesselation of Delaunay triangulation of the domain. Convective fluxes are evaluated by means of an Osher's approximate Riemann solver [26] with a second-order accurate MUSCL approach. Diffusive fluxes computation is performed using a finite element approach on the domain triangulation. Time

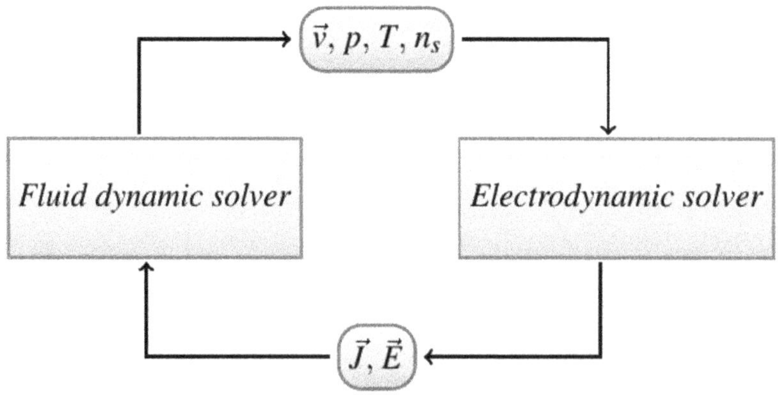

Figure 7.1. Coupling of the solvers in a low Re_m MHD numerical scheme.

integration is performed utilizing an explicit scheme, based on a fourth-order Runge–Kutta method.

Electrodynamics is solved by means of a finite element approach. The details of the formulation are given in [27, 28] and [29]. Defining $\boldsymbol{N}$ as the set of the shape functions on the given domain triangulation, we can approximate the unknown φ with a piecewise polynomial representation:

$$\tilde{\varphi} = \boldsymbol{N}^T \boldsymbol{\phi} \tag{7.72}$$

Introducing (7.72) in (7.71) and applying a weighted residual approach we obtain the weak formulation of the electrodynamic problem:

$$\int_\Omega \nabla W \cdot (\vec{\vec{\Sigma}} \cdot \nabla \tilde{\varphi})\, d\Omega = \int_\Omega \nabla W \cdot \left[\vec{\vec{\Sigma}} \cdot (\vec{v} \times \vec{B})\right] d\Omega - \oint_{\delta\Omega} W\, J_n\, d(\delta\Omega) \tag{7.73}$$

A linear algebraic system is obtained using the shape functions in $\boldsymbol{N}$ as a weighting function W:

$$\left[\int_\Omega [\nabla N]^T\, [\boldsymbol{\Sigma}][\nabla N]\, d\Omega\right]\boldsymbol{\phi} = \int_\Omega [\nabla N]^T\, [\boldsymbol{\Sigma}](\vec{v} \times \vec{B})\, d\Omega - \oint_{\delta\Omega} N\, J_n\, d(\delta\Omega) \tag{7.74}$$

The system (7.74) has to be defined and solved at each time step. Great attention has therefore to be given to optimizing the solution of this system, as poor performances of the electrodynamic solver would dramatically impair the overall efficiency of the whole code. We can observe in (7.27) that the Hall parameter determines the anisotropy of the conductivity tensor $\vec{\vec{\Sigma}}$. Thus the anisotropy of the electrodynamics are more pronounced for high values of the electron mobility. These conditions may be expected in plasmas at relatively low pressure and density, as in MHD aerospace applications in hypersonic regimes of interest. An ILU-type preconditioned GMRES solver has been adopted for the solution of (7.74). The convergence of solver can become very critical for high values of the Hall parameters in the plasmas, and high quality preconditioning is needed in order to enhance the performances. A quite straightforward way to obtain a higher accuracy in preconditioning is to perform a convenient numbering of the mesh nodes. In [29] is shown a reverse Cuthill–McKee ordering weighted on the local direction of the magnetic induction field with respect to the topology of the triangular mesh utilized.

7.3 Applications

The low Re_m numerical model described in the previous section will now be applied for the analysis of the MHD interaction around a body immersed in a hypersonic flow. The discussion is aimed at aerospace applications related to the re-entry in a planetary atmosphere. As already mentioned, the fundamentals of the MHD interaction have been investigated by many authors in the aerospace community, both experimentally [30–33] and numerically [34–37], and the efforts to develop reliable numerical models have gained great momentum.

The low Re_m approximation can be verified computing the MHD regime around a body immersed in a hypersonic flow and comparing the applied magnetic field with the field produced by the current density due to the MHD interaction. We consider then a axisymmetric body immersed in a Mach 15 argon flow (see figure 7.2). The applied magnetic induction field B_e, shown in figure 7.2, is assumed to be produced by a 56.5 kA electric current in a circular loop embedded in the body. The value of the electrical current is chosen so as to generate a 1 T induction field at the stagnation point. For our current purpose, a scalar conductivity is adequate, and thus the Hall parameter β_e is forcibly set to zero. In this case the only component of the current density is the azimuthal one, and the cross product between the velocity and the B field times the conductivity can be promptly evaluated. In figure 7.3 are shown some results of the calculation using the low Re_m model described in the previous section with a constant conductivity $\sigma = 100$ S m^{-1} in the shock layer. The azimuthal current distribution in shown in figure 7.3. It is worth noting that in the case of scalar conductivity, the azimuthal component of the electric field is equal to zero due to the axial symmetry of the problem. If not, there would exist a circular path centred on the symmetry axis along which the circulation of the electric field

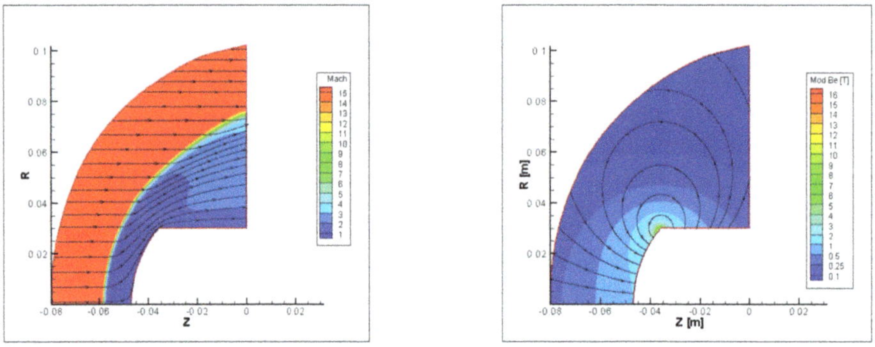

(a) Much number and stream lines

(b) Applied magnetic field

Figure 7.2. Applied magnetic field and Mach number no MHD.

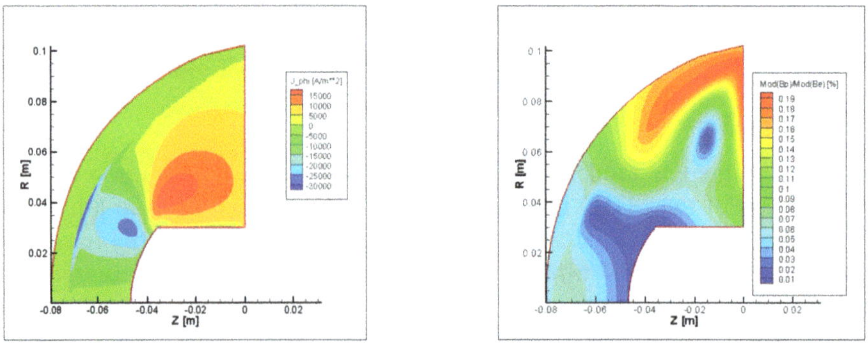

(a) Azimuthal current density

(b) ratio between induced and applied magnetic field

Figure 7.3. Calculation results for a scalar conductivity $\sigma = 100$ S m^{-1}.

would not be zero, violating condition in equation (7.69) which states that the electric field is conservative. As a result, the term $\vec{E} \cdot \vec{J}$ describing the contribution of the MHD interaction to the energy equation (7.58) is zero. This means that MHD interaction directly and locally converts kinetic energy into internal energy and vice versa leaving the total energy $\rho(e_i + v^2/2)$ unaffected. The magnetic induction field B_p generated by the current density in the shock layer can be evaluated using the Biot–Savart law. The percentage ratio of the induced field magnitude B_p to the applied field magnitude B_e is plotted in figure 7.3. As we can easily notice, in the given conditions the induced field B_p is well below 1 % of the applied field B_e.

In order to appreciate the impact that the tensorial characteristic has on the electrodynamics and, as a consequence, on the whole MHD regime, we now consider the same problem taking a constant scalar conductivity $\sigma = 100\ \mathrm{S\ m^{-1}}$ and a $100\ \mathrm{T^{-1}}$ electron mobility. In figure 7.4 some results of the calculations are shown. The modifications induced in the flow by the MHD interaction can be observed comparing the Mach number distribution and the streamlines in figure 7.4(a) with the same quantities in the unmodified case in figure 7.2. The increase of the shock stand-off distance caused by the MHD interaction is evident. One of the main consequences of the tensorial characteristic of the conductivity is

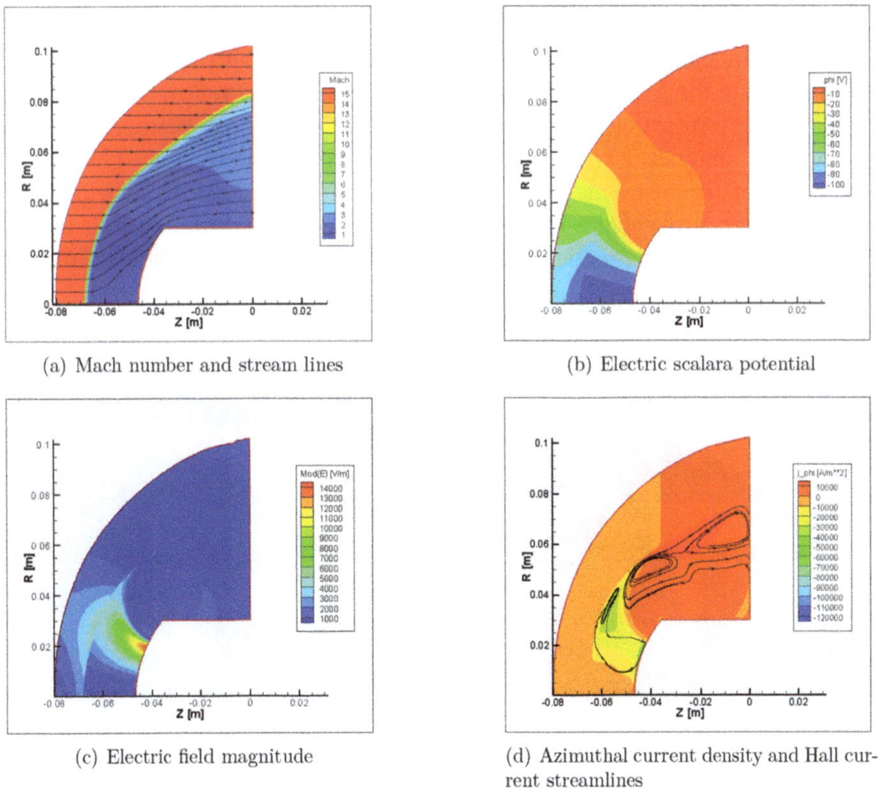

(a) Mach number and stream lines

(b) Electric scalara potential

(c) Electric field magnitude

(d) Azimuthal current density and Hall current streamlines

Figure 7.4. Calculation results for a tensorial conductivity ($\sigma = 100\ \mathrm{S\ m^{-1}}$, $\mu_e = 100\ \mathrm{T^{-1}}$).

the appearance of a Hall electric field, having direction approximately opposite to that of the velocity in the gas. This phenomenon is responsible for the electric potential distribution depicted in figure 7.4(b). A plot of the electric field magnitude is reported in 7.4(c). Since in high β_e regimes (i.e. in plasmas with low density and high electron mobility) the electric current density tends to be perpendicular to both the magnetic induction field $\vec{B}$ and the reduced electric field $\vec{E}'$, the Hall electric field is actually mainly responsible for the azimuthal current density, which in turn produces the body force $\vec{J} \times \vec{B}$. The azimuthal current density contour plot is shown in figure 7.4(d). In the same figure, the streamlines of the electric current density developing in the r–z plane due to the Hall effect are reported.

The formulation of models and the consequent implementation of numerical codes goes along with the experimental study of a given phenomenon. Numerical codes are useful in the process of designing an experiment, to assess the extent of the expected results. They are of course also essential tools for interpreting the experimental data and for achieving an insight on the phenomenon under investigation. On the other hand, the data produced by experiments are needed to carry out the validation of the numerical models and increase the confidence in the results that models provide. The investigations carried out in the field of MHD applied to hypersonic re-entry are no exception to this general rule, and many experiments in the last decades were accompanied by activities directed to the development of appropriate numerical tools.

The results in figure 7.6 report the numerical rebuilding of an experiment [38] aimed to evaluate the MHD effects in a Mach 6 ionized argon flow around a conical test body. The test article incorporates an arrangement of permanent magnets producing the magnetic induction field depicted in figure 7.5. The magnetic configuration of the test article is designed so as to have regions in which a high MHD interaction is expected. Indeed, in these regions, corresponding to the bright rings in figure 7.5, the magnetic induction field is roughly perpendicular to the flow velocity, as can be observed comparing the $\vec{B}$ field streamlines in figure 7.5 with the flow streamlines in figure 7.5(b). The electric transport properties for the calculations

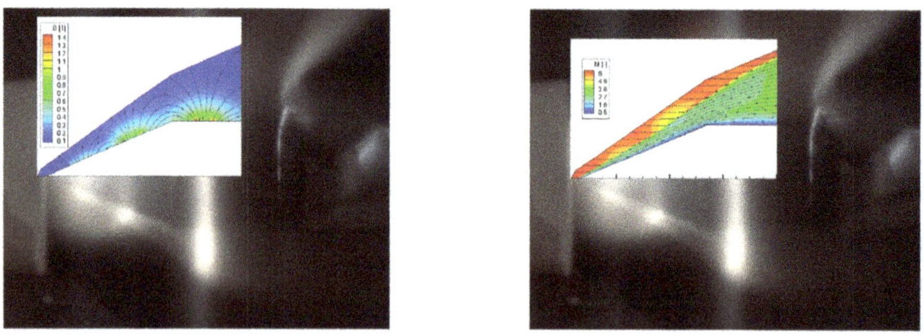

(a) Magnetic induction field distribution around the cone.

(b) Mach number ans flow streamlines around the cone

Figure 7.5. MHD experiment in an argon flow: the oblique shock is clearly visible. The bright rings around the cone corresponds to the region where the MHD interaction is stronger.

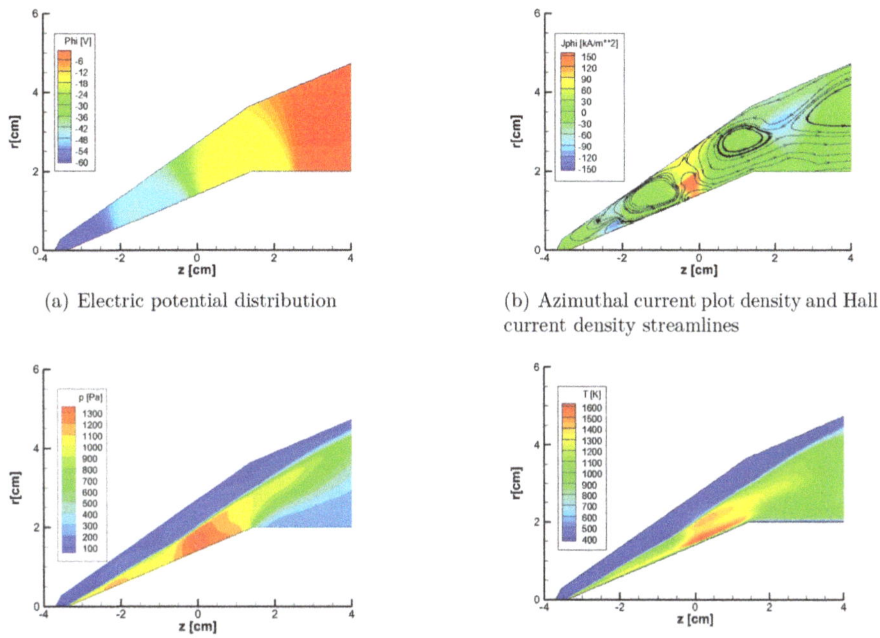

(a) Electric potential distribution

(b) Azimuthal current plot density and Hall current density streamlines

(c) Pressure distribution around the cone

(d) Pressure distribution around the cone

Figure 7.6. Numerical rebuilding of the MHD experiment.

have been evaluated utilizing the results of a model based on a two-terms approximation of the Boltzmann equation, in the assumption of a non-Maxwellian distribution of the electrons [39]. Due to the high electron mobility (μ_e ranges in the order of some tens of T^{-1}) the general behaviour of the phenomenon is similar to that seen in the previous illustrations. Also, in this case a Hall electric field roughly opposed to the direction of the flow is observed (see the electric potential plot in figure 7.6(a)). The azimuthal current is concentrated in the high MHD interactions regions, as shown in figure 7.6(b). Azimuthal current density, interacting with the magnetic induction field, produces the force per unit volume responsible for the modification of the flow field. Current densities, as well, are mainly responsible for the heating of the ionized gas flowing past the test article. These effects are well recognizable in figures 7.6(c) and 7.6(d), where the computed pressure field around the cone and the temperature are shown, respectively. In the interaction region, experimental measurements have shown a change in pressure induced by MHD above 10%. A good agreement has been observed between experimental data and numerical rebuilding [40, 41].

References

[1] Rosa R J, Krueger C H and Shioda S 1991 Plasmas in MHD power generation *IEEE Trans. Plasma Sci.* **19** 1180–90

[2] Rosa R J 1987 *Magnetohydrodynamic Energy Conversion* (New York: Hemisphere)

[3] Birat J and Chone J 1983 Electromagnetic stirring on billet, bloom, and slab continuous casters: state of the art in 1982 *Ironmaking Steelmaking* **10** 269–81
[4] Vives C and Ricou R 1985 Experimental study of continuous electromagnetic casting of aluminum alloys *Metall. Trans.* **16** 377–84
[5] Motora S, Takezawa S and Tamama H 1991 Development of the MHD ship YAMATO-1 *Oceans Conf. Record (IEEE)* vol 3 pp 1636–41
[6] Yohei S, Takezawa S, Sugawara Y and Kyotani Y 1995 The superconducting MHD-propelled ship YAMATO-1 *Proc. of the 4th Int. Conf. and Exhibition: World Congress on Superconductivity* vol 1 pp 167–76
[7] Jahn R G 1968 *Physics of Electric Propulsion* (New York: McGraw-Hill)
[8] Ziemer R 1959 Experimental investigation in magneto-aerodynamics Am *Rocket Soc.* J **29** 642–7
[9] Bityurin V, Lineberry J, Potebnia V, Alferov V, Kuranov A and Kuranov A 1997 Assessment of hypersonic MHD concepts *Meeting Papers of 28th Plasmadynamics and Lasers Conf.* Paper AIAA 1997-2393 https://doi.org/10.2514/6.1997-2393
[10] Bityurin V, Vatazhin A, Gus'kov O and Kopchenov V 2004 Hypersonic flow past the spherical nose of a body in the presence of a magnetic field *Fluid Dyn.* **39** 657–66
[11] Shang J 2001 Recent research in magneto-aerodynamics *Prog. Aerosp. Sci.* **37** 1–20
[12] Bityurin V, Bocharov A, Baranov D, Popov N, Alferov V and Tikhonov V 2011 Progress in MHD parachute concept study *49th AIAA Aerospace Sciences Meeting (Orlando, FL)* AIAA Paper 2011-902 https://doi.org/10.2514/6.2011-902
[13] Fomichev V and Yadrenkin M 2013 Experimental investigation of the magnetohydrodynamic parachute effect in a hypersonic air flow *Tech. Phys.* **58** 144–7
[14] Rybak J and Churchill R 1971 Progress in reentry communications *IEEE Trans. Aerosp. Electron. Syst.* **7** 879–94
[15] Mitchner M and Kruger C H J 1973 *Partially Ionized Gases* (New York: Wiley)
[16] Alfvén H and Fälthammar C G 1963 *Cosmical Electrodynamics* (Oxford: Oxford University Press)
[17] Alfvén H 1942 Existence of electromagnetic-hydrodynamic waves *Nature* **150** 405–6
[18] Goedbloed H P and Poedts S 2004 *Principles of Magnetohydrodynamics* (Cambridge: Cambridge University Press)
[19] Hosking R J and Dewar R L 2016 *Fundamental Fluid Mechanics and Magnetohydrodynamics* (Singapore: Springer)
[20] Brio M and Wu C 1988 An upwind differencing scheme for the equations of ideal magnetohydrodynamics *J. Comput. Phys.* **75** 400–22
[21] Cargo P 1997 Roe matrices for ideal MHD and systematic construction of Roe matrices for systems of conservation laws *J. Comput. Phys.* **136** 446–66
[22] Powell K, Roe P, Linde T, Gombosi T and De Zeeuw D 1999 A solution-adaptive upwind scheme for ideal magnetohydrodynamics *J. Comput. Phys.* **154** 284–309
[23] Janhunen P 2000 A positive conservative method for magnetohydrodynamics based on HLL and Roe methods *J. Comput. Phys.* **160** 649–61
[24] Gurski K 2004 An HLLC-type approximate Riemann solver for ideal magnetohydrodynamics *SIAM J. Sci. Comput.* **25** 2165–87
[25] Miyoshi T and Kusano K 2005 A multi-state HLL approximate Riemann solver for ideal magnetohydrodynamics *J. Comput. Phys.* **208** 315–44
[26] Harten A, Engquist B, Osher S and Chakravarthy S 1987 Uniformly high order accurate essentially non-oscillatory schemes, III *J. Comput. Phys.* **71** 231–303

[27] Borghi C, Carraro M and Cristofolini A 2003 Numerical modeling of MHD interaction in the boundary layer of hypersonic flows *IEEE Trans. Magn.* **39** 1507–10

[28] Borghi C, Carraro M and Cristofolini A 2005 Analysis of magnetoplasmadynamic interaction in the boundary layer of a hypersonic wedge *J. Spacecr. Rockets* **42** 45–50

[29] Cristofolini A, Latini C and Borghi C 2011 A weighted reverse Cuthill-McKee procedure for finite element method algorithms to solve strongly anisotropic electrodynamic problems *J. Appl. Phys.* **109** 033301

[30] Matsuda A, Wakatsuki K, Takizawa Y, Kawamura M, Otsu H, Konigorski D, Sato S and Abe T 2006 Shock layer enhancement by electro-magnetic effect for spherically blunt body *Meeting Papers of 37th Plasmadynamics and Lasers Conf.* AIAA Paper 2006-3573 https://doi.org/10.2514/6.2006-3573

[31] Gülhan A, Esser B, Koch U, Siebe F, Riehmer J, Giordano D and Konigorski D 2009 Experimental verification of heat-flux mitigation by electromagnetic fields in partially-ionized-argon flows *J. Spacecr. Rockets* **46** 274–83

[32] Cristofolini A, Borghi C, Neretti G, Passaro A, Fantoni G and Paganucci F 2008 Magnetohydrodynamics interaction over an axisymmetric body in a hypersonic flow *J. Spacecr. Rockets* **45** 438–44

[33] Cristofolini A, Borghi C, Neretti G, Schettino A, Trifoni E, Battista F, Passaro A and Baccarella D 2012 Experimental investigations on the magneto-hydro-dynamic interaction around a blunt body in a hypersonic unseeded air flow *J. Appl. Phys.* **112** 093304

[34] Poggie J and Gaitonde D 2000 Magnetic control of hypersonic blunt body flow *38th Aerospace Sciences Meeting and Exhibit* AIAA Paper 2000-0452 https://doi.org/10.2514/6.2000-452

[35] Gaitonde D and Poggie J 2002 Elements of a numerical procedure for 3-D MGD flow control analysis *38th Aerospace Sciences Meeting and Exhibit* AIAA Paper 2002-0198 https://doi.org/10.2514/6.2000-452

[36] MacCormack R 2002 Three-dimensional magneto-fluid dynamics algorithm development *40th Aerospace Sciences Meeting and Exhibit* AIAA Paper 2002-0197 https://doi.org/10.2514/6.2002-197

[37] D'Ambrosio D and Giodarno D 2007 Electromagnetic fluid dynamics for aerospace applications *J. Thermophys. Heat Transfer* **21** 284–302

[38] Cristofolini A, Borghi C, Carraro M, Neretti G, Passaro A, Fantoni G and Biagioni L 2008 Hypersonic MHD interaction on a conical test body with a Hall electrical connection *IEEE Trans. Plasma Sci.* **36** 530–41

[39] Colonna G and Capitelli M 2008 Boltzmann and master equations for MHD in weakly ionized gases *J. Thermophys. Heat Transfer* **22** 414–23

[40] Cristofolini A, Borghi C and Colonna G 2007 Numerical analysis of the experimental results of the MHD interaction around a sharp cone *Meeting Papers of the 38th AIAA Plasmadynamics and Lasers Conf.* AIAA Paper 2007-4252 https://doi.org/10.2514/6.2007-4252

[41] Cristofolini A, Neretti G and Borghi C 2012 Plasma parameters and electromagnetic forces induced by the magneto hydro dynamic interaction in a hypersonic argon flow experiment *J. Appl. Phys.* **112** 033302

IOP Publishing

Plasma Modeling (Second Edition)
Methods and applications
Gianpiero Colonna and Antonio D'Angola

Chapter 8

Drift-diffusion models and methods

Bernard Parent

Fluid models of plasmas can also be used to simulate non-neutral phenomena as would occur, for instance, within plasma sheaths. Sheaths are thin plasma regions juxtaposed to the interface between the plasma and the solid surface and that are inherently non-neutral. Sheaths play a critical role in gas discharges, lightning interactions with a surface, heat flux to the electrodes in MHD systems, etc. In this chapter, we will outline the basic equations governing the motion of the electrons and ions that are applicable not only within quasi-neutral regions but also within non-neutral sheaths. This will be done in the general case for a multicomponent plasma in the presence of a magnetic field. We will explain why such equations are especially difficult to simulate, and outline numerical methods that overcome the computational difficulties.

8.1 Drift-diffusion transport equations

Whether or not the plasma is quasi-neutral, the motion of the charged species with respect to the bulk can be found through the 'drift-diffusion' plasma model. Such can be used to find the charged species properties not only within the quasi-neutral plasma away from the surfaces but also within the non-neutral plasma sheaths juxtaposed to the surfaces.

In the drift-diffusion model, the mass conservation equation for the charged species can be written as:

$$\frac{\partial}{\partial t}N_k + \sum_i \frac{\partial}{\partial x_i} N_k \vec{V}_i^k = W_k \tag{8.1}$$

In the latter, N refers to the number density (i.e., number of particles per unit volume) and the super/subscript k refers to the kth species which can be either electrons or ions. The chemical source term W_k corresponds to the net rate of creation of the particles per unit volume by chemical reactions. Also, $\vec{V}^k$ is the

doi:10.1088/978-0-7503-3559-1ch8

velocity vector of the kth species which can be found from the momentum equation. Although the charged species mass conservation does not explicitly include diffusion processes, such are incorporated within the velocity $\vec{V}^k$ which is the sum of not only the convection but also the diffusion velocities.

The velocities of all charged species can be determined from the momentum equation. The momentum equation of the kth species includes the electromagnetic body forces, forces due to pressure gradients, and forces due to collisions as follows:

$$\rho_k \frac{\partial \vec{V}^k}{\partial t} + \rho_k \vec{V}^k \cdot \vec{\nabla} \vec{V}^k = \underbrace{q_k N_k (\vec{E} + \vec{V}^k \times \vec{B})}_{\text{electromagnetic force}} - \underbrace{m_k \nu_k N_k (\vec{V}^k - \vec{V})}_{\text{collision force with neutrals}} - \underbrace{\vec{\nabla} P_k}_{\text{pressure force}} \quad (8.2)$$

Let us go through the definition of all terms within the momentum equation. The charge of the kth species is denoted by q_k. Thus, q_k would be equal to the elementary charge e for singly-charged positive ions, to $-e$ for the electrons, to $-2e$ for the doubly-charged negative ions, etc. Also, $\vec{V}^k$ denotes the charged species velocity including both 'drift' and 'diffusion'. Thus, the velocity $\vec{V}^k$ is not limited to the convection of species k but also includes its diffusion due to density gradients or temperature gradients. We will discuss in more detail the drift and diffusion components of the velocity below. When the symbol $\vec{V}$ is used without any superscript, it refers to the velocity of the plasma mixture and thus constitutes an 'average' velocity of all electrons, ions, and neutrals present within the gas. Also, m corresponds to the mass of one particle and ν corresponds to the momentum transfer collision frequency between the charged species and the plasma bulk. The species partial pressure is denoted by the symbol P_k while the electric field and magnetic field vectors are denoted by $\vec{E}$ and $\vec{B}$, respectively.

In the momentum equation, it can be noticed that there are no terms on the RHS to account for forces originating from the shear stresses or from the collisions between charged species. This is because such forces are negligible compared to the forces originating from the collisions of the charged species with the neutrals when the plasma is *weakly ionized* (a plasma for which the electron or ion molar fraction is of order 10^{-4} or less). In that ionization range, it can be shown that the force due to collisions between charged species would amount to less than 1% of the force induced by ion–neutral or electron–neutral collisions, and that the force resulting from the shear stresses would typically amount to an even smaller quantity.

The drift-diffusion model further consists of assuming that the terms involving inertia gradients of the charged species in time or space are negligible compared to other terms This is a very good approximation for the electrons because their relatively small mass leads to their inertia forces being much smaller than the collision forces. This is also a good approximation for the ions as will be discussed later. After taking away the inertia terms from the momentum equation (8.2) the following can be obtained:

$$\vec{V}^k = \vec{V} + s_k\mu_k(\vec{E} + \vec{V}^k \times \vec{B}) - \frac{\mu_k}{|q_k|N_k}\vec{\nabla}P_k \tag{8.3}$$

In deriving the latter, we rewrote the momentum transfer collision frequency in terms of the mobility as follows:

$$\mu_k \equiv \frac{|q_k|}{m_k\nu_k} = \frac{s_k q_k}{m_k\nu_k} \tag{8.4}$$

where s_k is the sign of the charge of the species under consideration (-1 for electrons or negative ions, $+1$ for positive ions). The mobilities can be found from experiments measuring the drift velocity from the electric field. Empirical fits approximating experimental data of the mobilities of various ions moving in dry air are tabulated in table 8.1. Given the magnetic field and electric field distribution as well as the transport coefficients such as the mobility and diffusion coefficient, equation (8.3) gives one equation for each component of the velocity vector $\vec{V}^k$, and can thus be used to obtain the charged species velocity. Obtaining $\vec{V}^k$ from equation (8.3) is obviously trivial in the limit of no magnetic field. However, when the magnetic field is not zero, we need to rearrange the terms to yield the charged species velocity in a single step without resorting to a root solver. This can be done by regrouping the $\vec{V}^k - \vec{V}$ terms on the LHS of the momentum equation (8.3):

$$\vec{V}^k - \vec{V} - s_k\mu_k(\vec{V}^k - \vec{V}) \times \vec{B} = s_k\mu_k(\vec{E} + \vec{V} \times \vec{B}) - \frac{\mu_k}{|q_k|N_k}\vec{\nabla}P_k \tag{8.5}$$

Table 8.1. Ion and electron mobilities in dry air[a]. Reproduced with permission from reference [1]. Copyright 2022 Bernard Parent.

Species	Mobility μ, $m^2 \cdot V^{-1} \cdot s^{-1}$	Reference
Air^+	$N^{-1} \cdot \min(0.84 \cdot 10^{23} \cdot T^{-0.5},\ 2.35 \cdot 10^{12} \cdot (E^\star)^{-0.5})$	[2]
N_2^+	$N^{-1} \cdot \min(0.75 \cdot 10^{23} \cdot T^{-0.5},\ 2.03 \cdot 10^{12} \cdot (E^\star)^{-0.5})$	[3]
O_2^+	$N^{-1} \cdot \min(1.18 \cdot 10^{23} \cdot T^{-0.5},\ 3.61 \cdot 10^{12} \cdot (E^\star)^{-0.5})$	[3]
O_2^-	$N^{-1} \cdot \min(0.97 \cdot 10^{23} \cdot T^{-0.5},\ 3.56 \cdot 10^{19} \cdot (E^\star)^{-0.1})$	[4]
e^-	$\dfrac{1}{\frac{N_n}{3.74\cdot 10^{19}\cdot \exp(33.5/\sqrt{\ln T_e})} + \frac{N_i}{1.061\cdot 10^{16}\cdot T_e^{1.5}}}$	[2, 5], pp 180–1[b]

[a]Notation and units: T_e is the electron temperature in Kelvin; T is the heavy particle translational temperature in Kelvin; N is the total number density of the plasma in $1/m^3$; N_i is the sum of the positive ion number densities in $1/m^3$; N_n is the sum of the neutral species number densities in $1/m^3$; N_i is the sum of the ion species number densities in $1/m^3$; $E^\star$ is the reduced effective electric field ($E^\star \equiv |\vec{E}|/N$) in units of $V\cdot m^2$.
[b]The electron mobility includes both electron–neutral and electron–ion collisions.

The latter can be readily recast in matrix form as:

$$\begin{bmatrix} 1 & -s_k\mu_k\vec{B}_3 & s_k\mu_k\vec{B}_2 \\ s_k\mu_k\vec{B}_3 & 1 & -s_k\mu_k\vec{B}_1 \\ -s_k\mu_k\vec{B}_2 & s_k\mu_k\vec{B}_1 & 1 \end{bmatrix} \begin{bmatrix} (\vec{V}^k - \vec{V})_1 \\ (\vec{V}^k - \vec{V})_2 \\ (\vec{V}^k - \vec{V})_3 \end{bmatrix} = s_k\mu_k \begin{bmatrix} (\vec{E} + \vec{V} \times \vec{B})_1 \\ (\vec{E} + \vec{V} \times \vec{B})_2 \\ (\vec{E} + \vec{V} \times \vec{B})_3 \end{bmatrix} - \frac{\mu_k}{|q_k|N_k} \begin{bmatrix} \partial P_k/\partial x_1 \\ \partial P_k/\partial x_2 \\ \partial P_k/\partial x_3 \end{bmatrix} \tag{8.6}$$

Then, by defining the tensor mobility $\tilde{\mu}^k$ as:

$$\begin{aligned} \tilde{\mu}^k &\equiv \mu_k \begin{bmatrix} 1 & -s_k\mu_k\vec{B}_3 & s_k\mu_k\vec{B}_2 \\ s_k\mu_k\vec{B}_3 & 1 & -s_k\mu_k\vec{B}_1 \\ -s_k\mu_k\vec{B}_2 & s_k\mu_k\vec{B}_1 & 1 \end{bmatrix}^{-1} \\ &= \frac{\mu_k}{1 + \mu_k^2|\vec{B}|^2} \begin{bmatrix} 1 + \mu_k^2\vec{B}_1^2 & \mu_k^2\vec{B}_1\vec{B}_2 + s_k\mu_k\vec{B}_3 & \mu_k^2\vec{B}_1\vec{B}_3 - s_k\mu_k\vec{B}_2 \\ \mu_k^2\vec{B}_1\vec{B}_2 - s_k\mu_k\vec{B}_3 & 1 + \mu_k^2\vec{B}_2^2 & \mu_k^2\vec{B}_2\vec{B}_3 + s_k\mu_k\vec{B}_1 \\ \mu_k^2\vec{B}_1\vec{B}_3 + s_k\mu_k\vec{B}_2 & \mu_k^2\vec{B}_2\vec{B}_3 - s_k\mu_k\vec{B}_1 & 1 + \mu_k^2\vec{B}_3^2 \end{bmatrix} \end{aligned} \tag{8.7}$$

and multiplying all terms in the momentum equation by the tensor mobility and then dividing through by the scalar mobility, we obtain the charged species momentum equation in tensor form:

$$\vec{V}_i^k = \vec{V}_i + \underbrace{\sum_j s_k\tilde{\mu}_{ij}^k(\vec{E} + \vec{V} \times \vec{B})_j}_{\text{drift}} - \underbrace{\sum_j \frac{\tilde{\mu}_{ij}^k}{|q_k|N_k}\frac{\partial P_k}{\partial x_j}}_{\text{diffusion}} \tag{8.8}$$

We have obtained an equation giving an expression for the difference between the charged species velocity and the bulk velocity of the mixture written in terms of 'drift' and 'diffusion' components. The 'drift' corresponds to the part of the velocity component that is driven by the electromagnetic magnetic fields. The 'diffusion' corresponds to the part of the velocity component driven by the pressure gradients.

It may be argued that the latter is limited in scope because it was obtained assuming that the pseudo forces from the change in inertia as well as the electron–ion collision forces are small compared to the electron–neutral collision forces. However, such assumptions can be shown to lead to minimal error when the plasma is weakly ionized. Further, even when the plasma is strongly ionized, we can still use equation (8.8) to obtain a very good approximation to the charged species velocity simply by redefining the mobility of the electrons so that they include the electron–ion collisions, as done in table 8.1. Similarly, the ion mobilities can be modified to include the ion–ion collisions.

8.1.1 Drift-diffusion model in the absence of magnetic field

It is useful to simplify the drift-diffusion model to the case of no magnetic field whether the magnetic field is induced by the current or externally applied on the plasma by a magnet. This can be done starting from the general form of the drift-diffusion model shown in equation (8.8) while setting $\vec{B}$ to zero and noting that the tensor mobility would collapse to the scalar mobility when the magnetic field vanishes. Then, the following velocity for the kth charged species can be obtained:

$$\vec{V}_i^k = \vec{V}_i + s_k\mu_k\vec{E}_i - \frac{\mu_k}{|q_k|N_k}\frac{\partial P_k}{\partial x_i} \tag{8.9}$$

We can further use the ideal gas law to express the pressure as a function of number density and temperature:

$$P_k = N_k k_\mathrm{B} T_k \tag{8.10}$$

with k_B the Boltzmann constant and T_k the translational temperature of the kth species. After substituting the latter in equation (8.9) and rewriting in vector form we get:

$$\vec{V}^k = \vec{V} + \underbrace{s_k\mu_k\vec{E}}_{\text{drift}} - \underbrace{\frac{D_k}{N_k}\vec{\nabla}N_k}_{\text{mass diffusion}} - \underbrace{\frac{D_k}{T_k}\vec{\nabla}T_k}_{\text{thermal diffusion}} \tag{8.11}$$

In obtaining the latter, some terms were simplified through the definition of the diffusion coefficient $D_k \equiv \mu_k k_\mathrm{B} T_k/|q_k|$ (the so-called Einstein–Smoluchowski relation). The difference between the charged species velocity and the mixture velocity can be seen to be composed of a drift velocity function of the electric field as well as of a mass diffusion velocity due to number density gradients and of a thermal diffusion velocity function of temperature gradients.

8.1.2 Boundary conditions at solid surfaces

At the solid surfaces, the gradient of the mass fraction of the neutral species can generally be set to zero when there is no surface catalysis. In contrast, the charged species cannot be given a non-catalytic boundary condition. Rather, some of the ions release an electron when impacting the surface (secondary electron emission) while the rest recombine with the electrons at the surface. Both effects can be represented through the following boundary condition for the charged species, as outlined in reference [6]:

$$\frac{\partial}{\partial\chi}N_+\vec{V}_\chi^+ = 0 \quad \text{and} \quad N_- = 0 \quad \text{and} \quad N_\mathrm{e} = \frac{\gamma}{\mu_\mathrm{e}}\sum_{k=1}^{n_\mathrm{s}} N_k\mu_k\beta_k^+ \quad \text{for } \vec{E}_\chi < 0 \tag{8.12}$$

$$N_+ = 0 \qquad \text{and} \qquad \frac{\partial}{\partial\chi}N_-\vec{V}_\chi^- = 0 \qquad \text{and} \qquad \frac{\partial}{\partial\chi}N_\mathrm{e}\vec{V}_\chi^\mathrm{e} = 0 \qquad \text{otherwise} \tag{8.13}$$

where the subscripts '+' and '−' refer to the positive and negative ions, respectively, and the subscript 'e' refers to the electrons, and β_k^+ is equal to 1 should the kth species be a positive ion and to zero otherwise. Also, n_s is the number of species, χ is a coordinate perpendicular to the surface and pointing towards the fluid, and γ is the secondary emission coefficient which is typically set to 0.1 but could be as high as 2–5. The value of γ needs to be determined through experiments. What makes the latter boundary condition convenient is that it can be deployed without modification to various surfaces whether they are electrodes or dielectrics.

8.2 Stiffness and why it needs to be overcome

The simulation of the drift-diffusion model often requires an excessive number of iterations to reach convergence. Such slow convergence is generally attributed to the large discrepancy between the wave speeds within the system of equations. The largest wave speed within the system typically is that of the electrons, which can reach a million meters per second or more in regions where the electric field is high (such as within the plasma sheaths near the surfaces for instance). By itself, the large wave speed of the electrons is not problematic: simply reducing the integration time step to a value close to the one prescribed by the Courant–Friedrichs–Lewy (CFL) condition is sufficient to lead to stable convergence. The problem arises when integrating the electron transport equations in coupled form with other transport equations involving ions and neutrals which have much lower wave speeds in the order of a few tens or hundreds of meters per second. The physical processes associated with these lower wave speeds typically require much higher time steps to reach convergence rapidly than the physical processes associated with the electron wave speeds. Simulating both these slow and fast processes in coupled form leads to excessively slow convergence because (i) large integration time steps are needed in order to integrate efficiently the slow processes, and (ii) it is not possible to use such large integration time steps or the electron transport equations would grow unstable. A system of equations that has such a large discrepancy between its wave speeds leading to excessively slow convergence is commonly referred to as a *stiff* system.

Although the discrepancy between the wave speeds can lead to slow convergence, such is not the only property of the system that can lead to poor computational efficiency. Thus, we here prefer to define a stiff system in the more general sense following Lambert [7]:

> If a numerical method with a finite region of absolute stability, applied to a system with any initial conditions, is forced to use in a certain interval of integration a step length which is excessively small in relation to the smoothness of the exact solution in that interval, then the system is said to be stiff in that interval.

Slow convergence due to stiffness is not unique to plasma flows and is also encountered in non-ionized fluid dynamics. For instance, within the Navier–Stokes equations, the wave speeds of the diffusion terms can be orders of magnitude higher than the wave speeds of the convection terms Also, within the chemically reacting

Euler equations, the time scales associated with the chemical reactions can be thousands of times smaller than the time scales associated with the acoustic or convection waves. Nonetheless, the stiffness arising from the diffusion terms or the chemical source terms in non-ionized fluid dynamics can be easily overcome through the use of implicit methods which relieve the time step restrictions and lead to fast convergence in few iterations. Unfortunately, this doesn't hold when obtaining solutions to the drift-diffusion model where implicit methods alone often do not help much if at all to overcome the stiffness of the equations.

Overcoming stiffness and, hence, achieving faster convergence rates, is critical when doing numerical simulations of complex problems that require large computing effort because it leads to less *numerical error*. Indeed, for equal CPU hours, a solver that converges faster will enable the use of finer meshes than a solver that converges slower. In turn, the use of finer meshes leads to a smaller difference between the converged solution and the exact solution and, therefore, to less numerical error. Numerical error should not be confused with physical error. The former is simply due to the use of too few grid cells while the latter is due to an incomplete physical model. When doing numerical simulations, it is important to minimize both sources of error so that their sum (the total simulation error) is as small as possible. It would indeed be self-defeating to attempt solving a very stiff and detailed physical model with low physical error if such leads to very slow convergence and, thus, to excessive amounts of numerical error.

The importance of minimizing both physical modeling error and numerical error is schematized in figure 8.1. So far, numerical simulations of plasmas have reduced the total simulation error by reducing the complexity of the physical model such that the numerical error originating from its stiffness is not excessive. For instance, the stiffness associated with a detailed magnetohydrodynamics (MHD) physical model is removed by assuming that the plasma is quasi-neutral throughout. Then, there is no need to simulate the mass and momentum equations of the electrons, hence

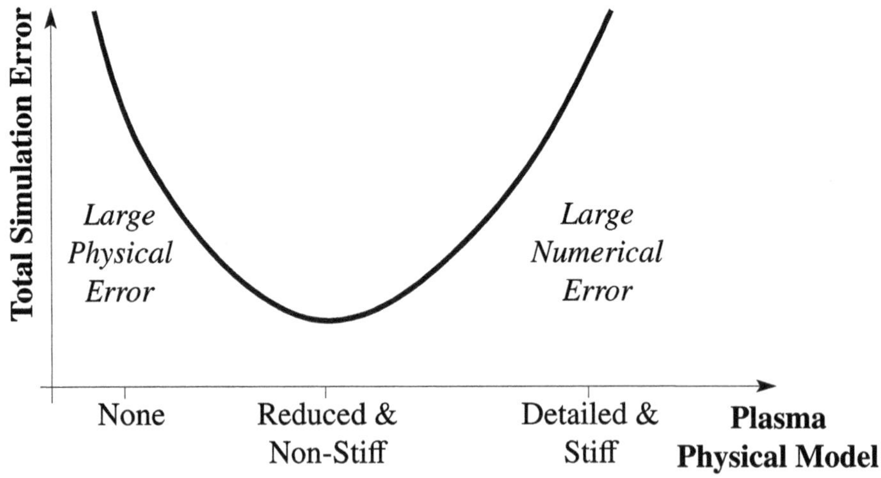

Figure 8.1. Effect of physical model accuracy on total simulation error.

permitting much larger integration time steps to be used, hence leading to much less numerical error. A similar strategy is employed when simulating planetary entry flows where non-neutral plasma effects are discarded in order to reduce the stiffness of the system. Of course, one caveat of doing so is that various non-neutral physical phenomena (such as sheaths for instance) are excluded from the physical model.

We will now outline strategies to overcome the stiffness within plasma equations without simplifying the physical model. It is emphasized that the methods outlined herein do not achieve fast integration by taking shortcuts. Rather, the methods presented here are strictly convergence acceleration techniques that get rid of the stiffness without reducing the physical model to a simpler form in any way. Put another way, the computational methods outlined below simply permit one to simulate more efficiently the drift-diffusion and thus this leads to less numerical error while not increasing the physical error.

8.3 Block-implicit schemes

Stiff physical systems of equations that include either chemical source terms or diffusion phenomena can be integrated much more efficiently using *implicit* schemes instead of *explicit* schemes. This can be attributed to the implicit schemes increasing the stability bounds and hence permitting much higher time steps to be used than explicit schemes. One class of implicit schemes that is especially efficient at integrating stiff systems of equations involving large variations of the time scales is the so-called 'block-implicit' approach. This has been used successfully to overcome stiffness in simulating chemically reacting fluid equations such as those encountered in re-entry flows for instance [8]. Let us here give an overview of such methods and how they can be deployed to the drift-diffusion model.

For illustration purposes, we will keep the physical model simple by limiting it to the drift-diffusion equations in the absence of magnetic field effects or unsteady effects. Thus, the species velocity can be obtained from equation (8.11). Also again for simplicity, let us consider a plasma with only three components: one type of positive ions, electrons, and one type of neutral species. Let us also keep the neutral species properties fixed and neglect temperature gradients. Let us further assume electrostatics and find the electric field from a potential ϕ obtained from Gauss's law. It is then convenient to express the steady-state mass conservation equations in matrix form as follows:

$$R = \sum_j \frac{\partial}{\partial x_j}(A_j U) - \sum_j \frac{\partial}{\partial x_j}\left(K\frac{\partial U}{\partial x_j}\right) - S \tag{8.14}$$

with R the residual vector which we seek to minimize ($R \to 0$) and the U, S, A, K matrices equal:

$$U = \begin{bmatrix} N_\mathrm{i} \\ N_\mathrm{e} \\ \phi \end{bmatrix} \quad S = \begin{bmatrix} W_\mathrm{i} \\ W_\mathrm{e} \\ \dfrac{e}{\epsilon_0}(N_\mathrm{i} - N_\mathrm{e}) \end{bmatrix} \quad A_j = \begin{bmatrix} \mu_\mathrm{i} E_j & 0 & 0 \\ 0 & -\mu_\mathrm{e} E_j & 0 \\ 0 & 0 & 0 \end{bmatrix} \quad K = \begin{bmatrix} D_\mathrm{i} & 0 & 0 \\ 0 & D_\mathrm{e} & 0 \\ 0 & 0 & 1 \end{bmatrix} \tag{8.15}$$

where the subscripts 'i' and 'e' stand for the positive ion species and the electron species, respectively. Before getting into the details of the implicit scheme, the governing equations shown in (8.14) need to be first discretized. We can indicate the discretization of a partial derivative along x with the symbol δ_x. Such a discretization operator does not point to a particular discretization stencil, but simply points to a discretized operator of the partial derivative ∂_x. Thus, the operator δ_x could refer to an upwind stencil or a central stencil, or any discretization operator that yields the partial derivative in the limit of a very small spacing between the nodes. Thus, after discretization, equation (8.14) becomes:

$$R_\Delta = \sum_j \delta_{x_j}(A_j U) - \sum_j \delta_{x_j}\big(K\delta_{x_j}(U)\big) - S \tag{8.16}$$

with R_Δ the discretized residual that needs to be minimized through an iterative technique. Also, the term $\delta_x(AU)$ is discretized using an upwind scheme with the direction of the upwind based on the eigenvalues of the matrix A while the term $\delta_x(K\delta_x(U))$ is discretized using central schemes. We can obtain a solution to the discretization equation (8.16) by implicit iterations. This can be done by first evaluating the discretization equation at iteration n, then evaluating it at iteration $n + 1$, and subtracting the latter from the former. This yields the following:

$$\begin{aligned}&\sum_j \delta_{x_j}(A_j U)^{n+1} - \sum_j \delta_{x_j}\big(K^{n+1}\delta_{x_j}(U^{n+1})\big) - S^{n+1}\\ &- \sum_j\Big(\delta_{x_j}(A_j U)^n - \sum_j \delta_{x_j}\big(K^n\delta_{x_j}(U^n)\big) - S^n\Big) = R_\Delta^{n+1} - R_\Delta^n\end{aligned} \tag{8.17}$$

We can simplify the latter by using the operator $\Delta^n(\cdot) = (\cdot)^{n+1} - (\cdot)^n$. Further, because we seek to reduce the residual to zero through the iterative process, stable convergence is obtained by setting the residual at the $n + 1$ iteration to zero. Then, the latter equation becomes:

$$\sum_j \delta_{x_j}\Delta^n(A_j U) - \sum_j \delta_{x_j}\Delta^n\big(K\delta_{x_j}(U)\big) - \Delta^n S = -R_\Delta^n \tag{8.18}$$

Because we aim to find in a single step the vector $\Delta^n U$, we need to assume that the matrices A and K remain constant between iteration n and $n + 1$. Also, the change in the source term S between iteration n and $n + 1$ can be approximated as $\Delta^n S \approx C^n \Delta^n U$ with C being the Jacobian of S equivalent to $\partial S/\partial U$. After making these simplifications in equation (8.18), we obtain:

$$\sum_j \delta_{x_j}(A_j^n \Delta^n U) - \sum_j \delta_{x_j}\big(K^n\delta_{x_j}(\Delta^n U)\big) - C^n\Delta^n U = -R_\Delta^n \tag{8.19}$$

To prevent divergence, it can be helpful to add a relaxation term as follows:

$$\Gamma^n\Delta^n U + \sum_j \delta_{x_j}(A_j^n \Delta^n U) - \sum_j \delta_{x_j}\big(K^n\delta_{x_j}(\Delta^n U)\big) - C^n\Delta^n U = -R_\Delta^n \tag{8.20}$$

with Γ the preconditioning matrix. There are various ways that the preconditioning matrix can be defined to yield faster convergence. One approach is to set it as a function of a pseudotime step which is either based on the CFL condition for the ion and electron transport equations or to a constant proportional to the grid spacing for the potential equation. Examples for the pseudotime step expressions for the preconditioning matrix can be found in [2]. Equation (8.20) is the so-called 'delta form' of the discretized equation. The change in the U vector can be found from the delta form in one step through a block-implicit algorithm. For instance, say that the system has only one dimension. Then, should the discretization stencils involve only three nodes, U at the next iteration can be found through a Tri-Diagonal-Matrix algorithm (TDMA) where each element of the TDMA is a square matrix.

Solving equation (8.20) in more than 1D is not as straightforward, however. One way this can be accomplished is through approximate factorization. Approximate factorization consists of first multiplying all terms by the inverse of the preconditioning matrix Γ:

$$\begin{aligned}&\Delta^n U + \sum_j (\Gamma^n)^{-1}\delta_{x_j}(A_j^n \Delta^n U) - \sum_j (\Gamma^n)^{-1}\delta_{x_j}\big(K^n \delta_{x_j}(\Delta^n U)\big) - (\Gamma^n)^{-1}C^n\Delta^n U \\ &= -(\Gamma^n)^{-1}R_\Delta^n\end{aligned} \tag{8.21}$$

Then, in 2D, the LHS of the latter equation can be approximately factored in the following manner:

$$\begin{aligned}&\big(I + (\Gamma^n)^{-1}\delta_{x_j}A_x^n - (\Gamma^n)^{-1}\delta_x(K^n\delta_x) - (\Gamma^n)^{-1}C^n\big) \\ &\times \big(I + (\Gamma^n)^{-1}\delta_y A_y^n - (\Gamma^n)^{-1}\delta_y\big(K^n\delta_y\big)\big)\Delta^n U = -(\Gamma^n)^{-1}R_\Delta^n\end{aligned} \tag{8.22}$$

with I the identity matrix. The latter would become equal to the former in the limit of a small magnitude of the pseudotime steps within the preconditioning matrix Γ. Indeed, recall that when using pseudotime stepping the diagonal elements of the preconditioning matrix are set to the inverse of the pseudotime steps. Thus, when the time step is small, the magnitude of the terms within Γ^{-1} would also be small. Therefore, when the time step is small, the terms on the LHS of (8.22) where Γ^{-1} appears twice or more would be negligible compared to the other terms, and the approximate factorization shown in (8.22) would collapse to the delta form shown in (8.21).

The approximately factored form can yield ΔU in two steps. First, we find $\Delta\tilde{U}$ using block TDMA sweeps in the first dimension as follows:

$$\big(I + (\Gamma^n)^{-1}\delta_{x_j}A_x^n - (\Gamma^n)^{-1}\delta_x(K^n\delta_x) - (\Gamma^n)^{-1}C^n\big)\Delta^n\tilde{U} = -(\Gamma^n)^{-1}R_\Delta^n \tag{8.23}$$

Knowing $\Delta\tilde{U}$, we find ΔU using block TDMA sweeps in the second dimension as follows:

$$\big(I + (\Gamma^n)^{-1}\delta_y A_y^n - (\Gamma^n)^{-1}\delta_y\big(K^n\delta_y\big)\big)\Delta^n U = \Delta\tilde{U} \tag{8.24}$$

It is noted that when the time step is large, the error induced by the approximate factorization would become more significant. However, approximate factorization error is often not a source of concern and large time steps can be imposed resulting in very fast convergence for many problems. Further, when using a block-implicit scheme, the error due to approximate factorization is not present for 1D problems (or multidimensional problems that are very close to 1D) because it originates from the factorization of the second and third dimension. Nonetheless, there are some multidimensional problems that are more difficult to integrate and that require a smaller factorization error. For these problems, the reader is referred to some other block-implicit schemes that are more robust than approximate factorization at higher time steps such as diagonally-dominant alternate direction implicit [9] or iterative modified approximate factorization [10].

8.4 Why the drift-diffusion system is particularly stiff

We will now assess how well the block-implicit scheme outlined above can integrate efficiently the drift-diffusion model. Integrating the drift-diffusion model using implicit methods can be particularly challenging and require excessive computational effort. As will be seen in this section, this can also be the case even for a very simple one-dimensional gas discharge problem between two electrodes, as illustrated in figure 8.2. Although we will use a 1D block-implicit scheme to integrate the equations to steady-state, we will see that its use alone does not overcome the stiffness of the drift-diffusion equations. An important goal of this section is thus to pinpoint why the block-implicit scheme (or other types of implicit scheme) is not efficient at integrating the drift-diffusion model.

In simulating this 1D gas discharge problem, the neutrals' densities, pressure, and temperature are not altered and remain fixed to the initial states specified in figure 8.2. Only the charged species' densities are allowed to vary. The air is here ionized through an electron beam acting in the direction perpendicular to the page, with the amount of ionization being a function of the electron beam power Q_{ebeam}.

It is convenient to choose e-beam ionization over thermal ionization in this case because it permits one to control the degree of ionization independently of the other properties within the plasma simply through the variation of the e-beam power. The

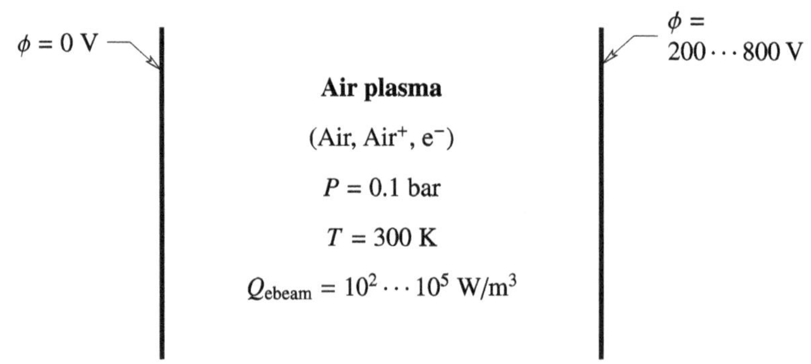

Figure 8.2. One-dimensional discharge in air between two electrodes ionized by an electron beam.

effect of the electron beam on the ionization rates can be modeled for convenience as an additional reaction. To keep the air ionization mechanisms as simple as possible we will limit here the other chemical reactions to Townsend ionization and electron attachment, with all reactions listed in table 8.2. The chemical reaction rates are then multiplied by the reactants concentrations to obtain the chemical source terms W that are added on the RHS of the charged species conservation equations, equation (8.1). The velocities in the conservation equations are obtained from the momentum equation (8.11) with the temperature gradients set here to zero for simplicity.

Because this is an electrostatic problem, the Maxwell equations lead to the curl of the electric field becoming zero. When the curl of a vector vanishes, a potential function exists. Thus, we can express the electric field as the negative of the gradient of a potential function as $\vec{E} = -\vec{\nabla}\phi$ with ϕ the electric field potential function. Substituting the latter in Gauss's law leads to the Gauss-based potential equation:

$$(\vec{\nabla} \cdot \vec{\nabla})\phi = -\frac{\rho_c}{\epsilon_0} \tag{8.25}$$

with ρ_c the net charge density which corresponds to $\sum_k N_k q_k$.

Let us first set the e-beam power deposited to 100 W m^{-3} and set the voltage difference between the electrodes to 200 Volts. As can be observed from the electron and ion density profiles shown in figure 8.3(a), this leads to a large cathode sheath on the left electrode and a quasi-neutral region of appreciable size in the middle of the domain. The equations were integrated to steady-state through the block-implicit method outlined in the previous section. Despite our best efforts to improve the convergence rates, the fastest convergence we were able to obtain for this problem is a residual decrease of 5 orders of magnitude in about 1 million iterations.

While such slow convergence to steady-state is perhaps not so surprising to researchers doing numerical simulations of the drift-diffusion model of plasma dynamics, it would appear peculiar to researchers doing numerical simulations of the Navier–Stokes equations of fluid dynamics. Even the most stiff problems simulated with the Navier–Stokes (as in hypersonic reacting flows for instance where the discrepancy between the chemical and convective time scales could reach one thousand times or higher) would not exhibit such slow convergence when using

Table 8.2. Ionization and recombination reactions taking place within a three-component e-beam-ionized air plasma.

No.	Reaction	Rate Coefficient[a]	References
1	$e^-+Air \rightarrow Air^++e^-+e^-$	$\exp(-0.010\,5031 \cdot \ln^2 E^\star - 2.409\,83 \cdot 10^{-75} \cdot \ln^{46} E^\star)$ cm^3 s^{-1}	[2]
2	$Air \rightarrow e^-+Air^+$	$1.84 \cdot 10^{17} \cdot Q_b/N$ s^{-1}	[2]
3	$e^-+Air^+ \rightarrow Air$	$2.24 \cdot 10^{-7} \cdot (300/T_e)^{0.5} + 0.4 \cdot 10^{-7} \cdot (300/T_e)^{0.7}$ cm^3 s^{-1}	[11]

[a]Notation and units: T_e is in Kelvin; T is in Kelvin; Q_b is the electron beam power deposited in W m^{-3}; N is the total number density of the plasma in 1 m^{-3}; $E^\star$ is the reduced electric field ($E^\star \equiv |E|/N$) in units of V·m^2.

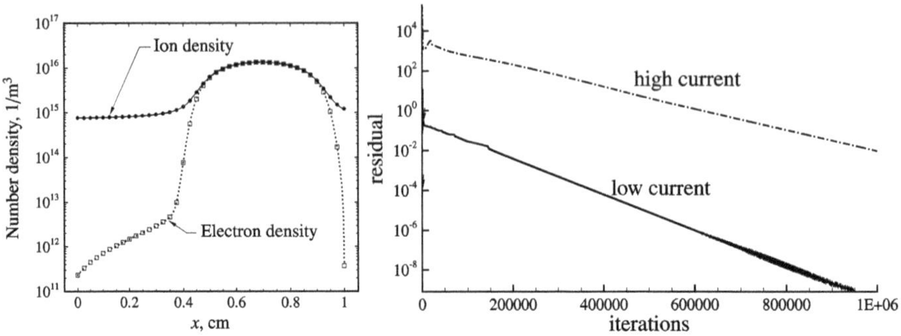

(a) Electron number density (b) Residual convergence

Figure 8.3. 1D discharge results obtained with the standard drift-diffusion model.

Table 8.3. Influence of voltage and e-beam power on number of iterations to reach 6 orders of magnitude of convergence for the 1D discharge case.

Case	$\Delta\phi$, V	Q_{ebeam}, W m^{-3}	Quasi-neutral region?	Iterations
Dielectrics	0	10^0	No	4701
Dielectrics	0	10^2	Yes	1 934 000
Dark discharge	800	10^2	No	25
Low J sheaths	200	10^2	Yes	947 000
High J sheaths	800	10^0	No	3117
Glow discharge	800	2×10^5	Yes	1 380 000

a block-implicit scheme. When used to integrate the 1D Navier–Stokes equations, the block-implicit strategy typically converges to steady-state in a few hundred iterations or less whether or not there is a high discrepancy of the time scales. Then, what is so special about the drift-diffusion model that leads to convergence requiring millions instead of hundreds of iterations?

Before attempting to answer why the drift-diffusion model is so stiff, let us first determine if this slow convergence always occurs when solving a discharge between two electrodes or if this is rather specific to this case in particular. Let us thus vary the voltage difference between the electrodes as well as the e-beam power to create various different types of discharges. For each case, we use the same physical model and the same block-implicit integrations strategy. Also, for a fair comparison we make sure that the user-specified convergence parameters (such as the CFL number) are chosen for each case such as to yield the fastest possible convergence. The convergence rates for various potential differences and e-beam power are tabulated in table 8.3. One striking finding is that the convergence rates tend to vary by *orders of magnitude* from case to case. For instance, for the 'dark discharge' case, only 25 iterations are needed to converge the residual by six orders of magnitude, or 40 000 times less computing than was needed to converge the 'glow discharge' case. The dark discharge case is not the only case to exhibit fast convergence. Several other cases could also be converged in a very respectable 3000-5000 iterations. What makes these very fast convergence rates intriguing is that they are obtained even though the discrepancy of the wave speeds is more or less the same as for the other much-slower-to-converge cases. Clearly, some cases are stiff and others are not and such stiffness cannot be attributed solely to the discrepancy of the wave speeds.

What property of the physical model other than wave speed discrepancy could cause slow convergence? Let us look carefully at the results shown in table 8.3. Intriguingly, the results show that the slow convergence only occurs when a quasi-neutral region forms within the domain. On the other hand, when the plasma is significantly non-neutral throughout, the convergence is typically 1000 times faster. What is so special about the quasi-neutral region that it leads to such slow convergence rates? The physics of the quasi-neutral region itself is not the culprit. Rather, the slow convergence is due to the way some terms are written within the potential equation which, when a quasi-neutral region is present, lead to divergence towards non-physical states unless very small time steps are used. This can best be illustrated by expanding the net charge density on the RHS of the potential equation (8.25) as a function of the electron and ion densities:

$$-\frac{\epsilon_0}{e}\nabla^2\phi = N_\mathrm{i} - N_\mathrm{e} \tag{8.26}$$

with N_i and N_e the ion and electron number densities, respectively. Because the electron and ion densities are obtained through separate mass transport equations, the absolute error associated with the RHS corresponds to the sum of the absolute error associated with N_i and the one associated with N_e:

$$\mathcal{E}_\mathrm{abs}(N_\mathrm{i} - N_\mathrm{e}) = \mathcal{E}_\mathrm{abs}(N_\mathrm{i}) + \mathcal{E}_\mathrm{abs}(N_\mathrm{e}) \tag{8.27}$$

with $\mathcal{E}_\mathrm{abs}(\cdot)$ the absolute error of a property due to integration process. But, the absolute error of a quantity can be written as the product between the relative error of such quantity and the quantity itself, as follows $\mathcal{E}_\mathrm{abs}(\cdot) = (\cdot)\mathcal{E}_\mathrm{rel}(\cdot)$. Using the latter, we can rewrite the absolute error associated with the RHS as a relative error as follows:

$$\mathcal{E}_\mathrm{rel}(N_\mathrm{i} - N_\mathrm{e}) = \frac{N_\mathrm{i}\mathcal{E}_\mathrm{rel}(N_\mathrm{i}) + N_\mathrm{e}\mathcal{E}_\mathrm{rel}(N_\mathrm{e})}{|N_\mathrm{i} - N_\mathrm{e}|} \tag{8.28}$$

From the latter, it can be seen that the relative error associated with $N_\mathrm{i} - N_\mathrm{e}$ scales with the relative error of N_i or N_e as expected. However, it also is inversely proportional to the difference between the ion and the electron densities. Thus, in quasi-neutral regions where N_i is very close to N_e, the error associated with N_e or N_i is amplified by a very large factor. Then, the integration of the potential equation can lead to divergence unless the relative error on either the ion or electron densities is kept to very small values. But such can only be accomplished by setting the integration time step to extremely small values, hence leading to an excessive amount of computational effort. When there is no quasi-neutral region, the error associated with the ion or electron densities is not amplified within the potential equation, and much larger time steps can thus be taken to integrate the charged species transport equations, leading to fast convergence.

As was demonstrated in this section through numerical examples in which a block-implicit method was used, the drift-diffusion equations do not always converge slowly. In many cases, the block-implicit scheme overcomes the stiffness and leads to very fast convergence to steady-state in a few hundred iterations. However, for some cases in which a quasi-neutral region is present, the stiffness of the system is not overcome by the block-implicit scheme. This was argued to be due to some terms within the Gauss-based potential equation amplifying the errors and forcing excessively small time steps to be used.

8.5 Overcoming the drift-diffusion stiffness

It was seen in the last section that the stiffness associated with the solution of the drift-diffusion model through block-implicit schemes is not solely due to the discrepancy of the wave speeds. The stiffness was seen to also originate from the Gauss-based potential equation amplifying the errors associated with the charged species densities when a quasi-neutral region forms. It follows that to overcome stiffness, we need to somehow avoid the solution of such terms that amplify error. Thus, we need to avoid obtaining the potential from Gauss's law.

8.5.1 Ohm-based potential equation

Another way that the electric field potential can be obtained is through Ohm's 'law'. Ohm's law is a misnomer in this case because it can be derived from the drift-diffusion transport equations and is thus not a law. For illustration purposes, we will here derive Ohm's law for a 1D plasma with only one type of ions and electrons and ignoring temperature gradients and magnetic field effects. This can be done starting from the mass conservation equation of each charged species as outlined in equation (8.1) and substituting the velocity by the expression presented in equation (8.11). We thus obtain:

$$\frac{\partial N_\mathrm{e}}{\partial t} + \frac{\partial}{\partial x}\left(-\mu_\mathrm{e} E N_\mathrm{e} - D_\mathrm{e}\frac{\partial N_\mathrm{e}}{\partial x}\right) = W_\mathrm{e} \tag{8.29}$$

$$\frac{\partial N_\mathrm{i}}{\partial t} + \frac{\partial}{\partial x}\left(\mu_\mathrm{i} E N_\mathrm{i} - D_\mathrm{i}\frac{\partial N_\mathrm{i}}{\partial x}\right) = W_\mathrm{i} \tag{8.30}$$

Then, multiply equation (8.29) by the electron charge $-e$, multiply equation (8.30) by the ion charge $+e$, and sum the two equations. We then note that no net charge can be created or destroyed through the chemical reactions, so that $W_\mathrm{e} = -W_\mathrm{i}$. After simplifications, and expressing the electric field as the negative of the gradient of the potential ϕ, we obtain the following Ohm-based potential equation:

$$\frac{\partial}{\partial t}e(N_\mathrm{i} - N_\mathrm{e}) + \frac{\partial}{\partial x}\left(-e(\mu_\mathrm{e} N_\mathrm{e} + \mu_\mathrm{i} N_\mathrm{i})\frac{\partial \phi}{\partial x} - eD_\mathrm{i}\frac{\partial N_\mathrm{i}}{\partial x} + eD_\mathrm{e}\frac{\partial N_\mathrm{e}}{\partial x}\right) = 0 \tag{8.31}$$

The latter can be used to find the electric field given the charged species densities. Contrarily to the Gauss-based potential equation, the Ohm-based potential equation does not include some terms that would lead to an amplification of the error associated with either the electron or the ion densities. Thus, the Ohm-based potential equation will not force excessively small time steps on the charged species transport equations that would lead to excessively slow convergence.

8.5.2 Modified ion transport equation

Although the use of an Ohm-based potential equation gets rid of the stiffness associated with the integration of the drift-diffusion model through block-implicit schemes, its use unfortunately leads to a system where one equation is redundant. For instance, finding a solution to the number densities through the electron and ion transport equations shown in equations (8.29)–(8.30) would automatically yield the correct solution to the Ohm-based potential equation independently of the value of the electric field. This is because the Ohm-based potential equation does not add any new physics that is not present within the ion and electron transport equations. Thus, a system composed of the Ohm-based potential and the charged species equations has one redundant equation and is thus under-specified.

In order to have three transport equations that do not lead to an under-specified system, we need to add new physics to the system. Some physics that is missing is the one related to the determination of the electromagnetic fields and, particularly, to Gauss's law. We can ensure that Gauss's law is satisfied by adding its terms to the charged species transport equations. Of course, this should be done in such a way that the added terms do not create stiffness. One approach that does not lead to stiffness is by first multiplying the terms within Gauss's law by the product of the ion mobility and the ion number density as follows:

$$-\mu_i N_i \frac{\partial E}{\partial x} = -\mu_i N_i \frac{e}{\epsilon_0}(N_i - N_e) \tag{8.32}$$

We can add the latter to the ion transport equation (8.30) and rearrange to obtain a *modified ion transport equation*:

$$\frac{\partial N_i}{\partial t} + E\frac{\partial}{\partial x}(\mu_i N_i) - \frac{\partial}{\partial x}\left(D_i \frac{\partial N_i}{\partial x}\right) = W_i - \mu_i N_i \frac{e}{\epsilon_0}(N_i - N_e) \tag{8.33}$$

Solving for a system of equations composed of the latter, the electron transport equation (8.29), and the Ohm-based potential equation (8.31), the solution of Gauss's law is guaranteed. This is because when the Ohm-based potential equation and the electron transport equation are solved conjunctly, the ion transport equation is automatically satisfied. Thus, solving the modified ion transport equation in which the Gauss's law terms are added necessarily ensures that Gauss's law is satisfied.

Let us now verify if indeed the modifications to the system of equations proposed above will get rid of the stiffness and lead to gains in computational efficiency. We will do so for the same 1D discharge problem tackled previously in section 8.4 and that was observed to be particularly stiff. For a fair comparison, both the Ohm-

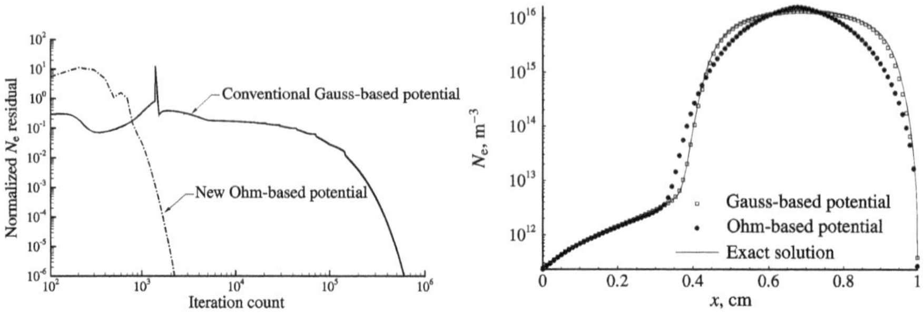

(a) Residual history (b) Electron density

Figure 8.4. Comparison between the Gauss-based and Ohm-based potential systems of equations.

based and the Gauss-based systems of equations will be integrated to steady-state using the same block-implicit scheme outlined in section 8.3. Also, the pseudotime steps were adjusted for each scheme to be optimal and thus to yield convergence in as few iterations as possible. As can be seen from figure 8.4(a), the new system of equations outlined in this section does result in a very significant improvement in convergence rates of several orders of magnitude. This confirms that the stiffness we observed previously with the drift-diffusion model was due to the terms on the RHS of the Gauss-based potential equation amplifying the error and thus preventing high time steps being used.

Obtaining the potential from Ohm's law instead of Gauss's law is not all beneficial, however. While it does lead to faster convergence, it also leads to a lower resolution of the converged solution. See for instance the profiles of electron density shown in figure 8.4(b). Both the Gauss-based and Ohm-based potential equations solve for exactly the same physical model and thus lead to the same exact solution as the number of grid points is increased to infinity. However, on a coarse mesh, the new system in which the potential is obtained from Ohm's law leads to a loss of resolution. The loss is so high that the mesh size would need to be multiplied by four times or more in every dimension in order to yield the same accuracy as the standard system. This fourfold decrease in resolution may not seem to be too concerning at first considering that the new system leads to more than a one-hundredfold decrease in the number of iterations. However, when simulating 3D problems, this fourfold decrease in resolution would offset the computational gains obtained through convergence acceleration. Indeed, a 3D problem typically requires a computational effort which scales with the number of mesh lines in one dimension to the fourth power. Thus, for a 3D problem, the new system would not yield a significant computational advantage over the standard system due to its much lower resolution.

8.5.3 Ambipolar form of the electron transport equation

Not all is lost however. We will see in this section that an additional transformation to the system can lead to spectacular gains in resolution while not affecting the fast convergence. This can be achieved by recasting the charged species transport equations in *ambipolar form* [12].

Before recasting the equations in ambipolar form, let us give a brief background of what 'ambipolar diffusion' is. In plasma physics, ambipolar diffusion refers to the diffusion of the electrons and ions impacting each other such that both the electrons and the ions diffuse at the same rate. Although electrons have a much higher diffusion coefficient than the ions and thus desire to diffuse at a faster rate, this cannot occur within a quasi-neutral plasma where the ion and electron densities must remain close to one another. To slow down the diffusion of the electrons and speed up the diffusion of the ions such that both diffuse at the same rate, an electric field forms in the opposite direction of the charged species density gradient. Because the electrons and ions have charges of opposite sign, this electric field slows down the electrons and accelerates the ions such that the plasma remains very close to neutral. Such an electric field is commonly referred to as the 'ambipolar electric field'.

Let us now find an expression for the ambipolar electric field for the case of a plasma composed of electrons and one type of positive ions. Let's also assume for simplicity that the plasma is one-dimensional and that there are no temperature gradients. We can do so starting from the expression of the current density which corresponds to the sum of each species current which itself is the product between the species velocity, charge, and density:

$$J = \sigma E - eD_{\mathrm{i}}\frac{\partial N_{\mathrm{i}}}{\partial x} + eD_{\mathrm{e}}\frac{\partial N_{\mathrm{e}}}{\partial x} \tag{8.34}$$

where σ is the plasma conductivity which in this case corresponds to $e(\mu_{\mathrm{e}}N_{\mathrm{e}} + \mu_{\mathrm{i}}N_{\mathrm{i}})$. When there is no externally applied electric field, the current within the ambipolar diffusion process will be close to zero because both positive ions and electrons diffuse at the same rate. Thus, we can find the ambipolar electric field simply by setting the current density to zero in the above and isolating the electric field. We thus obtain the following expression:

$$E' = \frac{eD_{\mathrm{i}}}{\sigma}\frac{\partial N_{\mathrm{i}}}{\partial x} - \frac{eD_{\mathrm{e}}}{\sigma}\frac{\partial N_{\mathrm{e}}}{\partial x} \tag{8.35}$$

where E' corresponds to the ambipolar electric field responsible for keeping the plasma neutral when electron and ion diffusion occurs.

Without loss of generality, we can add and subtract the ambipolar electric field E' within the electron transport equation (8.29):

$$\frac{\partial N_{\mathrm{e}}}{\partial t} + \frac{\partial}{\partial x}\left(-\mu_{\mathrm{e}}N_{\mathrm{e}}(E - E' + E') - D_{\mathrm{e}}\frac{\partial N_{\mathrm{e}}}{\partial x}\right) = W_{\mathrm{e}} \tag{8.36}$$

After substituting E' from the former and rearranging terms, the latter becomes the *ambipolar form* of the electron transport equation:

$$\frac{\partial N_{\mathrm{e}}}{\partial t} + \frac{\partial}{\partial x}\left(-\mu_{\mathrm{e}}N_{\mathrm{e}}(E - E') - \frac{e}{\sigma}\mu_{\mathrm{e}}N_{\mathrm{e}}D_{\mathrm{i}}\frac{\partial N_{\mathrm{i}}}{\partial x} - \frac{e}{\sigma}\mu_{\mathrm{i}}N_{\mathrm{i}}D_{\mathrm{e}}\frac{\partial N_{\mathrm{e}}}{\partial x}\right) = W_{\mathrm{e}} \tag{8.37}$$

At this stage, it is important to note that the latter does not change the drift-diffusion physical model in any way. We simply added and subtracted E' to the electric field. Thus, the integration of either the standard electron transport equation (8.29) or the ambipolar form shown in (8.37) will yield the same solution for a very fine mesh as long as the ambipolar electric field E' is taken from equation (8.35).

If the ambipolar form yields the same solution as the standard form, what is the advantage of doing this additional transformation? What makes the ambipolar form more computationally efficient is that each of its terms tends to be much smaller in magnitude than the corresponding terms within the standard form at least when the plasma is more or less neutral. Let us illustrate why this is by comparing one by one all terms of the ambipolar form to those of the standard form:

$$\frac{\partial N_e}{\partial t} + \frac{\partial}{\partial x}\left(\underbrace{-\mu_e N_e(E - E')}_{\ll \mu_e N_e E} - \underbrace{\frac{e}{\sigma}\mu_i N_i D_e \frac{\partial N_e}{\partial x}}_{\ll D_e \partial N_e/\partial x} - \underbrace{\frac{e}{\sigma}\mu_e N_e D_i \frac{\partial N_i}{\partial x}}_{\ll D_e \partial N_e/\partial x} \right) = W_e \tag{8.38}$$

Let us start with the convective wave speeds which are the terms within the first derivative multiplying the electron density. The convective wave speed within the ambipolar form corresponds to $\mu_e(E - E')$ and should be compared to the convective wave speed within the standard form, $\mu_e E$. In plasma regions that are quasi-neutral, the magnitude of $E - E'$ is generally less than the magnitude of the electric field E because most of the electric field is needed for ambipolar diffusion and only a small fraction is needed to carry current. Thus, the recast of the electron transport equation in ambipolar form generally leads to a reduction of the magnitude of its convective term.

We now proceed with the electron diffusion term and compare the magnitudes between the standard and ambipolar forms. The ambipolar form would reduce the magnitude of the diffusion term if the ambipolar electron diffusion coefficient $\frac{e}{\sigma}\mu_i N_i D_e$ is much less than the standard electron diffusion coefficient D_e as follows:

$$\frac{e}{\sigma}\mu_i N_i D_e \ll D_e \tag{8.39}$$

After substituting the conductivity in the latter and simplifying, the condition simplifies to:

$$1 + \frac{\mu_e N_e}{\mu_i N_i} \gg 1 \tag{8.40}$$

Noting that the ion mobility is several orders of magnitude lower than the electron mobility it is clear that the following condition would hold true when the plasma is close to being neutral (when the electron density approaches the ion density). Then, the electron diffusion coefficient within the ambipolar form will be much less than its counterpart within the standard form. Thus, again we see that the use of the ambipolar form leads to a reduction in the magnitude of the terms within the partial differential equation.

The last term we need to analyze is the one involving ion derivatives. Its magnitude will be small compared to the terms within the standard form if the following holds:

$$\frac{e}{\sigma}\mu_e N_e D_i \frac{\partial N_i}{\partial x} \ll D_e \frac{\partial N_e}{\partial x} \tag{8.41}$$

Consider a plasma close to neutrality. Then, the ion and electron densities can be taken as equal, yielding:

$$\frac{e}{\sigma}\mu_e N_e D_i \ll D_e \tag{8.42}$$

Substitute the conductivity by $e(\mu_e N_e + \mu_i N_i)$ and rearrange noting that we here focus on a quasi-neutral plasma with $N_i \approx N_e$:

$$\mu_e D_i \ll D_e(\mu_i + \mu_e) \tag{8.43}$$

The following condition is in general valid because the ion mobility is always considerably less than the electron mobility and because the ion diffusion coefficient is typically orders of magnitude less than the electron diffusion coefficient. Once more we find that the ambipolar form leads to a significant reduction in magnitude of the diffusion terms, at least in regions where the plasma is close to neutral.

We have thus shown that, in plasma regions that are close to neutral, all terms within the ambipolar form of the electron equation have a significantly smaller magnitude than their corresponding term within the standard form. We would thus expect the ambipolar form to yield a higher resolution because the discretization error associated with a certain stencil (whether a central stencil for the diffusion terms or some upwind stencil for the convective terms) will scale with the magnitude of the term. For instance, consider a standard second-order-accurate central discretization stencil used to discretize a diffusion term. Its numerical error (i.e., the difference between the discrete derivative and the partial derivative) can be found through Taylor series expansion to be proportional to the grid spacing to the second power and also proportional to the diffusion coefficient. Because the diffusion coefficients are significantly less within the ambipolar form of the electron transport equation than within the standard form, it follows that, for equal grid size, we would expect the ambipolar form to introduce much less numerical error than the standard form. Although this reduction in numerical error is expected to occur only in the plasma regions that are close to being neutral, this is not a source of concern because it is only in the quasi-neutral regions that the Ohm-based system was observed to lead to some resolution loss.

Let us now verify how much the ambipolar form can increase the resolution for the same 1D discharge test case investigated previously. For this purpose, we focus on the electron density profiles and compare various systems of equations: one being the standard system with the potential equation determined from Gauss's law; another being a system in which the potential is obtained from Ohm's law and the ion transport equation modified to ensure Gauss's law is satisfied; and the third

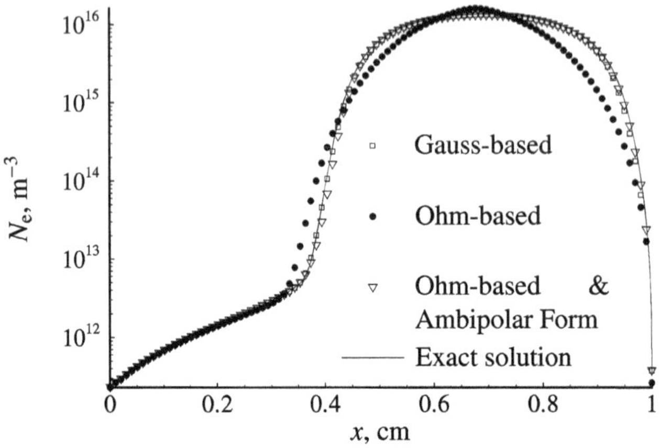

Figure 8.5. Electron density profiles of the 1D discharge obtained with the Gauss-based system, Ohm-based system, and Ohm-based system including the electron transport in ambipolar form.

being the same as the second but with the electron transport equation written in ambipolar form as specified above. For a fair comparison, the same stencils are used to discretize all systems. The results are plotted in figure 8.5. Clearly, rewriting the electron transport in ambipolar form leads to a remarkable improvement in resolution over the Ohm-based system. In fact, when using the ambipolar form, the Ohm-based system now has as high a resolution as the Gauss-based system. What makes this especially satisfying is that such gains in resolution are obtained while not sacrificing the convergence rates. Thus, the Ohm-based system with the electron transport in ambipolar form not only requires 100–1000 times fewer iterations to reach convergence than the standard Gauss-based system, but it does so while exhibiting as high a resolution of the converged solution.

We have here succeeded in recasting the drift-diffusion model differently to obtain a system of equations that overcomes the drift-diffusion stiffness, does not result in a loss of resolution, and is consequently more computationally efficient. For a typical discharge between two electrodes in which a quasi-neutral region forms, the recast of the system here proposed results in large gains in computational efficiency, approaching one thousand times. It is emphasized that such gains in computational efficiency were obtained without taking shortcuts and simplifying the physical model in any way. For a fine enough mesh, the new recast of the drift-diffusion system yields the same solution as the standard system.

8.6 Generalized recast of the drift-diffusion system

In previous sections, we have outlined an approach to overcome the stiffness of the drift-diffusion model. The approach consisted of recasting the system of equations differently by substituting one equation into another such as to avoid terms within the potential equation that lead to error amplification. However, for illustration purposes, such was done for a simplified case where the plasma has only three components, has no temperature gradients and is in the absence of magnetic field.

We now seek to extend this approach to the general case where the plasma has various types of positive ions, negative ions, temperature gradients, and is in the presence of a magnetic field. We can do so following the same steps as outlined in previous sections but starting from the general form of the drift-diffusion model outlined in equation (8.1) with the charged species velocities obtained from equation (8.8).

The first step consists of obtaining a potential equation that is not based on Gauss's law in order to avoid the terms that lead to error amplification. This can be obtained by multiplying the mass conservation equation (8.1) for each charged species by its respective charge, taking the sum over all charged species, and replacing the electric field by the negative of the gradient of the potential. After rearranging terms, we obtain the following *generalized* Ohm's law [13]:

$$\frac{\partial \rho_c}{\partial t} - \sum_{i=1}^{3} \frac{\partial}{\partial x_i}\left(\sigma \frac{\partial \phi}{\partial x_i}\right) + \sum_{k=1}^{n_s} \sum_{i=1}^{3} \frac{\partial}{\partial x_i} q_k N_k \left(\Delta \vec{V}_i^k + \vec{V}_i\right) - \sum_{k=1}^{n_s} \sum_{i=1}^{3} \frac{\partial}{\partial x_i}\left(s_k \mu_k \frac{\partial P_k}{\partial x_i}\right) = 0 \tag{8.44}$$

where σ is the plasma conductivity and is equal to:

$$\sigma \equiv \sum_{k=1}^{n_s} |q_k| N_k \mu_k \tag{8.45}$$

Also, the velocity difference $\Delta \vec{V}$ corresponds to the contribution to the charged species velocity due to the magnetic field and is equal to:

$$\Delta \vec{V}_i^k \equiv \sum_{j=1}^{3} s_k \tilde{\mu}_{ij}^k (\vec{E} + \vec{V} \times \vec{B})_j + \sum_{j=1}^{3} \left(\frac{\delta_{ij} \mu_k - \tilde{\mu}_{ij}^k}{|q_k| N_k}\right) \frac{\partial P_k}{\partial x_j} - s_k \mu_k \vec{E}_i \tag{8.46}$$

Noting that δ_{ij} is the Kronecker delta which is equal to 1 if i and j are equal and to 0 otherwise, it can be easily shown that $\Delta \vec{V}^k$ becomes zero when the magnetic field is zero and is thus the contribution to the velocity coming from the magnetic field.

The second step consists of adding Gauss's law to each positive ion mass conservation equation to ensure that Gauss's law is satisfied. We thus obtain the following modified positive ion conservation equation:

$$\begin{aligned} &\frac{\partial \rho_k}{\partial t} + \sum_{i=1}^{3} \frac{\partial}{\partial x_i}\left(\Delta \vec{V}_i^k + \vec{V}_i\right)\rho_k + \sum_{i=1}^{3} \vec{E}_i \frac{\partial}{\partial x_i} \mu_k \rho_k - \sum_{i=1}^{3} \frac{\partial}{\partial x_i}\left(\frac{\mu_k k_B T_k}{|q_k|} \frac{\partial \rho_k}{\partial x_i}\right) \\ &= m_k W_k - \mu_k \rho_k \left(\frac{\rho_c}{\epsilon_0}\right) + \sum_{i=1}^{3} \frac{\partial}{\partial x_i}\left(\frac{\mu_k k_B \rho_k}{|q_k|} \frac{\partial T_k}{\partial x_i}\right) \end{aligned} \tag{8.47}$$

where ρ_k is the partial density of the kth species.

The third step consists of rewriting in ambipolar form the mass conservation equations of the electrons and of the negative ions. This can be done following similar steps as outlined above in section 8.5.3. That is, we find the ambipolar electric field $\vec{E}'$ applicable to a multicomponent plasma in the presence of a magnetic field by setting the current density to zero and isolating the electric field. Then, we add and subtract $\vec{E}'$ to the electric field vector $\vec{E}$ within the charged species transport equations. After some manipulations (see reference [14] for the derivation details), the following is obtained:

$$\begin{aligned}&\sum_{r=1}^{n_s}\alpha_{kr}\frac{\partial\rho_r}{\partial t}+\sum_{i=1}^{3}\sum_{r=1}^{n_s}\frac{\partial}{\partial x_i}\alpha_{kr}(\Delta\vec{V}_i^{\,r}+\vec{V}_i)\rho_r-\sum_{i=1}^{3}\vec{J}_i\frac{\partial}{\partial x_i}\left(\frac{\mu_k\rho_k}{\sigma}\right)\\&-\sum_{i=1}^{3}\sum_{r=1}^{n_s}\frac{\partial}{\partial x_i}\left(\frac{\mu_r k_B T_r\alpha_{kr}}{|q_r|}\frac{\partial\rho_r}{\partial x_i}\right)=m_k W_k+\sum_{r=1}^{n_s}\sum_{i=1}^{3}\frac{\partial}{\partial x_i}\left(\frac{\mu_r k_B\rho_r\alpha_{kr}}{|q_r|}\frac{\partial T_r}{\partial x_i}\right)\end{aligned} \tag{8.48}$$

In the latter α is the non-dimensional ambipolar tensor which is defined as follows:

$$\alpha_{kr}\equiv\frac{m_k}{m_r}\left(\frac{\delta_{rk}\sigma+\beta_k^- q_r\mu_k N_k}{\sigma}\right) \tag{8.49}$$

where β_k^- is equal to 1 if the kth species is either composed of negative ions or of electrons, and to 0 otherwise.

Therefore, the generalized recast of the drift-diffusion system consists of obtaining the electric field potential from the generalized Ohm's law shown in equation (8.44), the positively-charged species densities from the modified mass conservation equation for the positive ions shown in equation (8.47) and the negatively-charged species densities from the ambipolar form shown in equation (8.48). It is emphasized that such a recast system does not add/remove any physics to/from the standard drift-diffusion system. Indeed, the recast system was obtained from the same transport equations as the standard system without dropping or adding terms in the process. Thus, the recast system solves for exactly the same physical model as the standard drift-diffusion model which is composed of the charged species mass conservation equations outlined in equation (8.1) and the Gauss-based electric field potential outlined in equation (8.25). This has been verified by simulating a multicomponent plasma with several types of positive ions, negative ions, and with a strong applied magnetic field. It was verified that, as the mesh is refined, the solution of the recast system converges towards the solution of the standard system. Further, on the finest meshes considered, no discernible difference could be observed with respect to electric field, species densities, or current density (see results in [14]). This shouldn't be surprising because both systems solve for the same physical model and should therefore yield the same solution for a fine enough mesh.

Despite solving the same physical model, the recast system and the standard system have very different convergence characteristics and computational efficiency. Because the recast system does not obtain the potential from Gauss's law, it is free of the error-amplifying terms that severely limit the time step size. Therefore, at least when using block-implicit schemes, the recast system overcomes the stiffness associated with the drift-diffusion model. This has been verified for a wide range of problems, including some with various types of positive ions, negative ions, imposed magnetic fields, etc. Further, because the electron transport equation within the recast system is written in ambipolar form, it leads to a high resolution of the converged solution, as high or higher than that of the standard system. Therefore, not only does the recast system lead to gains in computational efficiency by converging in much fewer iterations, but it also leads to gains in computational efficiency by giving a converged solution that has less numerical error. Various test cases indicate that for problems involving a large quasi-neutral region juxtaposed to sheaths, the use of the recast system shown here leads to gains in computational efficiency often exceeding one thousandfold.

References

[1] Parent B, Rajendran P T, Moses R W, Johnston C O, Cheatwood F, Macheret S O and Little J 2022 Effect of plasma sheaths on earth re-entry MHD processes *AIAA SCITECH 2022 Forum* p 0980

[2] Parent B, Shneider M N and Macheret S O 2013 Sheath Governing equations in computational weakly-ionized plasmadynamics *J. Comput. Phys.* **232** 234–51

[3] Sinnott G, Golden D E and Varney R N 1968 Positive-ion mobilities in dry air *Phys. Rev.* **170** 272–5

[4] Gosho Y and Harada A 1983 A new technique for measuring negative ion mobilities at atmospheric pressure *J. Phys.* D **16** 1159–66

[5] Chen F F 1984 *Introduction to Plasma Physics and Controlled Fusion Volume 1: Plasma Physics* (New York: Plenum)

[6] Parent B, Macheret S O and Shneider M N 2014 Electron and ion transport equations in computational weakly-ionized plasmadynamics *J. Comput. Phys.* **259** 51–69

[7] Lambert J D 1992 *Numerical Methods for Ordinary Differential Systems* (New York: Wiley)

[8] MacCormack R 2011 Implicit methods for fluid dynamics *Comput. Fluids* **41** 72–81

[9] Bardina J E and Lombard C K 1987 Three dimensional hypersonic flow simulations with the CSCM implicit upwind Navier-Stokes method *Proc. of the 8th AIAA Computational Fluid Dynamics Conf.* AIAA Paper 87-1114

[10] MacCormack R W 2001 Iterative modified approximate factorization *Comput. Fluids* **30** 917–25

[11] Kossyi A, Kostinsky A Y, Matveyev A A and Silakov V P 1992 Kinetic scheme of the non-equilibrium discharge in nitrogen-oxygen mixtures *Plasma Sources Sci. Technol.* **1** 207–20

[12] Parent B, Macheret S O and Shneider M N 2011 Ambipolar diffusion and drift in computational weakly-ionized plasmadynamics *J. Comput. Phys.* **230** 8010–27

[13] Parent B, Shneider M N and Macheret S O 2011 Generalized Ohm's law and potential equation in computational weakly-ionized plasmadynamics *J. Comput. Phys.* **230** 1439–53

[14] Parent B, Macheret S O and Shneider M N 2015 Modeling weakly-ionized plasmas in magnetic field: a new computationally-efficient approach *J. Comput. Phys.* **300** 779–99

IOP Publishing

Plasma Modeling (Second Edition)

Methods and applications

Gianpiero Colonna and Antonio D'Angola

Chapter 9

Self-consistent kinetics

Gianpiero Colonna, Mario Capitelli and Lucia Daniela Pietanza

An important aspect of modeling weakly ionized gases is to describe the chemical transformations occurring in the system. Thermal and non-thermal plasmas are commonly used in material processing [1], etching [2], deposition [3], waste treatment [4–6], combustion [7–10], pollutant reduction [11], medicine [12], agriculture [13], negative ion production [14], and so on [15], applications where the chemistry plays a fundamental role [16].

In non-thermal plasmas [17], the understanding of chemical kinetics is not a trivial issue, because of non-equilibrium conditions: commonly, the gas temperature is low, usually $T_g < 1000$ K, while the electron temperature, T_e, is above 10000 K. The population of internal states is determined, in a first approximation, by the balance between excitation induced by free electron collisions, quenching by heavy-particle interaction and radiative decay [18]. Another contribution comes from chemical reactions which can have a preferential path through excited states [10, 19, 20]. On the other hand, the electron energy distribution is strongly influenced by the population of excited levels, in particular vibrational states and long-living metastables, not decaying by radiative emission [21–23].

This scenario puts in evidence the complex interaction among the different plasma components, in particular the interplay between heavy-particle internal levels and free electrons [24, 25]. To describe these features, the *self-consistent* approach must be used, determining, at the same time, the chemical composition, the internal level and the free electron energy distributions [26], the latter calculated by solving the Boltzmann equation, as described in chapters 2, 3, and 13. The main difficulty of this approach is the size of the chemical problem. For diatomic molecules, as N_2, the number of vibrational levels in the ground electronic state is of the order of one hundred, while for triatomic molecules, as CO_2, this number reaches the order of tens of thousands [27].

Level kinetics should also be coupled with the radiative transport equation [28] (see also chapter 16). Part of the emitted photons are reabsorbed, exciting atoms and molecules, or inducing stimulated emission and the speed of these processes depends linearly on the density of photons. Usually, collisional-radiative models avoid the solution of the radiative transport equation by introducing the *escape factor* [29–31], the fraction of emitted photons reabsorbed locally. This approach cannot account for some interesting phenomena observed in non-homogeneous plasmas, such as the modification of the line shape[1], a consequence of the reabsorption of photons far from the emission zone, introducing *non-local effects* that can be described only by solving the radiative transport equation [33–35]. This aspect is relevant not only to optical emission spectroscopy, but also to vehicles entering planetary atmospheres, where part of the global heat flux to the vehicle surface is due to radiation, especially in regions where the conductive and convective fluxes are small [36, 37].

The *line-by-line* solution of the photon transport equation is a high-demanding computational problem due to the large number of wavelength sampling points to be considered. To reduce the computational load to calculate radiative properties, a *multi-group* approach can be used [38], considering a limited number of spectral intervals where emissivity and absorbance are estimated.

The complexity of the StS model is limited only by the capabilities of the computers to manage a large number of levels. It could be of interest to consider the vibrational ladder of electronically excited states. This is the case of modeling the emission spectra of N_2 [39] in recombining plasma or splitting global reactions in elementary processes [40] such as E–V, one vibrational excitation mechanism in H_2 that is the combination of

$$e + H_2(v) \rightarrow H_2^*(v') + e \text{ and } H_2^*(v') \rightarrow H_2(w) + \text{photon}.$$

This chapter introduces the state-to-state approach, giving at the same time the fundamentals of the self-consistent coupling and showing its application in modeling discharges and high-enthalpy flows.

This chapter has been enriched with the application of the self-consistent model to CO_2 plasma, a field that, in the last years, attracted the interest of the research community to investigate CO_2 activation (dissociation) in non-equilibrium plasmas [41], both by developing advanced simulation modeling [42–50] and by performing accurate experimental investigations [51–59].

The interest in CO_2 plasmas has a long history starting from the 1960s with different applications, such as CO_2 lasers [60, 61], astrophysical observations [62, 63], surface treatment processes [64, 65], or the design of the spacecraft shield for planetary (i.e. Mars) atmosphere entry [66–68]. The idea to use non-equilibrium plasma to improve CO_2 conversion energy efficiency started in the 1970s, with the works of Russian [16, 69] and Italian [70, 71] research groups. In these first studies, it was already clear that the presence of energetic electrons in plasma can activate the CO_2 gas by electron impact excitation (vibrational and electronic), increasing CO_2 reactive channels.

[1] As an extreme condition of line shape modification there is the double peak line [32], due to emission from hot plasma and absorption in a cold region.

9.1 The state-to-state approach

The state-to-state (StS) kinetics was born together with computer modeling of cold plasmas [72–76], when it was understood that some behaviors could be explained only by considering non-Boltzmann internal level distributions. State-to-state approach was also applied in modeling the high enthalpy flows [77–80]. The peculiarity of the StS approach is the inclusion in the kinetic model of those equations describing the evolution of internal levels, considered as independent chemical species, under the action of internal transitions[2] and chemical reactions. To extract global information on an internal degree of freedom from its level distribution χ_v, it is sufficient to calculate the weighted mean of a given state-specific quantity Q_v as

$$<Q> = \sum_{v=0}^{v_M} Q_v \chi_v \tag{9.1}$$

where $v = 0$ indicates the ground state, with the normalization condition

$$\sum_{v=0}^{v_M} \chi_v = 1 \tag{9.2}$$

where v_M is the last level of the considered internal degree of freedom. As an example, equation (9.1) can be used to calculate the internal energy by putting $Q_v = \varepsilon_v$, the energy of the vth level.

For non-Boltzmann distributions, the internal temperature loses its meaning, even if it remains one of the measured quantities in plasma experiments. In non-equilibrium conditions, different definitions of temperature are possible. The most commonly used is the so called *first-level temperature* $T_{1,0}$, calculated as the temperature of the Boltzmann function connecting the ground and the first excited states

$$T_{1,0} = \frac{\varepsilon_1 - \varepsilon_0}{\ln\left(\dfrac{\chi_0/g_0}{\chi_1/g_1}\right)} \tag{9.3}$$

This definition is based on the fact that low energy levels usually follow a Boltzmann distribution. Equation (9.3) can be generalized for any couple of levels by defining $T_{v,\omega}$ as the temperature of the Boltzmann distribution connecting vth and ωth levels [17]. An improvement of the previous definition can be obtained by modeling a group of levels with a Boltzmann distribution and considering the internal temperature as a fitting parameter. The definitions of temperature given above are adequate to compare numerical calculations with values measured by spectroscopy. In some experiments temperature is obtained from the measurement

[2] In the framework of StS kinetics, internal transitions are equalized to chemical reactions.

of thermodynamic quantities[3], in this case a suitable value can be estimated by solving the following algebraic non-linear equation

$$E_{\text{int}} = \sum_{v=0}^{v_M} \varepsilon_v \chi_v = \sum_{v=0}^{v_M} \varepsilon_v g_v e^{-\frac{\varepsilon_v - \varepsilon_0}{T_{\text{int}}}} \Bigg/ \sum_{v=0}^{v_M} g_v e^{-\frac{\varepsilon_v - \varepsilon_0}{T_{\text{int}}}} \tag{9.4}$$

where E_{int} is the mean energy of the non-equilibrium distribution and T_{int} is the temperature corresponding to a Boltzmann distribution having the same mean energy as the actual distribution. This definition gives a different value for the internal temperature than equation (9.3). If the energy of excited states is very high, as in noble gases or atomic hydrogen, the temperature calculated by equation (9.4) is weakly sensitive to the upper level population, because the contribution of these states to the mean energy is very small except at very high temperature. On the other hand, equation (9.3) can give negative temperatures in the case of population inversion, and therefore without the possibility of relating internal energy, always positive, to the corresponding temperature.

Another useful quantity connected to data measured in experiments is the global rate of a given processes, calculated extending equation (9.1) to a generic reaction

$$R_1(v_1) + \cdots + R_n(v_n) \xrightarrow{K_{v_1,\ldots,v_n}^{\omega_1,\ldots,\omega_m}} P_1(\omega_1) + \cdots + P_m(\omega_m) \tag{9.5}$$

It should be remarked that the above reaction in fact represents a whole family of processes obtained varying the internal levels $v_1, \ldots, v_n$ and $\omega_1, \ldots, \omega_m$ of the reactants and products. All these elementary processes contribute to the same global chemical reaction

$$R_1 + \cdots + R_n \xrightarrow{K_p} P_1 + \cdots + P_m \tag{9.6}$$

whose global rate is obtained weighting over the reactant distributions and summing over the states of the products

$$K_p = \sum_{v_1} \cdots \sum_{v_n} \chi_{R_1,v_1} \cdots \chi_{R_n,v_n} \sum_{\omega_1} \cdots \sum_{\omega_m} K_{v_1,\ldots,v_n}^{\omega_1,\ldots,\omega_m} \tag{9.7}$$

The same expression can be adapted to calculate the rate of internal energy variation of the Sth chemical species involved in the set of elementary processes described by equation (9.5)

$$K_{p,S}^{\text{en}} = z_S \sum_{v_1} \cdots \sum_{v_n} \chi_{R_1,v_1} \cdots \chi_{R_n,v_n} \sum_{\omega_1} \cdots \sum_{\omega_m} K_{v_1,\ldots,v_n}^{\omega_1,\ldots,\omega_m} \varepsilon_{S,v_S} \tag{9.8}$$

where v_S is the index for internal state and $z_s = -1$ for reactants and $z_s = 1$ for products.

[3] Shadowgraphy, based on the variation of density, can be considered a thermodynamic measurement.

9.2 Collisional-radiative model

The first example of StS kinetics is the *collisional-radiative* (CR) model, first developed by Bates in the 1960s [81, 82] to describe the spectra of atomic hydrogen and helium, in conditions where the Saha equation [83] fails. The simplest version of CR model considers a plasma containing an atomic species $\mathcal{A}$, its parent ion $\mathcal{A}^+$ and electrons e^-. Only electron impact collisional processes are considered, in particular: excitation/de-excitation

$$e^- + \mathcal{A}(i) \underset{K_{j,i}}{\overset{K_{i,j}}{\rightleftharpoons}} e^- + \mathcal{A}(j) \tag{9.9}$$

and ionization/three-body recombination

$$e^- + \mathcal{A}(i) \underset{K_i^{\mathrm{rec}}}{\overset{K_i^{\mathrm{ion}}}{\rightleftharpoons}} e^- + \mathcal{A}^+ + e^- \tag{9.10}$$

The collisional processes must fulfill the detailed balance principle, therefore the stationary solution is the Boltzmann distribution. However, by including radiative decay

$$\mathcal{A}(i) \xrightarrow{A'_{i,j}} \mathcal{A}(j) + h\nu \tag{9.11}$$

a source of non-equilibrium is introduced, since it is not balanced by the reverse process of radiation reabsorption, resulting in an irreversible energy loss outside the boundaries of the system. The usual approximation to model the reabsorption of radiation is to introduce an *escape factor* $\vartheta_{i,j}$ for each transition, with $0 \leqslant \theta_{ij} \leqslant 1$

$$A'_{i,j} = A_{i,j}\vartheta_{i,j} \tag{9.12}$$

which reduces the spontaneous emission Einstein coefficients $A_{i,j}$ by the fraction of photons reabsorbed very close to the emission position. For each transition, the $\vartheta_{i,j}$ coefficient is a function of the emission line shape, the source geometry, the optical depth and the space distribution of emitting atoms [84–87]. Another important process to be included in the CR model is the radiative-recombination

$$\mathcal{A}^+ + e^- \xrightarrow{\beta_i} \mathcal{A}(i) + h\nu \tag{9.13}$$

where β_i is corrected by its own escape factor, as in equation (9.12). Radiative recombination is responsible for the continuum spectrum, together with the bremsstrahlung radiation. In the traditional approach to CR model, two assumptions are considered

1. The electron energy distribution is Maxwellian;
2. The level distribution relaxes faster than the electron density, therefore the quasi-steady state (QSS) distribution can be considered, and the time evolution is calculated only for ground state and electron (and ion) densities.

As a consequence of these hypotheses, the electron impact rate coefficients are a function of T_e, while the electron density can be considered constant for the level kinetics.

In a two-level system, i.e. ground and first excited level, undergoing only electron-impact induced and radiative processes equations (9.9) and (9.11), the kinetic equations for the atomic level population is given in matrix form

$$\frac{d}{dt}\begin{pmatrix} n_1 \\ n_2 \end{pmatrix} = \begin{vmatrix} -K_{1,2}n_e & K_{2,1}n_e + A'_{2,1} \\ K_{1,2}n_e & -K_{2,1}n_e - A'_{2,1} \end{vmatrix} \begin{pmatrix} n_1 \\ n_2 \end{pmatrix} \tag{9.14}$$

where n_1, n_2, n_e are particle densities of ground state (1), excited level (2) and electrons (e), respectively, while $K_{1,2}$ and $K_{2,1}$ are the rate coefficients for the $1 \to 2$ and $2 \to 1$ electron impact induced transitions, function of T_e. Assuming constant n_e and T_e, it becomes a linear problem that can be solved by calculating the eigenvalues (λ) and eigenvectors ($\vec{v}$) of the matrix. With simple math, we obtain the following sets:

$$\begin{aligned} \lambda_1 &= 0 \\ \vec{v}_1 &= \frac{1}{K_{1,2}n_e + K_{2,1}n_e + A'_{2,1}} \begin{pmatrix} K_{2,1}n_e + A'_{2,1} \\ K_{1,2}n_e \end{pmatrix} \end{aligned} \tag{9.15}$$

$$\begin{aligned} \lambda_2 &= -(K_{1,2}n_e + K_{2,1}n_e + A'_{2,1}) \\ \vec{v}_2 &= \begin{pmatrix} 1 \\ -1 \end{pmatrix} \end{aligned} \tag{9.16}$$

which give the general solution of the differential equation

$$\begin{pmatrix} n_1 \\ n_2 \end{pmatrix} = C_1\vec{v}_1 e^{\lambda_1 t} + C_2\vec{v}_2 e^{\lambda_2 t} = C_1\vec{v}_1 + C_2\vec{v}_2 e^{-(K_{1,2}n_e + K_{2,1}n_e + A'_{2,1})t} \tag{9.17}$$

where C_1 and C_2 are constants that should be determined from the initial composition. From this simple problem some general properties of the solution can be deduced, extendable to multilevel atoms:

1. There is always the null eigenvalue and its corresponding eigenvector is the stationary solution. This property ensures the existence of a stationary value.
2. All the other eigenvalues are negative. As a consequence, the system will always evolve to the stationary solution.
3. The sum of the elements of the eigenvectors corresponding to negative eigenvalue is null. This means that the sum of levels is constant along the whole evolution.

These properties have also consequences on the numerical solution of the general problem. However, for an atom with n_{lev} levels, there are $n_{\text{lev}} - 1$ negative eigenvalues, usually different by orders of magnitude. Therefore, the numerical solution should be done using an algorithm adequate for problems with large stiffness.

In the case of the two-level system, we can understand if the stationary solution is an equilibrium Boltzmann distribution or not. This can be done by calculating the ratio between the two elements of the first eigenvector, i.e.

$$\frac{n_1}{n_2} = \frac{K_{2,1}n_e + A'_{2,1}}{K_{1,2}n_e} = \frac{K_{2,1}}{K_{1,2}} + \frac{A'_{2,1}}{K_{1,2}n_e} \tag{9.18}$$

The first fraction on the right-hand side is the ratio between the rates of direct and reverse electron-induced processes. Due to the detailed balance principle

$$\frac{K_{2,1}}{K_{1,2}} = \frac{g_1}{g_2}e^{(\varepsilon_2-\varepsilon_1)/T_e} = \frac{n_1^B(T_e)}{n_2^B(T_e)} \tag{9.19}$$

where ε_i is the energy of the ith level, g_i the corresponding statistical weight and n_i^B the Boltzmann distribution. The second fraction on the right-hand side of equation (9.18) represents the deviation from the equilibrium distribution due to the radiation emission, that reduces the population of the excited state, i.e. $n_2 < n_2^B$. If the plasma is thick ($A'_{2,1} \approx 0$) or if the electron density is high enough ($n_e \gg A'_{2,1}/K_{1,2}$) the radiative loss is negligible with respect to the up-pumping of the excited level and therefore the equilibrium condition is reached.

In the previous example, we have neglected ionization/recombination (9.10) processes, as well as radiative recombination (equation (9.13)). In the QSS approximation, we do not need the equation for the ground state, but only that for the excited state, i.e.

$$\frac{dn_2}{dt} = -(K_{2,1}n_e + A'_{2,1} + K_2^{\text{ion}}n_e)n_2 + K_{1,2}n_1n_e + (K_2^{\text{rec}}n_e + \beta_2)n_en_{A^+} \tag{9.20}$$

where, by considering n_1, n_e and n_{A^+} constant, the following general solution is obtained

$$n_2(t) = Ce^{-(K_{2,1}n_e+A'_{2,1}+K_2^{\text{ion}}n_e)t} + \frac{K_{1,2}n_1n_e + (K_2^{\text{rec}}n_e + \beta_2)n_en_{A^+}}{K_{2,1}n_e + A'_{2,1} + K_2^{\text{ion}}n_e} \tag{9.21}$$

with C being a constant determined by the initial condition. The stationary solution is given by

$$n_2 = \frac{K_{1,2}n_1n_e + (K_2^{\text{rec}}n_e + \beta_2)n_en_{A^+}}{K_{2,1}n_e + A'_{2,1} + K_2^{\text{ion}}n_e} \tag{9.22}$$

Let us analyze the solution in the case where n_e is much lower than equilibrium value, so that the recombination terms can be neglected. We obtain

$$\frac{n_1}{n_2} = \frac{K_{2,1}n_e + A'_{2,1} + K_2^{\text{ion}}n_e}{K_{1,2}n_e} = \frac{n_1^B}{n_2^B} + \frac{A'_{2,1}}{K_{1,2}n_e} + \frac{K_2^{\text{ion}}}{K_{1,2}} \tag{9.23}$$

showing that the excited state population is further reduced with respect to equation (9.18), by a term proportional to the ionization rate from the excited state.

When the electron density is high enough to neglect the radiative terms, considering the detailed balance for the ionization equilibrium

$$K_i^{\text{rec}} = \frac{K_i^{\text{ion}}}{K'_{\text{eq}}} \frac{n_2^B}{n_1^B} \tag{9.24}$$

where K'_{eq} is the equilibrium constant for the state-to-state model[4], from equation (9.22) we have that the population of the excited state is given by

$$\frac{n_2}{n_1} = \frac{n_2^B}{n_1^B}\left[1 + \frac{K_2^{\text{ion}} n_2^B}{n_1^B K_{1,2} + K_2^{\text{ion}} n_2^B}\left(\frac{n_e n_{A^+}}{n_1 K'_{\text{eq}}} - 1\right)\right] \tag{9.25}$$

According to this equation, the internal distribution shows a deviation from the Boltzmann one, whose sign depends on the departure from equilibrium of the ionization. If the electron density is higher than the equilibrium value ($n_e n_{A^+} > n_1 K'_{\text{eq}}$) the excited state is overpopulated with respect to the Boltzmann distribution, on the other hand, if electron density is lower than equilibrium value ($n_e n_{A^+} < n_1 K'_{\text{eq}}$) the excited state is underpopulated. These two conditions are defined as ionization and recombination regimes, respectively. Equation (9.25) can be easily extended by adding the radiative terms

$$\frac{n_2}{n_1} = \frac{n_2^B}{n_1^B}\left[1 + \frac{K_2^{\text{ion}} n_2^B\left(\frac{n_e n_{A^+}}{n_1 K'_{\text{eq}}} - 1\right) + \beta_2 n_{A^+} - \frac{A'_{2,1} n_2^B}{n_e n_1^B}}{n_1^B K_{1,2} + K_2^{\text{ion}} n_2^B + \frac{A'_{2,1} n_2^B}{n_e n_1^B}}\right] \tag{9.26}$$

Therefore, the radiative recombination gives a positive contribution, while that of spontaneous emission is negative, confirming the result obtained in equation (9.23).

In a multilevel system, the kinetic equations are more complex, because all the transitions between the levels give a contribution. However, the general behavior discussed in the two-level system is still valid, because the main contribution usually comes from the direct interaction with the ground state and the ionization. Therefore, this picture can be considered as a first-order approximation of the CR problem.

It should be pointed out that the deviation from the Boltzmann distribution is not the same for all the levels, depending on the state-specific rates or cross-sections. Fixing T_e, n_e and n_0, we have a linear set of equations [88].

As an example, let us consider the relaxation of the level distribution of hydrogen atoms in ionization and recombination regimes displayed in figure 9.1. In an ionization regime, the distribution tails are depleted with respect to a Boltzmann distribution, while a plateau appears in the recombination regime.

9.3 Vibrational kinetics

State-to-state approach has been used also to investigate the evolution of the vibrational distribution of diatomic molecules. For homonuclear molecules, such

[4] This is not the equilibrium constant obtained from statistical thermodynamics for the macroscopic kinetics. This aspect will be discussed in the next section.

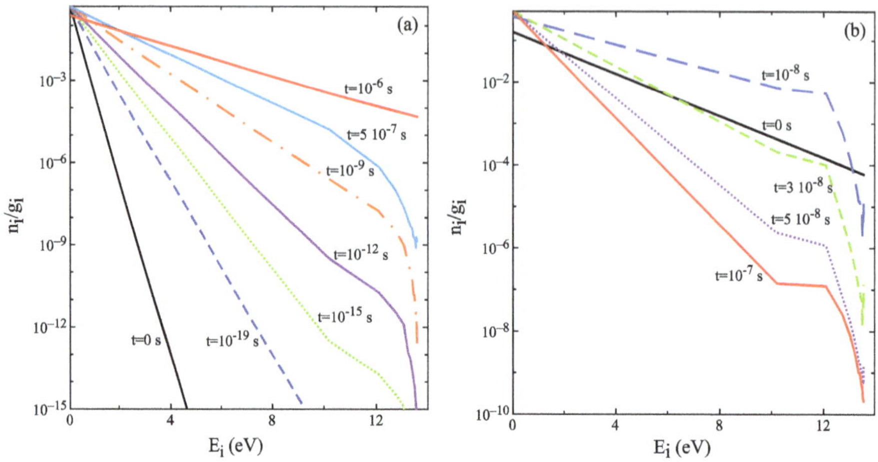

Figure 9.1. Time evolution of H level distribution in ionization (a) and recombination (b) regimes. Conditions are reported in table 9.1.

Table 9.1. Conditions for the calculations reported in figure 9.1.

	case (a)	case (b)
χ_e	10^{-3}	10^{-3}
T_e	20000 K	1000 K
$T_H(0)$	1000 K	20000 K

as N_2, O_2 and H_2, vibrational kinetics allows to determine theoretically the vibrational distribution of the ground electronic state, experimentally observed only by sophisticate techniques [89], with the radiative vibrational transitions optically-forbidden. State-to-state models were used to predict the behavior of high enthalpy flows in hypersonic wind tunnels [90, 91], shock tubes [92–94] or during the entry of space vehicles in planetary atmospheres [95–100] (see also chapter 10). In these conditions, the relaxation of vibrational levels to equilibrium passes through states characterized by non-Boltzmann distributions. If N_2/N mixture is considered, all internal transitions

$$N_2(v) + N_2(\omega) \rightleftharpoons N_2(v') + N_2(\omega') \tag{9.27}$$

are allowed and any quadruplet (v, ω, v', ω') possible. However, many of these processes have a very small probability and only monoquantum transitions are usually considered[5]:

[5] In *VTm* and *DRm* (equation (9.31)) processes, the rate coefficient is usually considered independent of ω and therefore, the state of the second molecule is omitted.

$$N_2(v) + N_2(\omega - 1) \rightleftharpoons N_2(v - 1) + N_2(\omega) \quad VV \tag{9.28}$$

$$N_2(v) + N_2(\omega) \rightleftharpoons N_2(v - 1) + N_2(\omega) \quad VTm \tag{9.29}$$

VV rate coefficients are written as $K_{v,\,v-1}^{\omega-1,\omega}(T)$, being a function of the temperature, with $v < \omega$, while for $v > \omega$ the detailed balance principle is applied to obtain the rate $K_{\omega-1,\omega}^{v,v-1}(T)$. The case with $v = \omega$, known as resonant transfer, has no global effect, with final and initial states indistinguishable. In the presence of atoms, the vibrational deactivation/excitation in atom–diatom collisions is also considered

$$N_2(v) + N \rightleftharpoons N_2(v - \Delta v) + N \quad VTa \tag{9.30}$$

Differently from *VTm*, *VTa* process includes also multi-quantum transitions. The kinetic scheme is completed with the dissociation/recombinations processes[5]

$$\begin{array}{ll} N_2(v) + N_2(\omega) \rightleftharpoons 2N + N_2(\omega) & DRm \\ N_2(v) + N \rightleftharpoons 2N + N & DRa \end{array} \tag{9.31}$$

It should be remarked that, varying v, ω over the vibrational ladder, the above reactions represent a few thousand elementary processes!

There are some aspects that should be considered in developing an StS kinetic model.

- The vibrational levels follow an anharmonic law, with the energy quanta decreasing as the levels approach the dissociation limit.
- The construction of a reliable, accurate and consistent database of rate coefficients and cross-sections for the huge number of reactive channels usually included in the StS vibrational kinetics is, even nowadays, a challenging problem. Semiclassical approaches are often used, with quantum calculations performed in a few cases for an accuracy check. In some cases, heuristic analytical expressions are used, with limited validity[6].
- Another problem is the matching of data coming from different sources. The rate coefficients are reported as a function of the vibrational quantum number. Researchers often neglect that the energy of the level, corresponding to a given vibrational quantum number, is not the same for the different sources, using the data without any adaption. Scaling relations should be used to adapt the different energy level sets. The most common procedure is to interpolate the data from one set to the other.

A sufficient condition to have a Boltzmann at any time of the evolution is:

1. the vibrational ladder is given by the truncated harmonic oscillator;
2. only monoquantum transitions are included;
3. VT rates grows linearly with the vibrational quantum number, i.e. $K_{v\to v-1}^{\mathrm{VT}} = vK_{1\to 0}^{\mathrm{VT}}(T)$;

These conditions are not fulfilled by real molecules.

[6] A discussion about dynamical data can be found in chapter 23, for processes induced by heavy-particle collisions, and in chapter 22, for electron–molecule collision cross-sections.

VV transitions, negligible for harmonic oscillators, are an important source of non-equilibrium for the anharmonic ladder. This process conserves the number of quanta but not the total energy. For a cooling gas, in the presence of *VV* only, the final solution is a distribution with the same number of quanta as the initial one, but with the minimum total vibrational energy. This behavior can be explained easily by considering the detailed balance principle for *VV* in equation (9.28),

$$K_{v,\,v-1}^{\omega-1,\omega} = K_{\omega,\,\omega-1}^{v-1,v} e^{-\frac{\varepsilon_\omega+\varepsilon_{v-1}-\varepsilon_{\omega-1}-\varepsilon_v}{T}} = K_{\omega,\,\omega-1}^{v-1,v} e^{-\frac{\Delta_\omega-\Delta_v}{T}} \tag{9.32}$$

where $\Delta_v = \varepsilon_v - \varepsilon_{v-1}$. For a harmonic oscillator, $\Delta_v = \Delta_\omega$ with the consequence that direct and reverse rates are always equal, independently on the temperature, giving the Boltzmann distribution as stationary solution. On the other hand, for real molecules, being Δ_v a decreasing function of v, when $\Delta_v > \Delta_\omega$ the process $v \to v - 1$, $\omega - 1 \to \omega$ is preferred with respect to its reverse, resulting in a parabolic-like distribution, first proposed by Treanor [101]. When both *VV* and *VT* are considered, during the time evolution, the distribution is characterized by an *S* shape, generated by the interplay between the two processes, each one dominating in a different region of the vibrational ladder.

In figure 9.2, the cooling of the N_2 vibrational distribution under the action of *VV* (a) and *VV* + *VTm* (b) is shown. From the inspection of the figure, inclusion of *VTm* processes drives the final distribution toward the Boltzmann one at the gas temperature, in a time scale of seconds.

Dissociation and recombination processes are another cause of non-equilibrium, these processes being favored from highly excited states, which are underpopulated in the dissociation regime and overpopulated in the recombination one [102].

As an example, let us estimate the behavior of the tail of the vibrational distribution in recombination regime. It should be noted that the gas in equilibrium

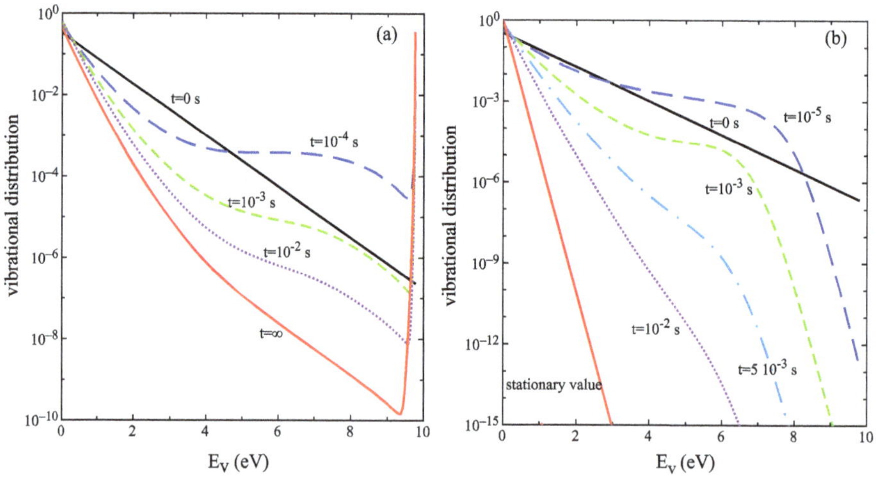

Figure 9.2. Time evolution of N_2 vibrational distribution, with initial vibrational temperature $T_v(0) = 8000$ K, for $P = 1$ Atm and $T_{gas} = 1000$ K calculated by considering only VV (a) and by including also *VTm* (b) processes.

at temperature T_0, instantaneously changed to $T < T_0$. For each vibrational level, the following evolution equation can be written:

$$\frac{dN_v}{dt} = G_v - L_v \tag{9.33}$$

where N_v is the concentration of molecules in the v state and G and L are, respectively, total gains and losses. As a general assumption, the VV processes (equation (9.28)) can be neglected [102]. Moreover, we will consider the evolution of the vibrational distribution at fixed dissociation degree, supposing that the distribution tail relaxes faster than the dissociation degree. The system is cooling and the pumping from lower states can be neglected. The pumping from higher states can be neglected as well, because the level population is usually lower. As a consequence, the *gain* term is approximately due only to recombination

$$G_v = N_{\mathrm{N}}^2(K_v^{Rm}N_{\mathrm{N}_2} + K_v^{Ra}N_{\mathrm{N}}) = K_v^r N_{\mathrm{tot}} N_{\mathrm{N}}^2 \tag{9.34}$$

where N_S is the global concentration of the Sth species. Losses come from all the processes, but VV, starting from the v level

$$L_v = N_v\left[\left(K_v^{Dm} + \sum_{\omega \neq v} K_{v,\omega}^{VTm}\right)N_{\mathrm{N}_2} + \left(K_v^{Da} + \sum_{\omega \neq v} K_{v,\omega}^{VTa}\right)N_{\mathrm{N}}\right] = K_v^d N_{\mathrm{tot}} N_v \tag{9.35}$$

K_v^r and K_v^d are global constant that depends on the temperature and on the composition, not on the distribution, and N_{tot} is the total gas density. Substituting in equation (9.33) the vibrational distribution and the dissociation degree

$$\chi_v = \frac{N_v}{N_{\mathrm{N}_2}} \quad ; \quad \alpha_d = \frac{N_{\mathrm{N}}}{N_{\mathrm{N}} + N_{\mathrm{N}_2}} = \frac{N_{\mathrm{N}}}{N_{\mathrm{tot}}} \quad ; \quad \frac{N_{\mathrm{N}_2}}{N_{\mathrm{tot}}} = 1 - \alpha_d \tag{9.36}$$

we obtain

$$\frac{d\chi_v}{dt} = N_{\mathrm{tot}}^2 \frac{\alpha_d^2}{1 - \alpha_d} K_v^r - N_{\mathrm{tot}} K_v^d \chi_v = a - b\chi_v \tag{9.37}$$

with a, b constants. The solution of this equation is

$$\chi_v(t) = \left(\chi_v^0 - \frac{a}{b}\right)e^{-bt} + \frac{a}{b} = (\chi_v^0 - \chi_v^\infty)e^{-bt} + \chi_v^\infty \tag{9.38}$$

where χ_v^0 is the initial distribution and χ_v^∞ is the QSS solution

$$\chi_v^\infty = \frac{a}{b} = N_{\mathrm{tot}} \frac{\alpha_d^2}{1 - \alpha_d} \frac{K_v^r}{K_v^d} \tag{9.39}$$

It should be noted that K_v^r and K_v^d are linear functions of α_d and therefore the QSS value for a given level is a rational function of the type $\chi_v^\infty \propto P_3(\alpha_d)/P_2(\alpha_d)$ where P_n is a generic polynomial of degree n. Differently from the collisional-radiative model, the QSS value of equation (9.39) could not be considered as a global solution because the approximations considered are not valid for all the levels, but only for those with high υ. In particular, the equation is very accurate for the last vibrational level υ_M. Global dissociation rates, as well as those of endothermic processes, are favored from highly excited states, and therefore they are not related to the vibrational temperature, but to the population of the last vibrational level. From these considerations, a simplified model, named *two-level distribution* (TLD) has been applied to pure nitrogen [103] and air [104]. Alternately, the *multi-internal temperature* model was proposed [105, 106], which considers the vibrational distribution as a given number of Boltzmann distributions, with their own temperature, each one describing the population in a given energy interval. These models can be regarded as the evolution of the traditional multi-temperature approach [107, 108], which considers a single temperature for each internal degree of freedom, and the chemical rate coefficients as an analytical function of the gas and relevant internal temperatures.

One difficulty of the StS model is the lack of state-specific rates, especially in molecule–molecule interaction. To overcome this problem, some heuristic approaches are used, such as SSH theory [109] or forced harmonic oscillator (FHO) [110] (see chapter 23). An interesting example is the dissociation (and the relative recombination) by molecule–molecule collisions, whose state-selective rates are unknown. Commonly, these rates are estimated by the so called *ladder-climbing* approximation, considering the dissociation as a virtual level ($\upsilon_d = \upsilon_M + 1$) instantaneously dissociating

$$N_2(\upsilon) + N_2(\omega) \xrightarrow{slow} N_2(\upsilon') + N_2(\upsilon_d) \xrightarrow{fast} N_2(\upsilon') + 2N \tag{9.40}$$

The rate is estimated extending the dependence of the vibrational excitations to the virtual level. This approach underestimates the dissociation rates by orders of magnitude [111]. Another approach assumes the same dependence on T and υ of the dissociation induced by atomic collisions, multiplied by a factor depending on the temperature to reproduce the experimental thermal rate [103].

9.4 The self-consistent approach

The previous sections offer only a partial view of the possible applications of the StS models. To characterize realistic systems, vibrational kinetics for molecules and level kinetics for atoms must be included in the same chemical model [112–114], together with electronically excited states of molecular species [21, 26], single and multiple ionized atoms and molecules [115], polyatomic species [41, 46–48, 116–121], interacting through relevant processes. Among them, electron induced processes play an important role, not only during discharges, but also in post-discharge and in high-enthalpy flows. Electron collisions with atoms and molecules in excited states can strongly affect the kinetics, introducing different mechanisms, such as

- ladder-climbing mechanism for dissociation and ionization, with successive excitation from one level to the next up to the highest level which can be easily dissociated or ionized [116, 117, 122];

- the excited level gives its energy to the impinging electron that acquires enough energy to induce high-threshold chemical processes [22, 123, 124];

This last class of processes (called superelastic or second kind collisions [22, 23]) produces non-equilibrium electron energy distributions, with bumps and plateaux. These examples illustrate how complex can be the link between electrons and excited states in affecting the chemistry of a plasma and its relaxation during the post-discharge [120]. These mechanisms are not accounted for when the E/N approach [125–128] is used. In this case the electron-impact rate coefficients are pre-calculated for different values of the reduced electric field from the electron energy distribution obtained by solving the stationary Boltzmann equation [129, 130] and neglecting the contribution of excited states [131–133]. According to which quantity is used as an independent variable for the electron-gas properties, this approach[7] is known as *local field approximation* (LFA, when considering E/N) or as *local energy approximation* (LEA, when considering electron mean energy or temperature). The ground state model is justified under the assumption that in the presence of strong electric fields, electron energy distribution reaches a steady state in a very short time and the contribution of superelastic collisions is negligible. This approximation is sometimes extended also to post-discharge conditions[8], though it is well-assessed that superelastic collisions significantly warm up the distribution of electrons when the electric field is switched off [88].

To properly account for the synergy between electron-impact and excited states, the self-consistent approach must be used. The general scheme is sketched in figure 9.3. What is usually considered as self-consistent kinetics is represented by the red-arrow path. The electron energy distribution function (EEDF) is calculated solving the Boltzmann equation (see chapter 2), which needs the gas composition and the internal distributions. From EEDF, electron impact rate coefficients are calculated, input data for the system of master equations, whose solution updates composition and internal distributions for the next cycle. This procedure should be followed at any time step, and, depending on the solution algorithm, also more than one time.

Many authors have investigated plasma kinetics in different systems using SC approach [134–141], however, in spite of the relevance of excited states in the kinetics, the most used approach remains the ground state model. In recent papers [23, 121] the SC model has been investigated also during the high pressure pulsed discharges [25], whose electric field temporal profile was represented in figure 9.4. Generally, in modeling vibrational kinetics in the presence of a discharge, only electron induced processes starting from the ground state are considered. In this example [25], a pure hydrogen discharge was considered, comparing results obtained with an electron impact cross-section set accounting for all possible vibrational excitation transitions (full model, $v = all$) in the following process

[7] This approach is also known as *cold gas approximation* or *ground state model* because molecular excitation is neglected.

[8] In the post-discharge, with $E/N = 0$ electron induced processes are neglected.

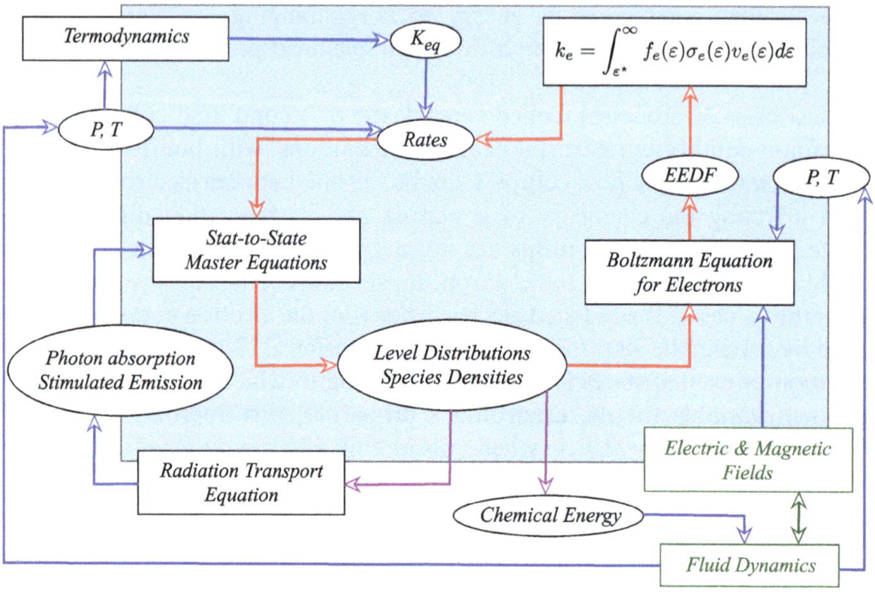

Figure 9.3. Scheme of the self-consistent approach. The self-consistent cycle is in red. Blue lines indicates information coming from external modules. Purple lines information passed to external modules.

$$H_2(v) + e \rightleftharpoons H_2(w > v) + e \tag{9.41}$$

with those obtained only with transitions from $v = 0$. The results present relevant differences in concentrations and distributions, as can be observed in figure 9.5. The full model predicts VDF with much higher tails than in the ground model, and, as a consequence, a higher dissociation degree and lower ionization. When the field is growing, the EEDFs are the same in both cases, while some differences are obtained during the decreasing phase of the electric field.

In ground kinetic models, however, superelastic collisions are neglected only in the electron Boltzmann equation, but are included in the heavy-particle kinetics. Moreover, the stationary EEDF is considered. Comparing directly LFA and SC-StS cannot discriminate if the differences in the results are due to the stationary approximation or to the missing of superelastic collisions. Therefore, a SC-StS code has been adapted mimicking LFA by neglecting superelastic collisions for electron kinetics and including in the level evolution equation, calculating the time dependent EEDF. This approach, has been applied to N_2/H_2 mixture [121] to investigate the formation of ammonia in a multi-pulse discharge with the temporal profile reported in figure 9.4. To put in evidence the contribution of superelastic collisions, an example is the time evolution of the electron density profile and of the particle density of NH and NH_3 in figure 9.6. The *full* model gives higher electron density at the peak of the electric field, because of superelastic collisions, that, as the density of the excited states increase, reach the equilibrium with the corresponding inelastic processes, and, as a consequence, the internal transitions become transparent to electrons, that can transfer more energy to ionization. The plasma

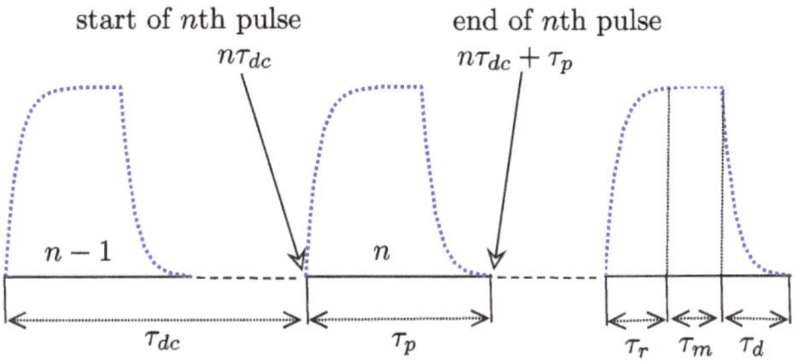

Figure 9.4. Scheme of the multi-pulse shape function and variable definition. The profile is normalized to the maximum field E_0/N. Reproduced with permission from [121], copyright IOP Publishing, all rights reserved.

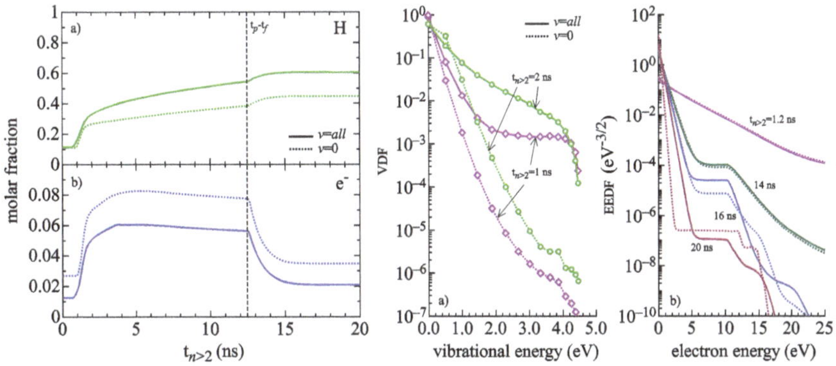

Figure 9.5. Atom and electron molar fractions (left), VDF (center) and EEDF (right): comparison between a full set of cross-sections (v = *all*) and a reduced model, including only transition with the ground state ($v = 0$). Field parameters (see figure 9.4): $\tau_r = \tau_d = 7.5$ ns, $\tau_p = 20$ ns, $\tau_{dc} = 25$ μs, $E_0/N = 200$ Td. Reproduced with permission from [25], copyright IOP Publishing, all rights reserved.

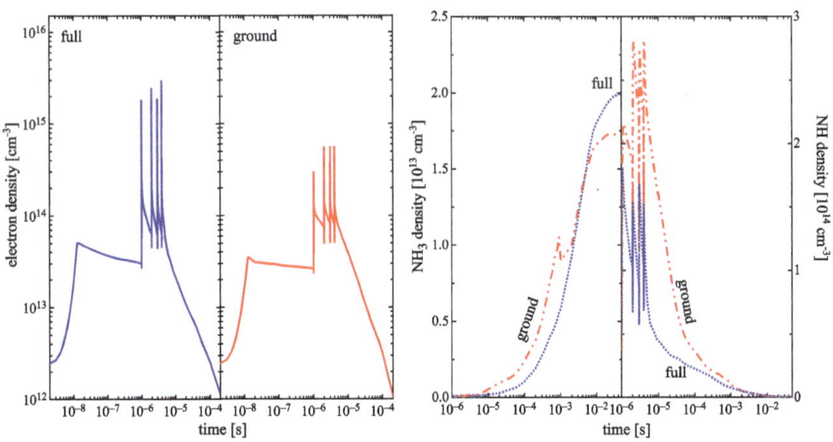

Figure 9.6. Time evolution of electron density (left) and NH and NH_3 (right) using the SC-StS (*full*) and *ground* models. Differently from figure 9.5 only 5 pulses have been considered. Field parameters (see figure 9.4): $\tau_r = \tau_d = 7.5$ ns, $\tau_p = 20$ ns, $\tau_{dc} = 1$ μs, $E_0/N = 100$ Td. Reproduced with permission from reference [121], copyright IOP Publishing, all rights reserved.

activation induce the formation of NH also in the post-discharge condition and the *full* model results, in the long time, in larger density. In contrast, it is the ground model that predicts larger ammonia formation probably because the large electron density increases the dissociation of NH_3 induced by electron collisions.

A step farther in understanding the role of superelastic collisions is to consider a system where the electric field is not imposed from outside, but it depends on the plasma composition. This condition is obtained when the plasma is inserted in a more complex circuit, including resistance and capacitance. A simple configuration can be to consider an external resistance in series with the plasma [23]. The current in the circuit induces a voltage drop in the external resistance, determining the plasma potential and therefore the ionization degree. On the other hand, the current is limited by the plasma conductivity that depends on the electron density N_e and mobility μ_e. Considering the plasma between parallel plate electrodes of surface S, separated by a distance d and connected to a generator, that in the absence of current produces an electric field $E_0 = V_g/d$, with a resistance R, the Ohm law can be written as [23]

$$E_0 d = E_p d + J_p S R \Rightarrow E_0 = E_p + N_e \mu_e E_p \hat{R} \Rightarrow E_p = \frac{E_0}{1 + N_e \mu_e \hat{R}} \tag{9.42}$$

where $J_p = N_e \mu_e E_p$ is the plasma (and circuit) current, E_p is the electric field in the plasma and $\hat{R} = RS/d$. Therefore, if the plasma is powered by a circuit, the applied field depends on the electron density, and as the electron density increases, the plasma field decreases. Including also a capacitance in the circuit, it is possible to describe also dielectric barrier discharges (DBDs) [120, 142].

Another possibility is to consider the power density w_p transferred to the plasma, a quantity that can be measured, giving for the plasma field [23]

$$w_p = E_p J_p = N_e \mu_e E_p^2 \Rightarrow E_p = \sqrt{\frac{w_p}{N_e \mu_e}}. \tag{9.43}$$

In both cases the plasma field decreases as the electron density increases, stabilizing the discharge.

The same approach used to investigate N_2/H_2 plasma has been applied to argon discharges considering a power source given by the equation as a function of time t

$$w_p = w_0 \left(\frac{t}{t_0} \right)^2 e^{-t/t_0} \tag{9.44}$$

with $w_0 = 10$ W cm^{-3} and $t_0 = 1$ ms [23]. The power supplied increases up to a maximum of $w_p \approx 6$ W cm^{-3} at $t \approx 2$ ms. The model includes ground and metastable states of argon, responsible for the superelastic collisions, and single ionized atoms. Also in this case, results obtained with the self-consistent model (*SC – StS*) and that where superelastic collisions are neglected in the Boltzmann equation for free electrons but included in the metastable state kinetics (*NoSup*) have been compared. In figure 9.7 the species concentration and electric field time

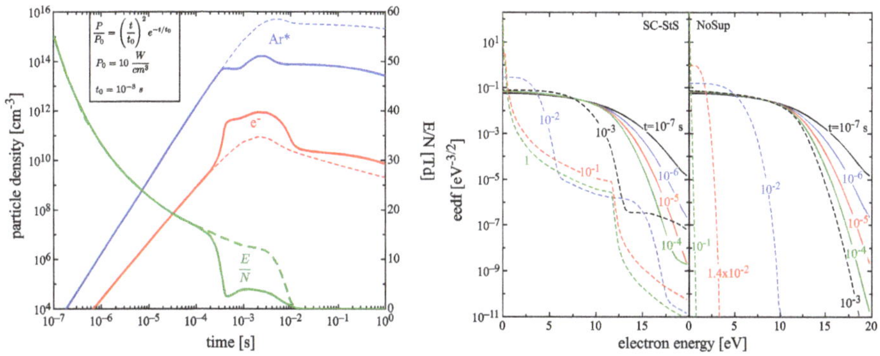

Figure 9.7. Temporal profile of species concentration and electric field (left) and the evolution of the EEDF (right) in argon plasma. Comparison between self-consistent (*SC – STS*) approach and calculation where the electron kinetics neglect the superelastic collisions (*NoSup*). Right viewgraph reproduced with permission from [23], copyright IOP Publishing, all rights reserved.

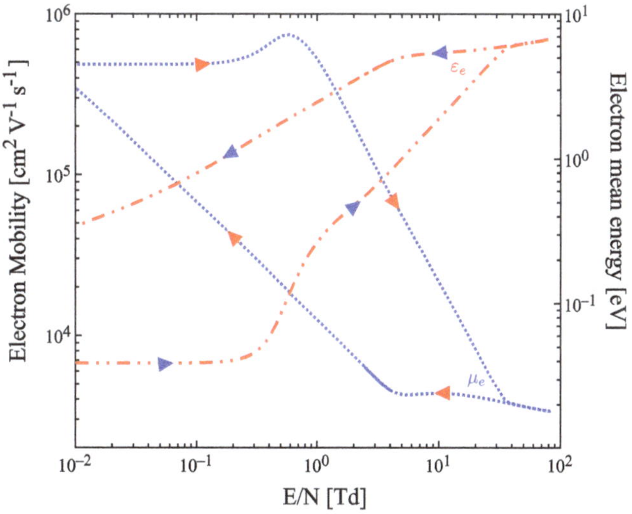

Figure 9.8. Electron mobility μ_e and mean energy ε_e as a function of the electric field. Reproduced with permission from [23], copyright IOP Publishing, all rights reserved.

profiles have been reported, together with the EEDFs. For $t < 0.1$ ms, the two models give the same profile for the field, concentrations and EEDF, but as the time proceeds, the results depart. E/N in the *SC – StS* is much slower, a consequence of the higher ionization degree, and lower metastable density are obtained. At the same time the EEDF starts to show the effects of superelastic collisions, and the distribution tails are quite high also after the power becomes null, at $t = 1$ s, while in the *NoSup* case the distribution thermalizes at 300 K already at $t = 0.1$ s.

The limits of the LFA are evinced from figure 9.8 where swarm parameters are reported as a function of the electric field. Both electron mean energy and mobility show a hysteresis loop, and at fixed electric field these quantity have two values, contradicting the assumption that all these quantities are only a function of E/N.

9.5 High enthalpy ionized flows

A further application of the self-consistent kinetics is the modeling of ionization dynamics in high enthalpy flows [21, 26, 143, 144]. At the same time, the radiation transport equation, which in principle is a Boltzmann equation for photons, is solved to determine the rates of processes involving photons such as stimulated emission, absorption, photo-ionization and photo-dissociation [35]. To determine bremsstrahlung radiation and radiative recombination, the radiation transport equation needs also the electron energy distribution function [30].

Shock tubes have been used since the beginning of space exploration to simulate the extreme conditions space vehicles experience when entering the atmosphere of a planet [145–147]. Some calculations have been performed for hydrogen [93, 94] and a helium–hydrogen mixture [112] in order to simulate Jupiter entry conditions, coupling the StS self-consistent model with the fluid dynamic equations for the shock tube and with the equation of radiation transport. The chemical model [93] includes the vibrational kinetics for molecules, as discussed in section 9.3, the collisional-radiative model, section 9.2, completed by processes involving electronically excited states of molecules, dissociation and ionization induced by electron collisions. Let us analyze the evolution of species and distributions relatively to the shock front at Mach 20. A fundamental role in initiating the ionization is played by atom–atom collisions in the region just after the shock, where the temperature is very high and the electron and ion densities are still very low. Figure 9.9 reports the spatial profiles of the plasma composition and temperatures as a function of the distance from the shock front. Vibrational temperature of molecules and electronic temperature of atoms are calculated from internal distributions using equation (9.3), while the electron temperature is obtained from the mean energy (equation (2.12)) by $T_e = \frac{2}{3}\varepsilon_e$. Just after the shock front, the pressure suddenly increases from about 100 Pa to 0.5

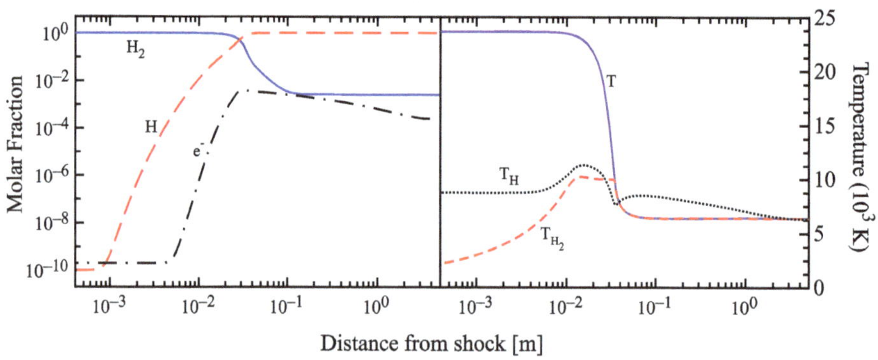

Figure 9.9. Spatial profiles of molar fractions and temperatures as a function of the distance from the shock for free stream velocity Mach = 20. T_{H_2} is the vibrational temperature of molecules, T_{H_2} the electronic temperature of atoms and T is the gas temperature.

Bar, while the gas temperature T reaches about 24 000 K, mainly heating the translational degree of freedom, the internal distribution and the chemical composition remaining those of the free stream condition.

As the distance from the shock front increases, collisions between molecules promote firstly vibrational excitation and after dissociation, reducing the gas temperature, as can be deduced from the evolution of the vibrational temperature of H_2 and from atomic molar fraction. At the same time, also free electron molar fraction increases, activating processes induced by electron collisions. The electron molar fraction presents a maximum, a behavior that can be reproduced only by using a state-to-state approach. The energy stored in the internal degrees of freedom are released slowly and therefore the ionization process can proceed, even if the gas temperature is not sufficient to sustain it. When the inertia due to internal states is exhausted, electrons and ions recombine. Another anomalous behavior can be observed in the internal temperature of atoms, which presents two maxima and a minimum. The first maximum and minimum are produced by internal collisions trying to equilibrate with the translational temperature.

The internal temperature remains a little lower than T because of the radiative losses. However, at about 3 cm from the shock front the gas temperature is still decreasing, while T_H starts increasing and then decreases again to equilibrate with T. This behavior, put in evidence in figure 9.10, can be explained considering that the recombination $H^+ + e^-$ is favored on excited states, increasing internal temperature.

The evolution of the gas temperature and compositions is reflected on the EEDF and internal distributions (figure 9.11). During the relaxation, the EEDF presents a markedly non-Maxwellian shape, due to superelastic collisions with the first excited state of atomic hydrogen, showing a bump ($x = 2$ cm) and a plateau ($x = 10$ cm) at energy of about 10 eV. The level distribution presents the characteristic shape observed in ionization regime up to $x = 2$ cm and recombination regime for larger distances. Comparing this result with that reported in figure 9.1 for the collisional-radiative model, a similar trend can be observed.

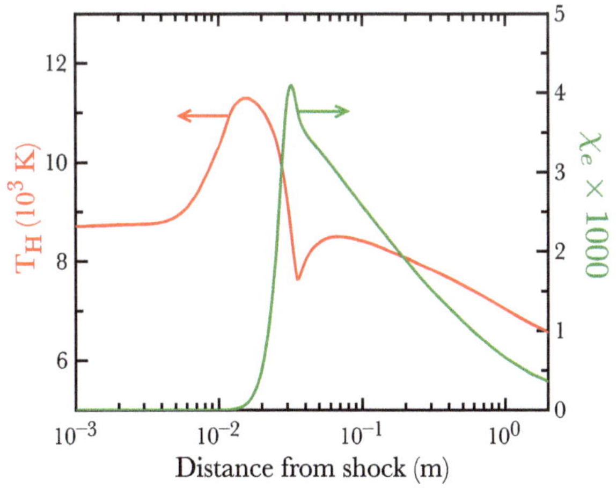

Figure 9.10. Spatial profiles of T_H and electron molar fraction for the free stream speed Mach = 20.

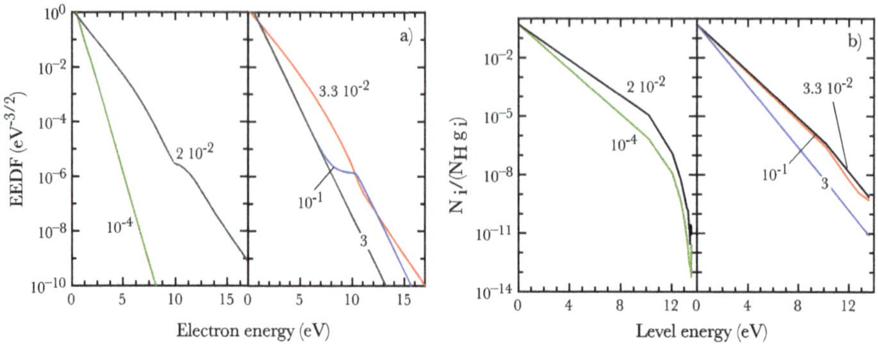

Figure 9.11. (a) EEDF and (b) atomic level distribution $N_i/(g_iN_H)$ at different distances from the shock front.

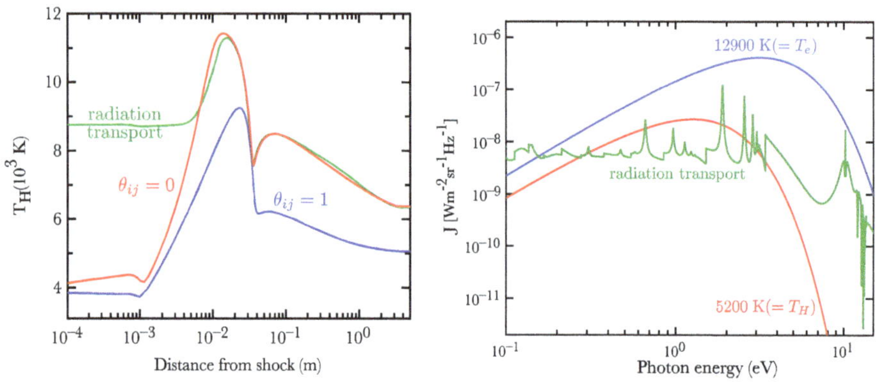

Figure 9.12. Left: Spatial distribution of T_H in the thin ($\theta_{ij} = 1$), thick ($\theta_{ij} = 0$) and radiation-coupled calculations for free stream velocity of Mach 20. Right: Monochromatic average J_ν (solid lines) and the local Planck distribution $B_\nu(T_e)$ (dotted lines) as a function of the photon energy at three different space locations in the post-shock region just behind the shock wave (0.1 nm from the shock).

Focusing again on figures 9.9, 9.10, we should note that the temperature of atomic hydrogen just after the shock front is quite high, about 9000 K, a result that cannot be explained with the collisional-radiative models, but only as non-local effects due to radiation emitted in regions where the atomic hydrogen has a large internal temperature, and absorbed in the region just after the shock. Therefore, the plasma behaves as if it were thick, i.e. the radiation emitted is reabsorbed. In figure 9.12 T_H calculated with the self-consistent approach (including radiation transport) is compared with the thin and thick cases, as described in section 9.2. The temperature in the self-consistent case behaves as in the thick case but for short distances from the shock, where the temperature is much higher than thin and thick cases. In this system, the optical depth is long enough to make inapplicable the escape factor approach, and only when solving the radiation transport equation can the observed behavior be simulated.

These non-local effects can be observed also in the spectra close to the shock front (see figure 9.12 (Right), where absorbing lines are observed for photons above 10 eV

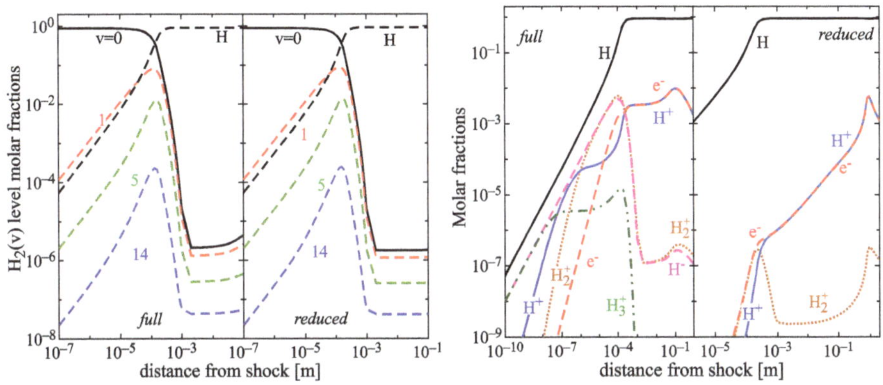

Figure 9.13. Profile of vibrational level (left) and chemical species (right) molar fractions. Comparison between kinetic model including (full) and neglecting (reduced) alternate reaction paths. Data from reference [94].

(Lyman lines). Another important feature observed in the spectra is the departure from the Plank distributions calculated at the local electron and atomic hydrogen temperatures.

If the temperature becomes sufficiently high so that ionization degree reaches relatively high values, new reactions paths must be taken into account to explain the growth of electron density in a shock tube [94]. The crucial point is to determine how the first electrons are produced because they rapidly equilibrate with the gas temperature and then electron-impact ionization rapidly dominates. The common idea is that the ionization happens in the following steps: just after the shock front, molecular collisions increase vibrational energy and when it becomes large enough, also dissociation is activated, speeding up as the atom density grows because $H_2 + H$ dissociation is more efficient than $H_2 + H_2$. Then electrons are produced via ionizing recombination

$$H + H \rightleftharpoons H_2^+ + e^- \tag{9.45}$$

However, these mechanisms are not sufficiently fast being the heating of the vibrational ladder a bottleneck. Alternate paths have been proposed [94] producing electrons faster. The first step is the formation of negative ions through the reaction

$$H_2 + H_2 \rightleftharpoons H_3^+ + H^- \tag{9.46}$$

where H^- releases the electron by detachment due to heavy species collisions

$$H^- + X \rightleftharpoons H + e^- + X \tag{9.47}$$

With this kinetic scheme (marked as *full* in figure 9.13) the role of vibrational kinetics becomes less relevant with respect to the model that neglects H_3^+ and H^- (marked as *reduced* in figure 9.13) as can be observed in figure 9.13. In fact, the evolution of vibrational states and atomic hydrogen[9] proceeds similarly in the two

[9] It is worth noting in figure 9.13 (left) the change of slope in the atom molar fraction, indication that dissociation induced by atom collisions is more effective than molecular induced dissociation.

models, contrarily to species molar fractions. In the *full* model, the electron density already overcomes 10^{-3} at $x = 0.1$ ms, while in the *reduced* case it is just above 10^{-7}. The maximum of electron molar fraction is achieved in the *full* case ≈ 0.1 m from the shock front, almost an order of magnitude before the *reduced* case. The *full* model was capable of reproducing with good accuracy some experimental data [94].

9.6 The self-consistent approach for CO_2 plasmas

In this section the SC-StS model will be applied to CO_2. The aim is to understand how CO_2 can be efficiently dissociate into CO and O and eventually converted into specific value-added chemicals by using plasma discharges in different conditions of pressure and frequencies. Experimental investigations showed that CO_2 dissociation can be promoted by selectively pumping energy into the lowest vibrational levels through electron–molecule collisions, followed by VV processes, populating upper vibrational levels and ending in dissociation. The power input for dissociation by vibrational excitation is that needed to excite by electron impact the first vibrational levels, a quantity lower than the power requested for thermal or direct electron impact dissociation. Therefore, the proposed mechanism considerably increases CO_2 conversion energy efficiency, attracting the interest of the scientific community to the study of CO_2 vibrational excitation in non-equilibrium plasmas. Moreover, due to the key role of the electrons, the importance of implementing the self-consistent approach has been recognized also for the investigation of CO_2 conversion by plasma discharges [41, 116].

In this section, CO_2 vibrational structure and kinetics will be presented, discussing most recent results obtained by an advanced SC-StS model applied to a CO_2 plasma mixture under different discharge and post-discharge conditions [46–48, 116–119].

9.6.1 CO_2 vibrational levels

CO_2 is a triatomic molecule with a linear geometry, with the carbon atom placed in the center, covalently double bonded with the oxygen atoms. It has three normal vibrational modes: the symmetric stretching mode, the doubly degenerate bending mode and the asymmetric stretching mode, characterized by the quantum numbers v_1, v_2, v_3, respectively. The double degenerate bending mode is also described by an additional quantum number l_2, the angular momentum of the quasi rotation of the linear molecule around its principal axis, which assumes the values $l_2 = v_2, v_2 - 2, v_2 - 4, \ldots,$ 1 or 0 depending on whether v_2 is odd or even. Therefore, CO_2 vibrational levels are denoted as $CO_2(v_1, v_2^{l_2}, v_3)$. According to quantum mechanical considerations, the energy of a level, defined by the set of quantum numbers (v_1, v_2, l_2, v_3), is given by the equation [148, 149]

$$E/hc = \sum_{k=1}^{3} \omega_k \left(v_k + \frac{d_k}{2} \right) + \sum_{k=1}^{3} \sum_{j \leqslant k}^{3} x_{jk} \left(v_k + \frac{d_k}{2} \right) \left(v_j + \frac{d_j}{2} \right) + g_{22} l_2^2 \qquad (9.48)$$

where ω_k is the normal mode frequencies, x_{kj} the anharmonicity coefficients, $d_1 = d_3 = 1$, $d_2 = 2$ and the coefficient g_{22} describes the polarization of the double

Table 9.2. Spectroscopic constant in cm^{-1} for calculating CO_2 vibrational level energies taken from [148].

Constant	ω_1	ω_2	ω_3	x_{11}	x_{22}	x_{33}	x_{12}	x_{13}	x_{23}	g_{22}
Value	1351.2	672.2	2396.4	−0.3	−1.3	−12.5	5.7	−21.9	−11.0	1.7

degenerate symmetric deformation vibrations[10]. The corresponding values, deduced from CO_2 experimental spectra, are reported in table 9.2.

Furthermore, since $v_1 \cong 2v_2$, several vibrational levels of the two symmetric modes v_1 and v_2 have nearly the same energy. This accidental degeneration is called Fermi resonance. The two vibrational levels with nearly the same energy repel each other and equation (9.48) cannot predict their correct energy values: one level is shifted up and the other down and the energy separation between them is much larger than expected. In addition to the energy condition, for symmetry reason, the Fermi resonance occurs for the level couples (v_1, $v_2^{l_2}$, v_3) and (v_1-1, v_2+2 ^{l_2}, v_3)[11]. As a consequence of the perturbation, the two wave functions are linearly combined and the Fermi levels cannot be designated separately. For example, the Fermi levels $(0,2^0,0)$ and $(1,0^0,0)$, which represent the Q-branches in CO_2 at 1285 cm^{-1} and 1388 cm^{-1} are indexed by placing both sets of symbols within the same parenthesis with a ranking number, i.e. $(0,2^0,0;\,1,0^0,0)_I$ and $(0,2^0,0;1,0^0,0)_{II}$. To correctly calculate the energy of Fermi levels, a complex procedure based on perturbation theory is necessary (see for example [150, 151]). The energy of some Fermi levels of CO_2 can be found in [152, 153].

9.6.2 CO_2 state-to-state kinetics

To model the StS vibrational kinetics of CO_2 in plasma discharges, beside the intra-mode transitions, as those introduced in section 9.3 for diatomic molecules, and electron induced processes, as discussed in section 9.4, also inter-mode energy transfer must be taken into account. Furthermore, an important role is played by electronically excited states. Therefore, the corresponding StS master equations describing the vibrational kinetics can be written by using the rate balance equations for the population N_v of the vth vibrational level

$$\begin{aligned}\frac{dN_v}{dt} &= \left(\frac{dN_v}{dt}\right)_{\text{eV}} + \left(\frac{dN_v}{dt}\right)_{\text{VT}} + \left(\frac{dN_v}{dt}\right)_{\text{VV}} + \left(\frac{dN_v}{dt}\right)_{\text{VV}'} \\ &+ \left(\frac{dN_v}{dt}\right)_{\text{VE}} + \left(\frac{dN_v}{dt}\right)_{\text{SE}} + \left(\frac{dN_v}{dt}\right)_{\text{React}}\end{aligned}. \tag{9.49}$$

where the terms on the right-hand side represents the contributions of the following processes:

- vibrational excitation by electron impact (eV);

[10] The conversion factor of energy from cm^{-1} to eV is $K_{cm^{-1}\to eV} = 1.239\,842\,447 \cdot 10^{-4}$.

[11] Note that only levels of the same l_2 can be shifted by the Fermi resonance.

- vibrational–translational energy exchange processes (VT) by collisions with molecules (VTm) and atoms (VTa);
- vibration–vibration (VV) energy exchange between the same mode (intra-mode);
- vibration–vibration (VV′) energy exchange among the different modes (inter-mode);
- vibrational-to-electronic energy exchange processes (VE);
- spontaneous emission processes (SE);
- any relevant reactive process by electron impact and/or by heavy-particle collisions, in particular, dissociation/recombination and/or ionization/recombination.

For the population N_i of the ith electronically excited state, the StS master equations is given by

$$\frac{dN_i}{dt} = \left(\frac{dN_i}{dt}\right)_{\mathrm{eE}} + \left(\frac{dN_i}{dt}\right)_{\mathrm{SE}} + \left(\frac{dN_i}{dt}\right)_{\mathrm{Quench}} \tag{9.50}$$

where the terms on the right-hand side corresponds to:

- excitation/de-excitation by electron impact (eE);
- spontaneous emission (SE) ;
- quenching by heavy-particle collisions (Quench).

In Fridman's book [16], it has been shown that the major contribution to the population of highly vibrationally excited states of CO_2 and, hence, to dissociation, is related to the excitation of the asymmetric vibrational modes. As a matter of fact, the asymmetric mode excitation is favored by the following conditions:

1. large eV rates: the asymmetric mode is activated by collisions with electrons at temperatures $T_e = 1 \div 3$ eV;
2. small VT rates: largest amount of energy is accumulated in the asymmetric mode, having smaller VT rates than symmetric ones;
3. large VV rates: VV exchange along the asymmetric mode is several orders of magnitude faster than along symmetric ones.

Based on these considerations, most of the vibrational kinetic models describing CO_2 dissociation in electrical discharges are focused essentially on the description of the asymmetric mode kinetics [42–48, 116–119]. In particular, in [46–48] all the pure asymmetric mode levels of the kind $(0,0,v)$[12] up to the dissociation energy[13] are included in the vibrational kinetics, together with the first bending mode, $v_{b_1} = (0,1,0)$, and three Fermi levels, $v_{F_1} = (0,2,0;\ 1,0,0)$, $v_{F_2} = (0,3,0;\ 1,1,0)$ and $v_{F_3} = (0,4,0;\ 1,2,0;\ 2,0,0)$ and one CO_2 electronic excited state at $\varepsilon = 10.5$ eV, treated as a metastable state.

---

[12] Note that, from now on, the vibrational quantum number of the asymmetric mode will be labeled as v instead of v_3 and only the $l_2 = 0$ quantum number will be considered.

[13] The dissociation energy of CO_2 is $\varepsilon_d = 5.5$ eV, corresponding to 21 levels.

An attempt to include more symmetric and bending levels for the CO_2 system have been done in [49, 50], accounting for the first 70 individual vibrational levels of the kind (v_1, v_2, v_3). This model has been validated by an experimental study applied to the description of a DC glow discharge in low excitation/dissociation conditions. A similar approach has been developed by Bultel [154], including individual levels up to a threshold energy of 0.8 eV, and considering independent modes for higher energy, i.e. (v_1,0,0), (0,v_2,0) and (0,0,v_3).

A more complete vibrational energy scheme, which takes into account possible combinations (v_1, v_2, v_3) up to a threshold energy of 3 eV, has been proposed in [155–159] and applied to the investigation of CO_2 non-equilibrium vibrational kinetics in the hypersonic boundary layer of bodies entering the Mars atmosphere. The application of this full kinetic scheme to the description of CO_2 dissociation by plasma discharges is, however, limited to the lack of electron impact cross-sections data for transitions involving all the vibrational level included in the model.

The StS approach of equations (9.49) and (9.50) needs an accurate estimation of the corresponding reaction rates, both for heavy-particle and electron impact collisions, and this is of fundamental importance for building realistic StS kinetic models. Complete and consistent sets of vibrational-state-resolved electron impact cross-sections for CO_2 and for its dissociation products such as CO, O_2 and O, describing energy transfer among internal states and reactive processes, are needed. Different open-access databases are available, among them, the PHYS4ENTRY DB [160] provides state-resolved cross-sections on vibrational energy levels and the LXCat [161] contains several swarm-derived complete cross-section sets, the most popular are those calculated by Phelps [162, 163]. Recently, a new complete and consistent cross-section set [164, 165], essentially based on the Phelps data, has been derived from a swarm analysis, and completed by Itikawa ionization cross-sections [166] and some electron impact resonant vibrational excitation cross-sections [167–169].

A crucial role is played by the direct electron impact dissociation mechanism of CO_2 [170] (DEM)

$$e + CO_2(v) \rightarrow e + CO + O \tag{9.51}$$

for which several models are available. As an example, for dissociation cross-section, Phelps [163] proposed to consider the data for the excitation with threshold at $\varepsilon^\star = 7$ eV while Cosby [171] suggested the experimental cross-section with $\varepsilon^\star \approx 12.8$ eV. The Phelps cross-section seems to lead to a reasonable estimation of CO_2 dissociation conversion rate [170]. However, very recently, also the cross-section of Polak and Slovetsky [172] has been considered, better describing CO_2 dissociation in the range E/N 45 ÷ 105 Td [173].

The electron impact cross-sections for vibrational excitation (eV), dissociation and ionization for CO_2 are available only for $v = 0$ level and to extend them to higher vibrational levels, approximative scaling laws can be used. For the vibrational excitation of the pure asymmetric levels of the kind (0,0,v), the Fridman scaling law [16]

$$\sigma_{v\omega}(\epsilon) = \exp\left(-\frac{\delta(\omega - v - 1)}{1 + \beta v}\right)\sigma_{01}(\epsilon + \varepsilon_{01} - \varepsilon_{\omega v}) \tag{9.52}$$

is recommended. According to equation (9.52), the cross-section for transitions $(0, 0, v) \rightarrow (0, 0, w)$ can be calculated from the cross-section σ_{01} of the transition $(0, 0, 0) \rightarrow (0, 0, 1)$ with threshold ε_{01} by shifting it by the difference of the energy levels ε_{vw} and by scaling its magnitude assuming $\delta = 0.5$ and $\beta = 0.0$.

On the other hand, for dissociation and ionization cross-sections, a simple threshold shifting by the ε_v vibrational energy is generally used, according to

$$\sigma_v(\varepsilon) = \sigma_0(\varepsilon + \varepsilon_0 - \varepsilon_v) \tag{9.53}$$

where σ_0 and ε_0 are the ground state cross-section and its threshold energy, respectively.

For the calculation of VV, VT and VV′ rate coefficients, several theoretical approaches can be used. As a first attempt, the first-order perturbation theory developed by Schwartz, Slawsky and Herzfeld (SSH) [109] for diatomic molecules was extended also to CO_2 and approximate scaling laws were obtained, calculating higher vibrational levels rate coefficients starting from available ones, experimental or theoretical, involving the lowest states [153, 174]. This approach was used in [42–50, 155–159]. Very recently, the FHO method [175–177] was extended to CO_2 providing a sets of vibrational-state resolved rate coefficients, including multi-quanta jumps [178]. Finally, some VV and VT transition rates of CO_2 were calculated by using the quasi-classical-trajectory (QCT) method with a semi-empirical PES [179, 180]. These calculations, however, are quite computationally expensive, preventing this method to be extensively used for all possible transitions.

The relevant heavy-particle chemical processes are in general described by Arrhenius rate coefficients [42–50, 155–159].

In particular, for CO_2 dissociation induced by heavy particle, two different process are taken into account:

$$CO_2(0, 0, v) + M \rightarrow CO + O + M \tag{9.54}$$

$$CO_2(0, 0, v) + O \rightarrow CO + O_2 \tag{9.55}$$

In general, the corresponding rate coefficients for higher vibrational levels, i.e. $k_D(v)$, are calculated from the ground state Arrhenius rate coefficients (k_0) by using the Fridman-Macheret α-model [16]

$$k_D(v, T_{gas}) = k_0 \exp\left(-\frac{E_a - \alpha\varepsilon_v}{KT_{gas}}\right) \tag{9.56}$$

where E_a is the activation energy and α a coefficient in the range $0 \leqslant \alpha \leqslant 1$.

According to this model, the vibrational energy lowers the activation energy of the chemical reactions, greatly enhancing the corresponding rate coefficient. The α parameter entering equation (9.56) determines the efficiency of lowering the reaction barrier by vibrational excitation and, in general, it is close to 1 for strongly endothermic reactions, such CO_2 dissociation, and close to zero for exothermic ones.

9.6.3 Results

The first results obtained by implementing a self-consistent model for CO_2 plasma are shown in [46] and have been obtained in low CO_2 dissociation conditions, in which the prevalent specie is CO_2 with low concentration ($\chi < 10^{-4}$) of its dissociation products, such as CO, O_2, O. In this way, only the effect of CO_2 vibrational and electronic excited state kinetics in the overall scheme and in the EEDF can be selectively taken into account. These conditions, in typical MW discharge test cases characterized by moderate values of pressure (P = 20 torr) and of E/N electric field values (E/N = 50 Td), can be obtained by appropriately shortening the electric discharge pulse duration from typical ms to μs values (i.e. $t_{\text{pulse}} \simeq 40\ \mu$s). Figure 9.14(a) shows the CO_2 vibrational distribution function (VDF) of the asymmetric mode levels (0,0,v) calculated at the end of the discharge pulse, at different gas temperatures (in the range 400–3000 K) and α values ($\alpha = 1$ and $\alpha = 0.8$) (see equation (9.56)) for the rate coefficient characterizing the heavy-particle dissociation process of equation (9.54). According to equation (9.56), the closer to 1 the α value, the stronger the dissociation from higher vibrational levels. Figure 9.14(b) shows the corresponding rate coefficients as a function of T_{gas} for heavy-particle dissociation process of equation (9.54), labeled as PVM (i.e. pure vibrational mechanism), and for the direct electron impact process of equation (9.51), labeled as DEM, which are calculated by

$$PVM = \sum_{v=0}^{v_{\max}} k_d(v) f(v) \tag{9.57}$$

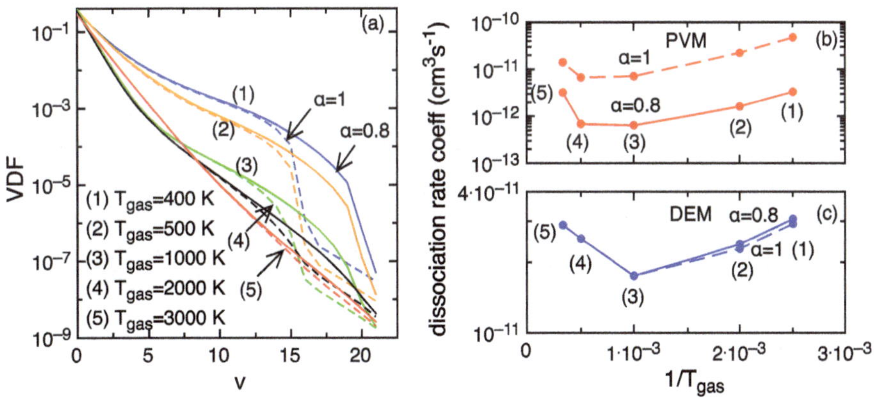

Figure 9.14. (a) VDF calculated at the end of the pulse ($t_{\text{pulse}} = 40\ \mu$s) at different T_{gas} in the range 400–3000 K, at two different α values; (b) PVM and (c) DEM dissociation rate coefficients as a function of $1/T_{\text{gas}}$ with $\alpha = 0.8$ (full lines) and $\alpha = 1$ (dashed lines) for a pure CO_2 plasma ($P = 20$ torr, E/N = 50 Td, $\chi_e(t = 0) = 10^{-6}$). Reproduced with permission from reference [46], copyright IOP Publishing, all rights reserved.

$$DEM = \sum_{v=0}^{v_{max}} k_{de}(v)f(v) \tag{9.58}$$

Where $k_d(v)$ and $k_{de}(v)$ are the corresponding vibrational-state resolved rate coefficients and $f(v)$ the CO_2 VDF.

By looking to figure 9.14(a), it is clear that the CO_2 VDF is characterized by an evident plateau due to VV processes, depleted at higher v by dissociation and VT deactivation processes. The VV plateau is shorter for $\alpha = 1$ as a consequence of a stronger dissociation from higher vibrational levels, as predicted by equation (9.56), and tends to disappear with the increase of T_{gas} due to the increase of VT deactivation thermalization process. The corresponding total PVM and DEM rates show a very singular behavior as a function of T_{gas}. In the low gas temperature range, both the rates increase as T_{gas} decreases. This effect corresponds to a non-Arrhenius behavior of the dissociation rates as a function of the gas temperature. The reason for this unexpected behavior is linked to the characteristic non-equilibrium trend of the CO_2 VDF: as the gas temperature decreases, VV plateaux are more pronounced due to a less efficient VT deactivation, increasing the CO_2 dissociation rates. At higher gas temperature, i.e. for T_{gas}>2000 K, the CO_2 VDF is thermal and the dissociation rate behavior follows the Arrhenius trend, i.e. the rate increases with the increase of T_{gas}. Figure 9.15 shows the corresponding time dependent EEDF in (a) discharge and (b) post-discharge conditions, characterized by a non-equilibrium behavior.

During the discharge, the EEDF is heated due to the effect of the electric field and of superelastic vibrational and electronic collisions of the kind, respectively

$$e(\epsilon) + CO_2(v_i) \rightarrow e(\epsilon + \epsilon_{v_i} - \epsilon_{v_j}) + CO_2(v_j) \tag{9.59}$$

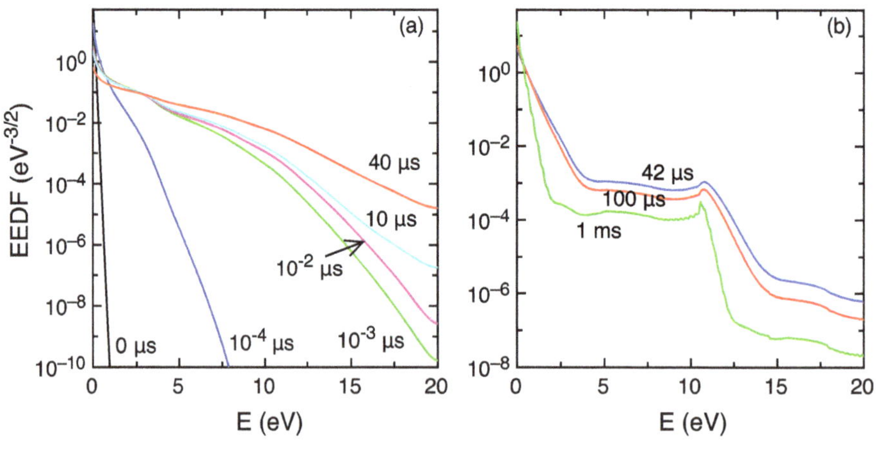

Figure 9.15. Time dependent EEDF as a function of the electron energy (a) during the discharge and (b) in the post-discharge for a pure CO_2 plasma with T_{gas} = 400 K, P = 20 torr, t_{pulse} = 40 μs, E/N = 50 Td, $\chi_e(t = 0) = 10^{-6}$. Reproduced with permission from reference [46], copyright IOP Publishing, all rights reserved.

$$e(\epsilon) + CO_2(10.5\ eV) \rightarrow e(\epsilon + 10.5\ eV) + CO_2 \tag{9.60}$$

Both previous processes create electrons at higher energies, near the threshold energies and their multiples, globally heating the EEDF. The EEDF is strongly affected by the electronic excited states included in the kinetics especially in the post-discharge. A well-defined peak at 10.5 eV emerges due to the dominant effect of the superelastic electronic collisions involving the CO_2 10.5 eV state (equation (9.60)). The peak, proportional to the concentration of the 10.5 eV electronic state, is spread over $0 < \epsilon < 10.5$ eV through inelastic, elastic and electron–electron collisions, forming a long plateau and decreasing its height during the post-discharge.

To investigate higher dissociation conditions, the SC-StS CO_2 kinetic model was successively improved with the inclusion of accurate StS kinetics for the main CO_2 dissociation products, i.e. CO, O_2, O and C (see [47, 48]), giving details about their vibrational and electronic excited state kinetics. Figure 9.16 shows, respectively, the calculated time dependent CO_2, CO and O_2 VDFs in discharge (a, c, e) and post-discharge (b, d, f) conditions in the case study: $T_{gas} = 300$ K, $P = 20$ Torr, $P_d = 80$ W cm^{-3}, $t_{pulse} = 50$ ms and $t_{pd} = 100$ ms [47].

As can be seen, transient and stationary VDFs are characterized by a non-equilibrium trend due to the interplay of eV, VV, VT and dissociation processes. In particular, during the discharge, eV processes heat up the VDFs, pumping energy predominantly to lower vibrational levels. This is more evident for CO_2 and CO, while for O_2, the lower energy part of the VDF remain essentially a Boltzmann at T_{gas} due to lower eV rates. VV processes redistribute these energies especially in the intermediate vibrational energy range, creating long non-equilibrium plateaux for CO_2 and CO. These plateaux enhance the vibrational excitation dissociation channel both for CO_2 and CO. In the high energy part these plateaux are strongly depleted by dissociation and VT deactivation processes. For the O_2 case, instead, the observed plateau is due to the three-body recombination of O atoms

$$O + O + O \rightarrow O_2(v) + O \tag{9.61}$$

which overpopulate preferentially the higher vibrational levels.

Note that for CO, a peak at $v = 27$ emerges (see figure 9.16 (c) and (d)) due to an important quenching process introduced for describing the kinetics of the first CO electronic excited state, i.e. $CO(a^3\Pi)$ at 6 eV

$$CO(a^3\Pi, \omega = 0) + CO \rightarrow CO(v = 27) + CO \tag{9.62}$$

This process was introduced by Porshnev [181] for matching calculated CO VDFs with measured ones in optically pumped experiments.

In the post-discharge, VT deactivation processes cool down the VDFs, pushing the low energy part of the distribution towards a Boltzmann distribution at the gas temperature ($T_{gas} = 300$ K). However, non-equilibrium trends are still present in the

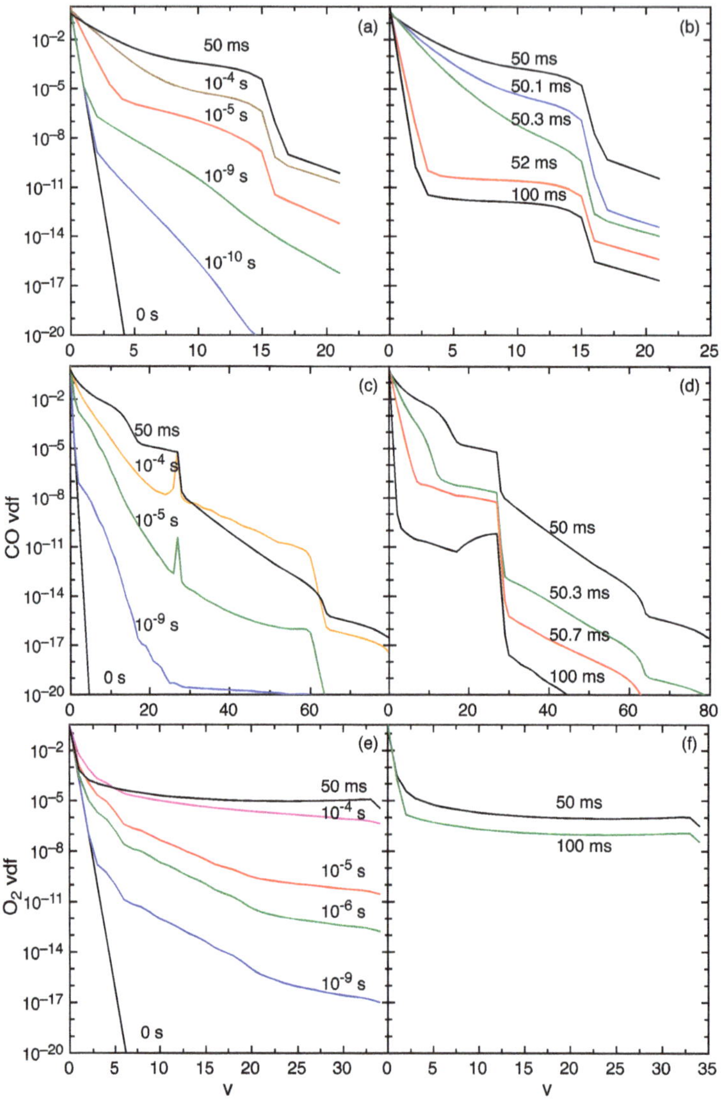

Figure 9.16. CO_2, CO and O_2 VDFs in discharge (a, c, e) and post-discharge (b, d, f) conditions for a MW discharge with $T_{gas} = 300$ K, $P = 20$ Torr, $t_{pulse} = 50$ ms and $P_d = 80$ W cm^{-3} in a CO_2–CO–O_2–O–C plasma. Data from [47].

high-energy part of the distributions. Figure 9.17 shows the corresponding EEDF in (a) discharge and (b) post-discharge conditions. The behavior is similar to the case of the pure CO_2 gas (figure 9.15), but this time, in the post-discharge, the EEDF is characterized by a more structured peak shape due to the presence of the electronic excited states of CO, O_2 and O and C [47]. In the post-discharge, the EEDF shape depends also on the presence of optically thin and thick plasma conditions for the electronic excited state kinetics. These two conditions correspond to the case in

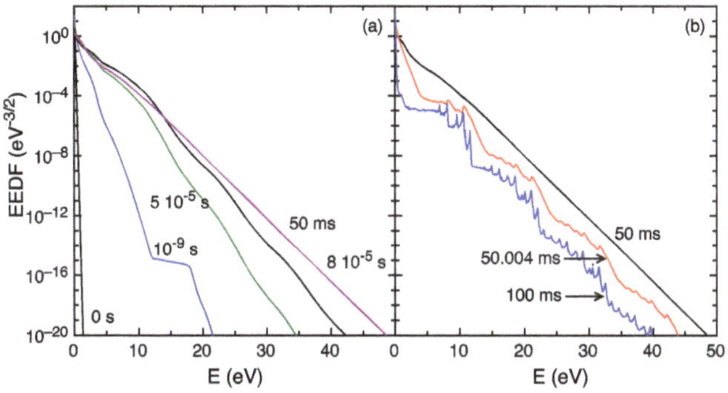

Figure 9.17. Time dependent EEDF as a function of the electron energy in (a) discharge and (b) post-discharge conditions for a MW discharge with T_{gas} = 300 K, P = 20 Torr, t_{pulse} = 50 ms and P_d = 80 W cm^{-3} in a CO_2–CO–O_2–O–C plasma. Data from [47].

which all the radiation emitted by the electronic excited states is assumed totally trapped or emitted by the plasma, respectively. As a consequence, in the thin case, the EEDF loses all the peaks corresponding to the emitting electronic excited states as shown in [47].

References

[1] Lieberman M and Lichtenberg A J 2005 *Principles of Plasma Discharges and Materials Processing* (New York: Wiley) https://doi.org/10.1002/0471724254

[2] Manos D and Flamm D (ed) 1989 *Plasma Etching: An Introduction* (New York: Academic)

[3] Bruno G, Capezzuto P and Madan A (ed) 1995 *Plasma Deposition of Amorphous Silicon-Based Materials* (New York: Academic) https://doi.org/10.1016/B978-0-12-137940-7.X5000-1

[4] Murphy A B and McAllister T 2001 Modeling of the physics and chemistry of thermal plasma waste destruction *Phys. Plasmas* **8** 2565–71

[5] Heberlein J and Murphy A B 2008 Thermal plasma waste treatment *J. Phys. D: Appl. Phys* **41** 053001

[6] Ying C D 2001 Cold plasma and its application in treatment of waste gas *Environ. Protect. Chem. Ind.* **21** 136–9

[7] Starikovskaia S M 2006 Plasma assisted ignition and combustion *J. Phys. D: Appl. Phys.* **39** R265

[8] Starikovskiy A and Aleksandrov N 2013 Plasma-assisted ignition and combustion *Prog. Energy Combust. Sci.* **39** 61–110

[9] Adamovich I V, Lempert W R, Rich J W, Utkin Y G and Nishihara M 2008 Repetitively pulsed nonequilibrium plasmas for magnetohydrodynamic flow control and plasma-assisted combustion *J. Propuls. Power* **24** 1198–215

[10] Starikovskaia S, Lacoste D A and Colonna G 2021 Non-equilibrium plasma for ignition and combustion enhancement *Eur. Phys. J.* D **75** 1–25

[11] Kim H-H 2004 Nonthermal plasma processing for air-pollution control: a historical review, current issues, and future prospects *Plasma Process. Polym.* **1** 91–110

[12] Fridman A A and Friedman G G 2013 *Plasma Medicine* (Chichester: Wiley)

[13] Šimek M and Homola T 2021 Plasma-assisted agriculture: history, presence, and prospects–a review *Eur. Phys. J.* D **75** 1–31

[14] Taccogna F *et al* 2021 Latest experimental and theoretical advances in the production of negative ions in caesium-free plasmas *Eur. Phys. J.* D **75** 1–34

[15] Colonna G, Capitelli M and Laricchiuta A 2021 Cold plasmas: fundamentals and applications *Eur. Phys. J.* D **75** 1–3

[16] Fridman A 2008 *Plasma Chemistry* (Cambridge: Cambridge University Press) https://doi.org/10.1017/CBO9780511546075

[17] Colonna G *et al* 2021 Theoretical and experimental aspects of non-equilibrium plasmas in different regimes: fundamentals and selected applications *Eur. Phys. J.* D **75** 1–35

[18] Smirnov B M 2008 *Physics of Ionized Gases* (New York: Wiley)

[19] Esposito F and Capitelli M 1999 Quasiclassical molecular dynamic calculations of vibrationally and rotationally state selected dissociation cross-sections: N+ N_2 (v, j) → 3N *Chem. Phys. Lett.* **302** 49–54

[20] Bose D and Candler G V 1996 Thermal rate constants of the N_2+ O → NO + N reaction using ab initio 3 Å and 3 Å potential energy surfaces *J. Chem. Phys.* **104** 2825–33

[21] Colonna G and Capitelli M 2001 The influence of atomic and molecular metastable states in high-enthalpy nozzle expansion nitrogen *J. Phys.* D **34** 1812–8

[22] D'Ammando G, Colonna G, Capitelli M and Laricchiuta A 2015 Superelastic collisions under low temperature plasma and afterglow conditions: a golden rule to estimate their quantitative effects *Phys. Plasmas* **22** 034501

[23] Colonna G 2020 On the relevance of superelastic collisions in argon and nitrogen discharges *Plasma Sources Sci. Technol.* **29** 065008

[24] Colonna G, Laporta V, Celiberto R, Capitelli M and Tennyson J 2015 Non-equilibrium vibrational and electron energy distributions functions in atmospheric nitrogen ns pulsed discharges and μs post-discharges: the role of electron molecule vibrational excitation scaling-laws *Plasma Sources Sci. Technol.* **24** 035004

[25] Colonna G, D'Ammando G and Pietanza L 2016 The role of molecular vibration in nanosecond repetitively pulsed discharges and in DBDs in hydrogen plasmas *Plasma Sources Sci. Technol.* **25** 054001

[26] Colonna G and Capitelli M 2001 Self-consistent model of chemical, vibrational, electron kinetics in nozzle expansion *J. Thermophys. Heat Transfer* **15** 308–16

[27] Armenise I and Kustova E 2013 State-to-state models for CO_2 molecules: from the theory to an application to hypersonic boundary layers *Chem. Phys.* **415** 269–81

[28] Surzhikov S 2019 Calculation of nonequilibrium radiation of shock waves in the air using two models *Fluid Dyn.* **54** 98–113

[29] Drawin H W 1971 *Reaction Under Plasma Condition* vol 1 (New York: Wiley) ch 3 53–238

[30] Griem H R 1997 *Principles of Plasma Spectroscopy* (Cambridge: Cambridge University Press) https://doi.org/10.1017/CBO9780511524578

[31] Bogaerts A, Gijbels R and Vlcek J 1998 Collisional-radiative model for an argon glow discharge *J. Appl. Phys.* **84** 121–36

[32] Ozaki N 1971 Temperature distribution of the high-pressure sodium vapour discharge plasma *J. Quant. Spectrosc. Radiat. Transfer* **11** 1111–23

[33] Colonna G, Pietanza L and D'Ammando G 2012 Self-consistent collisional-radiative model for hydrogen atoms: atom-atom interaction and radiation transport *Chem. Phys.* **398** 37–45

[34] Colonna G, Pietanza L D, D'Ammando G and Capitelli M 2012 Self-consistent coupling of chemical, electron and radiation models for shock wave in Jupiter atmosphere *Conf. Proc. AIP* (New York: AIP) 1400–7

[35] D'Ammando G, Pietanza L, Colonna G and Capitelli M 2012 Coupling of radiation transfer, level and electron kinetics under high-speed shock wave conditions in an H_2/He mixture *ESA Spec. Publ.* **714** 8

[36] Shang J and Surzhikov S 2011 Simulating stardust earth reentry with radiation heat transfer *J. Spacecr. Rockets* **48** 385–96

[37] Shang J and Surzhikov S 2012 Nonequilibrium radiative hypersonic flow simulation *Prog. Aerosp. Sci.* **53** 46–65

[38] Surzhikov S, Capitalli M, Colonna G and Defilippis F 2002 Comparative investigation of the multigroup optical models for the radiation characteristics prediction in the arc and laser plasma generators *Fourth Sym. on Aerothermodynamics for Space Vehicles* 487 p 153

[39] Laux C O, Pierrot L and Gessman R J 2012 State-to-state modeling of a recombining nitrogen plasma experiment *Chem. Phys.* **398** 46–55

[40] Colonna G, Pietanza L D, D'Ammando G, Celiberto R, Capitelli M and Laricchiuta A 2017 Vibrational kinetics of electronically excited states in H_2 discharges *Eur. Phys. J.* D **71** 1–8

[41] Pietanza L D *et al* 2021 Advances in non-equilibrium CO_2 plasma kinetics: a theoretical and experimental review *Eur. Phys. J.* D **75** 237

[42] Kozák T and Bogaerts A 2014 Splitting of CO_2 by vibrational excitation in non-equilibrium plasmas: a reaction kinetics model *Plasma Sources Sci. Technol.* **23** 045004

[43] Kozák T and Bogaerts A 2014 Evaluation of the energy efficiency of CO_2 conversion in microwave discharges using a reaction kinetics model *Plasma Sources Sci. Technol.* **24** 015024

[44] Berthelot A and Bogaerts A 2016 Modeling of plasma-based CO_2 conversion: lumping of the vibrational levels *Plasma Sources Sci. Technol.* **25** 045022

[45] Berthelot A and Bogaerts A 2017 Modeling of CO_2 splitting in a microwave plasma: how to improve the conversion and energy efficiency *J. Phys. Chem.* C **121** 8236–51

[46] Capitelli M, Colonna G, D'Ammando G and Pietanza L D 2017 Self-consistent time dependent vibrational and free electron kinetics for CO_2 dissociation and ionization in cold plasmas *Plasma Sources Sci. Technol.* **26** 055009

[47] Pietanza L D, Colonna G and Capitelli M 2020 Self-consistent electron energy distribution functions, vibrational distributions, electronic excited state kinetics in reacting microwave CO_2 plasma: an advanced model *Phys. Plasmas* **27** 023513

[48] Pietanza L D, Colonna G and Capitelli M 2020 Kinetics versus thermodynamics on CO_2 dissociation in high temperature microwave discharges *Plasma Sources Sci. Technol.* **29** 035022

[49] Silva T, Grofulović M, Klarenaar B L M, Morillo-Candas A S, Guaitella O, Engeln R, Pintassilgo C D and Guerra V 2018 Kinetic study of low-temperature CO_2 plasmas under non-equilibrium conditions. I. Relaxation of vibrational energy *Plasma Sources Sci. Technol.* **27** 015019

[50] Grofulović M, Silva T, Klarenaar B L M, Morillo-Candas A S, Guaitella O, Engeln R, Pintassilgo C D and Guerra V 2018 Kinetic study of CO_2 plasmas under non-equilibrium conditions. II. Input of vibrational energy *Plasma Sources Sci. Technol.* **27** 115009

[51] Snoeckx R and Bogaerts A 2017 Plasma technology–a novel solution for CO_2 conversion? *Chem. Soc. Rev.* **46** 5805–63
[52] Silva T, Britun N, Godfroid T and Snyders R 2014 Optical characterization of a microwave pulsed discharge used for dissociation of CO_2 *Plasma Sources Sci. Technol.* **23** 025009
[53] den Harder N *et al* 2017 Homogeneous CO_2 conversion by microwave plasma: Wave propagation and diagnostics *Plasma Process. Polym.* **14** 1600120
[54] Bongers W *et al* 2017 Plasma-driven dissociation of CO_2 for fuel synthesis *Plasma Process. Polym.* **14** e1600126
[55] Aerts R, Snoeckx R and Bogaerts A 2014 In-situ chemical trapping of oxygen in the splitting of carbon dioxide by plasma *Plasma Process. Polym.* **11** 985–92
[56] Brehmer F, Welzel S, van de Sanden M C M and Engeln R 2014 CO and byproduct formation during CO_2 reduction in dielectric barrier discharges *J. Appl. Phys.* **116** 123303
[57] Engeln R, Klarenaar B, Morillo-Candas A, Grofulović M, van de Sanden M and Guaitella O 2018 Diagnostics to study the vibrational excitation kinetics of CO_2 for renewable energy storage *Light, Energy and the Environment 2018 (E2, FTS, HISE, SOLAR, SSL)* EM2A.6 (Washington, DC: Optical Society of America)
[58] van de Steeg A W, Butterworth T, van den Bekerom D C M, Silva A F, van de Sanden M C M and van Rooij G J 2020 Plasma activation of N_2, CH_4 and CO_2: an assessment of the vibrational non-equilibrium time window *Plasma Sources Sci. Technol.* **29** 115001
[59] Klarenaar B L M, Grofulović M, Morillo-Candas A S, van den Bekerom D C M, Damen M A, van de Sanden M C M, Guaitella O and Engeln R 2018 A rotational Raman study under non-thermal conditions in a pulsed CO_2 glow discharge *Plasma Sources Sci. Technol.* **27** 045009
[60] Witteman W J 2013 *The CO_2 Laser* vol 53 (Berlin: Springer)
[61] Cenian A, Chernukho A and Borodin V 1994 Modeling of plasma-chemical reactions in gas mixture of CO_2 lasers I. Gas decomposition in pure CO_2 glow discharge *Contrib. Plasma Phys.* **34** 25–37
[62] d'Hendecourt L and Jourdain de Muizon M 1989 The discovery of interstellar carbon dioxide *Astron. Astrophys.* **223** L5–8 https://adsabs.harvard.edu/full/1989A&A...223L...5D
[63] Wyckoff S and Theobald J 1989 Molecular ions in comets *Adv. Space Res.* **9** 157–61
[64] Bousquet A, Cartry G and Granier A 2007 Investigation of O-atom kinetics in O_2, CO_2, H_2O and O_2/HMDSO low pressure radiofrequency pulsed plasmas by time-resolved optical emission spectroscopy *Plasma Sources Sci. Technol.* **16** 597–605
[65] Babu D J, Yadav S, Heinlein T, Cherkashinin G and Schneider J J 2014 Carbon dioxide plasma as a versatile medium for purification and functionalization of vertically aligned carbon nanotubes *J. Phys. Chem.* C **118** 12028–34
[66] Premathilake D, Outlaw R A, Quinlan R A and Byvik C E 2019 Oxygen generation by carbon dioxide glow discharge and separation by permeation through ultrathin silver membranes *Earth Space Sci.* **6** 557–64
[67] Marieu V, Reynier P, Marraffa L, Vennemann D, De Filippis F and Caristia S 2007 Evaluation of SCIROCCO plasma wind-tunnel capabilities for entry simulations in CO_2 atmospheres *Acta Astronaut.* **61** 604–16
[68] Kolesnikov A F, Pershin I S, Vasil'evskii S A and Yakushin M I 2000 Study of quartz surface catalycity in dissociated carbon dioxide subsonic flows *J. Spacecr. Rockets* **37** 573–9
[69] Rusanov V D, Fridman A A and Sholin G V 1981 The physics of a chemically active plasma with nonequilibrium vibrational excitation of molecules *Sov. Phys. - Usp.* **24** 447–74

[70] Capitelli M and Molinari E 1980 Kinetics of dissociation processes in plasmas in the low and intermediate pressure range *Plasma Chemistry II. Topics in Current Chemistry* (Berlin: Springer) **90** 59–109
[71] Capezzuto P, Cramarossa F, D'Agostino R and Molinari E 1976 Contribution of vibrational excitation to the rate of carbon dioxide dissociation in electrical discharges *J. Phys. Chem.* **80** 882–8
[72] Rich J W 1971 Kinetic modeling of the high-power carbon monoxide laser *J. Appl. Phys.* **42** 2719–30
[73] Capitelli M and Dilonardo M 1977 Nonequilibrium vibrational populations of diatomic species in electrical discharges: effects on the dissociation rates *Chem. Phys.* **24** 417–27
[74] Cartwright D C 1978 Vibrational populations of the excited states of N_2 under auroral conditions *J. Geophys. Res. Space Phys* **83** 517–31
[75] Gordiets B F, Osipov A I and Shelepin L A 1980 Kinetic processes in gases and molecular lasers *Moscow Izdatel Nauka* **1**
[76] Aleksandrov N, Konchakov A and Son E 1979 Electron energy distribution and kinetic coefficients of a CO plasma. II. Vibrationally excited molecules *Sov. Phys. - Tech. Phys.* **24** 661–6
[77] Rich J W and Treanor C E 1970 Vibrational relaxation in gasdynamic flows *Annu. Rev. Fluid Mech.* **2** 355–96
[78] Kewley D 1975 Numerical study of anharmonic diatomic relaxation rates in shock waves and nozzles *J. Phys. B: At. Mol. Opt. Phys.* **8** 2564–79
[79] Josyula E and Bailey W F 2001 Vibration-dissociation coupling using master equations in nonequilibrium hypersonic blunt-body flow *J. Thermophys. Heat Transfer* **15** 157–67
[80] Colonna G 2014 Problems and perspectives of state-to-state kinetics for high enthalpy plasma flows *AIP Conf. Proc.* **1628** 1176
[81] Bates D R and Kingston A E 1962 Recombination between electrons and atomic ions. I Optically thin plasmas *Proc. R. Soc.* **267** 685–9
[82] Bates D R and Kingston A E 1964 Recombination and energy balance in a decaying plasma. 2. He-He^+-e plasma *Proc. R. Soc.* A **279** 32–8
[83] Capitelli M, Colonna G and D'Angola A 2011 Fundamental Aspects of Plasma *Chemical Physics: Thermodynamics* vol 66 (Berlin: Springer) https://doi.org/10.1007/978-1-4419-8182-0
[84] Irons F E 1979 The escape factor in plasma spectroscopy–I. The escape factor defined and evaluated *J. Quant. Spectrosc. Radiat. Transfer* **22** 1–20
[85] Fill E E 1988 The radiation escape factor *J. Quant. Spectrosc. Radiat. Transfer* **39** 489–91
[86] Pestehe S J and Tallents G J 2002 Escape factors for laser-plasmas *J. Quant. Spectrosc. Radiat. Transfer* **72** 853–78
[87] Benilov M and Naidis G 2005 Asymptotic calculation of escape factor in atomic plasmas *J. Phys. D: Appl. Phys.* **38** 3599
[88] Capitelli M, Celiberto R, Colonna G, Esposito F, Gorse C, Hassouni K, Laricchiuta A and Longo S 2015 *Fundamental Aspects of Plasma Chemical Physics: Kinetics* vol 85 (Berlin: Springer) https://doi.org/10.1007/978-1-4419-8185-1
[89] Sharma S P, Ruffin S M, Gillespie W D and Meyer S A 1993 Vibrational relaxation measurements in an expanding flow using spontaneous Raman scattering *J. Thermophys. Heat Transfer* **7** 697–703

[90] Colonna G, Tuttafesta M, Capitelli M and Giordano D 1999 Non-Arrhenius NO formation rate in one-dimensional nozzle airflow *J. Thermophys. Heat Transfer* **13** 372–5

[91] Cutrone L, Tuttafesta M, Capitelli M, Schettino A, Pascazio G and Colonna G 2014 3D nozzle flow simulations including state-to-state kinetics calculation *AIP Conf. Proc.* **1628** **1154**

[92] Lordet F, Meolans J, Chauvin A and Brun R 1995 Nonequilibrium vibration-dissociation phenomena behind a propagating shock wave: vibrational population calculation *Shock Waves* **4** 299–312

[93] Colonna G, Pietanza L D and D'Ammando G 2012 Self-consistent collisional-radiative model for hydrogen atoms: atom-atom interaction and radiation transport *Chem. Phys.* **398** 37–45

[94] Colonna G, Pietanza L D and Laricchiuta A 2020 Ionization kinetic model for Hydrogen-Helium atmospheres in hypersonic shock tubes *Int. J. Heat Mass Transfer* **156** 119916

[95] Armenise I, Capitelli M, Colonna G, Kudriavtsev N and Smetanin V 1995 Nonequilibrium vibrational kinetics during hypersonic flow of a solid body in nitrogen and its influence on the surface heat flux *Plasma Chem. Plasma Process.* **15** 501–28

[96] Armenise I, Capitelli M, Colonna G and Gorse C 1996 Nonequilibrium vibrational kinetics in the boundary layer of re-entering bodies *J. Thermophys. Heat Transfer* **10** 397–405

[97] Giordano D, Bellucci V, Colonna G, Capitelli M, Armenise I and Bruno C 1997 Vibrational relaxing flow of N_2 past an infinite cylinder *J. Thermophys. Heat Transfer* **11** 27–35

[98] Tuttafesta M, Colonna G and Pascazio G 2013 Computing unsteady compressible flows using Roe's flux-difference splitting scheme on GPUs *Comput. Phys. Commun.* **184** 1497–510

[99] Bonelli F, Tuttafesta M, Colonna G, Cutrone L and Pascazio G 2017 An MPI-CUDA approach for hypersonic flows with detailed state-to-state air kinetics using a GPU cluster *Comput. Phys. Commun.* **219** 178–95

[100] Bonelli F, Pascazio G and Colonna G 2021 Effect of finite-rate catalysis on wall heat flux prediction in hypersonic flow *Phys. Rev. Fluids* **6** 033201

[101] Treanor C E, Rich J and Rehm R 1968 Vibrational relaxation of anharmonic oscillators with exchange-dominated collisions *J. Chem. Phys.* **48** 1798–807

[102] Colonna G, Armenise I, Bruno D and Capitelli M 2006 Reduction of state-to-state kinetics to macroscopic models in hypersonic flows *J. Thermophys. Heat Transfer* **20** 477–86

[103] Colonna G, Pietanza L D and Capitelli M 2008 Recombination assisted nitrogen dissociation rates under nonequilibrium conditions *J. Thermophys. Heat Transfer* **22** 399–406

[104] Colonna G, Pietanza L D and Capitelli 2011 Reduced two-level approach for air kinetics in recombination regime *AIP Conf. Proc* **1333** 1365

[105] Bourdon A, Annaloro J, Bultel A, Capitelli M, Colonna G, Guy A, Magin T, Munafó A, Perrin M and Pietanza L 2014 Reduction of State-to-State to macroscopic models for hypersonics *Open Plasma Phys. J.* **7** M4

[106] Guy A, Bourdon A and Perrin M-Y 2015 Consistent multi-internal-temperature models for vibrational and electronic nonequilibrium in hypersonic nitrogen plasma flows *Phys. Plasmas* **22** 043507

[107] Park C 1993 Review of chemical-kinetic problems of future NASA missions, I: Earth entries *J. Thermophys. Heat Transfer* **7** 385–98

[108] Chernyi G G, Losev S A, Macheret S O and Potapkin B V 2002 *Physical and Chemical Processes in Gas Dynamics: Cross Sections and Rate Constants for Physical and Chemical Processes* I (Reston, VA: American Institute of Aeronautics and Astronautics) https://doi.org/10.2514/4.866661
[109] Schwartz R, Slawsky Z and Herzfeld K 1952 Calculation of vibrational relaxation times in gases *J. Chem. Phys.* **20** 1591–9
[110] Adamovich I V, MacHeret S O, Rich J W and Treanor C E 1998 Vibrational energy transfer rates using a forced harmonic oscillator model *J. Thermophys. Heat Transfer* **12** 57–65
[111] Capitelli M, Colonna G and Esposito F 2004 On the coupling of vibrational relaxation with the dissociation-recombination kinetics: from dynamics to aerospace applications *J. Phys. Chem.* A **108** 8930–4
[112] Colonna G, D'Ammando G, Pietanza L and Capitelli M 2014 Excited-state kinetics and radiation transport in low-temperature plasmas *Plasma Phys. Control. Fusion* **57** 014009
[113] D'Ammando G, Capitelli M, Esposito F, Laricchiuta A, Pietanza L D and Colonna G 2014 The role of radiative reabsorption on the electron energy distribution functions in H_2/He plasma expansion through a tapered nozzle *Phys. Plasmas* **21** 093508
[114] Perrin M, Colonna G, D'Ammando G, Pietanza L, Riviere P, Soufani A and Surzhikov S 2014 Radiation models and radiation transfer in hypersonics *Open Plasma Phys. J.* **7** 114–26
[115] Pietanza L, Colonna G, De Giacomo A and Capitelli M 2010 Kinetic processes for laser induced plasma diagnostic: a collisional-radiative model approach *Spectrochim. Acta* **65** 616–26
[116] Pietanza L D, Colonna G, D'Ammando G, Laricchiuta A and Capitelli M 2015 Vibrational excitation and dissociation mechanisms of CO_2 under non-equilibrium discharge and post-discharge conditions *Plasma Sources Sci. Technol.* **24** 042002
[117] Pietanza L, Colonna G, D'Ammando G, Laricchiuta A and Capitelli M 2016 Electron energy distribution functions and fractional power transfer in "cold" and excited CO_2 discharge and post discharge conditions *Phys. Plasmas* **23** 013515
[118] Pietanza L, Colonna G, D'Ammando G, Laricchiuta A and Capitelli M 2016 Non equilibrium vibrational assisted dissociation and ionization mechanisms in cold CO_2 plasmas *Chem. Phys.* **468** 44–52
[119] Pietanza L D, Colonna G, D'Ammando G and Capitelli M 2016 Time-dependent coupling of electron energy distribution function, vibrational kinetics of the asymmetric mode of CO_2 and dissociation, ionization and electronic excitation kinetics under discharge and post-discharge conditions *Plasma Phys. Control. Fusion* **59** 014035
[120] Colonna G, D'Ammando G and Pietanza L 2016 The role of molecular vibration in nanosecond repetitively pulsed discharges and in DBD's in hydrogen plasmas *Plasma Sources Sci. Technol.* **25** 054001
[121] Colonna G, Laricchiuta A and Pietanza L 2019 Time dependent selfconsistent electron energy distribution functions during nano-second repetitive discharges in reacting N_2/H_2 mixtures *Plasma Phys. Control. Fusion* **62** 014003
[122] Capitelli M, Colonna G, D'Ammando G, Laporta V and Laricchiuta A 2014 Nonequilibrium dissociation mechanisms in low temperature nitrogen and carbon monoxide plasmas *Chem. Phys.* **438** 31–6

[123] Colonna G, Capitelli M, DeBenedictis S, Gorse C and Paniccia F 1991 Electron energy distribution functions in CO_2 laser mixture: the effects of second kind collisions from metastable electronic states Contrib *Plasma Phys* **31** 575–9

[124] Capriati G, Colonna G, Gorse C and Capitelli M 1992 A parametric study of electron energy distribution functions and rate and transport coefficients in nonequilibrium helium plasmas *Plasma Chem. Plasma Process.* **12** 237–60

[125] Rassou S, Packan D and Labaune J 2017 Numerical modeling of repetitive nanosecond discharge in air *Phys. Plasmas* **24** 100704

[126] Levaton J, Klein A N, de Amorim J and Binder C 2019 Temporal electron collisions and adjunct ionization mechanisms in the positive column of a flowing N_2 DC discharge Plasma Res *Express* **1** 045004

[127] Chen X, Lan L, Lu H, Wang Y, Wen X, Du X and He W 2017 Numerical simulation of Trichel pulses of negative DC corona discharge based on a plasma chemical model *J. Phys. D: Appl. Phys* **50** 395202

[128] Peters C J, Shneider M N and Miles R B 2019 Kinetics model of femtosecond laser ionization in nitrogen and comparison to experiment *J. Appl. Phys.* **125** 243301

[129] Hagelaar G J M and Pitchford L C 2005 Solving the Boltzmann equation to obtain electron transport coefficients and rate coefficients for fluid models *Plasma Sources Sci. Technol.* **14** 722

[130] Tejero-del Caz A, Guerra V, Gonçalves D, da Silva M L, Marques L, Pinhao N, Pintassilgo C and Alves L 2019 The LisbOn KInetics Boltzmann solver *Plasma Sources Sci. Technol.* **28** 043001

[131] Aerts R, Martens T and Bogaerts A 2012 Influence of vibrational states on CO_2 splitting by dielectric barrier discharges *J. Phys. Chem.* C **116** 23257–73

[132] Nighan W L 1970 Electron energy distributions and collision rates in electrically excited N_2, CO, and CO_2 *Phys. Rev.* A **2** 1989–2000

[133] Silva T, Britun N, Godfroid T and Snyders R 2014 Optical characterization of a microwave pulsed discharge used for dissociation of CO_2 *Plasma Sources Sci. Technol.* **23** 025009

[134] Dyatko N, Ionikh Y Z, Meshchanov A and Napartovich A 2018 Influence of a nitrogen admixture on the anomalous memory effect in the breakdown of low-pressure argon in a long discharge tube *Plasma Phys. Rep.* **44** 334–44

[135] Hannemann M, Hardt P, Loffhagen D, Schmidt M and Winkler R 2000 The electron kinetics in the cathode region of H_2/Ar/N_2 discharges *Plasma Sources Sci. Technol.* **9** 387

[136] Hübner M, Gortschakow S, Guaitella O, Marinov D, Rousseau A, Röpcke J and Loffhagen D 2016 Kinetic studies of NO formation in pulsed air-like low-pressure dc plasmas *Plasma Sources Sci. Technol.* **25** 035005

[137] Loffhagen D, Becker M M, Czerny A K, Philipp J and Klages C-P 2018 Impact of hexamethyldisiloxane admixtures on the discharge characteristics of a dielectric barrier discharge in argon for thin film deposition *Contrib. Plasma Phys.* **58** 337–52

[138] Pintassilgo C D and Guerra V 2017 Modelling of the temporal evolution of the gas temperature in N_2 discharges *Plasma Sources Sci. Technol.* **26** 055001

[139] Pintassilgo C D and Guerra V 2018 Kinetic mechanisms in air plasmas *Plasma Phys. Control. Fusion* **61** 014026

[140] Yang W, Zhou Q and Dong Z 2016 Simulation study on nitrogen vibrational kinetics in a single nanosecond pulse high voltage air discharge *AIP Adv.* **6** 055209

[141] Yang W, Zhou Q, Dong Z and Yan E 2019 Time-dependent modeling on initiation of pre-ionized laser filament in atmospheric air under an external field far below conventional threshold *Phys. Plasmas* **26** 073512

[142] Neretti G, Taglioli M, Colonna G and Borghi C 2017 Characterization of a dielectric barrier discharge in contact with liquid and producing a plasma activated water *Plasma Sources Sci. Technol.* **26** 015013

[143] Colonna G and Capitelli M 2008 Boltzmann and master equations for MHD in weakly ionized gases *J. Thermophys. Heat Transfer* **22** 414–23

[144] Colonna G, Capitelli M, Pietanza L D, Munafà A and Panesi M 2019 High-enthalpy ionized flows *Hypersonic Meteoroid Entry Physics* (Bristol: IOP Publishing) ch 8 https://doi.org/10.1088/2053-2563/aae894ch8

[145] Nerem R M and Stickford G H 1965 Shock-tube studies of equilibrium air radiation *AIAA J.* **3** 1011–8

[146] Livingston F R and Poon P T Y 1976 Relaxation distance and equilibrium electron density measurements in hydrogen-helium plasmas *AIAA J.* **14** 1335–7

[147] Johnson C O, Brandis A M and Bose D 2013 Radiative heating uncertainty for hyperbolic earth entry, Part 3: Comparisons with electric arc shock-tube measurements *J. Spacecr. Rockets* **50** 48–55

[148] Herzberg G 1945 *Molecular Spectra and Molecular Structure, Infrared and Raman Spectra of Polyatomic Molecules* (Princeton, NJ: Van Nostrand-Reinhold)

[149] Fridman A and Kennedy L A 2011 *Plasma Physics and Engineering* (Boca Raton, FL: CRC Press) https://doi.org/10.1201/9781482293630

[150] Dennison D M 1932 The vibrational levels of linear symmetrical triatomic molecules *Phys. Rev.* **41** 304–12

[151] Dennison D M 1940 The infra-red spectra of polyatomic molecules. Part II *Rev. Mod. Phys.* **12** 175–214

[152] Adel A and Dennison D M 1933 The infrared spectrum of carbon dioxide. Part I *Phys. Rev.* **43** 716–23

[153] Blauer J and Nickerson G 1974 A survey of vibrational relaxation rate data for processes important to CO_2-N_2-H_2O infrared plume radiation *7th Fluid and Plasma Dynamics Conf.* 536

[154] Annaloro J and Bultel A 2019 Vibrational and electronic collisional-radiative model in CO_2-N_2-Ar mixtures for Mars entry problems *Phys. Plasmas* **26** 103505

[155] Armenise I and Kustova E 2013 State-to-state models for CO_2 molecules: from the theory to an application to hypersonic boundary layers *Chem. Phys.* **415** 269–81

[156] Armenise I and Kustova E 2014 On different contributions to the heat flux and diffusion in non-equilibrium flows *Chem. Phys.* **428** 90–104

[157] Armenise I and Kustova E 2018 Mechanisms of coupled vibrational relaxation and dissociation in carbon dioxide *J. Phys. Chem.* A **122** 5107–20

[158] Kustova E, Savelev A and Armenise I 2019 State-resolved dissociation and exchange reactions in CO_2 flows *J. Phys. Chem.* A **123** 10529–42

[159] Armenise I and Kustova E 2018 Effect of asymmetric mode on CO_2 state-to-state vibrational-chemical kinetics *J. Phys. Chem.* A **122** 8709–21

[160] phys4entry Database (http://phys4entrydb.ba.imip.cnr.it/Phys4EntryDB/) retrieved January 2021

[161] The LXCat open-access website (http://lxcat.org/home/)

[162] Lowke J J, Phelps A V and Irwin B W 1973 Predicted electron transport coefficients and operating characteristics of CO_2-N_2-He laser mixtures *J. Appl. Phys.* **44** 4464
[163] Phelps Database (http://www.lxcat.net) retrieved on October 5, 2020
[164] IST-Lisbon Database (http://www.lxcat.net) retrieved on October 5, 2020
[165] Grofulović M, Alves L L and Guerra V 2016 Electron-neutral scattering cross sections for CO_2: a complete and consistent set and an assessment of dissociation *J. Phys. D: Appl. Phys.* **49** 395207
[166] Itikawa Y 2002 Cross sections for electron collisions with carbon dioxide *J. Phys. Chem. Ref. Data* **31** 749–67
[167] Celiberto R, Laporta V, Laricchiuta A, Tennyson J and Wadehra J 2014 Molecular physics of elementary processes relevant to hypersonics: electron-molecule collisions *Open Plasma Phys.* J **7** 33–47
[168] Celiberto R *et al* 2016 Atomic and molecular data for spacecraft re-entry plasmas *Plasma Sources Sci. Technol.* **25** 033004
[169] Laporta V, Tennyson J and Celiberto R 2016 Calculated low-energy electron-impact vibrational excitation cross sections for CO_2 molecule *Plasma Sources Sci. Technol.* **25** 06LT2
[170] Bogaerts A, Wang W, Berthelot A and Guerra V 2016 Modeling plasma-based CO_2 conversion: crucial role of the dissociation cross section *Plasma Sources Sci. Technol.* **25** 055016
[171] Cosby P C and Helm H 1993 Dissociation rates of diatomic molecules *Report no. AD-A266 464 WL-TR-93_2004* Wright-Patterson Airforce Base, Dayton, OH
[172] Polak L S and Slovetsky D I 1976 Electron impact induced electronic excitation and molecular dissociation *Int. J. Radiat. Phys. Chem.* **8** 257–82
[173] Morillo-Candas A S, Silva T, Klarenaar B L M, Grofulović M, Guerra V and Guaitella O 2020 Electron impact dissociation of CO_2/O_2 *Plasma Sources Sci. Technol.* **29** 01LT01
[174] Taylor R L and Bitterman S 1969 Survey of vibrational relaxation data for processes important in the CO_2-N_2 laser system *Rev. Mod. Phys.* **41** 26–47
[175] Treanor C E 1965 Vibrational energy transfer in high-energy collisions *J. Chem. Phys.* **43** 532–8
[176] Zelechow A, Rapp D and Sharp T E 1968 Vibrational-vibrational-translational energy transfer between two diatomic molecules *J. Chem. Phys.* **49** 286–99
[177] Adamovich I V, Macheret S O, Rich J W and Treanor C E 1998 Vibrational energy transfer rates using a forced harmonic oscillator model *J. Thermophys. Heat Transfer* **12** 57–65
[178] Vargas J, Lopez B and Lino da Silva M 2021 Heavy particle impact vibrational excitation and dissociation processes in CO_2 *J. Phys. Chem.* A **125** 493–512
[179] Lombardi A, Faginas-Lago N, Pacifici L and Grossi G 2015 Energy transfer upon collision of selectively excited CO_2 molecules: state-to-state cross sections and probabilities for modeling of atmospheres and gaseous flows *J. Chem. Phys.* **143** 034307
[180] Bartolomei M, Pirani F, Laganà A and Lombardi A 2012 A full dimensional grid empowered simulation of the $CO_2 + CO_2$ processes *J. Comput. Chem.* **33** 1806–19
[181] Porshnev P I, Wallaart H L, Perrin M-Y and Martin J-P 1996 Modeling of optical pumping experiments in CO. I. Time-resolved experiments *Chem. Phys.* **213** 111–22

Chapter 10

Hypersonic flows with detailed state-to-state kinetics using a GPU cluster

Giuseppe Pascazio, Davide Ninni, Francesco Bonelli and Gianpiero Colonna

A critical phase in space exploration missions is the entry into planetary atmospheres. The large kinetic energy of a space vehicle must be dissipated, reduced to values for safe landing. At a given altitude, when the pressure becomes sufficiently large, the formation of a strong bow shock transfers the kinetic energy to heat, activating the internal degrees of freedom and chemical processes such as dissociation and ionization, that by radiation, is partially dispersed in the environment.

A body entering a planetary atmosphere travels at high Mach number regimes [1]. At very high altitudes, the gas is very rarefied, and molecular collisions are negligible (molecular regime). As the altitude decreases, collisions become relevant, with the consequent formation of a shock wave in the front of the entering vehicle (continuum regime). Crossing the shock front, the translational temperature reaches high values while internal degrees of freedom and chemical composition remain unchanged. Then, inside the shock wave, the system relaxes towards equilibrium, each degree of freedom with its own relaxation time τ:

$$\tau_{\rm tr} < \tau_{\rm rot} < \tau_{\rm vib} < \tau_{\rm chem} < \tau_{\rm el} < \tau_{\rm ion} \tag{10.1}$$

where subscripts stand for translational (tr), rotational (rot), vibrational (vib), chemical (chem), electronic (el) and ionization (ion). Comparing the relaxation time τ with the flow characteristic time ($\tau_{\rm flow}$), it is possible to estimate the degree of non-equilibrium of the flow. All the degrees of freedom having $\tau \gtrsim \tau_{\rm flow}$ are in non-equilibrium, while those with $\tau \gg \tau_{\rm flow}$ are in equilibrium at the translational temperature [2]. If all the degrees of freedom have $\tau \ll \tau_{\rm flow}$ the flow is frozen and if $\tau \gg \tau_{\rm flow}$ the flow is in non-equilibrium.

The vehicle surface must be cooled to limit the temperature in order to avoid damage, breakage and interior overheating. Therefore, in the proximity of the wall,

doi:10.1088/978-0-7503-3559-1ch10

due to diffusion and thermal conductivity, a boundary layer is formed, showing again non-equilibrium conditions [3–5]. Some materials cool the surface by ablation, introducing further complexity in the model [6].

A numerical tool capable of modeling non-equilibrium should find a compromise between the accuracy of the thermochemical model and demand for computational resources. The *multi-temperature* (MT) approach (see chapter 16), that assigns an independent temperature to each degree of freedom and assuming the translational and rotational modes at equilibrium [7], is commonly used to model non-equilibrium because it leads to only a slight increase of the computational cost. However, assuming a Boltzmann distribution for internal degrees of freedom, multi-temperature models are not enough detailed to catch all the physical behaviors of non-equilibrium conditions.

This chapter presents an alternative strategy to accomplish these requirements, namely the state-to-state (StS) model [8, 9], that adds a mass continuity equation for each internal state, considered as pseudo-species. Comparing with the MT model, that adds one equation for the vibrational temperature of molecules (just three in the case of the air), the commonly used StS model of air considers about 100 species and more than 10 000 reactions.

The huge computational cost required by the StS model has limited its application to 1D configurations [3, 4, 10–12] with very few exceptions [13], with 2D numerical simulations being very onerous [14, 15] and 3D configurations exceptional [16].

In the view of reducing the computational cost, graphic processor units (GPUs) represent a powerful tool to speed up the algorithm. A GPU contains thousands of small computational units that can perform the same operation independently. This characteristic limits the applications of GPUs, requiring the use of ad hoc procedures to reach high computational efficiency. Over the years, programming language extensions have been developed for general-purpose computing on graphics processing units (GPGPUs) [17, 18]. The combination of Compute Unified Device Architecture (CUDA) with message passing interface (MPI) libraries allows one to employ these methods on GPU clusters, whose importance has been already highlighted recently [14, 19–21].

10.1 Physical model

10.1.1 Governing equations

The governing equations of hypersonic flows are given by the 3D compressible Navier–Stokes equations for a mixture:

$$\int_{V_0} \frac{\partial}{\partial t}\mathbf{U}\, dV + \oint_{S_0} \mathbf{F} \cdot \mathbf{n}\, dS = \int_{V_0} \mathbf{W}\, dV \tag{10.2}$$

where $\mathbf{U}$ and $\mathbf{W}$ are the vectors of conservative variables and source terms, whereas $\mathbf{F}$ is the fluxes tensor. They read:

$$\mathbf{U} = \begin{bmatrix} \rho_{1,1} \\ \vdots \\ \rho_{1,V_1} \\ \vdots \\ \rho_{N_s,V_{N_s}} \\ \rho\, u \\ \rho\, v \\ \rho\, w \\ \rho\, e \\ \rho_{N_2}\, \varepsilon_{vib,N_2} \\ \rho_{O_2}\, \varepsilon_{vib,O_2} \\ \rho_{NO}\, \varepsilon_{vib,NO} \end{bmatrix} \tag{10.3a}$$

$$\mathbf{F} = (\mathbf{F}_E - \mathbf{F}_V,\, \mathbf{G}_E - \mathbf{G}_V,\, \mathbf{H}_E - \mathbf{H}_V) \tag{10.3b}$$

$$\mathbf{F}_E = \begin{bmatrix} \rho_{1,1}u \\ \vdots \\ \rho_{1,V_1}u \\ \vdots \\ \rho_{N_s,V_{N_s}}u \\ \rho\, u^2 + p \\ \rho\, u\, v \\ \rho\, u\, w \\ (\rho\, e + p)\, u \\ \rho_{N_2}\, \varepsilon_{vib,N_2}\, u \\ \rho_{O_2}\, \varepsilon_{vib,O_2}\, u \\ \rho_{NO}\, \varepsilon_{vib,NO}\, u \end{bmatrix} \quad \mathbf{G}_E = \begin{bmatrix} \rho_{1,1}v \\ \vdots \\ \rho_{1,V_1}v \\ \vdots \\ \rho_{N_s,V_{N_s}}v \\ \rho\, u\, v \\ \rho\, v^2 + p \\ \rho\, v\, w \\ (\rho\, e + p)\, v \\ \rho_{N_2}\, \varepsilon_{vib,N_2}\, v \\ \rho_{O_2}\, \varepsilon_{vib,O_2}\, v \\ \rho_{NO}\, \varepsilon_{vib,NO}\, v \end{bmatrix} \quad \mathbf{H}_E = \begin{bmatrix} \rho_{1,1}w \\ \vdots \\ \rho_{1,V_1}w \\ \vdots \\ \rho_{N_s,V_{N_s}}w \\ \rho\, u\, w \\ \rho\, v\, w \\ \rho\, w^2 + p \\ (\rho\, e + p)\, w \\ \rho_{N_2}\, \varepsilon_{vib,N_2}\, w \\ \rho_{O_2}\, \varepsilon_{vib,O_2}\, w \\ \rho_{NO}\, \varepsilon_{vib,NO}\, w \end{bmatrix} \tag{10.3c}$$

$$(\mathbf{F}_V, \mathbf{G}_V, \mathbf{H}_V) = \begin{bmatrix} -\rho_{1,1}\mathbf{u}^D_{1,\,1} \\ \vdots \\ -\rho_{1,V_1}\mathbf{u}^D_{1,\,V_1} \\ \vdots \\ -\rho_{N_s,V_{N_s}}\mathbf{u}^D_{N_s,\,V_{N_s}} \\ \boldsymbol{\sigma} \\ \mathbf{u}\cdot\boldsymbol{\sigma} - \mathbf{q} \\ -\mathbf{q}_{vib,N_2} \\ -\mathbf{q}_{vib,O_2} \\ -\mathbf{q}_{vib,NO} \end{bmatrix} \tag{10.3d}$$

$$\mathbf{W} = \left[\dot{\omega}_{1,1},\, \ldots,\, \dot{\omega}_{1,V_1},\, \ldots,\, \dot{\omega}_{N_s,V_{N_s}},\, 0{,}0,\, 0{,}0,\, \dot{\omega}_{vib,N_2},\, \dot{\omega}_{vib,O_2},\, \dot{\omega}_{vib,NO}\right]^T \tag{10.3e}$$

where u, v and w are the velocity Cartesian components, p is the pressure, e is the total energy per unit mass. **F**, **G** and **H** are the flux vectors along the x, y, and z directions of the Cartesian reference frame, and the subscripts E and V stand for inviscid and viscous fluxes, respectively. Here, the continuity equation is substituted by a series of species continuity equations: for the case of air considered here, $N_s = 5$ is the number of species and V_s is the number of vibrational levels considered for the species s. For the Park model only the ground state is considered for vibrational modes ($V_s = 1$), leading to five continuity equations; for the StS model $V_{N_2} = 68$, $V_{O_2} = 47$ and $V_{NO} = V_N = V_O = 1$, for a total of 118 continuity equations since each vibrational level is treated as a singular species. On the other hand, in order to take into account non-equilibrium phenomena the multi-temperature approach solves three additional vibrational energy equations for the molecules, with $\varepsilon_{\text{vib},m}$ being the vibrational energy of the molecule m; these equations are not solved in the StS model. Other quantities in the governing equations are the viscous stress tensor ($\boldsymbol{\sigma}$) and the total and vibrational heat fluxes ($\mathbf{q}$ and $\mathbf{q}_{\text{vib}}$), evaluated as follows:

$$\boldsymbol{\sigma} = \mu[\nabla \mathbf{u} + (\nabla \mathbf{u})^T] - \frac{2}{3}\mu(\nabla \cdot \mathbf{u})\mathbf{I} \tag{10.4a}$$

$$\mathbf{q} = -\lambda_t \nabla T - \sum_{m=1}^{N_m} \lambda_{\text{vib},m} \nabla T_{\text{vib},m} + \sum_{s=1}^{N_s} \sum_{l=1}^{V_s} h_{s,l} \rho_{s,l} \mathbf{u}_{s,l}^D \tag{10.4b}$$

$$\mathbf{q}_{\text{vib},m} = -\lambda_{\text{vib},m} \nabla T_{\text{vib},m} + \varepsilon_{\text{vib},m} \rho_m \mathbf{u}_m^D \tag{10.4c}$$

with $N_m = 3$ being the number of molecules. Here, $T_{\text{vib},m}$ is the vibrational temperature of molecule m, the enthalpy per unit mass of each species s in the level l is indicated with $h_{s,l}$, while the transport properties are indicated with μ (mixture viscosity), λ_t (translational conductivity of the mixture) and $\lambda_{\text{vib},m}$ (vibrational conductivity of each molecule): their treatment is presented in next subsection. Finally, one can evaluate the diffusive velocities $\mathbf{u}_{s,l}^D$ of the species s in the level l by solving Fick's law:

$$\rho_{s,l} \mathbf{u}_{s,l}^D = -\rho \mathcal{D}_s \nabla Y_{s,l} + \mathbf{V}_c \tag{10.5}$$

where $\rho_{s,l}$ and $Y_{s,l}$ indicate the density and the mass fraction of the species s in the level l, respectively. The density of each species is computed as the sum of the densities of the single levels, thus $\rho_s = \sum_l \rho_{s,l}$ and obviously the density of the mixture is $\rho = \sum_s \rho_s$. $\mathcal{D}_s$ is the diffusion coefficient of species s in the mixture: its evaluation is presented in the next subsection. Note that $\mathbf{V}_c = \sum_{s=1}^{N_s} \sum_{l=1}^{V_s} \rho \mathcal{D}_s \nabla Y_{s,l}$ is a correction velocity to force the mass conservation [22]. Lastly, the system is closed by solving the perfect gas equation of state:

$$p = \frac{\rho \mathcal{R} T}{\mathcal{M}} = \rho R T \tag{10.6}$$

In the equation above $\mathcal{R}$ and $\mathcal{M}$ indicate the universal gas constant and the mixture molar mass, respectively, and the relation between p and e reads [23]:

$$p = (\bar{\gamma} - 1)\left[\rho e - \rho(\varepsilon_{\text{vib}} + \varepsilon_{\text{chem}}) - \rho\frac{u^2 + v^2 + w^2}{2}\right], \tag{10.7}$$

where $\bar{\gamma}$ is the specific heats ratio of the gas mixture, whereas ε_{vib} and $\varepsilon_{\text{chem}}$ are the total specific vibrational energy and the total specific chemical energy, respectively. The expressions of ε_{vib} depend on the thermochemical non-equilibrium model employed; thus, they will be given in the sections concerning the StS and the Park models. The total specific chemical term is given by

$$\varepsilon_{\text{chem}} = 1/\rho \sum_{s=1}^{N_s} \rho_s h_s^f, \tag{10.8}$$

where h_s^f is the specific formation enthalpy of the component s.

We compute the gas mixture specific heat at constant pressure by considering the degrees of freedom in equilibrium at the translational temperature

$$\bar{c}_p = \alpha\ R$$

where R is the specific gas constant and α is given by

$$\alpha = \sum_s \chi_s \alpha_s \tag{10.9}$$

where χ_s is the molar fraction of the component s and α_s is equal to 5/2 for monoatomic and 7/2 for diatomic components, respectively. Finally, by considering the Mayer relation, $R = \bar{c}_p - \bar{c}_v$, where $\bar{c}_v$ is the mixture specific heat at constant volume, we can write the specific heats ratio of the gas mixture as

$$\bar{\gamma} = \frac{\bar{c}_p}{\bar{c}_v} = \frac{\alpha}{\alpha - 1}. \tag{10.10}$$

10.1.2 Transport properties

The evaluation of the transport properties is here discussed, following classical methodologies available in literature [24–26]. We start from the single-species viscosity, evaluated through Gupta's curve fits [27]. They read:

$$\mu_s = \left[\exp\left(C_{\mu_s}\right)\right] T^{\left(A_{\mu_s} \ln T + B_{\mu_s}\right)} \tag{10.11}$$

where the coefficients for air species are given in table 10.1.

Table 10.1. Coefficients to compute the viscosity (obtained in g (cm s)$^{-1}$).

Species	A_{μ_s}	B_{μ_s}	C_{μ_s}
N_2	0.020 3	0.432 9	−11.815 3
O_2	0.048 4	0.145 5	−8.923 1
NO	0.045 2	−0.060 9	−9.459 6
N	0.012	0.593	−12.380 5
O	0.020 5	0.425 7	−11.580 3

Table 10.2. Coefficients to compute the viscosity for $T < 1000$ K (obtained in g (cm s)$^{-1}$).

Species	A_1	A_2	A_3	A_4
N_2	−16.261 728 437 2817	2.251 740 453 876 13	−0.213 834 089 369 937	0.009 477 823 154 448 48
O_2	−16.810 801 108 7253	2.522 528 725 238 04	−0.249 071 279 824 062	0.011 006 158 064 4724
NO	−16.227 369 563 0321	2.251 740 453 876 18	−0.213 834 089 369 944	0.009 477 823 154 448 81
N	−14.666 420 894 9590	1.577 763 512 732 12	−0.125 286 191 113 798	0.005 601 056 214 898 95
O	−14.815 635 915 8030	1.801 396 425 547 43	−0.154 910 226 483 785	0.006 908 759 721 670 36

Actually these expressions are valid in the range 1000–30 000 K, whereas polynomial expressions are used for low temperature (<1000 K) [28]:

$$\ln \mu_s = \sum_{n=1}^{N} A_{n,s} (\ln T)^{n-1} \tag{10.12}$$

where $N = 4$ to ensure a third-order polynomial expression, with the corresponding coefficients provided in table 10.2.

Once the single-species viscosity are computed, they are linked each other to compute the mixture viscosity through Wilke's mixing rule [24]:

$$\mu = \sum_{s=1}^{N_s} \frac{\chi_s \mu_s}{\sum_{r=1}^{N_s} \chi_r \phi_{sr}} \tag{10.13}$$

where

$$\phi_{sr} = \frac{1}{\sqrt{8}}\left(1 + \frac{\mathcal{M}_s}{\mathcal{M}_r}\right)^{-1/2}\left[1 + \left(\frac{\mu_s}{\mu_r}\right)^{1/2}\left(\frac{\mathcal{M}_r}{\mathcal{M}_s}\right)^{1/4}\right]^2 \tag{10.14}$$

For the evaluation of the conductivity, Eucken's formula is employed [29]

$$\lambda_s = \mu_s\left(\frac{5}{4} + \alpha_s\right)\frac{\mathcal{R}}{\mathcal{M}_s} \tag{10.15}$$

Wilke's mixture rule for viscosity is adapted for the conductivity, with an empirical correction [25]:

$$\lambda = \sum_{s=1}^{N_s} \frac{\chi_s\, \mu_s}{\sum_{r=1}^{N_s} 1.065 \cdot \chi_r\, \phi_{sr}} \tag{10.16}$$

For the vibrational conductivity, Eucken's formula is employed:

$$\lambda_{m,vib} = \mu_m c_{v,vib} \tag{10.17}$$

The last transport property is the diffusion. As shown in [30], in order to find the diffusion velocities, one should solve the following system:

$$\nabla\chi_s = \sum_{r=1}^{N_s} \frac{\chi_s\chi_r}{\mathcal{D}_{sr}}(\mathbf{u}_r^D - \mathbf{u}_s^D) + (Y_s - \chi_s)\frac{\nabla p}{p} + \frac{\rho}{s}\sum_{r=1}^{N_s} Y_s Y_r(\mathbf{f}_s - \mathbf{f}_r), \tag{10.18}$$

where $\mathbf{f}$ represents the volume force field, and $\mathcal{D}_{sr} = \mathcal{D}_{rs}$ is the binary diffusion coefficient. This system has a size of N_s^2 and it should be solved in each point of the discretized space domain at each time step: it is immediate to understand that this task is prohibitive. For this reason, the approximation by Hirschfelder and Curtiss can be employed [31]. It reads:

$$\mathbf{u}_s^D Y_s = -\mathcal{D}_s \nabla Y_s + \mathbf{V}_c \tag{10.19}$$

in which $\mathcal{D}_s = \frac{1 - \chi_s}{\sum_{r \neq s} \chi_r / \mathcal{D}_{sr}}$ is an equivalent diffusion coefficient, representing the diffusion of the species s into the entire mixture. In this way, only N_s diffusive velocities have to be computed, each one representing the diffusion of species s. Note that equation (10.19) is the same as equation (10.5) written in terms of mass fractions (here $\mathbf{V}_c = \sum_{s=1}^{N_s}\sum_{l=1}^{V_s} Y_{s,l}\mathcal{D}_s \nabla Y_{s,l}$). In the present work, the binary diffusion coefficient is computed through the Chapman–Enskog approximation [31] $\mathcal{D}_{sr} = \bar{\mathcal{D}}_{sr}/p$ where p is the pressure in atmospheres. In order to compute $\bar{\mathcal{D}}_{sr}$, as for the viscosity, Gupta's curve fitting is employed:

$$\bar{\mathcal{D}}_{sr} = \exp(D_{\bar{\mathcal{D}}_{sr}})\; T^{\left[A_{\mathcal{D}_{sr}} (\ln T)^2 + B_{\mathcal{D}_{sr}} (\ln T) + C_{\mathcal{D}_{sr}}\right]} \tag{10.20}$$

whose coefficients are collected in table 10.3. As for the viscosity, this evaluation is valid for high temperatures (>1000 K), whereas for low temperatures a third-order polynomial fitting ($N = 4$) is used, namely:

$$\ln \bar{\mathcal{D}}_{sr} = \sum_{n=1}^{N} D_n (\ln T)^{n-1} \tag{10.21}$$

whose coefficients are listed in table 10.4.

Table 10.3. Coefficients to compute the diffusion coefficients (obtained in $cm^2 \cdot atms^{-1}$).

Species	$A_{\mathcal{D}_{sr}}$	$B_{\mathcal{D}_{sr}}$	$C_{\mathcal{D}_{sr}}$	$D_{\mathcal{D}_{sr}}$
$N_2 - N_2$	0.0	0.011 2	1.618 2	−11.309 1
$N_2 - O_2$	0.0	0.046 5	0.927 1	−8.113 70
N_2 − NO	0.0	0.029 1	1.267 6	−9.687 80
N_2 − N	0.0	0.019 5	1.488 0	−10.365 4
N_2 − O	0.0	0.014 0	1.582 4	−10.881 9
$O_2 - O_2$	0.0	0.041 0	1.002 3	−8.359 70
O_2 − NO	0.0	0.043 8	0.964 7	−8.238 00
O_2 − N	0.0	0.017 9	1.484 8	−10.281 0
O_2 − O	0.0	0.022 6	1.370 0	−9.663 10
NO-NO	0.0	0.036 4	1.117 6	−8.969 50
NO-N	0.0	0.018 5	1.488 2	−10.330 1
NO-O	0.0	0.017 9	1.484 8	−10.315 5
N-N	0.0	0.003 3	1.557 2	−11.161 6
N-O	0.0	−0.004 8	1.919 5	−11.926 1
O-O	0.0	0.003 4	1.557 2	−11.172 9

10.1.3 Multi-temperature Park model

The multi-temperature Park model for air without ionization accounts for a five species mixture and considers the following dissociation reactions:

$$N_2 + X \leftrightarrow 2N + X \tag{10.22a}$$

$$O_2 + X \leftrightarrow 2O + X \tag{10.22b}$$

$$NO + X \leftrightarrow N + O + X \tag{10.22c}$$

and two NO exchange (Zeldovich) reactions:

$$NO + O \leftrightarrow N + O_2 \tag{10.23a}$$

$$N_2 + O \leftrightarrow NO + N \tag{10.23b}$$

Since each species can interact with any of the other ones, X represents a generic partner in the mixture, leading to 17 reactions. For the evaluation of chemical source terms $\dot{\omega}_s$, the law of mass action is employed. It reads:

$$\dot{\omega}_s = \mathcal{M}_s \sum_{i=1}^{N_r} \nu_{is} \, RR_i \tag{10.24}$$

where N_r is the number of chemical reactions, $\mathcal{M}_s$ is the molar weight of species s, ν_{is} is the difference between product and reactant stoichiometric coefficients of the species s in the reaction i. Rate coefficients of each reaction i, RR_i, are evaluated as:

Table 10.4. Coefficients to compute the diffusion coefficients for $T < 1000$ K (obtained in cm^2atms^{-1}).

Species	D_1	D_2	D_3	D_4
$N_2 - N_2$	−15.214 153 832 7568	3.348 053 449 783 54	−0.223 365 726 041 766	0.009 817 787 279 110 16
$N_2 - O_2$	−15.444 092 035 0701	3.434 913 447 661 11	−0.233 997 710 214 864	0.010 250 333 590 0079
$N_2 - NO$	−15.232 871 598 8803	3.348 822 234 893 93	−0.223 470 939 336 837	0.009 822 569 164 846 91
$N_2 - N$	−14.196 345 265 8321	3.084 392 753 918 00	−0.190 058 218 177 093	0.008 414 615 865 196 94
$N_2 - O$	−14.426 625 088 8514	3.216 422 191 096 48	−0.206 956 131 557 607	0.009 135 288 598 631 68
$O_2 - O_2$	−15.791 696 756 4624	3.572 143 437 285 50	−0.251 846 982 846 213	0.011 025 333 315 9281
$O_2 - NO$	−15.458 691 370 9497	3.433 490 032 521 87	−0.233 804 784 243 045	0.010 241 646 761 9844
$O_2 - N$	−14.468 559 371 8154	3.196 105 707 055 57	−0.204 296 979 798 262	0.009 019 114 102 561 01
$O_2 - O$	−14.743 638 324 8890	3.350 090 901 251 31	−0.224 538 224 379 586	0.009 906 528 312 288 99
NO-NO	−15.248 512 707 0064	3.348 053 449 783 49	−0.223 365 726 041 759	0.009 817 787 279 109 82
NO-N	−14.231 417 863 1530	3.094 550 770 723 80	−0.191 458 795 972 500	0.008 478 715 811 882 17
NO-O	−14.468 597 494 3429	3.228 900 405 620 62	−0.208 656 835 058 384	0.009 212 292 164 403 43
N-N	−12.989 609 352 8909	2.709 173 171 626 86	−0.140 472 900 804 897	0.006 229 986 628 462 49
N-O	−13.148 285 973 0107	2.823 021 206 300 66	−0.155 308 213 943 692	0.006 874 572 014 926 05
O-O	−13.295 037 783 9835	2.938 989 816 956 74	−0.170 601 246 501 854	0.007 547 358 484 820 40

$$RR_i = k_{f,i} \prod_{s=1}^{5} C_s^{\nu_{is}'} - k_{b,i} \prod_{s=1}^{5} C_s^{\nu_{is}''} \tag{10.25}$$

Here, ν_{is}' and ν_{is}'' are reactant and product stoichiometric coefficients, respectively, k_f and k_b are forward and backward rate coefficients and C_s is the concentration of species s. Following [7], Arrhenius formula is employed to evaluate forward coefficients, thus:

$$k_{f,i} = A_i\, T_a^{n_i}\, e^{-T_{d_i}/T_a} \tag{10.26}$$

where $T_a = \sqrt{T_v\, T}$ for dissociation reactions and $T_a = T$ for NO exchange reactions. Parameters A_i, n_i and T_{d_i} are given in [7] or [32]. The following expression is used to evaluate equilibrium constant as a function of $Z = 10\,000/T_a$:

$$\ln K_{eq,i} = \frac{A_1}{Z} + A_2 + A_3 \ln Z + A_4\, Z + A_5\, Z^2 \tag{10.27}$$

whose coefficients are given in [7]. Finally, backward rate coefficients are computed as follows:

$$k_{b,i} = k_{f,i}/K_{eq,i} \tag{10.28}$$

Concerning the vibrational source terms of molecules $\dot{\omega}_{\text{vib},m}$, two contributions are considered. Specifically, $\dot{\omega}_{\text{vib},m}$ can be decomposed in the collisional ($\dot{\omega}_{\text{LT},m}$) and chemical ($\dot{\omega}_{\text{chem},m}$) parts.

The collisional contribution $\dot{\omega}_{\text{LT},m}$ represents the energy transfer between translational and vibrational modes, and is modeled by the Landau–Teller equation:

$$\dot{\omega}_{\text{LT},m} = \rho_m \frac{\varepsilon_{\text{vib},m}(T) - \varepsilon_{\text{vib},m}(T_v)}{\tau_m} \tag{10.29}$$

This term is equal to zero in the case of thermal equilibrium ($T = T_v$). Here, τ_m is the relaxation time. As first approximation, it is evaluated through the Millikan–White expression [7, 32], that reads:

$$\tau_{m,\,X}^{\text{MW}} = \frac{p_{\text{atm}}}{p} \exp\left[A_{m,X}(T^{-1/3} - B_{m,X}) - 18.42\right] \tag{10.30}$$

It accounts for the collision of the molecule m with a generic partner X. The coefficients $A_{m,X}$ and $B_{m,X}$ can be evaluated following simple expressions given in [33]: specific values of this work are reported in [32]. Nevertheless, for temperatures above 5000 K, the prediction of the relaxation time obtained by using the Millikan–White expression does not well reproduce experimental results [7, 32, 34], since it does not account for the limits in collision cross-sections. To overcome this drawback, Park proposed a correction in the evaluation of τ_m [7], taking into account the effective excitation cross-section, that reads:

$$\tau_{m,\,X}^{c} = \frac{1}{n_m \sigma \sqrt{\dfrac{8\mathcal{R}T}{\pi \mu_{m,X}}}} \tag{10.31}$$

n_m being the number density of molecule m, σ the effective excitation cross-section equal to $3 \cdot 10^{-17}(50\,000/T^2)$ cm^2 [32] and $\mu_{m,X} = \mathcal{M}_m\mathcal{M}_X/(\mathcal{M}_m + \mathcal{M}_X)$ the equivalent molecular weight of two colliding particles.

Hence, the relaxation time for the collision between a molecule and a generic partner X is given by $\tau_{m,X} = \tau_{m,X}^{\mathrm{MW}} + \tau_{m,X}^{c}$ and its mean value is evaluated through a weighted harmonic average:

$$\frac{1}{\tau_m} = \frac{1}{n_t}\sum_{s=1}^{N_s}\frac{n_s}{\tau_{m,X}} \tag{10.32}$$

Here n_t is the total number density, while n_s is the number density of the species s.

The second contribution of the vibrational source terms is $\dot{\omega}_{\mathrm{chem},m}$, that takes into account the influence of the dissociation on the vibrational energy. In this work, the harmonic oscillator model is employed, thus the energy exchanged in dissociation process is equally divided into vibrational and translational degrees of freedom [7]:

$$\dot{\omega}_{\mathrm{chem},m} = \frac{D_m}{2}\dot{\omega}_m \tag{10.33}$$

where D_m is the dissociation energy per unit mass of molecule m. Finally, vibrational temperatures are evaluated as follows:

$$\varepsilon_{\mathrm{vib},m} = \frac{R_m\theta_m}{\exp(\theta_m/T_v) - 1} \tag{10.34}$$

where R_m is the gas constant of molecule m, θ_m is the characteristic vibrational temperature of molecule m. Following [7], these are equal to 3393 K, 2273 K and 2739 K for N_2, O_2 and NO respectively.

10.1.4 State-to-state model

The assumption of a Boltzmann distribution for vibrational levels considered in the MT approach is not suitable and it cannot explain the large deviations of the rate coefficients from the Arrhenius law observed in some conditions [8, 14].

In this scenario, an StS approach can be employed to provide the distribution of all internal states of the mixture. Nevertheless, this approach is time consuming since, for an air mixture, it accounts for all the vibrational levels of N_2 (68) and O_2 (47), whereas only the ground state is considered for NO, N and O. For this reason, the StS model is considered prohibitive for 2D/3D CFD problems.

The elementary processes contributing to the vibrational relaxation are classified as vibration–vibration (VV) and vibration–translation by molecules (VTm) and by atoms (VTa) energy transfer, together with dissociation-recombination by molecules (DRm) and by atoms (DRa) and other chemical reactions. The full set of processes is here summarized:

$$N_2(v) + N_2(w) \leftrightarrow N_2(v-1) + N_2(w+1) \quad (\mathrm{VV}) \tag{10.35a}$$

$$O_2(v) + O_2(w) \leftrightarrow O_2(v-1) + O_2(w+1) \quad (\mathrm{VV}) \tag{10.35b}$$

$$O_2(v) + N_2(w-1) \leftrightarrow O_2(v-2) + N_2(w) \;\; \text{(VV)} \tag{10.35c}$$

$$N_2(v) + N_2 \leftrightarrow N_2(v-1) + N_2 \;\; \text{(VTm)} \tag{10.35d}$$

$$O_2(v) + O_2 \leftrightarrow O_2(v-1) + O_2 \;\; \text{(VTm)} \tag{10.35e}$$

$$N_2(v) + O_2 \leftrightarrow N_2(v-1) + O_2 \;\; \text{(VTm)} \tag{10.35f}$$

$$O_2(v) + N_2 \leftrightarrow O_2(v-1) + N_2 \;\; \text{(VTm)} \tag{10.35g}$$

$$N_2(v) + N \leftrightarrow N_2(v-\Delta v) + N \;\; \text{(VTa)} \tag{10.35h}$$

$$O_2(v) + O \leftrightarrow O_2(v-\Delta v) + O \;\; \text{(VTa)} \tag{10.35i}$$

$$N_2(v) + O \leftrightarrow N_2(v-1) + O \;\; \text{(VTa)} \tag{10.35j}$$

$$O_2(v) + N \leftrightarrow O_2(v-1) + N \;\; \text{(VTa)} \tag{10.35k}$$

$$N_2(v) + N_2 \leftrightarrow 2N + N_2 \;\; \text{(DRm)} \tag{10.35l}$$

$$O_2(v) + O_2 \leftrightarrow 2O + O_2 \;\; \text{(DRm)} \tag{10.35m}$$

$$N_2(v) + N \leftrightarrow 2N + N \;\; \text{(DRa)} \tag{10.35n}$$

$$O_2(v) + O \leftrightarrow 2O + O \;\; \text{(DRa)} \tag{10.35o}$$

$$N_2(v_{\max}) + O_2 \leftrightarrow N_2(v_{\max}+1) + O_2 \equiv 2N + O_2 \;\; \text{(LCm)} \tag{10.35p}$$

$$O_2(v_{\max}) + N_2 \leftrightarrow O_2(v_{\max}+1) + N_2 \equiv 2O + N_2 \;\; \text{(LCm)} \tag{10.35q}$$

$$N_2(v_{\max}) + O \leftrightarrow N_2(v_{\max}+1) + O \equiv 2N + O \;\; \text{(LCa)} \tag{10.35r}$$

$$O_2(v_{\max}) + N \leftrightarrow O_2(v_{\max}+1) + N \equiv 2O + N \;\; \text{(LCa)} \tag{10.35s}$$

$$NO + X \leftrightarrow N + O + X \tag{10.35t}$$

$$N_2(v) + O \leftrightarrow NO + N \tag{10.35u}$$

$$O_2(v) + N \leftrightarrow NO + O \tag{10.35v}$$

with X being a generic partner, whereas v and w are vibrational levels indexes. The total vibrational energy per unit mass is defined as follows:

$$\varepsilon_{\text{vib}} = \frac{1}{\rho}\sum_{s=1}^{N_s}\sum_{l=1}^{V_s}\rho_{s,l}\varepsilon_{s,l} \tag{10.36}$$

with $\rho_{s,l}$ and $\varepsilon_{s,l}$ being the density and the vibrational energy per unit mass of the species s in the vibrational level l (V_s is the total number of levels of the species s).

For non-equilibrium distributions, the vibrational temperature loses its meaning. Nevertheless, low energy levels approximately follow a Boltzmann distribution, thus providing the vibrational temperature (from the first two levels) as:

$$T_{v,m} = \frac{\mathcal{M}_m(\varepsilon_{m,2} - \varepsilon_{m,1})}{\mathcal{R}\ln\left(\dfrac{\rho_{m,1}}{\rho_{m,2}}\right)} \tag{10.37}$$

Obviously, other definitions of the internal temperature are possible [35], such as the temperature of the Boltzmann distribution with the same mean energy as the actual one:

$$\frac{1}{\mathcal{Q}_m(T_{v,m})} = \sum_{l=1}^{V_m} e^{-\varepsilon_{m,l}/T_{v,m}} \varepsilon_{m,l} = \frac{1}{\rho_m} \sum_{l=1}^{V_m} \rho_{m,l}\varepsilon_{m,l} \tag{10.38}$$

where

$$\mathcal{Q}_m(T_{v,m}) = \sum_{l=1}^{V_m} e^{-\varepsilon_{m,l}/T_{v,m}} \tag{10.39}$$

and ρ_m is the total mass density of the mth molecule.

10.2 Numerical scheme

10.2.1 Finite-volume approach

The governing equations are discretized in space using a cell-centered finite-volume approach once the computational domain is discretized into a finite number of volumes (surfaces in 2D configurations). In this work, multi-block structured meshes are employed, each cell within a block being uniquely identified by three integers (i, j, k). Equation (10.2) can be written in discrete form as:

$$V_{i,j,k}\frac{d\mathbf{U}_{i,j,k}}{dt} + \sum_{f=1}^{N_f}(\mathbf{F}\cdot\mathbf{n}\Delta S)_f = V_{i,j,k}\mathbf{W}_{i,j,k} \tag{10.40}$$

where $V_{i,j,k}$ is the volume of the cell, N_f is the number of faces (4 in 2D and 6 in 3D) and $\mathbf{n}$ and ΔS are the outward unit normal and the surface of face f. The evaluation of the conservative variables stored in $\mathbf{U}$ is performed at the center of each cell, whereas the fluxes $\mathbf{F}$ are evaluated on the cell's surfaces, as shown in figure 10.1. For the purpose of fluxes formulation, it has to be pointed out that a vector $\tilde{\mathbf{U}}$ to store primitive variables is needed. It reads:

$$\tilde{\mathbf{U}} = \left[\rho_{1,1}, \ldots, \rho_{1,V_1}, \ldots, \rho_{N_s,V_{N_s}}, u, v, w, p, \varepsilon_{\text{vib},N_2}, \varepsilon_{\text{vib},O_2}, \varepsilon_{\text{vib},NO}\right] \tag{10.41}$$

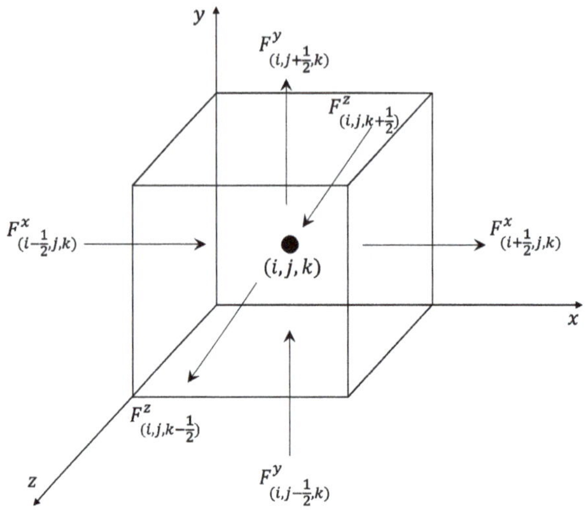

Figure 10.1. Computational cell and fluxes.

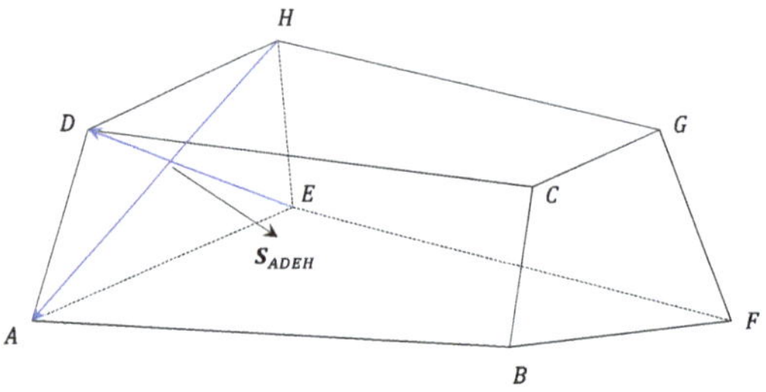

Figure 10.2. Sketch of a cell with surface vector.

To be consistent with the notation in section 10.1.1, one has to keep in mind that in figure 10.1 $\mathbf{F}^x = \mathbf{F}_E - \mathbf{F}_V$, $\mathbf{F}^y = \mathbf{G}_E - \mathbf{G}_V$ and $\mathbf{F}^z = \mathbf{H}_E - \mathbf{H}_V$.

The finite-volume approach requires the computation of the volume of each cell, along with the surface normal unit vector and area. In the following, the general procedure for a 3D configuration is presented, with the 2D configuration being a particular and easier case. Following the notation in [36], let us consider each cell as a six-sided hexahedron, as shown in figure 10.2. As an approximation, each edge of a surface is a straight-line segment so that one can divide the surface into two triangles. The surface vector is computed as the cross product of the two diagonals. As an example, considering the face ADHE, one has the surface vector:

$$\mathbf{S}_{\mathrm{ADHE}} = \frac{1}{2}(\mathbf{r}_{\mathrm{ED}} \times \mathbf{r}_{\mathrm{HA}}) \tag{10.42}$$

whose magnitude gives the area of the surface, also used to obtain the unit normal vector by normalization of the components of $\mathbf{S}_{\mathrm{ADHE}}$. Performing the same procedure on surfaces ABFE and EFGH, one can compute $\mathbf{S}_{\mathrm{ABFE}}$ and $\mathbf{S}_{\mathrm{EFGH}}$.

Finally, the volume of the cell is computed by subdividing the hexahedron into three pyramids that share the same diagonal EC, thus:

$$V = \frac{1}{3}(\mathbf{S}_{\mathrm{ADHE}} + \mathbf{S}_{\mathrm{ABFE}} + \mathbf{S}_{\mathrm{EFGH}}) \cdot (\mathbf{r}_{\mathrm{EC}}) \tag{10.43}$$

10.2.2 Convective fluxes discretization

The use of upwind schemes is a typical way to ensure stability when dealing with supersonic flows, in which shock waves propagate in the flow. Dealing with complex fluid flows, the solution presents discontinuities, that in the context of finite-volume methods require the solution of an approximate Riemann problem at each interface: the solution is represented by an arbitrary order polynomial function, that reduces to an average in case of a first-order scheme. Despite this, many reasons for using upwind schemes exist. For instance, in the case of implicit algorithms the use of centered schemes forces to the inversion of tridiagonal or pentadiagonal matrices. It can be a very tedious operation if compared with the treatment of triangular matrices when employing one-sided (upwind) schemes. Moreover, they often present more stable properties with respect to centered schemes. In the case of hyperbolic systems of equations, one has to evaluate the eignenvalues of the Jacobian matrices of inviscid fluxes; these define the speed and direction of the signals propagation that, also in the case of mixed sign eigenvalues, allow one to properly apply the upwinding. For this reason, a splitting approach for the fluxes formulation is an effective way for applying upwind methods.

Here, the flux vector splitting approach is employed, following the formulation proposed by Steger and Warming [37, 38]. The main idea is the splitting of the flux vector into two contributions: one linked to the positive eigenvalues (+) and the other linked to the negative eigenvalues (−). In this way, one can formulate the fluxes following the effective propagation direction of the characteristics, assumed to be normal to the cell face. This ensures stability also when dealing with subsonic flows (mixed signs eigenvalues). The total flux vector will be the sum of the positive and negative contributions. As an example, referring to figure 10.3, consider the 1D Euler's equations given by the following systems of partial differential equations (PDEs):

$$\frac{\partial \mathbf{U}_i}{\partial t} + \frac{\partial \mathbf{F}}{\partial x} = 0 \tag{10.44}$$

where $\mathbf{U} = [\rho, \rho u, \rho e]$ and $\mathbf{F} = [\rho u, \rho u^2 + p, (\rho e + p)u]$. Once the spatial discretization is performed, one has the solve the system of ordinary differential equations (ODEs):

$$\frac{d\mathbf{U}_i}{dt} = -\frac{\mathbf{F}_{i+1/2} - \mathbf{F}_{i-1/2}}{\Delta x} \tag{10.45}$$

Here, $\mathbf{F}_{i+1/2}$ and $\mathbf{F}_{i-1/2}$ are the total fluxes contributions, computed as follows:

$$\mathbf{F}_{i\pm1/2} = \mathcal{F}^{+}_{i\pm1/2} + \mathcal{F}^{-}_{i\pm1/2} \tag{10.46}$$

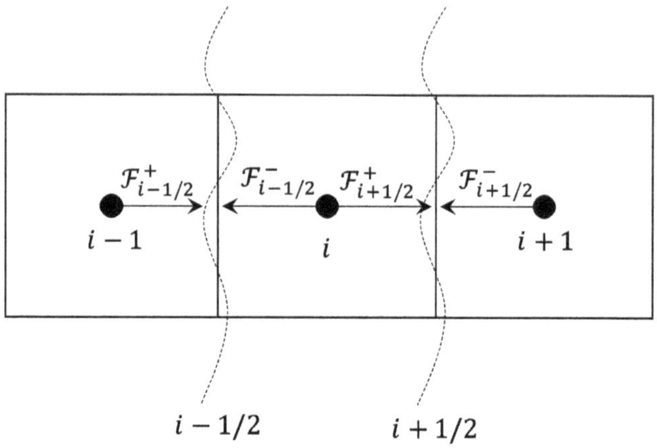

Figure 10.3. Interpretation of the flux vector splitting.

For the 1D gas dynamics equation, the expressions of $\mathcal{F}$ are given in [37]. Indicating with c the speed of sound, the three-dimensional generalized form of the splitting is:

$$\mathcal{F}^{\pm}=\frac{\rho}{2\bar{\gamma}}\begin{bmatrix} \beta \\ \beta\, u + c(\lambda_2^{\pm} - \lambda_3^{\pm})\hat{n}_x \\ \beta\, v + c(\lambda_2^{\pm} - \lambda_3^{\pm})\hat{n}_y \\ \beta\, w + c(\lambda_2^{\pm} - \lambda_3^{\pm})\hat{n}_z \\ \beta\left(\frac{u^2+v^2+w^2}{2} + \frac{\varepsilon_{\text{int}}}{\rho}\right) + c\, v_n(\lambda_2^{\pm} - \lambda_3^{\pm}) + c^2\frac{\lambda_2^{\pm}+\lambda_3^{\pm}}{\bar{\gamma}-1} \\ \beta\, \varepsilon_{\text{vib},N_2} \\ \beta\, \varepsilon_{\text{vib},O_2} \\ \beta\, \varepsilon_{\text{vib,NO}} \end{bmatrix} \tag{10.47}$$

where $\beta = 2(\bar{\gamma}-1)\lambda_1^{\pm} + \lambda_2^{\pm} + \lambda_3^{\pm}$, the eigenvalues are $(\lambda_1^{\pm}, \lambda_2^{\pm}, \lambda_3^{\pm}) = (v_n^{\pm}, v_n^{\pm}+c, v_n^{\pm}-c)$, and $v_n = u\,\hat{n}_x + v\,\hat{n}_y + w\,\hat{n}_z$, with $\left(\hat{n}_x, \hat{n}_y, \hat{n}_z\right)$ being the components of the unit normal vector to the cell face. It has to be pointed out that the internal energy ε_{int} is the sum of the chemical (formation) and vibrational contributions. More mathematical details are given in [37, 38]. For simplicity, in equation (10.47) symbols + and − have been reported only for the eigenvalues, to distinguish positive and negative ones. Nevertheless, these eigenvalues are in turn linked to the reconstruction of the solution in a certain direction.

To increase the order (and so, the accuracy) of the spatial reconstruction, a MUSCL (*monotone upstream-centered schemes for conservation laws*) scheme is employed [39]. Following the notation of the 1D configuration given in figure 10.4, let us consider the reconstructions of a generic variable u at the interface $i + 1/2$:

$$u^r_{i+1/2} = u_{i+1} - \frac{1+\kappa}{4}\delta u_{i+1/2} - \frac{1-\kappa}{4}\delta u_{i+3/2} \tag{10.48a}$$

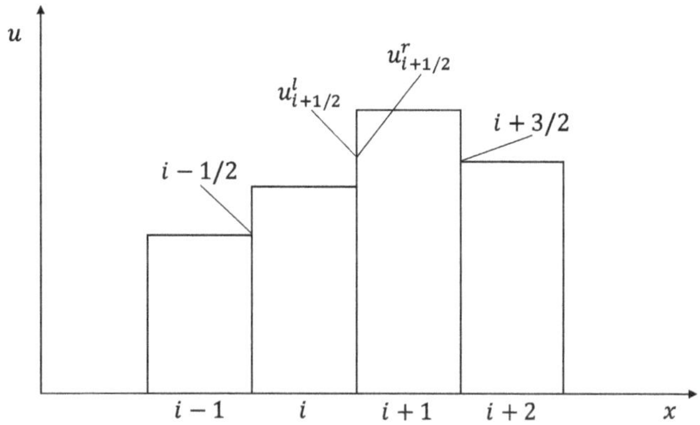

Figure 10.4. MUSCL reconstruction for spatial discretization.

$$u^l_{i+1/2} = u_i + \frac{1-\kappa}{4}\delta u_{i-1/2} + \frac{1+\kappa}{4}\delta u_{i+1/2} \tag{10.48b}$$

in which

$$\delta u_{i-1/2} = u_i - u_{i-1}, \quad \delta u_{i+1/2} = u_{i+1} - u_i, \quad \delta u_{i+3/2} = u_{i+2} - u_{i+1}$$

and κ is an integer parameter, whose value leads to different schemes, as follows:

$$\kappa = \begin{cases} 0 \longrightarrow \text{second order centered scheme} \\ -1 \longrightarrow \text{second order fully upwind scheme} \end{cases}$$

These reconstructions of the solution can be used for the flux formulation, keeping in mind that the positive flux contribution is linked to the left reconstruction, whereas the negative flux contribution is linked to the right reconstruction.

Despite the higher accuracy of a second-order scheme with respect to a first-order one, it is known from Godunov's theorem that linear high order schemes are not monotone, producing spurious oscillations near discontinuities [40]. In order to obtain a TVD (*total variation diminishing*) scheme, *flux limiters* functions are introduced in the reconstruction of the solution, with the purpose of combining first-order and higher order schemes: where the solution is continuous, the scheme follows a second-order formulation, whereas in the presence of discontinuities the scheme readapts itself to a first-order formulation. Hence, equation (10.48) becomes:

$$u^r_{i+1/2} = u_{i+1} - \frac{1+\kappa}{4}\phi(r_1)\delta u_{i+1/2} - \frac{1-\kappa}{4}\phi(r_2)\delta u_{i+3/2} \tag{10.49a}$$

$$u^l_{i+1/2} = u_i + \frac{1-\kappa}{4}\phi(r_3)\delta u_{i-1/2} + \frac{1+\kappa}{4}\phi(r_4)\delta u_{i+1/2} \tag{10.49b}$$

where

$$r_1 = \frac{\delta u_{i+3/2}}{\delta u_{i+1/2}}, \quad r_2 = \frac{\delta u_{i+1/2}}{\delta u_{i+3/2}}, \quad r_3 = \frac{\delta u_{i+1/2}}{\delta u_{i-1/2}}, \quad r_4 = \frac{\delta u_{i-1/2}}{\delta u_{i+1/2}}$$

with ϕ being the limiter function, that can assume different forms. It is $\phi(r) = \frac{r^2+r}{1+r^2}$ in the case of *van Albada* formulation or $\phi(r) = \max(0, \min(1, r))$ in the case of *minmod* approach.

In conclusion, the Steger–Warming flux vector splitting, combined with a second-order MUSCL reconstruction and flux limiters, provides high accuracy and numerical stability at the same time; specifically, dealing with hypersonic flows and shock waves, the scheme turns out to be stable also in the case of mixed sign eigenvalues, preserving the upwind nature of the fluxes formulation.

10.2.3 Diffusive fluxes discretization

The discrete representation of the diffusive terms requires the local evaluation of the flow gradients. These are mainly linked to the computation of stress tensor (velocity gradients) and heat flux (temperature gradients). These gradients can be evaluated by means of the generalized Gauss theorem:

$$\int_V \nabla\phi \, dV = \int_S \phi \, \mathbf{n} \, dS \tag{10.50}$$

where ϕ is a generic flow variable. Here, the main idea is to compute the gradients on the faces of each control volume. Consider two adjacent control volumes (in the computational domain) sharing face f (see 10.5), where the gradient of ϕ must be computed. Here index i is kept constant. The following procedure is valid for a 3D configuration (with the 2D formulation a particular case). By discretizing equation (10.50), one obtains:

$$\nabla\phi = \frac{1}{V_G} \sum_{f=1}^{N_f} \phi_f \bar{\mathbf{S}}_f \tag{10.51}$$

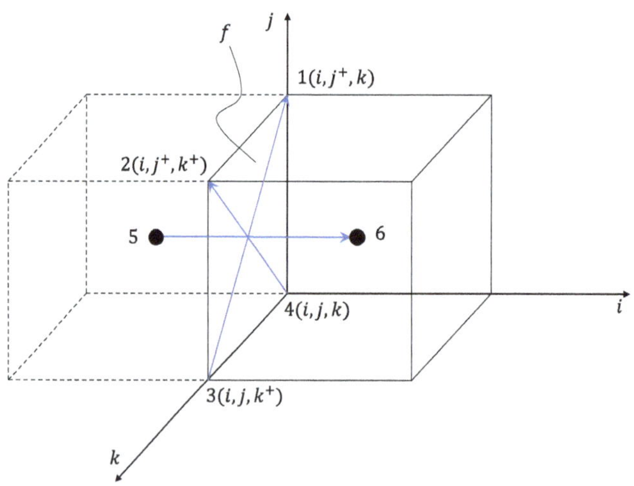

Figure 10.5. Sketch of numerical domain to perform Gauss theorem. Symbol + indicates the general increment of the index.

where V_G and $\bar{\mathbf{S}}_f$ are the volume and surface normal vectors of a fake cell, with N_f faces: we will refer to this cell as *Gauss cell* (subscript G stands for Gauss). For instance, in order to formulate the fluxes in i direction, one has to analyze the jk plane, as shown in figure 10.6: the dashed line identifies the Gauss cell of volume V_G, that extends from point 5 to point 6. Surface normal vectors can be computed through cross product of diagonals, as follows:

$$\bar{\mathbf{S}}_1 = \mathbf{r}_{42} \times \mathbf{r}_{56} \quad \bar{\mathbf{S}}_2 = \mathbf{r}_{56} \times \mathbf{r}_{31} \quad \bar{\mathbf{S}}_6 = \mathbf{r}_{31} \times \mathbf{r}_{42} \tag{10.52}$$

and obviously $\bar{\mathbf{S}}_3 = -\bar{\mathbf{S}}_1$, $\bar{\mathbf{S}}_4 = -\bar{\mathbf{S}}_2$ and $\bar{\mathbf{S}}_5 = -\bar{\mathbf{S}}_6$. In equation (10.51), ϕ_f represents the value of the flow variable ϕ in the point f. Since the solver is volume-based (variables computed in each cell center), ϕ_5 and ϕ_6 are obtained directly from the solution of the equations, whereas its values in the nodes (ϕ_1, ϕ_2, ϕ_3 and ϕ_4) are evaluated through a linear interpolation between values in the adjacent cells.

Once the gradients are computed, the fluxes can be assembled. While doing this, since they are evaluated on the interface between two contiguous cells, the values of the variables needed for the calculation (e.g. viscosity μ to compute $\boldsymbol{\sigma}$) have to be projected on the face using a linear interpolation (average) between points 5 and 6.

10.2.4 Time integration

Dealing with hypersonic entry flows, the problem formulation is stiff. The fluid dynamics time scale is comparable to the chemical one, but the treatment of the chemical and vibrational source terms is not carried out with the same temporal discretization. Indeed,

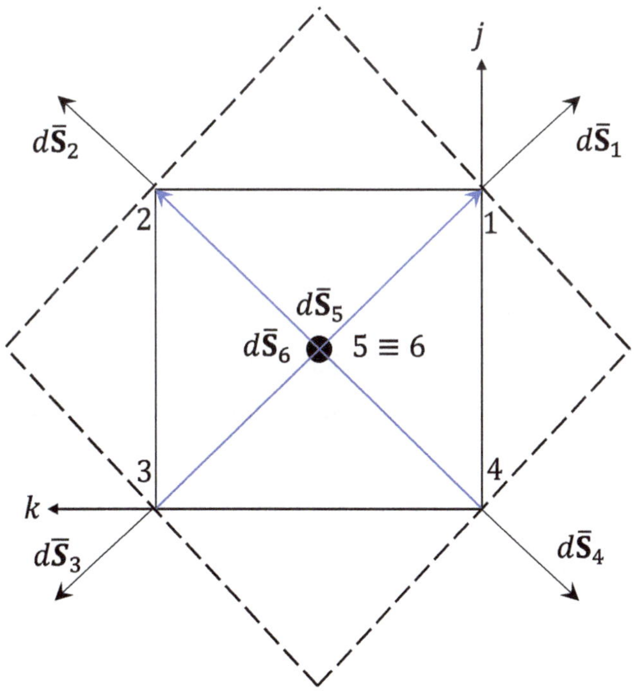

Figure 10.6. Interpretation of viscous fluxes in i direction.

to ensure stability a splitting approach is used: as shown in the next subsection, the source terms are evaluated at the end of the fluid dynamic time integration step. In other words, in the first step the homogeneous equations are solved (frozen step). So, once the spatial discretization is performed, equation (10.40) is written as:

$$\frac{d\mathbf{U}_{i,j,k}}{dt} = -\frac{1}{V_{i,j,k}}\sum_{f=1}^{N_f}(\mathbf{F}\cdot\mathbf{n}\Delta S)_f \tag{10.53}$$

The solution of this system of ordinary differential equations (ODEs) is performed through a third-order Runge–Kutta scheme [41], as follows:

$$\mathbf{U}^{(1)} = \mathbf{U}^n + \Delta t\ \mathcal{L}(\mathbf{U}^n) \tag{10.54a}$$

$$\mathbf{U}^{(2)} = \frac{3}{4}\mathbf{U}^n + \frac{1}{4}\mathbf{U}^{(1)} + \frac{1}{4}\Delta t\ \mathcal{L}(\mathbf{U}^{(1)}) \tag{10.54b}$$

$$\mathbf{U}^{n+1} = \frac{1}{3}\mathbf{U}^n + \frac{2}{3}\mathbf{U}^{(2)} + \frac{2}{3}\Delta t\ \mathcal{L}(\mathbf{U}^{(2)}) \tag{10.54c}$$

where $\mathcal{L}(\mathbf{U})$ is the spatial discretization operator (spatial indices i, j and k are omitted for simplicity); n and $n+1$ are, respectively, the old and new time step, whereas indexes (1) and (2) indicate the first and second Runge–Kutta sub-steps. The time step size is evaluated on the basis of the stability condition. According to [42] and considering a constant value of the CFL number:

$$\Delta t_{i,j,k} = \frac{\mathrm{CFL}\ V_{i,j,k}}{\left|\mathbf{u}_{i,j,k}\cdot\bar{\mathbf{S}}_i\right| + \left|\mathbf{u}_{i,j,k}\cdot\bar{\mathbf{S}}_j\right| + \left|\mathbf{u}_{i,j,k}\cdot\bar{\mathbf{S}}_k\right| + c_{i,j,k}\left(\bar{\mathrm{S}}_i + \bar{\mathrm{S}}_j + \bar{\mathrm{S}}_k\right)} \tag{10.55}$$

where $c_{i,j,k}$ and $\mathbf{u}_{i,j,k}$ are the speed of sound and the flow velocity in the cell (i, j, k), respectively; $\bar{\cdot}$ stands for face average value. Through this formulation, the time step size is calculated in each cell. At this stage, one can choose to perform the numerical simulation with a *global time stepping* approach, reducing the time step size to the minimum among all the cells. This approach ensures the transient state capturing. In contrast, if one is interested in a fast convergence to steady state, a *local time stepping* approach can be employed. In this case, each cell advances with its own time step size.

10.2.5 Evaluation of source terms: splitting approach

As anticipated, one has to account for the stiffness of the source terms when solving the system of equations for a reacting mixture. The strategy adopted is a splitting scheme, that first solves the homogeneous equations, and then updates the densities of each species in the mixture through an iterative Gauss–Seidel scheme. This is done by advancing in time with a time step size that is a fraction of the fluid dynamic one given in equation (10.55), such that:

$$\sum_{v=1}^{n}\Delta t^{(v)} = \Delta t \tag{10.56}$$

where n is the number of sub-steps. In general, it is reasonable to set $\Delta t^{(v)} = \Delta t/n$, with Δt evaluated through the CFL condition. In order to compute actual densities in the mixture, the species continuity equations are written as:

$$\frac{d\mathbf{y}}{dt} = \mathbf{P} - \mathbf{L} \cdot \mathbf{y} \tag{10.57}$$

where $\mathbf{y} = [\rho_1, \cdots, \rho_{N_L}]$ is the solution vector containing the densities of the species in the mixture (for simplicity, here N_L represents the total number of vibrational levels, namely five for the MT and 118 for the StS model in the case of air mixture). **P** and **L** are, respectively, a vector and a diagonal matrix whose components are non-negative and represent production and loss terms for a certain species. After discretizing equation (10.57), one obtains:

$$(\mathbf{I} + \Delta t^{(v)}\mathbf{L}) \cdot \mathbf{y}(t + \Delta t^{(v)}) = \Delta t^{(v)}\mathbf{P} + \mathbf{y}(t) \tag{10.58}$$

Despite the linear form of equation (10.58), it is non linear since **P** and **L** depend on **y**. By the way, matrix inversion is not needed since these represent uncoupled equations; indeed, the density of the generic species (vibrational level) s at the inner iteration k can be obtained through the following expression:

$$y_s^k(t + \Delta t^{(v)}) = \frac{\Delta t^{(v)}\, P_s(\mathbf{y}^{k-1}) + y_s(t)}{1 + \Delta t^{(v)}\, L_s(\mathbf{y}^{k-1})} \tag{10.59}$$

The parallelization of this problem is straightforward.

10.3 GPU clustering

10.3.1 CUDA environment

In principle, the main goal of the GPUs was video rendering. With technology progress, these hardware components have become more and more powerful. One of the best international companies is NVIDIA Corporation, that took (and takes) care of the improvement of the user interface on many laptops, mobile devices and gaming consoles.

Nevertheless, in the last few years GPUs have become an important computational resource that provides a non-negligible increment of performance in scientific computing. Thanks to a language interface, today several programming languages such as Fortran, C or C++ take advantage of the incredible performances of the last generation GPUs. Indeed, they are composed of thousands of micro-cores with a very large bandwidth for data transfer, so that parallel tasks can be performed with no need of manual handle in cores communications.

One of the most developed GPU programming environments is CUDA, that contains several extensions for basic programming languages. It allows one to develop GPU application to drastically reduce the computational cost. The most important benefit is the simplicity in the implementation of a function. For a better understanding, we will refer to a GPU function as *device kernel* (or simply *kernel*), whereas the part of the algorithm assigned to the central processing unit (CPU) is called *host*. This lets the reader understand that, although there is the possibility of assigning a task to a GPU, the overall algorithm is developed and run on conventional CPUs as shown in figure 10.7. A pure GPU code cannot exist since the initialization of the algorithm is always performed by the host.

In the following, a general overview of a kernel development is given, with particular attention to the definition of *blocks* and *threads*. For more programming details, the reader can refer to [18].

10.3.2 Kernel development

Before going ahead, it is necessary to introduce some definition of GPUs' hierarchy. Referring to figure 10.8, let us introduce some fundamental definition.

- *Thread*: the lowest level in the hierarchy, a thread represents a programming abstraction to which a certain operation is assigned (for instance, the sum of two elements).
- *Block*: it is a group of threads that can communicate with each other. Currently, a block can contain up to 1024 threads, but it depends on the hardware.
- *Grid*: ensemble of blocks to which the kernel invocation is assigned. Currently, a grid can contain up to 65 535 blocks.

Each thread and each block are identified by an index (`threadIdx` and `blockIdx`); the total number of threads in a block defines its dimension (`blockDim`), whereas the total number of blocks in a grid is referred to as `gridDim`. These are built-in functions useful to define the loop counter.

As an example, let us consider the code below. The arrays devoted to the computation (namely `a`, `b` and `c`) are declared and initialized on the host memory, and then copied on device through the command `cudaMemcpy`. Often the reverse operation can be useful (copying from device to host). A typical situation is the generation of an output file, since some function such as `fopen()` cannot be run on GPUs. In that case, it is affordable to perform the computations on GPUs and then copy device variables in host memory for file printing. After the declaration and initialization of the arrays, function calls are performed.

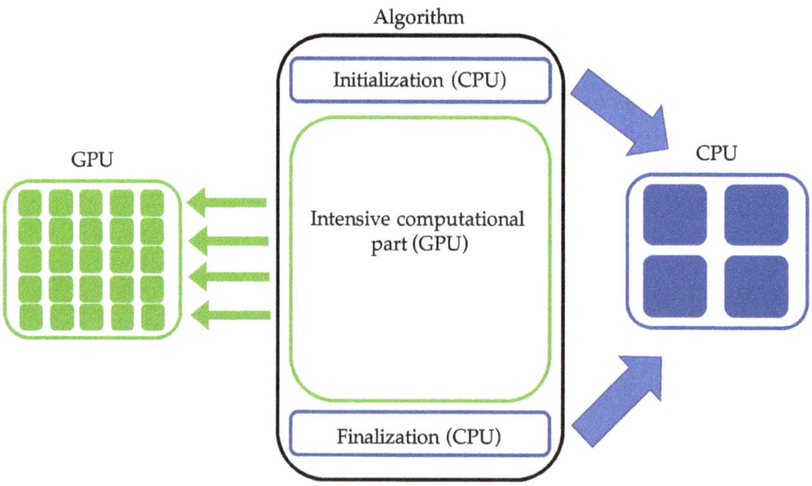

Figure 10.7. Assignment of the tasks to GPU and CPU.

```
#include <stdio.h>
#include <stdlib.h>

__global__
void sum_GPU(int *a, int *b, int *c, int N) {
    int i = threadIdx.x;
    if(i < N){
        c[i] = a[i] + b[i];
    }
}

void sum_CPU(int *a, int *b, int *c, int N) {
    int i = 0;
    while(i < N){
        c[i] = a[i] + b[i];
        i += 1;
    }
}

int main(void) {

    // Declaration of the arrays
    const int N = 1000;                 // This will be the size
    int a[N], b[N], c[N];               // Host variables (CPU)
    int *a_dev, *b_dev, *c_dev;         // Device variables (GPU)
    int B=1, T=N;                       // Blocks and Threads for kernel execution

    // Initialize host variables
    for (int i = 0; i < N; ++i)
    {
        a[i] = i + 1;
        b[i] = i + 2;
        c[i] = 0;
    }

    // Allocate device variables and copy from host
    cudaMalloc((void**)&a_dev,(N)*sizeof(int));
    cudaMalloc((void**)&b_dev,(N)*sizeof(int));
    cudaMalloc((void**)&c_dev,(N)*sizeof(int));
    cudaMemcpy(a_dev,a,(N)*sizeof(int),cudaMemcpyHostToDevice);
    cudaMemcpy(b_dev,b,(N)*sizeof(int),cudaMemcpyHostToDevice);
    cudaMemcpy(c_dev,c,(N)*sizeof(int),cudaMemcpyHostToDevice);

    // Perform the sum
    sum_CPU(a,b,c,N);
    sum_GPU<<<B,T>>>(a_dev,b_dev,c_dev,N);

    cudaFree(a_dev);
    cudaFree(b_dev);
    cudaFree(c_dev);

    return 0;

}
```

Listing 10.1. Example code: sum of two arrays

The differences in terms of implementation between a CPU function and a device kernel are minimal. The lettering `__global__` identifies the beginning of a kernel: the form of the loop is the same in the two cases, apart from the counter variable. In a CPU function it is an integer whose value is increased to span all the elements of the array, whereas in the case of a kernel it identifies each thread in a block. As is illustrated in figure 10.8, the organization of threads and blocks follows a multi-dimensional structure. Many ways to identify a certain thread exist [18]: a simple one is the 1D indexing, in which each thread is uniquely defined by an integer (line 6), that will refer to a specific element of the array. So, when calling a kernel, it will be run by a certain number of threads and blocks that are specified in the invocation. This is the second important difference: in the lettering `<<<B,T>>>` (line 46), the first integer is the number of blocks and the second one is the number of threads per block that will execute the kernel. Their product will give the total number of threads involved in the computation. These parameters depend on the hardware components of the machine and affect the efficiency of the computation: normally, the number of threads must not exceed the parameter `maxThreadsPerBlock` of the device

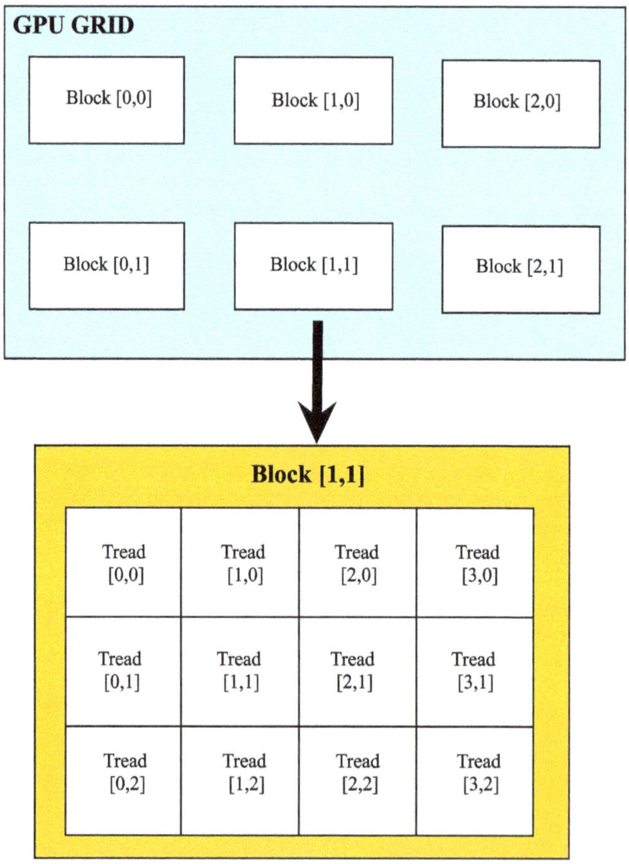

Figure 10.8. Grid structure in kernel invocation.

properties. Unfortunately, this is just a very simple example in which the array `c` contains 1000 elements and it is quite easy to find a suitable kernel configuration. Instead, when dealing with complex fluid flows, the number of elements is generic. In the view of algorithm scalability, one has to generalize the kernel configuration, taking care of the fact that enough threads are involved in the computation. If, for example, the array `c` contains 7137 elements (`N` =7137), it is not immediate to find an opportune combination of threads and blocks. So, a common procedure is to fix the number of threads, for instance `T` =128. At this point, the user must be sure to execute the kernel with enough threads (⩾`N`): the number of blocks needed can be computed as `B` =(`N` + `T` − 1)/`T`, that for this example is ≈56. This operation ensures launching the kernel with a suitable number of threads. In this case, the counter in the kernel must be defined differently. Also, a question arises: what happens when dealing with huge size arrays? Let us analyze this in detail.

In general, it might happen that the total number of threads involved in the computation does not cover the entire array. Let us suppose that we have set `T` =128. Since the maximum `gridDim` is 65 535, we would get an error when the size of our arrays exceeds 65 535 × 128 = 83 884 80. Even if this seems a very large number, there are many situations in which an array has a bigger size (for instance, when solving the flow equations employing the StS model, in which 118 densities must be stored in a grid with 512 × 256 cells). So, first of all we must compute the number of blocks as `B` =min(65 535, (`N` + `T` − 1)/`T`); also, we need to modify the counter in the kernel loop.

In this case, the counter at line 6 is defined as a combination of thread and block

```
1 __global__
2 void sum_GPU(int *a, int *b, int *c, int N) {
3     int i = threadIdx.x + blockIdx.x * blockDim.x;
4     while(i < N){
5         c[i] = a[i] + b[i];
6         i += blockDim.x * gridDim.x;
7     }
8 }
```

Listing 10.2. Modification of the previous kernel when the threads are less than the size of the array

index. If the total number of threads involved is less than the size of the array, an increment of the counter is needed. This allows one to use the blocks more than once. In this case, the 1D indexing follows a hierarchy as shown in figure 10.9.

10.3.3 MPI-CUDA environment

MPI environment (message passing interface) allows one to execute an algorithm on multiple CPUs to reduce the computational cost. MPI libraries contain built-in functions for data transfer between processors. Moreover, they allow one to take advantage of cluster supercomputers for multiple nodes executions. In this view, an

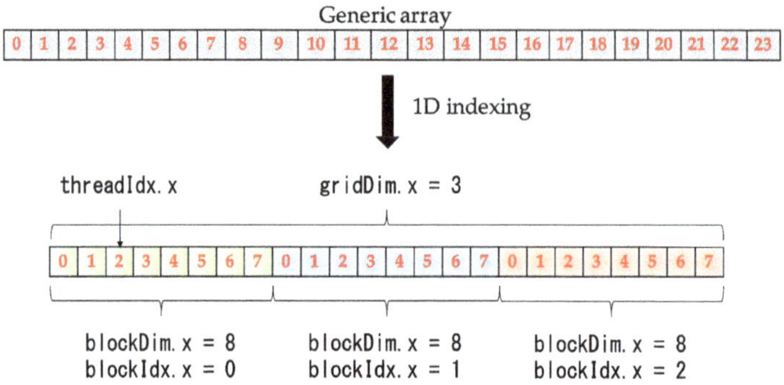

Figure 10.9. 1D indexing for kernel loop.

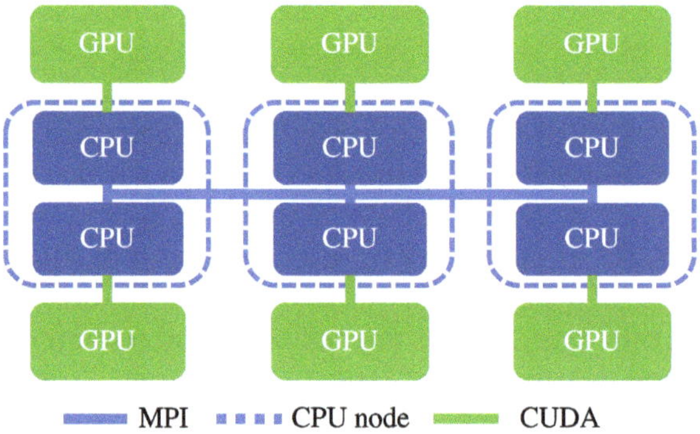

Figure 10.10. Scheme of MPI-CUDA parallelization on GPU clusters.

hybrid parallelization procedure can drastically increase the computational performance. A sketch of the strategy is given in figure 10.10.

The initialization is performed by the CPUs, responsible of the creation of a Cartesian topology of the domain through the well-known MPI libraries `MPI_ Cart_ create`, `MPI_ Cart_ coords` and `MPI_ Cart_ shift`. Within each cluster's node, the total number of GPUs is found through `cudaGetDeviceCount ()`; each of them is associated with a specific MPI process through `cudaSetDevice`. After the initialization, the main process (MASTER) reads the input data and broadcasts them to all the other processors, so that the computations can be performed in all the portions of the domain. Communications at the boundaries are ensured through the classical functions `MPI_ Send` and `MPI_ Recv`. Normally, GPUs communications require a re-adaptation of the transfer variables; once a data structure is created, the variables are copied from device to host to perform the data transfer, and then copied for the last time to device, before starting the new iteration. This tricky procedure can be by-passed thanks to CUDA-Aware MPI, that allows GPU–GPU communications.

10.3.4 Kernel examples

In this section, some kernels are shown as examples. For simplicity, a 2D configuration is considered.

MUSCL reconstruction

The first one is the kernel to reconstruct the primitive variables in space (*i* direction, 2D). For a better understanding of this part, let us refer to figure 10.11 (recall that C starts counting from 0). Here, a very simple 2D mesh is sketched.

```
__global__
void zz_muscl_i(int IMAX, int JMAX, REAL *qph, REAL *qp, REAL *qm, REAL
    rkap, int iphi)
{
   int i,j,l,k,kk,N;
   int jm_imax = (JMAX-1)*IMAX;
   REAL rkp,rkm,erre,phi1,phi2,phi3,phi4,d1,d2,d1m,d2m;

   N = NVAR*jm_imax;
   k = threadIdx.x + blockIdx.x * blockDim.x;

   rkp = 1.0 + rkap;
   rkm = 1.0 - rkap;

   while(k < N) {

      l  = k/jm_imax;     // Eq. index: jump plane of jm_imax faces
      kk = k%jm_imax;     // Useful to compute (i,j) of the cells
      j  = kk/IMAX;
      i  = kk%IMAX;

      // Compute delta
      d1  = Qph(i,j,l)   - Qph(i-1,j,l);
      d1m = Qph(i-1,j,l) - Qph(i-2,j,l);
      d2  = Qph(i+1,j,l) - Qph(i,j,l);
      d2m = Qph(i,j,l)   - Qph(i-1,j,l);

      // Minmod limiter
      erre = d2/(d1+1e-20);
      phi1 = MAX(0.0, MIN(1.0,erre));
      erre = d1/(d2+1e-20);
      phi2 = MAX(0.0, MIN(1.0,erre));
      erre = d1/(d1m+1e-20);
      phi3 = MAX(0.0, MIN(1.0,erre));
      erre = d1m/(d1+1e-20);
      phi4 = MAX(0.0, MIN(1.0,erre));

      // Reconstruct the solution
      Qp(i,j,l) = Qph(i,j,l)   - 0.25*(rkp*phi1*d1  + rkm*phi2*d2);
      Qm(i,j,l) = Qph(i-1,j,l) + 0.25*(rkm*phi3*d1m + rkp*phi4*d2m);

      k += blockDim.x * gridDim.x;

   }

}
```

Listing 10.3. Example code: MUSCL reconstruction in *i* direction

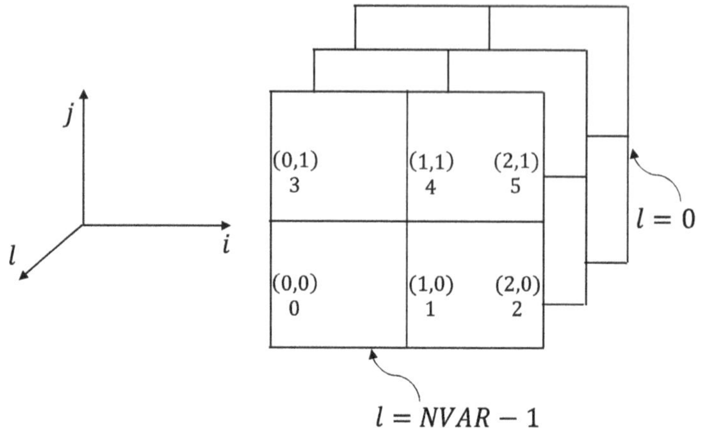

Figure 10.11. Interpretation of solution arrays.

In this kernel `IMAX` and `JMAX` represent the total number of nodes in each direction (in this case, three for each direction), useful to compute the total number of cells (`(IMAX-1)×(JMAX-1)`=4) and faces (`IMAX×(JMAX-1)+(IMAX-1)×JMAX` =12); `Qph` represents the macro function of the one-dimensional array of primitive variables (cell center), whereas `Qp` and `Qm` represent their right and left reconstructions with respect to the faces, respectively. At line 5, we define the total number of faces where the solution must be reconstructed: for the *i* direction it is `IMAX×(JMAX-1)`=6 and this value is stored in `jm_ imax` (the corresponding *j* direction would handle `(IMAX-1)×JMAX` faces). Since the reconstruction must be performed on all the primitive variables (equations), the loop will work on `N=NVAR×jm_ imax` elements (line 8), with `NVAR` =7 for a 2D configuration.

The most delicate part is the calculation of the integers (`i,j,l`) to identify the variable to be reconstructed (`l`) and the face where the reconstruction must be performed (`i,j`). Normally, a CPU function would take advantage of three `for()` loops, spanning all the spatial points for every variable. Nevertheless, this would be very inefficient for a device kernel. So, as anticipated above, we will build our variables by using a 1D indexing. The loop counter is `k` and identifies each thread. With the search for programming compactness, all the variables are stored in the same array, for example `Qp`: hence, `Qp` can be interpreted as a series of planes (specifically, `NVAR` planes), each linked to a specific variable (equation). In this simple case, `Qp` would contain 6 × `NVAR` elements: the first 6 elements (from 0 to 5) will represent the first primitive variable (e.g. ρ), elements from 6 to 11 will represent the second variable (e.g. u) and so on. So before performing the reconstruction, the equation index `l` must be computed, as is done at line 16. This ratio will give 0 for the first 5 faces: hence, the reconstruction will be performed on the first variable (`l` =0); in the case of element 6 the reconstruction would be performed on the second variable (`l` =1), and so on. When the equation index `l` is calculated, the second task is the identification of the face where the reconstruction will be performed. For this purpose, the remainder between `k` and `jm_ imax` is computed, so that the displacement among the elements of the array, linked to variable `l`, is unique. For example, for face number 4, `kk` =4 (when `k<jm_ imax`, the remainder is equal to `k`). At this stage, the computation of the couple of integers (`i,j`) is trivial. The reader can verify that in the case of face 4 (`i,j`)=(1, 1).

Even if this procedure seems very easy, some remarks are necessary:

- The calculation of index `l` depends on the definition of the macro functions. In the previous example, `Qp` is defined as follows

```
Qp(i,j,l) qp[(l)*(IMAX)*(JMAX)+(IMAX)*(j)+i].
```

- If the calculation involves ghost cells, they must be accounted for.
- For 3D configurations, this indexing is no longer valid. The 'planes' of `Qp` become 'volumes' because of the third dimension and the procedure must be readapted accordingly to the topology implemented.

To conclude this section, another example of kernel is provided.

Calculation of pressure

This kernel is different from the previous one: the MUSCL reconstruction is performed on the faces, whereas this kernel computes the pressure in the cell centers.

```
__global__
void zz_compute_pressure(int IMAX, int JMAX, REAL *q,  REAL *pr, REAL *Gamma, REAL *Eint, int ic)
{
  int i,j,k,N;
  int imax3 = IMAX + 3;
  int jmax3 = JMAX + 3;

  N = imax3*jmax3;
  k = threadIdx.x + blockIdx.x * blockDim.x;

  while(k < N) {

    i = k%imax3 - 1;
    j = k/imax3 - 1;

    PR(i,j) = (GAMMA(i,j)-1.0)*(Q(i,j,3)-EINT(i,j)-0.5*(Q(i,j,1)*Q(i,j,1)+Q(i,j,2)*Q(i,j,2))/Q(i,j,0));

    if( PR(i,j) <= 0 || isnan(PR(i,j)) || isinf(PR(i,j)) ){
      printf("\nNegative or nan pressure i = %i, j = %i at ic= %i\n",i,j,ic);
      printf("PR(%i,%i)      = %g\n",i,j,PR(i,j));
      printf("Q(%i,%i,3)     = %g\n",i,j,Q(i,j,3));
      printf("EINT(%i,%i)    = %g\n",i,j,EINT(i,j));
      printf("GAMMA(%i,%i)   = %g\n",i,j,GAMMA(i,j));
      printf("rho*u^2/2      = %g\n",0.5*(Q(i,j,1)*Q(i,j,1)+Q(i,j,2)*Q(i,j,2))/Q(i,j,0) );
      asm("trap;");
    }

    k += blockDim.x * gridDim.x;

  }

}
```

Listing 10.4. Example code: calculation of the pressure

As before, `IMAX` and `JMAX` represent the number of nodes in the two directions, so the internal cells are `(IMAX-1)` and `(JMAX-1)`, along *i* and *j* direction, respectively. Since the spatial discretization algorithm needs two ghost cells on each boundary (4 cells in each direction), the total number of cells where the pressure must be computed is `(IMAX+3)` along *i* direction and `(JMAX+3)` along *j* direction. These values are stored in `imax3` and `jmax3` (lines 5 and 6); their product defines the total number of elements to be handled by the kernel (`N` at line 8). The topology of the cells follows an indexing that starts from –1 and ends at `(IMAX+1)`: for this reason, when computing spatial indices `(i,j)`, the result must be reduced by 1 to force the loop to start counting from -1. Once the pressure is computed, a check is needed to ensure physically acceptable results (positive pressure); if it is not the case, the kernel execution is killed by mean of `asm("trap;")`.

10.4 Results

In this section, some results obtained with this GPU-based algorithm are presented. Specifically, a comparison between the multi-temperature model and the StS model is discussed for a 2D double-wedge configuration. In order to proof the capability of the code in capturing non-stationary flows, an unsteady test case is presented. Finally, a scalability performance analysis is illustrated.

10.4.1 High enthalpy flow over a double-wedge

The flow over a double-wedge is here considered for the geometry reported in figure 10.12, whose geometrical parameters are summarized in table 10.5. The analysis of hypersonic flows over this kind of geometry is interesting since it represents a simplified model of complex aerodynamics components, such as wings or fuselage [43–48].

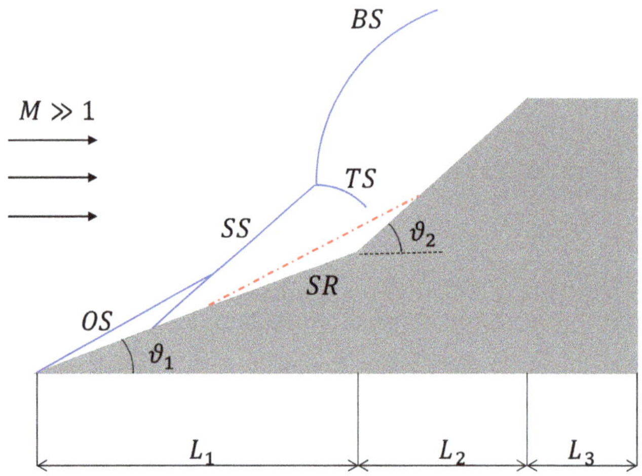

Figure 10.12. Sketch of the geometry (in grey) and of the shock structure (in blue) of a double-wedge: separation region (SR, in red), oblique shock (OS), separation shock (SS), transmitted shock (TS), bow shock (BS).

Table 10.5. Reference values for the double-wedge geometry: lengths in meters and angles in degrees.

θ_1	θ_2	L_1	L_2	L_3
30	55	0.043 99	0.014 58	0.010 82

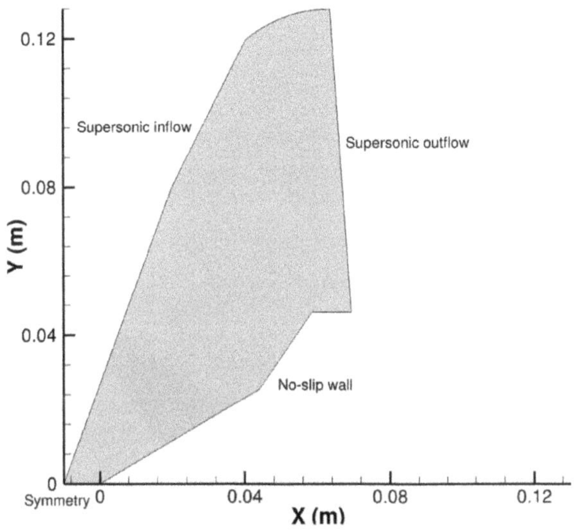

Figure 10.13. Computational domain with boundary conditions.

Such flows are characterized by a complex shock wave/boundary layer interaction (SWBLI). When a high Mach number flow impinges on a deflected wall (first wedge, with inclination θ_1), an oblique shock (OS) generates near the leading edge whose shape is given by classical gas dynamics laws. Since the second wedge presents a higher inclination angle θ_2, the flow cannot remain attached, giving rise to a bow shock (BS). While developing, the BS interacts with the OS near the compression corner, so a transmitted shock (TS) and a separation shock (SS) generate. In particular, the last one moves along the first wedge and it is responsible for the boundary layer separation, identified by a dashed-dot red line in figure 10.12. In this scenario, one of the important aspects is unsteadiness of the flow: in a recent work, Durna and Celik [49] found this geometrical configuration induces unsteadiness under certain free stream conditions. Specifically, for a low enthalpy regime the flow becomes periodic. Figure 10.13 illustrates the computational domain with the boundary conditions employed, whereas table 10.6 summarizes the free stream conditions. The wall is isothermal and non-catalytic, and the initial mixture is composed of molecular nitrogen and molecular oxygen, with fractions 78% and 22% in mass, respectively.

The flow rapidly reaches a stationary state. The shock structure is coherent with theoretical expectations, with vortex developing in the compression corner, as shown in figure 10.14. Non-equilibrium is relevant and, because of the high temperature,

Table 10.6. Flow conditions. τ_c is the characteristic time as defined in reference [45].

Quantity	
Mach number	7.14
Stagnation enthalpy [MJ kg^{-1}]	8.0
u_∞ [m s^{-1}]	3812
p_∞ [Pa]	780
T_∞ [K]	710
ρ_∞ [kg m^{-3}]	0.003 8
τ_c [μ s]	15.7
T_w [K]	300

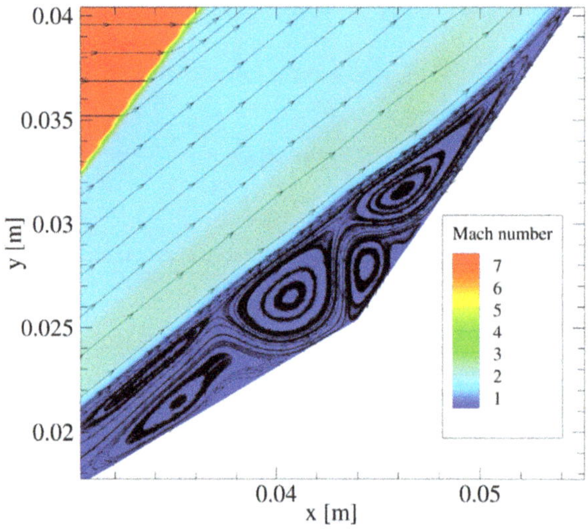

Figure 10.14. Zoom of the Mach number contour in the compression corner.

vibrational excitation and dissociation become effective. The results obtained through multi-temperature and StS models are discussed and compared with those reported by Hao *et al* [50]. Figure 10.15 shows the pressure and the heat flux distributions at $t = 242\,\mu$s, in terms of pressure coefficient and Stanton number, respectively, defined as:

$$C_p = \frac{p - p_\infty}{0.5\rho_\infty u_\infty^2} \qquad St = \frac{q_w}{0.5\rho_\infty u_\infty c_p(T_0 - T_w)} \tag{10.60}$$

where the subscript ∞ indicates free stream quantities, c_p is the mixture specific heat at constant pressure, T_0 and T_w represent the total free stream temperature and the wall temperature, respectively, whereas q_w represents the wall total heat flux.

Even if the data are comparable with the literature with both models, StS shows a larger extension of the separation region. Indeed, the separation point (identified by

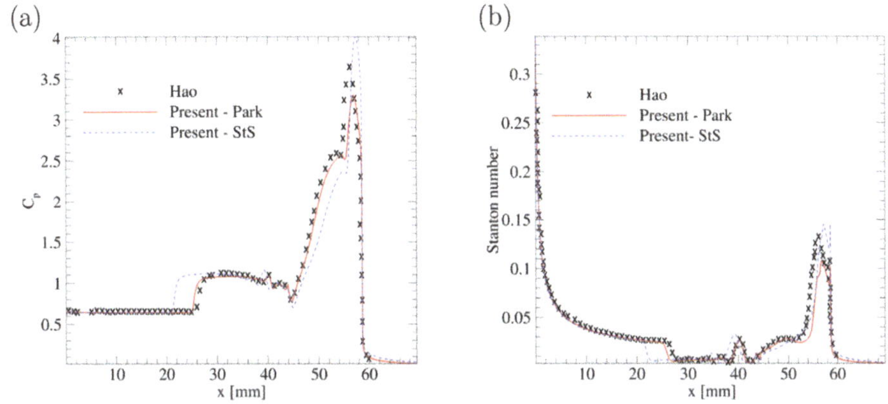

Figure 10.15. Wall distributions (present work and reference [50]) at $t = 242\ \mu s$ in terms of pressure coefficient (a) and Stanton number (b).

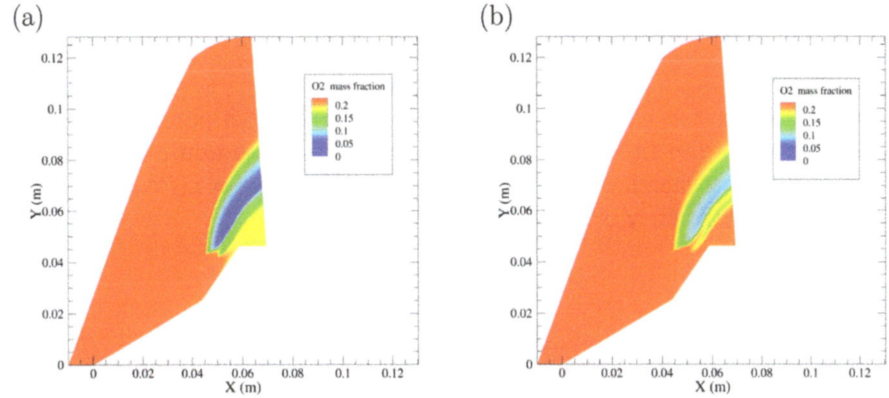

Figure 10.16. Contours of mass fraction of O_2 at $t = 242\ \mu s$: Park (a) and StS (b).

the first pressure jump) is placed more upstream when non-equilibrium is modeled by the StS kinetics. This feature might be linked to the stronger intensity of the shocks: the pressure peak (≈55 mm) is higher for the StS model, giving rise to faster waves. The same behavior can be observed for the heat flux. For these reasons, important differences have been observed on the dissociation, as illustrated in figure 10.16, that shows the contours of the mass fractions of O_2 at $t = 242\ \mu s$: indeed, the MT model predicts a smaller fraction of oxygen atoms than the StS model.

10.4.2 Scalability performance

An interesting aspect is to investigate the performances of the algorithm on different machines.

In order to measure the parallel performance of a computer code two approaches can be followed: the strong and the weak scaling [51]. Scaling requires observing how the execution time of a parallel code changes when changing the number of processors (GPUs in this case). Strong scaling describes the reduction of the execution time (wall-clock time) by increasing the number of processors while keeping constant the problem size. For example, once the computational model is fixed, in terms of equations and mesh size, the execution time when running on two processors should be halved with respect to the serial computation and so on. Therefore, speed-up (with respect to the serial case) is defined as

$$\text{speed-up}=\frac{T(N,\,1)}{T(N,\,P)} \tag{10.61}$$

where T is the execution time, N the problem size and P the number of processors. Obviously, the ideal speed-up (with respect to the serial case) is

$$\text{speed-up}_{\text{id}} = \frac{P}{1} \tag{10.62}$$

and efficiency

$$\eta_s = \frac{\text{speed-up}}{\text{speed-up}_{\text{id}}} = \frac{T(N,\,1)}{P \times T(N,\,P)}. \tag{10.63}$$

On the other hand, weak scaling measures the capability of a parallel code to keep constant the execution time when the problem size and the number of processors increase at the same rate, i.e., $T(N,1) = T(2N,2) = T(3N,3).... = T(PN,P)$. Therefore, for weak scaling efficiency is defined as

$$\eta_w = \frac{T(N,\,1)}{T(PN,\,P)}. \tag{10.64}$$

Strong and weak scalability have been tested on two different facilities: the hybrid partition of Piz Daint located at CSCS in Switzerland and Marconi100 at Cineca in Italy. The hybrid partition of Piz Daint includes 5704 Cray XC50 nodes each of them equipped with an Intel Xeon E5-2690 v3 2.60 GHz (12 cores, 64 GB RAM) and a NVIDIA Tesla P100 with 16 GB of dedicated RAM [52], whereas Marconi100 is a cluster based on POWER9 chips accelerated by Volta NVIDIA GPUs. It includes 980 nodes equipped with two 16 cores IBM POWER9 AC922 at 2.6 (3.1) GHz and four NVIDIA Volta V100 GPUs with 16GB of dedicated RAM [53].

On both supercomputers a hypersonic flow past a sphere has been considered by solving a five species neutral air mixture.

On the Piz Daint hybrid partition the reactive Euler equations in a 2D axisymmetric configuration have been solved by using the StS model with the chemical equation advanced in time by a single sub-step and eight Gauss–Seidel inner iterations.

On the other hand, on Marconi100 the Navier–Stokes equations have been solved for a 3D configuration with the Park model. The solution has been advanced for 100 time steps each including two sub-steps and eight Gauss–Seidel inner iterations for the chemical equation. Tables 10.7 and 10.8 provide the details of simulations, in terms of number of cells for the strong scalability and in terms of cells for GPU in the case of the weak scalability, for the Piz Daint and Marconi100 analysis, respectively. Figures 10.17 and 10.18 show the strong scalability results (speed-up and efficiency) for 2D StS (Piz Daint) and 3D Park (Marconi100) simulations, respectively. Speed-up values have been evaluated with respect to the smallest number of GPUs employed for each mesh resolution. In the case of 2D StS simulations (Piz Daint), the computational model scales very well even running on 2400 GPUs and it shows a superlinear behaviour in many occurrences. On the other hand, when the 3D Park model is employed on Marconi100 a good speed-up is obtained up to 512 GPU. Figures 10.19 and 10.20 show the weak scalability results in terms of wall-clock time and efficiency. By keeping constant the number of cells per GPU the time to perform the same number of iterations should be constant. Indeed, very good results have been obtained up to 1280 GPUs in the case of 2D StS (Piz Daint) and up to 512 for 3D Park (Marconi100) simulations.

Table 10.7. Strong and weak scalability test cases on Piz Daint (2D StS).

Strong scalability (2D cells mesh size)	Weak scalability (cells/GPU)
1024 × 512	13107.2
1024 × 1024	26214.4
2048 × 1024	52428.8
4096 × 2048	104857.6
8192 × 4096	
16 384 × 8192	

Table 10.8. Strong and weak scalability test cases on Marconi100 (3D Park).

Strong scalability (3D cells mesh size)	Weak scalability (cells/GPU)
512 × 256 × 256	524 288
1024 × 512 × 256	1048 576
1024 × 512 × 512	
1024 × 1024 × 512	

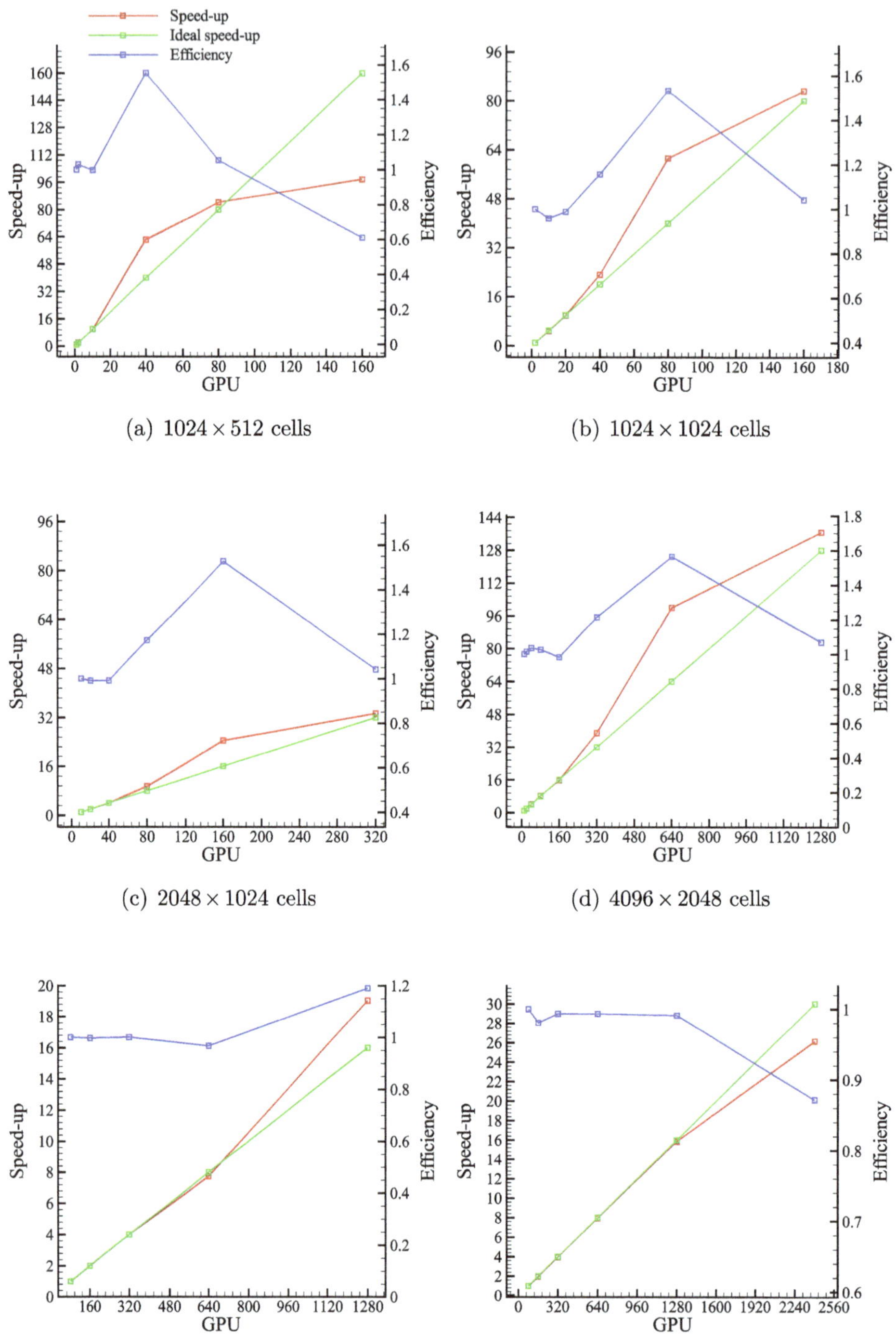

(a) 1024×512 cells

(b) 1024×1024 cells

(c) 2048×1024 cells

(d) 4096×2048 cells

(e) 8192×4096 cells

(f) 16384×8192 cells

Figure 10.17. 2D StS Piz Daint strong scaling: speed-up and ideal speed-up (with respect to one GPU (a), two GPUs (b), 10 GPUs (c) and (d), and 80 GPUs (e) and (f)) on the left axis and efficiency on the right axis.

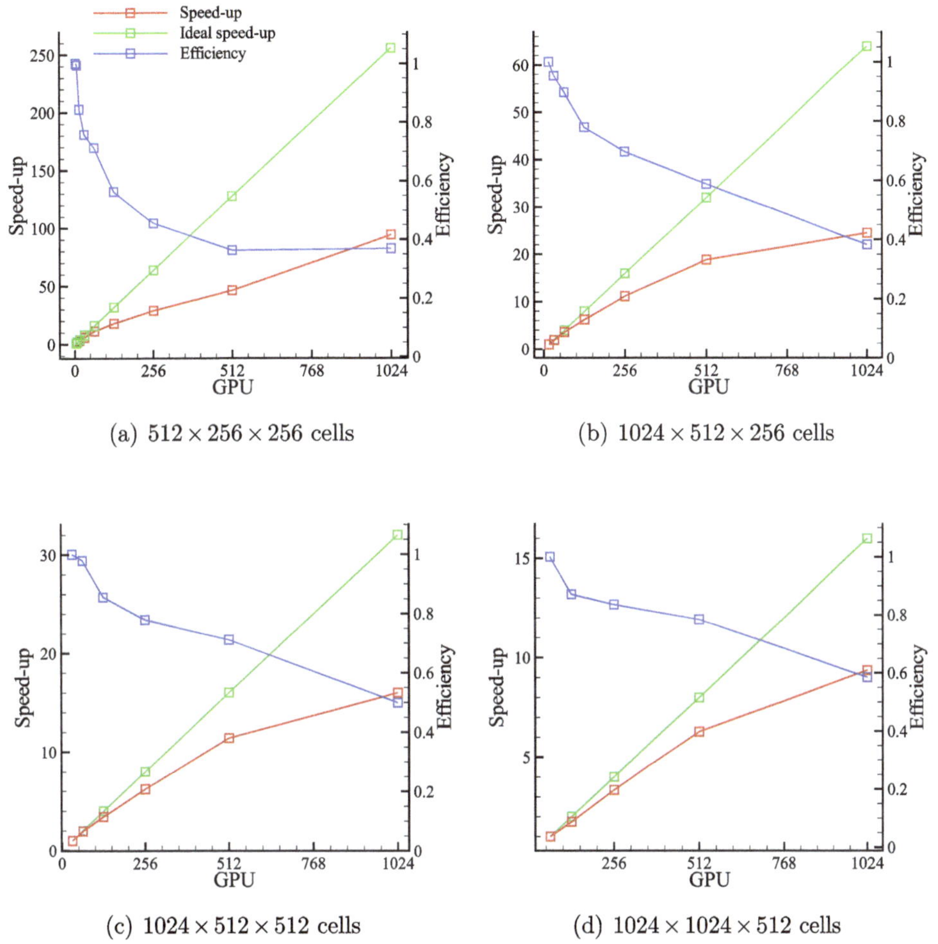

Figure 10.18. 3D Park Marconi100 strong scaling: speed-up and ideal speed-up (with respect to four GPUs (a), 16 GPUs (b), 32 GPUs (c) and 64 GPUs (d)) on the left axis and efficiency on the right axis.

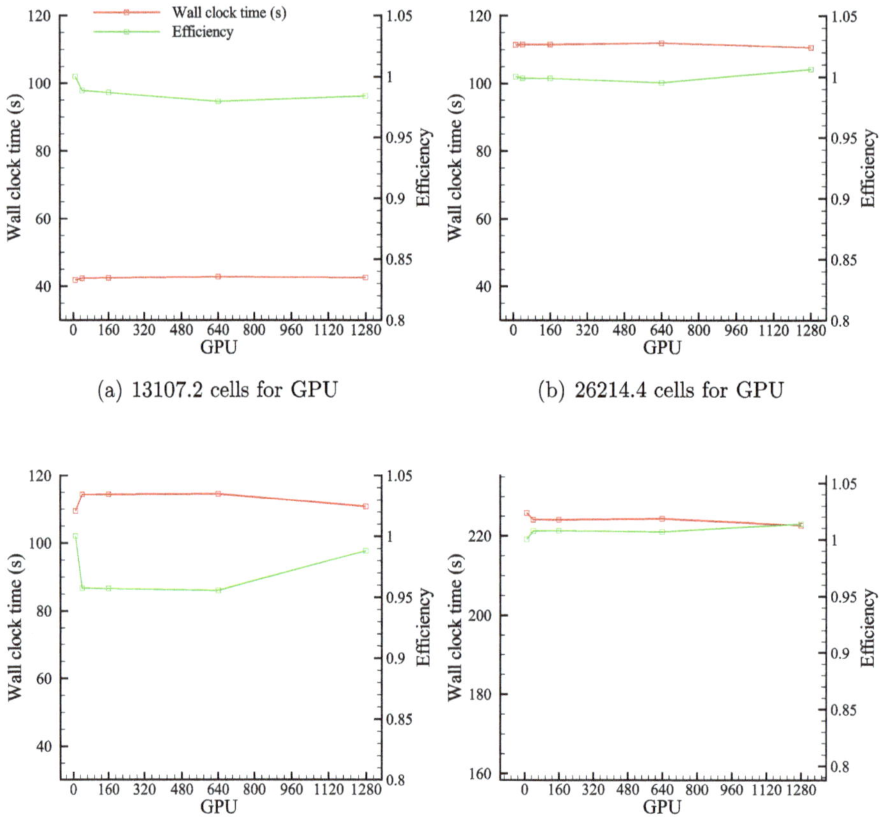

(a) 13107.2 cells for GPU (b) 26214.4 cells for GPU

(c) 52428.8 cells for GPU (d) 104857.6 cells for GPU

Figure 10.19. 2D StS Piz Daint weak scaling: wall-clock time (50 time steps (a) and (b), 21 time steps (c) and (d)) on the left axis and efficiency on the right axis.

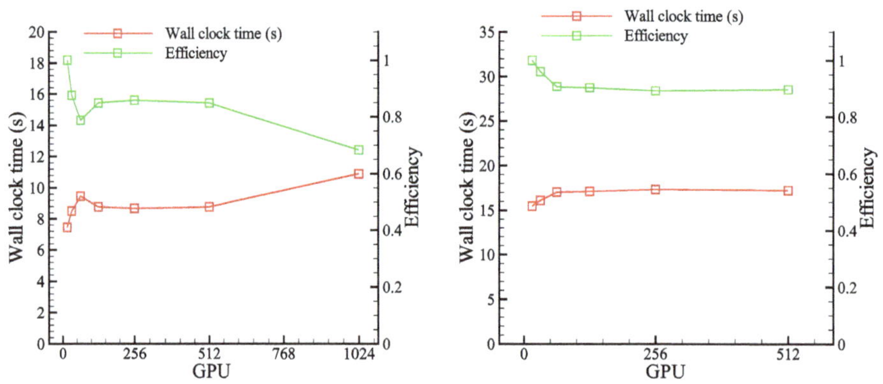

(a) 524288 cells for GPU (b) 1048576 cells for GPU

Figure 10.20. 3D Park Marconi100 weak scaling: wall-clock time (100 time steps) on the left axis and efficiency on the right axis.

References

[1] Bertin J J and Cummings R M 2003 Fifty years of hypersonics: where we've been, where we're going *Prog. Aerosp. Sci.* **39** 511–36

[2] Anderson J D Jr 2006 *Hypersonic and High-Temperature Gas Dynamics* (Reston, VA: American Institute of Aeronautics and Astronautics)

[3] Armenise I, Capitelli M, Colonna G, Koudriavtsev N and Smetanin V 1995 Nonequilibrium vibrational kinetics during hypersonic flow of a solid body in nitrogen and its influence on the surface heat-flux *Plasma Chem. Plasma Process.* **15** 501–28

[4] Colonna G and Capitelli M 1996 Electron and vibrational kinetics in the boundary layer of hypersonic flow *J. Thermophys. Heat Transfer* **10** 406–12

[5] Armenise I 2017 Excitation of the lowest CO_2 vibrational states by electrons in hypersonic boundary layers *Chem. Phys.* **491** 11–24

[6] Schrooyen P 2015 Numerical simulation of aerothermal flows through ablative thermal protection systems *PhD Thesis* UCL

[7] Park C 1989 *Nonequilibrium Hypersonic Aerothermodynamics* (New York: Wiley)

[8] Colonna G, Tuttafesta M, Capitelli M and Giordano D 1999 Non-Arrhenius NO formation rate in one-dimensional nozzle airflow *J. Thermophys. Heat Transfer* **13** 372–5

[9] Capitelli M, Celiberto R, Colonna G, Esposito F, Gorse C, Hassouni K, Laricchiuta A and Longo S 2015 *Fundamental Aspects of Plasma Chemical Physics: Kinetics* vol 85 (Berlin: Springer)

[10] Colonna G and Capitelli M 2001 The influence of atomic and molecular metastable states in high-enthalpy nozzle expansion nitrogen flows *J. Phys. D: Appl. Phys* **34** 1812

[11] Colonna G, Capitelli M, Pietanza L D, Munafò A and Panesi M 2019 High-enthalpy ionized flows *Hypersonic Meteoroid Entry Physics* (Bristol: IOP Publishing) ch 8 https://doi.org/10.1088/2053-2563/aae894ch8

[12] Colonna G, Pietanza L D and Laricchiuta A 2020 Ionization kinetic model for hydrogen-helium atmospheres in hypersonic shock tubes *Int. J. Heat Mass Transfer* **156** 119916

[13] Giordano D, Bellucci V, Colonna G, Capitelli M, Armenise I and Bruno C 1997 Vibrationally relaxing flow of N past an infinite cylinder *J. Thermophys. Heat Transfer* **11** 27–35

[14] Colonna G, Bonelli F and Pascazio G 2019 Impact of fundamental molecular kinetics on macroscopic properties of high-enthalpy flows: the case of hypersonic atmospheric entry *Phys. Rev. Fluids* **4** 033404

[15] Bonelli F, Pascazio G and Colonna G 2021 Effect of finite-rate catalysis on wall heat flux prediction in hypersonic flow *Phys. Rev. Fluids* **6** 033201

[16] Cutrone L, Tuttafesta M, Capitelli M, Schettino A, Pascazio G and Colonna G 2014 3D nozzle flow simulations including state-to-state kinetics calculation *AIP Conf. Proc.* vol 1628 pp 1154–61

[17] Guide D 2013 CUDA C programming guide *NVIDIA* **31** (July 29) 31

[18] Sanders J and Kandrot E 2010 *CUDA by Example: An Introduction to General-Purpose GPU Programming* (London: Addison-Wesley)

[19] Bonelli F, Tuttafesta M, Colonna G, Cutrone L and Pascazio G 2017 An MPI-CUDA approach for hypersonic flows with detailed state-to-state air kinetics using a GPU cluster *Comput. Phys. Commun.* **219** 178–95

[20] Tuttafesta M, Colonna G and Pascazio G 2013 Computing unsteady compressible flows using Roe's flux-difference splitting scheme on GPUs *Comput. Phys. Commun.* **184** 1497–510

[21] Tuttafesta M, Pascazio G and Colonna G 2016 Multi-GPU unsteady 2D flow simulation coupled with a state-to-state chemical kinetics *Comput. Phys. Commun.* **207** 243–57
[22] K-y K K and Acharya R 2012 *Fundamentals of Turbulent and Multiphase Combustion* (New York: Wiley)
[23] Capitelli M, Colonna G and D'Angola A 2011 *Fundamental Aspects of Plasma Chemical Physics: Thermodynamics* (Springer Series on Atomic, Optical, and Plasma Physics vol 66) 1st edn (New York: Springer)
[24] Wilke C 1950 A viscosity equation for gas mixtures *J. Chem. Phys.* **18** 517–9
[25] Mason E and Saxena S 1958 Approximate formula for the thermal conductivity of gas mixtures *Phys. fluids* **1** 361–9
[26] Sarma G 2000 Physico-chemical modelling in hypersonic flow simulation *Prog. Aerosp. Sci.* **36** 281–349
[27] Gupta R N, Yos J M, Thompson R A and Lee K-P 1990 A review of reaction rates and thermodynamic and transport properties for an 11-species air model for chemical and thermal nonequilibrium calculations to 30000 K *NASA STI/Recon Technical Report N 90* 27064
[28] Design R 2015 *ANSYS Chemkin Theory Manual* **17** 0
[29] Bird R B, Stewart W E and Lightfoot E N 2006 *Transport Phenomena* revised 2nd edn (New York: Wiley)
[30] Poinsot T and Veynante D 2005 *Theoretical and Numerical Combustion* (Philadelphia, PA: RT Edwards, Inc.)
[31] Hirschfelder J O, Curtiss C F and Bird R B 1964 *Molecular Theory of Gases and Liquids* (New York: Wiley)
[32] Park C 1993 Review of chemical-kinetic problems of future NASA missions. I–Earth entries *J. Thermophys. Heat Transfer* **7** 385–98
[33] Millikan R C and White D R 1963 Systematics of vibrational relaxation *J. Chem. Phys.* **39** 3209–13
[34] Park C 1988 Two-temperature interpretation of dissociation rate data for N_2 and O_2 (AIAA) 1988–458
[35] Colonna G *et al* 2021 Theoretical and experimental aspects of non-equilibrium plasmas in different regimes: fundamentals and selected applications *Eur. Phys. J. D* **75** 1–35
[36] Pletcher R H, Tannehill J C and Anderson D 2012 *Computational Fluid Mechanics and Heat Transfer* (Boca Raton, FL: CRC Press)
[37] Steger J L and Warming R 1981 Flux vector splitting of the inviscid gasdynamic equations with application to finite-difference methods *J. Comput. Phys.* **40** 263–93
[38] Hirsch C 1990 *Numerical Computation of Internal and External Flows, Volume 2: Computational Methods for Inviscid and Viscous Flows* vol 2 (New York: Wiley)
[39] Van Leer B 1979 Towards the ultimate conservative difference scheme. V. A second-order sequel to Godunov's method *J. Comput. Phys.* **32** 101–36
[40] Harten A 1997 High resolution schemes for hyperbolic conservation laws *J. Comput. Phys.* **135** 260–78
[41] Jiang G-S and Shu C-W 1996 Efficient implementation of weighted ENO schemes *J. Comput. Phys.* **126** 202–28
[42] Hirsch C 2007 *Numerical Computation of Internal and External Flows, Volume 1: The Fundamentals of Computational Fluid Dynamics* (Amsterdam: Elsevier)

[43] Reinert J D, Candler G V and Komives J R 2020 Simulations of unsteady three-dimensional hypersonic double-wedge flow experiments *AIAA J.* **58** 4055–67

[44] Swantek A and Austin J 2015 Flowfield establishment in hypervelocity shock-wave/boundary-layer interactions *AIAA J.* **53** 311–20

[45] Knight D *et al* 2017 Assessment of predictive capabilities for aerodynamic heating in hypersonic flow *Prog. Aerosp. Sci.* **90** 39–53

[46] de Rosa D, Pezzella G, Donelli R S and Viviani A 2016 Flap effectiveness appraisal for winged re-entry vehicles *Acta Astronaut.* **122** 175–84

[47] Shoev G, Oblapenko G, Kunova O, Mekhonoshina M and Kustova E 2018 Validation of vibration-dissociation coupling models in hypersonic non-equilibrium separated flows *Acta Astronaut.* **144** 147–59

[48] Zuo F-Y and Hu S-L 2021 Thermochemical non-equilibrium effects on aerothermodynamic prediction of laminar double-cone flow *Acta Astronaut.* **182** 179–88

[49] Durna A and Celik B 2019 Time-periodic shock interaction mechanisms over double wedges at Mach 7 *Shock Waves* **29** 381–99

[50] Hao J, Wen C-Y and Wang J 2019 Numerical investigation of hypervelocity shock-wave/boundary-layer interactions over a double-wedge configuration *Int. J. Heat Mass Transfer* **138** 277–92

[51] Pacheco P S 2011 *An Introduction to Parallel Programming* (Burlington, MA: Morgan Kaufmann Publishers)

[52] https://www.cscs.ch/computers/piz-daint/

[53] https://wiki.u-gov.it/confluence/display/SCAIUS/UG3.2

IOP Publishing

Plasma Modeling (Second Edition)
Methods and applications
Gianpiero Colonna and Antonio D'Angola

Chapter 11

Hybrid models

Francesco Taccogna and Pierpaolo Minelli

A computational approach extensively used in plasma computational physics is the hybrid method [1, 2]. It consists of solving the fluid laws (chapters 6 and 7) for the electrons using a kinetic approach (usually discrete-particle particle-in-cell with Monte Carlo collision (PIC-MCC) methods [3], described in chapter 4) to represent the evolution of the heavy species. Thus, the hybrid code combines a Lagrangian and Eulerian approach. It is suitable to study low frequency/high wave number phenomena that are not captured by full fluid theory because ion kinetic effects (ion beam instability, shock by backstreaming ions, ion-neutral charge exchange, etc) are at the origin of the phenomenon itself. On the other hand, a full kinetic particle code must resolve the electron behavior at scales much smaller than for ions, studying smaller regions for shorter times. Moreover, full particle techniques require much more computer memory. Therefore, the hybrid approach allows one to save computational time and, as a result, it has gained considerable popularity. The first one-dimensional hybrid codes appeared in the seventies thanks to the pioneering works of Mason [4], Chodura [5] and Sgro–Nielsen [6] on fusion plasmas. The extension to two-dimensional models dates back to the early eighties by Hewett [7] and Harned [8]. Finally, the first full three-dimensional representations can be considered the works of Brecht–Thomas [9] and Moore–Thomas–McComas [10] presented around the nineties.

This chapter is organized as follows: the next section is devoted to the description of the physics laws and hypothesis behind the hybrid description. Fluid electron and ion kinetic equations will be discussed. In section 11.2 the general numerical procedure necessary for the implementation of a hybrid code will be presented. Finally, in the last section 11.3 some numerical variants of the general scheme will be presented for selected space and laboratory applications of low-temperature plasmas. The corresponding results will be discussed.

doi:10.1088/978-0-7503-3559-1ch11

11.1 Basic assumptions and governing equations

Hybrid models describe phenomena occurring on space and time scales much larger than those typical of electrons; they are given to simulate and resolve processes that occur on ion gyroradius or ion inertial spatial scale, and inverse ion gyro-frequency or ion transit time. As a consequence, the space scales. are much larger than Debye length and electron Larmor radius and time scales are much larger than plasma and electron gyration periods.

The hybrid approximation treats the electrons as a fluid while the ions are modeled kinetically. The most used kinetic formulation is based on particle representation (PIC-MCC scheme [3], see chapter 4). It is to represent the ion sub-system as a collection of macro-ions (cloud of a fixed number of real ions) whose position $\overrightarrow{x_p}$ and velocity $\overrightarrow{v_p}$ are updated solving the usual equations of motion:

$$m_i \frac{d\overrightarrow{v_p}}{dt} = q_i\left(\vec{E} + \overrightarrow{v_p} \times \vec{B}\right) \tag{11.1}$$

$$\frac{d\overrightarrow{x_p}}{dt} = \overrightarrow{v_p} \tag{11.2}$$

where $\vec{E}$ and $\vec{B}$ (defined on a spatial grid through the coupling with electron sub-system) have been interpolated to the macro-particle location. The ion distribution function f_i and the corresponding generic momenta $\mathcal{M}_i$ (charge density ρ_i, bulk flow velocity $\vec{u}_i$, current $\vec{j}_i$, temperature T_i, etc) are reconstructed on the grid points through an inverse interpolation process using the same weighting function used for the gathering procedure in order to avoid self-forces on the particles [3]:

$$\mathcal{M}_i \leftarrow \left(\overrightarrow{x_p};\, \overrightarrow{v_p}\right). \tag{11.3}$$

Concerning the electron fluid description under hybrid approximation it can be shown that:

1. quasi-neutrality is assumed

$$\rho_e = \rho_i = \rho \tag{11.4}$$

so that given the ion density from the ion PIC module, the electron density is specified (here no negative ion species are assumed for simplicity);

2. neglecting the displacement currents in Ampère's law (Darwin approximation) $\vec{j} = \nabla \times \vec{B}/\mu_0$ (μ_0 is the vacuum permeability) and the electron mass in the electron momentum equation, it results in an equation of state (generalized Ohm's law) for the electric field:

$$\vec{E} = -\vec{u}_i \times \vec{B} + \frac{(\nabla \times \vec{B}) \times \vec{B}}{\mu_0 \rho} - \frac{\nabla p_e}{\rho} \tag{11.5}$$

once the electron pressure (considered isotropic) is known. Through the ideal gas law $p_e = n_e \kappa T_e$, it corresponds to handle the electron temperature and the following different options are available:

(a) isothermal (electrons distribute their energy fast in the whole domain because of high thermal conductivity):

$$T_e = \text{const}; \tag{11.6}$$

(b) adiabatic (small collision frequency):

$$T_e \propto n_e^{\gamma-1} \tag{11.7}$$

where γ is the polytropic index;

(c) solve for the massless energy equation:

$$\frac{\partial T_e}{\partial t} + \vec{u}_e \cdot \nabla T_e + \frac{3}{2}\nabla \cdot \vec{u}_e = \frac{2}{3}\frac{\eta|J|^2}{n_e}. \tag{11.8}$$

where $\vec{u}_e$ is the electron velocity.

In the collisional case, the right-hand side of equation (11.5) also contains a force term $C_{\text{coll}} = \eta\nabla \times \vec{B}/\mu_0$ due to electron–ion, electron–neutral and anomalous (i.e. wave-particle interaction) collisions (η is the plasma resistivity, considered for simplicity as scalar). Similarly, the electric field in the ion equation of motion equation (11.1) becomes $\vec{E}' = \vec{E} - \eta\nabla \times \vec{B}$ unless an explicit Monte Carlo module is used to take into account collisions.

It should be noted that under Darwin approximation there is no equation for the time evolution of the electric field, but it is rather determined by the requirement that the electric force balances the other forces acting on the massless electron fluid: the electric field is a state function of the ion moments ρ and $\overline{u_i}$, the magnetic field $\vec{B}$ and the electron temperature T_e.

Finally, in the electromagnetic case, the electric field is used to advance the magnetic field in time through Faraday's law $\frac{\partial \vec{B}}{\partial t} = -\nabla \times \vec{E}$ that substituting equation (11.5) becomes:

$$\frac{\partial \vec{B}}{\partial t} = \nabla \times (\overline{u_i} \times \vec{B}) - \nabla \times \frac{(\nabla \times \vec{B}) \times \vec{B}}{\mu_0 \rho}. \tag{11.9}$$

The first term on the right-hand side describes the induction, while the second represents the dispersion. Note that the electron pressure term of equation (11.5) does not influence the magnetic field evolution. Under previous assumptions the other's Maxwell equations

$$\nabla \cdot \vec{E} = -\frac{\rho}{\epsilon_0} \tag{11.10}$$

$$\nabla \cdot \vec{B} = 0 \tag{11.11}$$

are automatically satisfied.

11.2 Numerical implementation

In this section the implementation of the general hybrid approach will be presented. The different variations of the general scheme will be presented in the next section 11.3 for a number of applications ranging from plasma-based technological devices to astrophysical problems.

The governing laws presented in the previous section are solved on a computational grid discretizing the time. This requires defining a time step Δt and cell size Δx characteristics of the typical case studied. Usually the time step Δt is defined by ion plasma or cyclotron frequencies, while the cell size Δx is chosen such as to resolve ion gradients or to verify Courant–Friedrichs–Lewy (CFL) condition: $v_{i,\max}\,\Delta t < \Delta x$, where $v_{i,\max}$ represents the maximum speed of the ions. The computational grid can be uniform (Cartesian, spherical or even in general orthogonal curvilinear coordinates) or non-uniform (usually in Cartesian coordinates). In the latter case, hierarchical mesh or adaptive mesh refinement [11, 12] is used to generate a non-uniform grid.

Plasma in a hybrid scheme has two time-dependent components: positions/velocities of ion macroparticles and electromagnetic fields specified at the nodes of the computational grid. The bridge between the two kinds of quantities is carried by ion charge density ρ and ion flow velocity $\vec{u}_i$ interpolated from the particle location to the grid points. The magnetic field $\vec{B}$ is located on the grid (for example the cell centers). The electric fields $\vec{E}$, the plasma density ρ, the ion current densities $\vec{u}_i$ and the electron temperature T_e are all staggered from the locations of $\vec{B}$ (such as at the vertices of the cell).

11.2.1 Time-advance algorithm

Here we follow the implementation discussed in [11, 13, 14] and shown to be of second-order accuracy. It is an explicit time-centered leap-frog algorithm for both particles and fields, consisting of the following computational cycle (repeated until reaching stationarity):

1. The electromagnetic fields and particle positions are defined at full time steps $\vec{E}^{(n)}$, $\vec{B}^{(n)}$ and $\vec{x}_p^{(n)}$, as well as the density $\rho^{(n)}$ gathered from the particle positions:

$$\rho^{(n)} \leftarrow \left(\vec{x}_p^{(n)}\right). \tag{11.12}$$

Usually linear weighting are used for particle-to-grid location interpolation although more complex functions are used for cylindrical or spherical metrics [15–17].

2. The particle velocities defined at half time steps are advanced by solving equation (11.1)

$$\begin{aligned} \vec{v}_p^{(n+1/2)} &= \vec{v}_p^{(n-1/2)}\left(1 - \frac{h^2}{2}\vec{B}^{(n)} \cdot \vec{B}^{(n)}\right) + h\left(\vec{E}^{(n)} + \vec{v}_p^{(n-1/2)} \times \vec{B}^{(n)}\right) \\ &+ \frac{h^2}{2}\left(\vec{E}^{(n)} \times \vec{B}^{(n)}\right) + \frac{h^2}{2}\left(\vec{v}^{(n-1/2)} \cdot \vec{B}^{(n)}\right)\vec{B}^{(n)} \end{aligned} \tag{11.13}$$

where $h = q_i \Delta t / m_i$. For unmagnetized ions, this equation is greatly simplified.

3. Introduce the ion velocity $\vec{u}_i^{(-)}$ gathered from full time steps position $\vec{x}_p^{(n)}$ and half time steps velocity $\vec{v}_p^{(n+1/2)}$:

$$\vec{u}_i^{(-)} \leftarrow \left(\vec{x}_p^{(n)}; \vec{v}_p^{(n+1/2)}\right). \tag{11.14}$$

4. Advance $\vec{x}_p$ by solving equation (11.2)

$$\vec{x}_p^{(n+1)} = \vec{x}_p^{(n)} + \vec{v}^{(n+1/2)} \Delta t. \tag{11.15}$$

5. Gather $\rho^{(n+1)}$ and $\vec{u}_i^{(+)}$ from $\vec{x}_p^{(n+1)}$ and $\vec{v}_p^{(n+1/2)}$:

$$\rho^{(n+1)} \leftarrow \left(\vec{x}_p^{(n+1)}; \vec{v}_p^{(n+1/2)}\right) \tag{11.16}$$

$$\vec{u}_i^{(+)} \leftarrow \left(\vec{x}_p^{(n+1)}; \vec{v}_p^{(n+1/2)}\right) \tag{11.17}$$

in order to calculate the following averaged density and velocity centered at half time steps

$$\rho^{(n+1/2)} = \frac{\rho^{(n+1)} + \rho^{(n)}}{2} \tag{11.18}$$

$$\vec{u}_i^{(n+1/2)} = \frac{\vec{u}_i^{(+)} + \vec{u}_i^{(-)}}{2}. \tag{11.19}$$

6. Advance $\vec{B}^{(n+1)}$ by solving equation (11.9) using the previous averaged quantities $\rho^{(n+1/2)}$ and $\vec{u}_i^{(n+1/2)}$. This task is done by using the cyclic leap-frog technique: it consists in dividing the PIC cycle in m sub-time step $\Delta t_B = \Delta t_{\text{PIC}} / m$. With the notation $\vec{B}^{(p,\, n)}$ as $\vec{B}^{(n)}$ calculated at the pth sub-time step and $\vec{E}^{(p,\, n)}$ as the corresponding electric field obtained from equation (11.5) using the current value $\vec{B}^{(p,\, n)}$, we have the following iterative rule:

$$\vec{B}^{(1,\, n)} = \vec{B}^{(0,\, n)} - \Delta t_B \nabla \times \vec{E}^{(0,\, n)} \tag{11.20}$$

$$\vec{B}^{(2,\, n)} = \vec{B}^{(0,\, n)} - 2\Delta t_B \nabla \times \vec{E}^{(1,\, n)} \tag{11.21}$$

$$\vdots \tag{11.22}$$

$$\vec{B}^{(p+1,\, n)} = \vec{B}^{(p-1,\, n)} - 2\Delta t_B \nabla \times \vec{E}^{(p,\, n)} \tag{11.23}$$

$$\vdots \tag{11.24}$$

$$\vec{B}^{(m,\,n)} = \vec{B}^{(m-2,\,n)} - 2\Delta t_B \nabla \times \vec{E}^{(m-1,\,n)} \tag{11.25}$$

$$\overline{\vec{B}^{*}}^{(m,\,n)} = \vec{B}^{(m-1,\,n)} - \Delta t_B \nabla \times \vec{E}^{(m,\,n)} \tag{11.26}$$

$$\vec{B}^{(n+1)} = \frac{1}{2}\left(\vec{B}^{(m,\,n)} + \overline{\vec{B}^{*}}^{(m,\,n)}\right). \tag{11.27}$$

In addition, to stabilize the magnetic field integration against the influence of perturbations at short wavelengths, different smoothing procedures can be applied.

7. Extrapolate from $\vec{u}_i^{(n+1/2)}$ to $\vec{u}_i^{(n+1)}$. It can be done in several ways. First, keeping $\vec{u}_i^{(n-1/2)}$ as

$$\vec{u}_i^{(n+1)} = \frac{3\vec{u}_i^{(n+1/2)} - \vec{u}_i^{(n-1/2)}}{2} \tag{11.28}$$

or keeping $\vec{u}_i^{(n-3/2)}$ and improving the accuracy, as in the fourth order Bashforth–Adams extrapolation:

$$\vec{u}_i^{(n+1)} = \frac{4\vec{u}_i^{(n+1/2)} - 3\vec{u}_i^{(n-1/2)} + \vec{u}_i^{(n-3/2)}}{2}. \tag{11.29}$$

Alternatively, the velocity is calculated following the current advancement method (CAM) described by Matthews [13]. It consists of doing an extra half time step push using a mixed level evaluation of the electric field:

$$\vec{u}_i^{(n+1)} = \vec{u}_i^{(n+1)} + \frac{q_i \Delta t}{2m_i}\left(\vec{E}^{*} + \vec{u}_i^{(n+1)} \times \vec{B}^{(n+1)}\right) \tag{11.30}$$

where $\vec{E}^{(n+1)}$ is calculated from equation (11.5) using $\vec{B}^{(n+1)}$, $\vec{u}_i^{(n+1/2)}$ and $\rho^{(n+1)}$.

8. Calculate $\vec{E}^{(n+1)}$ from equation (11.5) using $\vec{B}^{(n+1)}$, $\vec{u}_i^{(n+1)}$ and $\rho^{(n+1)}$. In order to avoid numerical instability due to the term $(\nabla \times \vec{B}) \times \vec{B}/\mu_0\rho$ (the electric field becomes very large at low density) a minimum threshold density is defined below in which the field is set equal to zero [18].

Alternative time-advance algorithms are predictor–corrector [9], Horowitz [19] and a combination of velocity extrapolation and predictor–corrector [20].

11.2.2 Initialization and boundary conditions

The starting point of a simulation is always a sensitive issue. Initial conditions are specified by giving magnetic field $\vec{B}$, plasma density ρ and ion flow velocity $\vec{u}_i$ on the grid points. These latter two quantities are directly calculated from the initial ion distribution or loading/injection (only in few cases can one start from scratch) particles and the number of particles per cell plays a relevant role defining the resolution of phase space.

As regards the boundaries, two different kinds of elements can be distinguished: external surface domain and internal obstacle. The external boundaries can represent inflow, outflow, periodic, axis of symmetry or wall condition, while the obstacle can be insulator or metallic material and having its own intrinsic magnetic field. For all the cases, conditions have to be set for the particles and electromagnetic fields.

General rules have to be fulfilled. External boundaries should be as far as possible from the obstacle in order to reduce all possible perturbations coming from the obstacle itself. In addition, for periodic conditions, keep in mind that size of the imposed periodicity can induce an artificial scale or dump high wave numbers fluctuations.

11.3 Applications

11.3.1 Electrostatic case: plasma plume expansion and Langmuir probes

In the absence of an external magnetic field and for negligible induced magnetic field, the problem is purely electrostatic. Ohm's law equation (11.5) reduces to the the well-known Boltzmann relation:

$$n_e = n_0 \exp(e\phi/\kappa T_e) \tag{11.31}$$

where n_0 is the equilibrium density and ϕ the electric potential $\nabla\phi = -\vec{E}$. Equation (11.31) can be directly used to derive the electric field $\vec{E}$ from ion density through quasi-neutrality condition (11.4). Alternatively, without assuming quasi-neutrality, Poisson equation

$$\nabla^2\phi = -\frac{e}{\epsilon_0}(n_i - n_0 \exp(e\phi/\kappa T_e)) \tag{11.32}$$

has to be solved when space charge effects are important (small simulated regions, big object immersed in plasma, etc). In this case, equation (11.32) leads to a nonlinear elliptic differential equation which can be quickly solved by the iterative Newton–Raphson method [21]: the electric potential ϕ at the iteration $(n + 1)$ is a solution of a linear equation whose coefficients are functions of the solution at the previous iteration (n):

$$\left[\nabla^2 - \frac{e^2 n_0}{\epsilon_0 \kappa T_e} \exp(e\phi^{(n)}/\kappa T_e)\right]\phi^{(n+1)} = \left[1 - \frac{e}{\kappa T_e}\phi^{(n)}\right]\frac{e n_0}{\epsilon_0} \exp(e\phi^{(n)}/\kappa T_e) - \frac{e n_i}{\epsilon_0}. \tag{11.33}$$

The disadvantage with the Newton–Raphson method is that the equations have variable coefficients and direct use cannot be made of rapid elliptic solvers. Electron temperature T_e is considered fixed (isothermal) or coupled thought the adiabatic or energy conserving equation.

The electrostatic hybrid technique has been extensively used to study problems such as plume expansion from electric thrusters [22–24] and from laser-induced ablation [25] or in the calculation of ion current collection by electrostatic probe [26–29].

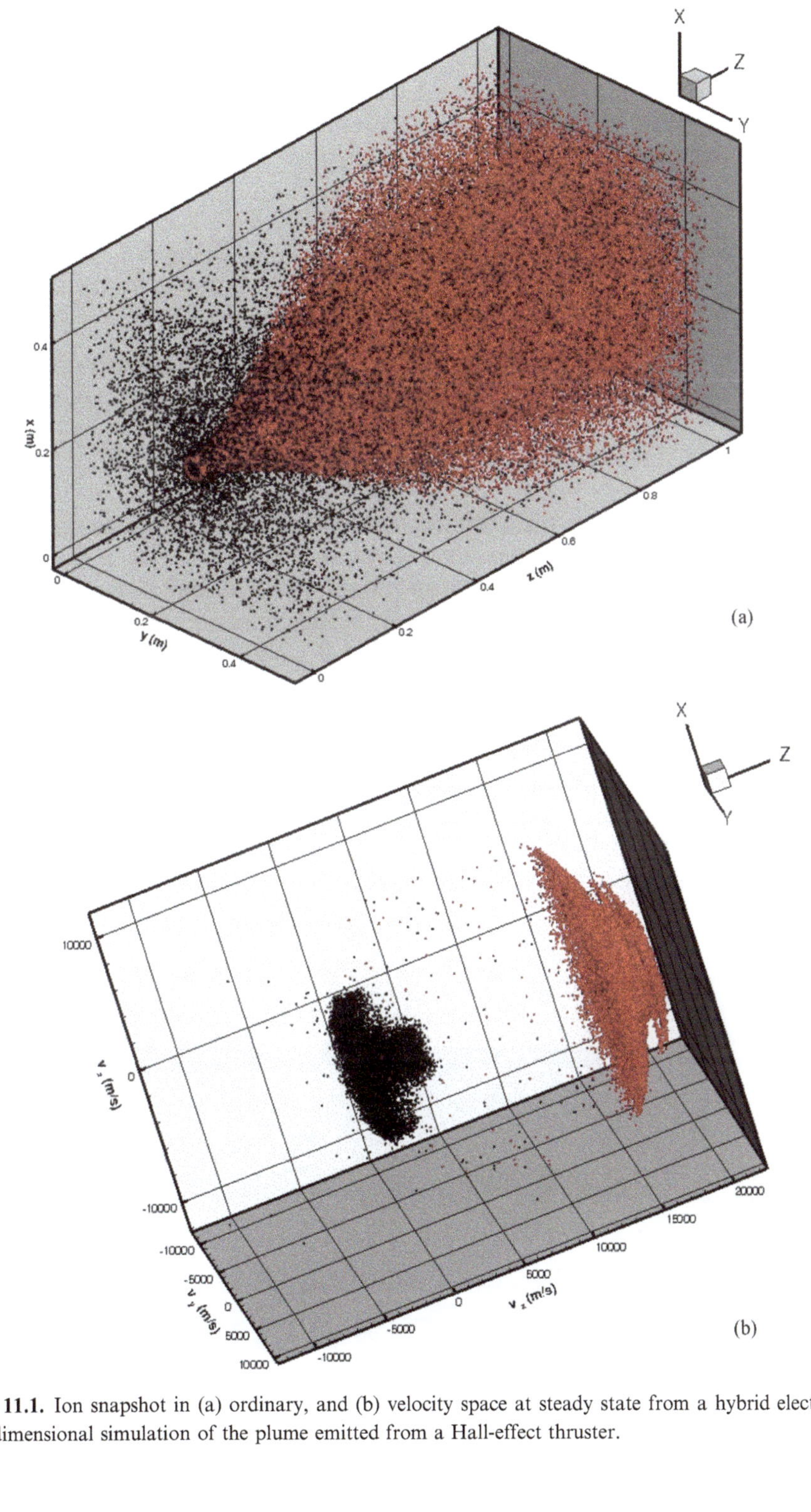

Figure 11.1. Ion snapshot in (a) ordinary, and (b) velocity space at steady state from a hybrid electrostatic three-dimensional simulation of the plume emitted from a Hall-effect thruster.

In figure 11.1 an ion swarm in (a) ordinary, and (b) velocity space from an electrostatic three-dimensional hybrid simulation of the plume emitted by a Hall-effect thruster (HET) [22–24] is reported. The model solves the nonlinear Poisson's equation (11.33) coupled with an adiabatic expansion equation (11.7) where the value $\gamma = 1.3$ has been chosen for the polytropic exponent to fit experimental measurements. The important effects of the production of slow ions (black spots in figures 11.1, while the red ones are beam ions) by resonant charge exchange CX collisions with neutral unionized propellant (where an electron is exchanged)

$$A_{\text{fast}}^{+} + A_{\text{slow}}^{0} \rightarrow A_{\text{fast}}^{0} + A_{\text{slow}}^{+} \tag{11.34}$$

is highlighted by the plume beam divergence: the fraction of CX ions is larger for higher angles, while in the 40 degrees (half-angle) cone the ion population is dominated by beam ions. In this way the production of slow CX ions reduces the thruster efficiency and is dangerous for the sensitive parts of the satellite figure 11.2.

The second example of hybrid electrostatic model represents the dynamics of ion collection by a Langmuir probe. Electrostatic probes are widely used to determine local values of plasma parameters, as electron and ion number densities, electron temperature, electron energy distribution function and plasma potential, from the experimentally measured current–voltage *IV* probe characteristic. For a negatively biased probe (ion collection regime) and in a middle pressure discharge, the commonly used orbital-motion-limited (OML) theory [30] does not explain experimental results and in particular, the fact that ion current reaches a maximum as a function of neutral gas pressure due to the destruction of the ion orbital motion around the probe (see figure 11.3.(a) for the collisionless case). The problem can be reduced to one-dimensional radial with ions moving in a central field of force under the effective potential defined as the sum of electrostatic attracting and centrifugal repelling contributions:

$$U_{\text{eff}}(r) = e\phi(r) + \frac{L^2}{2M_i r^2} \tag{11.35}$$

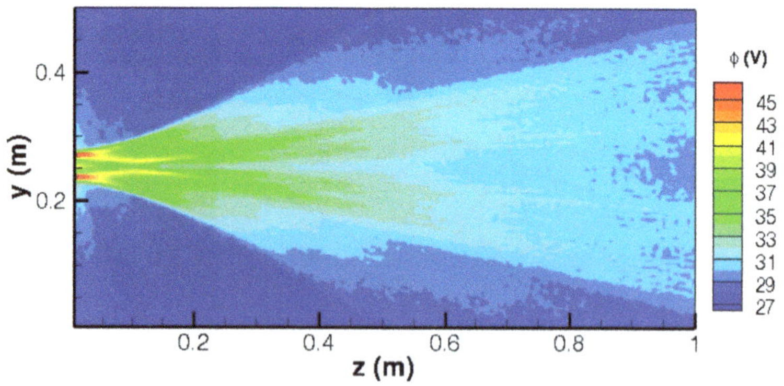

Figure 11.2. Two-dimensional (y,z) maps of electric potential ϕ (V) at the symmetric plane. The coaxial channel and plume divergency are evident.

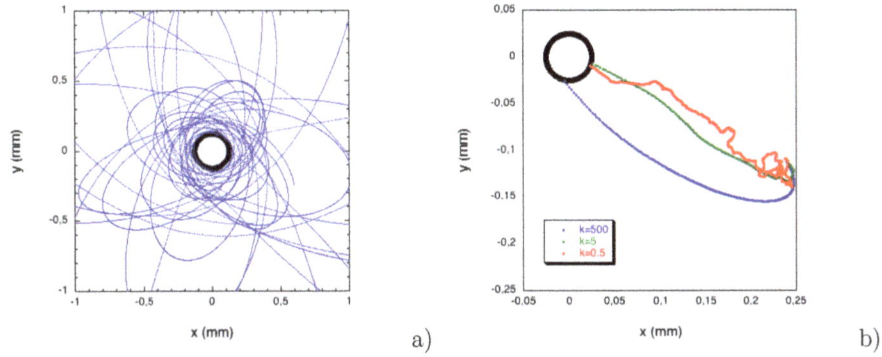

Figure 11.3. Ion trajectories around an electrostatic probe in (a) collisionless and (b) different collisional cases (k is the Knudsen number).

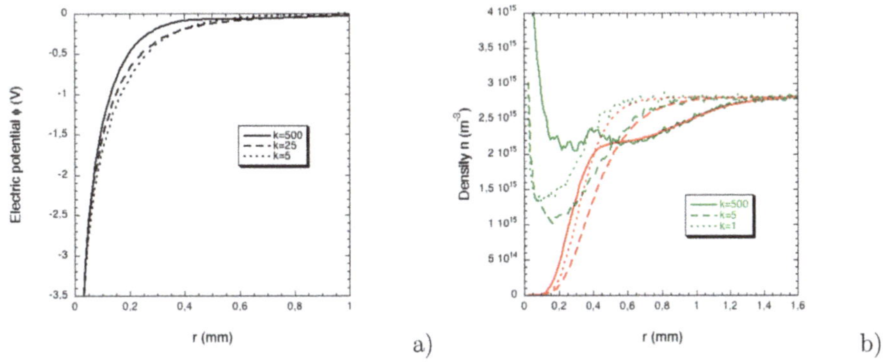

Figure 11.4. Radial profiles of (a) electric potential and (b) electron (red curves)/ion (green curves) density for different Knudsen numbers k.

where L is the angular momentum and ϕ is calculated by solving the nonlinear Poisson's equation (11.33) and assuming electrons as isothermal. Figure 11.3.(b) shows the typical ion trajectory for different Knudsen numbers k under the effect of the self-consistent electric potential reported in figure 11.4(a). The effect of ion-neutral collisions is to create a strong potential and consequently to reduce the sheath thickness. This effect is well understood and is caused by the increase of the viscous drag force that reduces the ion velocity in the collisional sheath. To satisfy the conservation of ion flux, the reduced ion velocity requires an increase in the ion density. Via Poisson equation, this increase leads to a stronger gradient in the electric field. A larger gradient implies a smaller scale length, i.e. a smaller sheath thickness. This is confirmed in figure 11.4(b) where ion (green curves) and electron (red curves) number density versus radial position at various k have been reported. Notice how the space charged region around the probe has been resolved without assuming quasi-neutrality $n_e \neq n_i$.

11.3.2 Magnetostatic case: $E \times B$ field devices

Electron confinement with magnetic field is often used in $E \times B$ configuration for different reasons. It reduces plasma diffusion to the wall, thus increasing plasma uniformity and its density at the same discharge power. Non-uniform magnetic field confines fast electrons better than cold ones, thus performing a filter between hot and cold electrons. Finally, electron magnetic confinement is often used in $E \times B$ configuration (HET, magnetron, ECR and helicon discharges) in order to increase ionization efficiency and create a strong impedance with a virtual cathode to accelerate the generated ions.

Hybrid models have been often developed for the simulation of $E \times B$ field devices. One of the first models to follow this approach was developed by Fife [31] for the simulation of the acceleration channel of a HET. Based on this model different variants have been developed during the years [32–35].

Here, the original Fife's numerical procedure is followed. The simulation is reduced to two-dimensional (r,z) cylindrical geometry using the axial-symmetry of the HET, schematically reported in figure 11.5(a). It is characterized by a DC glow discharge with a dielectric annular chamber: a magnetic circuit generates an axial-symmetric and quasi-radial magnetic field between the inner and outer poles (see figure 11.5(b)), while an electrical discharge is established between the anode (deep inside the channel) acting also as gas propellant distributor, and the external cathode, used also as electron emitter. In this configuration, cathode electrons are drawn to the positively charged anode, but the radial magnetic field creates a strong impedance, trapping the electrons in cyclotron motion which follows a closed drift path inside the annular chamber along the azimuthal θ direction. The trapped and

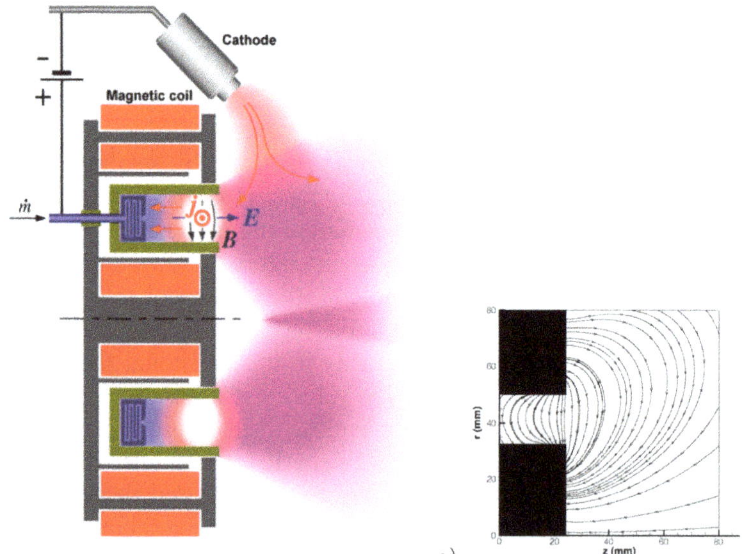

a) b)

Figure 11.5. (a) Cross-sectional cut view of an HET with (b) the axial-symmetric static magnetic field.

energized (by Joule heating) electrons act as a volumetric zone of ionization for atoms and as a virtual cathode to accelerate the generated ions which are not affected by the magnetic field due to their larger Larmor radii. The thruster is relieved of space charge limitations by the neutralizing effect of the trapped electrons, and HETs are therefore capable of providing higher thrust densities than gridded ion engines. The resulting external jet composed by the high-speed ion beam is subsequently neutralized by part of the electrons coming from the external cathode-compensator (see previous paragraph 11.3.1).

The system can be considered magnetostatic: the magnetic field variation due to plasma currents and changing electric fields is assumed to be small compared to the externally generated magnetic field. Ampère's law reduces to $\nabla \times \vec{B} = 0$ which allows one to define a magnetic potential function $\vec{B} = \nabla\sigma$ satisfying the Laplace equation

$$\nabla^2\sigma = 0. \tag{11.36}$$

Also, since $\nabla \cdot \vec{B} = 0$, it is possible to define a magnetic stream function λ whose gradient is everywhere orthogonal to $\vec{B}$:

$$\frac{\partial\lambda}{\partial z} = r\frac{\partial\sigma}{\partial r} = rB_r \tag{11.37}$$

$$\frac{\partial\lambda}{\partial r} = -r\frac{\partial\sigma}{\partial z} = -rB_z. \tag{11.38}$$

The function λ defined by these equations is constant along magnetic field lines $\vec{B} \cdot \nabla\lambda = 0$ and increases monotonically from anode to cathode. The cross-field gradient (gradient normal to magnetic field lines) of any quantity Q can be expressed in terms of λ as

$$\nabla_{\perp}Q = -rB\frac{\partial Q}{\partial\lambda}. \tag{11.39}$$

This equation allows us to write spatial derivatives across magnetic field lines as derivatives with respect to values invariant along the lines. This is due to overcoming the numerical errors associated with the extreme anisotropy of the electron transport: there is a large disparity (more than two orders of magnitude in regions with high values of the magnetic field) between perpendicular $\perp$ and parallel $||$ electron transport. Therefore, the fluid equations are solved on a computational mesh that is aligned with the applied magnetic field: magnetic field-aligned meshes (MFAM).

Along the magnetic field lines the problem reduces to electrostatic: as shown in the previous paragraph, equation (11.31), it is possible to define a quantity $\phi^*(\lambda)$ (called thermalized potential) constant along the magnetic field line

$$\phi^*(\lambda) = \phi - \frac{\kappa T_e}{e}\ln(n_e). \tag{11.40}$$

Differently, electron transport across the magnetic field lines obeys the generalized form of Ohm's law equation (11.5) in drift-diffusion approximation (the electron

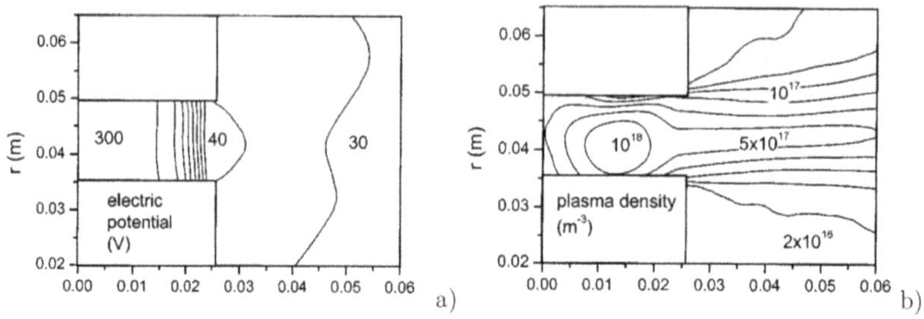

Figure 11.6. Contour plot of (a) electric potential and (b) plasma density in (r, z) plane of an HET. Reproduced with permission from [32], copyright 2002, AIP Publishing LLC.

collisional term is included); replacing the expression of the cross-field gradient equation (11.39) and using the definition of thermalized potential equation (11.40), it gives:

$$u_{e,\perp} = \mu_{e,\perp}\left[-rB\frac{\partial\phi^*}{\partial\lambda} - rB\frac{\kappa}{e}(\ln(n_e) - 1)\frac{\partial T_e}{\partial\lambda}\right] \quad (11.41)$$

where $\mu_{e,\perp} = \frac{\mu_{e,\|}}{1+\beta^2}$ represents the cross-field mobility ($\mu_{e,\|} = e/\nu_{en}m_e$ is the electron mobility and $\beta = \omega_e/\nu_{en}$ the Hall parameter). In this equation, the gradient of thermalized potential and electron temperature along the magnetic stream function are obtained from current and energy conservation equation (11.8), respectively. It should be noted that an artificial increased electron collision frequency $\nu_{en}^* = \nu_{en} + \nu_{ew} + \nu_{ano}$ is used to fit experimental measurements: the electron current collected at the anode becomes larger than obtained by classical collisional transport: electron–wall collision ν_{ew} and azimuthal fluctuation ν_{ano} contribute to increase the electron cross-field mobility. Figure 11.6 shows (a) electric potential and (b) plasma density maps in (r,z) plane as resulting from the magnetostatic hybrid model in [32]. The electric potential lines deviate significantly from the magnetic field lines (figure 11.5(b)) due to the second term of equation (11.40). Most of the electric field is concentrated near the exhaust, where the neutral density is so low that the electron mobility is strongly reduced. Outside the channel the larger fluctuation-induced mobility ν_{ano} imposed prevents a strong electric field. The plasma density is highest at the center of the channel and decreases at the exhaust due to the increasing ion velocity by the effect of the axial electric field.

MFAM method is strongly sensitive to small changes in the magnetic field strength or geometry and for this reason it has to be avoided in cases of strong magnetic field gradients or curvatures (as near ring cusps). In these cases non-MFAM approaches, such as simple orthogonal mesh or transverse-flux (TF) method [36], are suggested.

11.3.3 Electromagnetic case: fusion and space plasmas

Hybrid models have been developed to study plasma–wall transition regions in thermonuclear fusion reactors (scrape off layer (SOL) and divertor) [37] and instabilities driven by fusion-born ions [38]. Very commonly, hybrid models are used to simulate the interaction of the solar wind with planets, moons, comets, asteroids and local interstellar medium showing the different forming plasma structures: bow shock, magnetopause, magnetotail, etc. Here, as an example, the three-dimensional global simulation of the Martian plasma environment is reported. Mars represents a very challenging case for numerical simulation: the presence of multiple ion species in the ionosphere, the presence of a strong non-homogeneous crustal magnetic field at the surface and the fact that Mars radius and shock stand-offs are smaller than ion gyroradius makes the motivation very strong increased by the question about the escape of planetary oxygen ions as a phenomenon at the base of the loss of water from Mars [39].

Following [40, 41], the hybrid model treats six ion species: solar wind H^+ and He^{++}, planetary H^+, O^+, O_2^+ and CO_2^+. Although the density of alpha particles is smaller than the proton density, they carry out a significant fraction of the solar wind mass and momentum (20%), and thus can influence the overall dynamics. The electron fluid is assumed to be adiabatic with a polytropic index $\gamma = 2$ (equation (11.7)), with two specified temperatures for the undisturbed solar wind plasma and the ionospheric plasma. Planetary ions are produced by three processes: ionization by solar extreme ultraviolet (EUV) photons, resonant and non-resonant charge exchange, and electron impact ionization. The production rate of each ion species is computed from prescribed neutral reservoirs of atomic oxygen, hydrogen, and carbon dioxide. The model is implemented on a uniform Cartesian grid with cubic cells, extending from -2.4 to $3R_M$ in x and from -5 to $5R_M$ in y and z, where R_M is the Martian radii. The coordinate system is defined as follows: the x direction is aligned with the solar wind bulk flow velocity $\vec{u}_{sw}$ and points towards the Sun, z is the direction of the motional electric field of the solar wind $\vec{E}_{sw} = -\vec{u}_{sw} \times \vec{B}_{IMF}$, where $\vec{B}_{IMF}$ is the interplanetary magnetic field (IMF) conveyed by the solar wind and imposed to be directed along the z direction.

Figures 11.7 shows two-dimensional cuts of solar wind H^+ (left) and planetary O^+ (right) density (cm^{-3}) in the noon–midnight plane $y = 0$ containing $\vec{u}_{sw}$ and the interplanetary magnetic field $\vec{B}_{IMF}$. These figures show a first steep variation (at $x = 1.7R_M$ along the Sun–Mars line $z = 0$) clear signature of the bow shock (BS). The second steep increase (around $x = 1.3R_M$ along the Sun–Mars line $z = 0$) corresponding to the maximal value reached by the magnetic field intensity in the simulation domain, represents the magnetic pile-up boundary (MPB). It can also be considered as the boundary of the induced Martian magnetosphere. Three regions can be easily distinguished: the upstream region where the solar wind plasma density fluctuates around the undisturbed value $n_{H^+} = 2.3\ cm^{-3}$ (see figure 11.7.(a)), the downstream region between BS and MPB and finally, the region below MPB where planetary ions (see figure 11.7.(b)) dominate. The maps show a z-asymmetry, a consequence of the action of the motional electric field of the incoming plasma on the planetary ions which are accelerated mainly along the direction of the convection electric field, i.e. in the positive z direction, during a fraction of their first gyroperiod. Since their gyroradii are of the order of the size of the Martian obstacle, this initial acceleration breaks the symmetry between the two sides of the planet. Finally, the

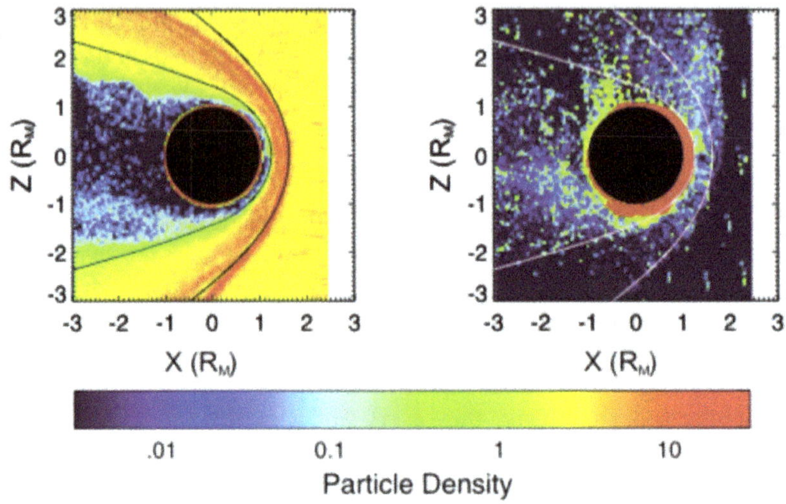

Figure 11.7. Two-dimensional cuts of H^+ (left) and O^+ (right) density (cm^{-3}) in the noon–midnight ($y = 0$) plane from [40, 41]. The nominal locations of the bow shock and MPB are indicated in black in each panel. Curves reported indicate location of the bow shock (BS), first from the right, and MPB, second from the right. Reproduced from [41], copyright 2010, with permission from Elsevier

plot of O^+ density (figure 11.7.(b)) indicates the dominant pathways for escaping planetary ions: due to the absence of a global intrinsic magnetic field, the Martian atmosphere is exposed to a continuous erosion process because the ionized component can be efficiently removed by ion pickup mechanisms. The escaping proton flux comes from charge exchange while photoionization is the main process which contributes to the escape of O^+ ions.

11.3.4 Spatially hybrid simulation: streamers and laser–plasma interaction

A variant of the hybrid approach is to resolve the electron sub-system as PIC-MCC in low density/high-field regions while keeping the fluid approach in high density/low field regions. This spatially hybrid simulation is well suited for cases characterized by high gradients, where plasma density spans several orders of magnitude, such as streamer discharges [42] or ignition laser fusion [43]. A spatially hybrid approach allows an improvement in computational speed in excess of 40× for a one-dimensional case and >500× in two dimensions.

The spatially hybrid model contains two important distinct features: the calculation of transport and reaction coefficients for the fluid model taken from the particle model to keep the two models consistent and the construction of the interface between particle and fluid regions. The model interface can be set at the position where the electron density reaches $n_e = \eta n_{e,\max}$ or where the electric field is $E = \xi E_{\max}$; here $n_{e,\max}$ and $E_{\max}$ stand for the maximal electron density and electric field, and η and ξ are real numbers between (0, 1), and chosen in such a manner that the calculation is efficient and the error is small. Different options for building the particle–fluid interface are reported in [42].

Usually, spatial hybrid simulations begin with a few electron–ion pairs followed by the full particle model. As new electrons are generated, the number of particles eventually reaches a given threshold, after which the simulation switches from full particle to hybrid approach. A this point the typical spatially hybrid cycle is repeated consisting of the following steps:

1. the particle–fluid boundary is determined and, in order to ensure a reasonable interaction between the particle and fluid approaches, buffer regions are built by extending the particle regions into the fluid region;
2. the movement of electrons is first followed in the particle and then in the buffer regions; the number of electrons crossing the model interface is recorded during this time step;
3. the electron fluxes on the interface are used as the boundary condition to update the densities in the fluid region;
4. the electron and ion densities are now known in both particle and fluid regions, and the new electric field is then calculated; with the updated electric field and the particle densities, the new model interface is determined.

Here, the example of a streamer discharge is reported as a spatial hybrid model application. Streamers are ionized fingers that penetrate into a previously non-ionized region under the action of a strong electric field (larger than the breakdown threshold); the resulting electron avalanche leaves behind positive ions, so in time more and more space charge is building up. If the electric field from the space charge becomes comparable to the background electric field, new avalanches predominantly grow in the high-field regions, so a self-propagating structure, known as streamer, can emerge. In figure 11.8 the three-dimensional structure of a negative streamer in air is reported as result of a spatially hybrid simulation [42]. The columns show from left to right: electron density isosurface at $n_e = 10^{13}\ \mathrm{cm}^{-3}$, electric field on two orthogonal planes and electron macroparticles with energy larger than 200 eV. The position of the particle–fluid interface used in the simulation is indicated in red in the middle column. It should be noted how hybrid simulation is able to trace the electrons with high energies that have the potential to run away.

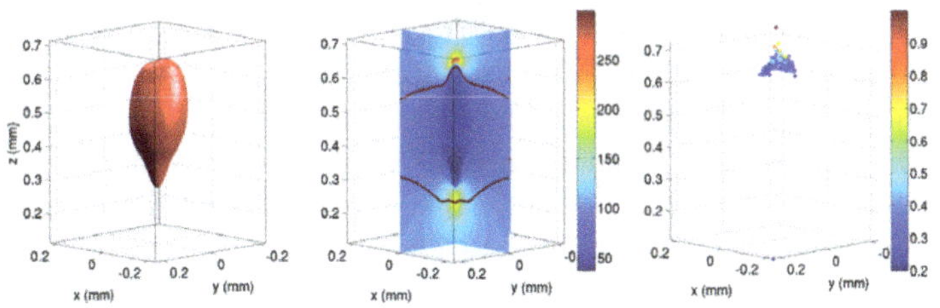

Figure 11.8. 3D simulation results of the 3D hybrid model for a negative streamer in air without photo-ionization developing in a background field of 100 kV cm^{-1}. The columns show from left to right: bulk electron density isosurface at $n_e = 10^{13}\ \mathrm{cm}^{-3}$, electric field on two orthogonal planes and electrons with energy larger than 200 eV. The position of the particle–fluid interface is indicated in red in the middle column. Reproduced from [42], copyright 2012, with permission from Elsevier.

References

[1] Brackbill J and Cohen B 1985 *Multiple Time Scales* (San Diego, CA: Academic)

[2] Lipatov A S 2002 *The Hybrid Multiscale Simulation Technology* (Berlin: Springer)

[3] Tskhakaya D, Matyash K, Schneider R and Taccogna F 2007 The particle-in-cell method *Contrib. Plasma Phys.* **47** 563–94

[4] Mason R J 1971 Computer simulation of ion-acoustic shocks. The diaphragm problem *Phys. Fluids* **14** 1943–58

[5] Chodura R 1975 A hybrid fluid-particle model of ion heating in high-Mach-number shock waves *Nucl. Fusion* **15** 55

[6] Sgro A G and Nielson C W 1976 Hybrid model studies of ion dynamics and magnetic field diffusion during pinch implosions *Phys. Fluids* **19** 126–33

[7] Hewett D 1980 A global method of solving the electron-field equations in a zero-inertia-electron-hybrid plasma simulation code *J. Comput. Phys.* **38** 378–95

[8] Harned D S 1982 Quasineutral hybrid simulation of macroscopic plasma phenomena *J. Comput. Phys.* **47** 452–62

[9] Brecht S H and Thomas V A 1988 Multidimensional simulations using hybrid particles codes *Comput. Phys. Commun.* **48** 135–43

[10] Moore K R, Thomas V A and McComas D J 1991 Global hybrid simulation of the solar wind interaction with the dayside of Venus *J. Geophys. Res. Space Phys.* **96** 7779–91

[11] Müller J, Simon S, Motschmann U, Schüle J, Glassmeier K-H and Pringle G J 2011 A.I.K. E.F.: Adaptive hybrid model for space plasma simulations *Comput. Phys. Commun.* **182** 946–66

[12] Leclercq L, Modolo R, Leblanc F, Hess S and Mancini M 2016 3D magnetospheric parallel hybrid multi-grid method applied to planet-plasma interactions *J. Comput. Phys.* **309** 295–313

[13] Matthews A P 1994 Current advance method and cyclic leapfrog for 2D multispecies hybrid plasma simulations *J. Comput. Phys.* **112** 102–16

[14] Büchner J, Dum C and Scholer M 2003 *Space Plasma Simulation*vol 615 (Berlin: Springer)

[15] Ruyten W M 1993 Density-conserving shape factors for particle simulations in cylindrical and spherical coordinates *J. Comput. Phys.* **105** 224–32

[16] Verboncoeur J 2001 Symmetric spline weighting for charge and current density in particle simulation *J. Comput. Phys.* **174** 421–7

[17] Cornet C and Kwok D T 2007 A new algorithm for charge deposition for multiple-grid method for PIC simulations in r-z cylindrical coordinates *J. Comput. Phys.* **225** 808–28

[18] Amano T, Higashimori K and Shirakawa K 2014 A robust method for handling low density regions in hybrid simulations for collisionless plasmas *J. Comput. Phys.* **275** 197–212

[19] Horowitz E J, Shumaker D E and Anderson D V 1989 QN3D: A three-dimensional quasi-neutral hybrid particle-in-cell code with applications to the tilt mode instability in field reversed configurations *J. Comput. Phys.* **84** 279–310

[20] Terasawa T, Hoshino M, Sakai J-I and Hada T 1986 Decay instability of finite-amplitude circularly polarized Alfven waves: a numerical simulation of stimulated Brillouin scattering *J. Geophys. Res. Space Phys.* **91** 4171–87

[21] Hockney R W and Eastwood J W 1989 *Computer Simulation using Particles* (Bristol: IOP Publishing)

[22] Taccogna F, Longo S and Capitelli M 2002 Particle-in-cell with Monte Carlo simulation of SPT-100 exhaust plumes *J. Spacecr. Rockets* **39** 409–19

[23] Taccogna F, Longo S and Capitelli M 2004 Very-near-field plume simulation of a stationary plasma thruster *Eur. Phys. J. Appl. Phys.* **28** 113–22

[24] Taccogna F, Pagano D, Scortecci F and Garulli A 2014 Three-dimensional plume simulation of multi-channel thruster configuration *Plasma Sources Sci. Technol.* **23** 065034

[25] Ellegaard O, Nedelea T, Schou J and Urbassek H 2002 Plume expansion of a laser-induced plasma studied with the particle-in-cell method *Appl. Surf. Sci.* 229–38

[26] Taccogna F, Longo S and Capitelli M 2003 A particle-in-cell/Monte Carlo model of the Ar^+ ion collection in He gas by a cylindrical Langmuir probe in the transition regime *Eur. Phys. J. Appl. Phys.* **22** 29–39

[27] Taccogna F, Longo S and Capitelli M 2004 PIC model of the ion collection by a Langmuir probe *Contrib. Plasma Phys.* **44** 594–600

[28] Taccogna F, Longo S and Capitelli M 2006 Ion orbits in a cylindrical Langmuir probe *Phys. Plasmas* **13** 043501

[29] Taccogna F, Longo S and Capitelli M 2008 Ion-neutral collision effects in Langmuir probe theory *Contrib. Plasma Phys.* **48** 509–14

[30] Lampe M 2001 Limits of validity for orbital-motion-limited theory for a small floating collector *J. Plasma Phys.* **65** 171–80

[31] Fife J and Martinez-Sanchez M 1995 Two-dimensional hybrid particle-in-cell (PIC) modeling of Hall thrusters *Proc. of the 24th Int. Electric Propulsion Conf., IEPC95-240, Electric Rocket Propulsion Society* pp 1213–24

[32] Hagelaar G J M, Bareilles J, Garrigues L and Boeuf J P 2002 Two-dimensional model of a stationary plasma thruster *J. Appl. Phys.* **91** 5592–8

[33] Parra F, Ahedo E, Fife J and Martinez-Sanchez M 2006 A two-dimensional hybrid model of the Hall thruster discharge *J. Appl. Phys.* **100** 023304

[34] Sommier E, Scharfe M K, Gascon N, Cappelli M A and Fernandez E 2007 Simulating plasma-induced Hall thruster wall erosion with a two-dimensional hybrid model *IEEE Trans. Plasma Sci.* **35** 1379–87

[35] Mikellides I G and Katz I 2012 Numerical simulations of Hall-effect plasma accelerators on a magnetic-field-aligned mesh *Phys. Rev.* E **86** 046703

[36] Hagelaar G J M 2007 Modelling electron transport in magnetized low-temperature discharge plasmas *Plasma Sources Sci. Technol.* **16** S57

[37] Furubayashi M, Hoshino K, Toma M, Hatayama A, Coster D, Schneider R, Bonnin X, Kawashima H, Asakura N and Suzuki Y 2009 Comparison of kinetic and fluid neutral models for attached and detached state *Proc. of the 18th Int. Conf. on Plasma -Surface Interactions in Controlled Fusion Device*

[38] Dendy R O and McClements K G 2015 Ion cyclotron emission from fusion-born ions in large tokamak plasmas: a brief review from JET and TFTR to ITER *Plasma Phys. Control. Fusion* **57** 044002

[39] Brecht S H and Ledvina S A 2010 The loss of water from Mars: numerical results and challenges *Icarus* **206** 164–73 (Special issue: Solar Wind Interactions with Mars)

[40] Modolo R, Chanteur G M, Dubinin E and Matthews A P 2005 Influence of the solar EUV flux on the Martian plasma environment *Ann. Geophys.* **23** 433–44

[41] Brain D *et al* 2010 A comparison of global models for the solar wind interaction with Mars *Icarus* **206** 139–51 (Special issue: Solar Wind Interactions with Mars)

[42] Li C, Ebert U and Hundsdorfer W 2012 Spatially hybrid computations for streamer discharges: II. Fully 3D simulations *J. Comput. Phys.* **231** 1020–50 (Special Issue: Computational Plasma Physics)

[43] Cohen B I, Kemp A and Divol L 2010 Simulation of laser-plasma interactions and fast-electron transport in inhomogeneous plasma *J. Comput. Phys.* **229** 4591–612

Part III

Applications

IOP Publishing

Plasma Modeling (Second Edition)
Methods and applications
Gianpiero Colonna and Antonio D'Angola

Chapter 12

On the coupling of vibrational and electronic kinetics with the electron energy distribution functions: past and present

Mario Capitelli

Non-equilibrium plasma kinetics is a topic of large interest for many applications in different fields such as plasma chemistry, plasma and laser physics, hypersonics and shock wave flows. Particular attention is paid to the development of kinetic models which couple the time dependent Boltzmann equation for the electron energy distribution function (EEDF) with the state-to-state vibrational kinetics and the radiative-collisional models of electronic excited states. This approach becomes essential when the chemistry at the basis of the relevant applications is dependent on the high lying vibrational levels of the considered molecules [1–4].

The following equations govern the time dependent behavior of the different quantities operating in plasmas (see chapters 2 and 9):

1. the electron Boltzmann equation describing the EEDF;
2. master equations describing the vibrational kinetics of molecules;
3. master equations describing the electronic excited state kinetics of atoms and molecules;
4. the plasma chemistry equations describing the concentrations of the different species (neutral and ionized) existing in the reactive plasma.

Previous equations are strictly correlated and should be solved with a self-consistent approach. The calculation of the EEDF is necessary to get the rate coefficients for electron-heavy particle collisions entering in the master and in the plasma chemistry equations, while the vibrational distribution functions of molecules, the electronic excited state concentrations and the plasma composition are necessary to solve the electron Boltzmann equation. In the following sections, a focus on the main experimental and theoretical investigations of some relevant molecular plasmas is provided.

doi:10.1088/978-0-7503-3559-1ch12

12.1 H_2 plasma

The H_2 system is one of the most studied systems including, for example: (1) the cracking of hydrocarbons in the presence of an H_2 plasma [5, 6]; (2) the role of atomic H species in the formation of diamond-like systems [7]; and (3) the role of H_2 for the formation of H^- negative ions to be used for injecting energy in nuclear fusion reactors [8]. The first work on the dissociation of H_2 in electrical discharges was published many years ago by Capitelli *et al* [1], parametrizing the electron temperature, the electron density, the gas temperature and pressure. The electron–molecule dissociation cross-sections, integrated by a Maxwell energy distribution function for free electrons, are used to calculate the rate of dissociation by direct electron impact (DEM). At the same time, electron-vibrational (e-V) processes, whose rates are calculated with the same approximation, are responsible of the introduction of energy quanta in the vibrational ladder of the molecules. The state-to-state vibrational kinetics including e-V, VV (vibration–vibration) and VT (vibration–translation) energy exchange processes, can be considered the starting point of non-equilibrium kinetics of molecules in plasmas. Beside the DEM dissociation mechanism, the concept of dissociation of the molecules by vibrational excitation, i.e. pure vibrational mechanism (PVM), has been introduced. In fact, the energy quanta injected in the first vibrational levels are redistributed in the whole ladder by VT and VV processes, reaching the last bound level and a pseudo state above the dissociation limit $v' = 14 + 1$ accounting for the PVM dissociation mechanism. Depending on electron temperature and electron density, the PVM dissociation can be more important than the DEM mechanism.

The presence of vibrational and electronic excited states modifies the EEDF through the action of superelastic vibrational and electronic collisions. The plasma kinetic description (see chapters 2 and 9) requires as input data a complete set of cross-sections. From the first simple electron impact H_2 cross-sections derived in the past by classical and semi-classical methods [9–12] and quasi-classical trajectory (QCT) approach [13, 14] for atom–molecule interaction, more recent and refined cross-sections [15–18] and new experimental observations [19, 20] have been deduced for H_2, providing an accurate description of H_2 plasma kinetics. Figure 12.1 reports the H_2 vibrational distribution function (VDF) and the EEDF in a pure hydrogen discharge calculated by Pagano *et al* [21]. These results correspond to the first source of negative ions appearing in the literature, composed by a filament emitting electrons and accelerated by an applied voltage [22–26]. Other configurations are under study with the aim of increasing the concentration of negative ions to be ejected as a beam, which, once formed, is accelerated and neutralized for entering the tokamak, creating a very hot plasma to be ready for the fusion [8].

The form of EEDF reported in figure 12.1 is in line with theoretical data by Gorse *et al* [22–24] and by the experimental determination of EEDF performed in reference [27]. The VDF and EEDF show non-equilibrium trends, characterized by plateaux, being at the basis of the formation of negative ions. These high-energy electrons entering the reactor containing H_2 (or D_2) form an EEDF composed by hot

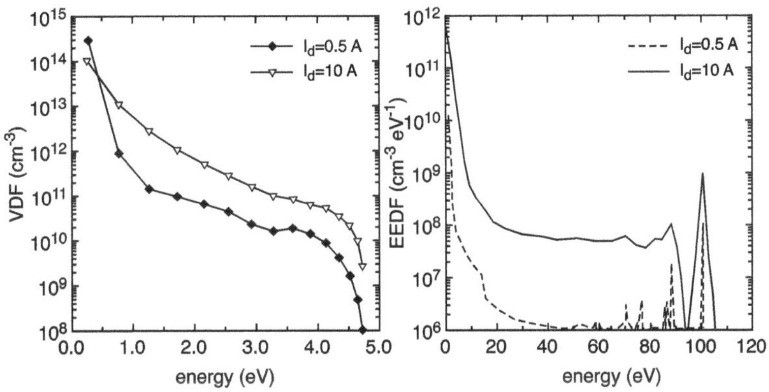

Figure 12.1. Stationary H_2 VDF and EEDF for different discharge currents for P = 1.5 Pa and V_d = 100 V. Data from reference [21].

electrons, which excite the Rydberg states of molecules and low energy electrons, forming the negative ions acting on the existing Rydberg states. All relevant quantities calculated by Pagano *et al* [21], i.e. the electron density and the vibrational distribution, are in satisfactory agreement with the modeling results of Mosbach [25], except for the concentration of negative ions, probably due to different cross-sections entering the two models. Finally, the recombination on the walls of atomic hydrogen can increase the concentration of H^- through the process (see Pagano *et al* [28])

$$H + H + \text{walls} \rightarrow H_2(v) + \text{walls} \tag{12.1}$$

a very important aspect, that has triggered the investigation on surface materials for the optimization of negative ion production.

12.2 N_2 plasma

A large number of theoretical and experimental papers focussed on N_2 plasmas appeared in the literature since 1978. At that time, the evolution of vibrational distribution of N_2 obtained by parametrizing the electron and the gas temperatures [2] has been simulated. By comparing the calculated VDF at three different gas temperatures (T_{gas} = 500–1000–2000 K), the VDF at 500 K showed a more extended plateau with respect to the one at 2000 K (see figure 12.2). In particular, the population densities of higher vibrational levels (i.e. v > 30) decreased with increasing gas temperature. Two processes are responsible of this behavior:

1. VT, which tend to depopulate the higher vibrational levels, increase their importance with increasing gas temperature;
2. the esothermic VV which, at low gas temperature, present rate coefficients much larger than the reverse ones, being the Boltzmann factor (see equation (9.32)) small at low gas temperature.

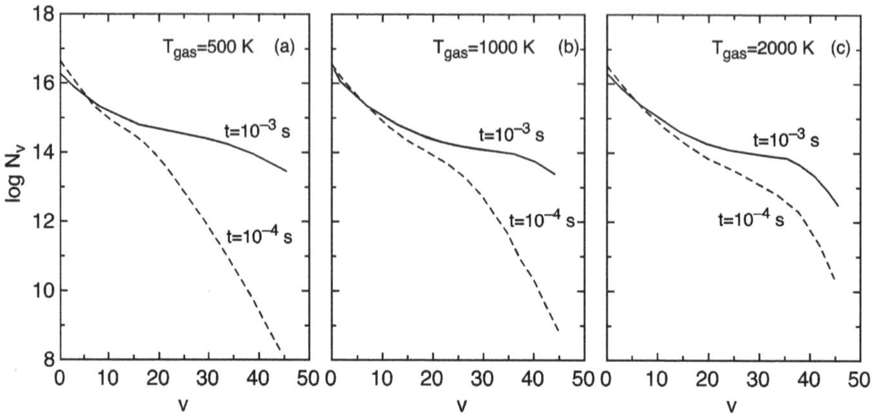

Figure 12.2. N_2 population densities (log N_v) as a function of vibrational quantum number at two different times and gas temperatures (T_e = 20 000 K and $n_e = 10^{12}$ cm^{-3}). Data from reference [2].

These results were used to rationalize the experimental CARS distributions obtained by Massabieaux *et al* [29], as well as the spectroscopic experimental results by Massabieaux *et al* [30].

Similar results were obtained more recently in [31], where the N_2 VDF was calculated at 500 K and 1000 K in a N_2 discharge characterized by a reduced electric field of E/N = 60 Td with a fixed electron molar fraction of 10^{-6}. The corresponding N_2 VDF showed a series of VV plateaux, more evident at low temperature (500 K) [3].

The role of vibrational levels in a N_2 glow discharge in affecting the dissociation rates was investigated in [31]. The temporal evolution of the dissociation rates from $v = 0$ and from all the vibrational levels was calculated in a discharge characterized by E/N = 60 Td and T_{gas} = 1000 K. The calculated results have shown that the two quantities are approximately the same up to 10^{-7} s, after that, they start diverging due to the contribution of vibrationally excited molecules in affecting the dissociation process, reaching a value of two orders of magnitude higher than the $v = 0$ rate. Another important aspect emerging in the study of a N_2 plasma in afterglow conditions was the importance of superelastic collisions of N_2 metastable electronic excited states in affecting the EEDF [32]. In this study, as initial condition, a high pressure (300 torr) nitrogen discharge with current density of about 100 A cm^{-2} sustained for approximately 200 ns is assumed [32]. The following concentrations of N_2 metastable states were estimated at the end of the discharge, together with the electron number density, reaching T_v = 400 K:

species	density [cm^{-3}]
$N_2(A)$	5 10^{14}
$N_2(B)$	10^{14}
$N_2(C)$	5 10^{13}
n_e	10^{12}

In these conditions, the vibrational excitation is negligible and the post discharge is dominated by superelastic electronic collisions involving the electronically excited states, especially the $A^3\Sigma$ metastable. The presence of superelastic electronic collisions creates, in the post discharge, a well structured EEDF, while such structures disappear when these superelastic processes are neglected (see figure 12.3). In both cases, the EEDF presents a hole around 2 eV due to the lack of energy exchange (inelastic) processes in nitrogen in this energy range.

Similar results due to superelastic vibrational collisions were obtained in [33], in which the relaxation of the EEDF in N_2 post-discharge regime was studied by solving the Boltzmann equation. Calculations were performed by including or neglecting the effect of superelastic vibrational collisions. In the first case, the EEDF relaxes toward nearly stationary values rich in electrons in the energy range 2–4 eV, while in the absence of superelastic vibrational collisions, the EEDF presents a hole in this energy range. This affects the electron-impact rate calculations: in the presence of superelastic vibrational collisions, the rate coefficients of the vibrational excitation and of excitation of the metastable nitrogen atoms $N(^2P, {}^2D)$ can conserve high values also in the post discharge regime.

More recently, a macroscopic kinetic model which accounts for non-Boltzmann vibrational distributions was proposed to describe the dissociation of N_2/N mixture in recombination regime [34]. This model was derived by the state-to-state approach, by taking into account the kinetic equation describing the last N_2 vibrational level and by relating the global dissociation rates to the population of this level instead of the vibrational temperature. The results obtained by this interesting approach were accurately validated by comparison with the state-to-state results.

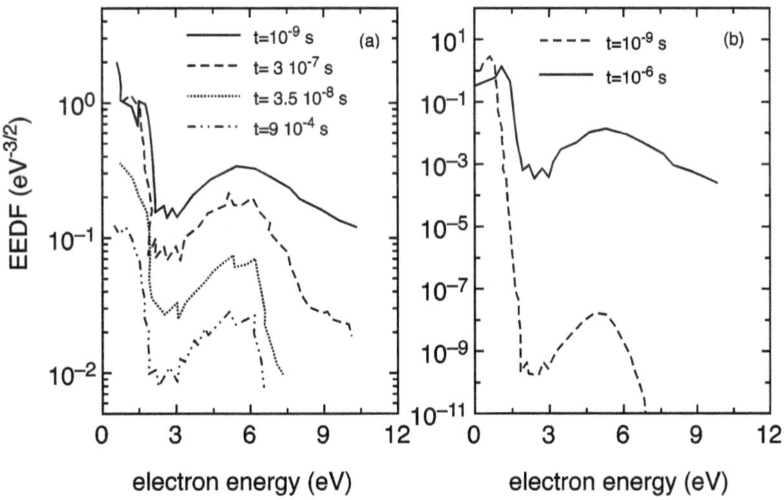

Figure 12.3. EEDF under post-discharge conditions with (a) and without (b) superelastic collisions. Data from reference [32].

12.3 O_2 plasma

The O_2 plasma is one of the most studied systems due to its importance in many applications. The first theoretical paper, in which the VDF of molecular oxygen were calculated in electric discharge conditions was published by Cacciatore *et al* [35], using the same method developed for H_2. In this case, the VT rates involving $O_2(v)$ and atomic oxygen O are very large so that the plateaux in the vibrational distribution function are expected to play a minor role. However, in the reacting O_2/O system important plateaux in the VDF can be formed through the three-body recombination of O atoms [35], i.e.

$$O + O + M \rightarrow O_2(v) + M \tag{12.2}$$

In the last few years, many papers appeared providing new electron impact cross-sections and heavy particle collision rate coefficients for the O_2 reacting plasma. In particular, complete state-to-state rate coefficients for the dissociation of $O_2(v)$ by atomic oxygen as well as a complete set of VT rates for O–$O_2(v)$ processes were calculated in [36, 37]. Moreover, a complete set of resonant e-V energy exchange collisions as well as the resonant dissociation [38] and dissociative attachment [39] collisions were recently calculated. Also, electron impact direct dissociation and ionization cross-sections calculated by classical and semi-classical methods as a function of vibrational quantum number are available for O_2 [40, 41]. On the basis of these new developments, the O_2 kinetics has been reconsidered and new calculations have been reported by Pietanza *et al* [42] by using a self-consistent model based on the coupling of the different time dependent kinetics equations reported in chapter 9. Recent advanced kinetic models for O_2 plasmas were developed also by Annusova *et al* [43] and Šimek *et al* [44]. The model results of Pietanza *et al* [42] showed long plateaux in the O_2 VDF created by the reaction in equation (12.2) at moderate pressure (i.e. 20 Torr) and by multi-quantum e-V processes at lower pressure (i.e. 20 mTorr). Moreover, the O_2 vibrational temperature remains close to the gas temperature due to the prevalence of VT processes over e-V ones. In the moderate pressure case, the EEDF in post-discharge conditions are characterized by declining plateaux due to superelastic collisions involving the electronic excited states of atomic oxygen. Low-lying metastable states of O_2 and O form structures in the EEDF in the energy range 0–3 eV. In the lower pressure case, the reached higher electron density induces a more Maxwellian behavior of the EEDF. Figure 12.4 shows a comparison of calculated and measured O_2 VDF by Pietanza *et al* [42] and Annusova *et al* [43] in two different pressure and gas temperature conditions. The reported VDF by Pietanza *et al* [42] have been calculated by applying a power density of $P_d = 0.02$ W cm^{-3} and a residence time of $t_r = 3$ ms and compared to stationary calculated and experimental VDF results by Annusova *et al* [43]. The comparison should be considered rather qualitative since Pietanza *et al* performed a time dependent solution of the equations, while Annusova *et al* reported stationary results without specifying the time of stationarity. Despite this, figure 12.4 shows a satisfactory agreement of the results. A difference indeed does exist in the model results from $v > 30$, showing a strong

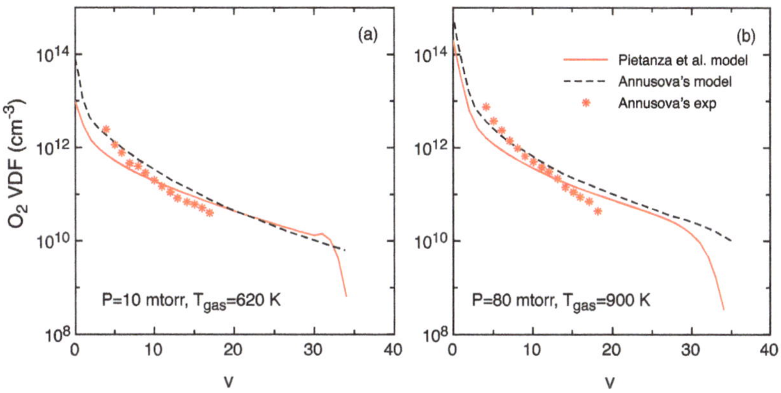

Figure 12.4. O_2 VDF comparison of Pietanza *et al* [42] and Annusova *et al* [43] results.

decay for Pietanza's ones and a flat behavior for Annusova's ones, probably due to differences in the used input data rates for processes involving higher vibrational levels.

12.4 CO plasma

One of the first observation of a VV pumped non-equilibrium VDF for the CO molecule was performed by Rich *et al* [45], in a steady state direct optical pumping experiment, in which a CO laser was used to excite a flowing mixture of CO. The measured VDF was a non-Boltzmann one with an overpopulation of the higher vibrational states, as predicted also by the theoretical treatments of the kinetic equations previously performed by Treanor, Rich and Rehm [46]. Soon after, CO VDFs were also directly measured by Fisher and Lightman [47] from the first overtone spontaneous emission spectra of CO in a liquid-N_2-cooled laser plasma and in the presence of varying amounts of O_2 (see figure 12.5). The experimental results were also quantitatively in agreement with calculated CO VDFs [47, 48].

The observed distributions showed again overpopulated tails, with an increasing deactivation as the O_2 concentration was increased. This effect was the result of the higher efficiency of the CO + O VT deactivation rates with the increase of the O_2 and O concentration in the plasma.

From a theoretical point of view, an important step forward in the description of CO non-equilibrium plasma discharges was made by Gorse *et al* [49, 50] with the development of a self-consistent kinetic model, coupling the vibrational master equations to the plasma chemistry and to the Boltzmann equation for the EEDF. This model showed the strong coupling between the vibrational distribution, the EEDF and the plasma composition, and the importance of the vibrational mechanism in dissociating CO, i.e. PVM, which could in some conditions predominate over DEM. The model was also extended to CO–N_2 mixtures to describe the non-equilibrium vibrational kinetics of CO pumped by vibrationally excited N_2 [51]. The pumping, the redistribution and the loss of vibrational quanta introduced by the

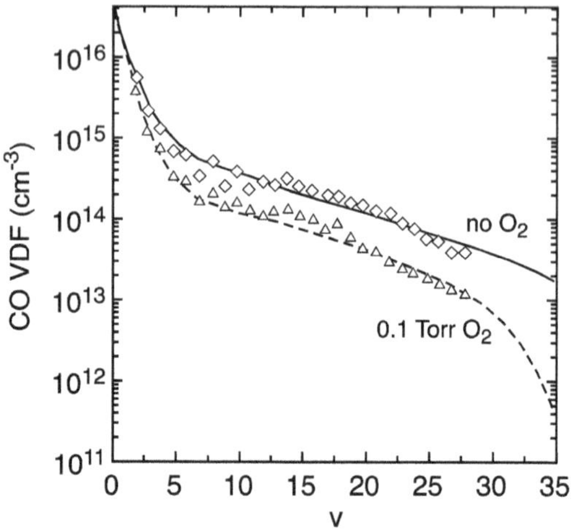

Figure 12.5. Measured (points) and calculated (lines) CO VDF for a 1.5/12.5 Torr CO/He mixture at 155 K with (0.1 Torr O_2, electron density = 2 10^{11} cm^{-3}, E/N = 1.02 10^{-16}, current = 10 mA) and without O_2 (electron density = 10^{10} cm −3, E/N = 1.68 10^{-16}, current = 10 mA). Data from reference [47].

N_2 system in the CO one were investigated. The calculated CO VDFs were also compared to experimental ones measured by infrared emission spectroscopy, showing a satisfactory agreement [52]. More recently, Essenigh *et al* [53] provided experimental measurements of CO VDFs in optically pumped CO–Ar–He mixtures at different He and CO partial pressures, by using FTIR spectroscopy and detecting vibrational levels as high as $v = 37$. The corresponding VDFs showed the characteristic VV plateaux, which were depleted only in the high-energy part ($v > 15$) by increasing the He partial pressure, as an effect of the deactivation of CO vibrational excitation by collisions with He atoms. They also showed that, in an optically pumped CO mixture, the CO_2 formation is controlled by the high CO vibrational levels populations, through the Boudouard reaction of the kind $CO(v) + CO(w) \rightarrow CO_2 + C$. The corresponding reaction rate constant and activation energy (i.e. 11.6 eV) were also evaluated. Very recently, a new estimation of this activation energy (i.e. 8.1 eV) was provided by Barreto *et al* [54] by using the transition state theory and a quantum chemistry approach.

A re-examination of the whole CO kinetics has been done very recently by Pietanza *et al* [55–57], with the implementation of an advanced self-consistent kinetic model applied to the description of CO plasma mixture in different discharge and post-discharge conditions. Particular attention was dedicated to the inclusion in the vibrational kinetic model of a complete set of resonant electron–molecule cross-sections, recently calculated by Laporta *et al* [58, 59], linking all the CO vibrational energy levels, and to the investigation of the role of the Boudouard reaction and CO metastable electronic excited states in affecting the kinetics. The following figure (figure 12.6) shows the time dependent CO VDF and EEDF in discharge (a)–(c) and post-discharge (b)–(d) conditions calculated by Pietanza *et al* [60] in a MW test case (P = 5 torr, T_{gas} = 500 K, t_{pulse} = 2.5 ms E/N = 60 Td).

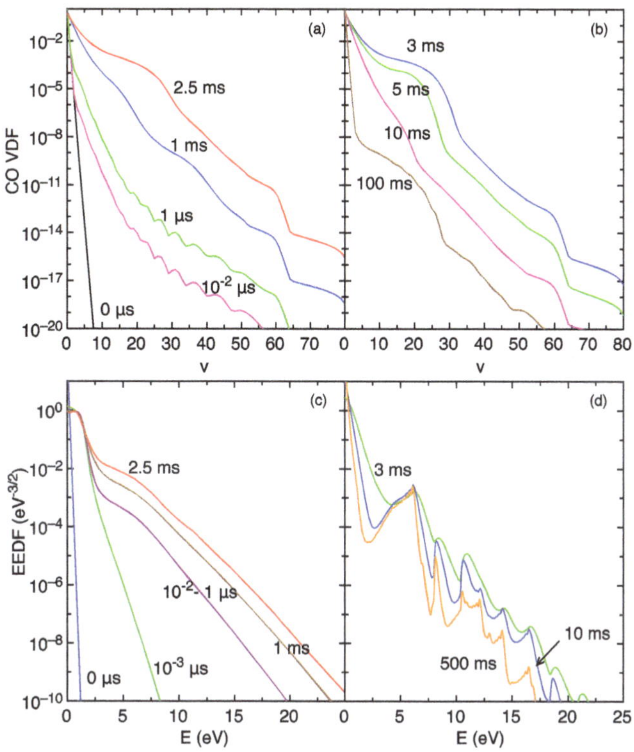

Figure 12.6. CO VDF (a, b) and EEDF (c, d) in discharge (a, b) and post-discharge (b–d) conditions in a MW discharge of CO plasma mixture in the following conditions $P = 5$ torr, $T_{gas} = 500$ K, $t_{pulse} = 2.5$ ms, E/N = 60 Td. Data from reference [60].

As it can be seen in figure 12.6, both the distributions are out of equilibrium. In particular, the CO VDF shows the characteristic plateau due to VV collisions up to nearly $v = 30$. This plateau is maintained also during the post-discharge with a progressive deactivation due to VT processes. The length of the plateau depends on the competition between the VV up pumping mechanism and to the two main deactivation processes, the Boudouard one and the VT deactivation by C and O atoms, which depopulate the VDF tail. The EEDF in the post-discharge is characterized by several peaks created by the superelastic processes involving the electronically excited states. The peaks appear at the corresponding electronic excited state energies (for example, $CO(a^3\Pi) = 6$ eV) and at multiples of them. The overpopulation of the CO VDF tail promotes CO dissociation by vibrational excitation (PVM). By performing an investigation on the dependence on T_{gas} of the CO VDF and dissociation rates, Pietanza *et al* [55] showed that the CO VDF have a more extended non-equilibrium plateau with the decrease of the T_{gas}, with the consequence of a non-Arrhenius behavior of the dissociation rates both by vibrational excitation and by electron impact in the low gas temperature range. The non-Arrhenius behavior consists in the increase of the rate with the decrease of T_{gas}. This effect is more evident for the dissociation rates by vibrational excitation due to the

$CO(v) + M \rightarrow C + O + M$ process (PVM_1) and the Boudouard process, which directly depend on the CO VDF, while it occurs to a minor extent for the electron impact dissociation process. At higher gas temperature ($T_{gas} > 2000$ K), the CO VDF becomes thermal and the dissociation rates present an Arrhenius behavior. The CO kinetic model by Pietanza *et al* [55–57] has been successively inserted in a more complex self-consistent kinetic model describing a CO_2 plasma mixtures, characterized also by all CO_2 main dissociation products, i.e. CO, O_2, C, O, etc for the purpose of investigating the role of vibrational excitation in increasing CO_2 conversion and energy efficiency rates in non-equilibrium plasmas [60–65] (see also chapter 9, section 9.6).

12.5 Nozzle flows

The nozzle flows are ideal tools to rapidly decrease the gas temperature opening the possibility of obtaining non-equilibrium vibrational distributions and non-equilibrium reactive rates. An interesting example is represented by the dissociation of N_2 in the nozzle investigated by Shizgal and Lordet [66], where the Euler equations are solved for the fluid dynamics [67]. We present in this section the dissociation of N_2 by using the QCT (quasi-classical trajectory) and the ladder climbing (LC) methods [68]. Figure 12.7 (a) reports the temperature profile (T/T_0) in the nozzle outlet for QCT and LC models ($T_0 = 8000$ K). We can see that the temperature profile T/T_0 strongly decreases as a function of nozzle position being practically independent passing from QCT to LC models. In contrast, the dissociation of N_2 is strongly dependent on QCT and LC models as can be seen in figure 12.7(b), which shows not only the differences of the two methods but also the non-Arrhenius of dissociation of N_2. In this case, the LC model presents a non-Arrhenius behavior more important than the QCT due to the differences in the

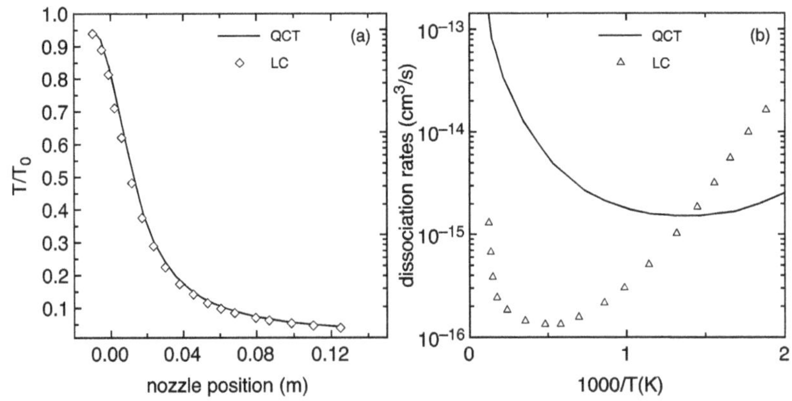

Figure 12.7. (a) Temperature profile in the nozzle outlet for QCT and LC models; (b) Arrhenius plot of the global dissociation rate for process $N_2(v) + N \rightarrow 2N + N$ for QCT and LC models ($T_0 = 8000$ K, $P_0 = 1$ atm). Data from reference [68].

selection of cross-sections. Nowadays, the QCT cross-sections together with a more accurate dynamic are used for many problems of hypersonic.

Another example is the reaction of N_2 and O_2 in a 1 m parabolic nozzle having cylindrical symmetry [67]. Also in this case, we can observe non-equilibrium behavior of N_2 and O_2 VDFs as well as a non-Arrhenius character of the rate of NO formation by the reaction $N_2 + O \rightarrow NO + N$. The method has been modified adding to the N_2 neutral species the ionization kinetics. In this case, one can observe also the non-equilibrium EEDF in different nozzle conditions. These differences being more important when one neglects the electron–electron collisions [69].

Finally, we want to briefly discuss the role of radiation during the supersonic expansion through a converging-diverging nozzle [70]. Figure 12.8 shows the EEDF calculated in [70] at different positions along the nozzle axis in two extreme conditions of optically thin (figure 12.8(a)) and thick (figure 12.8(b)) plasma for the Jupiter atmosphere (89% H_2, 11% He) with an inlet pressure of 1 atm and gas temperature of 8000 K.

Very structured EEDF are created, showing long plateaux and humps due to superelastic collisions of low energy electrons and electronically excited atomic hydrogen. As an example, in figure 12.8 two peaks at 10.2 eV and 20.4 eV and a hump at 12.3 eV are evidently due to the following processes

$$e(\epsilon \approx 0) + H(n = 2) \rightarrow e(\epsilon = 10.2\ \text{eV}) + H(n = 1) \tag{12.3}$$

$$e(\epsilon \approx 10.2\ \text{eV}) + H(n = 2) \rightarrow e(\epsilon = 20.4\ \text{eV}) + H(n = 1) \tag{12.4}$$

$$e(\epsilon \approx 0) + H(n = 3) \rightarrow e(\epsilon = 12.3\ \text{eV}) + H(n = 1) \tag{12.5}$$

The peaks and the humps are subsequently spread in plateaux by elastic collisions. The magnitudes of the plateaux are much higher in the thick case due to higher concentrations of the states involved in the superelastic collisions [70].

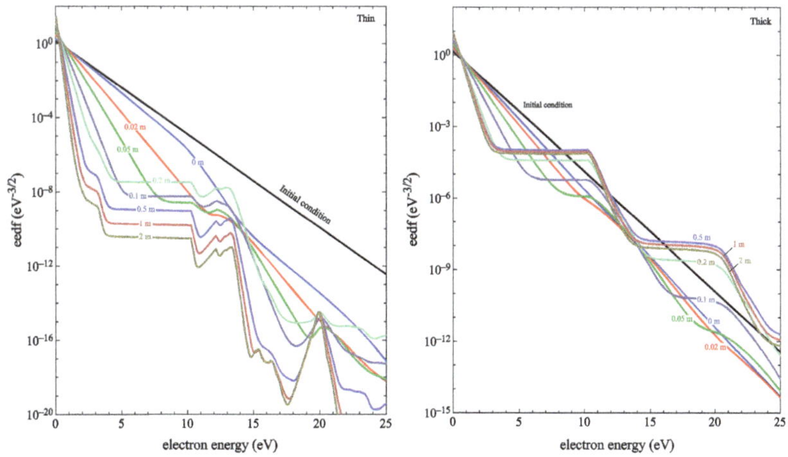

Figure 12.8. EEDF at different positions along the nozzle axis ($P_0 = 1$ atm, $T_0 = 8000$ K) in the thin and thick case. Reproduced from with permission from [70], copyright IOP Publishing, all rights reserved.

12.6 Conclusions

Old and new results on the application of a time dependent coupling of EEDF and VDF have been presented. In particular, we have discussed the progress made in the last years on different systems including H_2, N_2, O_2 and CO plasmas. In all of these systems we have discussed the plateaux of VDF and EEDF due to the vibrational and electronic superelastic collisions. An interesting case is the O_2 reacting plasma where the plateaux in the post discharge are due to the three-body recombination processes of atomic oxygen. In this case, a satisfactory agreement between recent modeling by Pietanza *et al* [42] and experimental results by Annusova *et al* [43] has been obtained. Another point is the discussion about the dissociation of N_2 by atomic N showing the differences between the LC and QCT theories. The results report the differences of the two methods, emphasizing the possibility of non-Arrhenius behavior of dissociation rates in both cases. Interesting results are also shown by studying the nozzle expansion through the state-to-state models. In this case, the non-equilibrium behavior of EEDF and VDF along the nozzle axis is well evident. Nowadays, the state-to-state approach is widely used in the kinetic study of hypersonic flows, including vibrationally-resolved rate coefficients for heavy particle collisions inducing energy transfer and dissociation. These rates can be obtained by the QCT method [71] integrating particle trajectories on accurate potential energy surfaces. QCT can also release state-specific cross sections needed in DSMC (direct simulation Monte Carlo) approach [72]. An evolution of DSMC is the DMS (direct molecular simulation) method which determines the output of each collisional event on the base of a trajectory calculation on the corresponding PES (potential energy surface) [73–75]. The above discussion refers to systems where free electrons are neglected. For higher temperatures electron-induced processes become relevant to the kinetics, requiring an extension of the model with improved ionization mechanisms, as shown by Levaton *et al* [76] for the nitrogen plasma.

References

[1] Capitelli M and Dilonardo M 1977 Nonequilibrium vibrational populations of diatomic species in electrical discharges: effects on the dissociation rates *Chem. Phys.* **24** 417–27

[2] Capitelli M and Dilonardo M 1978 Non-equilibrium dissociation of nitrogen *Rev. Phys. Appl.* **13** 115–23

[3] Capitelli M, Colonna G, D'Ammando G, Laporta V and Laricchiuta A 2013 The role of electron scattering with vibrationally excited nitrogen molecules on non-equilibrium plasma kinetics *Phys. Plasmas* **20** 101609

[4] Capitelli M and Pietanza L D 2019 Past and present aspects of Italian plasma chemistry *Rend. Lincei, Sci. Fis. Nat.* **30** 31–48

[5] Capezzuto P, Cramarossa F, D'Agostino R and Molinari E 1977 The decomposition of methane, ethane, ethylene and n-butane in electrical discharges of moderate pressures *Beiträge aus der Plasmaphysik* **17** 205–20

[6] Capitelli M and Molinari E 1980 Kinetics of dissociation processes in plasmas in the low and intermediate pressure range *Plasma Chemistry II(Topics in Current Chemistry)* vol 90 (Berlin: Springer)

[7] Hassouni K, Gicquel A, Capitelli M and Loureiro J 1999 Chemical kinetics and energy transfer in moderate pressure H_2 plasmas used in diamond MPACVD processes *Plasma Sources Sci. Technol.* **8** 494–512

[8] Taccogna F *et al* 2021 Latest experimental and theoretical advances in the production of negative ions in caesium-free plasmas *Eur. Phys. J.* D **75** 227

[9] Laricchiuta A, Celiberto R, Esposito F and Capitelli M 2006 State-to-state cross sections for H_2 and its isotopic variants *Plasma Sources Sci. Technol.* **15** S62

[10] Laricchiuta A, Celiberto R and Janev R 2004 Electron-impact-induced allowed transitions between triplet states of H_2 *Phys. Rev.* A **69** 22706

[11] Celiberto R, Janev R K, Laricchiuta A, Capitelli M, Wadehra J M and Atems D E 2001 Cross section data for electron-impact inelastic processes of vibrationally excited molecules of hydrogen and its isotopes *At. Data Nucl. Data tables* **77** 161–213

[12] Celiberto R, Capitelli M and Lamanna U 1994 Vibrational excitation of H_2 ($X^1\Sigma_g^+$, v)/D_2 ($X^1\Sigma_g^+$, v) through excitation of electronically excited singlet states and radiative cascade *Chem. Phys.* **183** 101–6

[13] Esposito F, Gorse C and Capitelli M 1999 Quasi-classical dynamics calculations and state-selected rate coefficients for H+ $H_2(v, j) \rightarrow$ 3H processes: application to the global dissociation rate under thermal conditions *Chem. Phys. Lett.* **303** 636–40

[14] Esposito F and Capitelli M 2009 Selective vibrational pumping of molecular hydrogen via gas phase atomic recombination *J. Phys. Chem.* A **113** 15307–14

[15] Scarlett L H, Tapley J K, Savage J S, Fursa D V, Zammit M C and Bray I 2019 Vibrational excitation of the state via electron-impact excitation and radiative cascade *Plasma Sources Sci. Technol.* **28** 025004

[16] Tapley J K, Scarlett L H, Savage J S, Zammit M C, Fursa D V and Bray I 2018 Vibrationally resolved electron-impact excitation cross sections for singlet states of molecular hydrogen *J. Phys. B: At. Mol. Opt. Phys.* **51** 144007

[17] Scarlett L H, Tapley J K, Fursa D V, Zammit M C, Savage J S and Bray I 2018 Electron-impact dissociation of molecular hydrogen into neutral fragments *Eur. Phys. J.* D **72** 34

[18] Celiberto R, Janev R, Wadehra J and Tennyson J 2012 Dissociative electron attachment to vibrationally excited H_2 molecules involving the resonant Rydberg electronic state *Chem. Phys.* **398** 206–13

[19] Zawadzki M *et al* 2018 Time-of-flight electron scattering from molecular hydrogen: benchmark cross sections for excitation of the $X^1\Sigma_g^+ \rightarrow b^3\Sigma_u^+$ transition *Phys. Rev.* A **97** 050702

[20] Pinnaduwage L A, Ding W X, McCorkle D L, Lin S H, Mebel A M and Garscadden A 1999 Enhanced electron attachment to Rydberg states in molecular hydrogen volume discharges *J. Appl. Phys.* **85** 7064–9

[21] Pagano D, Gorse C and Capitelli M 2007 Modeling multicusp negative-ion sources *IEEE Trans. Plasma Sci.* **35** 1247–59

[22] Gorse C, Capitelli M, Bretagne J and Bacal M 1985 Vibrational excitation and negative-ion production in magnetic multicusp hydrogen discharges *Chem. Phys.* **93** 1–12

[23] Gorse C, Capitelli M, Bacal M, Bretagne J and Laganà A 1987 Progress in the non-equilibrium vibrational kinetics of hydrogen in magnetic multicusp H^- ion sources *Chem. Phys.* **117** 177–95

[24] Bretagne J, Delouya G, Gorse C, Capitelli M and Bacal M 1985 Electron energy distribution functions in electron-beam-sustained discharges: application to magnetic multicusp hydrogen discharges *J. Phys. D: Appl. Phys.* **18** 811–25

[25] Mosbach T 2005 Population dynamics of molecular hydrogen and formation of negative hydrogen ions in a magnetically confined low temperature plasma *Plasma Sources Sci. Technol.* **14** 610–22

[26] Pagano D, Gorse C and Capitelli M 2005 Hydrogen plasmas for negative ion production *Proc. AIP Conf. 24th Int. Symp. on Rarefied Gas Dynamics* vol 762 ed M Capitelli 1155–60

[27] Bretagne J, Graham W G and Hopkins M B 1991 A comparison of experimental and theoretical electron energy distribution functions in a multicusp ion source *J. Phys. D: Appl. Phys.* **24** 668–71

[28] Pagano D, Gorse C and Capitelli M 2006 Atomic wall recombination and volume negative ion production *Rev. Sci. Instrum.* **77** 03A505

[29] Massabieaux B, Gousset G, Lefebvre M and Pealat M 1987 Determination of $N_2(X)$ vibrational level populations and rotational temperatures using CARS in a D.C. low pressure discharge *J. Physique* **48** 1939–49

[30] Massabieaux B, Plain A, Ricard A, Capitelli M and Gorse C 1983 Excitation of vibrational and electronic states in a glow discharge column in flowing N_2 *J. Phys. B: At. Mol. Phys.* **16** 1863–74

[31] Capitelli M, Colonna G, D'Ammando G, Laporta V and Laricchiuta A 2014 Nonequilibrium dissociation mechanisms in low temperature nitrogen and carbon monoxide plasmas *Chem. Phys.* **438** 31–6

[32] Gorse C and Capitelli M 1987 Coupled electron and excited-state kinetics in a nitrogen afterglow *J. Appl. Phys.* **62** 4072–6

[33] Capitelli M, Gorse C and Ricard A 1981 Influence of superelastic vibrational collisions on the relaxation of the electron energy distribution function in N_2 post discharge regimes *J. Phys. Lett.* **42** 469–72

[34] Colonna G, Pietanza L D and Capitelli M 2008 Recombination-assisted nitrogen dissociation rates under nonequilibrium conditions *J. Thermophys. Heat Transfer* **22** 399–406

[35] Cacciatore M, Capitelli M and Dilonardo M 1978 Non equilibrium vibrational population and dissociation rates of oxygen in electrical discharges: the role of atoms and of the recombination process *Beiträge aus der Plasmaphysik* **18** 279–99

[36] Esposito F and Capitelli M 2007 The relaxation of vibrationally excited O_2 molecules by atomic oxygen *Chem. Phys. Lett.* **443** 222–6

[37] Esposito F, Armenise I, Capitta G and Capitelli M 2008 O–O_2 state-to-state vibrational relaxation and dissociation rates based on quasiclassical calculations *Chem. Phys.* **351** 91–8

[38] Laporta V, Celiberto R and Tennyson J 2013 Resonant vibrational-excitation cross sections and rate constants for low-energy electron scattering by molecular oxygen *Plasma Sources Sci. Technol.* **22** 025001

[39] Laporta V, Celiberto R and Tennyson J 2015 Dissociative electron attachment and electron-impact resonant dissociation of vibrationally excited O_2 molecules *Phys. Rev.* A **91** 012701

[40] Laricchiuta A, Celiberto R and Capitelli M 2000 Electron impact cross-sections for electronic excitation of vibrationally excited O_2 to $B^3\Sigma_u^-$ state *Chem. Phys. Lett.* **329** 526–32

[41] Kosarim A, Smirnov B, Capitelli M, Laricchiuta A and Paniccia F 2006 Electron impact ionization cross sections of vibrationally and electronically excited oxygen molecules *Chem. Phys. Lett.* **422** 513–7

[42] Pietanza L D, Colonna G and Capitelli M 2020 Extended plateaux in the vibrational and electron distribution functions of O_2/O reacting plasmas in discharge and post-discharge conditions *Phys. Plasmas* **27** 093510

[43] Annušová A, Marinov D, Booth J-P, Sirse N, da Silva M L, Lopez B and Guerra V 2018 Kinetics of highly vibrationally excited $O_2(X)$ molecules in inductively-coupled oxygen plasmas *Plasma Sources Sci. Technol.* **27** 045006

[44] Šimek M and Bonaventura Z 2018 Non-equilibrium kinetics of the ground and excited states in N_2–O_2 under nanosecond discharge conditions: extended scheme and comparison with available experimental observations *J. Phys. D: Appl. Phys.* **51** 504004

[45] Rich J W, Bergman R C and Raymonda J W 1975 Vibration-vibration pumping of carbon monoxide initiated by an optical source *Appl. Phys. Lett.* **27** 656–8

[46] Treanor C E, Rich J W and Rehm R G 1968 Vibrational relaxation of anharmonic oscillators with exchange-dominated collisions *J. Chem. Phys.* **48** 1798–807

[47] Fisher E R and Lightman A J 1978 CO vibrational distributions in the presence of oxygen *J. Appl. Phys.* **49** 530–2

[48] Morgan W L and Fisher E R 1977 Effects of O_2 on low-pressure CO-laser discharges *Phys. Rev.* A **16** 1186–98

[49] Gorse C, Cacciatore M and Capitelli M 1984 Kinetic processes in non-equilibrium carbon monoxide discharges. I. Vibrational kinetics and dissociation rates *Chem. Phys.* **85** 165–76

[50] Gorse C and Capitelli M 1984 Kinetic processes in non-equilibrium carbon monoxide discharges. II. Self-consistent electron energy distribution functions *Chem. Phys.* **85** 177–87

[51] Gorse C, Billing G, Cacciatore M, Capitelli M and de Benedictis S 1987 Non-equilibrium vibrational kinetics of CO pumped by vibrationally excited nitrogen molecules: general theoretical considerations *Chem. Phys.* **111** 351–60

[52] De Benedictis S, Capitelli M, Cramarossa F and Gorce C 1987 Non-equilibrium vibrational kinetics of co pumped by vibrationally excited nitrogen molecules: a comparison between theory and experiment *Chem. Phys.* **111** 361–70

[53] Essenhigh K A, Utkin Y G, Bernard C, Adamovich I V and Rich J W 2006 Gas-phase Boudouard disproportionation reaction between highly vibrationally excited CO molecules *Chem. Phys.* **330** 506–14

[54] Barreto P R P, Euclides H d O, Albernaz A F, Aquilanti V, Capitelli M, Grossi G, Lombardi A, Macheret S and Palazzetti F 2017 Gas phase Boudouard reactions involving singlet–singlet and singlet–triplet CO vibrationally excited states: implications for the non-equilibrium vibrational kinetics of CO/CO_2 plasmas *Eur. Phys. J.* D **71** 259

[55] Pietanza L D, Colonna G and Capitelli M 2017 Non-equilibrium plasma kinetics of reacting CO: an improved state to state approach *Plasma Sources Sci. Technol.* **26** 125007

[56] Pietanza L D, Colonna G and Capitelli M 2018 Non-equilibrium electron and vibrational distributions under nanosecond repetitively pulsed CO discharges and afterglows: I. optically thick plasmas *Plasma Sources Sci. Technol.* **27** 095004

[57] Pietanza L D, Colonna G, Laricchiuta A and Capitelli M 2018 Non-equilibrium electron and vibrational distributions under nanosecond repetitively pulsed CO discharges and afterglows: II. The role of radiative and quenching processes *Plasma Sources Sci. Technol.* **27** 095003

[58] Laporta V, Tennyson J and Celiberto R 2016 Carbon monoxide dissociative attachment and resonant dissociation by electron-impact *Plasma Sources Sci. Technol.* **25** 01LT04

[59] Laporta V, Cassidy C M, Tennyson J and Celiberto R 2012 Electron-impact resonant vibration excitation cross sections and rate coefficients for carbon monoxide *Plasma Sources Sci. Technol.* **21** 045005

[60] Pietanza L D, Colonna G and Capitelli M 2020 Self-consistent electron energy distribution functions, vibrational distributions, electronic excited state kinetics in reacting microwave CO_2 plasma: An advanced model *Phys. Plasmas* **27** 023513

[61] Pietanza L D, Colonna G and Capitelli M 2020 Kinetics versus thermodynamics on CO_2 dissociation in high temperature microwave discharges *Plasma Sources Sci. Technol.* **29** 035022

[62] Pietanza L D *et al* 2021 Advances in non-equilibrium CO_2 plasma kinetics: a theoretical and experimental review *Eur. Phys. J.* D **75** 237

[63] Pietanza L D, Colonna G, D'Ammando G, Laricchiuta A and Capitelli M 2015 Vibrational excitation and dissociation mechanisms of CO_2 under non-equilibrium discharge and post-discharge conditions *Plasma Sources Sci. Technol.* **24** 042002

[64] Pietanza L D, Colonna G, D'Ammando G, Laricchiuta A and Capitelli M 2016 Electron energy distribution functions and fractional power transfer in 'cold' and excited CO_2 discharge and post discharge conditions *Phys. Plasmas* **23** 013515

[65] Pietanza L, Colonna G, D'Ammando G, Laricchiuta A and Capitelli M 2016 Non equilibrium vibrational assisted dissociation and ionization mechanisms in cold CO_2 plasmas *Chem. Phys.* **468** 44–52

[66] Shizgal B D and Lordet F 1996 Vibrational nonequilibrium in a supersonic expansion with reaction: application to O_2–O *J. Chem. Phys.* **104** 3579–97

[67] Colonna G, Tuttafesta M and Giordano D 2001 Numerical methods to solve Euler equations in one-dimensional steady nozzle flow *Comput. Phys. Commun.* **138** 213–21

[68] Capitelli M, Colonna G and Esposito F 2004 On the coupling of vibrational relaxation with the dissociation-recombination kinetics: from dynamics to aerospace applications *J. Phys. Chem.* A **108** 8930–4

[69] Colonna G and Capitelli M 2001 Self-consistent model of chemical, vibrational, electron kinetics in nozzle expansion *J. Thermophys. Heat Transfer* **15** 308–16

[70] Colonna G, D'Ammando G, Pietanza L D and Capitelli M 2014 Excited-state kinetics and radiation transport in low-temperature plasmas *Plasma Phys. Control. Fusion* **57** 014009

[71] Esposito F, Capitelli M and Gorse C 2000 Quasi-classical dynamics and vibrational kinetics of N + N_2(v) system *Chem. Phys.* **257** 193–202

[72] Bruno D, Capitelli M, Esposito F, Longo S and Minelli P 2002 Direct simulation of non-equilibrium kinetics under shock conditions in nitrogen *Chem. Phys. Lett.* **360** 31–7

[73] Grover M S, Torres E and Schwartzentruber T E 2019 Direct molecular simulation of internal energy relaxation and dissociation in oxygen *Phys. Fluids* **31** 076107

[74] Grover M S and Valentini P 2021 Ab initio simulation of hypersonic flows past a cylinder based on accurate potential energy surfaces *Phys. Fluids* **33** 051704

[75] Macdonald R L, Grover M S, Schwartzentruber T E and Panesi M 2018 Construction of a coarse-grain quasi-classical trajectory method. II. Comparison against the direct molecular simulation method *J. Chem. Phys.* **148** 054310

[76] Levaton J, Klein A N, Amorim J and Severo J H F 2021 Study of ionization and ion transfer rates in flowing N_2 DC discharges and post-discharges *J. Phys. D: Appl. Phys.* **54** 505205

IOP Publishing

Plasma Modeling (Second Edition)
Methods and applications
Gianpiero Colonna and Antonio D'Angola

Chapter 13

Atmospheric pressure plasmas operating in high frequency fields

Marco A Ridenti, Vasco Guerra and Jayr Amorim

The study of plasmas may be well fitted in the current area of complex systems, though the first of the scientific accomplishments from theoretical and experimental physicists dates from the seminal works from the 19th century, which developed into a quite elaborate science by the end of the first half of the 20th century. From the first laboratory production of electrical discharges by Sir William Crookes in 1879 [1, 2] and first attempts of modelling by Sir J J Thomson [3], plasma science has driven physicists to the exercise of formulating comprehensive models which required ingredients from many areas of physics, namely, statistical physics, kinetic theory, electromagnetic theory, fluid mechanics, atomic, molecular and nuclear physics, chemistry, and, in some cases, relativity theory. It not only required input from many fields of physics, but also the treatment of systems containing many kinds of interacting elements, such as electrons, ions, atoms and molecules, waves and radiation, which frequently must be treated simultaneously in order to guarantee the predictive power of the model. Another difficulty is the often found nonlinearities in the resultant system of equations, which may result in chaos, emergent phenomena, self organization, feedback loops, solitons, pattern formation and collective behaviour. Besides, plasmas are usually observed and modelled in macro-scales, but micro-scale phenomena must be studied and modelled to some degree, depending on the model purpose, in order to give correct predictions.

As stated by Weisskopf, two different trends in science research may be identified, one of which could be called 'intensive' and the other 'extensive'. The 'intensive research goes for the fundamental laws and the extensive research goes for the explanation of phenomena in terms of known fundamental laws' [4]. Plasma science is almost totally devoted to extensive research, since all the phenomena in this realm result from many interacting and non-interacting micro-scale physical entities which are governed by fundamental laws. The plasma research, as in other fields of

doi:10.1088/978-0-7503-3559-1ch13

'extensive' research, such as solid-state physics, was greatly inspired by the approach of starting from fundamental laws to derive the observed collective behaviour.

In this chapter, as far as possible we try to follow the 'extensive' approach mentioned above to give the best foundations to the model. As usually stated, the modelling process is strongly influenced by its purpose in the sense that some elements of the presumed reality of the system may be disregarded depending on the experimental conditions and measured quantities that one wishes to reproduce. For instance, if a measurement is designed to probe the electron density only, other elements of the reality, which only slightly influence electrons, such as radiation, may be completely disregarded. On the other hand, if one wishes to know the total radiative power delivered by a plasma discharge in a decontamination device, then radiation must be modelled, though its influence on the overall plasma behaviour could be minimal.

Herein we address a specific kind of discharge, known as surface-wave discharge, which may be found in many technological applications [5–7]. Surface-wave discharges all have in common the property of being sustained by high frequency electrical fields which also propagate in the medium. It may be operated at intermediate or low pressures, but here we focus on the atmospheric pressure case. Even in this specific case, so many differences may be found, depending on other elements, such as feeding gas, openness, flow regime, geometry, types of electromagnetic propagation, that it would be very difficult to offer all the necessary tools to build any model. Nonetheless, we try to raise as many issues as we can in order to cover the most common problems, giving the references to the specialized literature for further details. We also try to build general equations, which could be applied to as many cases as possible, leaving the most drastic simplifications in the worked example.

The phenomenon of plasma contraction, which is treated in the worked example, is still a matter of vigorous debate. Although some general agreement exists over the fundamental equations, its complexity almost always forces modellers to adopt some sort of approximation. Sometimes different approximations may explain the same collective phenomena, but in different ways, giving rise to such debates. Therefore, this is still an opened field, both experimentally and theoretically, even though the underlying fundamental laws may be well known.

Atmospheric pressure plasmas have some special features which should be pointed out, as some approximations may be applicable to shape their description into an adequate model. On the other hand, some types of reactions, which introduce nonlinearities, make it quite different from discharges at lower pressures. For instance, the hydrodynamic approximation, meaning that fluid equations may be used to describe the system macroscopic behaviour, is valid in most of the cases of interest. Furthermore, the pressure is not too high as to break the *molecular chaos* hypothesis underlying the kinetic theory which provides the transport parameters for the fluid model. However, three-body reactions may occur due to the elevated pressure, introducing additional non-linear terms in the system of differential equations describing the creation and destruction of particles through chemical reactions.

Microwave generated plasmas are especially interesting in technological applications for many reasons. Plasmas may be kept at *quasi* steady-state with power delivered continuously by microwaves because its frequency is usually higher than the plasma relaxation frequency. It means that the plasma system may be modelled with the usual steady-state fluid equations as in the DC case, with some modifications in the transport parameters, as pointed out in the text. From a technological point of view, microwave plasmas have much better power efficiency than radiofrequency or kilohertz discharges. Besides, they may be produced using an electrodeless configuration, which can completely eliminate the undesired residuals from plasma–electrode interaction. Such discharges are usually warmer than their low-frequency counterparts, but this can be overcome using pulsed microwave sources or gas flows with large flow rates.

The structure of this chapter is as follows. Section 13.1 deals with the general aspects of plasma modelling at atmospheric pressure based on the fluid equations and electron kinetic theory. We make a brief review of the classical electron kinetic theory giving special emphasis on the connection with fluid modelling. The continuity equation for the reactive neutral species and ions is written explicitly and the main creation and destruction terms are described in detail. We also give a brief description of the moment and energy equations of the gas mixture in a partially ionized gas, that would be needed in the most rigorous and comprehensive description of a non-magnetized laboratory plasma at atmospheric pressure. We also mention the usual method to account for space-charge fields using the ambipolar diffusion coefficient. Finally, we conclude this section describing the Maxwell equations for a plasma sustained by a high frequency field in TM_{00} mode. In section 13.2 we apply the theory described in the previous section to treat the problem of discharge contraction in a high frequency argon discharge at atmospheric pressure.

13.1 Atmospheric pressure plasmas modelling in high frequency fields

13.1.1 Transport properties of electrons in non-magnetized and partially ionized gases

Our goal is to derive the macroscopic equations that describe the density distributions of relevant species in the plasma, as well as some macroscopic physical quantities such as gas temperature and electric field. In this section, we discuss the connection between the kinetic theory of electrons and the fluid equations and show how electron transport parameters from two-term Boltzmann codes should be inserted into macroscopic equations.

To start, let us consider a system of electrons subject to a constant electric field, which can be non-uniform, and may collide with other electrons, neutral species or ions. No strong restrictions over the type of collisions are imposed, as inelastic collisions, ionization, attachment and superelastic collision can occur apart from elastic collisions with neutrals. As the molecular chaos hypothesis applies and correlation between particles is rapidly destroyed, the system may be fully described by means of a simple distribution function $f(\vec{r}, \vec{v}, t)$ subject to the condition

$\int f(\vec{r}, \vec{v}, t)\, v^2\, d^2\Omega\, dv = N_e(\vec{r}, t)$ where the integration is taken over all velocities and $N_e(\vec{r}, t)$ is just a function that gives the electron density at instant t and a small volume centred at position $\vec{r}$. Here we make use of the two-term distribution and the quasi-stationary approximation developed in chapter 2. Equations (2.31) and (2.36) are specially relevant in this context, because they are taken as the starting point in the next sections. Here we take $\hat{\Omega}$ in equation (2.31) as the sum of the collision frequencies due to elastic and inelastic processes, i.e., $\hat{\Omega} = \nu_{\text{col}} = \nu_{\text{el}} + \nu_{ei} + \sum_j^{N_0} \delta_j \sum_i^{M_j} \nu_{ji}$, ν_{el} being the elastic moment transfer collision frequency between electrons and neutrals, ν_{ei} the electron–ion collision frequency, N_0 the number of species, M_j the number of inelastic processes involving species j with fraction δ_j, and ν_{ji} the inelastic collision frequency of transition i. The effects of the magnetic field on the plasma can be neglected in the applications of interest in this chapter, so the tensor $\hat{\Omega}$ reduces to a scalar that will henceforth be designated as ν_{col}.

The expanded components of the distribution function $(f_0, \vec{f_1})$ depend on the position coordinate in a non-trivial manner unless some reasonable assumption may be made which greatly simplifies the problem. In the atmospheric pressure case, for instance, the electron mean free path ($\approx$1 μm) is usually much shorter than the scale at which strong density variations occur, so that the space dependence of $f(\vec{r}, \vec{v}, t)$ is approximately equal to the space dependence of $N_e(\vec{r}, t)$. For instance, the distribution function f_0 may be written as $f_0 = F_0(v, t)\, N_e(\vec{r}, t)$, where $4\pi \int F_0 v^2 dv = 1$. This approximation may be corrected in terms of a gradient expansion of the electron density, and we shall just take the first-order correction, $f_0 = F_0(v, t)\, N_e + \vec{\xi}_0(v) \cdot \nabla N_e$, where $\int \vec{\xi}_0 v^2 dv = 0$. The assumption underlying this approximation is usually referred to as hydrodynamic regime in the literature. In the hydrodynamic regime the fluid equations may be derived from equations (2.31) and (2.36), as indicated in the next section.

Electron fluid equations

In the quasi-stationary approximation $(\partial \vec{f_1}/\partial t \approx 0)$, the integration of equation (2.31) in the velocity space $(\int \{\ldots\} v^2 dv)$ after multiplication by $(4\pi/3)v$ gives the well-known drift–diffusion equation

$$\vec{\Gamma}_e = -\mathbf{D}_e \nabla(N_e(\vec{r}, t)) + \vec{v}_e N_e(\vec{r}, t), \tag{13.1}$$

where $\mathbf{D}_e$ is the diffusion tensor and $\vec{v}_e$ is the electron drift velocity, which may also be written as $\vec{v}_e = \mu_e \vec{E}$, μ_e being defined as the electron mobility. Their explicit expressions, given below, are especially important in the application of the fluid equations to model real problems.

The assumption $\partial \vec{f_1}/\partial t = 0$ is exact in the steady-state case and may be taken as a good approximation if $(\partial \vec{f_1}/\partial t)\hat{\Omega}^{-1} \ll \vec{f_1}$ after a small time comparable to $\hat{\Omega}^{-1}$. It was shown in chapter 2 that when the quasi-stationary approximation is assumed one may decouple the system of equations (2.31) and write an equation for the isotropic component of the distribution function only (see equation (2.36)). From this equation one may obtain the electron number conservation equation by integration in the velocity space ($\int \{\dots\} 4\pi v^2 dv$)

$$\frac{\partial N_e}{\partial t} + \vec{v}_e \cdot \nabla N_e - \mathbf{D}_e \nabla^2 N_e = (\tilde{\nu}_{\text{ion}} - \tilde{\nu}_{\text{att}})\, N_e, \tag{13.2}$$

where the electron drift velocity, diffusion tensor and mean ionization and attachment rates are given by the following formulas when the z-axis is taken as the acceleration direction

$$\vec{v}_e = -\frac{4\pi a_z}{3} \int_0^\infty \frac{v^3}{\hat{\Omega}} \frac{dF_0}{dv}\, dv\, \hat{z}, \tag{13.3}$$

$$D_{e,ij} = 0,\; i \neq j;\;\; D_{e,xx} = D_{e,yy} = D_{e,T} = \frac{4\pi}{3} \int_0^\infty \frac{v^4}{\nu_{\text{col}}} F_0\, dv, \tag{13.4}$$

$$D_{e,zz} = D_{e,L} = \frac{4\pi}{3} \int_0^\infty \frac{v^4}{\nu_i} F_0\, dv + \frac{4\pi a_z}{3} \int_0^\infty \frac{v^3}{\nu_{\text{col}}} \frac{d\xi_{0,z}}{dv}\, dv, \tag{13.5}$$

$$\tilde{\nu}_{\text{ion}} = N \sum_j^{N_0} \delta_j \int_0^\infty 4\pi\sigma_{j,ion}(v) F_0(v) v^3\, dv, \tag{13.6}$$

$$\tilde{\nu}_{\text{att}} = N \sum_j^{N_0} \delta_j \int_0^\infty 4\pi\sigma_{j,att}(v) F_0(v) v^3\, dv. \tag{13.7}$$

In the expressions above we immediately see that the electron diffusion tensor is diagonal, with equal diagonal components in the x and y directions, which are usually called the transversal diffusion coefficient, and the component in the z direction, which is usually called the longitudinal diffusion coefficient. In the expressions of the mean ionization and attachment rates we introduced the total particle density, N, the species fraction $\delta_j = N_j/N$ and the integral ionization or attachment cross-section $\sigma_{j,ion}$ and $\sigma_{j,att}$. Note that collision operator terms associated with conserving processes vanish after integration. Besides, the first-order correction of the hydrodynamic expansion was not taken into account on the ionization and attachment components of the collision operator, which would generate a correction on the electron drift velocity due to non-conserving processes. This correction is very important in some cases, where strongly attaching species are present, but it is usually very small in typical cases. If the second-order term was kept in the gradient expansion, corrections in the diffusion tensor would appear due to non-conserving transport. The explicit expression for these corrections may be found in [8], in the context of the two-term expansion.

The longitudinal and transverse diffusion coefficient may differ in magnitude depending on the gas species and the electron mean energy. For instance, a 3-D fluid description of electrons in a weakly ionized gas in a uniform electric field may be based on the following equation

$$\frac{\partial N_e}{\partial t} + v_{e,z}\frac{\partial N_e}{\partial z} - D_{e,L}\frac{\partial^2 N_e}{\partial z^2} - D_{e,T}\left(\frac{\partial^2 N_e}{\partial x^2} + \frac{\partial^2 N_e}{\partial y^2}\right) = \left(\tilde{\nu}_{\text{ion}} - \tilde{\nu}_{\text{att}}\right)N_e. \tag{13.8}$$

Finally, the electron energy equation may be obtained by multiplying equation (2.36) by $(m_e/2)v^2$, where m_e is the electron mass, and then integrating in the velocity space. The resulting equation may be written as

$$\frac{\partial(N_e\varepsilon_e)}{\partial t} + \nabla_r \cdot \vec{\Gamma}_{e,en} + e\vec{E} \cdot \vec{\Gamma}_e = P_{\text{coll}}, \tag{13.9}$$

where ε_e is the mean electron energy, defined as $\varepsilon_e = (2\pi m_e/N_e)\int_0^\infty f_0 v^4 dv$, P_{coll} is the net power loss per unit volume due to collisions, and $\vec{\Gamma}_{e,en}$ is the energy flux, which may be written as

$$\vec{\Gamma}_{e,en} = -\mathbf{G}\nabla n + \vec{\beta} n, \tag{13.10}$$

where $\mathbf{G}$ is the heat diffusion tensor and $\vec{\beta}$ is the thermoelectricity, defined as

$$G_{ij} = 0,\ i \neq j;\ \ G_{xx} = G_{yy} = G_T = 2\pi m_e \int_0^\infty \frac{F_0 v^6}{3\nu_{\text{col}}} dv, \tag{13.11}$$

$$G_{zz} = G_L = G_T + 2\pi m_e a_z \int_0^\infty \frac{v^5}{3\nu_{\text{col}}}\frac{d\xi_{0,z}}{dv}\, dv, \tag{13.12}$$

$$\vec{\beta} = -2\pi m_e a_z \int_0^\infty \frac{v^5}{3\nu_{\text{col}}}\frac{dF_0}{dv}\, dv\ \hat{z}. \tag{13.13}$$

In the literature, a different form of the electron energy flux is often found, which may be written as $\vec{\Gamma}_{e,en} = (5/3)N_e\varepsilon_e\vec{v}_d - (5/3)N_e D_T \nabla\varepsilon_e$. Although this is a simpler expression, other assumptions must be made, namely, a Maxwellian electron energy distribution function, a constant momentum transfer frequency (ν_{col} = cte) and constant kinetic pressure [9].

Collision operator and local field approximation

The collision operator gives the mathematical expression which determines how the collisional processes shape the electron energy distribution function (EEDF). The EEDF ($n(\varepsilon)$) is simply the isotropic velocity distribution function expressed in terms of the classical kinetic energy, i.e., $n(\varepsilon) = (4\pi/m_e)(\sqrt{2/m_e})f_0(v)/N_e$, with the normalization convention $1 = \int_0^\infty n(\varepsilon)\varepsilon^{1/2}d\varepsilon$. This is especially relevant when the modeller's efforts are directed towards the plasma chemistry, in applications where a good estimate of the reaction rates of electronic reactions is crucial. Here, we give a brief description of the formulas and their physical meaning, without going into the details of their derivation, which may be found in chapter 2 and/or in the specialized literature [10, 11].

In the previous sections, the collision operator was expanded in spherical harmonics. The isotropic term, which was designated by the symbol S_0 in chapter 2, accounts for different collisional processes, and contain corrections to the assumption of infinitely heavy species. It can describe the effect of several kinds of collisional processes, i.e.

$$S_0 = S_{0,el} + S_{0,in} + S_{0,ee} + S_{0,t} \tag{13.14}$$

where $S_{0,el}$ is the elastic term, which accounts for elastic collisions with heavy neutrals or ions with a Maxwellian distribution function, $S_{0,in}$ is the inelastic term, $S_{0,ee}$ is the electron–electron collision term and $S_{0,t}$ is the transformation term, which accounts for collisional processes that can create or destroy electrons, such as ionization, attachment or recombination ($S_{0,t} = S_{0,ion} + S_{0,rec} + S_{0,att}$).

Elastic collisions

The elastic collision term may be written as [10, 12] (see also section 2.6.1)

$$S_{0,el} = N\sum_{j}^{N_0} \delta_j \frac{m}{m_j}\frac{1}{v^2}\frac{\partial}{\partial v}\left[\nu_{g,j}v^3\left(f_0 + \frac{2k_bT_g}{m}\frac{\partial f_0}{\partial v^2}\right)\right] \tag{13.15}$$

where k_b is the Boltzmann constant, T_g is the equilibrium temperature of the heavy species and m_j is the mass of species j. The collision frequency ν_g is equal to the elastic moment transfer collision frequency if species j is neutral

$$\nu_g^{(n)} = n_j v \iint \left(1 - \cos\theta\right)\sigma_{\text{el}}\left(\theta, v\right)d^2\Omega. \tag{13.16}$$

If the scatterer is an ion one should write the electron–ion collision frequency instead

$$\nu_{g,j}^{(i)} = \frac{n_j}{v^3}4\pi\left(\frac{Ze^2}{4\pi\epsilon_0 m_e}\right)^2 \ln\Lambda, \tag{13.17}$$

where ϵ_0 is the vacuum permittivity and $\Lambda = 12\pi(\epsilon_0 k_b T_e)^{3/2}/(N_e e^3)$. The electron temperature, in the expression of Λ, is defined in terms of the mean energy, $T_e = (2/(3k_b))\varepsilon_e$. The electron–ion collision frequency is usually small in the case of chemical plasmas in which the degree of ionization is typically very small.

In the simple case of a discharge in a partially ionized gas, where there is only one dominant neutral species, it can be readily seen that the constant $\tau_e = (m_j/(m_e\nu_{g,j}))$ gives the relaxation time-scale of the isotropic component of f_0. The anisotropic component $\vec{f_1}$, on the other hand, relaxes with $\tau_p = 1/\nu_{\text{col}}$, therefore much faster than f_0.

Inelastic collisions

The general form for the inelastic term is given by the following formula [13] (see also section 2.6.2)

$$S_{0,in} = N\sum_{j=1}^{N_0}\sum_{i=1}^{M_j}\Big\{\delta_j\Big[\big(v + v_{ij}\big)\sigma_{ij}\big(v + v_{ij}\big)f_0\big(v + v_{ij}\big) - v\sigma_{ij}(v)f_0(v)\Big] \\ + \delta_{ji}\Big[\big(v - v_{ij}\big)\sigma'_{ij}\big(v - v_{ij}\big)f_0\big(v - v_{ij}\big) - v\sigma'_{ij}(v)f_0(v)\Big]\Big\}, \tag{13.18}$$

where the 'velocity' v_{ij} is equivalent to the energy threshold ($v_{ij} = \sqrt{(2/m)\varepsilon_{ij}}$) of the transition between the lower state of species j and its upper state i and σ_{ij} (σ'_{ij}) is the inelastic (or superelastic) cross-section. The newly introduced term δ_{ji} is the fractional population of level i for a given lower level species j. Note that the lower state j does not need to be on the ground state so that metastable or resonant species may be included as distinct species as long as only levels with higher energies are considered.

Rotational, vibrational and electronic inelastic collisions may be accounted for using equation (13.18), but the large difference between rotational thresholds and vibrational or electronic thresholds can make the numerical solution difficult or impossible. In the case of homonuclear diatomic molecules it has been proven in [14] that rotational collisions may be expressed by a continuous approximation

$$S_{0,rot} = N\sum_{j=1}^{N_0^{(r)}}\delta_j\frac{4N\sigma_0 B}{mv^2}\frac{\partial}{\partial v}\left[v^2\left(f_0 + \frac{2k_bT}{m}\frac{df_0}{dv^2}\right)\right], \tag{13.19}$$

where B is the rotational constant in units of energy and $\sigma_0 \equiv 8\pi Q^2 a_0^2/15$, with Q the quadrupole moment constant in ea_0^2 units and a_0 the Bohr radius.

Non-conserving collisions

The ionization can be seen as an inelastic processes which adds one electron per collision. It is usual to assume that this extra electron is created with zero energy, so that the collision operator may be written as [15]

$$S_{0,\,ion} = N\sum_{j=1}^{N_0}\delta_j\Big\{\Big[(v + v_{ion,\,j})\sigma_j^{(ion)}(v + v_{ion,\,j})f_0(v + v_{ion,\,j}) - v\sigma_j^{(ion)}(v)f_0(v)\Big] \\ + \frac{\delta(v)}{v^2}\int_0^\infty \sigma_j^{(ion)}(v')f_0(v')v'^3dv'\Big\}. \tag{13.20}$$

Other forms of this collision operator are possible, depending on the assumption over the energy partitioning between the two electrons after the collision. Unless the ionization collision rate is very high in comparison with other inelastic processes, which is unusual in most applications of plasma chemistry, then the electron energy partitioning choice make little difference in the final shape of the EEDF [16].

Attachment processes may be simply written as [15]

$$S_{0,att} = -\sum_j^{N_0} v\delta_j\sigma_j^{(att)}(v)f_0(v). \tag{13.21}$$

Both the ionization and recombination collision terms do not conserve the electron number density, as expected. In the cases in which ionization collisions occur at very low rates and attachment is negligible, the electron creation term (third term in the rhs of equation (13.20)) may be neglected in the numerical calculation of $F_0(v)$. Electron recombination by the processes $(e + A + \rightarrow A + h\nu)$ or $(e + e + A + \rightarrow A + e)$ are usually not taken into account in the collision term, though literature still lacks studies showing whether they could be relevant to the EEDF calculation in plasmas at atmospheric pressure.

Electron–electron collisions

Electron–electron collisions (*e–e* collisions) are important in the description of atmospheric pressure plasma as electron densities may be large enough to produce probable encounters between electrons. Collisions between elementary species of the same type do not produce power gain or loss, but *e–e* collisions force the EEDF towards a Maxwellian distribution, producing drastic changes in some reaction rates that strongly depend on the EEDF tail. The derivation of the $C_{0,ee}$ is somewhat cumbersome and usually takes as the starting point the Fokker–Planck equation [10]. It is given by the following equation

$$C_{0,ee} = \frac{Y}{v^2}\frac{\partial}{\partial v}\left[f_0 I_0^0 + \frac{mv^2}{3}\left(I_2^0 + J_{-1}^0\right)\frac{\partial f_0}{\partial \varepsilon}\right], \text{ where}$$

$$I_2^0 = \frac{4\pi}{v^2}\int_0^v f_0 v'^4 dv'; \; J_{-1}^0 = 4\pi v\int_v^{\infty} f_0 v' dv'; \; Y = 4\pi\left(\frac{Ze^2}{4\pi\epsilon_0 m_e}\right)^2 \ln\Lambda. \tag{13.22}$$

The above equation has some integral terms that need f_0 to be evaluated, which in turn is the desired solution. Therefore, if *e–e* collisions are included, then an iterative scheme is required for the numerical solution of the electron Boltzmann equation (2.36).

Local field approximation

Once the collision operator is defined and chosen to describe a particular electron discharge, the electron Boltzmann equation (2.35) may be solved numerically and P_{coll} in equation (13.9) may be evaluated. It is generally difficult to obtain simultaneously the velocity and space dependence of the electron distribution function, even in the hydrodynamic regime. In some cases, where the electron mean free path (l_c) is small when compared with the length in which strong densities variations occur, i.e., $l_c|\nabla n|/n \ll 1$, it is common to neglect all the terms with space derivatives in equation (2.36). In this approximation, the plasma can be viewed as a composition of small 'plasma volumes', each of them with uniform electron density. Besides, the electron energy distribution function is only determined by the average uniform field and collision processes in this small volume. The electron Boltzmann equation (equation (2.36)) and energy balance (equation (13.9)) could be simplified to

$$\frac{\partial f_0}{\partial t} - \frac{a_z^2}{v^2}\frac{\partial}{\partial v}\left(\frac{v^2}{3\nu_{\rm col}}\frac{\partial f_0}{\partial v}\right) = S_0 \tag{13.23}$$

and

$$\frac{\partial(N_e u_e)}{\partial t} + e\vec{E}\cdot(n\vec{v}_e) = P_{\rm coll}. \tag{13.24}$$

In the steady case, we could go even further, writing

$$-\frac{a_z^2}{v^2}\frac{\partial}{\partial v}\left(\frac{v^2}{3\nu_{\rm col}}\frac{\partial F_0}{\partial v}\right) = S_0 \tag{13.25}$$

and

$$e\vec{E}\cdot(N_e\vec{v}_e) = P_{\rm coll}. \tag{13.26}$$

In a practical application of this approximation, F_0 must be evaluated for each finite space domain of the numerical scheme and then applied to obtain the estimate of the transport parameters that are needed to evaluate the electron continuity equation (13.2). The local energy distribution is also needed to obtain the local electron reaction rates required to compute the continuity equations for neutrals and ions (see section 13.1.2).

Note that equation (13.26) has a very simple physical interpretation. The *lhs* side of the equation represents the power density due to the electric field energy gained by the electrons. It equates with the power loss and gain by collisions, which means that the field energy is totally dissipated by collisions. In this situation, it is usually said that electrons are in equilibrium with the electric field.

High frequency electric field

Let us consider now the problem of a discharge where the electric energy is provided by a high frequency field, e.g., microwaves. In the following analysis, we consider a linearly polarized field in the z direction, given by the expression $E_z = E_0 \cos \omega t$ (the phase is arbitrarily chosen to be zero). In general, even in the atmospheric pressure case, the energy relaxation frequency is much shorter than the frequency of a high frequency field $\omega \gg \nu_e$, so that the isotropic component of the EEDF may be considered almost constant in time. This is not so in the case of $\vec{f}_1$, as the varying field causes the electron mean drift velocity to alternate. The first anisotropic component $\vec{f}_1$ could be expanded in time harmonics, but since we are interested in the steady-state rather than the transient solution, it is enough to write $\vec{f}_1^{\,1} = \vec{f}_{0,1}^{\,1}\exp(j\omega t)$, where $\vec{f}_{0,1}^{\,1}$ is just a complex amplitude. The isotropic component, on the other hand, could be expanded as $f_0(t) = f_{0,cte} + f_{0,t}(t)$, where $f_{0,t}(t)$ is a small time-dependent perturbation. Writing the acceleration $\vec{a}$ as a complex number $\vec{a}^1 = \vec{a}_0 \exp j\omega t$, $\vec{a}_0$ being real since the phase is zero, then we may derive $\vec{f}_1^{\,1}$ from equation (2.31)

$$\vec{f}_1^1 = \left(-\frac{1}{\nu_{\text{col}} + j\omega}\right)\left[v\nabla_r f_0 + \vec{a}^1(t)\frac{\partial f_0}{\partial v}\right]. \tag{13.27}$$

Substitution of the above equation on equation (2.31) yields

$$\begin{aligned} &\frac{\partial f_0}{\partial t} + \frac{v}{3}\nabla_r \cdot \left[\frac{v\nabla_r f_0 + (\cos(\omega t) - \frac{\omega}{\nu_{\text{col}}}\sin(\omega t))\frac{\partial f_0}{\partial v}}{\nu_{\text{col}}(1 + \omega^2/\nu_{\text{col}}{}^2)}\right] \\ &- \vec{a}_0 \cdot \cos(\omega t)\frac{1}{3v^2}\frac{\partial}{\partial v}\left\{v^2\left[\frac{v\nabla_r f_0 + \vec{a}_0(\cos(\omega t) - \frac{\omega}{\nu_{\text{col}}}\sin(\omega t))\frac{\partial f_0}{\partial v}}{\nu_{\text{col}}(1 + \omega^2/\nu_{\text{col}}{}^2)}\right]\right\} = S_0 \end{aligned} \tag{13.28}$$

Note that only the real part of the solution was retained. Taking the time average over a period and neglecting the small time perturbation $f_{0,t}$, one arrives to the following equation for f_0

$$\frac{v^2\nabla^2 f_0}{3[\nu_{\text{col}}(1 + \omega^2/\nu_{\text{col}}{}^2)]} - \frac{a_0^2}{6v^2}\frac{\partial}{\partial v}\left[\frac{1}{\nu_{\text{col}}(1 + \omega^2/\nu_{\text{col}}{}^2)}\frac{\partial f_0}{\partial v}\right] = S_0. \tag{13.29}$$

A simple continuity equation may be obtained now, which writes as

$$-D_{\text{hf}}\nabla^2 N_e = (\tilde{\nu}_{ion} - \tilde{\nu}_{att})\, N_e, \tag{13.30}$$

where D_{hf} is the 'high frequency' diffusion coefficient defined as

$$D_{\text{hf}} = \frac{4\pi}{3}\int_0^\infty \frac{v^4}{\nu_{\text{col}}(1 + \omega^2/\nu_{\text{col}}{}^2)} F_0\, dv. \tag{13.31}$$

This equation shows that the higher the frequency ω, the lower will be the effective diffusion coefficient.

In the time averaging procedure we assumed a steady-state discharge, in the sense that the electron density is constant in time, though not uniform. This is possible if during one period of the field only few non-conserving collisions occur, $\tilde{\nu}_{ion,att} \ll \omega$. If this condition is not satisfied, then significant charge creation or destruction can occur during one period of the field and the full time differential equation has to be solved instead of the time averaged one, as f_0 could appreciably change over the period.

Let us look now at the power equation. We will take the averaged equation (13.28) and proceed with the energy integration, as we did before to derive the energy conservation equation in the DC case. If we limit ourselves to the local approximation, then the power balance equation may be written as

$$\frac{1}{2}eE_0^2N_e\mu_{\rm hf} = P_{\rm coll}, \tag{13.32}$$

where $\mu_{\rm hf}$ is the high frequency electron mobility, which may be written as

$$\mu_{\rm hf} = -\frac{4\pi}{3}\frac{e}{m}\int_0^\infty \frac{v^2}{\nu_{\rm col}(1 + \omega^2/\nu_{\rm col}{}^2)}\frac{dF_0}{dv}\,dv. \tag{13.33}$$

Equations (13.32) and (13.33) show that no power can be transferred to the electrons if collisions do not occur. In this case, the electrons simply move in quadrature with the field, so that the net work done by the field over a period on the moving particles is zero. In the other extreme, in which both the collision and the electron energy relaxation frequency are larger than the field frequency, the energy distribution function instantaneously adapts to the varying electron field and electrons 'swing' in phase with it. The intermediate case, where $\nu_e \approx \omega$ is more complicated, as f_0 will vary with time in a non-trivial manner.

13.1.2 Treatment of ions and neutral species

Ion and neutral species are assumed to be in equilibrium so that their energy distribution may be described by a Maxwellian distribution with a temperature T. This approximation is valid as long as the plasma is strongly collisional and the energy gained or lost in exothermic or endothermic reactions rapidly relaxes to the equilibrium temperature.

Both ions and neutrals can be described by the following continuity equation

$$\begin{aligned}\frac{\partial[N_j]}{\partial t} + \nabla\cdot\big([N_j]u\big) - D_j\nabla^2[N_j] = &-\sum_\alpha k_\alpha(m-p)[N_j]^m[N_i]^n \\ &-\sum_\beta k_\beta(m-p)[N_j]^m[N_i]^n[N_l]^s \\ &-\sum_{l<j}A_{jl}[N_j] + \sum_\zeta k_\zeta(p'-m')[N_j]^{m'}[N_i]^{n'}[N_l]^{s'} \\ &+\sum_\eta k_\eta p'[N_k]^{m'}[N_i]^{n'}[N_l]^{s'} + \sum_{l>j}A_{lj}[N_l] \\ &+\sum_\kappa \gamma_\kappa p_\kappa\Big(\nabla\cdot\vec{\Gamma}_\kappa\Big)_S - \gamma_j\Big(\nabla\cdot\vec{\Gamma}_j\Big)_S \\ &\text{where } p < m \text{ and } p' > m',\end{aligned} \tag{13.34}$$

where the square bracket is used to represent the particle density, in units of particles per cm^3. The velocity $\vec{u}$ is the fluid velocity of the mixture, which is mainly determined by the neutral species in a weakly ionized plasma and defined as $\vec{u} = \sum_j^{N_0}([N_j]m_j/\rho)\langle\vec{v}_j\rangle$, where ρ is the mass density, m_j the molecular mass and $\langle\vec{v}_j\rangle$ the mean velocity of species j. The parameter γ_κ is the probability of species conversion due to wall interaction and Γ_κ is the flux of species κ, given by $\vec{\Gamma}_\kappa = -D_\kappa\nabla n_\kappa + n_\kappa\vec{u}$. The divergence of the flux in equation (13.34) is taken over the vessel boundaries. All the terms in the *rhs* of the equation represent particle creation or loss due to chemical reactions, including electronic reactions and radiative processes. These terms do not exhaust all the possibilities, but include the most common types of reactions usually considered in global models.

The loss terms, represented in the first two sums in equation (13.34) with negative signal account for the destruction of species N_j by two-body and three-body reactions, respectively. These reactions may be represented explicitly as

$$mN_j + nN_i \xrightarrow{k_\alpha} pN_j + \sum_k q_k N_k, \tag{13.35}$$

$$\begin{gathered} mN_j + nN_i + sN_l \xrightarrow{k_\beta} pN_j + \sum_k q_k N_k, \\ \text{where } p < m, j \neq i, j \neq l, j \neq k, \forall k. \end{gathered} \tag{13.36}$$

The stoichiometric coefficients are represented by the letters m, n, p e q_k, and the reaction rate by k_α or k_β. Analogously, the gain reactions may be written explicitly as

$$m'N_j + n'N_i + s'N_l \xrightarrow{k_\zeta} p'N_j + \sum_k q'_k N_k, \tag{13.37}$$

$$\begin{gathered} m'N_k + n'N_i + s'N_l \xrightarrow{k_\eta} p'N_j + \sum_k q'_k N_k, \\ \text{where } p' > m', j \neq i, j \neq l, j \neq k, \forall k. \end{gathered} \tag{13.38}$$

Although both equations have three species in the *lhs*, the particular case of two-body reactions is included as long as one of the stoichiometric coefficients is zero.

The third sum in the *rhs* of equation (13.34) represents the loss of species N_j by radiative decaying, which may be expressed as

$$N_j \xrightarrow{A_{jl}} N_l + h\nu, \tag{13.39}$$

where A_{lj} is the Einstein coefficient of the transition of state j to l. On the other hand, the spontaneous emission from an upper energy state contributes to populate the state j, i.e.,

$$N_l \xrightarrow{A_{lj}} N_j + h\nu. \tag{13.40}$$

It is evident that the term corresponding to the radiative emission only occurs if species N_j is radiative. Besides, if the species is metastable, the radiative loss is usually negligible. Another effect that must be considered is the radiative imprisonment, which occurs due to radiative absorption by the species in the medium. The relevance of this effect depends on the plasma dimensions and the concentration of absorbing species. Usually, in weakly ionized plasmas, the ground-state species from the background gas or mixture are predominant and the main absorbers. The radiative imprisonment may be computed using Holstein's model [17, 18] which gives the formula to the correction coefficient g ($A_{\mathrm{lf}}^{(\mathrm{corr})} = g_{\mathrm{lf}} A_{\mathrm{lf}}$) depending on the plasma geometry and absorbing concentration.

The two last terms in the *rhs* of equation (13.34) represent the destruction of other species in the boundaries which may produce new molecules of species N_j, i.e.

$$N_\kappa \xrightarrow{\text{Wall}} pN_j. \tag{13.41}$$

or the destruction of species N_j in the wall.

$$N_j \xrightarrow{\text{Wall}} pN_l. \tag{13.42}$$

The first term in the *lhs* of equation (13.34) vanishes in the steady regime and the second term must only be evaluated if the convective flow is comparable with the reaction and diffusion losses. In the special case of a flow in a cylindrical container with negligible convective speed the diffusion term may be expressed as $D_j[N_j]/\Lambda^2$, where the constant Λ is the vessel radius R divided by the first zero $\lambda_0 \approx 2.405$ of the Bessel function $J_0(x)$.

One important observation must be made regarding the present form of equation (13.34). In the rigorous multicomponent transport theory [19], the mass flux vector contains the contribution due to the concentration gradients of all species, and also additional fluxes terms due to pressure and temperature gradients. In the present and simplified form of the theory, we assumed that all molecules have much lower densities than the background gas, so that all the D_j's are actually the binary diffusion coefficients of species j in the background gas. The density of the background gas is simply obtained using the equation of state of an ideal gas, i.e., $p = [N]k_bT \approx [N_{\mathrm{bg}}]k_bT$ where N_{bg} is the density of the background gas. If the background gas is a mixture with fixed composition, the D_j's of the species with low concentrations may be evaluated as the weighted mean of the binary diffusion coefficients [20]. We also note that the flux terms due to pressure and temperature gradients were neglected in equation (13.34).

The binary diffusion coefficients may be evaluated with aid of the kinetic theory [19] or, in some cases, extracted from theoretical and experimental tabulated data. In general, they vary with the gas temperature. This is also true for the reaction coefficients, which are usually expressed by the Arrhenius formula [21] and can often be found in the literature (e.g. reference [22]). The electron reaction rates, on the other hand, must be evaluated by means of the calculated electron distribution function, i.e., $k_{e,j} = 4\pi \int_0^\infty \sigma_j(v) F_0(v) v^3 dv$, although tabulated values or empirical formulas from literature may be found for many electronic reactions for the case of Maxwellian distributions. The parameter γ_k can also be found in the literature for some species and surfaces, though very often with large uncertainties.

With regard to the boundary conditions in the case of electrodeless discharges, simpler expressions can be defined because there is no need to account for the charge emission from electrodes. In the surfaces representing the vessel boundaries it is enough to impose the condition $N_j = 0$ for species that are totally destroyed in the walls; for species which only bounce back the boundary condition (b.c.) is $\hat{n} \cdot \vec{\Gamma}_j = 0$, for species which are partially destroyed the b.c. is $\hat{n} \cdot \vec{\Gamma}_j = \gamma_j \frac{1}{4} \langle v \rangle n_j$ and for species that may be created from wall interaction the b.c. is $\hat{n} \cdot \vec{\Gamma}_j = \sum_k (1 - \gamma_k) \frac{1}{4} p_k \langle v \rangle n_k$. In the case of a free-stream jet, for example, zero gradient or zero value condition may be imposed in the far away boundaries ($(\nabla \cdot N_j) \cdot \hat{n} = 0$ or $N_j = 0$) for charged and reactive species, and fixed values for the background gas.

A last observation concerns the *quasi*-neutrality hypothesis. If it is assumed that for every small volume of the plasma medium the negative charge density equals the positive one, then there is no need to solve the continuity equation for one of the charged species. The electron density is usually chosen as this species. Indeed, the *quasi*-neutrality hypothesis is valid in the plasma, except in the sheath region. In atmospheric pressure plasmas, the sheath dimensions are short compared with the bulk dimensions, except in the case of microplasmas. Unless the modeller's effort is directed towards sheath modelling, the application of the *quasi*-neutrality hypothesis to the entire space domain is valid.

13.1.3 Macroscopic equations for the weakly ionized gas flow

In the last section, we dropped some comments on the convective term in equation (13.34), but did not give the pathway to compute it. In the simplest case, the convective velocity may be considered constant if the flow can be modelled as a uniform and constant field. This would be the case in a laminar flow of an inviscid fluid through a tube. In this case, the velocity could be dropped out from the divergence term and convective losses would only be relevant in the regions where density gradients are present in the flow direction. In the general case of a weakly ionized plasma jet, the flow may be turbulent and in principle the flow field has to be determined from the Navier–Stokes equation for the background gas which is assumed to be totally unperturbed by the plasma

$$\frac{\partial \vec{u}}{\partial t} + \vec{u} \cdot \nabla \vec{u} = -\frac{1}{\rho} \nabla \cdot \mathbf{\Pi} + \vec{g} \tag{13.43}$$

where $\mathbf{\Pi}$ is the *pressure tensor*, given by [19]

$$\mathbf{\Pi} = -p\mathbf{I} + 2\eta\mathbf{S}, \quad \text{where } \mathbf{S} = \frac{1}{2}\left[\nabla \vec{u} + (\nabla \vec{u})^T - \frac{2}{3}(\nabla \cdot \vec{u})\mathbf{I}\right]. \tag{13.44}$$

In this particular form of the Navier–Stokes equation (13.43) the fluid is assumed to be compressible. In the derivation of equation (13.43) we use the fact that the overall mass conservation equation holds even when chemical reactions occur ($(\partial \rho)/(\partial t) + \nabla \cdot (\rho \vec{u}) = 0$). It also includes the gravity acceleration term ($\vec{g}$), which can be neglected in most cases. The bulk viscosity was also neglected. Although the charged species are supposed to have no influence on $\vec{u}$, the dynamic viscosity, represented by the symbol η, varies with temperature, which is determined mainly by the plasma heating. Similarly to the case of the diffusion coefficient, the dynamic viscosity may be determined from kinetic theory [19] or from parametrized formulas obtained from fittings of different sets of experimental data.

The energy equation of the flow closes the set of fluid equations, giving the gas temperature. In a weakly ionized plasma flow it assumes the following form[1]

$$\rho c_p\left[\frac{\partial T}{\partial t} + \vec{u}\cdot\nabla T\right] = \nabla\cdot(\lambda\nabla T) + e\vec{E}\cdot\vec{\Gamma}_e - \sum_i K_i^{(\mathrm{ch})}\delta h_i \\ - \sum_{i,j<i} A_{ij}g_{ij}h\nu_{ij}[N_i], \tag{13.45}$$

where c_p is the specific heat of the mixture at constant pressure, λ is the thermal conductivity of the system, K_i^{ch} and δh_i are the reaction rate and enthalpy variation per mol for reaction *i*, respectively. The energy losses due to friction were neglected, which is perfectly valid as the plasma input power is much larger. The second term in the *rhs* accounts for the energy transferred from the accelerated electrons to heavy particles. This is usually the most important source of energy in a chemical plasma. Note that energy transfer from ions to the medium was not considered, since their flux can be assumed to be very small in a weakly ionized plasma at atmospheric pressure. The third term accounts for the energy losses or gains in endothermic or exothermic chemical reactions. The last term is usually small and accounts for the radiative power losses due to electronic transitions of atoms (a similar but more complicated formula could be written in the molecular case).

The boundary conditions for $\vec{u}$ and T are deeply dependent on the particular problem. We could take, for example, a plasma enclosed in a tube with a background gas being fed at a constant mass flow rate. The velocity b.c. at the inlet is determined from the mass flow rate, while the outlet velocity boundary condition may be set as zero gradient in the outlet surface normal direction. The temperature at the inlet is usually fixed to a known value, while the outlet temperature boundary condition may be also set as zero gradient. At the tube walls the velocity component normal to the surface should be set to zero in the case of an inviscid flow, or zero velocity in the case of a viscous flow. If the temperature is kept constant at the walls by some means or is known by measurement then the boundary temperature at the walls is given by this value. If the value is not known, the temperature b.c. may be defined as zero gradient in the surface normal direction. If use is made of the axis of symmetry to reduce the space domain, the projection of the gradient of the temperature, particle fluxes and velocity in the symmetry surface should be set to zero.

13.1.4 Electrodynamics

Poison equation and ambipolar diffusion

In accordance with the *quasi*-neutrality hypothesis, any volume of the plasma is electrically neutral as long as the sampled volume dimensions are not smaller than the Debye length. If a small departure from neutrality occurs by a small charge

[1] The energy equation can be derived from the kinetic theory of dilute gases presented in reference [19] and the kinetic theory of electrons presented here.

separation due to some kind of transport effect, then an electric field will build up, pushing the particles with opposite charges back to neutrality, in accordance with Poison equation

$$\nabla \cdot \vec{E} = \frac{e}{\epsilon_0}\left(\sum_{j}^{N^{(+)}} Z_j n_j^{(+)} - \sum_{j}^{N^{(-)}} Z_j n_j^{(-)}\right). \tag{13.46}$$

where Z_j is the number of elementary charges of charge carrier j. For simplicity, let us consider a plasma with only electrons as negative charged particles and only one singly ionized positive species. Let us further suppose that a small departure from neutrality produces a local field $\vec{E_l}$, much smaller than the amplitude of an alternate microwave field. In this case, the electron energy distribution and transport parameters are determined by the alternating field and the local field influence on them could be neglected. The time averaged flow rate of ions and electrons are given by

$$\vec{\Gamma}_i = -D_i \nabla N_i + N_i \mu_i \vec{E_l} \text{ and } \vec{\Gamma}_e = D_e \nabla N_e - N_e \mu_e \vec{E_l}. \tag{13.47}$$

From the continuity equations, it can be readily found that $\nabla \cdot (\vec{\Gamma}_i - \vec{\Gamma}_e) = 0$. Removing the electric field from the above equations, using the *quasi*-neutrality condition and the latter result, it can be shown that

$$\nabla \cdot \vec{\Gamma}_i = \nabla \cdot \vec{\Gamma}_e = -D_a \nabla^2 n, \tag{13.48}$$

where D_a is the *ambipolar diffusion coefficient*

$$D_a = \frac{\mu_i D_e + \mu_e D_i}{\mu_i + \mu_e}. \tag{13.49}$$

Therefore, when the *quasi*-neutrality condition is assumed in a plasma—i.e., in a condition where space charges could have relevant effects—it is required that the diffusion coefficients of the charge carriers are replaced by the ambipolar diffusion coefficient D_a in the continuity equations. In the case of a plasma composed of many positive ions, the ambipolar diffusion coefficient must written as

$$D_{a,i} = D_i - \mu_i \frac{\sum_{j=1}^{n+1} n_j D_j}{\sum_{j=1}^{n+1} n_j \mu_j}, \quad i = 1, \ldots, n+1. \tag{13.50}$$

In this expression the index $n + 1$ represents the electrons. In the evaluation of the ambipolar diffusion coefficient the signal of the electronic density, n_{n+1}, and electron mobility, μ_{n+1}, must be negative in equation (13.50).

Note that the Poison equation must be solved in the study of the intermediate case where the plasma volume is comparable to the Debye sphere. It is also required when the study of the details of the sheath structure is desired. It is clear that in these cases D_a should not be replaced in the continuity equations.

Plasma sustaining by high frequency surface waves
Plasmas can be sustained by microwaves by several mechanisms. In principle, all of them can be well described by the solution of the Maxwell equations given the proper boundary and initial conditions and the plasma properties. The complexity of this problem lies in the fact that the electric field is in general strongly coupled to the plasma properties, which complicates the task of finding the solution.

In this section we treat the simple case of a plasma sustained by a propagating wave in magnetic transversal mode TM_{00}, which is the simplest model commonly applied to describe surface-wave discharges. These discharges are sustained by high frequency waves which in turn have the plasma as its propagating media.

In the case of a TM_{00} propagating wave, the electric and magnetic fields $\vec{E}$ and $\vec{B}$ can be written in cylindrical coordinates as $\vec{E} = (E_r, 0, E_z)$ and $\vec{B} = (0, B_\phi, 0)$. These components are assumed proportional to the factor $\exp(-j\omega t + j\int_z k(z')\,dz')$, where $k = \beta + j\alpha$, β is the propagation coefficient and α is the space damping factor (attenuation coefficient), and another factor which only depends on the radial coordinate. The relation between these components can be found using the Maxwell equations

$$\frac{\partial B_\phi}{\partial z} = j\frac{\omega}{c^2}\epsilon E_r;\ \frac{1}{r}\left[\frac{\partial}{\partial r}\left(rB_\phi\right)\right] = -j\frac{\omega}{c^2}\epsilon E_z;\ \frac{\partial E_r}{\partial z} - \frac{\partial E_z}{\partial r} = j\omega B_\phi \tag{13.51}$$

where ϵ is the medium relative dielectric constant. In the case of vacuum (or free space), the value is $\epsilon_0 = 1$; in the case of a dielectric, the value is ϵ_d (~4, if the medium material is quartz, for example); and in the case of a plasma the dielectric constant is given by [9]

$$\epsilon_p = 1 - \frac{\omega_p^2}{\omega^2 + {\nu_{\rm col}}^2} + j\frac{\nu_{\rm col}}{\omega}\frac{\omega_p^2}{\omega^2 + {\nu_{\rm col}}^2} = 1 - \xi + j\frac{\nu_{\rm col}}{\omega}\xi, \tag{13.52}$$

where $\omega_p = \left(\frac{N_e e^2}{\epsilon_0 m_e}\right)^{1/2}$ is the plasma oscillation frequency. After some algebraic manipulation of the equation (13.51), we arrive at the following expression for the component E_z

$$\frac{1}{r}\left[\frac{\partial}{\partial r}\left(r\frac{\partial E_z}{\partial r}\right)\right]\left[\frac{k^2}{\kappa_p^2} - 1\right] + \frac{\partial E_z}{\partial r}\frac{\partial}{\partial r}\left(\frac{k^2}{\kappa_p^2}\right) = \frac{\omega^2}{c^2}\epsilon E_z, \tag{13.53}$$

where $\kappa_p = k^2 - (\omega^2\epsilon)/c^2$. The radial component may be determined from E_z,

$$E_r = -j\frac{k}{\kappa_p^2}\frac{\partial E_z}{\partial r}. \tag{13.54}$$

We note that the solutions of the above equations are complex, so that the physical solution is obtained taking the real part. Note also that the factor $\exp(-j\omega t + j\int_z k(z')\,dz')$ may be cancelled out from equation (13.53) and equation (13.54) so that only the radial dependence is retained. Equation (13.53) assumes a simpler form if the plasma relative dielectric constant is assumed to be independent of the radial coordinate,

$$\frac{1}{r}\frac{\partial}{\partial r}\left(r\frac{\partial E_z}{\partial r}\right) + \zeta^2 E_z = 0, \tag{13.55}$$

where $\zeta^2 = -\kappa_p^2$. This equation may be written in the standard form as

$$r^2 E_z'' + rE_z' + \zeta^2 r^2 E_z = 0. \tag{13.56}$$

The general solution of this equation in a medium i with dielectric constant ϵ_i can be written with special complex functions [23]

$$E_{z,i}(r) = A_i J_0\left[\left(\frac{\omega^2}{c^2}\epsilon_i - k^2\right)^{1/2} r\right] + B_i H_0^{(1)}\left[\left(\frac{\omega^2}{c^2}\epsilon_i - k^2\right)^{1/2} r\right], \tag{13.57}$$

where J_0 is the Bessel function of first kind and order zero and $H_0^{(1)}$ is the Hankel function of first kind and first order. The constants A_i and B_i may be determined from the b.c.s in the interfaces between the different media and the imposed limits in $r = 0$ and $r \to \infty$. In a coaxial system plasma-dielectric-vacuum, we have $B_i = 0$ in the plasma and $A_i = 0$ in the vacuum from the requirement of finiteness in the limits $r = 0$ and $r \to \infty$. The signal of the arguments of the square-root functions were chosen so that the imaginary parts of the arguments of J_0 and $H_0^{(1)}$ were positive. This is necessary so that the function $H_0^{(1)}$ goes to zero when $r \to \infty$ [23].

The propagating and attenuation coefficients may be determined from the continuity condition applied to the components E_z and B_ϕ. The resulting values will also depend on the values of n_e, ω_p and ν_{col}. Dispersion curves for $\beta \times (\omega/\omega_p)$ and $\alpha \times (\omega/\omega_p)$ may be found in [23].

13.2 Application—contraction of an argon discharge

Let us consider the case of an argon discharge sustained by a surface-wave at atmospheric pressure. The discharge is supposed to be ignited inside a cylindrical tube made of a dielectric material. It expands into free space, but we shall ignore the diffusion of species from the environment into the discharge. Suppose also that the flow field of the background gas is always laminar and sufficiently small to be neglected. The axial gradients of the physical quantities are supposed to be small, so that at each axial position we may restrict the analysis to the radial description of the physical quantities. The axial mean or maximum values of the physical quantities could be determined from experiment (i.e., mean electron density or absorbed power) or fixed to a reasonable value in order to proceed with an exploratory study.

The continuity equation (13.34) may be written as

$$\begin{aligned}\frac{\partial[N_j]}{\partial t} &= \Gamma_j^{(+)}\left(T_e, T_g, [N_j], \ldots, [N_M]\right) - \Gamma_j^{(-)}\left(T_e, T_g, [N_j], \ldots, [N_M]\right)\\ &\quad + \frac{1}{r}\frac{\partial}{\partial r}\left(rD_j(r)\frac{\partial}{\partial r}[N_j]\right),\end{aligned} \tag{13.58}$$

where $\Gamma^{(+,-)}$ are the gain (+) and loss term (-) due to chemical reactions or spontaneous emission and M is the number of different species considered in the model. In this case, the reaction coefficients and transport parameters are functions of the space coordinates. The Boltzmann equation is solved for each point of the discretized space domain, so that the radial dependence of the transport parameters and electronic reaction coefficients can be determined. The local field approximation is assumed. As the EEDF depends on the electron density, which is given by the continuity equations in the *quasi*-neutrality approximation, an iterative procedure is required. Table 13.1 shows the electronic collisional processes which are included in

Table 13.1. Electronic cross-sections.

Collision[a]	Use[b]	Reference	Thr. (eV)
Elastic:			
$Ar(^1S_0) + e \rightarrow Ar(^1S_0) + e$	EB	IST	—
Excitation[c]:			
Excitation from $Ar(^1S_0)$:			
$Ar(^1S_0) + e \rightarrow Ar(^3P_1) + e$	EB, RC	IST	11.6
$Ar(^1S_0) + e \rightarrow Ar(^1P_1) + e$	EB, RC	IST	11.6
$Ar(^1S_0) + e \rightarrow Ar(^3P_2) + e$	EB, RC	IST	11.6
$Ar(^1S_0) + e \rightarrow Ar(^3P_0) + e$	EB, RC	IST	11.6
$Ar(^1S_0) + e \rightarrow Ar(4p) + e$	EB, RC	IST	13.0
$Ar(^1S_0) + e \rightarrow Ar^{**} + e$	EB, RC	IST	14.0
Excitation from 4*s*:			
$Ar(4s) + e \rightarrow Ar(4p) + e$	RC	IST	1.4
Excitation between states 4s:			
$Ar(4si) + e \rightarrow Ar(4sj) + e,\ i < j$	RC	IST	—
Ionization:			
Ionization from $Ar(^1S_0)$:			
$Ar(^1S_0) + e \rightarrow Ar^+ + 2e$	EB, RC	IST	15.6
Ionization from states 4*s*:			
$Ar(4s) + e \rightarrow Ar^+ + 2e$	RC	[28]	4.0
Ionization from states 4*p*:			
$Ar(4p) + e \rightarrow Ar^+ + 2e$	RC	[28]	2.6
Ionization from excimer Ar_2^*:			
$Ar_2^* + e \rightarrow Ar_2^+ + 2e$	RC	[28]	3.8
Superelastic processes:			
$Ar(4s) + e \rightarrow Ar(^1S_0) + e$	EB, RC	Klein–Rosseland	—
$Ar(4p) + e \rightarrow Ar(^1S_0) + e$	EB, RC	Klein–Rosseland	—
$Ar(4p) + e \rightarrow Ar(4s) + e$	RC	Klein–Rosseland	—
$Ar(4si) + e \rightarrow Ar(4sj) + e,\ i > j$	RC	IST	—

[a] The electronic states of argon are written in $L - S$ notation using capital letters. The lowercase letter, such as 4*s* e 4*p*, indicates the electronic configuration of the excited electron and it is used to group all the states with the same configuration.
[b] The cross-sections included in the Boltzmann code are indicated with the tag 'EB'; the cross-sections used in post-processing calculations are indicated with the tag 'RC'. Many cross-sections were used both in the Boltzmann equation evaluation and electronic reaction rate calculations.
[c] The 4*p* states were grouped in one 'effective' state. The four states in the 4*s* group were discriminated.

the computation of the electron Boltzmann equation. The corresponding cross-sections were extracted from the IST-Lisbon database which may be accessed in the web in the LXCat project's page [24]. Some processes were not considered in the computation of the EEDF, but the reaction rates were computed *a posteriori* to obtain the reaction coefficients. These mechanisms are also indicated in table 13.1.

Since *quasi*-neutrality is assumed, only the continuity equations for the positive ions have to be solved. In this model, only the ions Ar^+ and Ar_2^+ are considered. Although the electron continuity equation does not need to be computed, the electron diffusion coefficient and mobility which goes in the ambipolar diffusion formula, as well as the ionization rates, must be evaluated, since those quantities enter the ion continuity equations.

In total, the model considers seven neutral heavy species, ($Ar(^1S_0)$, $Ar(^3P_2)$, $Ar(^1P_1)$, $Ar(^3P_0)$, $Ar(^3P_1)$,$Ar(4p)$,Ar_2^*). The ten states that have the electronic configuration $4p$ are grouped together in a single state, indicated as $Ar(4p)$. The same applies to the excimer case Ar_2^*, which in reality has a more complex structure, including vibrational states [25]. Table 13.2 shows the reactions included in the model. The reaction coefficients were extracted from several sources and in order to abbreviate the extensive work of the bibliographical survey we refer to previous compilations [26, 27] except for the cross-sections which are not included in those works.

The solutions of the partial differential equation (13.58) are obtained using the boundary and initial conditions described next. In the walls, the concentration of the charged and reactive species were supposed to be zero, i.e. $[N_j]_{r=R_a} = 0$; from the requirement of symmetry, we have $(\partial[N_j]/\partial r)|_{r=0} = 0$, where R_a is the tube radius. The initial conditions are arbitrary, but one should choose one which leads to a feasible solution; we chose a Bessel function of the first kind and order zero to describe the initial profile of the charged particles, and a Gaussian profile for the neutral species. The initial density of the background gas (ground-state argon) is considered to be constant. The value was given by the state equation of an ideal gas at constant pressure. The temperature needed for the evaluation of the background gas density equation is determined from the heat equation

$$\frac{1}{r}\frac{d}{dr}\left(r\lambda(T_g(r))\frac{dT_g}{dr}\right) = -\sigma_e(r)E_0(r)^2, \tag{13.59}$$

where λ is the thermal conductivity of argon and σ is the electric conductivity ($\sigma_e = en_e\mu_e$). Note that the energy lost or gained due to radiation and chemical reactions is neglected, as electron collisions may be considered as the strongest heating source in the present case. The terms containing axial gradients were also neglected, in accordance with our earlier assumptions.

The thermal conductivity in equation (13.59) depends on the temperature, so an empirical relation for $\lambda(T)$ [36] had to be used to obtain a realistic solution. The electron conductivity $\sigma(r)$ was obtained from the solution of the electron Boltzmann equation. Rigorously, the electric field should be obtained from equation (13.53), but we used the approximate solution given by equation (13.55), using a similar

Table 13.2. Reaction between heavy species, spontaneous emission and reaction rates.

Reaction[a]	Reference	Reaction coefficient
Diffusion:		$[s^{-1}]$
Free diffusion		
$Ar(^3P_2)$, $Ar(^3P_0)$	[27]	D_f/Λ^2
Ambipolar diffusion		
Ar^+, Ar_2^+	[29]	D_a/Λ^2
Spontaneous emission		$[s^{-1}]$
$Ar(^3P_1) \rightarrow Ar(^1S_0) + h\nu$	[30]	$g \times (1.19 \cdot 10^8)$
$Ar(^1P_1) \rightarrow Ar(^1S_0) + h\nu$	[30]	$g \times (5.1 \cdot 10^8)$
$Ar(4p) \rightarrow Ar(^3P_2) + h\nu$	[30]	$1.43 \cdot 10^7$
$Ar(4p) \rightarrow Ar(^3P_1) + h\nu$	[30]	$7.3 \cdot 10^6$
$Ar(4p) \rightarrow Ar(^3P_0) + h\nu$	[30]	$3.5 \cdot 10^6$
$Ar(4p) \rightarrow Ar(^1P_1) + h\nu$	[30]	$9.83 \cdot 10^6$
$Ar_2(*) \rightarrow 2Ar(^1S_0) + h\nu$	[25]	$3.13 \cdot 10^5$
Molecular conversion:		$[cm^6\ s^{-1}]$
$Ar^+ + Ar(^1S_0) + Ar(^1S_0) \rightarrow Ar_2^+ + Ar(^1S_0)$	[31]	$2, 5 \cdot 10^{-31}$
Dissociative recombination[a]:		$[cm^3\ s^{-1}]$
$Ar_2^+ + e \rightarrow Ar(^1S_0) + Ar(^1S_0)$	[32]	$9.6 \cdot 10^{-7}\left(1 - \exp(-\frac{630}{T_g})\right)\chi\left(\frac{300}{T_e}\right)^{0.67}$
Penning Ionization		$[cm^3\ s^{-1}]$
$Ar(4si) + Ar(4sj) \rightarrow Ar(^1S_0) + Ar^+ + e$	[27]	$3.69 \cdot 10^{-11}T_g^{0.5}$
$Ar(4p) + Ar(4p) \rightarrow Ar^+ + Ar(^1S_0) + e$	[33]	$5 \cdot 10^{-10}$
$Ar(4p) + Ar(4s) \rightarrow Ar^+ + Ar(^1S_0) + e$	[33]	$5 \cdot 10^{-10}$
$Ar(4si) + Ar(4sj) \rightarrow Ar_2^+ + e$	[33]	$5 \cdot 10^{-10}$
Electronic impact dissociation		$[cm^3\ s^{-1}]$
$e + Ar_2^+ \rightarrow e + Ar(^1S_0) + Ar^+$	[34]	$1.11 \cdot 10^{-6}\exp\left(-\frac{(2.94 - 3(Tg[eV] - 0.026))}{(Te[eV])}\right)$
Atomic conversion		$[cm^3\ s^{-1}]$
$Ar_2^+ + Ar(^1S_0) \rightarrow Ar^+ + 2Ar(^1S_0)$	[34]	$5.22 \cdot 10^{-10}\frac{\exp(-\frac{1.304}{Tg[eV]})}{Tg[eV]}$
Three-body recombination		$[cm^6\ s^{-1}]$
$e + e + Ar^+ \rightarrow e + Ar(4p)$	[28]	$1.430\,3 \cdot 10^{-28}(T_e[eV])^{-3}$
$e + e + Ar^+ \rightarrow e + Ar(^3P_2)$	[28]	$6.143\,7 \cdot 10^{-29}(T_e[eV])^{-3}$
$e + e + Ar^+ \rightarrow e + Ar(^3P_1)$	[28]	$1.330\,27 \cdot 10^{-29}(T_e[eV])^{-3}$
$e + e + Ar^+ \rightarrow e + Ar(^3P_0)$	[28]	$3.812\,89 \cdot 10^{-29}(T_e[eV])^{-3}$
$e + e + Ar^+ \rightarrow e + Ar(^1P_1)$	[28]	$4.192\,07 \cdot 10^{-29}(T_e[eV])^{-3}$
Excimer formation		$[cm^6\ s^{-1}]$
$Ar(4s) + 2Ar(^1S_0) \rightarrow Ar_2(*) + Ar$	[26]	10^{-32}
Excimer extinction		$[cm^3\ s^{-1}]$
$Ar_2(*) + Ar_2(*) \rightarrow Ar_2^+ + 2Ar(^1S_0)$	[25]	$5 \cdot 10^{-10}$
Atomic three-body recombination		$[cm^6\ s^{-1}]$
$Ar^+ + Ar(^1S_0) + e \rightarrow 2Ar(^1S_0)$	[26]	$3.7 \cdot 10^{-29}T_e(eV)^{-1.5}T_g(K)^{-1}$

[a] Fast atoms produced after dissociative recombination usually lost their charge, so only a few percent of the collisional events effectively result in charge destruction [35]. We considered that just one percent ($\chi = 0.01$) of the collisional processes produce charge destruction. Other authors [34], adopted the value $\chi = 0.05$.

approach to that proposed in [37]. The general solution of the $E_z(r)$ component is given by the formula

$$F_p(r) = E_0 I_0\left(a_p r/R_a\right) = E_0 I_0\left[\left(\beta^2 - \frac{\omega^2}{c^2}\epsilon_p^2\right)^{1/2} r\right], \tag{13.60}$$

where R_a is the tube radius or the frontier between the plasma and the environment and I_0 is the modified Bessel function of the first kind. The dispersion relation is calculated using the explicit formulas derived by Zhelyazkov and Benova [37]. Equation (13.60) assumes a constant electron density, which was estimated as the radially averaged electron density

$$\langle N_e \rangle = \frac{\int_0^{2\pi}\int_0^{R_a} N_e(r) r \, dr d\theta}{\int_0^{2\pi}\int_0^{R_a} r \, dr d\theta} = \frac{2\int_0^{R_a} N_e(r) r \, dr}{R_a^2}. \tag{13.61}$$

This value was used to obtain the plasma permittivity, ϵ_p, and evaluate the dispersion relation [37].

The b.c.s of the heat equation are defined similarly to the continuity equations. From the requirement of symmetry we have $(\partial T/\partial r)|_{r=0} = 0$. The borders are assumed to have a constant temperature, $T|_{r=R_a} = T_0$, where T_0 is the laboratory temperature.

All the equations are coupled, since the continuity equations need the transport parameters and reaction rates from the Boltzmann equation, which in turn needs the electric field and electric density profiles. All the equations must be solved in an iterative and self-consistent fashion. Figure 13.1 shows the workflow diagram of the iterative procedure. One of the parameters, the electric amplitude, acts as an 'eigenvalue' of the problem, which should be adjusted in order to produce the electric density or another physical quantity (e.g. attenuation coefficient) given by experiment. In the present case, we set the electric density at the centre of the discharge, N_{e0}, as the fixed physical quantity, which in general is given by experiment. Here we adopted $N_e = 3.5 \cdot 10^{14}$ cm^{-3}, a typical value which may be found in this kind of discharge operating at a flow rate of 2.5 slm, applied power of 50 W and tube radius of 1.0 mm [38].

The solution converged to an electrical field amplitude ('eingenvalue') of $E_0 = 2505$ V m^{-1}. Figure 13.2 shows the normalized profile of the field, obtained from equation (13.60). It also shows the reduced field $E_{\mathrm{red}}(r)$ and the particle density, computed from $N(r) = p/(k_b T(r))$. The value of $N(r)$ grows faster than the electric field $E_z(r)$, resulting in a decreasing reduced electric field, which seems to be in contradiction with the experimental observation of an electron temperature increase towards the border [39].

Figure 13.4 shows two plots of the charged species profiles. From this graphic, it can be seen that the atomic ion in the discharge centre is predominant. In the borders, where the gas temperature and electric density are smaller, the molecular ion density becomes larger. These results agree qualitatively with experiment [40].

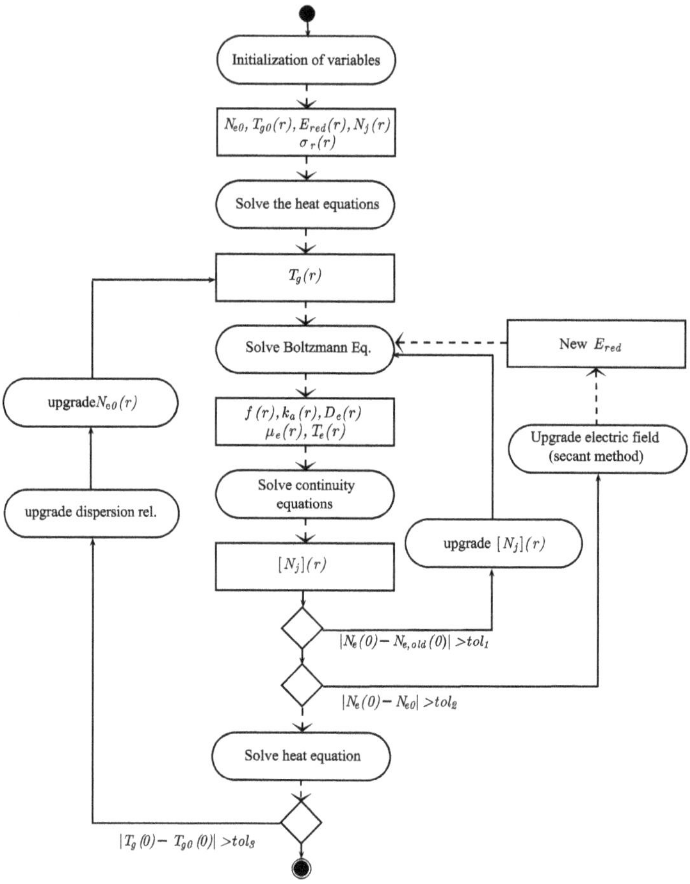

Figure 13.1. Workflow graph representing the iterative and auto-consistent procedure.

In reality, there are many factors which may contribute to the discharge contraction. These factors may be interrelated. The diffusion process naturally causes the density decrease towards the borders. Because of the dependence of the EEDF and ionization coefficient with the electron density (see figure 13.3), an abrupt fall in the ionization rate occurs. At the same time, the metastable densities decrease, which reinforces the decrease of the effective ionization rate due to the reduction of Penning ionization. On the other hand, the large relative concentration of the dimer Ar_2^+ in the discharge border guarantees an elevated recombination rate, which effectively contributes to negative charge destruction and electron density collapse. Note that the dimer creation rate increases when the temperature decreases, due to the dependence of the background gas with T_g, explaining why the relative concentration of the dimer increases towards the border. Another important effect particular to this model is the reduced electric field fall, which is also due to the increase of the background gas density towards the border (see figure 13.2). We call attention to the fact that the electric field was obtained assuming a constant electron density in the plasma medium; the detailed description

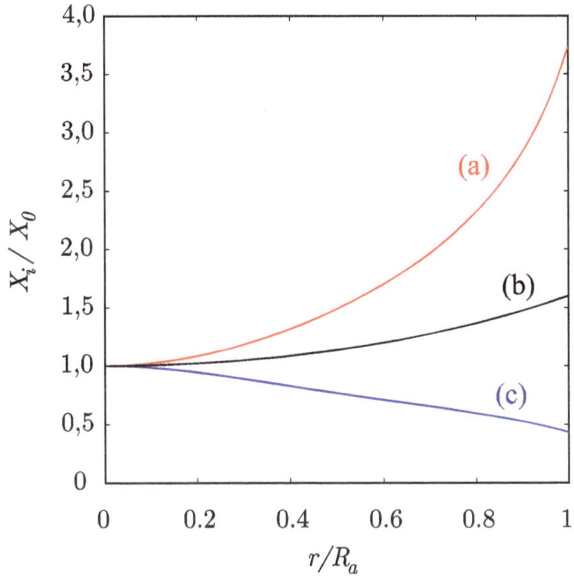

Figure 13.2. Normalized radial profile of particle density N, electric field $E_z(r)$ and reduced electric field $E_{\text{red}} = E_z(r)/N$. (a) $N(r)/N_0$, where $N_0 = 6.46 \cdot 10^{18}$ cm^{-3}; (b) $E_z(r)/E_0$, where $E_0 = 2505$ V m^{-1}; (c) $E_{\text{red}}(r)/E_{\text{red}0}$, where $E_{\text{red}0} = 0.39$ Td. The index 0 represents the value of a physical quantity in the column center.

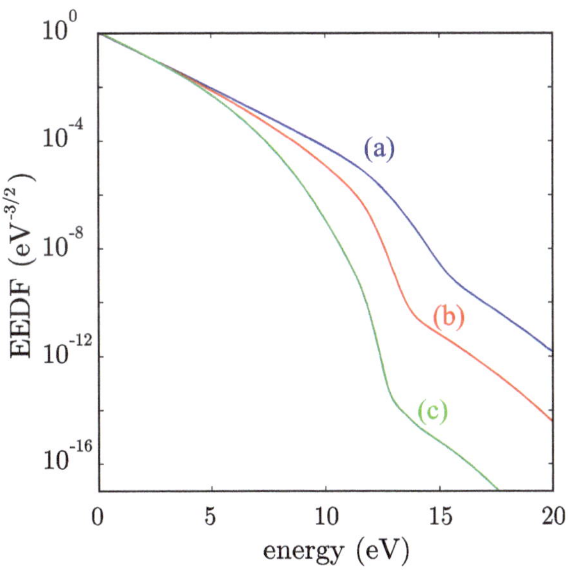

Figure 13.3. Comparison of the EEDFs at different positions in the plasma column, with $R_a = 1.0$ mm and $n_{e0} = 3.5 \cdot 10^{14}$ cm^{-3}. (a) $\rho = 0$, $T_e = 1.05$ eV; (b) $\rho = 0.1$, $T_e = 1.04$ eV; (c) $\rho = 0.2$, $T_e = 1.005$ eV.

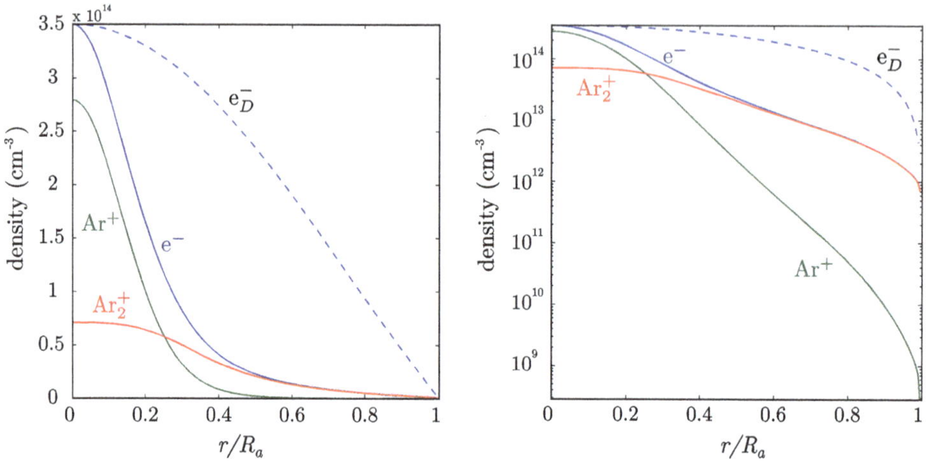

Figure 13.4. Radial profile in linear scale (*left*) or logarithmic scale (*right*) of the charged particles with $n_{e0} = 3.5 \cdot 10^{14}$ cm^{-3} in the column centre and $R_a = 1.0$ mm (e_D^- electron density in diffusion regime).

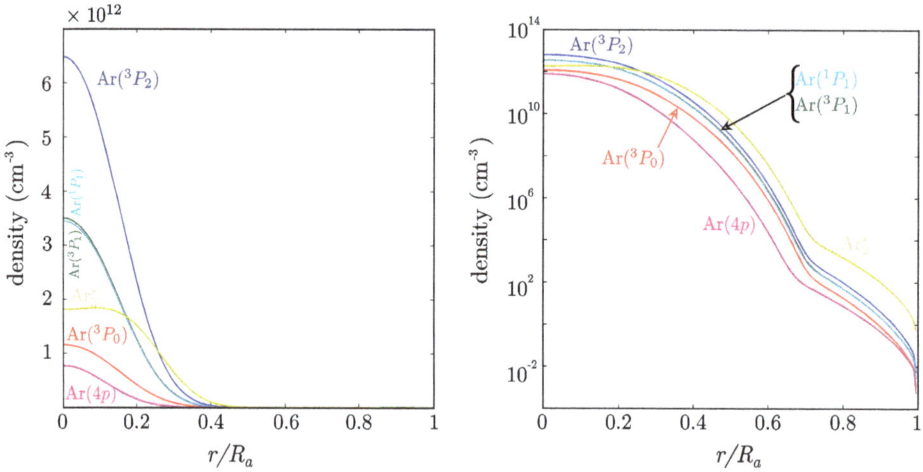

Figure 13.5. Radial profile in linear scale (*left*) or logarithmic scale (*right*) of the excited neutrals, with $n_e = 3.5 \cdot 10^{14}$ cm^{-3} and $R_a = 1.0$ mm.

of the field could possibly explain the experimental observation of an electric temperature increase towards the border [39]. In order to give a better idea of the contraction effect, the plot in figure 13.4 also shows the Bessel profile which is expected in the diffusion regime characteristic of low pressure conditions.

The profile of the excited neutral species is shown in figure 13.5. The concentration of this species also decreases towards the border, except in the excimer case. One of the destruction channels of this species is the ionization by electron impact and the only creation channel has a rate which depends quadratically on Ar($^1S_0$), which explains why the excimer maximum is displaced from the column center.

13.3 Conclusion

In this work we presented the theoretical foundations of plasma modelling for problems in high frequency fields and atmospheric pressure. We gave special emphasis to the classical electron kinetic theory and its connections with fluid modelling, since there are many evidences that non-equilibrium EEDFs play a very important role in many phenomena. The treatment of heavy species, flow dynamics, heat transport and electrodynamics was also discussed. In the last section, the theory was applied to the description of an argon discharge at atmospheric pressure and the mechanisms leading to contraction were identified and discussed.

We hope that this chapter may serve as a starting point for those interested in the design of more elaborate models, which would eventually contain more ingredients. As a matter of fact, a truly comprehensive model is very difficult to implement. Efforts in this direction may be found in the recent literature [26, 41–44], as the full understanding of the mechanisms of contraction clearly depends on the level of detail and completeness of the model. Besides, the description of plasma phenomena at atmospheric pressure in turbulent flows or plasmas sustained by TM fields at higher order modes, for instance, are poorly studied. In view of the many interesting applications of these discharges, further developments in this field are certainly worthwhile.

References

[1] Crookes W 1879 The Bakerian lecture: on the illumination of lines of molecular pressure, and the trajectory of molecules *Phil. Trans. R. Soc.* **170** 135–64

[2] Crookes W 1879 Contributions to molecular physics in high vacua. Magnetic deflection of molecular trajectory. Laws of magnetic rotation in high and low vacua. Phosphorogenic properties of molecular discharge *Phil. Trans. R. Soc.* **170** 641–62

[3] Thomson J 1899 XIX. On the theory of the conduction of electricity through gases by charged ions *Lond. Edinb. Dublin Philos. Mag. J. Sci.* **47** 253–68

[4] Weisskopf V F 1965 In defence of high energy physics *Nature of Matter: Purposes of High Energy Physics* (Long Island, NY: Brookhaven National Laboratory) pp 24–7

[5] Moisan M and Zakrzewski Z 1991 Plasma sources based on the propagation of electromagnetic surface waves *J. Phys. D: Appl. Phys.* **24** 1025

[6] Beenaker C 1977 Evaluation of a microwave-induced plasma in helium at atmospheric pressure as an element-selective detector for gas chromatography *Spectrochim. Acta* **32** 173–87

[7] Hopwood J, Hoskinson A R and Gregório J 2014 Microplasmas ignited and sustained by microwaves *Plasma Sources Sci. Technol.* **23** 064002

[8] Blevin H A and Fletcher J 1984 Electron transport and rate coefficients in Townsend discharges *Aust. J. Phys.* **37** 593–600

[9] Bittencourt J 2003 *Fundamentals of Plasma Physics* (Amsterdam: Elsevier) (INPE)

[10] Shkarofsky I, Johnston T and Bachynski M 1966 *The Particle Kinetics of Plasmas* (London: Addison-Wesley)

[11] Lin S, Robson R and Mason E 1979 Moment theory of electron drift and diffusion in neutral gases in an electrostatic field *J. Chem. Phys.* **71** 3483–98

[12] Loureiro J and Amorim J 2016 *Kinetics and Spectroscopy of Low Temperature Plasmas* 1st edn (Berlin: Springer)
[13] Holstein T 1946 Energy distribution of electrons in high frequency gas discharges *Phys. Rev.* **70** 367–84
[14] Ridenti M A, Alves L L, Guerra V and Amorim J 2015 The role of rotational mechanisms in electron swarm parameters at low reduced electric field in N_2, O_2 and H_2 *Plasma Sources Sci. Technol.* **24** 035002
[15] Hagelaar G J M and Pitchford L C 2005 Solving the Boltzmann equation to obtain electron transport coefficients and rate coefficients for fluid models *Plasma Sources Sci. Technol.* **14** 722–33
[16] White R D, Robson R, Schmidt B and Morrison M A 2003 Is the classical two-term approximation of electron kinetic theory satisfactory for swarms and plasmas? *J. Phys. D: Appl. Phys.* **36** 3125
[17] Holstein T 1947 Imprisonment of resonance radiation in gases *Phys. Rev.* **72** 1212–33
[18] Holstein T 1951 Imprisonment of resonance radiation in gases II *Phys. Rev.* **83** 1159–68
[19] Hirschfelder J O, Curtis C F and Bird R B 1954 *Molecular Theory of Gases and Liquids* (New York: Wiley)
[20] Drake G W 2006 *Springer Handbook of Atomic, Molecular, and Optical Physics* (Berlin: Springer)
[21] Laidler K 1987 *Chemical Kinetics* (New York: Harper and Row)
[22] Mallard W and National Institute of Standards and Technology (US) 1992 *NIST Chemical Kinetics Database* (National Institute of Standards and Technology)
[23] Glaude V M M, Moisan M, Pantel R, Leprince P and Marec J 1980 Axial electron density and wave power distributions along a plasma column sustained by the propagation of a surface microwave *J. Appl. Phys.* **51** 5693–8
[24] Alves L and Ferreira C 2014 *IST-Lisbon Database* http://www.lxcat.net, retrieved May 15, 2014
[25] Lam S K, Zheng C-E, Lo D, Dem'yanov A and Napartovich A P 2000 Kinetics of Ar_2^* in high-pressure pure argon *J. Phys. D: Appl. Phys.* **33** 242
[26] Petrov G M and Ferreira C M 2013 Numerical modelling of Ar glow discharge at intermediate and high pressures arXiv: 1308.2593
[27] Kutasi K, Guerra V and Sá P 2010 Theoretical insight into Ar–O_2 surface-wave microwave discharges *J. Phys. D: Appl. Phys.* **43** 175201
[28] Vriens L and Smeets A H M 1980 Cross-section and rate formulas for electron-impact ionization, excitation, deexcitation, and total depopulation of excited atoms *Phys. Rev.* A **22** 940–51
[29] Jonkers J, van de Sande M, Sola A, Gamero A and van der Mullen J 2003 On the differences between ionizing helium and argon plasmas at atmospheric pressure *Plasma Sources Sci. Technol.* **12** 30
[30] Wiese W L, Brault J W, Danzmann K, Helbig V and Kock M 1989 Unified set of atomic transition probabilities for neutral argon *Phys. Rev.* A **39** 2461–71
[31] Smith D, Dean A G and Plumb I C 1972 Three body conversion reactions in pure rare gases *J. Phys. B: At. Mol. Phys.* **5** 2134
[32] O'Malley T F, Cunningham A J and Hobson R M 1972 Dissociative recombination at elevated temperatures. II. Comparison between theory and experiment in neon and argon afterglows *J. Phys. B: At. Mol. Phys.* **5** 2126

[33] Brunet H, Lacour B, Rocca Serra J, Legentil M, Mizzi S, Pasquiers S and Puech V 1990 Theoretical and experimental studies of phototriggered discharges in argon and neon *J. Appl. Phys.* **68** 4474–80

[34] Jonkers J, van de Sande M, Sola A, Gamero A, Rodero A and van der Mullen J 2003 The role of molecular rare gas ions in plasmas operated at atmospheric pressure *Plasma Sources Sci. Technol.* **12** 464

[35] Ramos G, Schlamkowitz M, Sheldon J, Hardy K and Peterson J 1995 Observation of dissociative recombination of Ne_2^+ and Ar_2^+ directly to the ground state of the product atoms *Phys. Rev.* A **51** 2945

[36] Younglove B and Hanley H J 1986 The viscosity and thermal conductivity coefficients of gaseous and liquid argon *J. Phys. Chem. Ref. Data* **15** 1323–37

[37] Zhelyazkov I and Benova E 1989 Modeling of a plasma column produced and sustained by a traveling electromagnetic surface wave *J. Appl. Phys.* **66** 1641–50

[38] Ridenti M A, Souza-Corrêa J A and Amorim J 2014 Experimental study of unconfined surface wave discharges at atmospheric pressure by optical emission spectroscopy *J. Phys. D: Appl. Phys.* **47** 045204

[39] van Gessel A F H, Carbone E A D, Bruggeman P J and van der Mullen J J A M 2012 Laser scattering on an atmospheric pressure plasma jet: disentangling Rayleigh, Raman and Thomson scattering *Plasma Sources Sci. Technol.* **21** 015003

[40] Ridenti M, Spyrou N and Amorim J 2014 The crucial role of molecular ions in the radial contraction of argon microwave-sustained plasma jets at atmospheric pressure *Chem. Phys. Lett.* **595** 83–6

[41] Jimenez-Diaz M, Carbone E A D, van Dijk J and van der Mullen J J A M 2012 A two-dimensional Plasimo multiphysics model for the plasma-electromagnetic interaction in surface wave discharges: the Surfatron source *J. Phys. D: Appl. Phys.* **45** 335204

[42] Castaños Martinez E, Kabouzi Y, Makasheva K and Moisan M 2004 Modeling of microwave-sustained plasmas at atmospheric pressure with application to discharge contraction *Phys. Rev.* E **70** 066405

[43] Dyatko N A, Ionikh Y Z, Kochetov I V, Marinov D L, Meshchanov A V, Napartovich A P, Petrov F B and Starostin S A 2008 Experimental and theoretical study of the transition between diffuse and contracted forms of the glow discharge in argon *J. Phys. D: Appl. Phys.* **41** 055204

[44] Ridenti M A, de Amorim J, Dal Pino A, Guerra V and Petrov G 2018 Causes of plasma column contraction in surface-wave-driven discharges in argon at atmospheric pressure *Phys. Rev.* E **97** 013201

Chapter 14

Direct current microarcs at atmospheric pressure

Margarita Baeva and Markus M Becker

14.1 Introduction

Electric arcs are generated in gases and vapours at low, atmospheric and high pressure. They are characterized by low voltage, emission of high-intensity light and production of heat. They are widely utilized in welding, cutting, furnaces, lamps and other applications. Direct current (DC) electric arcs are now gaining even more importance due to the growing part of DC power applications spreading over household devices and renewable power sources.

Microarcs have a size of less than 1 mm in at least one direction. Arc plasmas of small size give an alternative to electron and laser beams [1] and are therefore considered for melting of materials in additive manufacturing, microwelding [2], repair of damaged surfaces of high value components [3], among others. Arcs of short lengths are even undesirably generated in switching devices during separation of contacts.

At moderate and high arc currents, most of the discharge gap is occupied by a plasma, which is close to the state of local thermodynamic equilibrium (LTE). Then, the LTE plasma core is quasi-neutral with equal temperatures of electrons and heavy particles, and a Maxwellian velocity distribution of the plasma species. Furthermore, the densities of the plasma species obey the Saha-Boltzmann equation [4–7]. Many arc models combining the equations of fluid dynamics and electromagnetics [8] have been developed. Deviations from thermal equilibrium can be taken into account by introducing a two-temperature description of the arc plasma [8], in which conservation equations for the energy of heavy species and electrons are solved separately. The deviations from thermal and chemical equilibrium, which occur in the vicinity of the electrodes [6, 9], and the formation of a space charge sheath adjacent to the electrodes due to the violation from quasi-neutrality, can be considered by coupling the LTE or two-temperature description of the core plasma with a non-equilibrium description of the boundary layer [7]. Alternatively, a fully non-equilibrium description of the quasi-neutral plasma core can be coupled with

doi:10.1088/978-0-7503-3559-1ch14

the space charge sheath [10, 11] to account for deviations from LTE in the arc core that can occur at low currents [12]. In these approaches, the boundary layers or space charge sheaths are introduced as infinitely thin interfaces between the plasma and the solid bodies of the electrodes and appropriate boundary conditions are formulated to account for the relevant phenomena.

In contrast, microarcs can deviate from LTE over their entire length. Thermal and ionization equilibria occur typically on the scale of several 10 μm to several 100 μm. The number of ionizing collisions that one electron undergoes decreases when the inter-electrode gap is shortened [13]. Fewer electrons in the microarc plasma are involved in a smaller number of elastic collisions with heavy particles, which in turn prevents the equilibration between electrons and heavy particles. Microarcs can even exist without a quasi-neutral arc column [14, 15]. A unified non-equilibrium description over the entire discharge gap is needed to provide the spatial resolution of the region(s) of space charge and the deviations from thermal and chemical equilibrium.

In what follows we consider the general backgrounds of a unified fluid modelling of microarcs that is presumably justified for microarcs at atmospheric pressure, where the mean free path of the electrons is expected to be the smallest characteristic length [16]. Fluid models of a gas discharge include equations for the moments of the Boltzmann equation as well as the Maxwell equations or the Poisson equation. The solution provides the number densities and energy of species along with the self-consistent electric field. A two-moment approach typically solves the equations of species conservation, conservation of electron energy, and the Poisson equation [17]. When the conservation of the gas energy is included, the dependence of plasma parameters on the gas temperature can be taken into account [18–20]. The Boltzmann equation for the electrons [21] provides the electron transport properties (mobility and diffusion) and the rate coefficients for processes of electron impact. These parameters can be calculated prior to the discharge modelling and prepared for use as lookup tables as a function of the mean electron energy according to the average energy approach (local mean energy approach) [22, 23]. The solution of the equations of electron momentum is commonly replaced by the classical drift-diffusion approximation, in which the electron particle flux and the flux of density of electron energy are expressed in terms of electric field and concentration gradients [24]. An alternative to the classical drift-diffusion approximation was introduced in [25] for the sake of improved reliability. This modified drift-diffusion approximation is considered below in more detail.

As in the general modelling of electric arcs, the coupling of the plasma with the electrodes is also essential for the modelling of microarcs. The interaction between the plasma and the electrodes controls their heat load and operation. While refractory cathodes can withstand high temperatures (e.g. above 3000 K if made of pure or doped tungsten), non-refractory cathodes start to melt before temperatures high enough for thermo-emission are reached. This on the one hand requires the account for thermo-field electron emission and on the other hand the release of metal atoms that enter the plasma. Metal atoms can also be released from the anode provided that the heat load leads to its melting and evaporation.

This chapter is organized as follows. In section 14.2, the equations of the unified fluid model of the arc are given. In section 14.3, attention is paid to transport quantities, thermodynamic and transport properties, and the description of the transport of electrons and electron energy is given in more detail. In section 14.4, the plasma chemistry is presented for the discharge in argon considered in the present contribution. Section 14.5 is devoted to the boundary conditions applied in the model. In section 14.6, selected results from the model are presented.

14.2 Unified fluid modelling of microarcs

A schematic view of the discharge arrangement is given in figure 14.1. The microdischarge burning between cylindrical electrodes is connected to a voltage source U_0 and a ballast resistor R_b.

The unified fluid modelling aims at describing the conservation of particles and energy of the plasma constituents coupled to the self-consistent electric field and the heat transfer in the electrodes. A convective motion of the plasma and self-induced magnetic field can be assumed to be negligible at low arc currents.

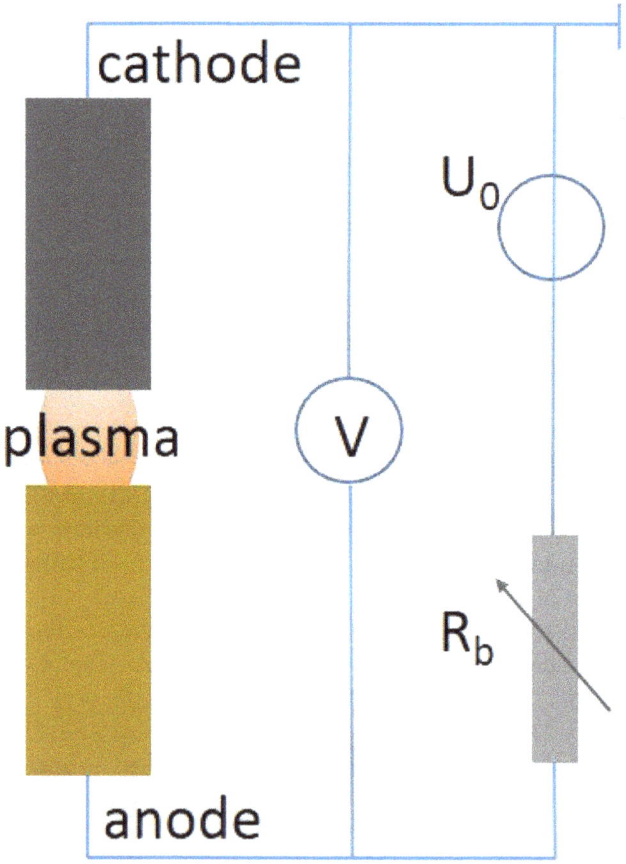

Figure 14.1. Schematic of the discharge arrangement.

For the sake of concreteness, we will consider argon gas at atmospheric pressure. The governing equations of the unified fluid model in their general form, however, are well applicable for other gases. We assume that the plasma behaves like a fluid, which is composed of electrons (e), ground state and excited atoms of argon (Ar, Ar*), atomic ions (Ar^+) and molecular ions (Ar_2^+), and dimers (Ar_2^*). The excited levels 4s and 4p with a statistical weight of 48 are lumped into the group Ar* [26].

The equations of particle and energy conservation of the electrons are written as

$$\frac{\partial n_e}{\partial t} + \nabla \cdot \mathbf{\Gamma}_e = S_e, \tag{14.1}$$

$$\frac{\partial n_\varepsilon}{\partial t} + \nabla \cdot \mathbf{\Gamma}_\varepsilon + e\mathbf{E} \cdot \mathbf{\Gamma}_e = S_\varepsilon - Q_h - Q_{rad}, \tag{14.2}$$

where n_e and n_ε denote, respectively, the electron number density and the density of electron energy, e is the elementary charge, $\mathbf{\Gamma}_e$ is the electron particle flux, and $\mathbf{\Gamma}_\varepsilon$ is the electron energy flux. These fluxes are expressed in terms of the drift-diffusion approximation, which will be considered in more detail in section 14.3. S_e and S_ε are source terms accounting, respectively, for the gain or loss of electrons and electron energy in inelastic processes. The terms Q_h and Q_{rad} describe, respectively, the loss of electron energy due to elastic collisions with heavy particles and volumetric radiation related to continuum contributions of electron–neutral and electron–ion transitions. We notice that line radiation is included in the term S_ε (see e.g. [27, 28]).

The transport of heavy particles (atoms, molecules and ions) is covered by equations of the type

$$\rho\frac{\partial Y_k}{\partial t} + \nabla \cdot \mathbf{J}_k = S_k, \tag{14.3}$$

where ρ, Y_k, $\mathbf{J}_k$, and S_k denote, respectively, the total mass density, the mass fraction, the mass flux, and the volumetric gain or loss of species of kind k.

The Poisson equation is solved for the self-consistent electric field $\mathbf{E}$

$$\nabla \cdot \mathbf{E} = \frac{\rho_q}{\varepsilon_0}, \tag{14.4}$$

where ρ_q is the space charge density, and ε_0 is the vacuum permittivity. The electric field $\mathbf{E}$ is related to the electric potential V by $\mathbf{E} = -\nabla V$.

Equations for the energy conservation of the heavy species in the plasma region and the heat transfer in the electrodes

$$A_s\rho C_p\frac{\partial T}{\partial t} + \nabla \cdot \mathbf{q} = A_s Q_h + A_s Q_{iJ}, \tag{14.5}$$

$$A_s\rho_s C_{ps}\frac{\partial T}{\partial t} + \nabla \cdot \mathbf{q}_s = 0 \tag{14.6}$$

are solved for the temperature T. In equations (14.5) and (14.6), C_p denotes the specific heat capacity at constant pressure, $\mathbf{q}$ is the heat flux due to heat conduction, and A_s is the area of the cross-section of the computational domain comprising the plasma region and the electrodes. The subscript s denotes the quantities corresponding to the solid. Further, the term Q_{iJ} in equation (14.5) represents the Joule heating of the ions.

The electric current density $\mathbf{j}$ is composed of the particle fluxes of electrons and ions, $\mathbf{j} = -e(\mathbf{\Gamma}_e - \Sigma_i \mathbf{\Gamma}_i)$. Its value is controlled by an external electric circuit including a voltage source U_0 and a ballast resistance R_b. Hence, an equation for the external circuit is added to equations (14.1)–(14.6)

$$V_a = U_0 - IR_b, \tag{14.7}$$

where $I = \int_S \mathbf{n} \cdot \mathbf{j} dS$ denotes the electric current collected on the electrode surface S, and V_a is the potential at the anode. The cathode is grounded.

The solution of the equation set (14.1)–(14.7) provides the knowledge of the physical behaviour of the microdischarge at currents of a few Ampere. The description of electric arcs at moderate and higher currents requires taking into account the self-induced magnetic field and the gas convection. The magnetic field gives rise to a volume force acting on the fluid (the Lorentz force) [8], which causes a magnetic pinch and the generation of a cathode jet [29]. The thickness of the regions of space charge progressively decreases with the increase of the arc current, while the spatial extent of the quasi-neutral core of the arc plasma increases [30]. This is related to an increase of the numerical effort and difficulties in solving equation (14.4). Instead of solving equation (14.4), the electric field is obtained by using the Ohm's law in its classic [5] or generalized [11] form and under the assumption of quasi-neutrality.

Time-dependent or steady-state solutions can be obtained by using commercial computation platforms such as COMSOL Multiphysics [31]. This provides the opportunity to combine various physical processes, e.g. transport of particles and energy with electromagnetics, as needed for arc plasma modelling and allows one to customize the governing equations. Additionally, COMSOL Multiphysics offers important features for generating geometries and computational meshes, a discretization of the system of partial differential equations using the finite element method for the space variables, powerful solvers that are fully parallelized and make use of multicore computing, and powerful visualization and post-processing of the results obtained. Time-dependent solvers can be used to achieve steady-state conditions, which can be easier than using stationary solvers under certain conditions.

14.3 Transport quantities, thermodynamic and transport properties

The equation set (14.1)–(14.7) involves fluxes concerning the transport of particles and energy, mass and charge. The electron particle flux $\mathbf{\Gamma}_e$ and the electron energy flux $\mathbf{\Gamma}_\varepsilon$, which appear in equations (14.1) and (14.2), can be described in terms of the drift-diffusion approximation in various forms. Due to the strong impact of the electron component on the overall behaviour of gas discharges, an accurate determination of these fluxes is crucial for the reliability of fluid models [32]. Therefore, approximations for the particle and energy flux of electrons are usually directly derived from the electron Boltzmann equation. The common drift-diffusion approximation

$$\mathbf{\Gamma}_{\mathrm{e}} = -\nabla(D_{\mathrm{e}} n_{\mathrm{e}}) - \mu_{\mathrm{e}} \mathbf{E} n_{\mathrm{e}}, \tag{14.8}$$

$$\mathbf{\Gamma}_{\varepsilon} = -\nabla(D_{\varepsilon} n_{\varepsilon}) - \mu_{\varepsilon} \mathbf{E} n_{\varepsilon} \tag{14.9}$$

is obtained by expanding the electron velocity distribution function (EVDF) for a reference system with preferential direction x in Legendre polynomials, neglecting the time derivative of the first anisotropy component of the expansion, f_1, and introducing the resulting expression into the equations

$$\Gamma_{\mathrm{e}}(x, t) = \frac{1}{3}\left(\frac{2}{m_{\mathrm{e}}}\right)^{\frac{1}{2}} \int_0^\infty U f_1(x, U, t) \mathrm{d}U, \tag{14.10}$$

$$\Gamma_{\varepsilon}(x, t) = \frac{1}{3}\left(\frac{2}{m_{\mathrm{e}}}\right)^{\frac{1}{2}} \int_0^\infty U^2 f_1(x, U, t) \mathrm{d}U \tag{14.11}$$

for the particle and energy flux, respectively [33, 34]. Here, U is the kinetic energy and m_{e} is the mass of electrons. The electron transport coefficients (mobility (μ_{e}, μ_{ε}) and diffusivity (D_{e}, D_{ε}) are defined then as energy integrals over expansion coefficients, which are determined prior to the fluid modelling by solving the stationary Boltzmann equations for the electrons in the framework of a two-term [33, 35] or a multi-term [36] approximation. It should be noted that common fluid models relying on the classical drift-diffusion approximation (14.8)–(14.9) use the approximations $D_{\varepsilon} = 5/3\, D_{\mathrm{e}}$ and $\mu_{\varepsilon} = 5/3\, \mu_{\mathrm{e}}$ rather than the consistent integral expressions for the energy transport coefficients. This is in favour of a better numerical stability of the system [25]. However, it also forces the particle and energy transport to occur on the same time scale, which is generally not the case.

An alternative consistent drift-diffusion approximation for the description of electrons within the framework of fluid models has been derived in [25]. Unlike the classical drift-diffusion approximation, this modified formulation is based on a hydrodynamic four-moment model and does not require a simplified approximation of the energy transport coefficients. The principal approach for derivation of the modified approximation of the particle and energy fluxes of electrons is the same as for the classical drift-diffusion model. The introduction of a multi-term expansion of the EVDF in Legendre polynomials into the spatially one-dimensional electron Boltzmann equation and the application of a proper integration over the electron kinetic energy lead to the two-moment equations of first and third order, respectively, [37]

$$\begin{aligned} &\frac{\partial}{\partial t}\Gamma_{\mathrm{e}}(x, t) + \frac{\partial}{\partial x}\left(\frac{2}{3m_{\mathrm{e}}} n_{\varepsilon}(x, t) + \frac{2}{5}\int_0^\infty U^{\frac{3}{2}} f_2(x, U, t)\, \mathrm{d}U\right) \\ &= -\frac{e}{m_{\mathrm{e}}} E(x, t) n_{\mathrm{e}}(x, t) - \frac{2}{3m_{\mathrm{e}}}\int_0^\infty \frac{U^{\frac{3}{2}}}{\lambda_{\mathrm{e}}(U)} f_1(x, U, t)\, \mathrm{d}U, \end{aligned} \tag{14.12}$$

$$\frac{\partial}{\partial t}\Gamma_\varepsilon(x, t) + \frac{\partial}{\partial x}\left(\frac{2}{3m_e}\int_0^\infty U^{\frac{5}{2}}\left(f_0(x, U, t) + \frac{2}{5}f_2(x, U, t)\right)\mathrm{d}U\right)$$
$$= -eE(x, t)\left(\frac{5}{3m_e}n_\varepsilon(x, t)\right) + \frac{4}{15m_e}\int_0^\infty U^{\frac{3}{2}}f_2(x, U, t)\,\mathrm{d}U \tag{14.13}$$
$$-\frac{2}{3m_e}\int_0^\infty \frac{U^{\frac{5}{2}}}{\lambda_e(U)}f_1(x, U, t)\,\mathrm{d}U,$$

where λ_e defines the mean free path of electrons. Together with the equations of particle and energy conservation (14.1) and (14.2), these time-dependent equations for the particle and energy flux of electrons represent a four-moment model for the electron component provided that transport coefficients are introduced to handle the integral expressions in equations (14.12)–(14.13). However, due to the small intrinsic time scales it is often highly time consuming to solve the fully time-dependent system. Therefore, as with the classical formulation, the time derivative is neglected here. This results in the modified drift-diffusion approximation [25, 32]

$$\mathbf{\Gamma}_e = -\frac{1}{m_e\nu_e}\nabla((\xi_0 + \xi_2)n_e) - \frac{e}{m_e\nu_e}\mathbf{E}n_e, \tag{14.14}$$

$$\mathbf{\Gamma}_\varepsilon = -\frac{1}{m_e\nu_\varepsilon}\nabla((\tilde{\xi}_0 + \tilde{\xi}_2)n_\varepsilon) - \frac{e}{m_e\nu_\varepsilon}\left(\frac{5}{3} + \frac{2}{3}\frac{\xi_2}{\xi_0}\right)\mathbf{E}n_\varepsilon, \tag{14.15}$$

which includes the momentum and energy flux dissipation frequencies ν_e and ν_ε as well as the transport coefficients ξ_0, ξ_2, $\tilde{\xi}_0$ and $\tilde{\xi}_2$. These properties are given as integrals of the isotropic part f_0 and the first two contributions f_1 and f_2 to the anisotropy of the EVDF over the electron kinetic energy U, respectively, according to

$$\nu_e = \frac{2}{3m_e\Gamma_e}\int_0^\infty \frac{U^{\frac{3}{2}}}{\lambda_e(U)}f_1(U)\,\mathrm{d}U, \tag{14.16}$$

$$\nu_\varepsilon = \frac{2}{3m_e\Gamma_\varepsilon}\int_0^\infty \frac{U^{\frac{5}{2}}}{\lambda_e(U)}f_1(U)\,\mathrm{d}U, \tag{14.17}$$

$$\xi_0 = \frac{2}{3n_e}\int_0^\infty U^{\frac{3}{2}}f_0(U)\,\mathrm{d}U, \tag{14.18}$$

$$\xi_2 = \frac{4}{15n_e}\int_0^\infty U^{\frac{3}{2}}f_2(U)\,\mathrm{d}U, \tag{14.19}$$

$$\tilde{\xi}_0 = \frac{2}{3n_e}\int_0^\infty U^{\frac{5}{2}}f_0(U)\,\mathrm{d}U, \tag{14.20}$$

$$\tilde{\xi}_2 = \frac{4}{15n_e}\int_0^\infty U^{\frac{5}{2}}f_2(U)\,\mathrm{d}U. \tag{14.21}$$

Note that the coefficients ξ_2 and $\tilde{\xi}_2$ are equal to zero if the two-term approximation is used for solving the electron Boltzmann equation. It has been shown for various applications that the modified drift-diffusion approximation (14.14)–(14.15) is broadly applicable and more reliable than the classical formulation [25, 27, 32].

In the case of a Maxwellian EVDF, the electron temperature is introduced as $T_e = 2\bar{\varepsilon}/(3k_B)$, where $\bar{\varepsilon} = n_\varepsilon/n_e$ denotes the mean electron energy, and k_B is the Boltzmann constant. Then, the electron transport parameters are written as $\mu_e = e/(m_e\nu_e)$ and $D_e = \mu_e k_B T_e/e$, where the frequency of momentum dissipation is approximated as $\nu_e = R_e/n_e$ with R_e being the total reaction rate for electron collision processes. The transport properties of electron energy are further written as $\mu_\varepsilon = 5\mu_e/3$ and $D_\varepsilon = 5D_e/3$.

The particle flux of heavy species of kind k, $\mathbf{\Gamma}_k$ and mass flux $\mathbf{J}_k$ (see equation (14.3)) are related through the mass of species k, m_k, i.e. $\mathbf{\Gamma}_k = \mathbf{J}_k/m_k$. The mass flux $\mathbf{J}_k$ accounts for mass diffusion and migration of charged species driven by the electric field $\mathbf{E}$ according to the expression

$$\mathbf{J}_k = -\rho Y_k\left(\frac{D_k}{Y_k}\nabla Y_k - \frac{D_k}{M}\nabla M + z_k\mu_k\mathbf{E}\right), \tag{14.22}$$

where D_k is the diffusion coefficient of species k in the mixture of species, z_k and μ_k are, respectively, the charge number and the mobility of species k, and M is the mean molar mass of the mixture. The diffusion coefficient D_k is expressed in terms of binary diffusion coefficients D_{kj}, which are calculated according to the gas kinetic theory [38, 39]. The species mobility μ_k is expressed by the Einstein relation with a common temperature T of all heavy species.

The heat fluxes due to heat conduction $\mathbf{q}$ and $\mathbf{q}_s$ in equations (14.5) and (14.6) are written as $\mathbf{q} = -A_s\lambda\nabla T$ and $\mathbf{q}_s = -A_s\lambda_s\nabla T$, respectively, with λ and λ_s denoting the thermal conductivity of the corresponding medium. The heat capacity at constant pressure and the thermal conductivity of the gas are computed using appropriate polynomial coefficients and tabulated data for the molar heat capacity of species [40, 41], and collisional integrals from the kinetic gas theory [42, 43]. The material properties of the electrode materials (tungsten, copper) are considered to depend on the temperature.

14.4 Plasma chemistry

The chemistry of the argon plasma can be considered with different levels of complexity regarding the excited electronic states of the argon atom. The level structure is treated in detail in global collisional-radiative models [44, 45], while in general, a reduced scheme is applied in fluid models [46] in order to limit the number of the followed species that corresponds to the number of the species equations to be solved. As mentioned in section 14.2, the considered unified model of the micro-discharge comprises electrons (e), ions (Ar^+, Ar_2^+), and neutral heavy species (Ar, Ar^*, Ar_2^*). Table 14.1 summarizes the reactions that are taken into account in the reaction kinetics model. The rate coefficients (RCs) for electron impact reactions considered by a cross section (CS) are calculated by integrals of the EVDF.

Table 14.1. Volume processes considered in the microarc model.

	Process	CS/RC	Reference
1	$e + Ar \rightarrow e + Ar$	CS	[47]
2	$e + Ar \rightarrow e + Ar^*$	CS	[47]
3	$e + Ar^* \rightarrow e + Ar$	CS	[47]
4	$e + Ar \rightarrow e + e + Ar^+$	CS	[47]
5	$e + Ar^* \rightarrow e + e + Ar^+$	CS	[48]
6	$e + e + Ar^+ \rightarrow e + Ar$	$8.75 \times 10^{-39} T_e^{-4.5}$	[13]
7	$Ar^* + Ar^* \rightarrow e + Ar^+ + Ar$	$1.625 \times 10^{-16} T^{0.5}$	[49]
8	$Ar^* + Ar \rightarrow Ar + Ar$	3×10^{-21}	[50]
9	$e + Ar_2^+ \rightarrow Ar^* + Ar$	$1.04 \times 10^{-12} \left(\frac{0.026}{Te}\right)^{0.67} \frac{1 - \exp(-418/T)}{1 - 0.31 \exp(-418/T)}$	[51]
10	$e + Ar_2^+ \rightarrow e + Ar + Ar^+$	$1.11 \times 10^{-12} \exp\left[-2.94 + 3\left(\frac{T}{11\,600} - \frac{0.026}{Te}\right)\right]$	[52]
11	$2Ar + Ar^+ \rightarrow Ar + Ar_2^+$	$7.5 \times 10^{-41}/T$	[53]
12	$Ar + Ar_2^+ \rightarrow 2Ar + Ar^+$	$6.06 \times 10^{-12} T^{-1} \exp\left(-\frac{1.258 \times 10^5}{R_g T}\right)$	[52]
13	$e + Ar_2^* \rightarrow e + e + Ar_2^+$	CS	[54]
14	$e + Ar_2^* \rightarrow e + 2Ar$	CS	[55, 56]
15	$2Ar_2^* \rightarrow e + 2Ar + Ar_2^+$	$1.624\,8 \times 10^{-16} T^{0.5}$	[53]
16	$2Ar + Ar^* \rightarrow Ar + Ar_2^*$	3.3×10^{-44}	[57]
17	$Ar^* \rightarrow Ar + h\nu$	3.145×10^5	[58, 59]
18	$Ar_2^* \rightarrow 2Ar + h\nu$	6×10^7	[57]

Remarks: The rate coefficients of the reactions for which collision cross-section data (CS) is used are calculated with a convolution of the cross-section with the corresponding EVDF. The rate coefficients of reactions (RC) are expressed in units of 1 s^{-1} for one-body reaction, $m^3\ s^{-1}$ in the case of two reactants, and in $m^6\ s^{-1}$ in the case of three reactants. T and T_e have units of K and eV, respectively. R_g is the gas constant.

In general, the data is obtained with EVDF as a solution of the space-independent Boltzmann equation for given reduced electric field and a multi-term approximation of the expansion in Legendre polynomials [36] with accounting for nonconservative electron collision processes. The RCs are then prepared in lookup tables as functions of the mean electron energy according to the local mean energy approach [23]. The other rate coefficients are taken from published data.

14.5 Boundary conditions

Appropriate boundary conditions complete the fluid model of the microdischarge. At the electrodes, the fluxes of electrons and electron energy, and the ion fluxes are written as [21]:

$$\mathbf{n} \cdot \mathbf{\Gamma}_e = \frac{1}{4} n_e v_{th,e} - \gamma \Sigma_i \mathbf{\Gamma}_i \cdot \mathbf{n} - \mathbf{\Gamma}_{th} \cdot \mathbf{n}, \tag{14.23}$$

$$\mathbf{n} \cdot \mathbf{\Gamma}_\epsilon = \frac{1}{4} n_e v_{th,e} \cdot 2k_B T_e - \gamma \Sigma_i \epsilon_{se,i} \mathbf{\Gamma}_i \cdot \mathbf{n} - \epsilon_{th} \mathbf{\Gamma}_{th} \cdot \mathbf{n}, \tag{14.24}$$

$$\mathbf{n} \cdot \mathbf{\Gamma}_i = \frac{1}{4} n_i v_{\text{th},i} + a \mu_i n_i \mathbf{E} \cdot \mathbf{n}. \tag{14.25}$$

In equations (14.23)–(14.25), $\mathbf{n}$ is the unit outward normal vector, $v_{\text{th,e}}$ and $v_{\text{th},i}$ denote the thermal velocity of electrons and ions, respectively, and γ is the coefficient of secondary emission of electrons caused by the flux $\mathbf{\Gamma}_i$ of ions of kind i onto the electrode. The parameter a takes the value of one, if the electric field is directed towards the electrode ($\mathbf{E} \cdot \mathbf{n} > 0$) and zero otherwise.

The flux of electrons due to emission from the cathode caused in general by thermo-field emission, $\mathbf{\Gamma}_{\text{th}}$, is related to the corresponding emission current density $j_{\text{em}} = e\mathbf{\Gamma}_{\text{th}}$. Secondary and thermo-field emission of electrons are considered to occur from the cathode. These processes are essential for the self-sustainment of the discharge. On the edge of transition to microarc, the thermo-field emission becomes the dominant mechanism. Depending on the material properties, cathodes are classified as refractory or non-refractory. While refractory cathodes (e.g. made of tungsten) can withstand temperatures above 3000 K so that thermionic emission can be strong, non-refractory cathodes (e.g. made of copper) start to melt before the temperature for appreciable thermionic emission is reached. Atoms released from the cathode due to evaporation enter the plasma and can be ionized easily since their ionization potential is well below that of the inert gas. The evaluation of the electric current density has to consider the general case of thermo-field emission [60]. For the sake of simplicity, we consider a refractory cathode at temperature T_c, for which the emission current density is given by the Richardson–Dushman–Schottky expression as

$$j_{\text{em}} = A_{\text{R}} T_{\text{c}}^2 \exp\left(-\frac{W - \Delta W}{k_{\text{B}} T_{\text{c}}} \right), \tag{14.26}$$

where A_{R} is the Richardson constant, W is the work function of the cathode material, and $\Delta W = e\sqrt{eE/4\pi\varepsilon_0}$ accounts for a lowering of the work function due to the electric field of strength E. Alternatively, the approach by Murphy and Good [61] can be applied.

The mean energy of the secondary electrons is expressed by the corresponding ionization energy $E_{\text{ion},i}$ and the work function as $\epsilon_{\text{se},i} = e(E_{\text{ion},i} - 2W)$. The average kinetic energy taken away by the electrons released due to thermionic emission is $\epsilon_{\text{th}} = 2k_{\text{B}}T_{\text{c}}$.

The flux boundary condition for the excited species reads

$$\mathbf{n} \cdot \mathbf{\Gamma}_* = \frac{1}{4} n_* v_{\text{th},*}. \tag{14.27}$$

Here, n_* is the number density of the excited species and $v_{\text{th},*}$ their thermal velocity.

The boundary conditions for the energy conservation of heavy particles and the heat transfer in the electrodes, which are governed by equations (14.5) and (14.6), employ a fixed temperature of 300 K for the anode and the end of the cathode, which is not in contact with the plasma. On the edge between the plasma and the cathode, a heat flux Q_{c} is introduced to account for contributions of ion bombardment and recombination of ions on the cathode, electron emission due to secondary and thermionic emission, plasma electrons reaching the cathode, and black-body radiation. It reads

$$\begin{aligned} Q_c = j_i(E_{\text{ion}} - W) - j_{\text{se}}(E_{\text{ion}} - W) - j_{\text{em}}(W + \frac{1}{e}2k_BT_c) \\ + \frac{1}{4}n_e v_{\text{th,e}}(2k_BT_e + eW) - \epsilon_W\sigma_{SB}T_c^4. \end{aligned} \tag{14.28}$$

A detailed description of the derivation of (14.28) is given in [30]. It represents a simplified form, which considers the presence of atomic ions in the vicinity of the cathode. In equation (14.28), ϵ_W is the temperature-dependent emissivity of the cathode material, and σ_{SB} is the Stefan–Boltzmann constant. The anode is kept at constant temperature, which corresponds to the case of a water-cooled anode. This is a simplification of the model, which spares us the solution of the heat transfer in the anode. Otherwise, a corresponding heat flux can be assigned on the edge between the anode and the plasma [62].

14.6 Realization and selected results

The realization of a multidimensional (in two or three dimensions) unified model is expected to provide a more realistic physical picture of the microdischarge. It necessarily faces the problem of an enormous computational effort, which is caused by the spatial resolution of the regions of space charge. The unified model in one dimension (1D) allows us to study the discharge properties with high spatial resolution and a larger level of plasma chemistry and to avoid the computational overhead. This is very important in cases, where a wide range of parameter variation has to be done. The 1D computational domain is typically resolved in 1000 to 3000 mesh elements of decreasing size towards the electrodes. Depending on the discharge regime, the computational time needed to cover a physical time of 10 s on a computer with 16CPUs@1.26 GHz varies between 40 s (glow microdischarge) and 15 min (microarc).

A broad spectrum of modelling that results from the unified model as realized on the computational platform COMSOL Multiphysics in one and two dimensions is presented in recent works [15, 27, 30, 60, 62, 63]. Some selected results are discussed here.

The discharge arrangement (figure 14.1) is considered with cylindrical electrodes with a radius of 2 mm. The cathode is made of tungsten ($W = 4.32$ V, $\gamma = 0.096$). The anode is made of copper and is water-cooled. It has a fixed temperature of 300 K, which is low enough to neglect a release of copper atoms. The discharge is ignited in argon gas at atmospheric pressure in the inter-electrode gap, which has a length from 20 to 500 μm.

Figure 14.2 shows the calculated voltage–current density characteristics of a microdischarge with a gap length of 400 μm for three approaches applied. They differ in the description of the electron transport. The modified drift-diffusion approximation (equations (14.14) and (14.15)) is combined with a solution of the

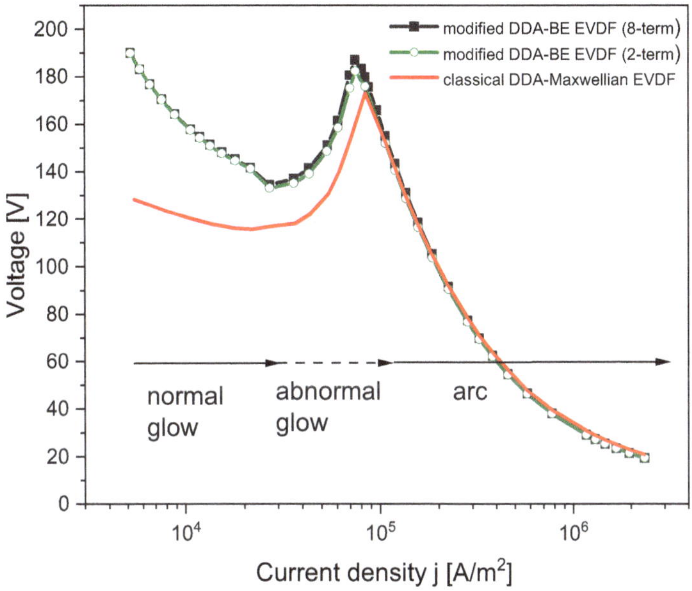

Figure 14.2. Voltage–current density characteristics for a gap of 400 μm obtained with the modified DDA-BE EVDF (two-term and eight-term expansion of the EVDF) and the classical DDA-Maxwellian EVDF approaches.

Boltzmann equation (BE) in the framework of a two-term and an eight-term expansion of the EVDF in Legendre polynomials (DDA-BE). The classical DDA (equations (14.8) and (14.9)) is applied assuming a Maxwellian EVDF (DDA-Maxwellian EVDF). The transition from a normal glow discharge to an abnormal glow discharge and further to an arc discharge is evident in the results of the three approaches. Both BE-based approaches provide results in good agreement to each other over the entire region of current density values. The approach 'classical DDA-Maxwellian EVDF' deviates from the 'modified DDA-BE EVDF' for normal and abnormal glow microdischarges. It agrees well for microarcs.

Figure 14.3 shows the predicted electron (a) and gas (b) temperatures for a current density of about 10^6 A m^{-2} over the discharge gap length of 400 μm. The positions of the anode and the cathode ends in contact with the plasma correspond to $x = 0$ and 400 μm, respectively. The considered current density corresponds to the arc regime of the microdischarge (figure 14.2). The arc regime occurs beyond current density of 1.1×10^5 A m^{-2}, where the heating of the refractory cathode by the plasma is strong enough to ensure that the current density due to field enhanced thermo-emission exceeds the current density due to secondary electron emission. The results obtained by the modified DDA-BE EVDF with a two-term and an eight-term expansion of the EVDF are practically identical (symbols are used to indicate the cases). The approach using the classical DDA with a Maxwellian EVDF yields in general lower temperature values for both the electron and gas temperature in the bulk of the microarc plasma. The peak value of the electron temperature reached close to the cathode is also lower than in the case of non-Maxwellian EVDF. However, the gas temperature on the edge

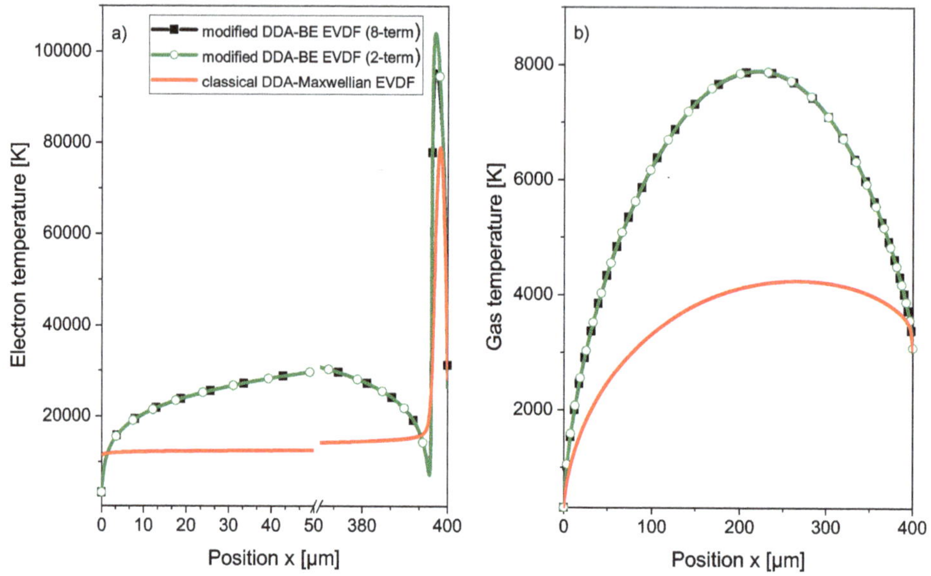

Figure 14.3. Distribution of electron temperature (a) and gas temperature (b) for a gap of 400 µm and a current density of 10^6 A m^{-2} obtained with the modified DDA-BE EVDF (two-term and eight-term expansion of the EVDF) and the classical DDA-Maxwellian EVDF approaches.

between the plasma and the cathode agrees well. This temperature determines the thermionic emission of the cathode. It is important for the prediction of the current density of the emitted electrons. It should be noticed that the solution of the Boltzmann equation for the EVDF did not account for electron–electron collisions that in general lead to a Maxwellization of the EVDF.

A detailed analysis of the influence of the description of the electron transport on the predicted plasma parameters is given in [27]. There, it has been shown that also densities of charge carriers and neutrals can vary depending on description of the electron transport at similar or almost equal discharge voltages. Related measurements of particle densities and temperatures are needed for appropriate validation of the different approaches for description of the electron transport. Since microarcs stay in the focus of this work, results of the 'classical DDA-Maxwellian EVDF' approach are considered below.

Figure 14.4 shows the distribution of the electron density (n_e and the total ion density (n_i) over the gap distance of 400 µm for a current density of 10^6 A m^{-2}. A quasi-neutral plasma bulk (figure 14.4b) occupies most of the inter-electrode region. Thin sheaths of positive charge occur adjacent to the electrodes (figure 14.4(a,c)). Their thicknesses can be estimated as the fractional part of 0.01 of the difference in the ion and electron density with respect to the ion density $(n_i - n_e)/n_i = 0.01$ [64, 65]. We obtain values $L_{as} \approx 2$ µm and $L_{cs} \approx 8$ µm, respectively, for the thickness of the space charge sheaths of the anode and the cathode.

From the positions of the sheath edges and figure 14.5 showing the distribution of the electric potential and electric field, we can now estimate the voltage drop in the

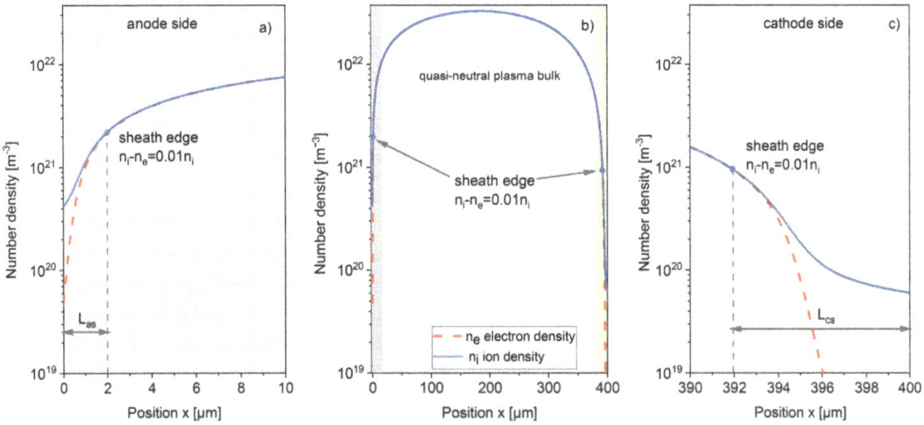

Figure 14.4. Distribution of charged particle densities for a gap of 400 μm obtained with the classical DDA-Maxwellian EVDF approach at a current density of 10^6 A m^{-2}.

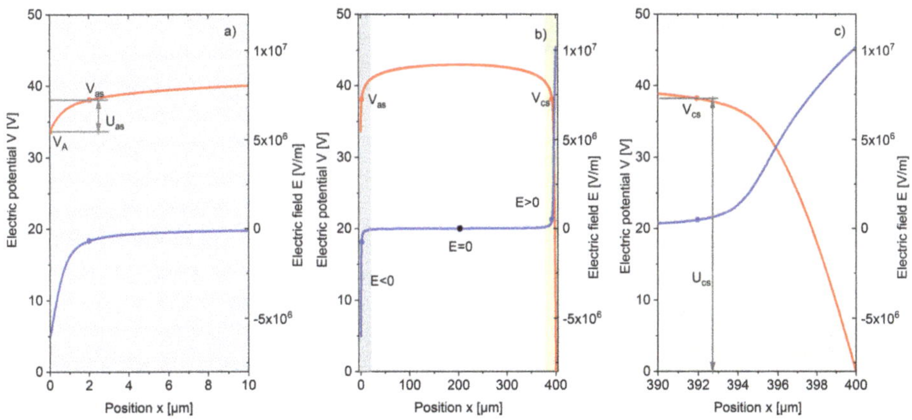

Figure 14.5. Electric potential and electric field across the discharge gap for the conditions in figure 14.4.

space charge sheaths. On the anode side, the voltage drop is $U_{\mathrm{as}} = V_{\mathrm{A}} - V_{\mathrm{as}} = -4.3$ V. Here, V_{A} denotes the potential on the anode, which is obtained from the solution of the equation set. The negative value of the voltage drop reveals a reversal of the electric field outside of the anode space charge sheath. Similarly, a fully non-equilibrium modelling of DC arcs in argon at atmospheric pressure [11] has predicted a negative anode fall. The voltage drop in the cathode space charge sheaths is $U_{\mathrm{cs}} \simeq 38$ V and appears larger than the discharge voltage of 33.7 V due to the negative anode fall.

The length of the quasi-neutral plasma bulk at given current density decreases with the decrease of the gap length (figure 14.6). Towards minuscule discharge gaps, the anode and cathode space charge sheaths merge and the microarc plasma remains positively charged. It is worth noting that the fluid modelling of microarcs of very short lengths becomes critical since the mean free path of electrons and the length of

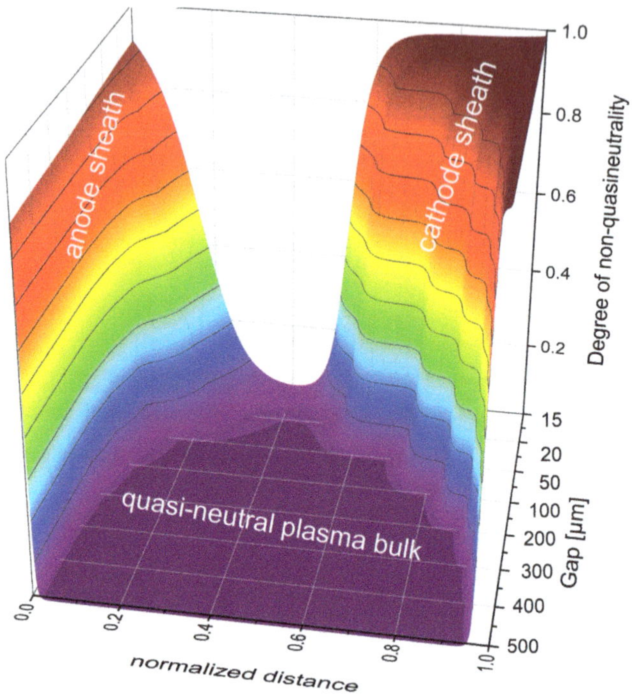

Figure 14.6. Spatial distribution of the degree of non-quasi-neutrality over the normalized gap distance for a current density of 10^6 A m^{-2}.

the inter-electrode gap become comparable. Furthermore, an increase of the current density leads to a decrease of the sheath thickness so that sheaths can become collision-free.

Figure 14.7 shows the distribution of the electric potential (a), the electron temperature (b) and the temperature of heavy particles (c) at a current density of 10^6 A m^{-2} for gap lengths of 20 μm (solid), 25 μm (dashed), 50 μm (dash-dotted), and 100 μm (short dashed). The results indicate the increase of the voltage and the maximum electron temperature, and the decrease of the maximum gas temperature when the gap length is shortened. An increase of the voltage was also observed experimentally [66]. A probable explanation is that at short gaps the number of ionizing collisions decreases compared to that at larger gaps and a stronger field is required to achieve densities of charge particles needed to sustain the discharge. The increase of the discharge voltage is related to the increase of the electron temperature in the entire gap and of its peak value on the cathode side. The heating of the gas slightly decreases because of the lower electron density and in turn lower frequency of elastic collisions with the neutrals.

The extension of the unified non-equilibrium model in two dimensions (2D) allows us to obtain radial dependence of the plasma properties and to explore the potentials and limitations of the 1D model with respect to 2D modelling [63]. The 1D model of the microarc presumes that the arc plasma fills the space between the

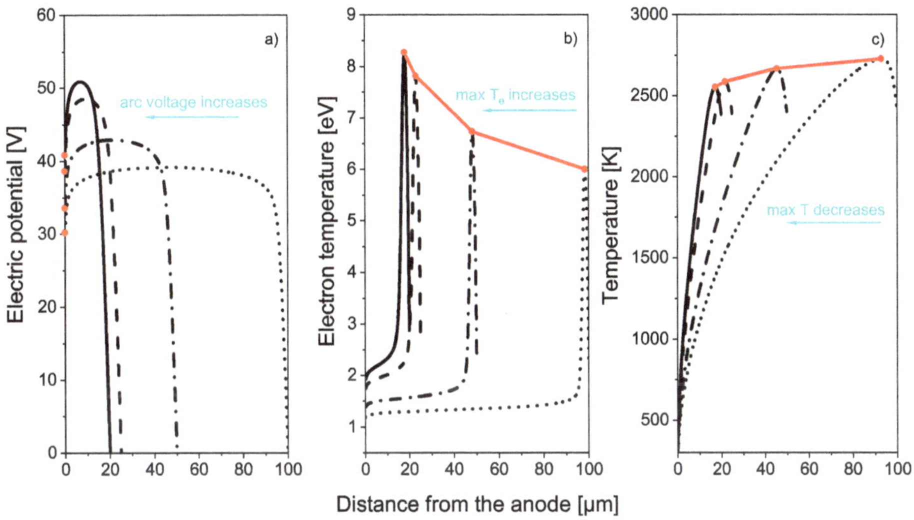

Figure 14.7. Behaviour of plasma parameters when shortening the discharge gap.

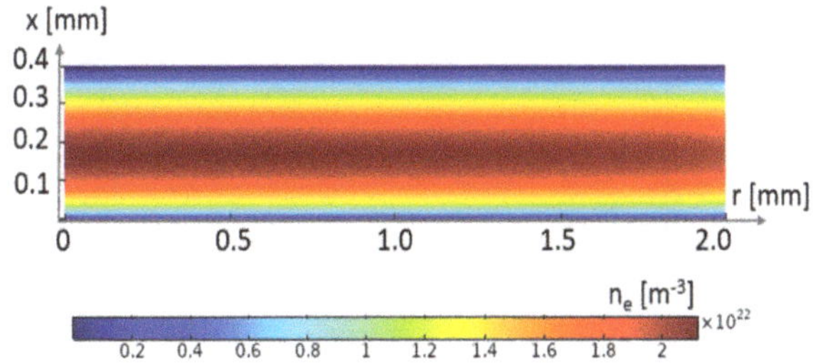

Figure 14.8. Electron density in the gap region obtained by the 2D model at an electric current of 5.3 A. Reproduced with permission from reference [63], copyright 2020 IOP Publishing, all rights reserved.

electrodes and the arc attachment to the electrodes occurs uniformly on the surfaces of the cathode and anode facing each other.

The electron density obtained by 2D modelling of the gap region of the microarc with electric current of 5.3 A is shown in figure 14.8. A well-extended plasma occupies the gap. For all radial positions, the electron density reaches a maximum approximately in the middle of the gap. The maximum value slightly decreases from the axis of symmetry towards the periphery.

The result of the 2D model is compared for several radial positions with that of the 1D model applied for a current density of 4.24×10^5 A m^{-2} in figure 14.9. The 1D distribution of n_e lies well between the 2D results. The maximum values reached on the axis and at $r = 2$ mm differ by about 4% only. This indicates a slightly stronger ionization towards the arc axis and is related to the radial profile of the current density shown in figure 14.10. Good agreement between the 1D model and the 2D model of the gap region is obtained also for the other plasma parameters.

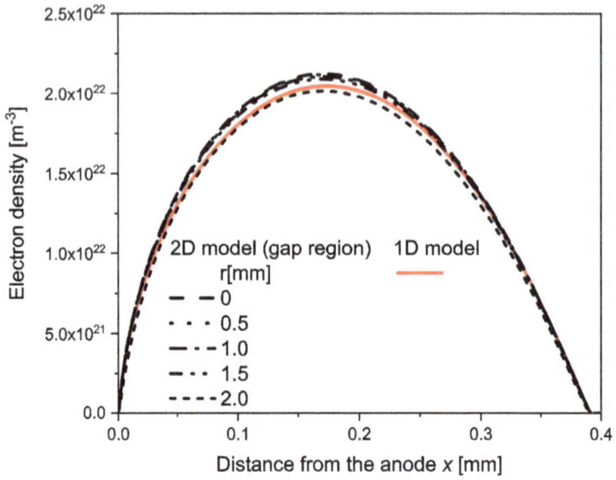

Figure 14.9. Electron density obtained by 1D modelling in comparison with that of 2D modelling for different radial positions. Reproduced with permission from reference [63], copyright 2020 IOP Publishing, all rights reserved.

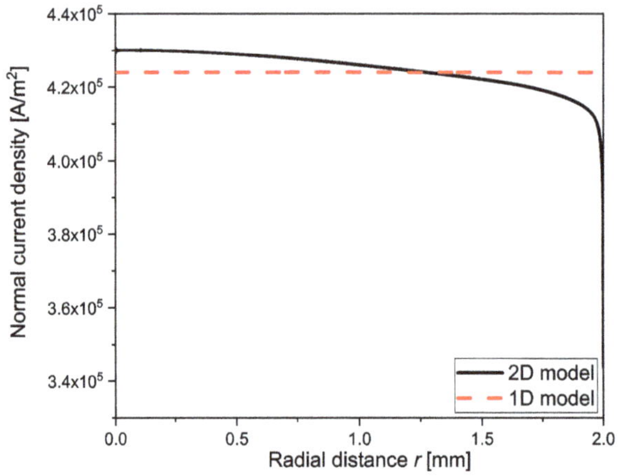

Figure 14.10. Current density on the cathode resulting from 1D and 2D modelling. Reproduced with permission from reference [63], copyright 2020 IOP Publishing, all rights reserved.

14.7 Conclusion

In this chapter, we present the physical background and results of recent self-consistent unified plasma modelling of microarcs at low currents in argon at atmospheric pressure. We consider the fluid description of the microarc plasma in the framework of the classical and modified drift-diffusion approximations, and show selected results obtained by one- and two-dimensional modelling.

It is shown that the results of the different drift-diffusion approximations for the current–voltage characteristics are in good agreement over the whole current range (from normal glow to arc regime) when the solution of the electron Boltzmann equation is taken into account for determination of the transport properties of electrons. Larger deviations are found in the normal and abnormal glow regime when considering a Maxwellian EVDF. All approaches predict generally the discharge voltage in good agreement in the arc regime.

A detailed analysis of the arc properties for different gap lengths is performed for a current density of 10^6 A m^{-2} and it is shown that the quasi-neutral plasma bulk vanishes when the gap length is reduced to less than 15 μm. By comparing results of spatially one- and two-dimensional modelling it can be concluded that the 1D model is well applicable for the considered conditions.

Further results and analysis of the plasma parameters can be found in the journal papers [15, 27, 30, 63]. The knowledge gained in the modelling of microarcs will be further extended towards short arcs occurring in low-voltage circuit breakers. We are convinced that the presented modelling approach will inspire a further development and exploration of microarcs and the phenomena in the regions of space charge of electric arcs, both experimentally and computationally.

References

[1] Fridman A, Gutsol A and Cho Y I 2007 Non-thermal atmospheric pressure plasma *Advances in Heat Transfer* ed A Fridman, Y Cho, A George and A B-C Greene (New York: Academic) pp 1–142

[2] Nowack M, Leidich S, Reuter D, Kurth S, Kuechler M, Bertz A and Gessner T 2012 Micro arc welding for electrode gap reduction of high aspect ratio microstructures *Sens. Actuators* A **188** 495–502

[3] Jhavar S, Paul C P and Jain N K 2016 Micro-plasma transferred arc additive manufacturing for die and mold surface remanufacturing *JOM* **68** 1801–9

[4] Boulos M I, Fauchais P and Pfender E 1994 *Thermal Plasmas, Fundamentals and Applications* vol 1 (Berlin: Springer)

[5] Gleizes A, Gonzalez J J and Freton P 2005 Thermal plasma modelling *J. Phys. D: Appl. Phys.* **38** R153–83

[6] Murphy A B and Uhrlandt D 2018 Foundations of high-pressure thermal plasmas *Plasma Sources Sci. Technol.* **27** 063001

[7] Benilov M S 2020 Modeling the physics of interaction of high-pressure arcs with their electrodes: advances and challenges *J. Phys. D: Appl. Phys.* **53** 013002

[8] Mitchner M and Kruger C 1973 *Partially Ionized Gases* (New York: Wiley)

[9] Benilov M S 2008 Understanding and modelling plasma-electrode interaction in high-pressure arc discharges: a review *J. Phys. D: Appl. Phys.* **41** 144001

[10] Baeva M, Kozakov R, Gorchakov S and Uhrlandt D 2012 Two-temperature chemically non-equilibrium modelling of transferred arcs *Plasma Sources Sci. Technol.* **21** 055027

[11] Baeva M, Benilov M S, Almeida N A and Uhrlandt D 2016 Novel non-equilibrium modelling of a DC electric arc in argon *J. Phys. D: Appl. Phys.* **49** 245205

[12] Baeva M 2017 A survey of chemical nonequilibrium in argon arc plasma *Plasma Chem. Plasma Process.* **37** 513–30

[13] Raizer Y P 1991 *Gas Discharge Physics* (New York: Springer)

[14] Dreizin E L 2004 Microarc discharge: mechanism and applications *Proceedings of 2nd International Workshop on Microplasmas* (Hoboken, NJ: Stevens Institute of Technology)

[15] Baeva M, Loffhagen D, Becker M M, Siewert E and Uhrlandt D 2020 Plasma parameters of microarcs towards minuscule discharge gap *Contrib. Plasma Phys.* **60** e202000033
[16] Kolobov V I and Arslanbekov R R 2006 Simulation of electron kinetics in gas discharges *IEEE Trans Plasma. Sci.* **34** 895–909
[17] Graves D B and Jensen K F 1986 A continuum model of DC and RF discharges *IEEE Trans. Plasma Sci.* **14** 78–91
[18] Arslanbekov R R 2000 Model of the cathode region and gas temperature of a DC glow discharge at high current density *J. Phys. D: Appl. Phys.* **33** 524–44
[19] Kushner M J 2004 Modelling of microdischarge devices: pyramidal structures *J. Appl. Phys.* **95** 846–59
[20] Farouk T, Farouk B, Gutsol A and Fridman A 2008 Atmospheric pressure radio frequency glow discharges in argon: effects of external matching circuit paramaters *Plasma Sources Sci. Technol.* **17** 035015
[21] Arslanbekov R R and Kolobov V I 2003 Two-dimensional simulations of the transition from Townsend to glow discharge and subnormal oscillations *J. Phys. D: Appl. Phys.* **36** 2986–94
[22] Boeuf J P and Pitchford L C 1995 Two-dimensional model of a capacitively coupled RF discharge and comparisons with experiments in the gaseous electronics conference reference reactor *Phys. Rev.* E **51** 1376–90
[23] Donko Z, Hartmann P and Kutasi K 2006 On the reliability of low pressure DC glow discharge modelling *Plasma Sources Sci. Technol.* **15** 178–86
[24] Hagelaar G J M and Pitchford L C 2005 Solving the Boltzmann equation to obtain electron transport coefficients and rate coefficients for fluid models *Plasma Sources Sci. Technol.* **14** 722–33
[25] Becker M M and Loffhagen D 2013 Enhanced reliability of drift-diffusion approximation for electrons in fluid models of nonthermal plasmas *AIP Adv.* **3** 012108
[26] Moore C 1971 *Atomic Energy Levels* (Washington DC: NSRDS-NBS 35)
[27] Baeva M, Loffhagen D, Becker M M and Uhrlandt D 2019 Fluid modelling of DC argon microplasmas: effects of the electron transport description *Plasma Chem. Plasma Process.* **39** 949–68
[28] Beulens J J, Milojevic D, Schram D C and Valinga M 1991 A two-dimensional non-equilibrium model of cascaded arc plasma flows *Phys. Fluids* **3** 2548–57
[29] Maecker H 1955 Plasmastroemungen in Lichtboegen infolge eigenmagnetischer Kompression *Z. Phys.* **141** 198–216
[30] Baeva M, Loffhagen D and Uhrlandt D 2019 Unified non-equilibrium modelling of tungsten-inert gas microarcs in atmospheric pressure argon *Plasma Chem. Plasma Process.* **39** 1359–78
[31] Multiphysics v 5.6 COMSOL (Stockholm: COMSOL)
[32] Becker M M, Kählert H, Sun A, Bonitz M and Loffhagen D 2017 Advanced fluid modeling and PIC/MCC simulations of low-pressure CCRF discharges *Plasma Sources Sci. Technol.* **26** 044001
[33] Hagelaar G J M and Pitchford L C 2005 Solving the Boltzmann equation to obtain electron transport coefficients and rate coefficients for fluid models *Plasma Sources Sci. Technol.* **14** 722–33
[34] Grubert G K, Becker M M and Loffhagen D 2009 Why the local-mean-energy approximation should be used in hydrodynamic plasma descriptions instead of the local-field approximation *Phys. Rev.* E **80** 036405
[35] Colonna G and D'Angola A 2016 The two-term Boltzmann equation in [67] pp 2–1–34

[36] Loffhagen D 2016 Multi-term and non-local electron Boltzmann equation in [67] pp 3–1–30
[37] Becker M M and Loffhagen D 2013 Derivation of moment equations for the theoretical description of electrons in nonthermal plasmas *Adv. Pure Math.* **3** 343–52
[38] Bird R R 2002 *Transport Phenomena* vol 1 (New York: Wiley)
[39] Surzhikov S T 2016 High-enthalpy radiating flows in aerophysics in [67] pp 12–1–30
[40] Chase M W J 1998 NIST-JANAF thermochemical tables, fourth edition *J. Phys. Chem. Ref. Data* **25** 551
[41] Gurvich L V, Veyts I V and Alcock C B 1989 *Thermodynamic Properties of Individual Substances, Part 2 tables* vol 1 (Washington, DC: Hemisphere)
[42] Neufeld P, Janzen A and Aziz R 1972 Empirical equations to calculate 16 of the transport collision integrals $\Omega^{(l,\,s)*}$ for the Lennard–Jones (12–6) potential *J. Chem. Phys.* **57** 1100–2
[43] Brokaw R S 1969 Predicting transport properties of dilute gases *Ind. Eng. Process. Des. Dev.* **8** 240–53
[44] Vlcek J 1989 A collisional-radiative model applicable to argon discharges over a wide range of conditions. I: Formulation and basic data *J. Phys. D: Appl. Phys.* **22** 623–31
[45] Bogaerts A, Gijbels R and Vlcek J 1998 Collisional-radiative model for an argon glow discharge *J. Appl. Phys.* **84** 121–36
[46] Ridenti M A, Guerra V and Amorim J 2016 Atmospheric pressure plasmas operating in high-frequency fields in [67] pp 11–1–30
[47] Hayashi M 2003 Bibliography of electron and photon cross sections with atoms and molecules published in the 20th century ArgonNIFS-DATA-72 Report
[48] Vriens L and Smeets A H M 1980 Cross-section and rate formulas for electron-impact ionization, excitation, deexcitation, and total depopulation of excited atoms *Phys. Rev.* A **22** 940–51
[49] Kolokolov N B, Kudrjavtsev A A and Blagoev A B 1994 Interaction processes with creation of fast electrons in the low temperature plasma *Phys. Scr.* **50** 371–402
[50] Lymberopoulos D P and Economou D J 1993 Fluid simulations of glow discharges: effect of metastable atoms in argon *J. Appl. Phys.* **73** 3668–79
[51] Cunningham A J, O'Malley T F and Hobson R M 1981 On the role of vibrational excitation in dissociative recombination *J. Phys. B: At. Mol. Phys.* **14** 773–82
[52] Marchenko V S 1983 Dissociation of homonuclear ions by electron impact *Sov. Phys. JETP* **58** 292–8
[53] Lam S, Zheng C-E, Lo D, Dem'ynov A and Napartovoch A P 2000 Kinetics of Ar_2^* in high-pressure pure argon *J. Phys. D: Appl. Phys.* **33** 242–51
[54] Flannery M R and McCann K J 1980 Cross sections for ionization of rare gas excimers by electron impact and atomic and molecular processes in excimer lasers *Technical Report* AFWAL-TR-80-2015
[55] Gregorio J, Leprince P, Bolsse-Laporte C and Alves L L 2012 Self-consistent modelling of atmospheric micro-plasmas produced by a microwave source *Plasma Sources Sci. Technol.* **21** 015013
[56] Bretagne J, Godart J and Puech V 1982 Low-energy electron distribution in an electron-beam generated argon plasma *J. Phys. D: Appl. Phys.* **15** 2205–25
[57] Kannari F, Suda S, Obara M and Fujioka T 1983 Theoretical simulation of electron-beam-excited xenonchloride (XeCl) lasers *IEEE J. Quant. Electr.* **19** 1587–600
[58] Kramida A, Ralchenko Y, Reader J and NIST ASD Team 2018 NIST Atomic Spectra Database

[59] Biberman L M, Vorob'ev V and Yakubov I T 2012 *Kinetics of Nonequilibrium Low-Temperature Plasmas* vol 1 (Berlin: Springer)
[60] Baeva M, Boretskij V F, Gonzalez D, Methling R, Murmantsev O, Uhrlandt D and Veklich A 2021 Unified modelling of low-current short-length arcs between copper electrodes *J. Phys. D: Appl. Phys.* **54** 025203
[61] Murphy E L and Good R H 1956 Thermionic emission, field emission, and the transition region *Phys. Rev.* **102** 1464–73
[62] Baeva M, Methling R and Uhrlandt D 2021 Unified modelling of TIG microarcs with evaporation from copper anode *Plas. Phys. Technol.* **8** 1–4
[63] Baeva M, Uhrlandt D and Loffhagen D 2020 Unified modelling of non-equilibrium microarcs in atmospheric pressure argon: potentials and limitations of one-dimensional models in comparison to two-dimensional models *Jap. J. Appl. Phys.* **59** SHHC05
[64] Tonks L and Langmuir I 1929 A general theory of the plasma of an arc *Phys. Rev.* **34** 876–922
[65] Franklin R N 2003 The plasma-sheath boundary: its history and Langmuir's definition of the sheath edge *J. Phys. D: Appl. Phys.* **36** R309–20
[66] Baeva M, Siewert E and Uhrlandt D 2019 Comparative studies of low-intensity short-length arcs *Plas. Phys. Technol.* **6** 56–9
[67] Colonna G *et al* 2016 *Plasma Modeling: Methods and Applications* 1st edn (Bristol: IOP Publishing)

IOP Publishing

Plasma Modeling (Second Edition)
Methods and applications
Gianpiero Colonna and Antonio D'Angola

Chapter 15

Multiscale phenomenona in a self-organized plasma jet

Florian Sigeneger

15.1 Introduction

Non-thermal plasma jets have become a widespread type of discharge in recent decades. The localized operation at atmospheric pressure and the limited thermal load enable various applications in the fields of, e.g., surface cleaning and activation, deposition of thin functional coatings and treatment of biological surfaces. Overviews on plasma jet types and their applications have been published in several review papers [1–3].

A large variety of plasma jet configurations can be found in the literature. The configurations differ with respect to the arrangement of capillaries, electrodes, the power supply frequency and the gas composition. From the physical point of view, the choice of all these parameters strongly influences the phenomena and the processes which are to be taken into account in a theoretical description [4]. Even the type and electrical properties of the substrate may influence the behaviour of the plasma jet which is directed to this substrate [5]. Model descriptions of plasma jets have to consider the various relevant processes and phenomena such as power deposition, plasma generation, species reactions and transport, gas flow, gas heating and surface reactions. The very different scales in time and space of these processes are often treated by a partition of the models in separate solution steps [6].

Most plasma jets have a cylindrical geometry. A tube guides the working gas from the inlet through the activation zone towards the outlet. Partly, a gas with a different composition is additionally guided to the outlet to shield the working gas [7] or provide precursor molecules [8]. The arrangement of electrodes may differ remarkably leading to very different mechanisms of power supply. Some setups use a pin in the centre of the gas flow as powered electrode leading to large electric fields in its vicinity [9]. In other arrangements, the exposition of electrodes to the gas is avoided using dielectric tubes which separate electrodes and gas [10]. Special

doi:10.1088/978-0-7503-3559-1ch15

arrangements with plane-parallel electrodes have been designed to make the physics as simple as possible and to enable direct investigations of basic mechanisms in the flowing discharge [11].

Various types of power supply are used in plasma jets which largely determine the dominant physical processes. Many plasma jets are driven by pulses in the kHz region. Each pulse ignites a discharge event which is coupled to the previous pulses only due to long-lived species. A series of papers have been published based on the nonPDPSim model consisting of several interacting modules which have been used to investigate, e.g., the formation of reactive oxygen and nitrogen species [12] or the spatio-temporal behaviour of bullets, occurring in such discharges [13].

Also, microwaves are used to supply power to the flowing gas in an electrodeless arrangement. High power densities and the continuous operation may lead to elevated gas temperatures in these applications. A fully coupled 2D model was able to reveal the complex interaction between gas flow, plasma, and electromagnetic field in a microwave driven argon plasma jet [14]. The plasma-induced flow instabilities in a power modulated atmospheric-pressure microwave plasma jet have been described by a self-consistent fluid model [15].

Plasma jets may also be driven by radio frequency (RF) voltages in the MHz range. Apart from few inductively coupled setups [16], RF plasma jets are mostly capacitively coupled devices. From the modelling point of view, these discharges are difficult to describe because of the wide range of time scales to be taken into account. Besides fast processes within the sheath of the RF discharge long-term processes mainly occurring in the bulk require a relaxation over many periods. Simplifications become possible if the setup allows a global model [17] or a one-dimensional description because of plane-parallel electrodes causing an electric field perpendicular to the gas flow [18].

An RF plasma jet is also the subject of this chapter. However, this plasma jet requires at least a two-dimensional model because of the longer distance between the electrodes which are separated in flow direction leading to a filamentary discharge. Further complications arise from self-organizing phenomena which concern stratification of the discharge filaments and a flow-discharge interaction leading to mode-locked phenomena as detailed below. The plasma jet operates in argon. Additionally, supplied precursor leads to deposition of a thin film on a substrate. The theoretical investigations of this plasma jet have been reported in a series of papers [19–24].

This chapter is organized as follows. In section 15.2, the setup of the plasma jet considered is explained and the different model approaches are briefly outlined. In section 15.3, the common equations for the model approaches are given and supplemented by a description of the argon chemistry. In section 15.4 two model approaches including selected results are presented.

15.2 Setup and discharge behaviour

The setup of the plasma jet is shown in figure 15.1 together with computational domains used in different model approaches. It consists of an outer quartz capillary which carries the lower grounded and the upper powered electrode and an inner

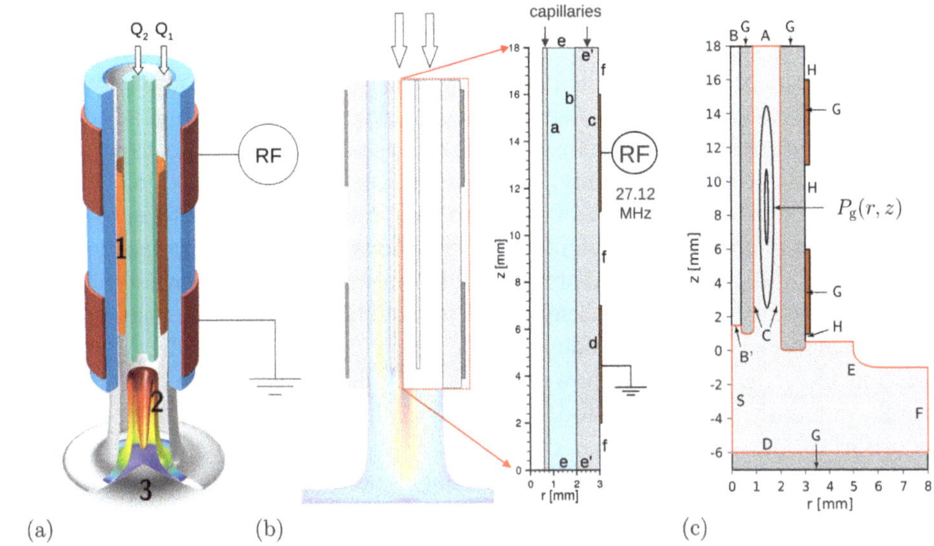

Figure 15.1. Configuration of the plasma jet: (a) overview with the active zone (1), mixing zone (2) and substrate (3); (b) solution domain in the single filament model (section 15.3.1); (c) solution domain in the overall jet model (section 15.3.2). Figures reproduced with permission from (a) reference [22], copyright 2017 John Wiley & Sons; (b) reference [19], copyright 2016 IOP Publishing; reference [20], copyright 2016 IOP Publishing.

ceramic capillary which guides the mixture of argon and precursor with gas flow Q_2 into the mixing zone (2) below the jet. Between inner and outer capillary, the working argon gas flows with the gas flow Q_1 from the inlet towards the mixing zone. It passes the active zone (1) where plasma is sustained due to the RF power supply. Active species are generated in this zone and transported with the gas flow towards the mixing zone and into the effluent. Reactions between the active argon species and the precursor molecules occur in the mixing zone which cause the generation of precursor fragments (PFs). These fragments are guided further with the gas flow to reach finally the substrate surface (3). Here, absorption and subsequent surface reactions lead to material deposition. Using hexamethyldisiloxan (HMDSO) as precursor, a film mostly consisting of SiO_x compounds is deposited on the substrate.

A special feature of the plasma jet considered here is the generation of filaments. While the generation of filaments is a phenomenon which occurs in most plasmas at atmospheric pressure, a special kind of self-organization is observed in this plasma jet [25]. The collective behaviour of the filaments between both electrodes organizes itself in different modes which change with increasing power and flow rate. These modes are characterized by distinct numbers of filaments which are distributed equidistantly to each other in azimuthal direction. Furthermore, stationary modes alternate with so-called locked modes where the filaments incline and move in azimuthal direction with constant velocity. Chaotic behaviour of the filaments is observed during mode transitions. The in-depth analysis of this self-organizing behaviour is beyond this chapter contribution. A partial aspect, namely the relation between the inclination and the azimuthal velocity has been investigated separately [22].

The discharge behaviour of the plasma jet incorporates a very complex interplay of various physical processes such as plasma generation including volume and surface charges, plasma chemistry, transport due to diffusion, convection and migration, gas flow, gas heating, constriction and self-organization. All these processes occur in a wide range of scales in time and space. Therefore, in view of the limitations of current computers, approximations are required, to reach a model description of this plasma jet. Moreover, different models have been established to describe selected aspects of the plasma jet in more detail while approximating others.

Two such models will be considered in the following sections. Model A, described in section 15.4.1, is devoted to power supply and plasma generation during the RF period. An axially symmetric model of a single filament is applied to describe the generation of the argon plasma in the active zone by a model which resolves temporally the RF period of the applied voltage and describes the formation of the filament due to constriction.

Model B, described in section 15.4.2, is devoted to the influence of the gas flow on the plasma generation and species transport. A period-averaged description of the plasma generation allows the extension of the geometry which includes all zones shown in figure 15.1. This becomes possible by using a period-averaged heating profile for electrons which is extracted from results of model A. The interplay of plasma generation, gas flow, precursor kinetics, transport of active species in the effluent and deposition is investigated in this model.

15.3 Model equations

Although partly different variants of the basic equations are used in both models, some common aspects will be presented in this section.

15.3.1 Gas dynamics

The argon gas flowing through the plasma jet is considered as a compressible continuum with the mass density ρ, the mass-averaged flow velocity $\mathbf{u}$ and the temperature T_g. The dynamic behaviour of this continuum can be described by the Navier–Stokes equations

$$\frac{\partial \rho}{\partial t} + \nabla \cdot (\rho \mathbf{u}) = 0, \tag{15.1}$$

$$\rho \frac{\partial \mathbf{u}}{\partial t} + \rho(\mathbf{u} \cdot \nabla)\mathbf{u} = -\nabla p + \nabla \cdot \left(\mu(\nabla \mathbf{u} + (\nabla \mathbf{u})^T) - \frac{2}{3}\mu(\nabla \cdot \mathbf{u})\mathbf{I}\right), \tag{15.2}$$

$$\rho\, c_p \frac{\partial T_g}{\partial t} + \rho\, c_p \mathbf{u} \cdot \nabla T_g + \nabla \cdot \mathbf{q} = S_T, \tag{15.3}$$

which arise from the conservation of mass (equation (15.1)), momentum (equation (15.2)) and energy (equation (15.3)). Here, p and $\mathbf{I}$ denote the pressure and the unity tensor, respectively. The second term on the right-hand side of equation (15.2) arises from the stress tensor assuming the gas to be an isotropic Newtonian viscous fluid with the dynamic viscosity μ. The energy equation (15.3) contains the specific heat capacity at constant pressure c_p, the heat flux $\mathbf{q} = -\lambda \nabla T_g$ with the heat conductivity λ and the energy source S_T. This term includes according to $S_T = S_\epsilon^{(el)} + S_\epsilon^{(h)}$ the heating of the gas due to elastic collisions of electrons with gas atoms and due to reactions between heavy particles as detailed below.

The required coefficients μ and λ for the argon gas are taken from [26] and a value of $\lambda = 1.4$ W m^{-1}K^{-1} [27] is used for the heat conductivity of the capillary quartz glass. To complete the system and relate the continuum with the species densities, the equation of state of the ideal gas $p = N_g k_B T_g$ with $\rho = N_g \sum_l \frac{n_l}{N_g} M_l$ is used where k_B is the Boltzmann constant and n_l an M_l denote density and particle mass of species l, respectively. The boundary conditions for the Navier–Stokes equations will be discussed in the respective subsections.

15.3.2 Plasma description

The plasma model is based on the assumption that the plasma consists of a mixture of m fluids corresponding to the number of species taken into account. In the case of heavy species, Maxwell distribution can be assumed due to the large collisionality at atmospheric pressure and the effective energy transfer in elastic collisions between these species. For the same reason, a common temperature T_g is assumed for all heavy species.

The electron component is not in equilibrium with the heavy species because of the small mass ratio and the resultant weak energy transfer in elastic collisions between electrons and heavy species. Electrons receive energy from the electric field and reach a mean energy which is much larger than that of the heavy species.

While kinetic methods are required to strictly account for the non-equilibrium behaviour of electrons, a hybrid approach became very successful during recent decades. According to this concept, the electron component is treated as a fluid solving macroscopic balance equations derived by integrating moments of the time and space-dependent Boltzmann equation. Transport and rate coefficients occurring in these equations are assumed to depend not directly on time and space but on the mean electron energy u_m and on the ionization degree. To obtain this dependence, the stationary space-independent Boltzmann equation is solved for a series of electric field strength and ionization degrees.

Thus, the system of equations describing the argon plasma comprises the continuity equations

$$\frac{\partial}{\partial t} n_l + \nabla \cdot \mathbf{\Gamma}_l = S_l \tag{15.4}$$

for all species densities n_l with the source term S_l, the electron energy balance equation

$$\frac{\partial}{\partial t} n_\epsilon + \nabla \cdot \mathbf{\Gamma}_\epsilon = e_0 \mathbf{\Gamma}_e \cdot \nabla \phi + S_\epsilon \tag{15.5}$$

for the electron mean energy density $n_\epsilon = u_\mathrm{m} n_\mathrm{e}$ with the energy source term S_ϵ and Poisson's equation

$$-\Delta(\varepsilon\phi) = \sum_l q_l n_l \tag{15.6}$$

for the electrostatic potential ϕ with the permittivity ε and the species volume charge q_l.

The particle fluxes $\mathbf{\Gamma}_l$ and the electron energy flux $\mathbf{\Gamma}_\epsilon$ are driven according to

$$\mathbf{\Gamma}_l = \mathrm{sgn}(q_l)\mu_l n_l \mathbf{E} - D_l \nabla n_l + n_l \mathbf{u} \tag{15.7}$$

$$\mathbf{\Gamma}_\epsilon = -\mu_\epsilon n_\epsilon \mathbf{E} - D_\epsilon \nabla n_\epsilon + n_\epsilon \mathbf{u} \tag{15.8}$$

by drift (first term), diffusion (second term) and convection (third term). The corresponding transport coefficients are the particle mobility μ_l and the diffusion coefficient D_l as well as the corresponding coefficients μ_ϵ and D_ϵ for the electron energy transport. The representations (15.7) and (15.8) are obtained from the momentum balance equation neglecting the inertia of the particles. This assumption is justified at atmospheric pressure.

The source terms S_l in equation (15.4) include the gain and loss rates due to electron impact and heavy particle collisions as well as radiation. The energy source term $S_\epsilon = S_\epsilon^{(2)} + S_\epsilon^{(3)} - S_\epsilon^{(\mathrm{el})}$ in equation (15.5) includes the energy change rate in two-body collisions $S_\epsilon^{(2)} = \sum_j \Delta\epsilon_j k_j n_{l_1(j)} n_{l_2(j)}$, in three-body collisions $S_\epsilon^{(3)} = \sum_j \Delta\epsilon_j k_j n_{l_1(j)} n_{l_2(j)} n_{l_3(j)}$ and the energy loss rate in elastic collisions $S_\epsilon^{(\mathrm{el})} = n_\mathrm{e} N_\mathrm{g} \tilde{k}_\mathrm{el}$. Here, $\Delta\epsilon_j$ and k_j denote the energy change and the collision rate coefficient of the collision process, respectively. Furthermore, $\tilde{k}_\mathrm{el}$ represents the rate coefficient for the energy loss in elastic electron collisions. The heat source due to reactions between heavy particles contributing to equation (15.3) is given by $S_\epsilon^{(\mathrm{h})} = \sum_j \Delta h_j$, where Δh_j denotes the enthalpy change in reaction j.

The rate coefficients of electron–atom collisions and the particle transport coefficients D_e and μ_e are determined by solving the electron Boltzmann equation as described above. The corresponding transport coefficients μ_ϵ and D_ϵ for the energy flux are evaluated approximately according to $\mu_\epsilon = \frac{5}{3}\mu_\mathrm{e}$ and $D_\epsilon = \frac{5}{3}D_\mathrm{e}$ from the particle transport coefficients [28].

A further equation is required to determine the surface charge density σ_s at the surface of the dielectric capillary which causes the jump $\varepsilon_1 E_1 - \varepsilon_2 E_2 = \sigma_s$ of the normal component of the electric field strength E from the plasma (1) to the dielectrics domain (2). The surface charge is obtained by integrating according to

$$\sigma_s(t) = \sum_l q_l \int_0^t (\mathbf{n} \cdot \mathbf{\Gamma}_l(\tilde{t}))d\tilde{t} \tag{15.9}$$

the flux of positive and negative charge carriers towards the dielectric surface.

15.3.3 Argon plasma chemistry

When selecting the list of heavy species, a compromise between the request for maximal information and the limitations caused by computing capabilities and acceptable calculation times must be found.

For argon such a collection of dominant species comprises atomic ions (Ar^+), atoms excited in metastable (Ar^m), resonance (Ar^r) and a higher lumped (Ar^p) state. Because of the atmospheric pressure also molecular species such as molecular ions (Ar_2^+) and excimers in triplet ($Ar_2(^3\Sigma_u^+)$) and singlet state ($Ar_2(^1\Sigma_u^+)$) must be taken into account. The reaction kinetics of these species is summarized in table 15.1. It includes elastic collisions, 12 exciting and de-exciting and 6 ionizing collision processes between electrons and neutral species. The rate coefficients of these processes are calculated in advance by solving the electron Boltzmann equation as

Table 15.1. Reactions included in the plasma model. The dimensions of the rate coefficients are [m^3/s] for two-body reactions, [m^6/s] for three-body reactions and [s^{-1}] for radiation processes. The electron temperature $T_e = (2/3)u_m$ and the heavy particle temperature T_g are specified in [K]. Reproduced with permission from [20]. Copyright 2016 IOP.

No.	Reaction	Rate coefficient	Reference
	Elastic collisions		
	$Ar + e \rightarrow Ar + e$	$f(u_m)$	[29]
	Excitation and de-excitation		
1	$Ar + e \rightarrow Ar^m + e$	$f(u_m)$	[30, 31]
2	$Ar^m + e \rightarrow Ar + e$	$f(u_m)$	[30, 31]
3	$Ar + e \rightarrow Ar^r + e$	$f(u_m)$	[30, 31]
4	$Ar^r + e \rightarrow Ar + e$	$f(u_m)$	[30, 31]
5	$Ar + e \rightarrow Ar^p + e$	$f(u_m)$	[30, 31]
6	$Ar^p + e \rightarrow Ar + e$	$f(u_m)$	[30, 31]
7	$Ar^m + e \rightarrow Ar^r + e$	$f(u_m)$	[30, 31]
8	$Ar^r + e \rightarrow Ar^m + e$	$f(u_m)$	[30, 31]
9	$Ar^m + e \rightarrow Ar^p + e$	$f(u_m)$	[30, 31]
10	$Ar^p + e \rightarrow Ar^m + e$	$f(u_m)$	[30, 31]
11	$Ar^r + e \rightarrow Ar^p + e$	$f(u_m)$	[30, 31]
12	$Ar^p + e \rightarrow Ar^r + e$	$f(u_m)$	[30, 31]
	Ionization		
13	$Ar + e \rightarrow Ar^+ + 2e$	$f(u_m)$	[32]
14-16	$Ar^i + e \rightarrow Ar^+ + 2e$, $i =$ m,r,p	$f(u_m)$	[33]
17	$Ar_2(^1\Sigma_u^+) + e \rightarrow Ar_2^+ + 2e$	$f(u_m)$	[34]
18	$Ar_2(^3\Sigma_u^+) + e \rightarrow Ar_2^+ + 2e$	$f(u_m)$	[34]
	Three-body recombination		
19	$Ar^+ + 2e \rightarrow Ar^p + e$	$f(u_m)$	[28]
	Ion conversion		
20	$Ar^+ + 2Ar \rightarrow Ar_2^+ + Ar$	$2.25 \times 10^{-43}(T_g/300)^{-0.4}$	[35]

(*Continued*)

Table 15.1. *(Continued)*

No.	Reaction	Rate coefficient	Reference
	Dissociative recombination		
21	$Ar_2^+ + e \rightarrow Ar^p + Ar$	$8.1 \times 10^{-13}(T_e/300)^{-0.64}$ $\times(T_g/300)^{-0.86}$	[36]
	Neutral association		
22	$Ar^m + 2Ar \rightarrow Ar_2(^3\Sigma_u^+) + Ar$	8.3×10^{-45}	[37]
23	$Ar^m + 2Ar \rightarrow Ar_2(^1\Sigma_u^+) + Ar$	2.5×10^{-44}	[37]
24	$Ar^r + 2Ar \rightarrow Ar_2(^3\Sigma_u^+) + Ar$	8.3×10^{-45}	[37]
25	$Ar^r + 2Ar \rightarrow Ar_2(^1\Sigma_u^+) + Ar$	2.5×10^{-44}	[37]
	Chemo ionization		
26	$Ar^m + Ar^m \rightarrow Ar^+ + Ar + e$	1.3×10^{-15}	[38]
27	$Ar^r + Ar^r \rightarrow Ar^+ + Ar + e$	5×10^{-16}	[38]
28	$Ar^r + Ar^m \rightarrow Ar^+ + Ar + e$	5×10^{-16}	[38]
29	$Ar^p + Ar^p \rightarrow Ar^+ + Ar + e$	5×10^{-16}	[39]
30	$Ar^p + Ar^m \rightarrow Ar^+ + Ar + e$	5×10^{-16}	[39]
31	$Ar^p + Ar^r \rightarrow Ar^+ + Ar + e$	5×10^{-16}	[39]
32-34	$Ar^i + Ar_2(^3\Sigma_u^+) \rightarrow Ar_2^+ + Ar + e$, i= m,r,p	5×10^{-16}	[39]
35	$Ar_2(^3\Sigma_u^+) + Ar_2(^3\Sigma_u^+) \rightarrow Ar_2^+ + 2Ar + e$	5×10^{-16}	[39]
	Spontaneous radiation		
36	$Ar^p \rightarrow Ar^m + h\nu$	1.91×10^7	[40, 41]
37	$Ar^p \rightarrow Ar^r + h\nu$	1.38×10^7	[40, 41]
38	$Ar^r \rightarrow Ar + h\nu$	2.01×10^6 (see text)	[40–42]
39	$Ar_2(^1\Sigma_u^+) \rightarrow 2Ar + h\nu$	2.38×10^8	[40, 41]
40	$Ar_2(^3\Sigma_u^+) \rightarrow 2Ar + h\nu$	3.13×10^5	[40, 41]

detailed above. Recombination in three-body collisions and dissociative recombination represent electron loss processes.

Molecular species are generated in collisions between heavy particles inducing ion conversion and neutral association. Chemo ionization reactions lead to the generation of atomic ions and the reaction enthalpies of these reactions contribute to the gain term S_ϵ in the electron energy balance equation (equation (15.5)).

The list of reactions contain five radiation processes from excited atomic and excimer states. The rate coefficients of these processes were taken from literature as indicated in table 15.1. For the radiation at $\lambda_0 = 106$ nm from resonance atoms Ar^r with a transition probability of $k_0 = 3.15 \times 10^8$ s^{-1}, radiation trapping must be taken into account. This is considered approximately by an effective lifetime according to Holstein's theory for a cylindrical plasma [42].

15.3.4 Solution method

Both parts of the model have been solved using Comsol Multiphysics [43]. Triangular grids with boundary layers near walls are used as mesh for the solution regions of the flow and plasma parts. The equation system of the flow part is solved applying a stationary solver and a temporal relaxation process is solved to determine the periodic state of the plasma equations in model A and the stationary state in model B. A logarithmic representation of the particle densities is used to avoid negative densities during the solution.

15.4 Plasma jet models

15.4.1 Single filament model

The aim of the single filament model is to describe the plasma generation in the active volume caused by the RF voltage between the powered and grounded electrodes. This task is separated from the overall modelling of the plasma jet as described in section 15.3.2 because of the large computational cost of the phase-resolved modelling of the RF discharge. The period-averaged spatial profile of the electron heating obtained from the single filament model is used to avoid the phase resolution in the overall jet model B.

Therefore, the solution domain is restricted to a cylindrical section of the plasma jet as shown in figure 15.1(b). Furthermore, the gas flow is neglected in the framework of this model. However, the heating of the gas is included to account for the considerably power supply.

The system of continuity equations (15.4), electron energy balance equation (15.5) and surface charge equation (15.9) is solved in the rectangular domain surrounded (in clockwise direction) by boundaries a,e,b,e. Furthermore, Poisson's equation (15.6) and the heat equation (15.3) are solved in the larger domain surrounded by boundaries a,e,e′,f,c,f,d,f,e′,e.

Appropriate conditions are required at the boundaries marked in figure 15.1(b). At the boundaries a, c, d and f, the fixed temperature $T_g = 450$ K is used and the zero gradient condition $(\partial T_g)/(\partial z) = 0$ is applied at the boundaries e and e′. The RF voltage $V = V_0 \sin(2\pi f t)$ with the amplitude $V_0 = 3$ kV and the frequency $f = 27.12$ MHz is prescribed at the powered electrode c while $V = 0$ is used at the grounded electrode d. The zero gradient condition $\mathbf{n} \cdot \nabla V = 0$ is applied along the remaining outer boundaries e, e′ and f.

Because of the demanding long-term calculation, the set of species and reaction is reduced. Thus, all excited argon species are lumped in one species Ar* and both excimer species are lumped in one species Ar_2^*, too. De-excitation is used at boundary condition for these species at the walls as well as recombination for the ions Ar^+ and Ar_2^+. Insulation boundary condition $\mathbf{n} \cdot \Gamma_l = 0$ is used at boundaries e for all species and similarly for the electron energy according to $\mathbf{n} \cdot \Gamma_\epsilon = 0$.

Selected results of this model are shown in figures 15.2–15.4. All these figures depict the establishment of a filament which carries the current between both electrodes and is characterized by constriction and stratification. Pronounced maxima and minima with modulation degree up to 90% can be observed along the trace of the filament. A distinct separation between the constricted stratified filament and the sheath region can be observed from the logarithmic plot of the electron density in figure 15.2 and the linear plots

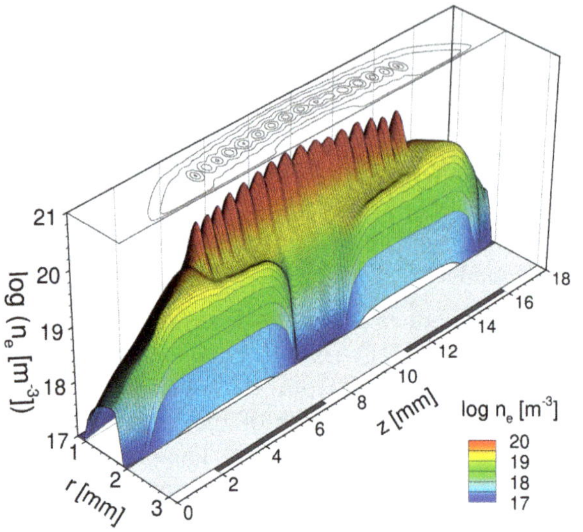

Figure 15.2. Period-averaged electron density in the filament. The dark grey rectangles mark the electrodes which are attached to the outer capillary, coloured in light grey. Reproduced with permission from [19]. Copyright 2016 IOP Publishing.

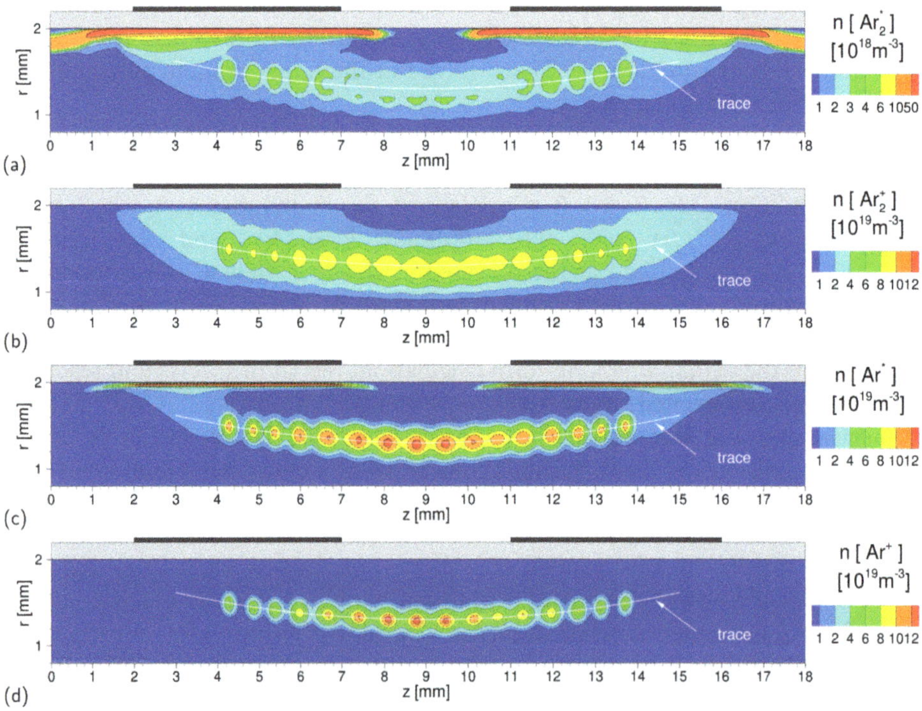

Figure 15.3. Period-averaged particle densities in the filament. Reproduced from [19]. Copyright 2016 IOP Publishing.

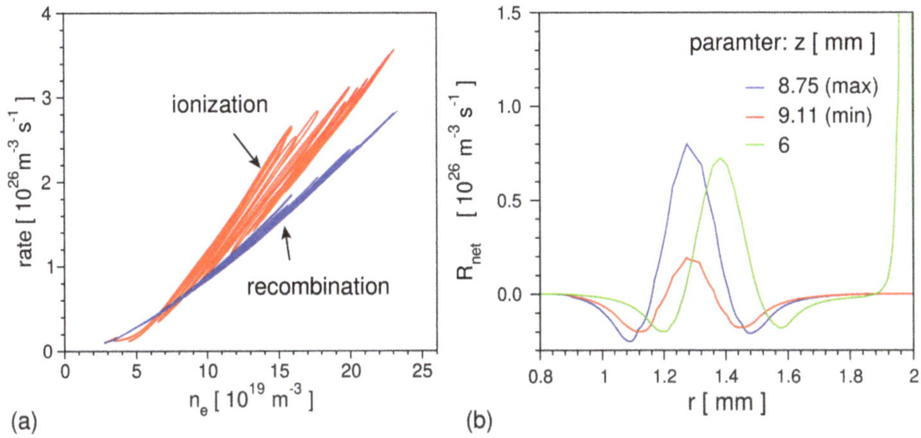

Figure 15.4. Ionization and recombination rates along the filament trace as function of the electron density (a) and radial profiles of the net ionization rate at striation maximum (8.75 mm) and minimum (9.11 mm) positions and at a position in front of the grounded electrode (6 mm) (b). Left frame reproduced with permission from [19]. Copyright 2016 IOP Publishing.

of the excited species densities in figure 15.3(c) and figure 15.3(d). Large densities of excited species are obtained in the vicinity of the electrodes because of the large electric field there.

The generation of striations in the filament is confirmed by experimental observations. This kind of self-organization is caused by the nonlinear dependence of the ionization frequency on the electron density as detailed in the following. Some information regarding this phenomenon is given in figure 15.4. Along the trace of the filament, the ionization (red) and recombination (blue) rates are plotted as functions of the electron density in figure 15.4(a). This figure illustrates the nonlinear dependence of the net ionization rate and the positive feedback. In figure 15.4(b), the radial profile of the net ionization rate is shown for selected axial positions. The net ionization rate in the striation maximum (z = 8.75 mm) is much larger than in the minimum (z = 9.11 mm) and is negative in the wings. This is again an indication for the mechanism of the striation generation: Due to the positive feedback, i.e. increased ionization with increased electron density, emerging inhomogeneities are enhanced. Furthermore, the negative net recombination (i.e. dominant recombination) in the wings enhances the radial constriction of the filament. This effect acts in a similar manner in axial and radial direction which leads to the pronounced structures around the ends of the filament, as shown in figure 15.3. The third curve in figure 15.4(b) shows the radial profile of the net ionization rate in front of the grounded electrode where the pronounced ionization rate in the sheath becomes visible.

Striations of this type should be distinguished from striations which are generated at lower pressures and currents. Under these conditions, a very different mechanism, namely the nonlocal response of electrons to disturbances of the electric field, leads to the generation of spatially periodic structures [44].

From the spatial profile of the period-averaged power density in the single filament the analytic representation

$$P^{\mathrm{pa}}(r, z) = P^{\mathrm{pa}}_{\max}\, \Omega(r)\Theta(z), \quad \Theta(z) = \left(1 + \cos\left(\frac{2\pi}{L_f}\delta(z)\right)\right)/2,$$
$$\delta(z) = \min(|z - z_0|, L_f/2), \quad \Omega(r) = (e^{-(r-r_0)^2/\sigma_1^2} + e^{-(r-r_0)^2/\sigma_2^2})/2 \tag{15.10}$$

with $L_f = 15$ mm, $r_c = 1.45$ mm, $z_c = 8.5$ mm is extracted to be used in the overall period-averaged plasma jet model as described in the following section.

15.4.2 Period-averaged plasma jet model

The spatial profile $P^{\mathrm{pa}}(r, z)$ of the period-averaged power density from the single filament model is used in the period-averaged plasma jet model to investigate the interplay between plasma generation, gas flow and heating and the reactions of active argon species with precursor molecules in the mixing zone. The overall solution domain as shown in figure 15.1(c) comprises the plasma jet, the mixing zone, the effluent and the substrate which is situated at a distance of 6 mm from the jet capillary.

The period-averaged plasma jet model can be divided into a flow and a plasma part. In the flow part, equations (15.1) and (15.2) are solved in the light grey shaded domain with boundaries S,B,C,A,C,E,F,D. Furthermore, the heat balance equation (15.3) is solved in the extended domain including the dark grey shaded domains with boundaries S,B,G,A,G,H,E,F,G. In addition, the transport of the precursor molecules is described. Through the inner capillary (boundary B), a mixture of argon and precursor (also called monomer) is supplied with a flow rate of Q_2. It reaches the mixing zone after passing boundary B′. As precursor (Pc), the organosilicon compound hexamethyldisiloxane (HMDSO) with the formula $Me_3SiOSiMe_3$ ($Me \equiv CH_3$) is used which is a common precursor for thin film deposition. The transport of the precursor molecules with the gas flow is described by the continuity equation

$$\frac{\partial n_{\mathrm{Pc}}}{\partial t} + \nabla \cdot (-D_{\mathrm{Pc}} \nabla n_{\mathrm{Pc}} + \mathbf{u}\, n_{\mathrm{Pc}}) = 0 \tag{15.11}$$

where n_{Pc} denotes the particle density of the precursor molecules. The diffusion coefficient of the precursor in the buffer gas argon $D_{\mathrm{Pc}} = 1.366 \times 10^{20}\ \mathrm{m}^{-1}\ \mathrm{s}^{-1}/N_g$ has been determined using Fuller's method [45].

An improvement is undertaken to approximately account for the string-like structure of the filament in the gas temperature equation (15.3) which would be otherwise neglected in the framework of the axially symmetric geometry. Therefore, according to $S_T = S_\epsilon^{(\mathrm{el})} + S_\epsilon^{(\mathrm{h})} + S_\epsilon^{(\mathrm{azi})}$, the action of the additional heat flux from the filament in azimuthal direction is included as a loss contribution $S_\epsilon^{(\mathrm{azi})} \approx -\lambda(\partial^2 T_g/\partial r^2)$ in the source term of (equation (15.5)) assuming that the profiles of temperature and heat flux in the vicinity of the filament centre are similar in radial and azimuthal directions.

In the plasma part, the continuity equations (15.4), electron energy balance equation (15.5), surface charge equation (15.9) and Poisson's equation (15.6) are solved in the domain which is marked in figure 15.1(c) by the red boundaries S,B′,C,A, C,E,F,D. The RF voltage as boundary condition at the powered electrode is omitted. Instead of that, the spatial profile $P^{\mathrm{pa}}(r, z)$ of the period-averaged power density according to (equation (15.10)) is inserted as a gain contribution in the source term S_ϵ of the electron energy balance equation (15.5) and the potential determined from solving Poisson's equation reflects the influence of space charges in the discharge.

The reactions which are initiated by collisions of HMDSO with active argon species in the mixing zone are highly branched and complex [46]. Several studies can be found in the literature which contribute to the understanding of the main reaction paths [47–49]. To limit the number of species and the difficulty of obtaining data, only a simplified model is applied here. The considered reactions are listed in table 15.2. Only lumped neutral and charged reaction products (RP, RP^+) are taken into account representing the large variety of generated species.

Because of the mostly unidirectional coupling between both parts, the flow part with equations (15.1)–(15.3) is solved as a stationary equation system using initially $P^{\mathrm{pa}}(r, z)$ from (equation (15.10)) as source in the heat balance equation (equation (15.3)). The flow part delivers the flow velocity $\mathbf{u}$, the gas temperature T_g and the precursor density n_{Pc}. These quantities are used in the plasma part to solve the time-dependent equation system equations (15.4)–(15.9) which delivers the species densities n_l, the mean electron energy u_m, the potential V and the surface charge density σ_s. Because of the slight influence via the energy source term $S_\epsilon^{(\mathrm{h})}$ of the plasma results to the flow part an iteration of both parts is required which is, however, converged after few iterations.

Table 15.2. Reactions between precursor molecules and active argon species. The electron temperature is specified in [K]. Reproduced with permission from [20]. Copyright 2016 IOP.

No.	Reaction	Rate coefficient [m^3/s]	Reference
	Collisions with ions		
41	$Pc + Ar^+ \rightarrow PF^+ + Ar$	2×10^{-15}	[50]
42	$Pc + Ar_2^+ \rightarrow PF^+ + 2Ar$	1.4×10^{-15}	[51]
	Collisions with excited atoms		
43	$Pc + Ar^m \rightarrow PF^+ + 2Ar + e$	2.5×10^{-16}	[52]
44	$Pc + Ar^r \rightarrow PF^+ + 2Ar + e$	2.5×10^{-16}	[52]
45	$Pc + Ar^p \rightarrow PF^+ + 2Ar + e$	2.5×10^{-16}	[52]
46	$Pc + Ar^m \rightarrow PF + Ar$	2.5×10^{-16}	[53]
47	$Pc + Ar^r \rightarrow PF + Ar$	2.5×10^{-16}	[53]
48	$Pc + Ar^p \rightarrow PF + Ar$	2.5×10^{-16}	[53]
	Collisions with excimers		
49	$Pc + Ar_2(^3\Sigma_u^+) \rightarrow PF + 2Ar$	5×10^{-16}	estimated
50	$Pc + Ar_2(^1\Sigma_u^+) \rightarrow PF + 2Ar$	5×10^{-16}	estimated
	Recombination		
51	$PF^+ + e \rightarrow PF$	$3.6 \times 10^{-12}(T_e/300)^{-0.7}$	[50]

Boundary conditions

Several boundary conditions are imposed. For all quantities, axial symmetry is applied at the symmetry axis (boundary S in figure 15.1(c)). With respect to the gas flow, two inlets are to be taken into account. The inflow of argon gas at boundary A is prescribed by the parabolic velocity profile $-v_z = 6s_1(1 - s_1)\langle v\rangle_1$ and the inflow of the argon HMDSO mixture at boundary B is prescribed by the parabolic velocity profile $-v_z = 2(1 - s_2^2)\langle v\rangle_2$ at boundary B with the averaged velocities $\langle v\rangle_1$ and $\langle v\rangle_2$ determined from the corresponding flow rates. Here, s_1 and s_2 denote normalized space coordinates running from zero to one across the inlet boundaries A and B. The no slip condition is used at all capillary walls and at the substrate. The pressure is fixed to atmospheric pressure $p_0 = 1.013 \times 10^5$ Pa at the outlet boundaries (E,F).

The gas temperature is fixed to $T_0 = 300$ K at all outer walls G and at the outlet boundaries (E,F). A zero gradient condition $\mathbf{n} \cdot \nabla T_g = 0$ is applied at the inlet boundaries A and B and ambient radiation is used as condition at the boundaries H according to $-\mathbf{n} \cdot \mathbf{q} = \varepsilon_s \sigma_{sb}\left(T_{amb}^4 - T_g^4\right)$ where ε_s and σ_{sb} denote the surface emissivity of quartz and the Stefan–Boltzmann constant, respectively. For the ambient temperature, a value of $T_{amb} = 300$ K is used. The precursor concentration is set equal to 1 % at boundary B.

In the plasma model (equations (15.4)–(15.6)), zero field and zero gradient conditions for the heavy particles and the electron particle and energy flux are applied at the inlet (A,B′) and outlet boundaries (E, F). The potential at the substrate surface (D) is grounded. The surface charge condition $\varepsilon\, \mathbf{n} \cdot \nabla\phi = \sigma_s$ is used at the walls of the capillaries.

With respect to plasma species, boundary conditions for the particle fluxes $\mathbf{\Gamma}_e$ according to [54]

$$\mathbf{n} \cdot \mathbf{\Gamma}_l = \frac{1 - r_l}{1 + r_l}\left(|\mathbf{n} \cdot \mathbf{v}_l| n_l + \frac{1}{2} v_l^{th} n_l\right) + \delta_{le}\mathbf{\Gamma}_{sec} \tag{15.12}$$

with the thermal velocity $v_l^{th} = \sqrt{8k_B T_l/(\pi m_l)}$ and the directed velocity $\mathbf{v}_l$ of the species l caused by migration and convection are applied. In the case of electrons, secondary electron emission flux $\mathbf{\Gamma}_{sec} = -\frac{2\gamma}{1 + r_e}\sum_i \max(\mathbf{\Gamma}_i \cdot \mathbf{n}, 0)$ due to the fluxes $\mathbf{\Gamma}_i$ of the ions Ar^+, Ar_2^+ and PF^+ onto the surfaces is taken into account. Values of 0.07 and 0.02 are used for the secondary emission coefficient γ at the grounded substrate [27, 32] and the dielectric surfaces [55], respectively. However, the secondary electron emission is not very important in the framework of the period-averaged plasma jet model because the plasma is mainly sustained here due to the power source P^{pa} provided to the electrons. For the electron reflection coefficient r_e the values 0.7 and 0.3 are used at the grounded substrate and the dielectric surfaces, respectively [56]. The analogous condition for the electron energy flux $\mathbf{\Gamma}_\epsilon$

$$\mathbf{n} \cdot \mathbf{\Gamma}_\epsilon = \frac{1 - r_e}{1 + r_e}\left(|\mathbf{n} \cdot \mathbf{v}_\epsilon| n_\epsilon + \frac{2}{3} v_e^{th} n_\epsilon\right) - u_m^{wall}\mathbf{\Gamma}_{sec} \tag{15.13}$$

is applied at the walls C and D. The energy of emitted electrons is set to $u_m^{wall} = 1$ eV.

Results

Results for the period-averaged plasma jet model will be shown for a flow rate of $Q_1 = 0.5$ slm of the main argon supply between inner and outer capillary which corresponds to a maximum velocity of 1.25 m s^{-1} at inlet A. The flow rate of the argon precursor mixture through the inner capillary amounts to $Q_2 = 0.25$ slm in the reference case which is related to a maximum velocity of about 16.6 m s^{-1} at inlet B. In addition to the reference case $Q_2/Q_1 = 0.5$, results are shown also for varying flow rate ratios $Q_2/Q_1 = 0.1$ and $Q_2/Q_1 = 1$ in the following figures. The maximal power density of the power profile (equation (15.10)) is set to $P^{\mathrm{pa}}_{\mathrm{max}} = 10^9$ Wm^{-3}. The concentration of HMDSO molecules in the inlet flow Q_1 amounts to 1%.

The gas temperature and the gas velocity for the reference case is shown in figure 15.5(a) and figure 15.6(a). The influence of the flow rate through the inner capillary on both quantities is depicted by comparisons of radial and axial cuts in figure 15.5(b,c) and figure 15.6(b,c). The gas temperature reaches its maximum of about $T_g = 1220$ K at an axial position which is somewhat shifted in downstream direction from the plasma centre due to convection. The effect of convection is more pronounced in the inner capillary where the larger velocity leads to a remarkable decrease of gas temperature in the centre due to convective cooling. This effect becomes visible in figure 15.5(b,c) comparing the results for different flow rates Q_2. Especially, the axial temperature profile at $r = 0$ is strongly affected by the flow through the inner capillary.

The strongly inhomogeneous gas velocity, shown in figure 15.6(a) is characterized by remarkable differences between the gas flow in the outer and inner capillary and by the action of the stagnation point at $r = 0$, $z = -6$ mm. The comparison of the cuts for the different cases in figure 15.6(b,c) illustrates the influence of the gas flow

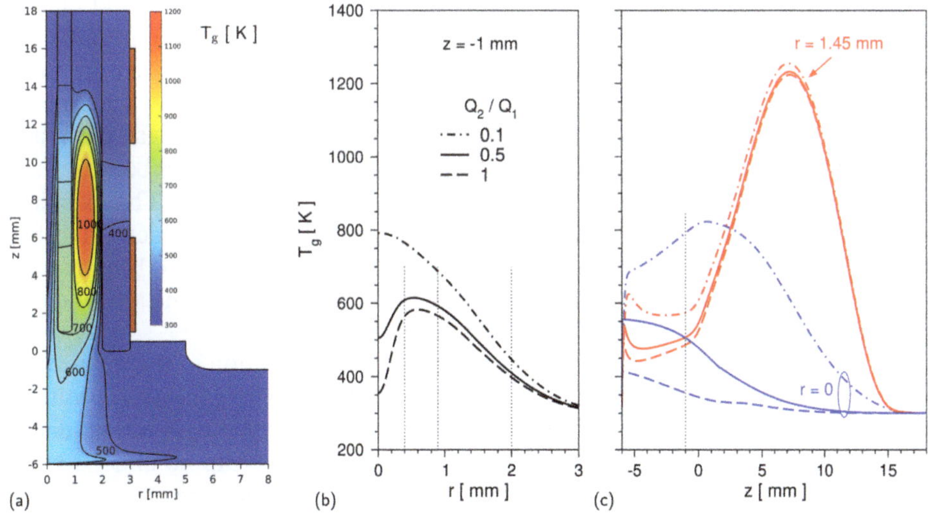

Figure 15.5. Gas temperature in the jet and effluent (a) and along radial (b) and axial (c) cuts. The results for the reference case $Q_2/Q_1 = 0.5$ are compared with those obtained with modified ratios $Q_2/Q_1 = 0.1$ and $Q_2/Q_1 = 1$.

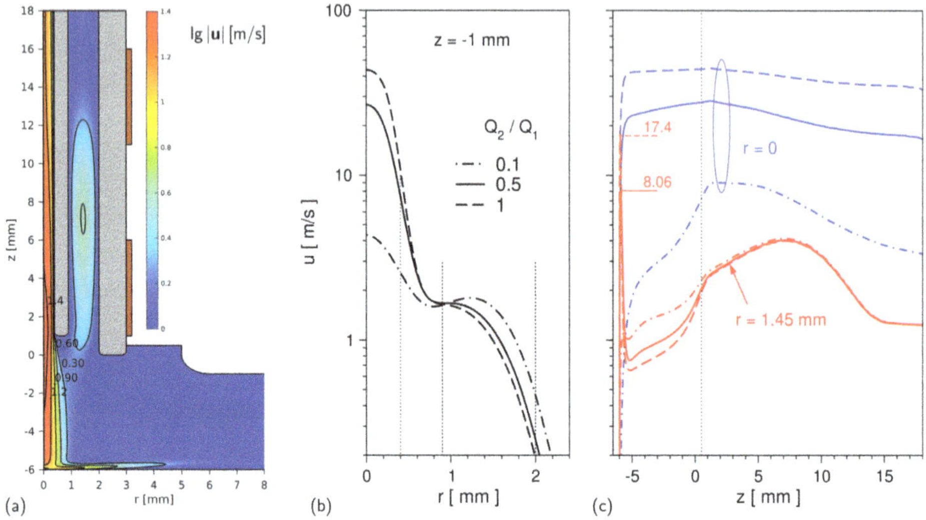

Figure 15.6. Gas velocity in the jet and effluent (a) and along radial (b) and axial (c) cuts. The results for the reference case $Q_2/Q_1 = 0.5$ are compared with those obtained with modified ratios $Q_2/Q_1 = 0.1$ and $Q_2/Q_1 = 1$.

rate Q_2. With increasing flow rate Q_2, the axial profile at $r = 0$ becomes more flat and the outward directed radial flow velocity from the stagnation point near the substrate surface increases reaching maximal values of $u_r = 8.06$ m s^{-1} and $u_r = 17.4$ m s^{-1} at $Q_2/Q_1 = 0.5$ and $Q_2/Q_1 = 1$, respectively. Despite the larger gas velocity at higher Q_2 values, the gas flow through both capillaries and in the effluent remains laminar with maximal Reynolds numbers below 200.

The particle densities of selected argon species and of the neutral and charged precursor fragments are shown for the reference case $Q_2/Q_1 = 0.5$ as contour plots in figures 15.7 and 15.9. Furthermore, axial and radial cuts of these densities are shown in figure 15.8. The dominant species in the active zone with maximal densities of about 4×10^{19} m^{-3} are electrons and Ar_2^+ ions. As typical for inert gas plasmas at atmospheric pressure, the molecular ions are the dominant ion species because of the rapid conversion from atomic ions by reaction 20 (see table 15.1). The calculated density of charge carriers is somewhat smaller than the values which have been obtained by spectroscopic methods [57] in the range between 2.2×10^{20} m^{-3} and 3.3×10^{20} m^{-3}. The deviation may be caused by the limitations of the model caused by the axially symmetric geometry.

The electrons and Ar_2^+ ions are also dominant in the transition to the mixing zone and in the effluent with densities up to 10^{16}m^{-3}. Contrary to this, the axial profiles of the excited species are mainly restricted to the active zone with rapid density gradients towards the mixing zone. This different behaviour can be understood by analysing the main reaction channels. The molecular ions are predominantly generated due to ionizing collisions of electrons with the excimer molecules $Ar_2(^3\Sigma_u^+)$. The total production rate of Ar_2^+ is larger than its loss rate due to dissociative recombination by a factor of about five. This net production is the

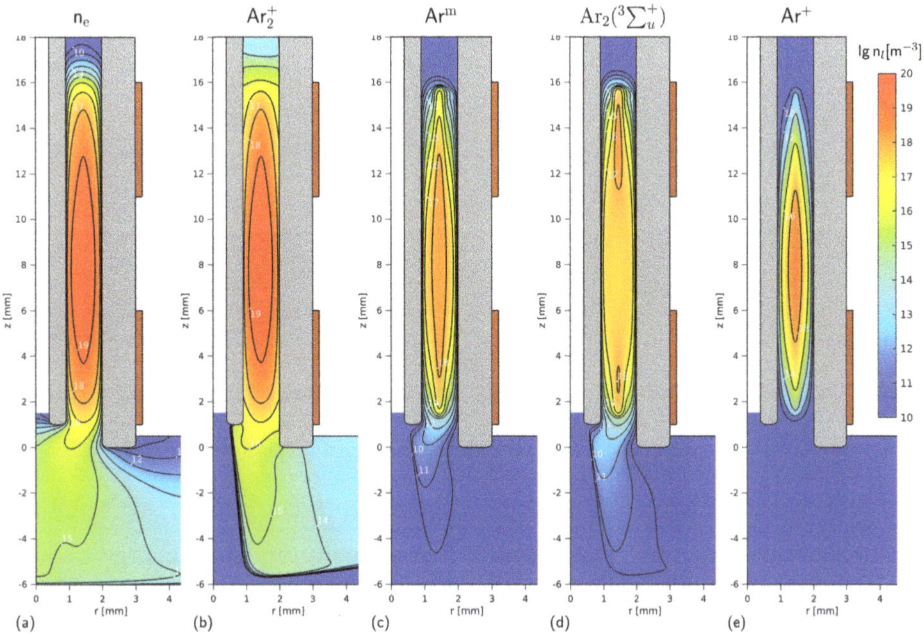

Figure 15.7. Particle densities of electrons (a), molecular ions (b), metastable atoms (c), excimers (d) and atomic ions (e) in the jet and effluent for the reference case $Q_2/Q_1 = 0.5$.

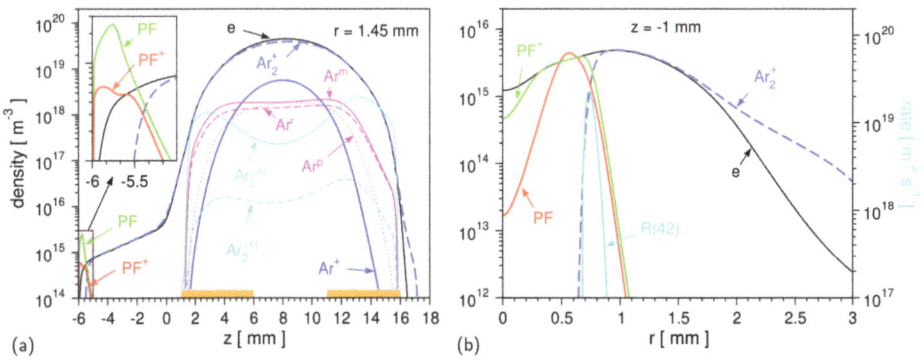

Figure 15.8. Particle densities along axial (a) and radial cuts (b). The bars in the bottom of the left figure mark the position of the electrodes.

reason for the outflow of Ar_2^+ and electrons into the effluent, whereas the atomic ions and the excited species are generated and destroyed more or less locally.

Radial profiles of particle densities in the effluent are shown in figure 15.8(b) with values reduced by about four orders of magnitude compared to figure 15.8(a). At about $r = 1$ mm, a pronounced separation between the inner region with dominant precursor fragment ions PF^+ and the outer region containing the molecular argon

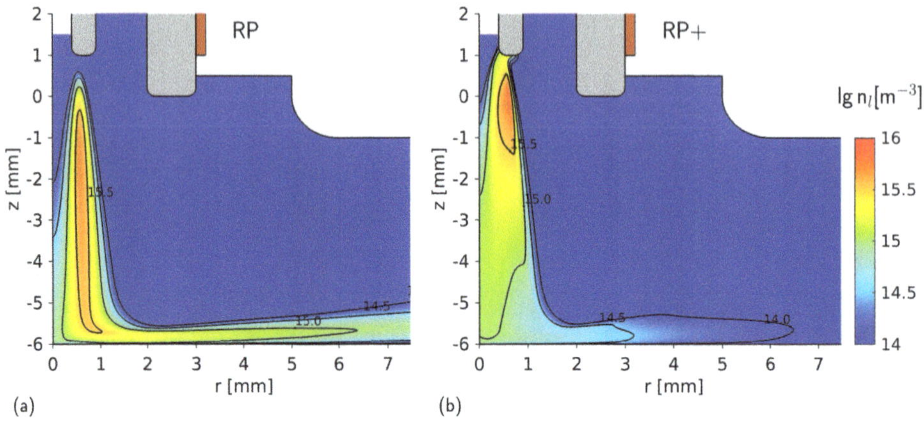

Figure 15.9. Particle densities of neutral (a) and charged (b) precursor fragments.

ions Ar_2^+ can be observed. The reason for this separation is the charge exchange reaction 42 (see table 15.2). This reaction is restricted to the small region where the radially strongly inhomogeneous profiles of Pc and Ar_2^+ overlap. Consequently, the rate of this reaction has a sharp radial profile with a pronounced maximum of about 2×10^{19} m^{-3} s^{-1}at $r \approx 0.75$ mm as shown in figure 15.8(b). The depopulation of Ar_2^+ ions due to this reaction in the inner part of the effluent becomes visible in figure 15.7(b), too. Similarly, depopulations of excited species caused by reactions 43–48 with precursor molecules can be observed in figure 15.7(c) and figure 15.7(d).

The transport of neutral and charged precursor fragments PF and PF^+ towards the substrate with the gas flow is represented in figure 15.9. Neutral precursor fragments PF (figure 15.9(a)) are mainly generated by the electron–ion recombination reaction 51. Consequently, the stream of PF starts at a position somewhat downstream from the creation maximum of PF^+ shown in figure 15.9(b). Finally, neutral PFs represent the dominant species reaching the substrate and forming a thin layer with a thickness of about 5 mm in front of the substrate. Corresponding density profiles at $r = 1.45$ mm are included in figure 15.8(a) and enlarged in a subfigure.

The influence of the flow rate Q_2 on the radial density profiles of dominant species in the effluent at $z = -1$ mm and $z = -3$ mm is shown in figure 15.10. The changing flow rate Q_2 directly modifies the radial extension of the HMDSO profile. This is caused by the varying relation between axial flow and radial diffusion of HMDSO in the argon gas. As a result, the radial separation between the inner region with dominant PF^+ ions and the outer region containing Ar_2^+ ions moves inwards in radial direction with increasing flow rate Q_2. Furthermore, the larger flow in the centre leads to a decrease of the PF^+ density and a remarkable depletion of PF in the centre. This effect is mainly caused by the reduced residence time of the reactants in the effluent which weakens the recombination reaction 51. The depletion of PF^+ and PF in the centre is more pronounced at $z = -1$ mm where the gas flow has crossed only a shorter distance from the beginning of the mixing zone than at $z = -3$ mm.

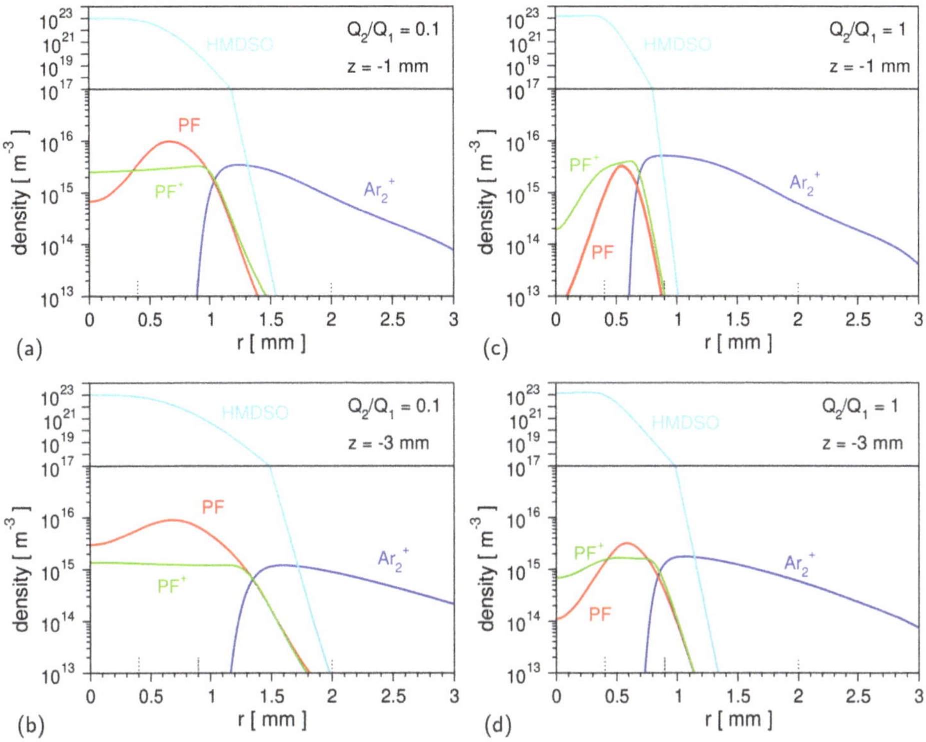

Figure 15.10. Radial density profiles of dominant species in the effluent at $z = -1$ mm (a,c) and $z = -3$ mm (b,d) for a flow rate ratio Q_2/Q_1 of 0.1 (a,b) and 1 (c,d).

Finally, the period-averaged model allows one to determine the radial profiles of the fluxes of neutral and charged precursor fragments onto the surface of the substrate. These profiles can be used as a measure of growth rate of the deposited film. Both flux profiles are shown in figure 15.11. The results of the reference case are compared with those obtained with smaller and larger flow rate Q_2 through the inner capillary. The most striking changes arise when comparing the transition from $Q_2/Q_1 = 0.1$ to $Q_2/Q_1 = 0.5$. They concern mostly the PF^+ flux around the centre ($r \lesssim 1$ mm). Here, the direct influence of the increasing flow rate Q_2 leads to an increased transport of charged PFs, PF^+, in the inner region towards the substrate. The influence of the inner flow rate Q_2 weakens in the outer region ($r \gtrsim 2$ mm), where the incoming PFs originate from the transition zone in the upstream effluent.

A structural modification of the flux profile of the neutral PFs can be observed. With increasing flow rate Q_2 the PF flux decreases in the outer region by a factor of about two. This change is related to the constriction of the radial HMDSO profile in the effluent. At the same time, a remarkable increase of the PF flux is found in the inner region despite the decrease of the PF density in the effluent (see figure 15.10). This change is caused by the pronounced structure of the gas flow arising at the higher flow rate Q_2 near the stagnation point (see figure 15.6a) which influences the density gradients near the substrate.

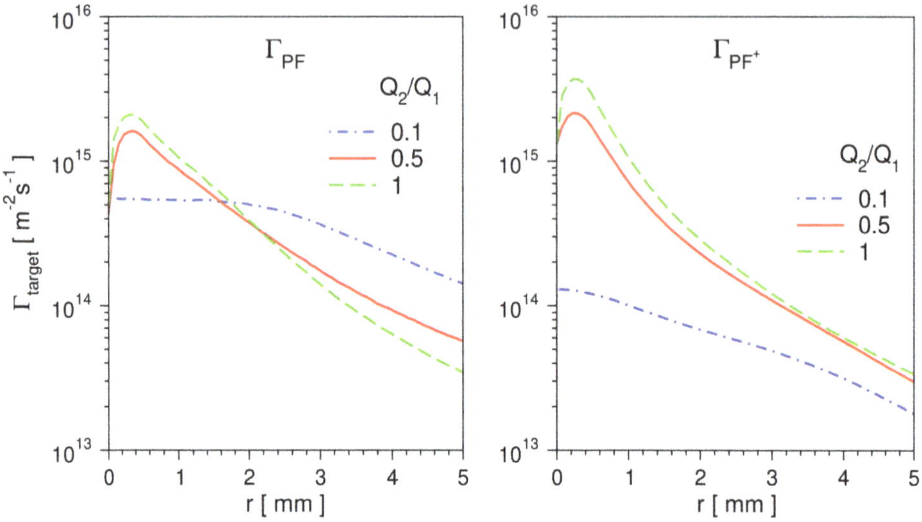

Figure 15.11. Flux of charged and neutral PFs toward the substrate for varying flow rate Q_2.

15.5 Concluding remarks

Plasma jets represent very diverse discharge configurations which combine a large manifold of physical processes. As one example of plasma jet configurations, a non-thermal RF plasma jet is considered in the current chapter. Because of the complexity of the physics involved, different models are used to describe selected aspects of the plasma jet in more detail.

A phase-resolved plasma model of the active zone reveals the mechanisms of plasma generation which are accompanied by constriction and stratification. This model is used to extract a power gain profile which is applied in a period-averaged model of the overall plasma jet including the gas flow of argon and the argon precursor mixture through the plasma jet and towards the substrate.

The period-averaged model delivers results on the interplay of gas flow, heating and reactions. Molecular argon ions are found as the dominant species which initiates reactions with the precursor molecules in the mixing zone. The subsequent transport of charged and neutral precursor fragments remarkably depends on the flow rates through the inner and outer capillaries resulting in different profiles of the fluxes onto the substrate which are related to the growth of the deposited film.

The self-organization of the plasma jet leading to multiple standing or rotating filaments in the active zone represents a further interesting aspect of the RF plasma jet. Due to the limitation of the axially symmetric geometry used in both models, this aspect cannot be considered in the framework of this chapter. The reader is referred to further literature concerning this subject [22, 25].

References

[1] Lu X, Laroussi M and Puech V 2012 On atmospheric-pressure non-equilibrium plasma jets and plasma bullets *Plasma Sources Sci. Technol.* **21** 034005

[2] Winter J, Brandenburg R and Weltmann K-D 2015 Atmospheric pressure plasma jets: an overview of devices and new directions *Plasma Sources Sci. Technol.* **24** 064001

[3] Penkov O V, Khadem M, Lim W-S and Kim D-E 2015 A review of recent applications of atmospheric pressure plasma jets for materials processing *J. Coat. Technol. Res.* **12** 225–35

[4] Lietz A M and Kushner M J 2018 Electrode configurations in atmospheric pressure plasma jets: production of reactive species *Plasma Sources Sci. Technol.* **27** 105020

[5] Viegas P, Hofmans M, van Rooij O, Obrusnik A, Klarenaar B L M, Bonaventura Z, Guaitella O, Sobota A and Bourdon A 2020 Interaction of an atmospheric pressure plasma jet with grounded and floating metallic targets: simulations and experiments *Plasma Sources Sci. Technol.* **29** 095011

[6] Semenov I L and Weltmann K-D 2020 Modelling of turbulent reacting flow for a cold atmospheric pressure argon plasma jet *Plasma Sources Sci. Technol.* **29** 055001

[7] Schmidt-Bleker A, Winter J, Bösel A, Reuter S and Weltmann K-D 2016 On the plasma chemistry of a cold atmospheric argon plasma jet with shielding gas device *Plasma Sources Sci. Technol.* **25** 015005

[8] Favaro M, Patelli A, Ceccato R, Dire S, Callone E, Fredi G and Quaranta A 2021 Thin films of plasma-polymerized n-hexane and ZnO nanoparticles co-deposited via atmospheric pressure plasma jet *Coatings* **11** 167

[9] Reuter S, von Woedtke T and Weltmann K-D 2018 The kINPen—a review on physics and chemistry of the atmospheric pressure plasma jet and its applications *J. Phys. D: Appl. Phys.* **51** 233001

[10] Breden D and Raja L L 2014 Computational study of the interaction of cold atmospheric helium plasma jets with surfaces *Plasma Sources Sci. Technol.* **23** 065020

[11] Golda J *et al* 2016 Concepts and characteristics of the 'cost reference microplasma jet' *J. Phys. D: Appl. Phys.* **49** 084003

[12] Norberg S A, Johnsen E and Kushner M J 2015 Formation of reactive oxygen and nitrogen species by repetitive negatively pulsed helium atmospheric pressure plasma jets propagating into humid air *Plasma Sources Sci. Technol.* **24** 035026

[13] Babaeva N Y, Naidis G V, Panov V A, Wang R, Zhang S, Zhang C and Shao T 2019 Plasma bullet propagation and reflection from metallic and dielectric targets *Plasma Sources Sci. Technol.* **28** 095006

[14] Baeva M, Bösel A, Ehlbeck J and Loffhagen D 2012 Modeling of microwave-induced plasma in argon at atmospheric pressure *Phys. Rev.* E **85** 56404

[15] Kubecka M, Snirer M, Obrusnik A, Kudrle V and Bonaventura Z 2020 Computational study of plasma-induced flow instabilities in power modulated atmospheric-pressure microwave plasma jet *Plasma Sources Sci. Technol.* **29** 075001

[16] Klute M, Porteanu H-E, Stefanovic I, Heinrich W, Awakowicz P and Brinkmann R P 2020 Theoretical investigation of a miniature microwave driven plasma jet *Plasma Sources Sci. Technol.* **29** 065018

[17] Petrova T B, Boris D R, Hinshelwood M, Johnson M J, Gillman E D and Walton S G 2020 Helium and oxygen excited states densities in a He-air RF-driven atmospheric pressure plasma jet *Phys. Plasmas* **27** 103512

[18] Golda J, Held J and Schulz-von der Gathen V 2020 Comparison of electron heating and energy loss mechanisms in an RF plasma jet operated in argon and helium *Plasma Sources Sci. Technol.* **29** 025014

[19] Sigeneger F and Loffhagen D 2016 Fluid model of a single striated filament in an RF plasma jet at atmospheric pressure *Plasma Sources Sci. Technol.* **25** 035020

[20] Sigeneger F, Becker M M, Foest R and Loffhagen D 2016 Modelling of plasma generation and thin film deposition by a non-thermal plasma jet at atmospheric pressure *J. Phys. D: Appl. Phys.* **49** 345202

[21] Sigeneger F, Schäfer J, Weltmann K-D, Foest R and Loffhagen D 2017 Modeling of a non-thermal RF plasma jet at atmospheric pressure *Plasma Process. Polym.* **14** 1600112

[22] Schäfer J, Sigeneger F, Sperka J, Rodenburg C and Foest R 2018 Searching for order in atmospheric pressure plasma jets *Plasma Phys. Control. Fusion* **60** 014038

[23] Sigeneger F, Schäfer J, Foest R and Loffhagen D 2019 Phase-resolved modeling of a filamentary argon plasma in an RF plasma jet *Plasma Sources Sci. Technol.* **28** 055004

[24] Schäfer J, Quade A, Abrams K J, Sigeneger F, Becker M M, Majewski C and Rodenburg C 2019 Helixjet: an innovative plasma source for next-generation additive manufacturing (3d printing) *Plasma Process. Polym.* **17** 1900099

[25] Schäfer J, Foest R, Ohl A and Weltmann K-D 2009 Miniaturized non-thermal atmospheric pressure plasma jet-characterization of self-organized regimes *Plasma Phys. Control. Fusion* **51** 124045

[26] Boulos M I, Fauchais P and Pfender E 1994 *Thermal Plasmas: Fundamentals and Applications* vol 1 (New York: Plenum)

[27] Lide D R (ed) 2010 *CRC Handbook of Chemistry and Physics* 90th edn (Boca Raton, FL: CRC Press)

[28] Raizer Y 1997 *Gas Discharge Physics* (Berlin: Springer)

[29] Yamabe C, Buckman S and Phelps A 1983 Measurement of free-free emission from low-energy-electron collisions with Ar *Phys. Rev.* A **27** 1345–52

[30] Zatsarinny O and Bartschat K 1999 B-spline Breit-Pauli R-matrix calculations for electron collisions with argon atoms *J. Phys. B: At. Mol. Opt. Phys.* **37** 4693–706

[31] Hayashi M 2003 Bibliography of electron and photon cross sections with atoms and molecules published in the 20th century—argon Tech. Rep. NIFS-DATA-72, National Institute for Fusion Science, Oroshi-cho, Japan

[32] Rapp D and Englander-Golden P 1965 Total cross sections for ionization and attachment in gases by electron impact. I. Positive ionization *J. Chem. Phys.* **43** 1464–79

[33] Deutsch H, Becker K, Matt S and Märk T 1999 Calculated cross sections for the electron-impact ionization of metastable atoms *J. Phys. B: At. Mol. Opt. Phys.* **32** 4249–59

[34] Flannery M and McCann K J 1980 Cross sections for ionization of rare gas excimers by electron impact and atomic and molecular processes in excimer lasers Tech. Rep. AFWAL-TR-80-2015, Georgia Institute of Technology, Atlanta, USA

[35] Jones J D C, Lister D G, Wareing D P and Twiddy N D 1980 The temperature dependence of the three-body reaction rate coefficient for some rare-gas atomic ion-atom reactions in the range 100–300 K *J. Phys. B: At. Mol. Opt. Phys.* **13** 3247–55

[36] Lukac P, Mikus O, Morva I, Zabudla Z, Trnovec J and Morvova M 2011 Electron and gas temperature dependences of the dissociative recombination coefficient of molecular ions Ar_2^+ with electrons *Plasma Sources Sci. Technol.* **20** 055012

[37] Eletskii A and Smirnov B 1986 Physical processes in gas lasers *J. Russ. Las. Res.* **7** 207–323

[38] Kolokolov N, Kudrjavtsev A and Blagoev A 1994 Interaction processes with creation of fast electrons in the low temperature plasma *Phys. Scr.* **50** 371–402

[39] Lam S K, Zheng C-E, Lo D, Dem'yanov A and Napartovich A P 2000 Kinetics of Ar_2^* in high-pressure pure argon *J. Phys. D: Appl. Phys.* **33** 242–51

[40] Wiese W, Braulth J, Danzmann K, Helbig V and Kock M 1989 Unified set of atomic transitions probabilities for neutral argon *Phys. Rev.* A **39** 2461–71

[41] Irimia A and Fischer C 2004 Breit-Pauli and Dirac-Hartree-Fock energy levels and transition probabilities in neutral argon *J. Phys. B: At. Mol. Opt. Phys.* **37** 1659–72

[42] Holstein T 1951 Imprisonment of resonance radiation in gases. II *Phys. Rev.* **83** 1159–68

[43] http://www.comsol.de

[44] Kolobov V I 2006 Striations in rare gas plasmas *J. Phys. D: Appl. Phys.* **39** R487–506

[45] Fuller E N, Ensley K and Giddings J C 1969 Diffusion of halogenated hydrocarbons in helium. The effect of structure on collision cross sections *J. Chem. Phys.* **73** 3679–85

[46] Creatore M, Barrell Y, Benedikt J and van de Sanden M C M 2006 On the hexamethyldisiloxane dissociation paths in a remote ar-fed expanding thermal plasma *Plasma Sources Sci. Technol.* **15** 421

[47] Enache I, Caquineau H, Gherardi N, Paulmier T, Maechler L and Massines F 2007 Transport phenomena in an atmospheric-pressure townsend discharge fed by N_2/N_2O/HMDSO mixtures *Plasma Process. Polym.* **4** 806–14

[48] Schäfer J, Foest R, Quade A, Ohl A and Weltmann K-D 2008 Local deposition of SiO_x plasma polymer films by a miniaturized atmospheric pressure plasma jet (APPJ) *J. Phys. D: Appl. Phys.* **41** 194010

[49] Trunec D, Zajickova L, Bursikova V, Studnicka F, Stahel P, Prysiazhnyi V, Perina V, Houdkova J, Navratil Z and Franta D 2010 Deposition of hard thin films from HMDSO in atmospheric pressure dielectric barrier discharge *J. Phys. D: Appl. Phys.* **43** 225403

[50] Carles S, Le Garrec J L and Mitchell J B A 2007 Electron and ion reactions with hexamethyldisiloxane and pentamethyldisiloxane *J. Chem. Phys.* **127** 144308

[51] Fridman A 2008 *Plasma Chemistry* (New York: Cambridge University Press)

[52] Schmidt M and Maass M 1989 Modelling of the positive column in an argon-hexamethyldisiloxane dc glow discharge *Contrib. Plasma Phys.* **29** 197–209

[53] Jauberteau J L and Jauberteau I 2012 Comparison of hexamethyldisiloxane dissociation processes in plasma *J. Phys. Chem.* A **116** 8840–50

[54] Hagelaar G J M, de Hoog F J and Kroesen G M W 2000 Boundary conditions in fluid models of gas discharges *Phys. Rev.* E **62** 1452–4

[55] Wagenaars E, Brandenburg R, Brok W J M, Bowden M D and Wagner H-E 2006 Experimental and modelling investigations of a dielectric barrier discharge in low-pressure argon *J. Phys. D: Appl. Phys.* **39** 700

[56] Phelps A V and Petrovic Z L 1999 Cold-cathode discharges and breakdown in argon: surface and gas phase production of secondary electrons *Plasma Sources Sci. Technol.* **8** R21–44

[57] Schäfer J, Sigeneger F, Foest R, Loffhagen D and Weltmann K-D 2010 On plasma parameters of a self-organized plasma jet at atmospheric pressure *Eur. Phys. J.* D **60** 531–8

IOP Publishing

Plasma Modeling (Second Edition)
Methods and applications
Gianpiero Colonna and Antonio D'Angola

Chapter 16

High-enthalpy radiating flows in aerophysics

Sergey T Surzhikov

The expression 'high enthalpy radiating flows' is often used to indicate gas flows, heated up to high temperature, with developed physico-chemical processes, such as dissociation, ionization and photon emission. Such high-enthalpy gas flows are very often encountered in nature (e.g. astrophysical plasmas [1–4]). Tasks of aerothermodynamics of space vehicles and of hypersonic aircraft, as well as of different high-energy devices, are directly related to the solution of the complex problem of the physics of high-enthalpy flows, which in turn are closely related with the problem of radiation heat transfer.

Unique features of hypersonic flight for space access are embedded in the phenomena of rarified gas dynamics, viscous–inviscid interaction, surface ablation, non-equilibrium thermodynamics and chemical reactions, as well as, radiative heat transfer [5–9]. The traditional computational simulation capability has been developed on the framework of kinetic theory of gases. In a high-temperature environment, the internal degrees of freedom of the gaseous medium become exited and the knowledge of internal structure of gas components is required for understanding the interactions at molecular and atomic scales [10, 11]. Therefore, the relatively poorly understood but most challenging topics for high enthalpy flow simulation can be identified as the physical-based models for the non-equilibrium thermodynamic state of the medium.

Flows in chemical and thermal non-equilibrium are often investigated by using conservation laws of interdisciplinary computational fluid dynamics (CFD) for high-enthalpy gases. In the framework of continuum mechanics, the bridge linking the Boltzmann equation and conservative laws of aerodynamics is the kinetic theory of gas through Chapman–Enskog expansion [10]. Since the classical kinetic theory does not include the internal structure of atoms and molecules, the higher degrees of excitation beyond the molecular translational motion must be solved by approximations. Traditionally the connection between molecular and atomic structure and thermodynamic behavior of high-temperature gas is described by statistical

doi:10.1088/978-0-7503-3559-1ch16

thermodynamics and augmented by the knowledge from quantum physics. The evaluation of finite-rate chemical reactions usually adopts chemical kinetic models through the law of mass action [12–15]. Some attempts have been recently undertaken, dedicated to the application of quantum chemistry and quantum statistics to modern aerophysics [16–18].

In a hypersonic atmospheric entry, the radiation heat transfer contributes to a substantial amount of energy transfer in addition to the conductive and convective contributions. For example, at the peak heating condition of the Stardust re-entry [19], the radiative heat transfer rate is 120 W cm^{-2} versus the convective–conductive heat transfer rate of 839 W cm^{-2}. Therefore, it is critically important to understand the basic heat transfer mechanisms developing accurate and efficient predictive methods.

All transitions involving radiation emission or absorption can be divided into bound–bound, bound–free, and free–free processes according to initial and final quantum state of atom or molecule, resulting in a continuum or discrete spectrum [20, 21]. In general, the thermal radiative heat transfer has a frequency spectrum from the far infrared to vacuum ultraviolet bands. High enthalpy gas mixtures in the shock layers and in the wakes of re-entering space vehicles, also with species coming from the ablative heat shield, usually contain optically active components as CO_2, H_2O, CH_4, N_2, O_2, NO, N_2^+, C_2, CO, etc. At a re-entry speed around 10 km s^{-1} strong absorption by C, N, O, NO, CO, O_2 has been detected in vacuum ultraviolet (80–200 nm) and ultraviolet spectra. Some of these transition processes are not fully understood but become critical mechanisms for radiative heat transfer [22, 23].

The computation of radiative heat transfer requires the development of effective numerical simulation methods not only for prediction of spectral radiation fluxes inside the radiating control volume but also for the evaluation of optical properties. The energy exchange mechanism of radiative heat transfer is fundamentally different from the convective or conductive ones. It is a phenomenon associated with quantum transitions of molecule and atom. Therefore, the electronic excitation is not ignorable in radiative calculation as can be done in convective–conductive heat transfer simulations in most Earth re-entry conditions.

The development of computational models and codes for description of aerothermodynamics of re-entry space vehicles, intended for investigation of planets of the Solar system and for return to Earth, is the important part of scientific and engineering programs which are currently realized in different space agencies in the world. It is also well known that the general method for verification and validation of such models and codes is the comparison with available experimental and flight data, still very scanty. Existing well-known flight experiments, such as RAM-C-II [24–26] and Fire-II [27], which are still referred to for the computational works [27–31], probed radiation properties in the shock layer, in different conditions, and from full non-equilibrium to local thermodynamic equilibrium (LTE). Ionization processes in air compressed layer at altitudes 61–81 km in the vicinity of the spherically blunted cone at orbital velocity were studied in RAM-C-II flight experiments, while the density of convective and radiative heat fluxes to the surface of the super-orbital spacecraft Fire-II has been detected, conditions under which non-equilibrium dissociation and ionization play an important role.

The highest entry velocity of a spacecraft into the Earth's atmosphere (12.4 km s^{-1}) was achieved for the spacecraft Stardust. In the first series of works focused on Stardust re-entry [19, 22, 23] predictions can be found on aerothermodynamics and ablation of the thermal protection system, based on flight data analysis after the return of the spacecraft [31] and remote observation [32].

In 2010, a special issue of *Journal of Spacecraft and Rockets* published a number of computational results performed using the new models created during the decade. It has been shown that predicted data in 1999 [19] have been confirmed with good accuracy. It is worth noting that the data from the flight experiments of Fire-II allowed the verification of the computer models used to describe the conditions of Stardust's entry with good reliability.

Non-equilibrium is a critical aspect for spacecraft entering the atmosphere with super-orbital velocity, where strong gas dynamics–radiation interaction is expected and large contribution to the radiative heat transfer comes from atomic and ionic transitions [28–30].

Particularly relevant is the problem of thermal non-equilibrium which actually means that there are no equilibrium conditions between Boltzmann distribution for populations of the vibrational and electronic excited states, which determine the intensity of the luminescence of the thermal radiation. To take into account the disequilibrium, a multi-temperature approach can be used, where the energy stored in each vibrational or electronically excited state can be estimated by the Boltzmann distribution with the introduction of a characteristic vibrational and electron temperatures, the latter is set equal to that of electronically excited states.

An alternative approach, called the *collisional-radiative* model (CR), consists in solving the master equation for each excited state level of atoms and molecules [33] (see also chapters 9 and 10).

Another challenging task which can be solved using models developed in the framework of high-enthalpy flows is the physics of meteoroids. It is well known that the determination of the concentration of atoms and molecules, and relative ions, is the main problem of the quantitative analysis of meteor spectra [34]. The models should allow the determination of the chemical composition and the types of materials making up meteoroids, and consequently, the quantitative abundances of the constituent elements, thus to establish to what degree the minor bodies of the Solar System are similar. This is a fundamental problem, being nowadays widely accepted that there exists a genetic interrelation between asteroids, comets and meteoroids. Different methods for quantitative analysis of meteor spectra have been developed and used by researchers, classified as the quantitative physical approach. However, other research activities are developing methodologies based on the computational physical gas dynamics of high-enthalpy flows. The emission and ablation of large meteoroids in the course of their motion through the Earth's atmosphere are considered in reference [35]. The radiative gas dynamics and emission of small meteoroids were considered in reference [36]. It was shown that the non-equilibrium effects of vibrational excitation of diatomic molecules N_2, O_2, and NO behind the leading shock wave play a significant role in forming the spectral signatures of bolides, as well as in ionization processes in shock layer and wake.

The radiative gas-dynamic model of high-enthalpy flows, which can be applied to investigation of most of the mentioned problems, will be described in this chapters.

The theory is based on the equations of viscous, heat conducting, chemically non-equilibrium, emitting and absorbing gas, and includes databases of thermodynamic, kinetic and radiation properties.

16.1 Fluid dynamic model

Most general radiative gas-dynamic models should contain Navier–Stokes or Euler continuity equations with mass conservation of chemical species, vibrational energy conservation, and the radiation heat transfer equation (in multi-group approximation). These are:

$$
\begin{aligned}
&\frac{\partial \rho}{\partial t} + \vec{\nabla} \cdot \rho \vec{v} = 0 && (a)\\
&\frac{\partial \rho v_k}{\partial t} + \vec{\nabla} \cdot \rho v_k \vec{v} = -\frac{\partial p}{\partial x_k} + F_{\mu,k} && (b)\ k = x, y, z\\
&\frac{\partial \rho_i}{\partial t} + \vec{\nabla} \cdot \rho_i \vec{v} = -\vec{\nabla} \cdot \vec{J}_i + \dot{\omega} && (c)\ i = 1 \ldots N_s\\
&\rho c_p \frac{\partial T}{\partial t} + \rho c_p \vec{v} \cdot \vec{\nabla} T = \vec{\nabla} \cdot (\lambda \vec{\nabla} T) + \frac{\partial p}{\partial t} + \vec{v} \cdot \vec{\nabla} p + \\
&+ \Phi_\mu + Q_{\text{vib}} - \vec{\nabla} \cdot \vec{q}_R - \sum_{i=1}^{N_s} \left[h_i \dot{\omega}_i + c_{p,i} \vec{J}_i \cdot \vec{\nabla} T \right] && (d)\\
&\frac{\partial \rho e_{v,m}}{\partial t} + \vec{\nabla} \cdot \rho \vec{v} e_{v,m} = \dot{e}_{v,m} && (e)\ i = 1 \ldots N_v\\
&\vec{\Omega} \cdot \vec{\nabla} J_\omega(\vec{r}, \vec{\Omega}) + k_\omega(\vec{r}) J_\omega(\vec{r}, \vec{\Omega}) = j_\omega(\vec{r}) && (f)
\end{aligned}
\tag{16.1}
$$

where for the Navier–Stokes equations

$$
\begin{aligned}
F_{\mu,x} &= -\frac{2}{3}\frac{\partial}{\partial x}(\mu \vec{\nabla} \cdot \vec{v}) + 2\frac{\partial}{\partial x}\mu\frac{\partial v_x}{\partial x} + \frac{\partial}{\partial y}\left[\mu\left(\frac{\partial v_y}{\partial x} + \frac{\partial v_x}{\partial y}\right)\right] + \frac{\partial}{\partial z}\left[\mu\left(\frac{\partial v_z}{\partial x} + \frac{\partial v_x}{\partial z}\right)\right]\\
F_{\mu,y} &= -\frac{2}{3}\frac{\partial}{\partial y}(\mu \vec{\nabla} \cdot \vec{v}) + 2\frac{\partial}{\partial y}\mu\frac{\partial v_y}{\partial y} + \frac{\partial}{\partial x}\left[\mu\left(\frac{\partial v_x}{\partial y} + \frac{\partial v_y}{\partial x}\right)\right] + \frac{\partial}{\partial z}\left[\mu\left(\frac{\partial v_z}{\partial y} + \frac{\partial v_y}{\partial z}\right)\right]\\
F_{\mu,z} &= -\frac{2}{3}\frac{\partial}{\partial z}(\mu \vec{\nabla} \cdot \vec{v}) + 2\frac{\partial}{\partial z}\mu\frac{\partial v_z}{\partial z} + \frac{\partial}{\partial x}\left[\mu\left(\frac{\partial v_x}{\partial z} + \frac{\partial v_z}{\partial x}\right)\right] + \frac{\partial}{\partial y}\left[\mu\left(\frac{\partial v_y}{\partial z} + \frac{\partial v_z}{\partial y}\right)\right]\\
\Phi_\mu &= \mu\left[2\left(\frac{\partial v_x}{\partial x}\right)^2 + 2\left(\frac{\partial v_y}{\partial y}\right)^2 + 2\left(\frac{\partial v_z}{\partial z}\right)^2 + \right.\\
&\left. + \left(\frac{\partial v_y}{\partial x} + \frac{\partial v_x}{\partial y}\right)^2 + \left(\frac{\partial v_z}{\partial x} + \frac{\partial v_x}{\partial z}\right)^2 + \left(\frac{\partial v_y}{\partial z} + \frac{\partial v_z}{\partial y}\right)^2 - \frac{2}{3}\left(\frac{\partial v_x}{\partial x} + \frac{\partial v_y}{\partial y} + \frac{\partial v_z}{\partial z}\right)^2\right]\\
\vec{J}_i &= -\rho D_i \vec{\nabla} Y_i\\
c_p &= \sum_{i=1}^{N_s} Y_i c_{p,i}
\end{aligned}
\tag{16.2}
$$

and for vibrational energy conservation equations:

$$\begin{aligned}\dot{e}_{v,i} &= \rho_{i(m)}\frac{e^{0}_{v,m} - e_{v,m}}{\tau_m} - e_{v,m}\dot{\omega}_{im}\\ e_{v,m} &= \frac{R_0\theta_m}{M_{i(m)}[e^{\theta_m/T_{v,m}} - 1]}\end{aligned} \tag{16.3}$$

This system is closed by the ideal gas equation of state

$$p = \rho\frac{R_0}{M}T = \rho R_0 T\sum_{i=1}^{N_s}\frac{Y_i}{M_i} \tag{16.4}$$

and the total enthalpy equation

$$h = \int_{T_0}^{T} c_p dT + h_0 \tag{16.5}$$

where t is the time; v_{ik} are the projection of velocity $\vec{v}$ on x, y, z axes; p is the pressure, ρ and ρ_i are the total and single species mass density, T is the translational temperature, M is the mean molar mass, μ and λ are viscosity and heat conductivity coefficients, c_p is the specific heat capacity of gas mixture. The symbol i is the index running over the N_s number of species having Y_i the mass fraction, $c_{p,i}$ is the constant pressure specific heat, h_i is the specific enthalpy, $\dot{\omega}_i$ production rate, D_i is the effective diffusion coefficient, $\vec{J}_i$ is the mass diffusion flux and M_i the molar mass. The index m runs over N_v vibrational modes where $i(m)$ is the index of species with mth vibrational mode and $e_{v,m}(T_{v,m})$ is the specific vibrational energy corresponding to vibrational temperature $T_{v,m}$ and $e^{0}_{v,m} = e_{v,m}(T_{v,m} = T)$ is the corresponding equilibrium energy. $R_0 = 8.314\ 10^7$ erg (K mole)$^{-1}$ is the universal gas constant. Other variables are defined below. Equation (16.1)(e) describes vibrational relaxation processes in non-equilibrium gas flows. The term Q_{vib} in equation (16.1)(d) is the energy production due to vibrational relaxation in multi-component gas mixture. For air $N_v = 3$ assigning $m = 1$ for N_2, $m = 2$ for O_2 and $m = 3$ for the NO.

The radiation heat transfer equation (see equation (16.1)(f)) is formulated in general form neglecting light scattering. Being solved, this equation allows one to calculate total radiation heat flux

$$\begin{aligned}\vec{q}_R &= \int_{\Delta\omega_{\text{tot}}}\vec{q}_{R,\omega}d\omega\\ q_R &= \vec{q}_R\cdot\vec{n}\\ \vec{q}_{R,\omega} &= \int_{4\pi}J_\omega\left(\vec{R},\vec{\Omega}\right)\vec{\Omega}d\Omega\\ q_{R,\omega} &= \vec{q}_{R,\omega}\cdot\vec{n}\end{aligned} \tag{16.6}$$

where J_ω is the spectral intensity of heat radiation; $k_\omega(\vec{r})$ is the spectral absorption coefficient; $j_\omega(\vec{r})$ is the spectral emission coefficient, which is calculated in the case of local thermodynamic equilibrium (LTE) by the Kirchhoff law

$$j_\omega(\vec{r}) = k_\omega(\vec{r})J_{b,\omega}(\vec{r}) \tag{16.7}$$

where $J_{b,\omega}(\vec{r})$ is the black body spectral intensity (the Planck function). The spectral intensity $J_\omega(\vec{r},\vec{\Omega})$ is a function of the wave number ω, of the space coordinate $\vec{r}$ and of the direction of propagation $\vec{\Omega}$. This means that equation (16.1)(f) is a differential equation in a space with six dimensions, considering that the direction vector is completely determined by two angles. The total spectral region of heat radiation $\Delta\omega_{\text{tot}}$ in the given case =1000—200 000 cm^{-1}.

A cumulative function, which will be used below, is defined as follows

$$Q_{\text{rad}}(\omega) = \int_{\omega_{\min}}^{\omega} q_{R,\omega'} d\omega' \qquad (16.8)$$

The cumulative function Q_{rad} gives a representation concerning a part of total radiation heat flux for the given spectral region $[\omega_{\min}, \omega_{\max}]$.

In the case under consideration thermodynamic properties of gas mixture cannot be described with the use of single temperature because there are relaxation zones with non-equilibrium (i.e. non-Boltzmann, see section 9.1) populations of internal degree of freedoms of molecules and atoms. Most simple and popular non-equilibrium thermodynamic models are based on introduction of various translational and vibrational temperatures. Such models are used for description of relaxation processes behind shock wave fronts and in expansion parts of gas flows. In such cases the Kirchhoff low is applied to averaged vibrational temperature or to the separate temperatures.

Using a symbolic formulation of the nth chemical reaction

$$\sum_{j=1}^{N_s} a_{j,n}[X_j] \underset{k_n^r}{\overset{k_n^f}{\rightleftharpoons}} \sum_{j=1}^{N_s} b_{j,n}[X_j] \qquad (16.9)$$

the rate of formation of ith species due to nth chemical reaction can be written as follows

$$\left(\frac{dX_i}{dt}\right)_n = (b_{i,n} - a_{i,n})\left(k_n^f \prod_{j=1}^{N_s} X_j^{a_{j,n}} - k_n^r \prod_{j=1}^{N_s} X_j^{b_{j,n}}\right) = (b_{i,n} - a_{i,n})(S_n^f - S_n^r) \qquad (16.10)$$

and for N_r reactions the equation for the time evolution is given by

$$\frac{dX_i}{dt} = \sum_{n=1}^{N_r}\left(\frac{dX_i}{dt}\right)_n = \sum_{n=1}^{N_r}(b_{i,n} - a_{i,n})(S_n^f - S_n^r) \qquad (16.11)$$

where $a_{i,n}$ and $b_{i,n}$ are the stoichiometric coefficients of the nth reaction; $[X_i]$ is the chemical symbol of ith species and X_i is its molar density; k_n^f and k_n^r are the rate constants of the forward and reverse reactions, respectively. Equation (16.11) is written for molar density. To have the equation for the mass density ρ_i, equation (16.11) must be multiplied for the molar mass M_i, i.e.

$$\dot{\omega}_i = M_i\frac{dX_i}{dt} = M_i \sum_{n=1}^{N_r}(b_{i,n} - a_{i,n})(S_n^f - S_n^r) \qquad (16.12)$$

where $\dot{\omega}_i$ is the chemical contribution in equation (16.1). In the framework of the Arrhenius theory of reactions, the forward and reverse constants for the nth reaction are given by

$$k_n^{f/r} = A_n^{f/r} T^{p_n^{f/r}} \exp\left(-\frac{E_n^{f/r}}{kT}\right) \tag{16.13}$$

where $A_n^{f/r}$, $p_n^{f/r}$ and $E_n^{f/r}$ are the approximation coefficients for the forward (f) and reverse (r) reaction rates. Note that the equilibrium constants for each nth chemical reaction are determined as

$$K_n^{\mathrm{eq}} = \frac{k_n^f}{k_n^r} \tag{16.14}$$

In turn, the equilibrium constant can be calculated directly from thermodynamic data of individual species. Different databases (see for example [37, 38]) give approximate expressions for thermodynamic properties of individual chemical species in the temperature range of 100–20 000 K. For example, the Gibbs function Φ_i and its derivatives are given by the following expression as a function of $x = T \cdot 10^{-4}$

$$\begin{aligned}
\Phi_i &= \phi_{1,i} + \phi_{2,i}\ln x + \phi_{3,i}x^{-2} + \phi_{4,i}x^{-1} + \phi_{5,i}x + \phi_{6,i}x^2 + \phi_{7,i}x^3 \\
\frac{d\Phi_i}{dx} &= \frac{1}{x}\left(\phi_{2,i} - 2\phi_{3,i}x^{-2} - \phi_{4,i}x^{-1} + \phi_{5,i}x + 2\phi_{6,i}x^2 + 3\phi_{7,i}x^3\right) \\
\frac{d^2\Phi_i}{dx^2} &= \frac{1}{x^2}\left(-\phi_{2,i} + 6\phi_{3,i}x^{-2} + 2\phi_{4,i}x^{-1} + 2\phi_{6,i}x^2 + 6\phi_{7,i}x^3\right)
\end{aligned} \tag{16.15}$$

from which is possible to determine, for each species, molar enthalpy H_i (in J mole^{-1}) and constant pressure heat capacity $C_{p,i}$ (in J mole^{-1} K^{-1})

$$H_i = xT\frac{d\Phi_i}{dx} + x^2\phi_{8,i} \tag{16.16}$$

$$C_{p,i} = 2x\frac{d\Phi_i}{dx} + \frac{d^2\Phi_i}{dx^2} \tag{16.17}$$

The equilibrium constants for partial pressures (with respect to the reference value $p_0 = 101\,325$ Pa $= 1$ atm) is given by

$$\ln K_n^{\mathrm{eq}} = \frac{1}{R_0 T}\sum_{j=0}^{N_s}(a_{j,n} - b_{j,n})(-T\Phi_j + \phi_{8,j} \cdot 10^3) \tag{16.18}$$

Transport coefficients such as viscosity (μ) and heat conductivity (λ) of gas mixtures are calculated by the following mixing rules [39, 40]

$$\mu = \left(\sum_{i=1}^{N_s} \frac{Y_i}{\mu_i} \right)^{-1} \tag{16.19}$$

$$\lambda = \frac{1}{2}\left[\sum_{i=1}^{N_s} x_i \lambda_i + \left(\sum_{i=1}^{N_s} \frac{x_i}{\lambda_i} \right)^{-1} \right] \tag{16.20}$$

where the single species quantities are calculated as

$$\mu_i(g\ \mathrm{cm}^{-1}\ \mathrm{s}^{-1}) = 2.67 \cdot 10^{-5} \frac{\sqrt{M_i T}}{\sigma_i^2 \Omega_i^{(2,\,2)\star}} \tag{16.21}$$

$$\lambda_i(\mathrm{erg\ cm}^{-1}\ K^{-1}) = 8330 \cdot 10^{-5} \sqrt{\frac{T}{M_i}} \frac{1}{\sigma_i^2 \Omega_i^{(2,\,2)\star}} \tag{16.22}$$

where x_i are the molar fractions, σ_i is the diameter of collision and $\Omega_i^{(2,\,2)\star}$ is the viscosity type reduced collision integral. Effective diffusion coefficient D_i for ith species is calculated from the binary diffusion coefficients $D_{i,j}$ by the Wilke formula [41]

$$D_i(\mathrm{cm}^2\ \mathrm{s}^{-1}) = (1 - x_i)\left(\sum_{j \neq i}^{N_s} \frac{x_j}{D_{i,j}} \right)^{-1} \tag{16.23}$$

where

$$D_{i,j}(\mathrm{cm}^2\ \mathrm{s}^{-1}) = 1.858 \cdot 10^{-3} \sqrt{T^3 \frac{M_i + M_j}{M_i M_j}} \frac{1}{p \sigma_{i,j}^2 \Omega_{i,j}^{(1,\,1)\star}} \tag{16.24}$$

where $\Omega_{i,j}^{(1,\,1)\star}$ is the diffusion type reduced collision integrals. The reduced collision integrals are calculated with the approximations suggested by Anfimov [42]

$$\begin{aligned} \Omega_i^{(2,\,2)\star} &= 1.157 T_i^{-0.1472} \\ \Omega_{i,j}^{(1,\,1)\star} &= 1.074 T_{i,j}^{-0.1604} \end{aligned} \tag{16.25}$$

where $T_i = kT/\varepsilon_i$, with ε_i the well depth of the attractive interaction potential and

$$\begin{aligned} T_{i,j} &= \frac{kT}{\varepsilon_{i,j}} \\ \varepsilon_{i,j} &= \sqrt{\varepsilon_i \varepsilon_j} \\ \sigma_{i,j} &= \frac{\sigma_i + \sigma_j}{2} \end{aligned} \tag{16.26}$$

The last aspect to consider is the interaction with the entry vehicle. A pseudo-catalytic surface was presumed, i.e. $Y_i(\mathrm{wall}) = Y_i(\infty)$ where $Y_i(\infty)$ are the mass fraction in the oncoming flow. Surface temperature is calculated with the assumption of equilibrium emitting surface

$$\varepsilon\sigma T_w^4 \tag{16.27}$$

where ε is the surface emissivity factor, $\sigma = 5.67 \cdot 10^{-12}$ W cm^2 K^4 is the Stephan–Boltzmann constant and Q_w is the total (convective plus radiative) heat flux density on the surface.

16.2 Radiative gas dynamics of re-entry space vehicles

Predictions of the shock layer radiative heating of entry vehicles and the coupled radiative gas-dynamic processes in shock layers at orbital and super-orbital return conditions have received renewed interest due to modern national plans for a return to lunar and neighboring planets of the Solar System exploration. Despite the fact that radiation heat transfer processes applied to re-entry space vehicles have been considered for more than 40 years, unanswered questions still remain in the existing planetary data. The scientific unknowns following from these questions have to be resolved to provide acceptable aerothermodynamic prediction of creating new-generation space vehicles.

One of the specific distinguishing features of the new generation of space vehicles is their large size in comparison with those of the previous generation. Space vehicles with midship diameters of about 5 meters are considered in the current plans. For example, the ORION project considers the possibility to use spacecraft with nose radius about 400 cm [43]. This concept has significant advantages, as it decreases the convective heat load. However, it is also necessary to take into account a significant disadvantage, namely the increase of radiative heating load. It can be estimated that typical stand-off distances for the large spacecraft of similar size is about 30–50 cm

To determine the radiation heating, atomic lines in the spectral range $\Delta\omega \approx 1000$–$150\,000$ cm^{-1} (≈ 10 µm–0.067 µm) must be considered, modeling with good accuracy the line shape, whose typical half-widths are in the range $\gamma \approx 0.01$–1.0 cm^{-1}. A significant challenge is to manage the large number of atomic lines (for air $\approx$ 2000-3000), which should be considered individually or by using special spectral methods.

The radiative gas-dynamics of different kinds of space vehicles will be considered below. These are: Fire-II [27–30], Stardust [22, 23, 31–33], RAM-C-II [24–26], ORION [43], Prospective Transport Vehicle (PTV) [44], and Mars Science Laboratory (MSL) [45, 46]. Their re-entry trajectories are shown in figure 16.1, and some of their trajectory parameters are presented in tables 16.1–16.6.

16.2.1 Fire-II

Flight data of the Fire-II experimental program have been used for validation of physical-chemical models and computer codes for the aerophysics of re-entry vehicles for more than 40 years [27–30]. Nevertheless, it is well known that there are several basic issues in aerophysics for the re-entry probe simulation that need further development:

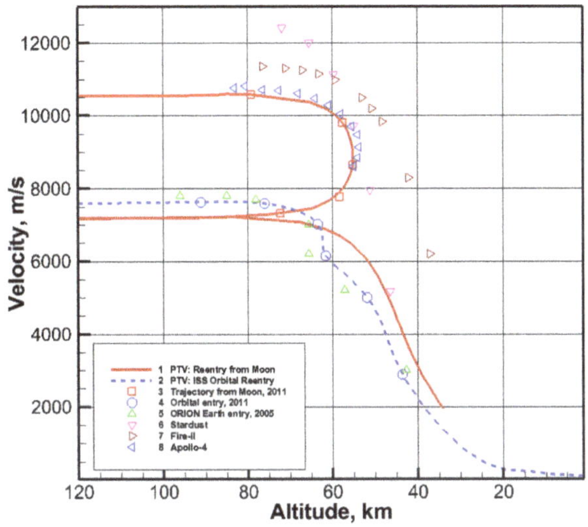

Figure 16.1. Typical trajectory parameters of Fire-II, Stardust, ORION and PTV.

Table 16.1. Trajectory parameters for Fire-II [27–30]. H is the altitude, t the trajectory time, ρ_∞, V_∞ and T_∞ gas density, velocity and temperature in the free stream, T_w the wall temperature, p_0 the stagnation pressure and R_n is the blunting radius.

t s	H km	$\rho_\infty \cdot 10^7$ g/cm^3	p_∞ erg/cm^3	V_∞ km/s	T_∞ K	T_w K	$p_0 \cdot 10^{-5}$ erg/cm^3	R_n cm
1634	76.42	0.372	20.8	11.36	195	615	0.48	
1636	71.04	0.857	51.7	11.31	210	810	1.10	
1637	67.05	1.47	96.2	11.25	228	1030	1.86	93.5
1639	63.11	2.41	167.4	11.14	242	1325	2.99	
1640	59.26	3.86	281.4	10.97	254	1560	4.65	
1643	53.04	7.80	617.9	10.48	276	640	8.57	
1644	50.67	10.2	822.6	10.19	281	1100	10.5	80.5
1645	48.37	13.2	1079.7	9.83	285	1520	12.8	
1648	42.14	30.0	2298.9	8.30	267	1560	20.7	70.2
1651	37.19	60.5	4393.0	6.19	253	1060	23.2	

Table 16.2. Trajectory parameters for Stardust. See table 16.1 for parameter definitions.

t s	H km	$\rho_\infty \cdot 10^8$ g cm^{-3}	p_∞ erg cm^{-3}	V_∞ km s^{-1}	T_∞ K	$p_0 \cdot 10^{-5}$ erg cm^{-3}
42	71.9	4.16	26.40	12.4	221.4	0.972
48	65.4	10.6	103.0	12.0	229.0	2.23
54	59.8	23.4	160.0	11.1	238.5	3.96
66	51.2	72.1	688.0	7.96	253.5	5.36
76	46.5	135	1 230.0	5.18	256.9	5.61

Table 16.3. Trajectory parameters for RAM-C-II. *Re* is the Reynolds number. For the other parameters see table 16.1.

H km	V_∞ km s^{-1}	p_∞ erg cm^{-3}	$\rho_\infty \cdot 10^8$ g cm^{-3}	T_∞ K	Mach	$Re \cdot 10^{-3}$
61	7.50	192.	27.3	244	23.9	19.5
71	7.64	44.8	7.20	217	25.9	6.28
81	7.95	8.89	1.57	197	28.3	1.59

Table 16.4. Conditions for the curve numbers in figure 16.5.

H [km] V_∞ [km s^{-1}]	61 7.5	71 7.64	81 7.95
LTE	1	2	3
Park dissociation model [12]	4	5	6
First dissociation model in reference [48]	7	8	9
Second dissociation model in reference [48] with $U = 6$	10	11	12
Second dissociation model in reference [48] with $U = 3$	13	14	15
Experimental data [24, 25]	16	17	18

Table 16.5. Trajectory parameters for ORION. See table 16.1 for parameter definitions.

trajectory points	t s	H km	$\rho_\infty \cdot 10^8$ g cm^{-3}	p_∞ erg cm^{-3}	V_∞ km s^{-1}	T_∞ K	$p_0 \cdot 10^{-4}$ erg cm^{-3}
1	150	83.0	0.630	3.37	7.7	187	0.374
2	200	78.2	2.45	14.2	7.7	202	1.45
3	300	65.6	15.1	100.	7.0	232	7.40
4	400	65.6	15.1	100.	6.2	232	5.82
5	500	57.1	44.3	324.	5.2	255	12.0
6	600	42.8	267	1980.	3.0	258	24.0

Table 16.6. PVT trajectory parameters (see tables 16.1 and 16.3 for definitions).

H km	$\rho_\infty \cdot 10^8$ g cm^{-3}	p_∞ erg cm^{-3}	T_∞ K	V_∞ km s^{-1}	$p_0 \cdot 10^{-4}$ erg cm^{-3}
50	103	798	247	9.9	100
60	31	220	247	9.5	27.7
80	1.85	10.5	199	8.5	1.33

- kinetic models for non-equilibrium dissociation, ionization, relaxation and radiation processes;
- efficient CFD tools for solving the governing equations on both structured and unstructured two- and three-dimensional grids;
- adequate models for turbulent flow and laminar-turbulent transition.

The uncertainty in numerical predictions for the convective and radiative heating of the re-entering capsule using different models of chemical and physical kinetics have been analyzed in several studies [27–30]. These calculations were performed with chemical models by Dunn and Kung [47] and Park [13, 14]. Several models of dissociation have also been used, among them the one-temperature model assuming LTE, and different models of non-equilibrium dissociation, such as that proposed by Treanor and Marrone [48, 49]. In this chapter, all the calculations are performed with and without coupling effects between radiation and gas-dynamics. The spectral optical properties of high temperature air are calculated using *ab initio*, quasi-classical and quantum mechanics approaches. The line-by-line spectral data calculations are performed with high accuracy for about 80 000 nodes of spectral points.

All computational results are compared with experiments and other calculations. For each trajectory point the following data have been analyzed: velocity components, pressure, density, translational and vibrational temperatures, as well as mass fractions along the stagnation streamline. Finally, the one-sided integral radiation heat fluxes, spectral distributions of radiation heat flux density, and the cumulative function are presented and delineated.

Two-dimensional fields of translational temperature and longitudinal velocity are shown in figure 16.2(a) for trajectory point t = 1639 s. The axial distributions of translational and vibrational temperatures for N_2, O_2 and NO molecules along the stagnation line are shown in figure 16.2(b). These results are obtained using two kinetic models [13, 47]. The two-dimensional fields of temperature and velocity were obtained with the Park kinetic model [13].

The density of spectral radiation heat fluxes and corresponding cumulative functions are shown in figure 16.2(c). These data allow us to demonstrate the contribution of spectral lines of atoms and ions into the total radiative heating. One can see that atomic lines give ≈ 100%–200% contribution to radiative heating of the frontal radiation heat shield.

16.2.2 Stardust

Calculation data for radiation-gas dynamics of Stardust will be considered in this section. The coupled radiative gas-dynamic problems are solved for the trajectory point t = 54 s (table 16.2).

Figure 16.3(a) shows the longitudinal velocity and translational temperature in the calculation domain, including the leading shock layer and flow-field in wake. These data are also important for the analysis of the spectral signature of the space vehicle entering the atmosphere, which can be compared with experimental measures [32].

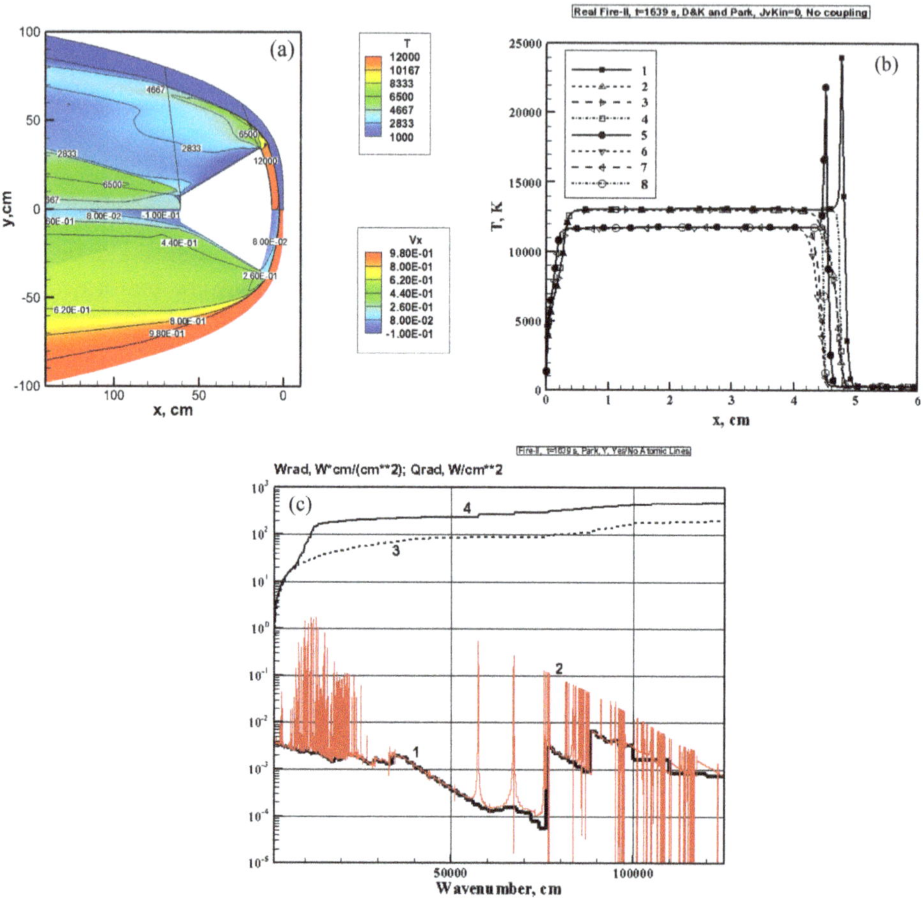

Figure 16.2. Space vehicle Fire-II at t = 1639 s. (a) Translational temperature, T in K (top) and longitude velocity normalized to V_∞ (bottom); (b) axial distribution of translational (1,5) and vibrational (2,6 - $T_v(N_2)$, 3,7 - $T_v(O_2)$, 4,8 - $T_v(NO)$) temperatures along the critical line. The Dunn–Kang (1–4) and the Park (5–8) kinetic models with the second model of Marrone–Treanor [49] at $U = 6$ were used. (c) Spectral radiation heat fluxes without (1) and with (2) atomic lines and corresponding cumulative functions (3,4) of radiation heat fluxes in stagnation point. Note that dimensionless parameter U is used in the model [49] to characterize the probability of dissociation from vibrationally excited states of diatomic molecules.

A representation of the distribution of translational and vibrational temperatures in leading shock layer can be obtained from figure 16.3(b). It can be observed that the translational temperature achieves values larger than 20 000 K, while vibrational temperatures are located below 12 000 K. As the altitude and velocity decrease the thermal equilibrium in shock layer becomes more obvious. The depth of the non-equilibrium zone at $t = 54$ s achieves about 0.3 cm, and, obviously, that temperature distribution in this region plays a significant role in the radiative heating. The spectral distribution of the surface radiation heat fluxes and the corresponding cumulative functions are presented in figure 16.3(c). Comparing the spectral

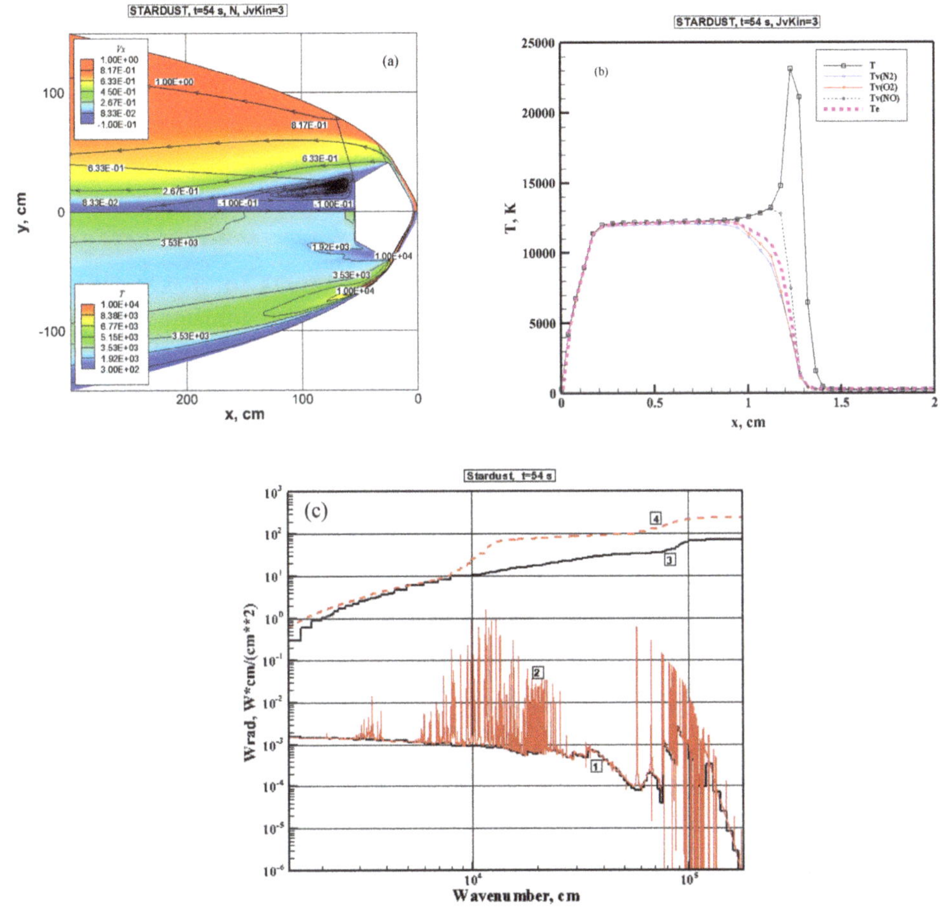

Figure 16.3. Space vehicle Stardust at t = 54 s. (a) Translational temperature, T in K (top) and longitude velocity normalized to V_∞ (bottom); (b) axial distribution of translational (1,5) and vibrational (2,6 - $T_v(N_2)$, 3,7 - $T_v(O_2)$, 4,8 - $T_v(NO)$) temperatures along the stagnation line of the flow field. The Dunn–Kang (1–4) and the Park (5–8) kinetic models with the second model of Marrone–Treanor [49] at $U = 6$ were used. (c) Spectral radiation heat fluxes without (1) and with (2) atomic lines and corresponding cumulative functions (3,4) of radiation heat fluxes in stagnation point.

distribution of the heat fluxes with the cumulative functions shows that the two significant jumps of the total heat flux in regions 10 000–20 000 cm^{-1} and 78 000–105 000 cm^{-1} are connected to free–free (radiative-recombination) and bound–bound transitions in atomic and ionic species. The main contribution in the spectral region 20 000–80 000 cm^{-1} is due to diatomic molecules neutral and ionized. One can see that molecular emission, located in the region between near infrared and near ultraviolet, gives a small contribution into integral radiation heat flux. However, the photon emission of electronic bands is used for spectral analysis, therefore, several electronic bands of heated air have attracted much attention from specialists in the aerophysics of return space vehicles, such as: the Vegard–Kaplan bands, the Birge–Hopfield, 1 and 2 bands, the first, second and the fourth positive systems of N_2, γ^-, β^-, δ^-, and ε^- systems of NO, namely the Meinel's auroral system, the first and fourth negative systems of N_2^+. The population of highly excited electronic states just behind the shock wave front reflects in full measure the differences in translational and electronic temperatures.

16.2.3 RAM-C-II

Examples of the 2D and 3D numerical simulation results for RAM-C-II space vehicle are presented in this section. A kinetic model of air ionization behind the shock front at speeds of $V_\infty < 9$ km s^{-1} was analyzed in references [26, 50]. The seven components of partially ionized air (N_2, O_2, N, O, NO, NO^+, e^-) and the following six chemical reactions were considered:

$$\begin{aligned}
(R1)\quad & N_2 + M \rightleftharpoons N + N + M \\
(R2)\quad & O_2 + M \rightleftharpoons O + O + M \\
(R3)\quad & NO + M \rightleftharpoons N + O + M \\
(R4)\quad & N_2 + O \rightleftharpoons NO + N \\
(R5)\quad & O_2 + N \rightleftharpoons NO + O \\
(R6)\quad & N + O \rightleftharpoons NO^+ + e^-
\end{aligned} \tag{16.28}$$

The first three reactions of this model are the dissociation and recombination in collisions with a third particle M. Bimolecular reactions ($R4$) and ($R5$) are important in the kinetics of formation of NO. In the model under consideration, the basic process with participation of charged particles is the associative ionization ($R6$) (and the reverse, the dissociative recombination). In reference [51] the role of electron impact ionization is discussed. Based on these results [26, 51] the conclusion that the main reaction is the associative ionization is reached. At higher speeds, and correspondingly at higher temperatures, collisional ionization and ion–molecule reactions become more significant.

Park's kinetic model [13] was chosen for the calculations in the approximation of LTE, together with various models of non-equilibrium dissociation [12, 48, 49]. It is well known that assuming an Arrhenius rate for the associative ionization reaction ($R6$) introduces some errors, because it proceeds at least in two steps. Firstly, N and O atoms form NO^* which thereafter decay into the NO^+ and an electron. This is the reason for the large differences between the rate constants used by different authors. A detailed description of this reaction is also given in reference [52]. The free stream conditions for the calculation here presented have been taken from the experimental values reported in table 16.3. For reference, characteristic Mach numbers and Reynolds numbers are also reported in the table. The distributions of translational temperature and the longitudinal velocity at altitude $H = 71$ km are shown in figure 16.4(a).

With increasing altitude, the degree of non-equilibrium in the stream grows. Figure 16.4(b) shows the axial distributions of translational and vibrational temperatures along the stagnation line. It can be observed that relevant differences between

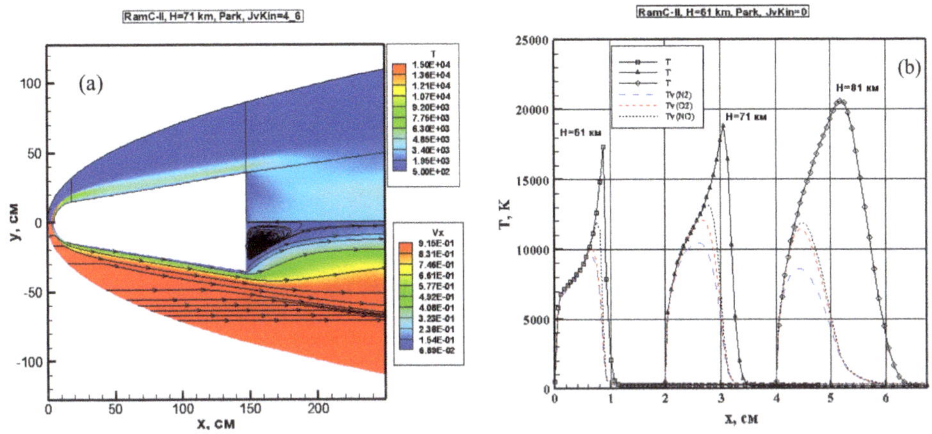

Figure 16.4. RAM-C-II: (a) Translational temperature, T in K (top) and the longitude velocity (bottom) at $H = 71$ km and $V_\infty = 7.64$ km s^{-1}; (b) axial distribution of translational and vibrational temperature, along the front of the critical line. Point of the trajectory $H = 61$ km and $V_\infty = 7.5$ km s^{-1}, $H = 71$ km and $V_\infty = 7.64$ km s^{-1}, $H = 81$ km and $V_\infty = 7.95$ km s^{-1}. Park's kinetic model with the second model of Marrone–Treanor [49] at $U = 6$. Temperature distribution for different heights are shifted to the right by 2 cm.

the temperatures are achieved at a relatively large height ($H = 81$ km). In this case, the calculation of the rate constants of dissociation was performed in LTE, i.e. the translational temperature is used in equation (16.13).

Figure 16.5 shows the maximum electronic density along the surface of the blunted cone at three different altitudes, calculated using different models of non-equilibrium dissociation as reported in table 16.4 and compared with the experimental data [24, 25]. Increasing the U from 3 to 6 results in faster dissociation from vibrational states with small vibrational quantum number. All the dissociation models show good agreement with experimental data for low altitudes (61 and 71 km), however differences larger than one order of magnitude are obtained. A better agreement is obtained by using the second Marrone–Treanor model [49], while LTE model overestimates electron density by more than a factor of 10.

Let us analyze the behaviors of the flow around the hypersonic vehicle RAM-C-II changing the angle of attack. A windward surface of the cone is exposed to higher pressure than in the axisymmetric case. This is clearly seen when comparing the temperature fields shown in figure 16.6. Even if a good accuracy was achieved in reproducing the flight test data of RAM-C-II in a wide range of altitudes, it should be pointed out that this result was actually possible only by selecting suitable physical and chemical models. Constructing realistic models for non-equilibrium conditions capable of predicting flight conditions still remains an open problem in modern aerophysics.

16.2.4 ORION

In this part we consider general elements of radiative gas dynamics of large-scale space vehicle ORION at its entry into Earth atmosphere with orbital velocity.

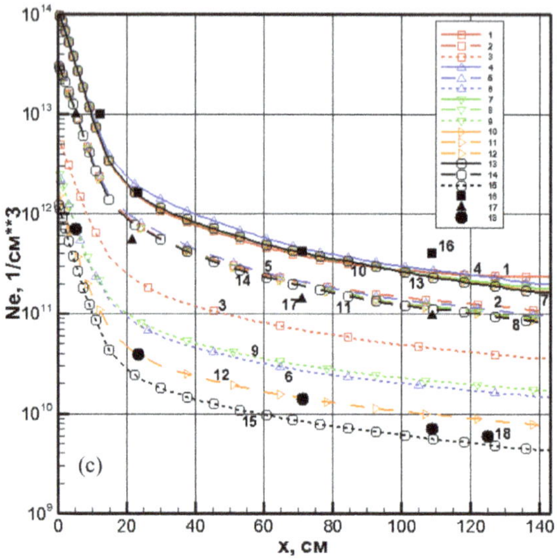

Figure 16.5. Distribution of the maximum electron density along x-axis of the experimental spacecraft. See table 16.4 for curve descriptions.

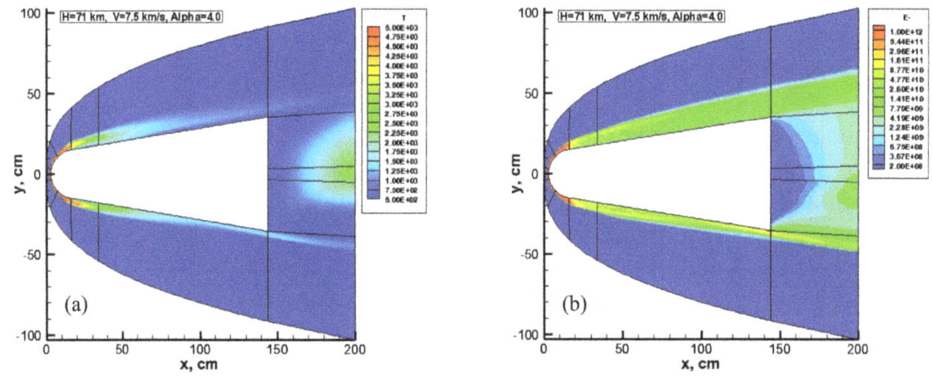

Figure 16.6. RAM-C-II: 3D numerical simulation results for translational temperature T (in K) and electron number density N_e (cm^{-3}) in the plane of symmetry at $H = 61$ km and $V_\infty = 7.5$ km s^{-1} at angle of attack $\alpha = 4°$.

Two-dimensional fields of the translational temperature and longitudinal velocity are shown in figure 16.7(a) while temperature distributions along stagnation line for trajectory point $t = 200$ s are shown in figure 16.7(b). The dotted lines correspond to calculations without coupling gas dynamics with radiation, while the solid lines correspond to translational and vibrational temperatures including radiation–gas-dynamics coupling. These temperature profiles indicate the non-equilibrium relaxation zone behind the shock wave front (larger than 2 cm). Total stand-off distance of the shock wave achieves $\approx$ 32 cm. This shock layer is a noticeable source of heat radiation. It is well known that increasing the blunt radius of a space vehicle leads to decrease of

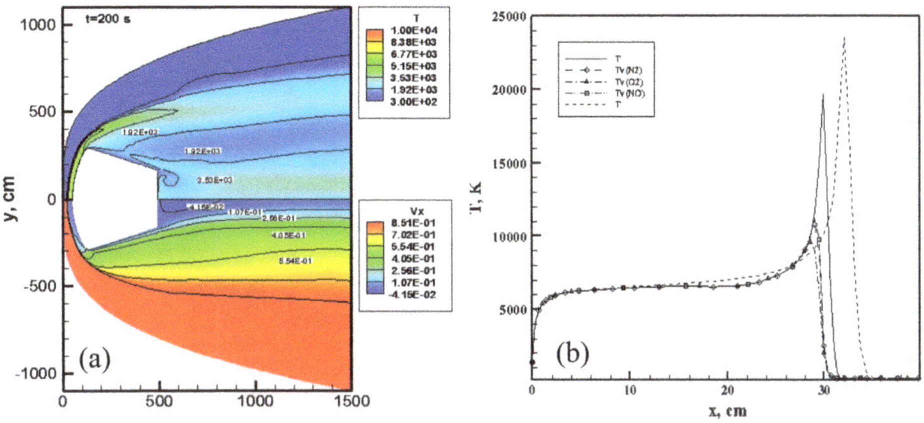

Figure 16.7. Space vehicle ORION: (a) field of longitudinal velocity (normalized to V_∞) and translational temperature (in K) at $t = 200$ s; (b) translational and vibrational temperatures along forward stagnation line. Solid and dotted lines without markers are the translational temperatures with and without radiation–gas-dynamics coupling correspondingly.

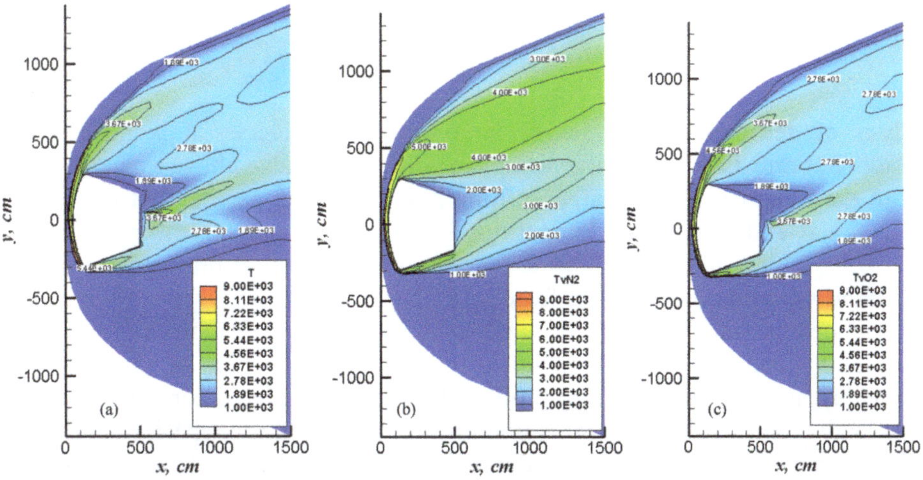

Figure 16.8. Space vehicle ORION: mass fractions of N_2 and O_2 at $t = 200$ s. Angle of attack $\alpha = 15°$.

convective heat fluxes, as well as to increasing the shock wave stand-off distance, and, as a consequence, radiative heat fluxes also increase. Therefore, it is expected that for some trajectory points one can find conditions where radiative heating overcomes the convective contribution. Such a situation dictates more attention to which relaxation kinetics should be used to model the shock formation (see table 16.5).

More realistic cases for this kind of space vehicle must consider entering conditions with given angle of attack. Therefore, three-dimensional radiative gas-dynamic models must be used just for these purposes. Numerical simulation results obtained for ORION spacecraft at angle of attack $\alpha = 15°$ at the first trajectory points ($t = 200$ s) are shown in figures 16.8 and 16.9. It should be noted that for

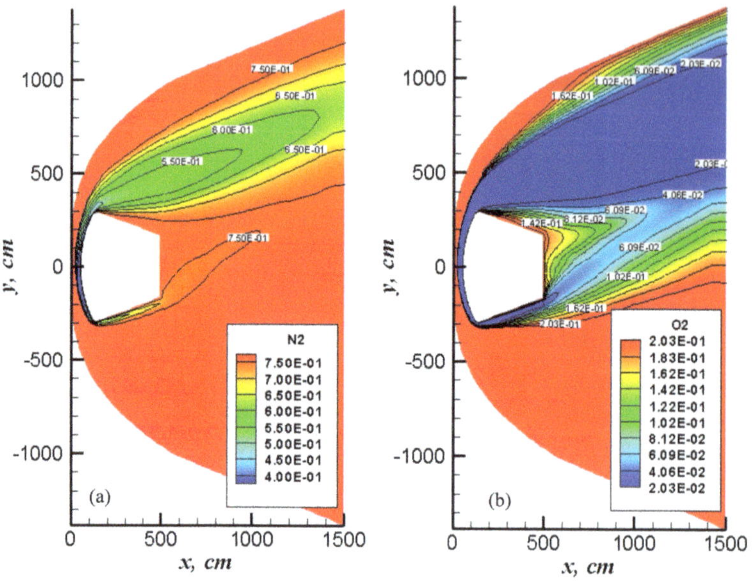

Figure 16.9. Space vehicle ORION: Translational and vibrational temperatures of N_2 and O_2 at t = 200 s. Angle of attack $\alpha = 15°$.

analysis of aerothermodynamics of large-size space vehicles it is necessary to consider not only flow field above windward surface, but also above leeward surface and in the wake because radiative heating has volumetric genesis. Let us focus on translational and vibrational temperatures in the near and far wake reported in figure 16.8. These figures adumbrate a degree of non-equilibrium between translational and vibrational degrees of freedom in N_2 and O_2.

The effective temperature of vibrational excitation of molecules of N_2 is significantly higher than other temperatures in the large part of the flow field. Noticeably, not only temperature profiles should be considered correctly to predict the radiative heating, but also mass fractions of optically active species (figure 16.9) and in particular N_2, O_2 and NO, the main emitters of heat radiation in the case under consideration. The distribution of species concentrations is strongly dishomogeneous not only along the leeward surface but also above the windward surface, which is accompanied by appreciable variation of the shock layer depth having a direct impact on the intensity of the heat radiation.

16.2.5 PTV

The radiating aerothermodynamics of Prospective Transport Vehicle (PTV) in two-dimensional formulation was investigated at six points of trajectory of super-orbital entry, which are presented and marked in table 16.6. These points correspond to non-equilibrium, quasi-equilibrium and actually fully equilibrium conditions in the shock layer. The calculations of flow fields around the space vehicle were carried out

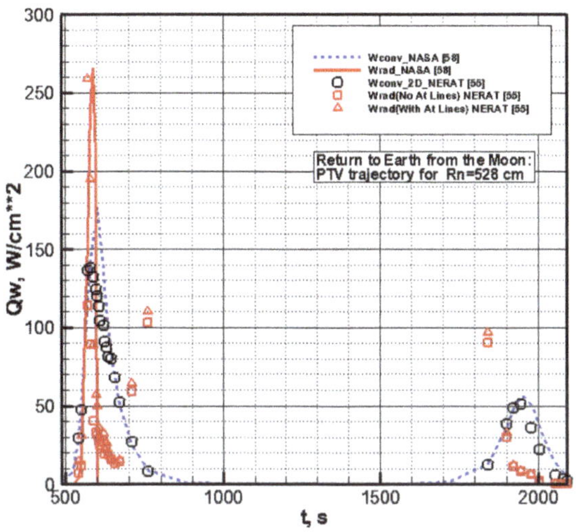

Figure 16.10. PTV: (a) p/p_0 (p_0 is the break pressure); (b) longitudinal and (c) radial velocities normalized to V_∞; (d) translational temperature and vibrational of (e) N_2 ($Tv1$) and (f) O_2 ($Tv2$) in K.

in three series of multi-block computational grids of different minuteness. All calculations were performed for pseudo-catalytic surface.

Figure 16.11 shows the comparison between the convective and radiative heating of super-orbital PTV versus time of flight. In the first part of the trajectory the radiative heating is dominant and the contribution of atomic lines is significant. Therefore, the conclusions concerning the influence of atomic lines on radiative heating, already discussed for Fire-II flight data, holds also for super-orbital large-size space vehicles.

Let us discuss the basic laws of radiative and convective heat transfer in a shock layer on the surface of a super-orbital spacecraft of large size considering the example of the first trajectory point in table 16.6. Figure 16.11 shows flow quantities in the shock layer and figure 16.12(a) reports the profile of the translational and vibrational temperatures along the stagnation line. Significant differences between the translational and vibrational temperatures can be observed, evidence of thermodynamic non-equilibrium behind the shock wave front. Distributions of mole fractions along the stagnation line are presented in figure 16.12(b). These profiles allow one to estimate the degree of dissociation and ionization in the shock layer. The spectral structure of the radiative heat fluxes in PTV's stagnation point is illustrated in figure 16.12(c), reporting the cumulative functions and corresponding spectral heat fluxes. Note that the value of the cumulative function at the maximal value of the wavenumbers $\omega_{max} = 120\,000$ cm^{-1} corresponds to the total integral flux to the stagnation point while the cumulative function as a function of ω is calculated according equation (16.8).

These results allow us to determine the characteristics of radiation-gas dynamics for the super-orbital re-entry of PTV with radius of blunting of the aerodynamic heat shield of 528 cm:

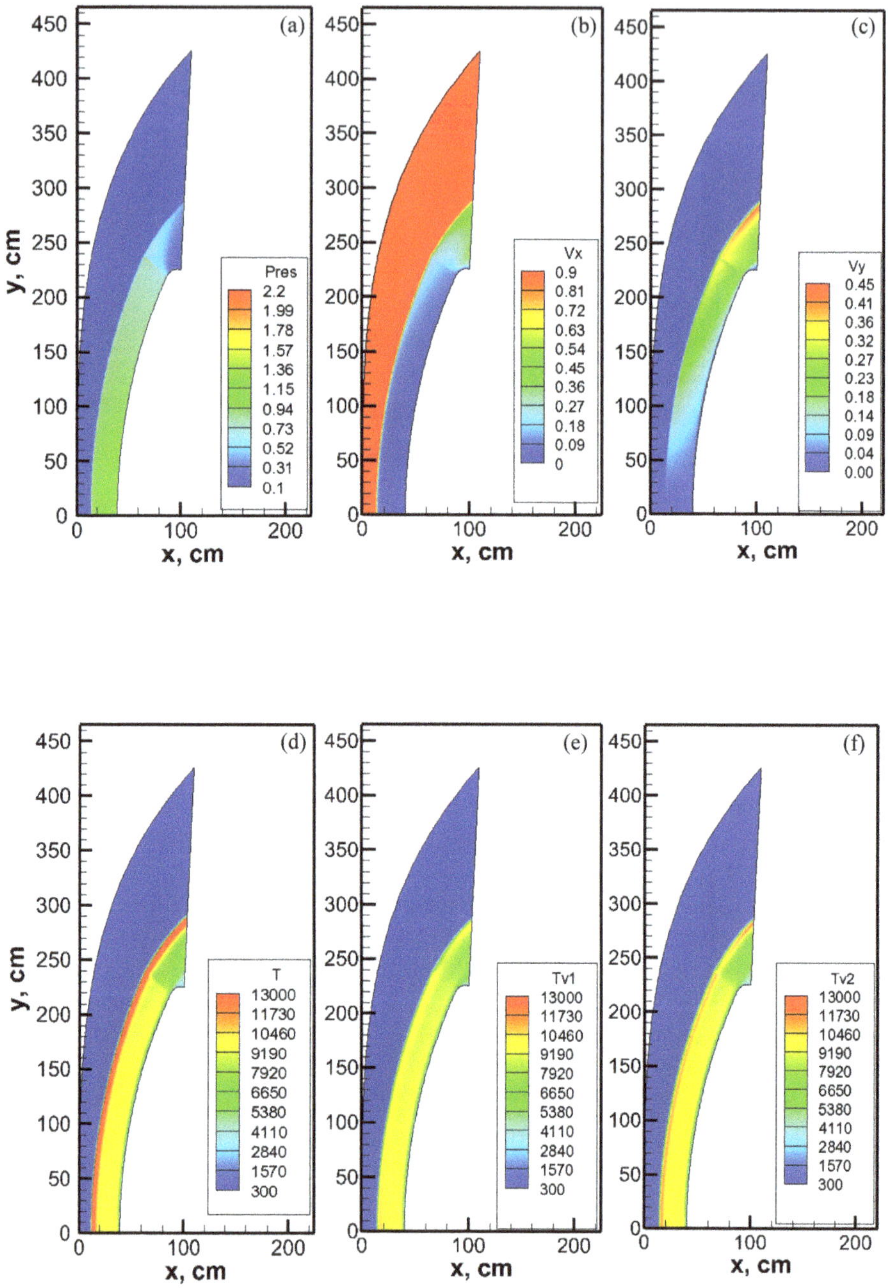

Figure 16.11. PVT: radiative and convective heating of super-orbital large-size PTV for two trajectory points. Solid red line and dotted blue line correspond to correlation [53]. Black circles are the convective heating, predicted by code NERAT [44]. Red squares and triangles are radiative heating without and with atomic lines, predicted by codes NERAT + ASTEROID [44].

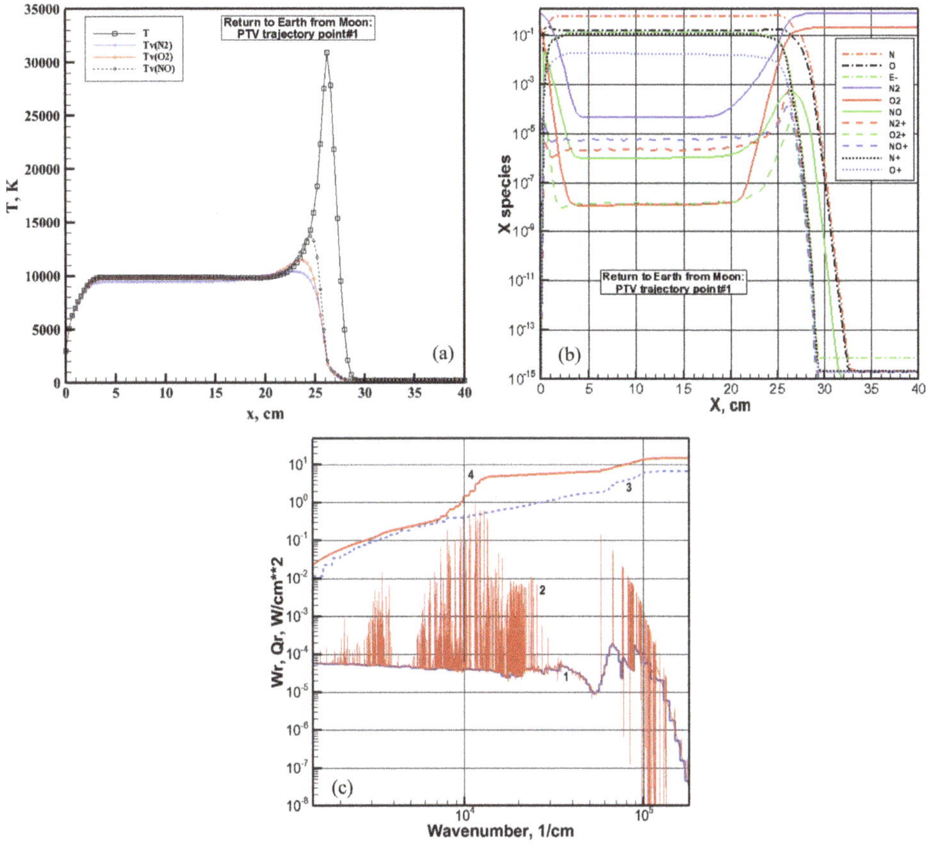

Figure 16.12. PTV: (a) axial distribution of translational and vibrational temperatures in K; (b) species molar fractions along stagnation line; (c) spectral radiation heat fluxes (1,2) and cumulative functions of radiation heat fluxes with (2,4) and without (1,3) atomic lines in stagnation point.

1. The thickness of the compressed layers at the front surface varies in the range $\delta = 22–35$ cm. The thickness of the relaxation zone behind the shock front varies in the range of $h = 1–10$ cm;
2. At the first entering of the spacecraft into Earth's atmosphere at velocities $V_\infty > 9.5$ km s^{-1}, the temperature in the compressed layer reaches $T = 10\,000$ K. The detachments of shocks from the surface and the thicknesses of the relaxation zone are: $\delta_1 = 23$ cm and $h_1 = 4$ cm.
3. The distributions of radiation fluxes towards the surface are of special practical interest. The first trajectory point corresponds to the beginning of the growth of radiant fluxes, reaching its maximum in the non-equilibrium relaxation zone behind the shock front. Compressed layers are regions where weakening and strengthening of radiation fluxes are observed. In the second trajectory point, corresponding to the maximum of radiative heating, radiation fluxes to the surface grow over the whole compressed layer. Only in the boundary layer near the surface there is a slight weakening. In the third

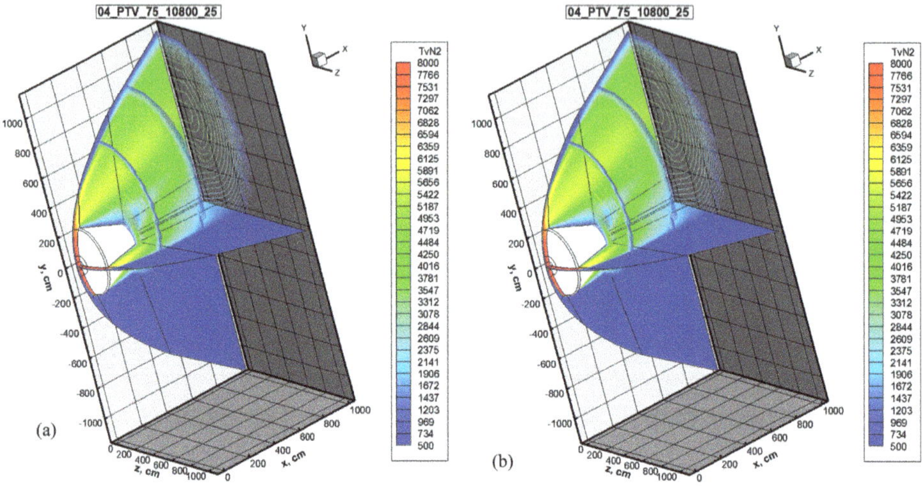

Figure 16.13. PTV: 3D field of (a) T and (b) $T_v(N_2)$ at H = 75 km, V_∞ = 10.8 km s^{-1}, α = 25°.

point of the trajectory, corresponding to the beginning of PTV's entry in the rarefied layers of atmosphere, there is a sharp increase in the radiative flux in the relaxation zone behind the shock front, monotonically increasing in the compressed layer. Thus, we can conclude that a significant portion of radiation flux to the surface is initiated in the non-equilibrium relaxation zone.

4. Analysis of the spectral distribution of the radiation fluxes and cumulative functions shows that the contribution of the atomic lines may exceed 100% in the conditions under investigation. By decreasing the velocity below 10 km s^{-1}, the contribution of atomic lines is reduced.
5. Distributions of the gas-dynamic functions in the compressed layer show a sufficiently high uniformity along the front surface of the aerodynamic shield. This justifies the use of the approximation of the locally one-dimensional layer to calculate radiative transfer. The increase of viscosity and thermal conductivity in the region of shock wave is notable. The degree of ionization decreases quite rapidly, departing from the critical line of the flow to the periphery of the aerodynamic shield.

The three-dimensional gas-dynamics fields around landing module of large size under an angle of attack of α = 25° are shown in figures 16.13 and 16.14. Parameters of investigated trajectory points are presented in table 16.1. Figure 16.13 reports translational temperature and vibrational temperatures of N_2. These figures are provided by the same color scale, in order to allow direct comparison.

The degree of non-equilibrium in the flow field can be appreciated, noticing that the vibrational temperature is much higher than the translational one. The high temperature in the shock layer and strong asymmetry of the flow field, a consequence of the angle of attack, are clearly seen here.

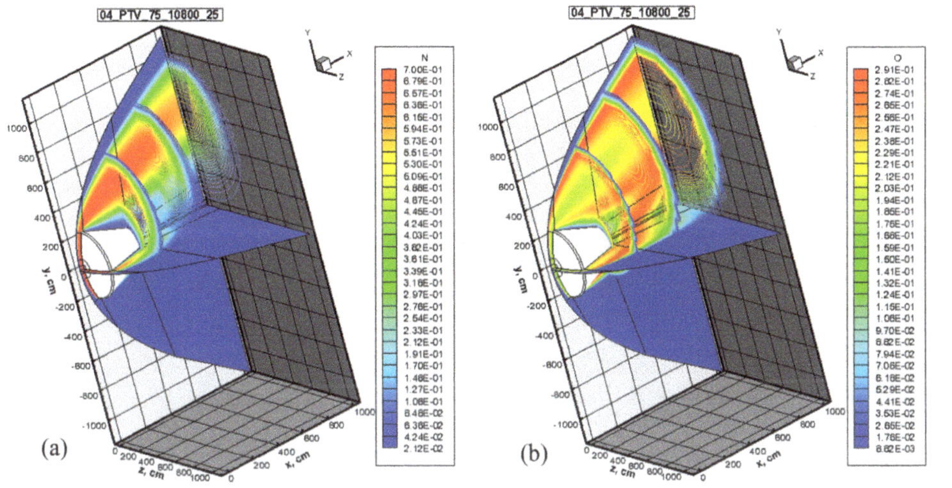

Figure 16.14. PTV: 3D field of molar fraction of (a) N and (b) O at $H = 75$ km, $V_\infty = 10.8$ km s^{-1}, $\alpha = 25°$.

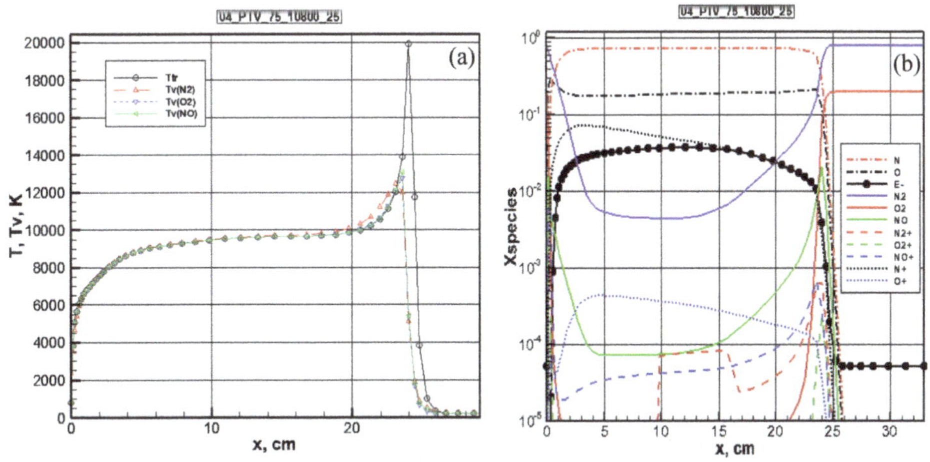

Figure 16.15. PTV: Axial distributions of translational and vibrational temperatures (a) and molar fractions (b) along the stagnation line at $H = 75$ km, $V_\infty = 10.8$ km s^{-1}, $\alpha = 25°$.

The molar fractions of atoms N and O, as well as of diatomic molecules N_2 and O_2 are presented in figure 16.14. These figures give general representation concerning degree of dissociation of diatomic molecules, which was calculated using models of non-equilibrium dissociation.

Axial distributions of the translational and vibrational temperatures, as well as molar. The fractions of the gas components along the stagnation line are shown in figure 16.15. These data give a full representation concerning the thermal conditions in the most high temperature region and degree of ionization and dissociation.

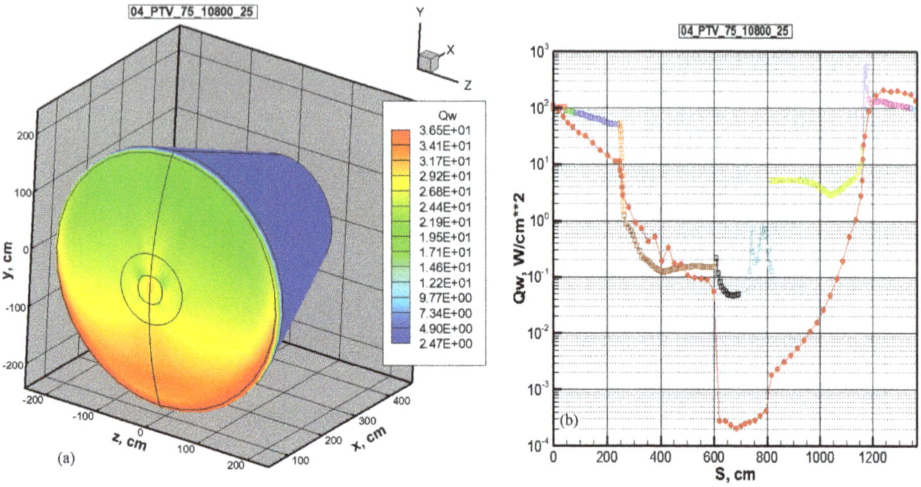

Figure 16.16. PTV: (a) distribution of convective heat flux density (in W cm^{-2}) on the front shield and (b) convective (squares) and radiative (circles) heating of PTV surface at angle of attack $\alpha = 25°$ and $H = 75$ km, $V_\infty = 10.8$ km s^{-1}. Coordinate S along the surface is counted off from the black point in the direction marked by the arrow in the left side of the figure.

The general results of the numerical simulations are presented in figure 16.16, where the convective heat flux on the front aerodynamic shield, distributions of the convective heat flux, and the integral radiation heat fluxes along the surface in meridional cross-sections are shown. In the case under consideration, the radiation heat flux exceeds the convective heat flux on the front shield surface, especially in the region of leeward part.

16.2.6 MSL

In the previous sections we considered radiative gas-dynamic problems of Earth's return hypersonic flights. The radiative gas dynamics of Martian space vehicles are also of practical interest. Recently obtained experimental data on heating of MSL during real descent trajectory [45], provide the possibility to validate computational data. Figure 16.17 shows the flow field around the space vehicle MSL with catalytic surface under angle of attack $\alpha = 16°$. The distinctive features of the flow field are:

- The predicted asymmetry of the flow field (in particular, of the translational temperature and longitudinal velocity) is formed due to angle of attack $\alpha = 16°$;
- An expansion of the shock layer above the windward surface from the direction of incident flow is observed, as well as the vortex structure above the top part of the leeward surface;
- High-mass fractions of CO_2 and low mass fractions of CO are observed for catalytic surface;

The distributions of convective and radiative heat fluxes along the meridional surface of MSL entry module are presented in figure 16.18(b). The starting position

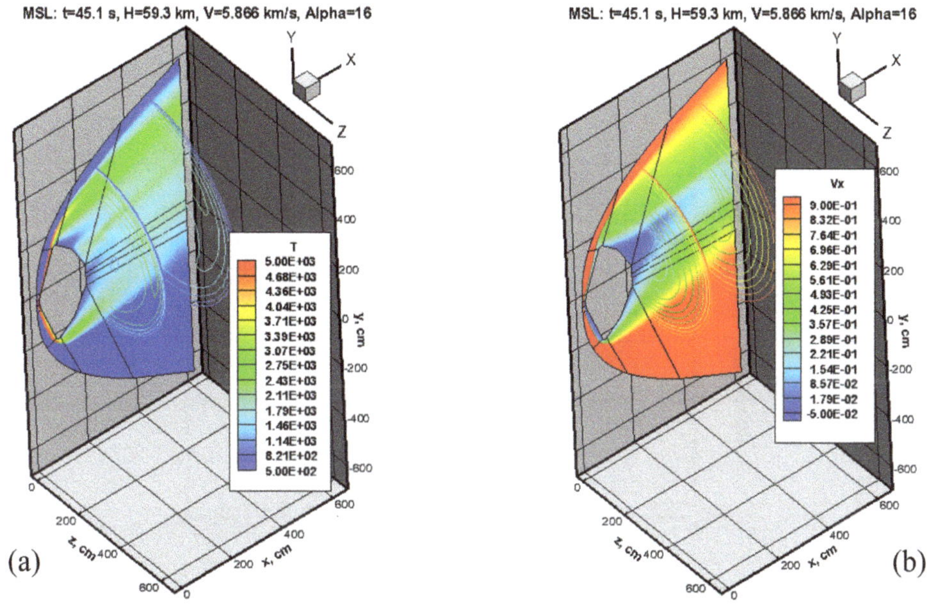

Figure 16.17. (a) Translational temperature and (b) longitudinal velocity normalized to V_∞ at t = 45.1 s, H = 59.3 km, $\rho_\infty = 2.67 \cdot 10^{-8}$ g cm^{-3}, p_∞ = 4.29 erg cm^{-3}, V_∞ = 5.866 km s^{-1}, T = 85.7 K, $\alpha = 16°$.

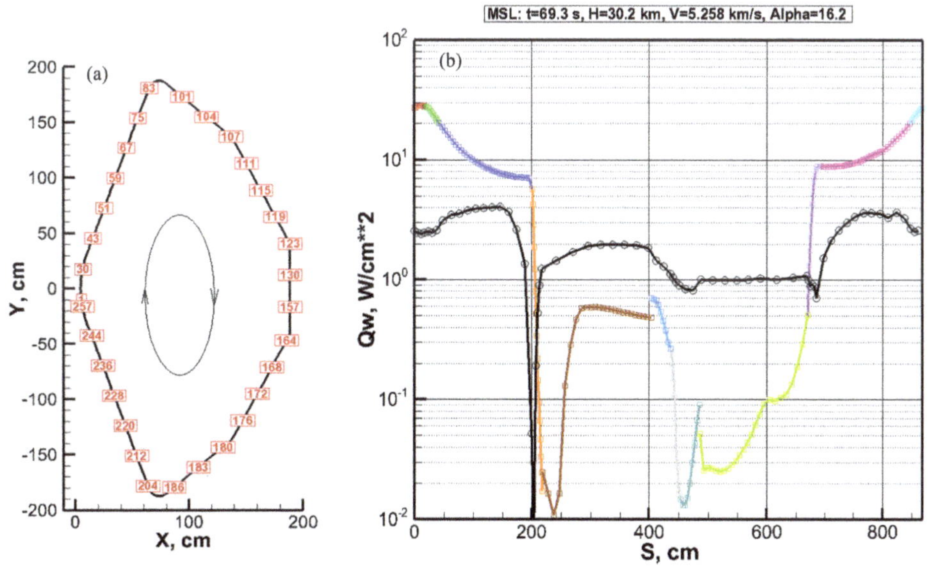

Figure 16.18. (a) Location of grid nodes in equatorial cross-section and (b) convective heat flux (color curve) and integral radiation heat flux (black line) along surface coordinate S in meridional plane. t = 69.3 s, H = 30.2 km, $\rho_\infty = 6.96 \cdot 10^{-7}$ g cm^{-3}, p_∞ = 222 erg cm^{-3}, V_∞ = 5.258 km s^{-1}, T = 168.5 K, $\alpha = 16.2°$.

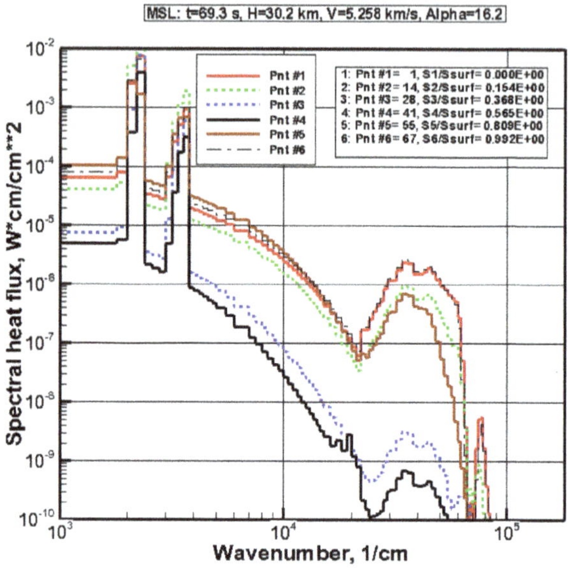

Figure 16.19. Spectral radiation heat fluxes at six points on the surface in meridional cross-section. $t = 69.3$ s, $H = 30.2$ km, $\rho_\infty = 6.96 \cdot 10^{-7}$ g cm^{-3}, $p_\infty = 222$ erg cm^{-3}, $V_\infty = 5.258$ km s^{-1}, $T = 168.5$ K, $\alpha = 16.2°$.

of the S coordinate is counted from the point shown in figure 16.18(a). It should be noted that radiative heating is significant for the back side of the landing module. The maximum value of radiative heat fluxes reaches 1.7 W cm^{-2}. Spectral radiation heat fluxes for separate points on the landing module surface are shown in figure 16.19, reporting the spectral radiation heat fluxes in six different points on the surface, corresponding to positions (1, 36, 95, 146, 209, 256) in figure 16.18(a). Two bands of CO_2 and CO contribute to surface illumination for all trajectory points. These are 2.7 μ and 4.3 μ bands, corresponding to wavenumbers $\omega \approx 3703$ cm^{-1} and $\omega \approx 2325$ cm^{-1}. Some electronic bands are observed on the front shield of the lander, whose contribution is practically small.

16.3 Conclusions

The modern theory of high-enthalpy radiating flows has been realized in computational models of the aerophysics of spacecraft. Numerical analysis of radiative aerothermodynamics of the Earth's and Martian space vehicles has been presented and analyzed. The distinguishing feature of the presented data is the comparison of convective and radiative heating of the landing module surfaces, as well as the comparison of obtained calculation data on convective heating with measured experimental data. Three-dimensional CFD and Radiative Gas Dynamics (CFD/RadGD) code NERAT(3D) has been used for the investigation of radiative gas dynamics of different landing modules under angles of attack. Three-dimensional calculations of convective and radiative heating have been performed. It has been shown that for some trajectory points the radiative heating of the leeward surface exceeds the convective heating. The data obtained on convective heating are in good agreement with real space mission data.

References

[1] Pringle G and King A 2007 *Astrophysical Flows* (Cambridge: Cambridge University Press) p 206

[2] LeVeque R J, Mihalas D, Dorfi E and Müller E 1998 *Comutational Methods for Astrophysical Fluid Flow: Saas-Fee Advanced Course* (Lecture Notes 1997 Swiss Society for Astrophysics and Astronomy vol 27) (Berlin: Springer) p 508

[3] Shu F 1992 *Physics of Astrophysics, Vol. II: Gas Dynamics* (Mill Valley, CA: University Science Books) p 476

[4] Rybicki G B and Lightman A P 2008 *Radiative Processes in Astrophysics* (Weinheim: Wiley) p 382

[5] Pal S-I 1966 *Radiation Gas Dynamics* (New York: Springer)

[6] Martin J 1966 *Atmospheric Entry–An Introduction to Its Science and Engineering* (Englewood Cliffs, NJ: Prentice-Hall)

[7] Bond J W, Watson K M and Welch J A 1965 *Atomic Theory of Gas Dynamics* (Reading, MA: Addison-Wesley)

[8] Siegel R and Howell J 1972 *Thermal Radiation Heat Transfer* (New York: McGraw-Hill)

[9] Öışık M N 1973 *Radiative Transfer and Interactions with Conduction and Convection* (New York: Wiley)

[10] Chapman S and Cowling T G 1965 *The Mathematical Theory of Non-uniform Gases* (Cambridge: Cambridge University Press) pp 134–50

[11] Clarke J F, McChesney M and Talbot L 1964 *The Dynamics of Real Gases* vol 17 (Washington, DC: Butterworth Inc.) p 76

[12] Park C 1990 *Nonequilibrium Hypersonic Aerothermodynamics* (New York: Wiley)

[13] Park C 1993 Review of chemical-kinetic problems of future NASA missions. I–Earth entries *J. Thermophys. Heat Transfer* **7** 385–98

[14] Park C, Jaffe R L and Partridge H 2001 Chemical-kinetic parameters of hyperbolic earth entry *J. Thermophys. Heat Transfer* **15** 76–90

[15] Tchuen G and Zeitoun D E 2009 Effects of chemistry in nonequilibrium hypersonic flow around blunt bodies *J. Thermophys. Heat Transfer* **23** 433–42

[16] Bird G 2011 The Q-K model for gas-phase chemical reaction rates *Phys. Fluids* **23** 106101

[17] Wagnild R and Gallis M A 2013 *Continuum Simulation of Hypersonic Flows using the Quantum-Kinetic Chemical Reaction Model* AIAA 2013-3144 https://doi.org/10.2514/6.2013-3144

[18] Liechty D S 2013 *State-to-State Internal Energy Relaxation Following the Quantum-Kinetic Model in DSMC* AIAA 2013-2901 https://doi.org/10.2514/6.2013-2901

[19] Olynick D, Chen Y-K and Tauber M E 1999 Aerothermodynamics of the Stardust sample return capsule *J. Spacecr. Rockets* **36** 442–62

[20] Zel'dovich Y B and Raizer Y P 2002 *Physics of Shock Waves and High-temperature Hydrodynamics Phenomena* (New York: Dover)

[21] Mihalas D and Weibel-Mihalas B 1984 *Foundations of Radiation Hydrodynamics* (New York: Oxford University Press) pp 309–402

[22] Shang J S and Surzhikov S T 2011 Simulating Stardust Earth reentry with radiation heat transfer *J. Spacecr. Rockets* **48** 385–96

[23] Liu Y, Prabhu D, Trumble K A, Saunders D and Jenniskens P 2010 Radiation modeling for the reentry of the Stardust sample return capsule *J. Spacecr. Rockets* **47** 741–52
[24] Akey N D and Cross A E 1970 *Radio blackout alleviation and plasma diagnostic results from a 25000 foot per second blunt-body reentry* Tech. Rep. TN D-5615, NASA
[25] Grantham W L 1970 *Flight results of a 25000-foot-per-second reentry experiment using microwave reflectometers to measure plasma electron density and standoff distance* Tech. Rep. TN D-6062, NASA
[26] Candler G V and MacCormack R W 1991 Computation of weakly ionized hypersonic flows in thermochemical nonequilibrium *J. Thermophys. Heat Transfer* **5** 266–73
[27] Olynick D, Henline W, Hartung L and Candler G 1994 *Comparison of Coupled Radiative Navier–Stokes Flow Solutions with the Project Fire-II Flight Data* AIAA 94-1955 https://doi.org/10.2514/3.712
[28] Johnston C O, Hollis B R and Sutton K 2008 Spectrum modeling for air shock-layer radiation at lunar-return conditions *J. Spacecr. Rockets* **45** 865–78
[29] Johnston C O, Hollis B R and Sutton K 2008 Nonequilibrium stagnation-line radiative heating for Fire-II *J. Spacecr. Rockets* **45** 1185–95
[30] Surzhikov S and Shang J 2012 Fire-II flight data simulations with different physical-chemical kinetics and radiation models *Frontiers Aerosp. Eng.* **1** 1–12
[31] Desai P N, Lyons D T, Tooley J and Kangas J 2008 Entry, descent, and landing operations analysis for the Stardust entry capsule *J. Spacecr. Rockets* **45** 1262–8
[32] Jenniskens P 2010 Observations of the Stardust Sample Return Capsule entry with a slitless echelle spectrograph *J. Spacecr. Rockets* **47** 718–35
[33] Panesi M, Magin T, Bourdon A, Bultel A and Chazot O 2011 Electronic excitation of atoms and molecules for the Fire-II flight experiment *J. Thermophys. Heat Transfer* **25** 361–74
[34] Trigo-Rodriguez J M, Rietmeijer F J, Llorca J, Llorca J and Janches D 2008 *Advances in Meteoroid and Meteor Science* (Berlin: Springer) p 561
[35] Kosarev I 2009 The optical properties of vapors of matter of cosmic bodies invading the Earth atmosphere *High Temp.* **47** 777–87
[36] Surzhikov S T 2014 *Non-Equilibrium Radiative Gas Dynamics of Small Meteor* AIAA 2014-2636 https://doi.org/10.2514/6.2014-2636
[37] Gurvich L, Veitc I and Medvedev V *et al* 1978 *Thermodynamic Properties of Individual Substances* (Moscow: Nauka)
[38] Chase M, Davies C A and Downey J R *et al* 1985 JANAF thermochemical tables *J. Phys. Chem. Ref. Data* **14** suppl. 1 p 1856
[39] Bird R, Stewart W and Lightfoot E 1965 *Transport Phenomena* (New York: Wiley)
[40] Hirschfelder J O, Curtiss C F, Bird R B and Mayer M G 1954 *Molecular Theory of Gases and liquids* vol 26 (New York: Wiley)
[41] Wilke C R 1950 Diffusional properties of multicomponent gases *Chem. Eng. Prog.* **46** 95–104
[42] Anfimov N 1962 The laminar boundary layer above chemically active surface *Proc. of the Academy of Sciences, Mechanics and Machine-building* vol 3
[43] Stanley D, Cook S, Connolly J, Hamaker J, Ivins M, Peterson W, Geffre J, Cirillo B, McClesky C and Hanley J *et al* 2005 *NASA's exploration systems architecture study* Tech. Rep. TM-2005-214062, NASA
[44] Surzhikov S 2016 *Radiative Gas Dynamics of Large Superorbital Space Vehicle at Angle of Attack* AIAA 2016-0741 https://doi.org/10.2514/6.2016-0741

[45] Edquist K T, Hollis B R, Johnston C O, Bose D, White T R and Mahzari M 2014 Mars science laboratory heat shield aerothermodynamics: design and reconstruction *J. Spacecr. Rockets* **51** 1106–24

[46] Edquist K T 2005 *Afterbody heating predictions for a Mars science laboratory entry vehicle* AIAA 2005-4817 https://doi.org/10.2514/6.2005-4817

[47] Dunn M G and Kang S 1973 *Theoretical and experimental studies of reentry plasmas* Tech. Rep. CR-2232, NASA

[48] Treanor C E and Marrone P V 1962 Effect of dissociation on the rate of vibrational relaxation *Phys. Fluids* **5** 1022–6

[49] Marrone P V and Treanor C E 1963 Chemical relaxation with preferential dissociation from excited vibrational levels *Phys. Fluids* **6** 1215–21

[50] Park C, Howe J T, Jaffe R L and Candler G V 1994 Review of chemical-kinetic problems of future NASA missions. II—Mars entries *J. Thermophys. Heat Transfer* **8** 9–23

[51] Josyula E, Bailey W F and Ruffin S M 2003 Reactive and nonreactive vibrational energy exchanges in nonequilibrium hypersonic flows *Phys. Fluids* **15** 3223–35

[52] Boyd I D 2007 Modeling of associative ionization reactions in hypersonic rarefied flows *Phys. Fluids* **19** 096102

[53] Brandis A M and Johnston C O 2014 *Characterization of Stagnation-Point Heat Flux for Earth Entry* AIAA 2014-2374 https://doi.org/10.2514/6.2014-2374

IOP Publishing

Plasma Modeling (Second Edition)
Methods and applications
Gianpiero Colonna and Antonio D'Angola

Chapter 17

Simulating plasma aerodynamics

Jonathan Poggie

For high-speed travel, the greatest expenditure of energy goes toward overcoming drag, the force impeding motion in air. Designers attempt to minimize drag through careful shaping of the vehicle. They try to avoid separation (detachment of the flow from the surface), and to delay transition from orderly, laminar flow to chaotic, turbulent flow. On aircraft, they try to avoid or weaken shock waves, which dissipate mechanical energy into heat.

Vehicle design involves many trade-offs, and the vehicle shape cannot always be optimized to minimize drag. For example, the long, tapered shape required to avoid separation on the back of an automobile would make access to the trunk difficult, and parking almost impossible. Thus aerodynamicists must turn to flow control to mitigate the effects of a suboptimal shape.

Flow control has its own complications. Fixed flow control devices may increase drag under conditions where their presence is unnecessary. The small vanes (vortex generators) used for separation (stall) resistance on an airplane wing fall into this category. Active flow control devices, such as synthetic jets, must respond at the characteristic time scales of the flow. As flow speed increases, the characteristic time scales become smaller, and mechanical actuators are unable to produce the needed control.

Electrical discharges and other plasma-based devices are an alternative means of flow control that show promise for a variety of aerodynamic applications [1]. Plasma actuators can be turned completely off when not in use. Furthermore, low-profile, flush-mounted electrodes have very low drag penalty when turned off. These devices offer the opportunity to actuate the flow at electrical time scales as short as nanoseconds, whereas mechanical devices are typically limited to millisecond time scales and above.

The use of plasma-based technology for aerodynamics applications has an approximately seventy-year history. Work during the Space Race era (1950s to 1970s) was motivated by the thought that atmospheric entry plasma could be

doi:10.1088/978-0-7503-3559-1ch17

manipulated for beneficial effects [2]. A resurgence of the field in the post Cold War era (1990s to 2000s) was related to the disclosure of the Soviet AJAX hypersonic concept vehicle [3] and pioneering work on plasma actuators [4, 5]. The term *plasma aerodynamics* was coined in that era to describe this field. Since then, the field has grown into an important component of aerodynamic research and development.

Plasma-based devices can interact with a flow by introducing new species, by introducing an electromagnetic body force, or by thermal interaction with the flow. The thermal interaction is usually in the form of Joule heating, in which part of the energy of an electric current running through a resistive medium is dissipated as heat. As will be discussed in the next section, each of these processes corresponds to a source term in the continuum conservation equations. Although the introduction of new species has many potential benefits for combustion applications, this chapter will focus on aerodynamic applications in which significant flow heating and body forces can used to beneficially manipulate the flow.

The target applications are ground transportation and flight in the troposphere and stratosphere, where pressures and densities are reasonable fractions of standard, sea-level values. To make the scope of this review reasonable, and to narrow the focus to continuum modeling, the important applications of near-space flight and electric spacecraft propulsion will be excluded here.

The objective of this chapter is to help researchers and designers select a suitable continuum model for plasma applications in which the aim is to alter the bulk gas flow. Given the disparity between the space and time scales of plasma phenomena and fluid dynamic phenomena, it is usually necessary to model the two separately, and use strong approximations to couple them. The aim of this chapter is to help practitioners make a good choice among the many options.

17.1 Background and levels of modeling

In aerodynamic applications, two classes of plasma are generally encountered [6]. One is a thermal plasma, generated through aerodynamic heating and concomitant thermal ionization. Thermal plasmas occur particularly in atmospheric entry flight from space, but can also occur in any hypersonic flight condition and in combustion applications and engine exhaust. Arc and spark discharges can also fall into this category. Thermal plasmas are characterized by hot gas, relatively close to thermochemical equilibrium, with temperatures on the order of 10^4 K (1 eV) and a relatively high fractional ionization of 10^{-4} to 10^{-3}.

Another category is cold plasma, which is in a highly nonequilibrium state. The electrons are hot (above 1 eV or 10^4 K), whereas the ions and the background gas are near ambient temperature. These plasmas are sustained by impact ionization from free electrons accelerated in a strong electric field. They are weakly ionized, with fractional ionization of 10^{-8} to 10^{-6}, and collisions are primarily between charged particles and the neutral background gas.

In the laboratory, electrical discharges incorporating both kinds of plasma are typically generated in a closed chamber at low pressure containing a pair of electrodes [6]. Perhaps the simplest electrical discharge is a DC discharge, in which the potential difference across the gap is held constant. Beyond a certain threshold voltage, the discharge can contain a self-sustaining plasma. Ionization is supplied through secondary emission at the cathode and the Townsend avalanche process. Analogous mechanisms support ionization in relatively low pressure discharges driven by sinusoidal or pulsed voltage waveforms At higher pressures, the mechanism for maintaining the plasma can come from streamers. Here, space charge at the head of the streamer maintains a strong electric field, which sustains electron impact ionization. (Examples of both kinds of discharge will be presented in section 17.4.) At higher currents, the ionization mechanism transitions to thermal, and the discharge enters the arc regime.

The same categories of discharge appear in aerodynamic applications, with the addition of bulk gas flow, and a variety of local conditions are encountered. Electrical discharges can resemble glow discharges, streamer discharges, arcs, and so on. Since aerodynamics is a continuum phenomenon, it is natural to attempt to describe plasma-flow interactions in terms of moment equations or continuum conservation laws.

To recap the material of earlier parts of this volume, ionized gases can be modeled at several levels, trading physical detail and computational cost. At the most detailed level (Part I), the motion of individual particles can be tracked, as in molecular dynamics simulations. This approach can be extremely computationally expensive for high-density conditions; an effective compromise is to track representative particles, as in particle-in-cell (PIC) and direct simulation Monte Carlo (DSMC) methods. If less fine-grained detail is required, moment methods and magnetohydrodynamics (Part II) can be employed. This chapter focuses on these latter continuum methods, attempting to make a practical compromise between physical fidelity and the need to capture both plasma and fluid dynamic phenomena. The discussion begins by considering appropriate levels of modeling, then goes on to consider flow control by heating, by magnetic body forces, and by electrical body forces.

Continuum conservation equations can be derived by taking the first few moments of the Boltzmann equation (chapter 6). The mass, momentum, and energy conservation equations for species-s derived in this way are:

$$\frac{\partial}{\partial t}(n_s m_s) + \nabla \cdot (n_s m_s \boldsymbol{v}_s) = S_s \tag{17.1}$$

$$\frac{\partial}{\partial t}(n_s m_s \boldsymbol{v}_s) + \nabla \cdot \left(n_s m_s \boldsymbol{v}_s \boldsymbol{v}_s + p_s \boldsymbol{I}\right) = \nabla \cdot \boldsymbol{\tau}_s + n_s m_s \boldsymbol{g} + n_s q_s (\boldsymbol{E} + \boldsymbol{v}_s \times \boldsymbol{B}) + \boldsymbol{A}_s \tag{17.2}$$

$$\begin{aligned} \frac{\partial}{\partial t}\left[n_s m_s\left(\epsilon_s + \frac{1}{2}v_s^2\right)\right] + \nabla \cdot \left[n_s m_s \boldsymbol{v}_s\left(\epsilon_s + \frac{1}{2}v_s^2\right) + p_s \boldsymbol{v}_s\right] &= \nabla \cdot \left[\boldsymbol{\tau}_s \cdot \boldsymbol{v}_s - \boldsymbol{Q}_s\right] \\ &+ n_s m_s \boldsymbol{v}_s \cdot \boldsymbol{g} + n_s q_s \boldsymbol{v}_s \cdot \boldsymbol{E} + M_s \end{aligned} \tag{17.3}$$

Here n_s is the number density, m_s is the mass per particle, q_s is the charge per particle, $\boldsymbol{v}_s$ is the local velocity, and ϵ_s is the thermal energy per unit mass. Further, $\boldsymbol{E}$ is the electric field, $\boldsymbol{B}$ is the magnetic field, and $\boldsymbol{g}$ is the local acceleration due to gravity. Forces in the continuum framework are divided into short-range (contact forces like pressure p_s and viscous stress $\boldsymbol{\tau}_s$) and long-range (electromagnetic and gravitational body forces). Species thermal energy is exchanged through space by the heat flux vector $\boldsymbol{Q}_s$.

Exchange of conserved quantities between species comes through the source terms These are the transfer of mass from one species to another through chemical reactions S_s, the exchange of momentum through elastic and inelastic collisions $\boldsymbol{A}_s$, and the corresponding total energy exchange in collisions M_s. Equations (17.1)–(17.3) are written in what is called, in the fluid dynamics literature, strong conservation form. The equations contain flux terms (terms with a divergence operator) that are continuous across shocks and other flow discontinuities, and thus well suited for numerical calculations. It can be helpful in thinking about various flows to convert to nonconservative form through manipulations employing the mass conservation equation (17.1). The nonconservative form of the momentum equation (17.2) is:

$$n_s m_s \frac{D_s \boldsymbol{v}_s}{Dt} = -\nabla p_s + \nabla \cdot \boldsymbol{\tau}_s + n_s m_s \boldsymbol{g} + n_s q_s (\boldsymbol{E} + \boldsymbol{v}_s \times \boldsymbol{B}) + \hat{\boldsymbol{A}}_s \tag{17.4}$$

where the source term is $\hat{\boldsymbol{A}}_s = \boldsymbol{A}_s - S_s \boldsymbol{v}_s$ and $D_s/Dt = \partial/\partial t + (\boldsymbol{v}_s \cdot \nabla)$ is the material derivative with respect to the motion of species-s. The acceleration of a material particle of species-s is seen to be altered by the pressure gradient, viscous forces, body forces, and a particular portion of the momentum source term due to collisions.

To derive the thermal energy equation, take the scalar product of the velocity with the momentum equation (17.4) and subtract that from the nonconservative form of the total energy equation (17.3). The result is:

$$n_s m_s \frac{D_s \epsilon_s}{Dt} = -p_s \nabla \cdot \boldsymbol{v}_s + \boldsymbol{\tau}_s : \nabla \boldsymbol{v}_s - \nabla \cdot \boldsymbol{Q}_s + \hat{M}_s \tag{17.5}$$

where the source term is $\hat{M}_s = M_s - S_s(\epsilon_s + \frac{1}{2}v_s^2) - \hat{\boldsymbol{A}}_s \cdot \boldsymbol{v}_s$. The colon operator in the dissipation term $\boldsymbol{\tau}_s : \nabla \boldsymbol{v}_s$ designates contraction on both indices of the two second-order tensors. The internal energy of a material particle of species-s is seen to be changed by compression work, viscous dissipation, heat flux, and a portion of the energy source term due to collisions. It is not directly affected by body forces. For this reason, it is somewhat misleading to speak of Joule heating for a single species; heating comes about through acceleration by the fields and subsequent randomization of that motion in collisions (represented by the term $\hat{M}_s$).

The various forms of the source terms are important because they are the means by which charged particles interact with the neutral bulk-gas flow. The path of energy exchange is of particular interest. For example, energy in an external circuit becomes energy in an electric field, which in turn increases the kinetic energy of ions, which collide with neutral atoms. Part of the energy flow is randomized in these collisions (heating) and part becomes directed bulk-gas motion.

Examining a relatively simple model for elastic collisions is educational. One such model assumes that particles coming out of collisions have completely randomized velocities [7]. Under this model, the components of the source terms due to elastic collisions are:

$$\boldsymbol{A}_s^E = -n_s \sum_t m_{st}\nu_{st}(\boldsymbol{v}_s - \boldsymbol{v}_t) \tag{17.6}$$

$$M_s^E = -n_s \sum_t \frac{m_{st}\nu_{st}}{m_s + m_t}[3k_B(T_s - T_t) + (\boldsymbol{v}_s - \boldsymbol{v}_t) \cdot (m_s\boldsymbol{v}_s + m_t\boldsymbol{v}_t)] \tag{17.7}$$

Here k_B is Boltzmann's constant, T_s is the species temperature, ν_{st} is the average momentum transfer collision frequency between particles of types s and t, and $m_{st} = m_s m_t/(m_s + m_t)$ is the reduced mass. The corresponding terms in equations (17.4)–(17.5) are $\hat{\boldsymbol{A}}_s^E = \boldsymbol{A}_s^E$ and

$$\hat{M}_s^E = n_s \sum_t \frac{m_{st}\nu_{st}}{m_s + m_t}[-3k_B(T_s - T_t) + m_t|\boldsymbol{v}_s - \boldsymbol{v}_t|^2] \tag{17.8}$$

This latter form illustrates how the source terms change the acceleration and thermal energy in the reference frame of a material particle. The velocity difference and temperature difference terms tend to equilibrate the state of all species. Nonetheless, the last term in equation (17.8) involving the square of the magnitude of the velocity difference is always positive. It is seen to have a dissipative nature, always tending to increase the internal energy, that is, converting directed mechanical energy into heat. These terms illustrate the two-stage process for heating of an individual species: acceleration by the fields, followed by randomization in collisions of some of that directed energy.

Conservation laws for the gas as whole can be obtained by summing equations (17.1)–(17.3) over all species, noting that conservation requires that the collision source terms sum to zero. The resulting global conservation laws for mass, momentum, and energy are as follows:

$$\frac{\partial \rho}{\partial t} + \nabla \cdot (\rho \boldsymbol{w}) = 0 \tag{17.9}$$

$$\frac{\partial}{\partial t}(\rho \boldsymbol{w}) + \nabla \cdot (\rho \boldsymbol{w}\boldsymbol{w} + p\boldsymbol{I}) = \nabla \cdot \boldsymbol{\tau} + \rho \boldsymbol{g} + \zeta \boldsymbol{E} + \boldsymbol{j} \times \boldsymbol{B} \tag{17.10}$$

$$\frac{\partial}{\partial t}\left[\rho(\epsilon + \frac{1}{2}w^2)\right] + \nabla \cdot \left[\rho \boldsymbol{w}(\epsilon + \frac{1}{2}w^2) + p\boldsymbol{w}\right] = \nabla \cdot [\boldsymbol{\tau} \cdot \boldsymbol{w} - \boldsymbol{Q}] + \rho \boldsymbol{w} \cdot \boldsymbol{g} + \boldsymbol{E} \cdot \boldsymbol{j} \tag{17.11}$$

where $\rho = \sum_s n_s m_s$ is the overall density, $\zeta = \sum_s n_s q_s$ is the space charge, $\boldsymbol{j} = \sum_s n_s q_s \boldsymbol{v}_s$ is the total current density, and $\boldsymbol{w} = \sum_s n_s m_s \boldsymbol{v}_s/\rho$ is the mass-weighted average velocity. The remaining terms are summations of the corresponding terms for the individual species, with (generally small) additional terms [8] involving the diffusion velocities that arise because of the change of reference frame from that of the individual species to that of the mass-averaged velocity. The electromagnetic body force is $\zeta \boldsymbol{E} + \boldsymbol{j} \times \boldsymbol{B}$. In equation (17.11), the term $\boldsymbol{E} \cdot \boldsymbol{j}$ represents the total power delivered to matter by the electromagnetic fields; it includes both the rate of work done by the body force and thermal interaction.

The corresponding thermal energy equation, derived in a manner analogous to equation (17.5), is:

$$\rho \frac{D\epsilon}{Dt} = -p\nabla \cdot \mathbf{w} + \boldsymbol{\tau} : \nabla \mathbf{w} - \nabla \cdot \mathbf{Q} + \mathbf{E}' \cdot \mathbf{j}' \tag{17.12}$$

where $\mathbf{E}' = \mathbf{E} + \mathbf{w} \times \mathbf{B}$ is the electric field and $\mathbf{j}' = \mathbf{j} - \zeta \mathbf{w}$ is the current, both in a reference frame moving with the mass-averaged velocity $\mathbf{w}$. Here $D/Dt = \partial/\partial t + (\boldsymbol{w} \cdot \nabla)$ is the material derivative with respect to the mass-averaged motion. The internal energy of the overall flow is seen to be changed by pressure work, viscous dissipation, and heat flux. The term $\mathbf{E}' \cdot \mathbf{j}' = \mathbf{E} \cdot \mathbf{j} - (\zeta \boldsymbol{E} + \boldsymbol{j} \times \boldsymbol{B}) \cdot \boldsymbol{w}$ represents thermal interaction, including the Joule heating effect; the rate of work done by the electromagnetic body force has been subtracted.

A further level of simplification of the particle motion model (17.1)–(17.3) can be obtained if particle inertia can be neglected. Neglecting acceleration terms and diffusion due to temperature gradients, and assuming that the gas is sufficiently weakly ionized that only collisions with neutrals are important in equation (17.6), the nonconservative momentum equation (17.4) can be simplified to obtain a drift–diffusion model, and introduced into the particle conservation equation (17.1):

$$\frac{\partial n_s}{\partial t} + \nabla \cdot \left\{ n_s \left[\mathbf{w} + s_s \mu_s \mathbf{M}^s \cdot (\mathbf{E} + \mathbf{w} \times \mathbf{B}) \right] \right\} = \nabla \cdot (D_s \mathbf{M}^s \cdot \nabla n_s) + \omega_s \tag{17.13}$$

$$M_{ij}^s = \frac{1}{1 + \mu_s^2 B^2} \left(\delta_{ij} + \mu_s^2 B_i B_j + s_s \mu_s \epsilon_{ijk} B_k \right) \tag{17.14}$$

where $s_s = q_s/|q_s|$ is the sign of the charge of species-s, μ_s is the mobility, D_s is the diffusion coefficient, $\omega_s = S_s/m_s$ is the net rate of production of species-s in reactions, and the dot product corresponds to summation on the second index of M_{ij}^s.

The conservation equations must be coupled to some form of Maxwell's equations to predict the electric and magnetic fields. The general form of the equations is:

$$\epsilon_0 \nabla \cdot \boldsymbol{E} = \zeta \tag{17.15}$$

$$\nabla \cdot \boldsymbol{B} = 0 \tag{17.16}$$

$$\nabla \times \boldsymbol{E} = -\frac{\partial \boldsymbol{B}}{\partial t} \tag{17.17}$$

$$\frac{1}{\mu_0} \nabla \times \boldsymbol{B} = \boldsymbol{j} + \epsilon_o \frac{\partial \boldsymbol{E}}{\partial t} \tag{17.18}$$

where ϵ_0 is the permittivity and μ_0 the permeability of vacuum. The equations are, respectively, Gauss' law, conservation of magnetic flux, Faraday's law, and the Ampère–Maxwell law. For low-frequency electromagnetic phenomena, the displacement current $\epsilon_0 \partial \boldsymbol{E}/\partial t$ can often be neglected.

In the magnetohydrodynamic approximation, the gas is quasi-neutral, $\zeta \approx 0$. The current is given by Ohm's law:

$$\boldsymbol{j} = \boldsymbol{\sigma} \cdot (\boldsymbol{E} + \boldsymbol{w} \times \boldsymbol{B}) \tag{17.19}$$

where $\boldsymbol{\sigma}$ is the conductivity tensor. Ohm's law (17.19) can be derived from the flux expression in the drift–diffusion approximation (17.13)–(17.14) by multiplying by the charge per particle and summing over all species. Using a scalar conductivity σ, instead, is a good approximation if the collision frequency is much greater than the cyclotron (gyro) frequency, as in a relatively dense gas. If a scalar conductivity applies, then the thermal energy exchange term in equation (17.12) becomes $\mathbf{E}' \cdot \mathbf{j}' = j^2/\sigma$, that is, simple Joule heating.

If the displacement current is neglected, equations (17.15)–(17.19) can be manipulated to eliminate the electric field, giving the magnetic induction equation:

$$\frac{\partial \boldsymbol{B}}{\partial t} = \nabla \times (\boldsymbol{w} \times \boldsymbol{B} - \boldsymbol{\eta} \cdot \nabla \times \boldsymbol{B}) \tag{17.20}$$

where $\boldsymbol{\eta} = \boldsymbol{\sigma}^{-1}/\mu_0$ is the magnetic diffusivity tensor, or the scalar diffusivity $\eta = 1/(\sigma\mu_0)$ in the case of scalar conductivity.

If distortion of the magnetic field by currents in the flow is small, the magnetic field can be considered to be the imposed field. In that case, in place of (17.20), the current continuity equation is solved for the electric field:

$$\nabla \cdot \boldsymbol{j} = \nabla \cdot [\boldsymbol{\sigma} \cdot (\boldsymbol{E} + \boldsymbol{w} \times \boldsymbol{B})] = 0 \tag{17.21}$$

where the current has been given by equation (17.19).

The assumption of quasi-neutrality is not valid for some conditions, such as an electrode sheath or a streamer head. In that case, Gauss' law (17.15) must be solved for the electric field. If the magnetic field is constant, or varies slowly in time, then Faraday's law (17.17) implies that the curl of the electric field is zero. Under those conditions, the electric field is derivable from a potential $\boldsymbol{E} = -\nabla\phi$. Equation (17.15) can then be converted to Poisson's equation:

$$\nabla^2\phi = -\zeta/\epsilon_0 \tag{17.22}$$

This is a standard model for calculations that include space charge and sheaths.

In subsequent sections, modeling of flow control actuators is explored at three levels of approximation. At the first level (section 17.2), the action of the plasma actuator is represented by specified source terms in the overall fluid conservation equations (17.9)–(17.11). This approach is useful when the fluid mechanics are the dominant consideration, and the plasma device is used to provide a perturbation to the flow, such as local heat release. The next level of approximation (section 17.3) is magnetohydrodynamics, combining either equation (17.20) or (17.21) with the overall conservation laws to capture the effects of an imposed magnetic field on an ionized gas flow. Finally, detailed modeling (section 17.4), combining the species conservation equations (17.1)–(17.3) or (17.13)–(17.14) with a solution of the Poisson equation (17.22), is applied to plasma flow control actuators based on the ion wind effect.

17.2 Flow control via plasma heating

Starting in the late 1990s, motivated by a number of promising experiments on the propagation of shocks through electrical discharges [9], there was a great deal of interest in using regions of plasma to control shock waves. Early analytical work focused on determining whether observed changes in shock waves were due to inherent plasma effects or thermal nonuniformities [10]. Thermal effects proved to be the main contributor to the observed flow behavior, and those effects were subsequently explored for flow control [11].

The heating mechanism comes through the source term in equation (17.12). If the conduction current is in the direction of the electric field, which is commonly the case except close to boundaries, then $\mathbf{E}' \cdot \mathbf{j}' > 0$, which represents Joule heating. A very complete model would solve for charged species motion and the electric field, but when the main physical mechanism of interest is heating, modeling using localized thermal energy deposition in a conventional computational fluid dynamics code can replicate most of the important physical effects [12].

The relevant nondimensional parameter [13] characterizing this process is the ratio in equation (17.11) of the deposited thermal power to the flow power:

$$\tilde{Q} = \frac{Q}{\rho U^3 L^2} \tag{17.23}$$

where $Q = \int \boldsymbol{E} \cdot \boldsymbol{j}\, dV$ is the total power deposition, ρ is density, U is velocity, and L is a characteristic length. In most applications, the relevant location for evaluating the density and velocity is the uniform freestream, but sometimes local values are appropriate. The relevant length is the scale over which flow modification is desired.

This approach has been of interest in very high speed applications, where conventional means of flow control are not viable. Unfortunately, the required levels of power for brute-force application of the concept are very high. For example, consider an aircraft flying at 19 km altitude at a speed of 1000 m s^{-1} (Mach 3.4). Here the air density is about 0.1 kg m^{-3}. Taking the characteristic length as 1 m, the required power to obtain $\tilde{Q} = 1$ is about 100 MW. It is more appropriate to employ such power levels in the propulsion system than use them for flow control.

Local changes in the flow, on the other hand, can be obtained at much lower power levels, since the length scale is much smaller. Sufficient thermal energy deposition can generate flow asymmetry, which in turn can generate forces and moments for flight control [13–15].

In the examples that follow, the body length was used as the characteristic length scale in (17.23); this choice results in low values of the nondimensional power deposition parameter. An even better form of the nondimensional power for these cases might be a representative boundary layer thickness.

A popular application of microwave or laser power deposition is the air-spike concept [16]. In this concept, a conventional physical aerodynamic spike is replaced by a region of power deposition ahead of the vehicle. This small region of plasma effectively turns a blunt body shape into a sharp one, with potential reductions in drag and heat transfer. With off-axis placement of the deposition point, flight control can be achieved. Experiments on this concept have been carried out using laser energy deposition, microwave energy deposition, and counterflow plasma jets [17].

Flight control could also be achieved, potentially at lower source power levels, using near-body energy addition. Bisek *et al* [13] carried out a study of the required power levels for hypersonic flight control with near-body energy addition, as with an electric discharge at the body surface. The power deposition was modeled as a modified Gaussian function with an ellipsoidal shape.

A generalized form [14] of the power deposition pattern is:

$$g(x, y, z) = \frac{1}{K}\exp\left[-\left|\frac{x - x_c}{a}\right|^m - \left|\frac{y - y_c}{b}\right|^n - \left|\frac{z - z_c}{c}\right|^p\right] \qquad (17.24)$$
$$K = 8\, abc\, \Gamma(1 + 1/m)\, \Gamma(1 + 1/n)\, \Gamma(1 + 1/p)$$

where the centroid of the deposition region is (x_c, y_c, z_c) and the characteristic length scales along each of the coordinate axes are a, b, c. These axes can be aligned with an arbitrary direction through a coordinate rotation (omitted for brevity; see [13]). The parameter K is used to scale $g(x, y, z)$ so that its integral over all space is unity. The Euler gamma function $\Gamma(x)$ arises in this context, and K is seen to have units of volume. The term added to the total energy equation (17.11) is:

$$S = Q\, g(x, y, z) \qquad (17.25)$$

where Q is the total power. A smooth distribution of power of this form is easy to resolve numerically and can have reasonable physical realism.

Figure 17.1 shows some of the results obtained by Bisek *et al* [13]. The configuration studied was a blunt-nosed elliptic cone, chosen as a generic hypersonic vehicle shape for which validation data were available for the baseline, no-control case. The freestream Mach number was $M = 14$, and small ($L = 0.21$ m), medium ($L = 0.62$ m), and large cones ($L = 3.00$ m) were considered. The temperature field with control applied is shown on the left side of figure 17.1 for the small cone configuration. The power deposition is centered near the surface close to the nose, and a hot wake is evident downstream. The corresponding asymmetric changes in pressure over the vehicle surface lead to net moments for flight control.

Control effectiveness was characterized by the nondimensional moment coefficient:

$$C_m = \frac{M}{\frac{1}{2}\rho U^2 L^2 d} \qquad (17.26)$$

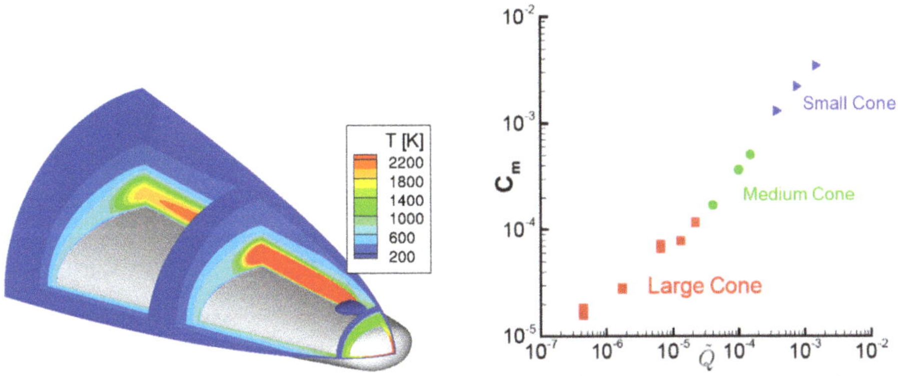

Figure 17.1. Effect of near-body power deposition on a hypersonic, blunt elliptic cone flow. Temperature field for $Q = 1$ kW and $L = 0.2$ m (left) and nondimensional moment coefficient C_m versus nondimensional power $\tilde{Q}$ (right). From Bisek *et al* [13].

where M is the moment induced by the distributed force at the body surface, d is the maximum spanwise length, L is the body length, and the remaining parameters are freestream values. The power deposition was found to generate a moment of the same order as a small mechanical flap for a deposition level of $\tilde{Q}{\sim}10^{-4}$. For a fixed volume, the general shape of the deposition region (sphere, oblate spheroid, prolate spheroid) did not have a strong effect. Including thermochemical nonequilibrium in the model was found to be important: endothermic reactions absorbed some of the energy that would otherwise go into translational modes altering the flow. Similarly, higher wall temperature diminished the flow control effect. As shown in the right part of figure 17.1, the predicted moment coefficients were found to collapse on a common curve when plotted against the nondimensional power parameter (17.23). These results suggest that feasible power levels can be obtained for high altitude and small bodies moving at relatively low speed.

In a follow-on study, Atkinson *et al* [14] carried out a similar study of flow control for flight control at hypersonic speed (Mach 14.5) and very high angle-of-attack, up to 60 deg, characteristic of certain atmospheric entry flows. The model was somewhat more detailed, including both a magnetohydrodynamic body force and power deposition. Envisioning small magnetohydrodynamic actuators, flush-mounted with the surface, the following source terms were employed:

$$\begin{aligned} \boldsymbol{f} &= \boldsymbol{I} \times \boldsymbol{B}\, \ell\, g(x, y, z) \\ S_t &= VI\, g(x, y, z) \\ S_v &= \eta_v(S_t - \boldsymbol{f} \cdot \boldsymbol{w}) \end{aligned} \tag{17.27}$$

Here $\boldsymbol{I}$ is the current expressed as a vector, $\boldsymbol{B}$ is the magnetic field, V is voltage across the electrodes, and ℓ is the length of the plasma column between the electrodes. The function $g(x, y, z)$ is given by equation (17.24). Each term is a source in the conservation equations: $\boldsymbol{f}$ is a body force in (17.10), S_t is the total power per unit

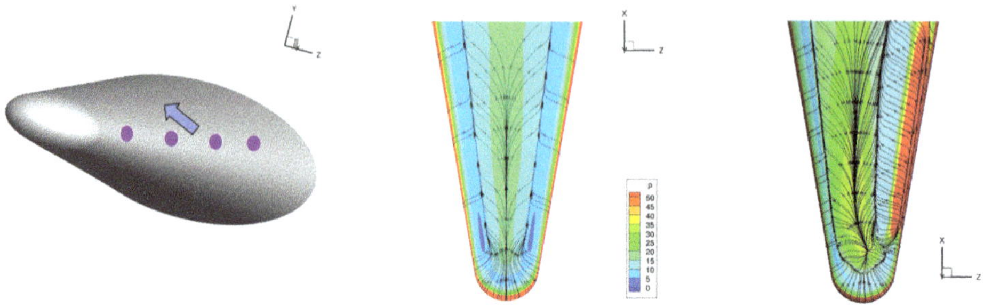

Figure 17.2. Control of hypersonic flow at high angle-of-attack. Plasma actuator configuration and direction of force (left). Limiting streamlines and pressure contours (Pa) on upper surface for baseline (middle) and actuation (right) cases. From Atkinson *et al* [14].

volume in the total energy equation (17.11), and S_v is the portion of the thermal power that goes into the vibrational energy of air molecules (equation omitted for brevity; see reference [14]). A typical value for the fraction transferred in this way (lost to flow control) is $\eta_v = 0.5$.

The parameters were chosen to be representative of laboratory experiments with magnetohydrodynamic actuators: $B = 3.0$ T, I = 35–250 mA, and $V = 3.7$ kV. Selected results are shown in figure 17.2 for an asymmetric actuator configuration (left image) designed to create moments for flight control. Images of the limiting (wall) streamlines and surface pressure on the upper (leeward) surface are also shown. The baseline case (middle image) is symmetric, and is a product of a complex crossflow separation pattern in this high angle-of-attack flow. Significant flow asymmetry is produced when actuation is applied (right image), leading to moments on the vehicle configuration. The model actuator was effective in altering the flow structure and creating moment coefficients of $C_m \sim 10^{-4}$ for nondimensional power inputs of $\tilde{Q} \sim 10^{-2}$.

In summary, a simplified physical model, in which a thermal source and a localized body force are the main additions to the conservation equations of fluid dynamics (17.9)–(17.11), can be very useful in exploring the effects of plasma-based flow control. These reduced-order physical models are appropriate when the fluid mechanics and heat release are the dominant physics of interest, and the main goal is to estimate the order of magnitude of the power required to achieve significant flow control effects. They can be highly computationally efficient, allowing optimization studies. Nonetheless, more detailed models are required to fully investigate actuator performance, particularly if electromagnetic body forces are a significant aspect of the fluid control scheme.

17.3 Flow control via magnetic forces

In the case where the primary flow control mechanism is an applied magnetic field, a magnetohydrodynamic model, assuming quasi-neutral plasma, is often appropriate. The conservation equations for the gas as a whole (17.9)–(17.11) are solved, along with electromagnetic equations to predict the fields. In magnetohydrodynamics, the displacement current is assumed to be negligible in the Ampère–Maxwell law (17.18), which leads to the magnetic induction equation (17.20).

Looking at the terms in the magnetic induction equation (17.20), the ratio of the order of magnitude of the convection and diffusion terms is given by the magnetic Reynolds number:

$$\mathrm{Re}_M = \sigma\mu_0 VL \tag{17.28}$$

In the high magnetic Reynolds number regime, the full magnetic induction equation (17.20) must be solved. In the low magnetic Reynolds number regime, the current continuity equation (17.21) is used to find the electric field, and the applied magnetic field is assumed to be undistorted by the flow. The magnetic Reynolds number is typically low in aerodynamic applications. For example, for $\sigma = 100\ (\Omega\mathrm{m})^{-1}$, $V = 8000\ \mathrm{m\ s^{-1}}$, and $L = 1$ m, the magnetic Reynolds number is about unity. These figures are typical of atmospheric entry flows with natural ionization.

To consider the influence of magnetohydrodynamic effects on the flow, take the ratio of the magnetic body force to the convective acceleration in (17.10), assuming that the electric field is of order $E \sim UB$. The result is the magnetic interaction parameter, or Stuart number:

$$S = \frac{\sigma B^2 L}{\rho U} \tag{17.29}$$

For values typical of atmospheric entry flows, $\sigma = 100\ (\Omega\mathrm{m})^{-1}$, $V = 8000\ \mathrm{m\ s^{-1}}$, $\rho = 0.01\ \mathrm{kg\ m^{-3}}$, $L = 1$ m, and $B = 1.0$ T, interaction parameters of order one are achievable.

Another important parameter for flight magnetohydrodynamics is the Hall parameter, defined as:

$$\beta_e = \frac{\sigma B}{e n_e} \tag{17.30}$$

The Hall parameter measures the tendency of electrons to spiral around magnetic field lines at high magnetic field strength, rather than follow the material frame field $\boldsymbol{E}' = \boldsymbol{E} + \boldsymbol{w} \times \boldsymbol{B}$. The Hall parameter has typical values of $\beta_e \sim 10$ for atmospheric entry flows, and thus the Hall effect complicates manipulation of these flows with magnetic forces.

A strong applied magnetic field tends to slow the flow in the shock layer ahead of a hypersonic blunt body, increasing shock standoff (minimum distance from shock to body). The reason for this effect is illustrated in figure 17.3. In the absence of a strong electric field, a current $\boldsymbol{j} = \sigma\, \boldsymbol{v} \times \boldsymbol{B}$ tends to flow in a direction perpendicular to both the local magnetic field and the local fluid velocity. For a dipole magnetic field, aligned with a blunt body flow, the corresponding current lines are circles around the axis of symmetry (see figure 17.6). The corresponding force $\boldsymbol{j} \times \boldsymbol{B}$ is perpendicular to both the current and the magnetic field. Considering the geometry, the magnetic force will lie in a plane with the magnetic field and the velocity, and will have a component that locally decelerates the flow. Mass conservation requires that

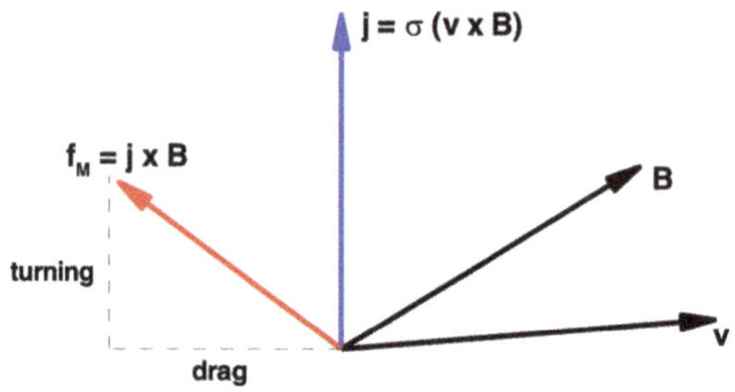

Figure 17.3. Magnetic force in the absence of strong electric field.

the shock move away from the body to pass the same mass flow at lower speed, and the corresponding reduction in stagnation point gradients yields a reduction in heat transfer.

In the mid-1950s, when researchers first realized that the strong shock waves and the high temperatures of atmospheric entry flight would be accompanied by significant flow ionization, a magnetohydrodynamic heat shield was proposed to protect the vehicle from thermal damage [2]. An increase in shock standoff and drag, and a reduction in heat transfer rate, have been confirmed experimentally [18, 19]. A fairly complete approximate analytical theory of the problem was developed by Bush [20, 21].

Poggie and Gaitonde [22] applied modern computational fluid dynamics methods to the problem, using a magnetohydrodynamic model in the low magnetic Reynolds number approximation. In agreement with the Bush theory [20, 21], an applied magnetic field was found to increase shock standoff and reduce heat transfer. Nonuniform electrical conductivity was found to reduce the effectiveness of control.

Figure 17.4 shows selected results for Mach 5, viscous flow over a hemisphere. The figures show pressure contours, and the flow is from left to right, with the dark region representing the solid hemisphere. A strong bow shock appears as the sharp, curved pressure jump upstream of the hemisphere. For the baseline case (left), the shock standoff is about 13% of the body radius, in agreement with standard experimental values. With the application of a strong magnetic field, corresponding to an interaction parameter of $S = 6$, the shock standoff increases to about 28% of the body radius. There is a corresponding drop in stagnation point heat transfer of about 26%.

The work of Poggie and Gaitonde used a simple correlation of electrical conductivity as a function of temperature and neglected the Hall effect. Their results compared well with the simplified theory of Bush, but they did not validate their results against experimental data.

Bisek *et al* [23] attempted to confirm this approach by predicting the experimental results of Kranc *et al* [18] for an ionized, Mach 4.75 argon flow over a hemispherical-nosed cylindrical body. They found a strong sensitivity of the shock standoff change to the electrical conductivity model. Interestingly, although the stagnation point heat flux was reduced by the magnetohydrodynamic heat shield, additional heating was observed on the downstream portion of the body.

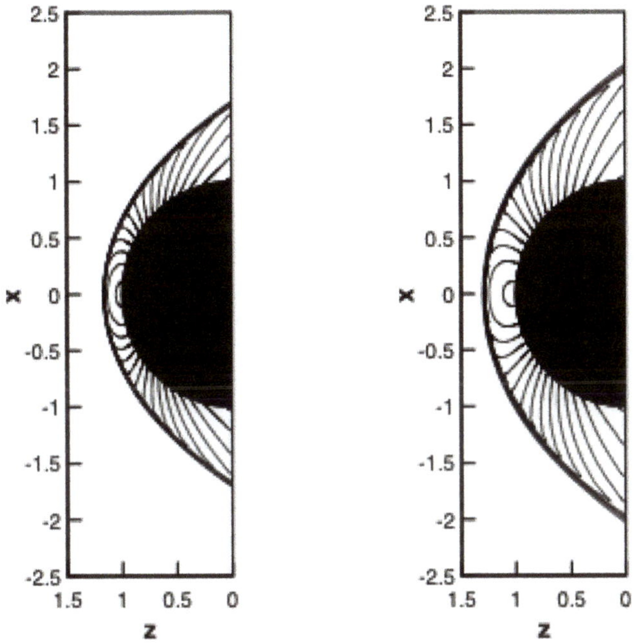

Figure 17.4. Effect of applied magnetic field on hypersonic flow over a hemisphere. Baseline case (left) and applied magnetic field with interaction parameter $S = 6$ (right). Pressure contours of interval $p/p_\infty = 2$. Reproduced with permission from Poggie and Gaitonde [22]. Copyright 2002 American Institute of Physics.

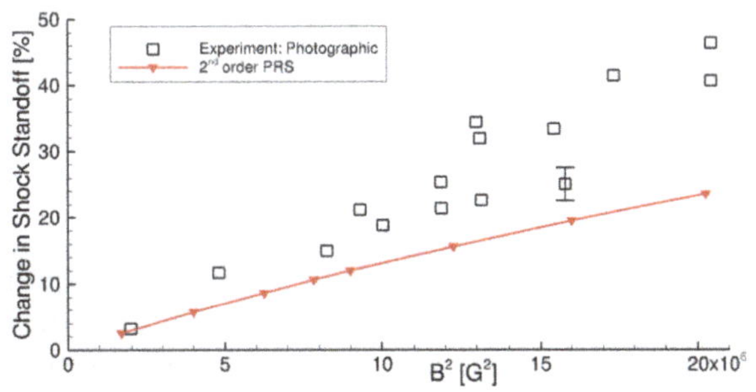

Figure 17.5. Effect of applied magnetic field on shock standoff in a hypersonic blunt body flow. From Bisek *et al* [23].

Figure 17.5 shows a comparison to experimental data of predictions with the most effective conductivity model, a curve fit to data obtained from Boltzmann solutions. The vertical axis is the percent change in shock standoff distance, and the horizontal axis shows magnetic field strength squared. The corresponding interaction parameter for the range of values on the horizontal axis is about $0 \leqslant S \leqslant 40$.

The level of disagreement between theory and experiment seen here is typical of attempts to predict plasma aerodynamics experiments. In this case, the experiments were carried out in an arc-heated facility, in which the freestream was already

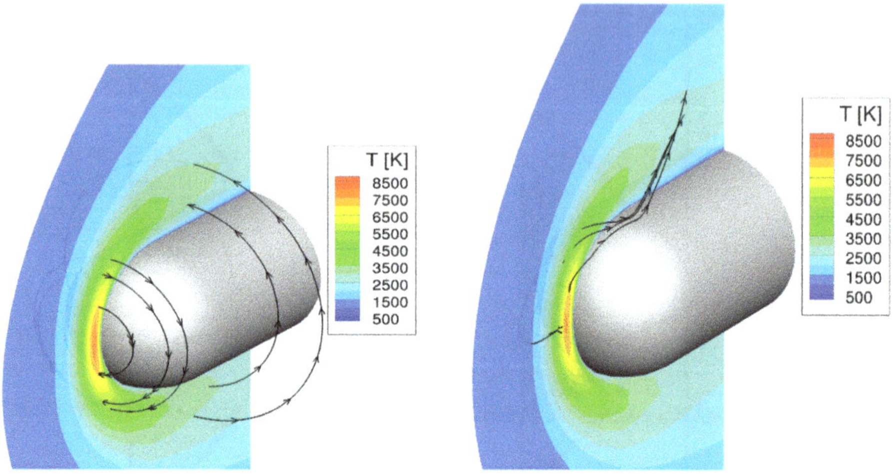

Figure 17.6. Temperature field with current lines for a Mach 4.75 argon flow over a blunt body with an applied magnetic field of 0.13 T. Without (left) and with (right) the effect of the Hall parameter. From Bisek *et al* [24].

ionized, not ionized after the shock. The uniformity of the freestream flow is not known, nor are some of the other details of the experimental conditions.

Subsequent work by Bisek *et al* [24] expanded the model to include the Hall and ion slip effects. As observed in early theoretical studies [25], the effectiveness of the magnetohydrodynamic heat shield at high values of the Hall parameter was found to be reduced.

Selected results are presented in figure 17.6, which shows the temperature field and current lines for two physical models. For the left image, Hall effect is neglected, and current flows in circles around the axis of symmetry. The right image shows the results for the same conditions, accounting for a Hall parameter of $\beta_e \approx 18$. Here, current lines follow a complicated path, and the effectiveness of the magnetohydrodynamic heat shield is much reduced.

Including the effect of finite magnetic Reynolds number requires the solution of the magnetic induction equation (17.20) to determine the distortion of the applied magnetic field by currents generated in the flow. The effectiveness of the magnetohydrodynamic heat shield approach is reduced for increasing magnetic Reynolds numbers, but the effect is not strong for the regime of practical applications [26]. Certain studies have included these effects [27], and distortion of the imposed field may become important for flows with high levels of ionization.

Despite the challenges in accurately predicting electrical conductivity and characterizing complex flow conditions, the magnetohydrodynamic model has been found to be a valuable tool in understanding and developing flow control approaches based on applied magnetic fields. Magnetohydrodynamic devices have many potential applications beyond the magnetohydrodynamic heat shield. Via electrodes, a magnetohydrodynamic generator can extract power from the flow, and a magnetohydrodynamic pump can inject power into the flow. Further, small-scale, local flow control avoids the need for heavy magnets and power conditioning equipment, and seems very promising for control of unsteady separation and shock-wave/boundary-layer interaction [28].

17.4 Flow control via electrical forces

Even a small fractional imbalance of charged particles in a gas produces such a large electrical force that excess charge is quickly eliminated by particle motion. Quasi-neutrality and negligible electric body forces are thus an excellent approximation in most plasma aerodynamics applications, as discussed in section 17.3. Significant electrical body forces only exist in the presence of significant space charge. This generally occurs only in the sheath at a solid surface or in a streamer head. These are the regions where electrical body forces can be generated for flow control [29].

The relative strength [30] of the electric body force in a cathode sheath is:

$$Z_E = \frac{en_i\Delta\phi}{\rho U^2} \tag{17.31}$$

Typical ion densities are 10^{14} m^{-3} in a corona discharge and 10^{17} m^{-3} in glow discharges. Typical potentials in a corona discharge are 10^4 V, and 10^2 V in a normal glow and 10^3 V in an abnormal glow. Thus, for near standard atmospheric conditions, $Z_E \sim 1$ for speeds of a few meters per second. This makes brute-force flow control with electrical forces unlikely for typical commercial aircraft speeds of more than 200 m s^{-1}, or indeed even for automobile speeds of 20 m s^{-1}.

More localized flow control, however, is feasible. Interest in this concept for flow control was stimulated by pioneering work in the 1980s and 1990s [4, 5]. Several comprehensive reviews of the application of electric forces to flow control are available [11, 31, 32].

For manipulation of the very low speed flow near the wall in a boundary layer, for example, the relevant parameter [30] is:

$$Z_\tau = \frac{en_i\Delta\phi}{\tau_w} = Z_E\frac{2}{c_f} \tag{17.32}$$

where τ_w is the wall shear stress. Typical values of the skin friction coefficient are $c_f = 2\tau_w/(\rho U^2) \sim 10^{-3}$, so $Z_\tau \sim 1$ might be obtained for flow speeds up to 50 m s^{-1}. This explains the relative success of flow control approaches based on plasma actuators [11, 31, 32].

As examples of discharges with significant space charge layers, drift–diffusion calculations, based on equations (17.13)–(17.14) and (17.22), are shown in 17.7. For simplicity, the calculations were carried out assuming two-dimensional, planar symmetry. The gas was nitrogen at 670 Pa. For the DC discharge, an external circuit with a 2 kV voltage source and resistance of 300 kΩ was employed. For the RF discharge, a 10 MHz sinusoidal voltage of amplitude 550 V was applied with a 5 kΩ external resistance. Both cases were chosen to lie in the normal glow regime. The normal current density effect, in which increasing discharge current is accommodated by increasing discharge area at nearly constant current density,

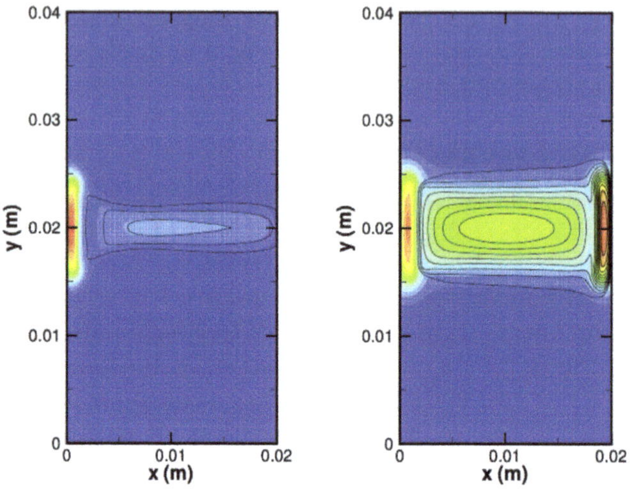

Figure 17.7. Electrical discharges in 670 Pa nitrogen computed with the drift–diffusion model. Number densities in a DC (left) and RF discharge (right). From Poggie [33].

has been observed experimentally in DC and RF glow discharges [6]. This effect was reproduced approximately in the numerical calculations [33]; it is evident in the figures that discharge in each case is limited to a finite cathode spot.

The images in figure 17.7 show number density distributions for positive ions (color contours) and electrons (black contour lines); the contour interval in each case is 1×10^{15} m^{-3}. The left image shows the DC discharge case. The cathode is represented by the left side of the domain ($x = 0$) and the anode by the right ($x = 0.02$ m). The cathode spot is quite evident as a peak in the ion number density at the left, and the quasi-neutral positive column can be seen as a region where ion and electron number density contours coincide near the center of the domain.

Corresponding data for the RF case are shown at the right for a time corresponding to the start of a cycle. The electrodes lie at the left (driven electrode) and right (grounded electrode) boundaries of the domain. The strong peak in the ion concentration occurring near the electrodes indicates that the discharge state corresponds to the high-current γ-mode, in which secondary emission plays a strong role. The sheath structure is similar to that of the cathode region of the DC case.

Classical low-pressure DC and RF discharges can generate flow control for certain applications [29], but much of the research on flow control via electrical forces has been done with dielectric barrier discharges (DBDs). An insulating barrier is placed over one (single DBD) or both (double DBD) metal electrodes, and the system is driven with an AC or pulsed power source. The idea is to limit the current through charge accumulation on the dielectric surface, allowing the discharge to be run at high pressure without transition to the arc regime [34]. The most common form of the device in aerodynamic applications is an asymmetric DBD, as shown schematically in figure 17.8. Here one electrode is exposed to the flow, whereas the other is buried under the dielectric layer. Various configurations and alignments of the electrodes have been explored.

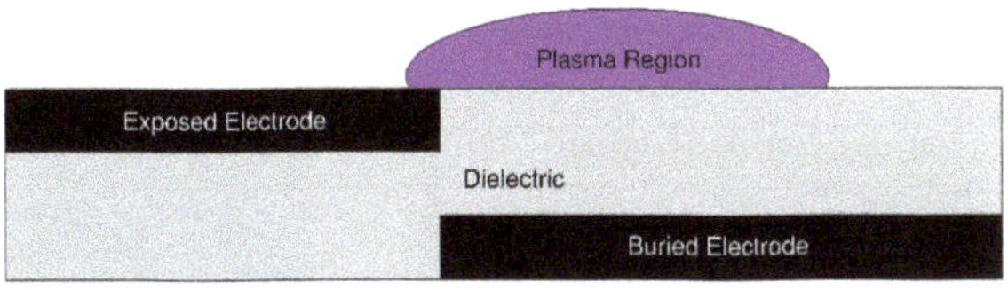

Figure 17.8. Diagram of an asymmetric dielectric barrier discharge plasma actuator.

There are two main categories of DBD discharge. The AC-DBD is typically driven by a sinusoidal waveform with amplitude on the order of 10 kV and frequency of 1 kHz to 10 kHz. AC-DBD actuators are thought to interact with the bulk flow primarily through electric body forces. The ns-DBD is driven by pulses of 1 kV to 10 kV amplitude and 1 ns to 100 ns width, with a repetition rate on the order of 10 kHz. The pulsed discharges have been suggested as more efficient sources of ionization [35] and more efficient production of electric body force [36], but ns-DBDs are believed to interact with the flow primarily through heat release [11]. The heat release occurs through a combination of Joule heating and a relaxation of different internal energy modes. It occurs over two time scales: a fast heat release in microseconds, and a slower heat release over milliseconds.

DBD plasma actuators produce relatively low flow speeds. Benard and Moreau [32] report that designs with a single actuator have produced air speeds of up to 7 m s^{-1} in the laboratory, and that multi-actuator designs have produced up to 11 m s^{-1}. The velocity profile near the actuator has the form of a wall jet, with a maximum about 1 mm from the surface. Because of the considerations discussed in conjunction with equation (17.32), DBD actuators have nonetheless been shown to be useful for many practical flow control applications. For example, demonstrations of closed-loop stall control have been carried out [37].

At atmospheric pressure, a DBD discharge consists alternately of groups of streamers and diffuse micro-discharges (corona regions), depending on the phase of the cycle [32]. It is not currently feasible to capture this three-dimensional, unsteady forest of small structures in routine numerical simulations. Nonetheless, capturing a single streamer is feasible, and aids in understanding the most important physics governing these devices.

An example from Likhanskii and Poggie [38] is shown in 17.9. The simulations were carried out for air, approximated as a mixture of nitrogen and oxygen, at standard atmospheric conditions. The lower (grounded) cathode was 0.02 mm thick and 1.6 mm long, the upper, exposed anode was 0.02 mm thick and 0.2 mm long, and they were separated by a 0.1 mm thick dielectric with relative permittivity $\epsilon_r = 5.2$. Positive Gaussian pulses of 5 kV amplitude and 4 ns duration were applied to the exposed anode.

In figure 17.9, the left image shows the electron number density at a simulated time of 3.30 ns, as predicted by a kinetic, PIC model. The exposed anode and buried cathode are labeled in the image. A streamer is seen traveling from left to right in the domain, originating at the anode-dielectric juncture. The streamer head (right side of the ionized region) is the site of significant space charge and extremely high electric

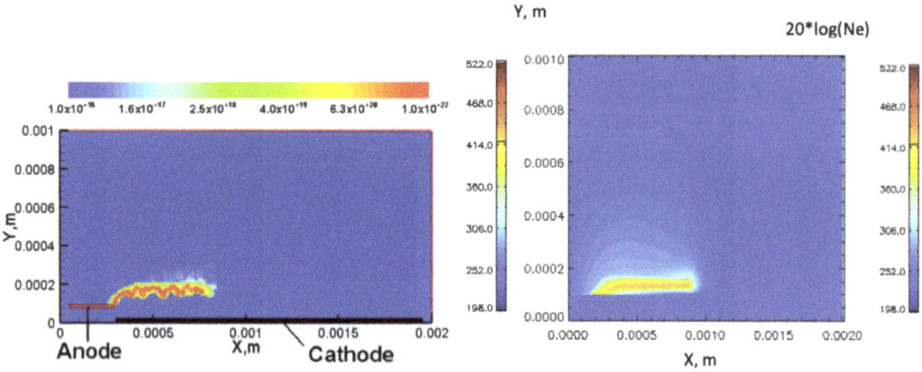

Figure 17.9. Numerical simulation of a single streamer in a DBD actuator configuration. Electron number density field for particle-in-cell simulation (left) and drift–diffusion simulation with photoionization (right). Reproduced with permission from Likhanskii and Poggie [38]. Copyright 2013 Likhanskii and Poggie.

field strength. The figure on the right shows the prediction of a drift–diffusion model at 3.15 ns, for the same configuration and conditions. The results are very similar, although the streamer propagation speed is somewhat different, possibly due to different rates of photoelectron production. For both cases, electric fields as high as 100 MV m^{-1} (about $E/N = 4000$ Td) were observed at the streamer head.

Such extreme fields seem to be a common feature of pulsed discharges. Values of the reduced electric field as high as $E/N = 10^4$ Td have been observed in pulsed discharge simulations [8]. An unexplored effect is the behavior of rate coefficients in this regime with very high electron temperature and velocity. The separation of time scales between the picosecond scale of electron motion and the millisecond scale required to achieve a periodic state with repetitive pulses, imposes a heavy computational cost [39].

Another concern is the validity of the drift–diffusion model under such extreme conditions. It is appropriate to look at the approximations used to reduce the full continuum conservation equations (17.1)–(17.3) to the drift–diffusion model (17.13)–(17.14) to explore whether the situation could be improved with changes to the model.

The relative importance of the inertia terms can be evaluated through nondimensional ratios of the elastic collision term and the electric body force term to the convective acceleration term in the species momentum equation (17.2).

$$\begin{aligned} R_1 &= \frac{m_s n_s \nu_s v_s}{m_s n_s v_s^2 / L} = \frac{\nu_s L}{v_s} \\ R_2 &= \frac{e n_s \, \Delta\phi / L}{m_s n_s v_s^2 / L} = \frac{e \, \Delta\phi}{m_s v_s^2} \end{aligned} \tag{17.33}$$

Here ν_s is the momentum-transfer collision frequency for collisions with neutrals, e is the unit charge, L is a characteristic width of the sheath, and $\Delta\phi$ is the potential drop across the sheath.

For the 400 Pa argon, 850 V, 3 ns pulse discharge experiments considered by Piskin *et al* [39], the drift–diffusion model is appropriate for the ions ($R_1 \sim 100$, $R_2 \sim 100$), but possibly less so for the electrons ($R_1 \sim 10$, $R_2 \sim 10$). In contrast, for simulations [8] of 5 kV, 5 ns pulse experiments [40] in 5 kPa air, particle acceleration plays an important role for electrons ($R_1 \sim 1$, $R_2 \sim 10$), and a fairly significant role for ions ($R_1 \sim 10$, $R_2 \sim 100$).

The latter calculation [8] implemented a relatively complete continuum model of the nanosecond-pulse air discharge, and provides an example of such calculations. A reduced plasma kinetic model (23 species and 50 processes) attempted to capture the dominant species and reactions for energy storage and thermalization in the discharge. It included electronically and vibrationally excited species, and several species of ions and ground state neutrals.

To capture the bulk gas flow that resulted from rapid thermalization, mass-averaged conservation equations (17.9)–(17.11) were solved for the gas as a whole. A five-moment model (17.1)–(17.3) was employed for the charged particles, including inertia terms and a temperature for each species. Diffusion equations (17.13) were used for the neutral species and the Poisson equation (17.22) was solved for the electric potential.

Sample results are presented in 17.10 at the time corresponding to the peak in the 5 kV, 5 ns (full width at half maximum) input voltage pulse. These conditions are extreme; the ions reach speeds and temperatures of around 4×10^3 m s^{-1} and 2 eV, and the electrons 3×10^6 m s^{-1} and 50 eV. Rapid bulk gas heating was observed near the sheath edge as a result of relaxation of excited molecules and Joule heating. The result was a temperature rise in the bulk gas and formation of acoustic waves, in agreement with experimental observation. A portion of the input energy was stored in vibrational excitation of nitrogen molecules. Due to the slow relaxation of the vibrational energy, this energy did not contribute to the formation of acoustic waves.

Multi-fluid modeling, coupled with a solution of the Poisson equation for the electric potential, is seen to be a useful tool for studying the effects in plasma actuator configurations of both the electric body force and rapid heat release. The disparate time scales of the most important physics range from picoseconds to milliseconds, making time-accurate simulations computationally expensive. Further, there is a continuing need for accurate rate data under the extreme conditions that exist in pulsed discharges.

17.5 Summary and paths forward

Several levels for computational modeling have been presented here, all practical under different circumstances for predicting and understanding the behavior of plasma actuators in flow control applications.

The computationally least expensive approach (section 17.2) is to employ a very reduced-order model of the plasma effects, implemented as source terms in the overall fluid conservation equations. This approach is appropriate when fluid dynamic effects are the most important consideration, and the plasma device serves mainly to perturb the flowfield. For example, Poggie [28] found that forcing a

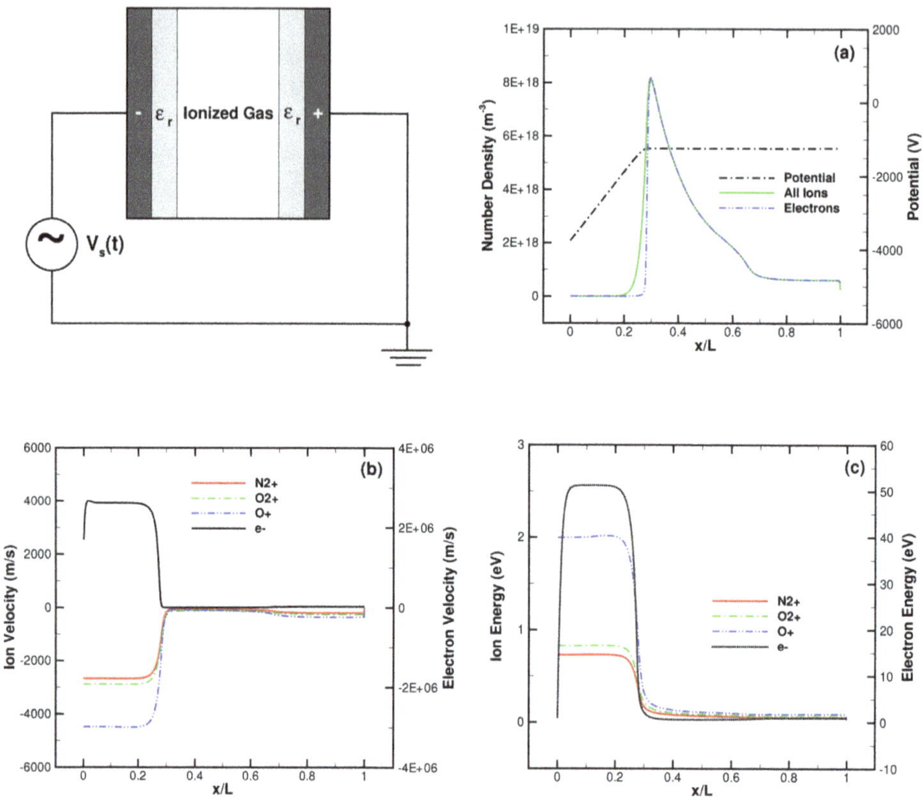

Figure 17.10. Discharge profiles at 12 ns in discharge driven by a 5 kV, 5 ns pulse. Configuration (top left). Number densities and potential (top right), velocities (bottom left), and temperatures (bottom right). Reproduced with permission from Poggie *et al* [8]. Copyright 2013 IOP Publishing.

separated flow with a body force profile tailored to a particular receptive mode, identified through conditional averaging, was able to lock in separation motion. The examples presented here show how such a simple model can also be used to estimate the order of magnitude of the power required for useful flow control. In other circumstances, however, the details of the plasma actuator behavior are an important consideration, and need to be modeled in detail. Thus additional models are considered.

The next level of modeling considered (section 17.3) is magnetohydrodynamics, in which quasi-neutrality is assumed and the magnetic body force and Joule heating are the most important effects on the flow. Via electrodes, magnetohydrodynamic devices can extract power from the flow (magnetohydrodynamic generator) or inject power into the flow (magnetohydrodynamic pump); passive magnetohydrodynamic devices without electrodes can dissipate flow power as heat. This latter situation describes the magnetohydrodynamic atmospheric entry heat shield device, which has been considered for protecting hypersonic vehicles from the extremely hot, ionized gas behind the strong bow shock. An important challenge to accurate flight magnetohydrodynamic calculations is the prediction of the electric conductivity of air under extreme, nonequilibrium conditions. This deficiency is being addressed by the hypersonic thermophysics community through systematic measurement and calculation of rate data.

The most detailed level of continuum modeling considered here (section 17.4) involves the solution of conservation equations for each significant species of charged particles. Modeling at this level of detail is necessary for understanding the process of rapid energy storage and release in ns-DBD actuators. The main challenges in this regime again involve the need for accurate rate data under extreme conditions. An additional difficulty lies in numerically capturing the disparate time scales that are important to ns-DBD physics.

Plasma-based devices have been shown experimentally to be promising for a variety of flow control applications, in which they perturb a flow by injecting heat or momentum. With a low profile and potential for rapid actuation, they can address flow control needs that are not accessible with mechanical actuators, like synthetic jets. Recent actuator development has been based, however, in a large part on trial and error, the so-called Edisonian approach.

With improved modeling of these devices, significant improvement and optimization may be achieved. Despite the challenges to accurate modeling listed here, improved computational modeling in the continuum formulation seems very feasible. With the introduction of accurate rate data for extreme conditions, and development of more efficient numerical formulations, continued progress will be achieved.

References

[1] Poggie J, McLaughlin T and Leonov S 2015 Plasma aerodynamics: current status and future directions *ONERA Aerosp. Lab J.* **10** AL10–01

[2] Kantrowitz A R 1955 A survey of physical phenomena occurring in flight at extreme speeds *Proc. of the Conf. on High-Speed Aeronautics* ed A Ferri, N J Hoff and P A Libby (New York: Polytechnic Institute of Brooklyn) pp 335–9

[3] Gurijanov E P and Harsha P T 1996 *AJAX: New Directions in Hypersonic Technology* (Reston, VA: American Institute of Aeronautics and Astronautics) AIAA Paper 1996-4609

[4] Malik M R, Weinstein L M and Hussaini M Y 1983 *Ion Wind Drag Reduction* (Reston, VA: American Institute of Aeronautics and Astronautics) AIAA Paper 83-0231

[5] Roth J R and Sherman D M 1998 *Boundary Layer Flow Control with a One Atmosphere Uniform Glow Discharge Surface Plasma* (Reston, VA: American Institute of Aeronautics and Astronautics) AIAA Paper 98-0328

[6] Raizer Y P 1991 *Gas Discharge Physics* (Berlin: Springer)

[7] Burgers J M 1969 *Flow equations for Composite Gases* (New York: Academic)

[8] Poggie J, Adamovich I, Bisek N and Nishihara M 2013 Numerical simulation of nanosecond-pulse electrical discharges *Plasma Sources Sci. Technol.* **22** 015001

[9] Ganguly B N, Bletzinger P and Garscadden A 1997 Shock Wave damping and dispersion in nonequilibrium low pressure argon plasmas *Phys. Lett.* A **230** 218–22

[10] Macheret S O, Ionikh Y Z, Chernysheva N V, Yalin A P, Martinelli L and Miles R B 2001 Shock wave propagation and dispersion in glow discharge plasmas *Phys. Fluids* **13** 2693–705

[11] Leonov S B, Adamovich I V and Soloviev V R 2016 Dynamics of near-surface electric discharges and mechanisms of their interaction with the airflow *Plasma Sources Sci. Technol.* **25** 063001
[12] Bisek N, Poggie J, Nishihara M and Adamovich I 2014 Hypersonic flow over a cylinder with a nanosecond pulse electrical discharge *J. Thermophys. Heat Transfer* **28** 18–26
[13] Bisek N, Boyd I and Poggie J 2009 Numerical study of plasma-assisted aerodynamic control for hypersonic vehicles *J. Spacecr. Rockets* **46** 568–76
[14] Atkinson M, Poggie J and Camberos J 2013 Control of high angle-of-attack reentry flow with plasma actuators *J. Spacecr. Rockets* **50** 337–46
[15] Gutiérrez D R and Poggie J 2018 Effects of power deposition on the aerodynamic forces on a slender body *AIAA J.* **56** 2911–7
[16] Myrabo L N and Raizer Y P 1994 *Laser-Induced Air Spike for Advanced Transatmospheric Vehicles* (Reston, VA: American Institute of Aeronautics and Astronautics) AIAA Paper 94-2451
[17] Knight D D 2019 *Energy Deposition for High-Speed Flow Control* (New York: Cambridge University Press)
[18] Kranc S, Yuen M C and Cambel A B 1969 *Experimental Investigation of Magnetoaerodynamic Flow around Blunt Bodies* (Washington, DC: NASA) Contractor Report NASA CR-1393 https://ntrs.nasa.gov/api/citations/19690025519/downloads/19690025519.pdf
[19] Gülhan A, Esser B, Koch U, Siebe F, Riehmer J, Giordano D and Konigorski D 2009 Experimental verification of heat-flux mitigation by electromagnetic fields in partially-ionized-argon flows *J. Spacecr. Rockets* **46** 274–83
[20] Bush W B 1958 Magnetohydrodynamics-hypersonic flow past a blunt body *J. Aerosp. Sci.* **25** 728
[21] Bush W B 1961 The stagnation-point boundary layer in the presence of an applied magnetic field *J. Aerosp. Sci.* **28** 630
[22] Poggie J and Gaitonde D V 2002 Magnetic control of flow past a blunt body: numerical validation and exploration *Phys. Fluids* **14** 1720–31
[23] Bisek N J, Boyd I D and Poggie J 2010 Numerical study of magnetoaerodynamic flow around a hemisphere *J. Spacecr. Rockets* **47** 816–26
[24] Bisek N J, Poggie J and Boyd I D 2010 *Numerical Study of a MHD-Heat Shield* (Reston, VA: American Institute of Aeronautics and Astronautics) AIAA Paper 2010-4487
[25] Porter R W and Cambel A B 1967 Hall effect in flight magnetogasdynamics *AIAA J.* **5** 2208–13
[26] Poggie J and Gaitonde D V 2000 *Magnetic Control of Hypersonic Blunt Body Flow* (Reston, VA: American Institute of Aeronautics and Astronautics) AIAA Paper 2000-0452
[27] Gaitonde D V and Poggie J 2003 Implicit technique for three-dimensional turbulent magnetoaerodynamics *AIAA J.* **41** 2179–91
[28] Poggie J 2019 Effect of forcing on a supersonic compression ramp flow *AIAA J.* **57** 3765–72
[29] Poggie J 2008 Numerical simulation of direct current glow discharges for high-speed flow control *J. Propuls. Power* **24** 916–22
[30] Macheret S O, Shneider M N and Miles R B 2004 Magnetohydrodynamic and electrohydrodynamic control of hypersonic flows of weakly ionized plasmas *AIAA J.* **42** 1378–87
[31] Corke T C, Enloe C L and Wilkinson S P 2010 Dielectric barrier discharge plasma actuators for flow control *Annu. Rev. Fluid Mech.* **42** 505–29

[32] Benard N and Moreau E 2014 Electrical and mechanical characteristics of surface AC dielectric barrier discharge plasma actuators applied to airflow control *Exp. Fluids* **55** 1846
[33] Poggie J 2007 *Numerical Simulation of DC and RF Glow Discharges* (Reston, VA: American Institute of Aeronautics and Astronautics) AIAA Paper 2007-632
[34] Kunhardt E E 2000 Generation of large-volume, atmospheric-pressure, nonequilibrium plasmas *IEEE Trans. Plasma Sci.* **28** 189–200
[35] Macheret S O, Shneider M N and Miles R B 2002 Modeling of air plasma generation by repetitive high-voltage nanosecond pulses *IEEE Trans. Plasma Sci.* **30** 1301–14
[36] Likhanskii A V, Shneider M N, Macheret S O and Miles R B 2007 Modeling of dielectric barrier discharge plasma actuators driven by repetitive nanosecond pulses *Phys. Plasmas* **14** 073501
[37] Poggie J, Tilmann C P, Flick P M, Silkey J S, Osborne B A, Ervin G, Maric D, Mangalam S and Mangalam A 2010 Closed-loop stall control system *J. Aircr.* **47** 1747–55
[38] Likhanskii A and Poggie J 2013 *On the validation of fluid plasma model for pulsed DBD plasma actuator simulations against full kinetic approach* (Reston, VA: American Institute of Aeronautics and Astronautics) AIAA Paper 2013-2494
[39] Piskin T, Podolsky V A, Macheret S O and Poggie J 2019 Challenges in numerical simulation of nanosecond-pulse discharges *J. Phys. D: Appl. Phys.* **52** 304002
[40] Nishihara M, Takashima K, Rich J W and Adamovich I V 2011 Mach 5 bow shock control by a nanosecond pulse surface dielectric barrier discharge *Phys. Fluids* **23** 066101

IOP Publishing

Plasma Modeling (Second Edition)
Methods and applications
Gianpiero Colonna and Antonio D'Angola

Chapter 18

Dust–plasma interaction: a review of dust charging theory and simulation

Gian Luca Delzanno and Xian-Zhu Tang

18.1 Introduction

The interaction of material objects with plasmas has been a major research area since the beginning of plasma physics, spanning a variety of applications in laboratory and space and astrophysics. In laboratory experiments, this includes the development of diagnostic techniques to probe the plasma properties [1], while experiments on solid particulate (dust) affecting streamer discharges were reported by Langmuir *et al* as early as 1924 [2]. Almost in parallel, remote observations indicated the presence of dust as a medium responsible for absorption, scattering and emission of light in an astrophysical environment [3].

With the advent of the space age and the beginning of *in situ* observations, it became clear that dust is also a ubiquitous component of the solar system [4, 5]. The Voyager observations of the mysterious Saturn spokes [6] indicated that the dust grains were charged by the plasma and gave impetus to the development of some of the theories of dust–plasma interaction. In the early 90s further interest was driven by the microelectronics industry, particularly with regard to plasma deposition and etching techniques, where the negative impact of the presence of dust particulates in the processing chamber was recognized [7, 8]. Finally, tremendous growth of the field of dusty plasmas was driven by the discovery of ordered structures (i.e. plasma crystals) [9, 10] and the possibility to investigate phase transitions in these systems. Indeed, the fact that dust particles can be diagnosed with simple optical techniques (and in fact can be normally seen with the naked eye) is one of the main advantages of dusty plasmas as it allows measurements at the kinetic level [11]. Today the field keeps growing with new emerging applications. For instance, dust created by plasma–material interaction at the chamber walls of tokamak devices poses important safety and operational concerns [12]. This has led to the development of sophisticated dust transport codes [13–18] and experimental campaigns [19–22] aimed at understanding the impact of dust in magnetic fusion energy applications. Furthermore, plasma medicine exploiting the interaction of plasmas with biological matter is a new research frontier that can have many aspects in common with dust–plasma interaction [23] and is showing much promise for the treatment of diseases for which traditional medicine has yet to find a definitive answer [24]. New laboratory experiments with high magnetic fields are also leading the development of the field of magnetized dusty plasmas [25].

A dusty plasma is loosely defined as a system containing electrons, ions, neutrals and dust particulates [26]. The dust is macroscopic relative to the size of plasma particles, atoms and molecules but still microscopic relative to the system size. There is indeed a great variety of dust sizes in the Universe. In astrophysical environments one can find macromolecules, micron-sized grains, planetesimals all the way to boulders [5]. In laboratory, the dust size can also vary widely. There are two characteristic regimes that depend on the relative ratios of the dust characteristic radius r_d, the plasma Debye length λ_D and the dust interparticle distance d [26]. Typically, the case $r_d < \lambda_D < d$ is referred to as 'dust-in-a-plasma' since the dust grains act individually as screened charged particles, while the case $r_d < d < \lambda_D$ is that of a 'dusty-plasma' since the dust particles can act collectively. Note that in space and astrophysics $r_d/\lambda_D \ll 1$ (see for instance references [3, 27] for some representative parameters), while in laboratory experiments parameters can vary widely and $r_d/\lambda_D \gg 1$ is also possible.

Dusty plasmas are quite different from conventional plasmas [28–30]. The dust particles are charged by absorption of plasma particles (and possibly also by other processes like electron emission) and are coupled to the plasma through the electromagnetic force. Since the dust charge-to-mass-ratio is much smaller than that of the plasma, the dust particles introduce new characteristic scales and dynamics into the system. In addition, the dust can significantly affect the plasma through the quasi-neutrality condition, $q_i n_i = e n_e - q_d n_d$, where n_d, n_e and n_i are the dust, electron and ion densities, respectively, q_d is the dust charge (assumed here to be negative), e is the elementary charge and q_i is the ion charge. Oftentimes the dust particles can accumulate $10^3 - 10^5$ electron charges, implying that even a small (relative to the plasma density) dust density can contribute appreciably to quasi-neutrality. Furthermore, dust particles can also act as a source of plasma through electron emission processes. Lastly, the dust charge is a dynamical variable that depends on the local dust and plasma conditions.

A dust charging theory is the building block of any study in dusty plasma physics and its review is the main objective of the present chapter. In section 18.2 we review the basics of the interaction of a single dust grain with a collisionless, unmagnetized plasma in the simplest case of primary charging. We derive the equations of the orbital-motion-limited (OML) theory, the most widely used charging theory in the dusty plasma community. In section 18.3 we briefly discuss the numerical approaches that can be used in the context of dust charging. In section 18.4 we extend the charging theory to the case of dust electron emission. Electron emission can significantly alter the heating of a dust grain in a plasma, playing a major role in dust survivability. With electron emission, we show that OML can become inaccurate in the regime where the dust grain is positively charged and the power collected by the dust from the background electrons can be significantly reduced relative to the OML value. We conclude by highlighting some emerging applications and related important effects where further developments of dust–plasma interaction theories are necessary.

18.2 Basics of dust–plasma interaction

Let us begin by considering a spherical, absorbing dust grain of radius r_d at rest in an unmagnetized, collisionless plasma. The grain gets charged by absorbing plasma particles, creating a sheath electric field near the grain. If the grain is at floating potential with respect to the plasma, the net current collected by the grain from the plasma vanishes as the electron and ion currents cancel each other. In this simple case, often denoted as primary charging, the grain is normally negatively charged due to the higher mobility of the electron relative to the ions.

We assume spherical symmetry, introduce a spherical reference system centered on the grain and seek the steady-state solution of this system. Such state is determined by the laws of Hamiltonian mechanics in a central force field (known as orbital motion (OM) theory), where the particle's motion is confined to a plane determined by its initial radial (v_r) and tangential (v_θ) velocities and the conservation of energy

$$\frac{m_\alpha}{2}(v_r^2 + v_\theta^2) + q_\alpha \phi(r) = E \tag{18.1}$$

and angular momentum

$$m_\alpha r v_\theta = J \tag{18.2}$$

apply to each particle. In equations (18.1) and (18.2), m_α (q_α) is the mass (charge) of particles of species α ($\alpha = e, i$ labels electrons or ions), E (J) is the particle energy (angular momentum) and $\phi(r)$ is the radially-symmetric electrostatic potential while r is the radial distance. From classical mechanics, the motion of a particle can be reduced to a one-dimensional (radial) motion in an effective potential field given by

$$U(r, J) = q_\alpha \phi(r) + \frac{1}{2}\frac{J^2}{m_\alpha r^2}. \tag{18.3}$$

Clearly the first term in equation (18.3) corresponds to the electrostatic force while the second is the centrifugal force.

In terms of boundary conditions, at infinity $lim_{r \longrightarrow +\infty}\phi(r) = 0$, the plasma is unperturbed and we will assume the plasma distribution function to be Maxwellian:

$$f_\alpha(\mathbf{v}) = n_\infty \left(\frac{m_\alpha}{2\pi T_\alpha}\right)^{3/2} \exp\left[-\frac{m_\alpha}{2T_\alpha}(v_r^2 + v_\theta^2)\right], \tag{18.4}$$

where n_∞ is the unperturbed plasma density and T_α is the temperature of species α expressed in eV. This implies that in equation (18.1), $E > 0$, a condition that must be taken into account when the distribution function is integrated to obtain plasma densities and currents. Note also that, by virtue of Liouville's theorem, the distribution function at any point in space remains Maxwellian and is given by

$$f_\alpha(\mathbf{v}) = n_\infty \left(\frac{m_\alpha}{2\pi T_\alpha} \right)^{3/2} \exp\left[-\frac{m_\alpha}{2T_\alpha}(v_r^2 + v_\theta^2) - \frac{q_\alpha \phi(r)}{T_\alpha} \right]. \tag{18.5}$$

It is important to emphasize that we are only considering *unbounded* particles, that is, particles whose orbits connect to infinity.

Furthermore, one must take into account the absorbing nature of the grain, namely the fact that particles on inbound orbits ($v_r < 0$) hitting the grain will not be present in the outbound orbit ($v_r > 0$). Again from equations (18.1) and (18.2), it follows that the condition for absorption is

$$v_r^2 + \left[1 - \left(\frac{r}{r_d} \right)^2 \right] v_\theta^2 + 2\frac{q_\alpha}{m_\alpha} \left[\phi(r) - \phi_d \right] \geqslant 0, \; v_r \leqslant 0, \tag{18.6}$$

where $\phi_d = \phi(r_d)$ is the dust potential. This implies that, in order to calculate the plasma density around the grain, the distribution function must be integrated with the condition

$$v_r^2 + \left[1 - \left(\frac{r}{r_d} \right)^2 \right] v_\theta^2 + 2\frac{q_\alpha}{m_\alpha} \left[\phi(r) - \phi_d \right] \leqslant 0, \; v_r \geqslant 0. \tag{18.7}$$

The density of the plasma can then be computed by integrating equation (18.5) over the *accessible* part of phase space:

$$n_\alpha(r) = 2\pi \int \mathrm{d}v_r \int \mathrm{d}v_\theta v_\theta f_\alpha(r, \mathbf{v}). \tag{18.8}$$

Similarly, the current reaching the dust grain is obtained by

$$I_\alpha = -q_\alpha 4\pi r_d^2 2\pi \int \mathrm{d}v_r v_r \int \mathrm{d}v_\theta v_\theta f_\alpha(r_d, \mathbf{v}), \tag{18.9}$$

and the power collected by the dust grain from species α is

$$h_\alpha = -4\pi r_d^2 2\pi \int \mathrm{d}v_r v_r \int \mathrm{d}v_\theta v_\theta \frac{m_\alpha}{2}(v_r^2 + v_\theta^2) f_\alpha(r_d, \mathbf{v}), \tag{18.10}$$

where the domain of integration is now restricted to incoming particles (i.e. $v_r < 0$) and, as in equation (18.8), again the limits of integration will be specified on a case-by-case basis. The factor $2\pi v_\theta$ in equations (18.8)–(18.10) comes from expressing the tangential velocity in terms of its magnitude v_θ and an isotropic angle that can be integrated away and leads to 2π. Consistently, $v_\theta > 0$.

18.2.1 Repelled species ($q_\alpha \phi_d > 0$)

Let us now proceed to treat the species that is repelled by the dust grain, that is electrons in the case of a negatively-charged grain. The effective potential $U(r)$ is always positive and monotonically decreasing with r, implying that all the particles are on unbounded orbits and that $E > U(r_d, J)$ is necessary to reach the dust grain. The distribution function must be integrated with the absorption condition (18.7), yielding

$$\frac{n_\alpha(z)}{n_\infty} = \frac{1}{2}\left\{1 + \mathrm{Erf}\left(\sqrt{\varphi_d - \varphi}\right) + \sqrt{1 - z^{-2}}\left[1 - \mathrm{Erf}\left(\sqrt{\frac{\varphi_d - \varphi}{1 - z^{-2}}}\right)\right]\right.$$
$$\left.\exp\left[\frac{\varphi_d - \varphi}{z^2 - 1}\right]\exp(-\varphi),\right\} \tag{18.11}$$

where we have introduced $z = r/r_d$ and $\varphi = q_\alpha\phi/T_\alpha$. Figure 18.1 shows the phase-space domain for the repelled species. Similarly, one can compute the current reaching the dust grain by integrating over the whole $v_r < 0$ domain. It yields

$$I_\alpha = q_\alpha 4\pi r_d^2 n_\infty \sqrt{\frac{T_\alpha}{2\pi m_\alpha}}\exp\left(-\frac{q_\alpha\phi_d}{T_\alpha}\right). \tag{18.12}$$

Similarly, the dust collected power is

$$h_\alpha = 2T_\alpha\frac{I_\alpha}{q_\alpha}. \tag{18.13}$$

18.2.2 Attracted species ($q_\alpha\phi_d < 0$)

For the attracted species, the situation is complicated by the fact that the electrostatic and centrifugal forces have now opposite signs and therefore the effective potential can be negative and/or non-monotonic. To illustrate this point, let us consider a Debye–Hückel potential (normalized to the temperature T_α) around the grain

$$\varphi(r) = \frac{q_\alpha\phi(r)}{T_\alpha} = \varphi_d\frac{r_d}{r}\exp\left(-\frac{r - r_d}{\lambda}\right), \tag{18.14}$$

where λ is the screening length. We choose $\varphi_d = -2.5$ and $\lambda/r_d = 2$ and show the effective potential for different values of the dimensionless parameter $\delta = J^2/(2m_\alpha r_d^2 T_\alpha)$ in figure 18.2. For $\delta = 0$, the effective potential is monotonically

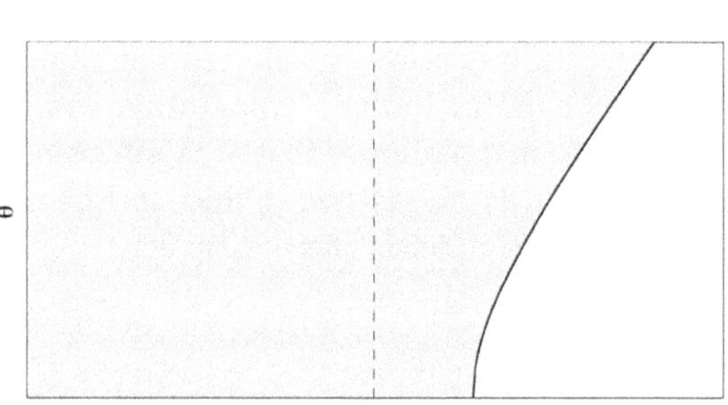

Figure 18.1. Cartoon of the phase-space integration domain to obtain the density of the repelled species.

increasing with r. For small values of δ, $\delta < \delta_1$ ($\delta_1 = -3\varphi_d/4 = 1.875$ for the parameters considered here) the effective potential is non-monotonic, with a maximum at a certain $r_m(J)$. For $\delta_1 \leqslant \delta \leqslant \delta_2$ ($\delta_2 \simeq 3.46$), the effective potential is still non-monotonic, but with a minimum close to r_d and a maximum away from r_d. For $\delta \geqslant \delta_2$, the effective potential is again monotonic, but decreasing with r. The fact that the effective potential can be non-monotonic has two important consequences:

1. There could be particles performing bounded motion, i.e. motion confined between $r_A \leqslant r \leqslant r_B$, where r_A and r_B depend on the particle energy and angular momentum and $r_A > r_d$. Bounded particles are typically neglected in theories of dust–plasma interaction and we will follow the same path here, but the reader should keep in mind that this might not always be the case.
2. Particles might be reflected before reaching the dust grain if the condition

$$U^{\max} = U(r_m(J),\, J) > E \tag{18.15}$$

holds, where r_m is the position of the maximum of U and, most importantly, is dependent on the particle's angular momentum J. This is known as *the absorption radius* effect [31], which might reflect away some particles of the attractive species even if $E > U(r_d,\, J)$ is satisfied. One would then need to integrate equation (18.5) with the condition (18.15) (instead of equation (18.7)) when $v_r > 0$. It yields an expression for the density in integral form and the interested reader is referred to references [32, 33] for the exact expressions.

Here we take an alternative path, known as the OML theory [34]. The simplifying assumption is to assume that the effective potential is monotonic or, in other words, that all the particles such that $E > U(r_d)$ (which of course is equivalent to equation

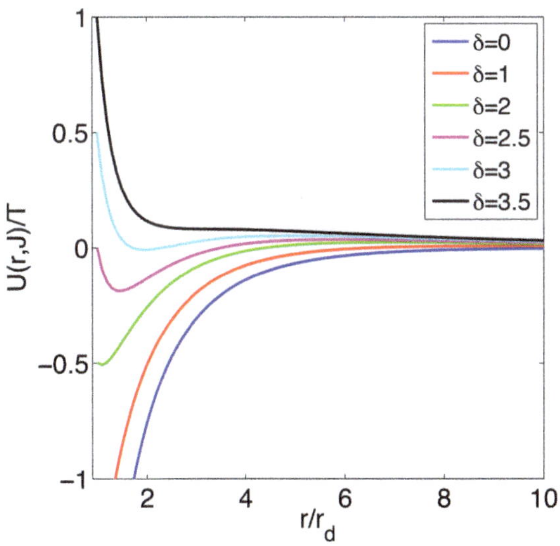

Figure 18.2. Profiles of the effective potential varying J via $\delta = J^2/(2m_\alpha r_d^2 T_\alpha)$ for the Debye–Hückel potential given in equation (18.14), illustrating the non-monotonic nature of the effective potential for particles of the attracted species in a certain range of J. Adapted with permission from [35]. Copyright 2014, AIP Publishing LLC.

(18.7)) can reach the dust grain. From figure 18.2, one expects that OML would overestimate the current and the density near the grain since it neglects that some particles can be reflected by (effective) potential barriers.

Let us now proceed to calculate the density of the attracted species in the OML framework. First, the distribution function has to be integrated with the condition $E > 0$, whose boundary is a circle in phase space

$$v_r^2 + v_\theta^2 \geqslant -\frac{2q_\alpha \phi(r)}{m_\alpha} \tag{18.16}$$

which intersects the $v_r = 0$ axis at

$$v_\theta^b = \sqrt{-2q_\alpha \phi(r)/m_\alpha}\,. \tag{18.17}$$

For particles moving towards the grain, $v_r < 0$, equation (18.16) is the only condition necessary to calculate the density. For $v_r > 0$, on the other hand, one must also consider equation (18.7), which corresponds to a parabola in phase space whose intersection with the $v_r = 0$ axis is

$$v_\theta^r = \sqrt{\frac{2q_\alpha}{m_\alpha}\frac{\phi - \phi_d}{z^2 - 1}}\,. \tag{18.18}$$

Thus, the domain of integration in the $v_r > 0$ space depends on the value of v_θ^b and v_θ^r. For the case $v_\theta^b > v_\theta^r$, which corresponds to $\phi(r) < \phi_d/z^2$, the domain of integration is

$$\begin{aligned} &v_r < 0,\ v_r^2 + v_\theta^2 \geqslant -2q_\alpha\phi(r)/m_\alpha, \\ &v_r \geqslant 0,\ v_r \in [0, v_r^c],\ v_\theta \in [\sqrt{-(2q_\alpha/m_\alpha)\phi - v_r^2}, \infty), \\ &v_r \geqslant 0,\ v_r \in [v_r^c, \infty),\ v_\theta \in \left[\sqrt{\frac{v_r^2 + \frac{2q_\alpha}{m_\alpha}(\phi - \phi_d)}{z^2 - 1}}, \infty\right), \end{aligned} \tag{18.19}$$

where v_r^c and v_θ^c correspond to $v_\theta^b = v_\theta^r$, i.e.

$$v_r^c = \sqrt{-\frac{2q_\alpha}{m_\alpha}\left(\phi - \frac{\phi_d}{z^2}\right)};\ v_\theta^c = \sqrt{-\frac{2q_\alpha}{m_\alpha}\frac{\phi_d}{z^2}} \tag{18.20}$$

See figure 18.3 (left). This yields

$$\begin{aligned} \frac{n_\alpha(z)}{n_\infty} = &\sqrt{-\frac{\varphi}{\pi}}\left[1 + \sqrt{1 - \frac{\varphi_d}{z^2\varphi}}\right] + \frac{e^{-\varphi}}{2}\left[1 - \text{Erf}\left(\sqrt{-\varphi}\right)\right] \\ &+ \frac{\sqrt{1 - z^{-2}}}{2}e^{-\tilde{\varphi}}\left[1 - \text{Erf}\left(\sqrt{-\tilde{\varphi}}\right)\right],\ \phi \leqslant \phi_d/z^2 \end{aligned} \tag{18.21}$$

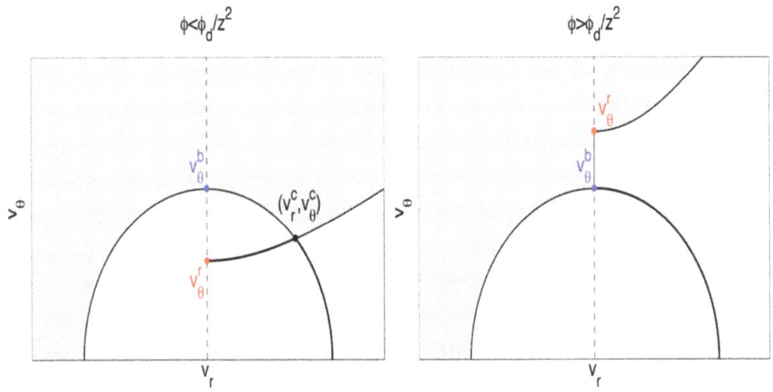

Figure 18.3. Cartoon of the phase-space integration domain for the attracted species. The intersection of the two curves, $E > 0$ and equation (18.7), gives rise to two different integration limits depending on the behavior of $\phi(r)$ versus ϕ_d/z^2. Adapted with permission from [35]. Copyright 2014, AIP Publishing LLC.

with

$$\tilde{\varphi} = \frac{z^2\varphi - \varphi_d}{z^2 - 1}. \tag{18.22}$$

For $v_\theta^b < v_\theta^r$, which corresponds to $\phi(r) > \phi_d/z^2$, the domain of integration becomes

$$\begin{aligned} &v_r < 0,\ v_r^2 + v_\theta^2 \geqslant -2q_\alpha\phi(r)/m_\alpha, \\ &v_r \geqslant 0,\ v_r \in [0, +\infty),\ v_\theta \in \left[\sqrt{\frac{v_r^2 + \frac{2q_\alpha}{m_\alpha}(\phi - \phi_d)}{z^2 - 1}}, \infty\right). \end{aligned} \tag{18.23}$$

See figure 18.3 (right). It yields

$$\begin{aligned} \frac{n_\alpha(z)}{n_\infty} = &\sqrt{-\frac{\varphi}{\pi}} + \frac{e^{-\varphi}}{2}\left[1 - \mathrm{Erf}\left(\sqrt{-\varphi}\right)\right] \\ &+ \frac{\sqrt{1 - z^{-2}}}{2}e^{-\tilde{\varphi}},\ \phi \geqslant \phi_d/z^2, \end{aligned} \tag{18.24}$$

We note that the different expressions for the density depending on the behavior of ϕ relative to ϕ_d/z^2 have only been derived fairly recently [35, 36]. The distinction between ϕ and ϕ_d/z^2 is not necessary to compute the current, which is obtained from equation (18.9) by integrating over $v_r < 0$ and $v_r^2 + v_\theta^2 \geqslant -2q_\alpha\phi_d/m_\alpha$. It leads to

$$I_\alpha = q_\alpha 4\pi r_d^2 n_\infty \sqrt{\frac{T_\alpha}{2\pi m_\alpha}} \left(1 - \frac{q_\alpha \phi_d}{T_\alpha}\right). \tag{18.25}$$

The dust collected power is

$$h_\alpha = T_\alpha \left(\frac{2 - \varphi_d}{1 - \varphi_d} - \varphi_d\right) \frac{I_\alpha}{q_\alpha}. \tag{18.26}$$

18.2.3 Summary of OML theory

In summary, the steady-state solution of the interaction between a spherical dust grain and a plasma in the OML framework can be obtained by the solution of the non-linear Poisson equation

$$\nabla^2 \phi = \frac{1}{r^2} \frac{d}{dr}\left(r^2 \frac{\mathrm{d}\phi}{\mathrm{d}r}\right) = -\frac{\sum_\alpha q_\alpha n_\alpha(\phi)}{\varepsilon_0}, \tag{18.27}$$

where ε_0 is vacuum permittivity and the densities are given by equations (18.11) and (18.21) for $\phi \leqslant \phi_d / z^2$ and (18.24) for $\phi \geqslant \phi_d / z^2$. Equation (18.27) is solved with the following boundary conditions

$$\text{Floating potential condition: } \sum_\alpha I_\alpha(\phi_d) = 0 \tag{18.28}$$

and

$$\lim_{r \longrightarrow +\infty} \phi(r) = 0. \tag{18.29}$$

From a numerical point of view, equation (18.29) is often substituted by $\phi(R) = 0$, where $R \gg r_D$ is the boundary of the computational domain. For a probe biased at ϕ_{bias}, equation (18.28) is substituted by the condition $\phi(r_d) = \phi_{\text{bias}}$.

The theory is based on the following assumptions:

1. Spherical symmetry.
2. Unbounded particles: the trajectory of all the plasma particles (either collected by the dust grain or reflected at some radius) connects to infinity.
3. Collisionless plasma: collisions between plasma particles or between particles and neutrals are neglected, implying that the theory is valid when r_d, $\lambda_D \ll \lambda_{\text{mfp}}$, where λ_D is the plasma Debye length and λ_{mfp} is the collisional mean-free-path.
4. Unmagnetized plasma: the effect of the magnetic field on the trajectory of the plasma particles is neglected, implying that the theory if valid when r_d, $\lambda_D \ll \rho_e$, where ρ_e is the electron gyroradius.
5. Potential barriers to the motion of the particles of the attracted species are neglected, namely all the attracted particles satisfying $E > U(r_d, J)$ are assumed to be collected by the dust grain. Relative to the assumptions listed above, this is the only one that differentiates the OML theory from the more complete (but rarely used) OM theory.

18.2.4 Some important considerations

We have shown that the OML approximation consists in neglecting that particles (of the attracted species) with angular momentum in a certain range will be reflected before reaching the dust grain if their absorption radius is bigger than the dust radius. Thus, OML can/will overestimate the collected current and the density near the grain. This quite innocent approximation has, in practice, remarkable consequences: one can obtain the dust potential by solving equation (18.28) and *without* solving Poisson's equation! Given its simplicity and considering that the dust potential is the quantity that determines currents, forces and heat fluxes to the grain, all necessary ingredients to study dust transport in a plasma, one can understand why OML is the most widely used dust charging theory.

One might ask the question of the validity of OML theory. The answer, of course, has to do with the particular form of the shielding potential and whether or not potential barriers to the particles motion are present. By studying the behavior of $r^3 dU/dr$ one can conclude that if $\phi(r)$ varies more slowly than $1/r^2$ (say like a Coulomb potential $\propto 1/r$), then the effective potential can only have a minimum or be monotonic, see references [32, 37]. For these conditions OML theory would be exact. Unfortunately, these are not the conditions that one encounters for dust–plasma interaction, where the potential decreases more slowly than $1/r^2$ (i.e. Coulomb) close to the grain, faster than $1/r^2$ in the Debye shielding region $r \sim \lambda_D$ and exactly as $1/r^2$ asymptotically for $r \gg \lambda_D$. Thus, OML can only be an approximation.

What is then the accuracy of OML theory? To answer this question one must compare OML with an exact solution obtained either by solving the OM theory or by simulations. The fact that OML reproduces OM in the limit of small dust radius relative to the Debye length has been investigated by several authors [33, 38–41] and is well established. Furthermore particle-in-cell (PIC) simulations of dust–plasma interaction were performed in references [40, 41] to investigate the limit of large dust radius. It was concluded that OML is a very good approximation for $r_d/\lambda_{\mathrm{De}}$ of order unity (with a dependence on the electron-to-ion temperature ratio [40]), with λ_{De} the electron Debye length. For $r_d \gg \lambda_{\mathrm{De}}$, the sheath near the dust grain approaches a planar sheath (with the ion density increasing monotonically with r) and the transition from a spherical to a planar sheath is not captured by OML. Nevertheless, from reference [41] we find that for $T_i/T_e = 1$, $m_i/m_e = 1836$ and up to $r_d/\lambda_{\mathrm{De}} = 10$, OML is still reasonable with ~20% accuracy on dust charge, currents and heat fluxes.

An important issue relates to the dust charge Q_d, a quantity critically important for dust transport since it is directly related to the Lorentz force. The dust charge is given by the electric field at the dust surface as dictated by Gauss' law:

$$Q_d = -4\pi\varepsilon_0 r_d^2 \frac{\mathrm{d}\phi}{\mathrm{d}rr}\bigg|_{r=r_d}, \tag{18.30}$$

where we are still taking advantage of spherical symmetry. For a Debye–Hückel potential (18.14), equation (18.30) becomes

$$Q_d = 4\pi\varepsilon_0 r_d\left(1 + \frac{r_d}{\lambda}\right)\phi_d, \tag{18.31}$$

where one recognizes the capacitance of a spherical conductor in vacuum, $C_d = 4\pi\varepsilon_0 r_d$. In the limit $r_d/\lambda \ll 1$, expression (18.31) simplifies to

$$Q_d = 4\pi\varepsilon_0 r_d \phi_d, \tag{18.32}$$

namely the dust charge is completely determined by the dust potential ϕ_d. However, equation (18.31) reveals that in general the dust charge depends on the space-charge distribution outside the grain, i.e. by the screening λ, and this requires the solution of the non-linear Poisson equation (18.27). Note also that the screening contribution to Q_d can in fact be dominant in regimes where $r_d/\lambda > 1$. For instance, reference [41] shows that for $T_i/T_e = 1$, $m_i/m_e = 1836$ and $r_d/\lambda_{\mathrm{De}} = 10$, the dust charge is six times higher than that given by equation (18.32).

18.3 A note on the numerical solution of dust–plasma interaction problems

Before we proceed with investigating further aspects of dust–plasma interaction, let us discuss how the problem can be treated numerically. There are essentially two approaches. Under certain assumptions such as those of OML or OM theories, the problem is reduced to the numerical solution of a non-linear Poisson equation. In order to do so, let us introduce a computational mesh for the radial coordinate, where mesh points are given by

$$r_i = r_d + (i - 1)\Delta r, \quad i = 1, \ldots, N + 1, \tag{18.33}$$

with N the number of mesh points and $\Delta r = (R - r_d)/N$ the mesh spacing (assumed uniform for simplicity). The non-linear Poisson equation can be discretized using for instance finite differences techniques and cast in residual form as

$$R[\phi_i] = R_i = \frac{r_{i+1/2}^2}{r_i^2}\frac{\phi_{i+1} - \phi_i}{\Delta r^2} - \frac{r_{i-1/2}^2}{r_i^2}\frac{\phi_i - \phi_{i-1}}{\Delta r^2} + \sum_\alpha \frac{q_\alpha}{\varepsilon_0} n_\alpha(\phi_i), \quad i = 2, \ldots, N, \tag{18.34}$$

where $\phi_i = \phi(r_i)$, together with the boundary conditions

$$\begin{aligned} R[\phi_1] &= R_1 = \sum_\alpha I_\alpha(\phi_1), \\ R[\phi_{N+1}] &= R_{N+1} = \phi_{N+1}. \end{aligned} \tag{18.35}$$

In matrix form, we write equations (18.34) and (18.35) as $\mathbf{R}(\phi) = 0$, with ϕ the column vector containing all the unknowns. The system of discrete equations can be solved numerically with standard techniques of numerical analysis, such as Newton's method [42]. The idea is to solve iteratively successive linear systems of the form

$$\mathbf{J}(\phi^n)\Delta\phi^n = -\mathbf{R}(\phi^n) \tag{18.36}$$

with the Jacobian matrix $\mathbf{J}$ defined as $J_{ij} = \frac{\partial \mathbf{R}_i(\phi)}{\partial \phi_j}$ (where $\mathbf{R}_i$ labels the ith row of $\mathbf{R}$), n labels the Newton iteration and $\phi^{n+1} = \phi^n + \Delta\phi^n$. Non-linear convergence is reached when

$$\| \mathbf{R}(\phi^n) \|_2 \leqslant \tau_a + \tau_r \| \mathbf{R}(\phi^0) \|_2, \tag{18.37}$$

with $\|\cdot\|$ the L_2-norm and τ_r and τ_a are relative and absolute tolerances that are provided by the user. A discussion of iterative methods for linear and non-linear equations can be found in reference [42] and a series of MATLAB codes that implement these solvers is available on the related website. We use a Jacobian-Free Newton–Krylov (JFNK) method for the solution of $\mathbf{R}(\phi) = 0$, which has the advantage that the Jacobian matrix is never formed explicitly and the user only has to provide the residual function.

An alternative approach is to solve directly the time-dependent Vlasov–Poisson equations. For plasma–material interaction problems, the most popular technique is the PIC method. In PIC the plasma is discretized with macroparticles, each carrying the statistical information (like the charge) corresponding to many particles of the physical system, moving through a computational mesh. At each time step, the plasma density is obtained by interpolating the particles' information on the mesh, Poisson's equation is solved on the mesh to construct the electrostatic field there and the electrostatic force is interpolated back to the particles' position. The time step is completed by moving the particles via Newton's equations. The reader interested in the details of the PIC algorithm is referred to chapter 4 and classic monographs [43, 44]. Note that the presence of a dust grain requires two adjustments to the classic PIC algorithm. First, plasma particles absorbed by the grain are removed from the simulation and their charge is accumulated onto the dust grain, translating in a boundary condition on the electric field given by Gauss' law. Second, since the dust grain is a sink of plasma particles, a source of plasma must be injected in the system (typically at the outer boundaries) to be able to reach a steady state.

Each of the two techniques has advantages and disadvantages. The Newton solver can deliver the solution within minutes on a conventional laptop (although convergence can be difficult depending on the level of nonlinearity of the densities and might require some user intervention to provide a good initial guess to the iterative procedure) but the non-linear Poisson's equation can only be derived under some quite restrictive assumptions. PIC, on the other hand, is a first-principle method that can be easily extended to more general cases, such as to include the effect of collisionality or of an external magnetic field, or can be used to study transients. The price to pay for this flexibility is that it might take a long time to obtain the steady-state solution of the problem and this might require significant computational resources depending on the system parameters. Nevertheless, several PIC codes are in use in the dusty plasma community [45–49].

In the next section we will use both techniques to explore the limitations of OML in the limit of electron-emitting dust grains.

18.4 Dust electron emission

Let us now investigate dust–plasma interaction for electron-emitting dust grains, a case of interest in many applications in nature and in the laboratory. Dust grains can emit electrons according to three main mechanisms: thermionic emission, photoemission and secondary emission. Thermionic emission occurs when a material is heated to high enough temperature that some electrons have enough kinetic energy to overcome the attractive force that holds them in the material (i.e. the material work function). Thermionic emission is important for the dynamics of dust particles created in magnetic fusion energy devices [14] or for meteoroids frictionally heated by the interaction with the Earth's atmosphere [50]. Electron emission can also be induced by the photo-electric effect, when incident photons have enough energy to release electrons from the material. This occurs commonly in space because of the UV field of nearby stars. Examples include dust transport in the solar system [51], planetary rings [52], cometary environments [53], the Earth's Moon [54] or spacecraft–plasma interaction when the spacecraft is exposed to sunlight [55]. Secondary electron emission, on the other hand, occurs when incident, energetic plasma particles eject electrons from materials [56]. Once again, secondary emission is important in many applications in the laboratory [14, 57] and space and astrophysical environments [55, 58, 59]. A recent review on secondary electron emission can be found in reference [60].

In what follows, we will consider Maxwellian emitted electrons

$$f_{\rm em}(r_d, v_r, v_\theta) = \frac{J_{\rm em}}{e2\pi v_{\rm th,\,em}^4}\exp\left[-\frac{m_e}{2T_{\rm em}}(v_r^2 + v_\theta^2)\right], \tag{18.38}$$

where $J_{\rm em}$ is total emitted current density and $v_{\rm th,em} = \sqrt{T_{\rm em}/m_e}$ is the thermal velocity of the emitted electrons with temperature $T_{\rm em}$. A Maxwellian distribution function is representative of thermionic emission, where $J_{\rm em}$ is given by the Richardson–Dushman current [61]

$$J_{\rm em} = e\frac{4\pi T_{\rm em}^2 m_e}{h^3}\exp\left(-\frac{W}{T_{\rm em}}\right), \tag{18.39}$$

with h Planck's constant and W the thermionic work function of the dust material. A Maxwellian distribution function is also sometimes used to model photoemission, with the current density $J_{\rm em}$ depending on properties of the material and of the incident light. For secondary electron emission, on the other hand, the current density depends nonlinearly on the plasma response to the charging process (via the electron background current to the dust grain) and changes with the polarity of the grain.

In order to treat the emitted electrons we can again use orbital motion. The procedure follows that of section 18.2, with the difference that now the particle energy is in the range $-e\phi_d < E < \infty$. The expressions for the emitted-electron densities, current and emitted power are equivalent to equations (18.8), (18.9) and

(18.10). The steady state of the system including electron emission amounts again to solving Poisson's equation with the emitted-electrons density $n_{\rm em}$:

$$\nabla^2\phi = -\frac{\sum_\alpha q_\alpha n_\alpha(\phi) - en_{\rm em}(\phi)}{\varepsilon_0}, \tag{18.40}$$

Similarly, the floating condition (18.28) is modified to include the emitted-electron current, $I_{\rm em}$:

$$I_e(\phi_d) + I_i(\phi_d) + I_{\rm em}(\phi_d) = 0. \tag{18.41}$$

18.4.1 The OML approach

Let us begin our investigation of electron emission effects by reviewing the OML approach, which assumes a monotonic shielding potential around the grain. We will treat the negatively-charged and positively-charged cases separately, depending on the sign of ϕ_d.

The negatively-charged case with monotonically increasing ϕ

Let us consider the case where the grain is negatively charged and the shielding potential is *negative and monotonically increasing*. Under these assumptions, the effective potential is always positive and monotonically decreasing. All the emitted electrons are accelerated away from the grain, since $E > U(r_d, J)$ holds. There are no potential barriers (as described above for the attracted species) to the motion of the emitted electrons and the expressions that are derived below are exact. The emitted-electron density is obtained by integrating the distribution function over the domain

$$v_r \in \left[+\sqrt{\frac{2e}{m_e}(\phi - \phi_d)}, +\infty\right), \; v_\theta \in \left[0, +\sqrt{\frac{v_r^2 - \frac{2e}{m_e}(\phi - \phi_d)}{z^2 - 1}}\right] \tag{18.42}$$

leading to

$$n_{\rm em}(r) = \sqrt{\frac{\pi}{2}}\frac{J_{\rm em}}{ev_{\rm th,em}}\exp\left(e\frac{\phi - \phi_d}{T_{\rm em}}\right)\left[1 - {\rm erf}\sqrt{e\frac{\phi - \phi_d}{T_{\rm em}}} - \exp\left[\frac{e(\phi - \phi_d)}{T_{\rm em}(z^2 - 1)}\right]\frac{\sqrt{z^2 - 1}}{z}\left(1 - {\rm erf}\sqrt{\frac{e(\phi - \phi_d)z^2}{T_{\rm em}(z^2 - 1)}}\right)\right], \tag{18.43}$$

while the emitted electron current is

$$I_{\rm em} = 4\pi r_d^2 J_{\rm em} \tag{18.44}$$

and the electron-emitted power is

$$h_{\mathrm{em}} = -2T_{\mathrm{em}}\frac{I_{\mathrm{em}}}{e}. \tag{18.45}$$

Since in this regime the only approximation in OML is to neglect potential barriers to the motion of the particles of the attracted species, as discussed in section 18.2, but the treatment of the emitted electrons is exact, one presumes that OML remains a good approximation for the negatively-charged case for $r_d/\lambda_{\mathrm{De}} \sim 1$. Simulations confirm that this is indeed true [62].

The positively-charged case with monotonically decreasing ϕ

As the emitted current is increased, the grain becomes positively charged. In the spirit of section 18.2, let us now assume a *positive, monotonically-decreasing* shielding potential. All the considerations performed in section 18.2 for the background plasma remain the same, with the obvious difference that now background electrons are the attracted species.

Let us then focus on the emitted electrons. An inspection of the effective potential quickly reveals that the contribution of the electrostatic force is negative while that from the centrifugal force is positive, exactly as for the attracted species in section 18.2. Thus, potential barriers can appear. Depending on the angular momentum, the slow emitted electrons can be attracted back to the grain, while the conditions for emitted electrons to escape the grain attraction are given by

$$E \geqslant U^{\max}(r_m(J), J), \tag{18.46}$$

whereas before r_m is the position of the maximum of U, given by:

$$-r_m^3\phi'(r_m) = \frac{J^2}{em_e}, \tag{18.47}$$

where $\phi' = d\phi/dr$. At this point the attentive reader will not be surprised to discover that, in the OML framework, a monotonic effective potential is assumed and potential barriers for the emitted electrons are neglected. Let us calculate the emitted-electron density. For particles moving away from the grain, the distribution function has to be integrated over the whole $v_r > 0$ space. For particles attracted back to the grain, i.e. moving with $v_r < 0$, one has to distinguish between negative and positive energy E. For $-e\phi_d \leqslant E \leqslant 0$, the effective potential is monotonically increasing and its maximum is at infinity, $U^{\max} = 0$. Thus, one has $E < U^{\max} = 0$, corresponding to

$$v_r^2 + v_\theta^2 \leqslant \frac{2e}{m_e}\phi. \tag{18.48}$$

For $0 \leqslant E \leqslant \infty$, the maximum of the effective potential is at $r = r_d$. The emitted electrons cannot reach infinity if $E < U^{\max}(r_d)$, which can be reformulated as

$$v_r^2 + v_\theta^2(1 - z^2) \leqslant \frac{2e}{me}(\phi - \phi_d), \tag{18.49}$$

whose intersection with the $v_r = 0$ axis corresponds to v_θ^r given by equation (18.18) (in this case $q_\alpha = -e$). Equation (18.49) is valid when $v_r^2 + v_\theta^2 \geqslant \frac{2e}{m_e}\phi$, whose intersection with the $v_r = 0$ axis corresponds to v_θ^b in equation (18.17). As for the attracted species in section 18.2, the limits of phase-space integration depend on the relative magnitude of v_θ^r and v_θ^b. For $v_\theta^r > v_\theta^b$, namely $\phi < \phi_d/z^2$, the phase-space domain of integration is

$$\begin{aligned}
&v_r > 0,\ v_r \in [0, +\infty),\ v_\theta \in [0, +\infty),\\
&v_r \leqslant 0,\ v_r^2 + v_\theta^2 \leqslant \frac{2e}{m_e}\phi,\\
&v_r \leqslant 0,\ v_r \in (-\infty, 0],\ v_\theta \in \left[\sqrt{\frac{v_r^2 - \frac{2e}{m_e}(\phi - \phi_d)}{z^2 - 1}},\ \infty\right),
\end{aligned} \tag{18.50}$$

leading to

$$\begin{aligned}
n_{\text{em}}(r) = \sqrt{\frac{\pi}{2}}\frac{J_{\text{em}}}{ev_{\text{th,em}}}\exp\left(e\frac{\phi - \phi_d}{T_{\text{em}}}\right)\Bigg[1 + \operatorname{erf}\sqrt{\frac{e\phi}{T_{\text{em}}}} - \frac{2}{\sqrt{\pi}}\sqrt{\frac{e\phi}{T_{\text{em}}}}\exp\left(-\frac{e\phi}{T_{\text{em}}}\right)\\
-\frac{\sqrt{z^2 - 1}}{z}\exp\left[\frac{e(\phi - \phi_d)}{T_{\text{em}}(z^2 - 1)}\right]\Bigg],\quad \phi \leqslant \phi_d/z^2.
\end{aligned} \tag{18.51}$$

For $v_\theta^r < v_\theta^b$, namely $\phi > \phi_d/z^2$, the phase-space domain of integration is

$$\begin{aligned}
&v_r > 0,\ v_r \in [0, +\infty),\ v_\theta \in [0, +\infty),\ v_r \leqslant 0,\\
&v_r \in \left[-\sqrt{\frac{2e}{m_e}\left(\phi - \frac{\phi_d}{z^2}\right)},\ 0\right],\ v_\theta \in [0, +\infty),\\
&v_r \leqslant 0,\ v_r \in \left[-\sqrt{\frac{2e}{m_e}\phi},\ -\sqrt{\frac{2e}{m_e}\left(\phi - \frac{\phi_d}{z^2}\right)}\right],\ v_r^2 + v_\theta^2 \leqslant \frac{2e}{m_e}\phi,\\
&v_r \leqslant 0,\ v_r \in \left(-\infty,\ -\sqrt{\frac{2e}{m_e}\left(\phi - \frac{\phi_d}{z^2}\right)}\right],\\
&\qquad v_\theta \in \left[\sqrt{\frac{v_r^2 - \frac{2e}{m_e}(\phi - \phi_d)}{z^2 - 1}},\ \infty\right),
\end{aligned} \tag{18.52}$$

leading to

$$n_{\mathrm{em}}(r) = \sqrt{\frac{\pi}{2}}\frac{J_{\mathrm{em}}}{ev_{\mathrm{th,em}}}\exp\left(e\frac{\phi - \phi_d}{T_{\mathrm{em}}}\right)\left[1 + \operatorname{erf}\sqrt{\frac{e\phi}{T_{\mathrm{em}}}} + \frac{2}{\sqrt{\pi}}\left[\sqrt{\frac{e}{T_{\mathrm{em}}}\left(\phi - \frac{\phi_d}{z^2}\right)} - \sqrt{\frac{e\phi}{T_{\mathrm{em}}}}\right]\exp\left(-\frac{e\phi}{T_{\mathrm{em}}}\right) - \frac{\sqrt{z^2 - 1}}{z}\exp\left[\frac{e(\phi - \phi_d)}{T_{\mathrm{em}}(z^2 - 1)}\right]\left[1 + \operatorname{erf}\sqrt{\frac{e}{T_{\mathrm{em}}}\frac{z^2\phi - \phi_d}{z^2 - 1}}\right]\right], \quad \phi \geqslant \phi_d/z^2. \tag{18.53}$$

The emitted-electron current is obtained by integrating over $v_r > 0$ and

$$v_r^2 + v_\theta^2 \geqslant 2e\phi_d/m_e. \tag{18.54}$$

It leads to

$$I_{\mathrm{em}} = 4\pi r_d^2 J_{\mathrm{em}}\left(1 + \frac{e\phi_d}{T_{\mathrm{em}}}\right)\exp\left(-\frac{e\phi_d}{T_{\mathrm{em}}}\right). \tag{18.55}$$

Once again, one can solve for the dust potential via the floating condition (18.41) without solving Poisson's equation. The emitted-electron power is

$$h_{\mathrm{em}} = -T_{\mathrm{em}}\left(\frac{2 + \dfrac{e\phi_d}{T_{\mathrm{em}}}}{1 + \dfrac{e\phi_d}{T_{\mathrm{em}}}} + \frac{e\phi_d}{T_{\mathrm{em}}}\right)\frac{I_{\mathrm{em}}}{e}. \tag{18.56}$$

18.4.2 Transition from negatively- to positively-charged states

It was shown in equation (18.31) that in the conventional OML theory for a non-emitting dust, the dust charge is linearly proportional to the dust potential. This would imply that dust charge vanishes when $\phi_d = 0$. For a dust that emits electrons, the transition from negative to positive dust charge occurs at a dust potential that is less than zero, which we denote as $\phi_d^* < 0$. In this section, we contrast the prediction of a modified OML theory, which includes the contribution of emitted electrons, with the first-principles kinetic simulations to elucidate the physics underlying ϕ_d^*.

PIC simulations of dust–plasma interaction with electron emission

We present some PIC simulations (which are equivalent to the OM theory) to contrast with the OML approximation described above [62]. We consider a system with a spherical dust grain of radius $r_d/\lambda_{\mathrm{De}} = 1$ [$\lambda_{\mathrm{De}} = \sqrt{\varepsilon_0 T_e/(n_\infty e^2)}$] interacting with a collisionless, unmagnetized hydrogen plasma ($m_i/m_e = 1836$). At time $t = 0$ the plasma is Maxwellian with $T_i/T_e = 1$ and the outer domain is at $R/\lambda_{\mathrm{De}} = 10$. The dust emits electrons thermionically (see equation (18.38)). We fix $T_{\mathrm{em}}/T_e = 0.03$ and vary $\hat{J}_{\mathrm{em}} = \frac{J_{\mathrm{em}}}{en_\infty\sqrt{T_e/m_e}}$ parametrically to investigate negatively- and positively-charged regimes. The smallest time step in the simulations is $\Delta t\omega_{\text{pe}} = 0.012\,5$ ($\omega_{\text{pe}} = \sqrt{e^2 n_\infty/(\varepsilon_0 m_e)}$ is the electron plasma frequency). Figure 18.4 (left) shows the steady-state dust potential from OML (i.e. solving equation (18.41)) and from PIC. For PIC, the dust potential is averaged over the last quarter of the simulation, once the system has reached a steady state, to reduce statistical noise. For small values of $\hat{J}_{\text{em}}$ the grain is negatively charged and the potential asymptotes to the value obtained from the expressions derived in section 18.2, $e\phi_d/T_e = -2.5$. As the emitted current is increased, the dust potential starts to decrease (in magnitude) significantly since the grain is less negatively charged. At a critical current the two curves bend, signaling that the dust is now positively charged and part of the emitted electron population is attracted back to the grain. One can see that OML and PIC agree very well in the negatively-charged regime. However, OML misses the transition to the positively-charged regime (which in OML corresponds to $\phi_d = 0$). For PIC, this transition occurs at a negative dust potential, $e\phi_d/T_e = -0.22$. Along the same line, figure 18.4 shows the power collected by the dust grain from the background electrons, indicating that OML overestimates this quantity by ~30% for the parameters considered. This is a particularly important result, since it is well known that in the positively charged regime the heating of the dust grain is dominated by the background electrons. This can quickly lead to dust destruction via a positive feedback [14], and might affect dust survival in applications such as magnetic fusion energy. The reason for the discrepancy between OML and PIC resides in the presence of a non-monotonic shielding potential (a potential well, also known as virtual cathode) [62, 63], which invalidates the assumption of OML. This can be seen in figure 18.5 which shows the steady-state shielding potential obtained by PIC for various values of $\hat{J}_{\text{em}}$. We note that the importance of potential-well effects increases if, for instance, one increases the dust radius (while the other parameters remain fixed). This can be seen in figure 18.6, again showing

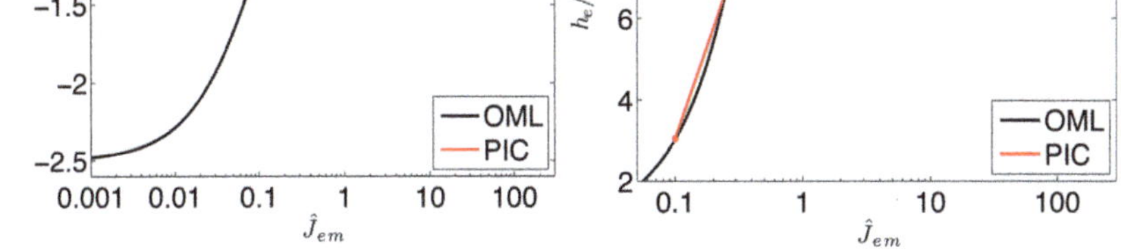

Figure 18.4. The dust potential (left) and power collected from the background electrons (right) from OML theory and PIC simulations as a function of the emitted-electron current $\hat{J}_{\text{em}}$. Adapted from reference [62], copyright 2014 by the American Physical Society.

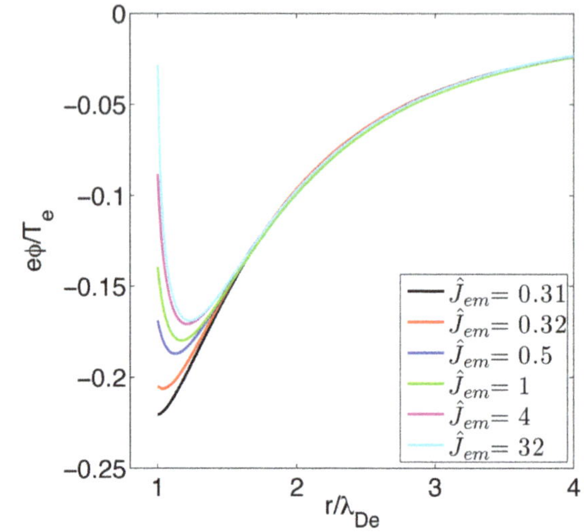

Figure 18.5. The shielding potential varying the emitted-electron current. Adapted from reference [62], copyright 2014 by the American Physical Society.

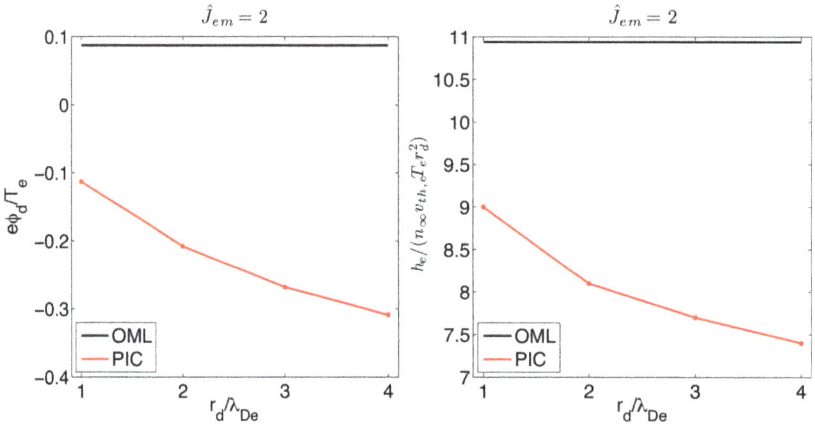

Figure 18.6. The dust potential (left) and power collected from the background electrons (right) from OML theory and PIC simulations as a function of dust radius ($\hat{J}_{\mathrm{em}} = 2$).

the dust potential (left) and the power collected from the background electrons (right) varying the dust radius for the case $\hat{J}_{\mathrm{em}} = 2$. While the OML prediction is independent of the dust radius, the discrepancy between PIC and OML widens as the dust radius increases. For instance, for $r_d/\lambda_{\mathrm{De}} = 4$ OML overestimates the power collected from the background electrons by $\sim$50%. Finally, the relation between the dust charge and potential given by the vacuum capacitance no longer holds, as the screening length is dominated by the emitted electrons.

Critical ϕ_d^* for transition from negatively- to positively-charged states
In the positively-charged regime the slowest emitted electrons can be attracted back to the grain, creating an electron population confined near the grain and the development of a non-monotonic shielding potential. Let us examine this case from the point of view of the OM theory, assuming that the shielding potential has a minimum at $r = r_{\min}$ which is labeled as $\phi_{\min}$. The emitted electrons experience potential barriers to their motion. The position of the extrema of U is still given by equation (18.47). Clearly, $\phi' < 0$ for $r_d \leqslant r_m \leqslant r_{\min}$ and there is only one solution to equation (18.47) giving $r_m = r_m(J)$ corresponding to a maximum of $U(r)$. Also, $J = 0$ corresponds to $r_m = r_{\min}$. No solution exists for $r > r_{\min}$, since $\phi' > 0$, corresponding to a monotonically decreasing effective potential and one can define a *critical angular momentum* above which all the emitted electrons leave the dust grain because of the strong centrifugal force:

$$J^* = \sqrt{-e r_d^3 m_e \phi'(r_d)}\,. \tag{18.57}$$

Again, the emitted electrons that are able to escape the grain attraction and contribute to the emitted current are those with enough energy to satisfy $E \geqslant U^{\max}(r_m(J), J)$, equation (18.46).

The implications of equation (18.46) can be understood if one considers the limit where $J^* \gg m_e r_d v_{\text{th,em}}$. In this case, since the emitted electrons are Maxwellian, most of them are characterized by $J \leqslant m_e r_d v_{\text{th,em}}$ and see the potential barrier located at $r_m \sim r_{\min}$. Thus one can reduce expression (18.46) to a somewhat more manageable expression:

$$E \geqslant U(r_{\min}, J) \longrightarrow v_r^2 + \left[1 - \left(\frac{r_d}{r_{\min}}\right)^2\right] v_\theta^2 \geqslant \frac{2e}{m_e}(\phi_d - \phi_{\min}), \tag{18.58}$$

corresponding to an ellipsoid in phase space with aspect ratio given by $1 - (r_d/r_{\min})^2$. When compared to the OML condition for the emitted-electron current (18.54), it is clear that if the shielding potential has a deep ($|\phi_d - \phi_{\min}| \gg |\phi_d|$) and localized (near the grain, $r_{\min} \sim r_d$) potential well, then the emitted electron current can be significantly different from that predicted by OML. Note that a full OM theory for non-monotonic shielding potentials (with $\phi_d > 0$) has been developed in reference [64].

Furthermore, a consequence of the presence of non-monotonic potentials is that the negatively-to-positively-charged transition $Q_d = 0$ does not occur for $\phi_d = 0$. In fact, in section 18.2 we have discussed that in general the dust charge depends on the space-charge distribution outside the grain, equations (18.30) and (18.31), and that this requires the solution of the non-linear Poisson equation. Motivated by the good agreement between PIC and OML in the negatively-charged regime, we solve the OML Poisson's equation in this regime to calculate the value of the critical current (or dust potential) corresponding to $Q_d = 0$. Thus, we solve equation (18.40) with the JFNK method discussed in section 18.3, with the emitted-electron density given by equation (18.43). The boundary conditions are given by the floating condition (18.41) at $r = r_d$ and by the condition $\phi'(R) = 0$. An inspection of the plasma densities, equations (18.11), (18.21), (18.24), and (18.43) reveals that the system depends on the following dimensionless parameters: $\frac{m_e}{m_i}$, $\frac{T_i}{T_e}$, $\frac{T_{\rm em}}{T_e}$, $\hat{J}_{\rm em}$, while the dependence on the dust radius comes through the floating condition (18.41). We fix $m_i/m_e = 1836$, consider four representative values of $T_i/T_e(= 0.001, 0.1, 1, 10)$, and vary parametrically $T_{\rm em}/T_e$ and $r_d/\lambda_{\rm De}$. The emitted-electron current density is obtained by imposing the additional condition $Q_d = 0$ and we label quantities corresponding to the negatively- to positively-charged transition with the superscript *. Once $\hat{J}^*_{\rm em}$ is obtained, ϕ^*_d follows from the floating condition (18.41).

The results are presented in figure 18.7, showing the normalized current density $\hat{J}^*_{\rm em}$, the dust potential ϕ^*_d and γ^*, the ratio between the emitted current and the background electron current $\gamma^*=\frac{I^*_{\rm em}}{I_e(\phi^*_d)}$, a quantity that is often used in the literature to characterize electron emission. One can see in figure 18.7 that $\hat{J}^*_{\rm em}$ varies between $0.2 - 0.4$, quite insensitive to T_i/T_e. Interestingly, γ^* has a rather weak dependence on $r_d/\lambda_{\rm De}$ and $T_{\rm em}/T_e$: it varies between $0.98 - 0.995$, $0.96 - 0.99$, $0.94 - 0.98$ and $0.86 - 0.93$ for $T_i/T_e = 0.01, 0.1, 1, 10$. Most importantly, the negatively- to positively-charged grain transition occurs with a negative dust potential. Only for $r_d/\lambda_{\rm De} \ll 1$ and for smaller values of T_i/T_e, the dust potential remains quite small and the OML approximation is reasonable. On the other hand, for large values of $r_d/\lambda_{\rm De}$ and small $T_{\rm em}/T_e$, figure 18.7 shows that $e\phi^*_d/T_e$ can be sizable, with values in the range -0.3 to -0.6 for $T_i/T_e \geqslant 0.1$. The dust potential is more negative when T_i/T_e is higher, consistent with the fact that higher T_i/T_e implies a higher ion current and therefore less emitted current is necessary to reach $Q_d = 0$. Note also that for the parameters of section 18.4.2, $r_d/\lambda_{\rm De} = 1$ and $T_{\rm em}/T_e = 0.03$, this procedure gives $e\phi^*_d/T_e = -0.22$ and $\hat{J}^*_{\rm em} = 0.31$, in excellent agreement with figure 18.4. These results confirm that, depending on the regime of parameters, the OML approximation can be inaccurate in capturing the negatively- to positively-charged transition. Most importantly, ϕ^*_d can be used as a proxy to signal when potential-well effects are important and the OML theory becomes inadequate.

18.5 Final remarks

In this chapter we have reviewed the basics of the theory of charging and shielding of a spherical object in a collisionless, unmagnetized plasma, one of the building blocks for studies of plasma–material interaction including dusty-plasma physics. We have derived the basic equations in the OML approximation and contrasted them with first-principle PIC simulations. While PIC simulations of dust–plasma interaction can be computationally expensive depending on the scale separation in the system, the beauty of OML is that it delivers a simple and fast solution where Poisson's equation does not even need to be solved. For this reason, the OML theory is the basis of all existing codes to study dust transport in a plasma. Although OML is traditionally accepted in the limit of $r_d/\lambda_D \ll 1$ (a regime which is very common in space and astrophysics), PIC simulations show that in the negatively-charged regime it can be quite accurate for $r_d/\lambda_D \gtrsim 1$ (a regime of interest to laboratory applications).

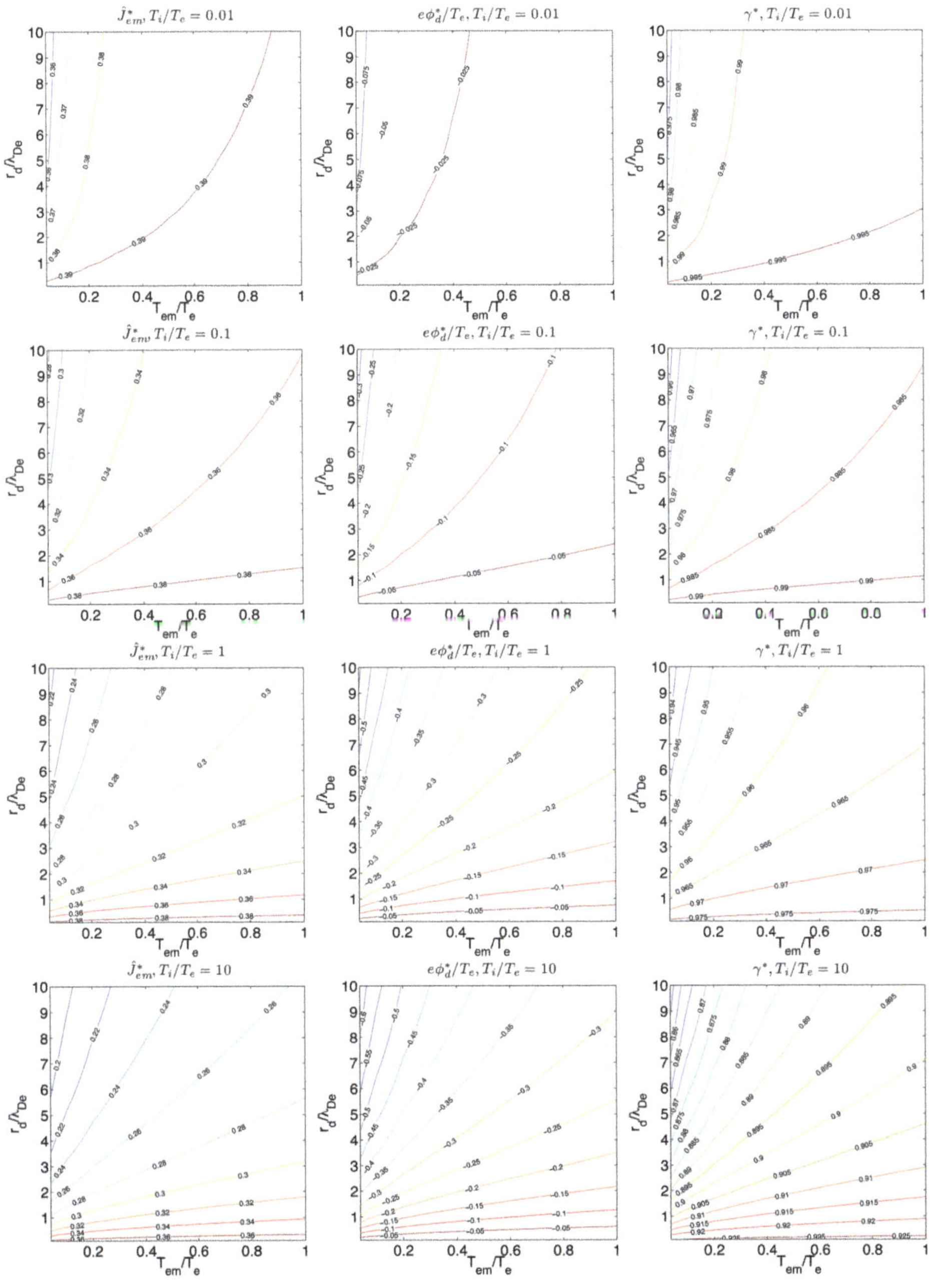

Figure 18.7. The case $Q_d = 0$ as a function of $T_{\rm em}/T_e$, $r_d/\lambda_{\rm De}$ and for four different values of T_i/T_e. The plots show the value of the emitted current $J^*_{\rm em}$, the dust potential ϕ^*_d and the ratio between the emitted current and the background electron current defined as γ^*. Reprinted with permission from [62], copyright 2014 by the American Physical Society.

In the positively-charged regime driven by dust electron-emission processes, on the other hand, OML can become very inaccurate when $r_d/\lambda_D \gtrsim 1$ because of the development of a non-monotonic shielding potential. This also implies that the dust collected power from the background electrons can be significantly reduced relative to OML and that electron emission can play a major role in determining dust survivability in a plasma.

As new emerging applications continue to drive the development of the field, it is clear that further developments in dust–plasma interaction theories are necessary. This is particularly true if one considers the necessity of bridging the microscopic scales associated with dust charging to the macroscopic scales of dust transport, which makes OML-like models very popular. For instance, in magnetic fusion energy applications and recent laboratory experiments with high magnetic fields, the electron gyroradius can become comparable to or smaller than the characteristic dust size, thus affecting the charging currents (particularly in the positively-charged regime). Furthermore, in many applications the dust grains are far from spherical: a variety of shapes arises depending on the dust formation processes. Whether OML-like models can be formulated to capture these non-OML effects (including the anisotropy induced by the presence of a directed ion flow) accurately is at this point an open question. Furthermore, the interaction of plasmas with biological matter, at the heart of the infant field of plasma medicine, is already showing great promise for the treatment of diseases like cancer, opening new perspectives and a host of opportunities at the intersection of plasma physics, atomic physics and chemistry.

In summary, this is a rather exciting time for researchers interested in plasma–material interaction problems, and in particular for dust–plasma interaction!

References

[1] Hutchinson I H 2005 *Principles of Plasma Diagnostics* (Cambridge: Cambridge University Press)

[2] Irving Langmuir A F D and Found C G 1924 A new type of electric discharge: the streamer discharge *Sci.* **60** 392–4

[3] Mendis D A and Rosenberg M 1994 Cosmic dusty plasma *Annu. Rev. Astron. Astrophys.* **32** 419–63

[4] Goertz C K 1989 Dusty plasmas in the solar system *Rev. Geophys.* **27** 271–92

[5] Verheest F 2000 *Waves in Dusty Space Plasmas* (Dordrecht: Kluwer)

[6] Smith B A *et al* 1982 A new look at the Saturn system—the Voyager 2 images *Science* **215** 504–37

[7] Selwyn G S, Haller K L and Patterson E F 1993 Trapping and behavior of particulates in a radio frequency magnetron plasma etching tool *J. Vac Sci. Technol.* A **11** 1132–5

[8] Bouchoule A 1999 *Dusty Plasmas: Physics, Chemistry, and Technological Impacts in Plasma Processing* (New York: Wiley)

[9] Chu J H and Lin I 1994 Direct observation of Coulomb crystals and liquids in strongly coupled rf dusty plasmas *Phys. Rev. Lett.* **72** 4009–12

[10] Thomas H, Morfill G E, Demmel V, Goree J, Feuerbacher B and Möhlmann D 1994 Plasma crystal: coulomb crystallization in a dusty plasma *Phys. Rev. Lett.* **73** 652–5

[11] Fortov V E, Khrapak A G, Khrapak S A, Molotkov V I and Petrov O F 2004 Dusty plasmas *Phys.-Usp.* **47** 447–92
[12] Krasheninnikov S I, Smirnov R D and Rudakov D L 2011 Dust in magnetic fusion devices *Plasma Phys. Control. Fusion* **53** 083001
[13] Ticos C M, Wang Z, Delzanno G L and Lapenta G 2006 Plasma dragged microparticles as a method to measure plasma flows *Phys. Plasmas* **13** 103501
[14] Smirnov R D, Pigarov A Y, Rosenberg M, Krasheninnikov S I and Mendis D A 2007 Modelling of dynamics and transport of carbon dust particles in tokamaks *Plasma Phys. Control. Fusion* **49** 347–71
[15] Bacharis M, Coppins M and Allen J E 2010 Critical issues for modeling dust transport in tokamaks *Phys. Rev.* E **82** 026403
[16] Ratynskaia S, Vignitchouk L, Tolias P, Bykov I, Bergsåker H, Litnovsky A, den Harder N and Lazzaro E 2013 Migration of tungsten dust in tokamaks: role of dust-wall collisions *Nucl. Fusion* **53** 123002
[17] Delzanno G L and Tang X Z 2014 Survivability of dust in tokamaks: dust transport in the divertor sheath *Phys. Plasmas* **21** 022502
[18] Vignitchouk L, Tolias P and Ratynskaia S 2014 Dust-wall and dust-plasma interaction in the MIGRAINe code *Plasma Phys. Control. Fusion* **56** 095005
[19] Rudakov D 2009 Dust studies in DIII-D and TEXTOR *Nucl. Fusion* **49** 085022
[20] Litnovsky A *et al* 2013 Dust investigations in TEXTOR: impact of dust on plasma–wall interactions and on plasma performance *J. Nucl. Mater.* **438** S126–32 (Proc. of the 20th Int. Conf. on Plasma-Surface Interactions in Controlled Fusion Devices)
[21] Ratynskaia S *et al* 2015 Elastic–plastic adhesive impacts of tungsten dust with metal surfaces in plasma environments *J. Nucl. Mater.* **463** 877–80
[22] Tolias P *et al* 2016 Dust remobilization in fusion plasmas under steady state conditions *Plasma Phys. Control. Fusion* **58** 025009
[23] Fridman A and Friedman G 2012 *Plasma Medicine* (New York: Wiley)
[24] Keidar M 2015 Plasma for cancer treatment *Plasma Sources Sci. Technol.* **24** 033001
[25] Thomas E Jr, Merlino R L and Rosenberg M 2012 Magnetized dusty plasmas: the next frontier for complex plasma research *Plasma Phys. Control. Fusion* **54** 124034
[26] Shukla P K and Mamun A A 2001 *Introduction to Dusty Plasma Physics* (Philadelphia: Institute of Physics Publishing)
[27] de Angelis U 1992 The physics of dusty plasmas *Phys. Scr.* **45** 465
[28] Tsytovich V N and de Angelis U 1999 Kinetic theory of dusty plasmas. I. General approach *Phys. Plasmas* **6** 1093–106
[29] Schram P P J M, Trigger S A and Zagorodny A G 2003 New microscopic and macroscopic variables in dusty plasmas *New J. Phys.* **5** 27
[30] Tolias P, Ratynskaia S and de Angelis U 2011 Kinetic models of partially ionized complex plasmas in the low frequency regime *Phys. Plasmas* **18** 073705
[31] Allen J 1992 Probe theory–the orbital motion approach *Phys. Scr.* **45** 497–503
[32] Al'pert Y L, Gurevich A V and Pitaevskii L P 1965 *Space Physics with Artificial Satellites* (New York: Plenum)
[33] Kennedy R and Allen J 2003 The floating potential of spherical probes and dust grains. II: Orbital motion theory *J. Plasma Phys.* **69** 485–506
[34] Mott-Smth H and Langmuir I 1926 The theory of collectors in gaseous discharges *Phys. Rev.* **28** 0727–63

[35] Tang X-Z and Delzanno G L 2014 Orbital-motion-limited theory of dust charging and plasma response *Phys. Plasmas* **21** 123708
[36] Bystrenko T and Zagorodny A 2002 Effects of bound states in the screening of dust particles in plasmas *Phys. Lett.* A **299** 383–91
[37] Allen J, Annaratone B and de Angelis U 2000 On the orbital motion limited theory for a small body at floating potential in a Maxwellian plasma *J. Plasma Phys.* **63** 299–309
[38] Daugherty J, Porteous R, Kilgore M D and Graves D 1992 Sheath structure around particles in low-pressure discharges *J. Appl. Phys.* **72** 3934–42
[39] Lampe M 2001 Limits of validity for orbital-motion-limited theory for a small floating collector *J. Plasma Phys.* **65** 17–80
[40] Willis C T N, Coppins M, Bacharis M and Allen J E 2010 The effect of dust grain size on the floating potential of dust in a collisionless plasma *Plasma Sources Sci. Technol.* **19** 065022
[41] Delzanno G L and Tang X-Z 2015 Comparison of dust charging between orbital-motion-limited theory and particle-in-cell simulations *Phys. Plasmas* **22** 113703
[42] Kelley C T 1995 *Iterative Methods for Linear and Nonlinear equations* (Philadelphia, PA: SIAM)
[43] Birdsall C K and Langdon A B 2004 *Plasma Physics Via Computer Simulation* (London: Taylor and Francis) https://doi.org/10.1201/9781315275048
[44] Hockney R and Eastwood J 1988 *Computer Simulation Using Particles* (London: Taylor and Francis) https://doi.org/10.1201/9780367806934
[45] Koziel S, Leifsson L, Lees M, Krzhizhanovskaya V V, Dongarra J, Sloot P M and Miloch W 2015 Simulations of several finite-sized objects in plasma *Procedia Comput. Sci.* **51** 1282–91
[46] Patacchini L and Hutchinson I H 2007 Ion-collecting sphere in a stationary, weakly magnetized plasma with finite shielding length *Plasma Phys. Control. Fusion* **49** 1719
[47] Lapenta G 2011 DEMOCRITUS: An adaptive particle in cell (PIC) code for object-plasma interactions *J. Comput. Phys.* **230** 4679–95
[48] Taccogna F 2012 Dust in plasma I. Particle size and ion-neutral collision effects *Contrib. Plasma Phys.* **52** 744–55
[49] Delzanno G L, Camporeale E, Moulton J D, Borovsky J E, MacDonald E A and Thomsen M 2013 CPIC: a curvilinear particle-in-cell code for plasma–material interaction studies *IEEE Trans. Plasma Sci.* **41** 3577
[50] Sorasio G, Mendis D and Rosenberg M 2001 The role of thermionic emission in meteor physics *Planet. Space Sci.* **49** 1257–64
[51] Kimura H and Mann I 1998 The electric charging of interstellar dust in the solar system and consequences for its dynamics *Astrophys. J.* **499** 454
[52] Horányi M and Cravens T E 1996 The structure and dynamics of Jupiter's ring *Nature* **381** 293–5
[53] Klumov B, Vladimirov S and Morfill G 2007 On the role of dust in the cometary plasma *JETP Lett.* **85** 478–82
[54] Poppe A, Halekas J S and Horányi M 2011 Negative potentials above the day-side lunar surface in the terrestrial plasma sheet: evidence of non-monotonic potentials *Geophys. Res. Lett.* **38** L02103
[55] Hastings D and Garrett H 2004 *Spacecraft-Environment Interactions* (Cambridge: Cambridge University Press)
[56] Bruining H 1962 *Physics and Applications of Secondary Electron Emission* (London: Pergamon)

[57] Wang X, Pilewskie J, Hsu H-W and Horanyi M 2016 Plasma potential in the sheaths of electron-emitting surfaces in space *Geophys. Res. Lett.* **43** 525–31
[58] MacDonald E A, Lynch K A, Widholm M, Arnoldy R, Kintner P M, Klatt E M, Samara M, LaBelle J and Lapenta G 2006 In situ measurement of thermal electrons on the SIERRA nightside auroral sounding rocket *J. Geophys. Res. Space Phys* **111** A12310
[59] Meyer-Vernet N 1982 Flip-flop of electric potential of dust grains in space *Astron. Astrophys.* **105** 98–106
[60] Tolias P 2014 On secondary electron emission and its semi-empirical description *Plasma Phys. Control. Fusion* **56** 123002–11
[61] Ashcroft N and Mermin N 1976 *Solid State Physics* (Philadelphia, PA: Saunders)
[62] Delzanno G L and Tang X Z 2014 Charging and heat collection by a positively charged dust grain in a plasma *Phys. Rev. Lett.* **113** 035002
[63] Delzanno G L, Lapenta G and Rosenberg M 2004 Attractive potential around a thermionically emitting microparticle *Phys. Rev. Lett.* **92** 350021–24
[64] Delzanno G L, Bruno A, Sorasio G and Lapenta G 2005 Exact orbital motion theory of the shielding potential around an emitting, spherical body *Phys. Plasmas* **12** 062102

IOP Publishing

Plasma Modeling (Second Edition)
Methods and applications
Gianpiero Colonna and Antonio D'Angola

Chapter 19

Magnetic confinement for thermonuclear energy production

Enzo Lazzaro

An energy generator is basically a device for the *conversion* of different forms of *potential* or stored energy, into eventually usable electrical or mechanical form, with a *gain* $Q = \frac{P_{\text{out}}}{P_{\text{ext}} + P_{\text{loss}}} > 1$, where *out*, *ext*, *loss* label the net power output, the input from external sources, and the irreversible losses, respectively. In the case of a thermonuclear reactor $P_{\text{out}} = P_\alpha$, where P_α refers to the power released by the α particles. An energy conversion device is a *machine* when it responds in a controlled and stable way to variation of the input parameters (in most applications with a *linear* response to controls). Furthermore such devices must be as *reliable* as possible, which means having a small probability of fault or malfunction, uniformly distributed among the essential components and must be *robust* with respect to minor perturbations. The goal of research on controlled thermonuclear fusion is the creation of an apparatus capable of performing a controlled conversion, with gain $Q \gg 1$ of the amount of energy released by fusion reactions. The practically most promising process involves D and T nuclei in the reaction

$$ {}_1D^2 + {}_1T^3 \longrightarrow {}_2He^4(3.5\ \text{MeV}) + {}_0n^1(14.7\ \text{MeV}) \tag{19.1}$$

The device must be brought to and maintained at its required working regime, by external sources and the α particles energy, producing an output consisting of the *kinetic energy* of neutron, eventually converted into heat to be used in conventional thermodynamic cycles. The power density generated by fusion reactions is given by the product of number of reactions per unit time and unit volume times the energy E_{rs} produced in a single reaction between nuclei of species r, s:

$$P_{rs} = c_{r,s} n_r n_s \langle \sigma_{rs} v_{rs} \rangle E_{rs} \tag{19.2}$$

where n_r, n_s are the densities of the reactants, $\langle \sigma_{rs} v_{rs} \rangle$ is the reactivity, a function of E_{rs} and of the plasma temperature, with $c_{r,s} = 1$ if $r \neq s$, and $c_{r,s} = 0.5$ if $r = s$ [1].

doi:10.1088/978-0-7503-3559-1ch19

Only charged particles can keep within the reactor the energy produced. If E_c is the energy of the charged species c, the fraction of confined power is then

$$P_c / P_{rs} = E_c / E_{rs} \tag{19.3}$$

For a typical DT plasma with values $n_D = n_T = 2.5 \cdot 10^{20}\ \text{m}^{-3}$ and $T \approx 10$ keV the power density produced by the fusion reaction is $P_{\text{DT}} = 2.1 \cdot 10^5\ \text{W m}^{-3}$, while the power density associated with the α particles confined in the magnetized plasma is $P_\alpha = 4.2 \cdot 10^4\ \text{W m}^{-3}$. This 20% of the total fusion power is used to keep up the plasma temperature. In the simplest approach, with DT reactions, the output power eventually would be transformed through *conventional* thermodynamic cycles (typically a Rankine or Brayton cycle [2, 3]) in consumer usable electricity. In order to clarify the objectives it is necessary to anticipate some consideration based on key parameters which will be fully spelled out in the following paragraphs. A fundamental relation for any magnetic fusion reactor design specifies the dependence of the volumetric fusion power density on the macroscopic parameters of the device. In a toroidal device plasma carrying a toroidal current I_t a key parameter for the quality of the confinement configuration is $\beta = 2\mu_0 \langle p \rangle / B^2$ and often used in the normalized form as $\beta_N = \beta\epsilon(RB_t/\mu_0 I_t)$, where $\epsilon = a/R_0$. Equilibrium and stability constraints set an upper limit to β, which, when expressed as a percentage [%] is $\beta_{\text{max}} = gI_t/aB_t$ (with I_t in MA, B_t in T, and a in m and $[g] = [\mu_0]$. The fusion power density can then be expressed as:

$$\frac{P_{\text{Th}}}{V} \propto \frac{\langle p \rangle^2}{4} \left(\frac{<\sigma v>}{T^2} \right)_{T \approx 15\ \text{keV}} \propto \frac{\beta_N^2 \epsilon^4 B_0^4 R^3}{q^2} \tag{19.4}$$

where $\langle p \rangle = \langle 2nT \rangle$ is the volume-averaged plasma pressure in MPa and $q \simeq 5(a^2/R)(B_t/I_t)$ $[m, T, MA]$ is the *field line winding number, or 'safety factor'* described later. From this scaling, two strategies appear possible to achieve the high fusion power density desirable for a nuclear fusion facility and required for economical fusion power: a high β or a high B. These two strategies are dramatically different. The quantity β is limited by the risk of onset of magnetohydrodynamic (MHD) instabilities and increasing the frequency of disruptions, which are an unacceptable liability in an industrial facility. Increasing B meets technological limits and costs. Actually in a fusion power plant the electric power density P_E eventually made available to the consumer, is a fraction η of the wall loading power *surface* density $P_w[Wm^{-2}]$ crossing a unit area of the wall. A high wall loading reduces the specific cost of the plant. In order to evaluate what is the β value needed to reach a certain P_w, technically tolerable by the wall material, one must consider that for a tokamak of effective cross-section radius $a = S/C_p$, where S is the plasma meridian area, C_p its contour and R the torus major radius, the wall loading is related to the thermonuclear power density $P_{\text{Th}}[Wm^{-3}]$ by the energy density flux balance condition and the relation to the total power $P_{\text{tot}}[W]$:

$$\pi a^2 P_{\text{Th}} = 2\pi a P_w \tag{19.5}$$

$$P_{\text{tot}} = 4\pi^2 a^2 \varepsilon^{-1} P_w \tag{19.6}$$

The radius a is eliminated from these relations, leading to

$$P_{\mathrm{Th}} = \frac{4\pi}{\sqrt{\varepsilon}} \frac{P_w^{3/2}}{P_{\mathrm{tot}}^{1/2}} [Wm^{-3}] \tag{19.7}$$

This relation establishes the link between the fusion power produced and the tolerable power load bearing on the reactor *wall*. A schematic outline of a fusion reactor based power plant is shown in figure 19.1.

It is apparent that a certain imbalance in the present concept exists, in coupling the extremely advanced and delicate science and technology of the reactor with the transformation of the output energy through conventional, thermodynamic cycles, with efficiency limited by Carnot's principle. This will be a challenge for the future designers of practical devices. Among the various paths tried to reach this goal, the experimental *Darwinian selection* has left just few machines, with expected characteristics of reliability and controllability, although still far from continuous performance. The most prominent are the tokamak (see figure 19.2) and the stellarator, for which still formidable problems must be solved, but have reached already significant values of temperature, density and confinement time of the reactants [4–7].

The invention and first experiments of the tokamak (Russian acronym for toroidal'naya kamera s magnitnymi katushkami) are due to a number of Russian scientists such as Tamm, Sakharov, Artsimovich, Shafranov, Kadomtsev, Yavlinsky [9–13]. The concept of magnetic confinement of a mixture of D and T nuclei having a temperature ($T_{\mathrm{fus}} \geqslant 10$ keV) that enables thermonuclear fusion, must be based on key considerations of the order of magnitude involved; containing in a vessel a plasma with density of $n \approx 10^{20}$ m^{-3} requires a high vacuum. The thermal expansion of a gas at the temperature T_{fus} would cool it down on a space scale of 1 m. The problem of containment, in a practical volume, by means of magnetic fields of the hot charged species constituting the plasma can be addressed looking for a magnetic field configuration which applies a force $\propto \nabla B^2$ acting perpendicularly to the plasma displacement. This is a characteristic of the Lorentz force applied on charged particles moving in a **B** field. The objective of the present text is to give an overview of the present status and ongoing progress of this rich field of research, with sufficient information and detailed elements for further study and help for applications, but just limited to the case of the equilibrium configuration of the *paradigmatic* tokamak device.

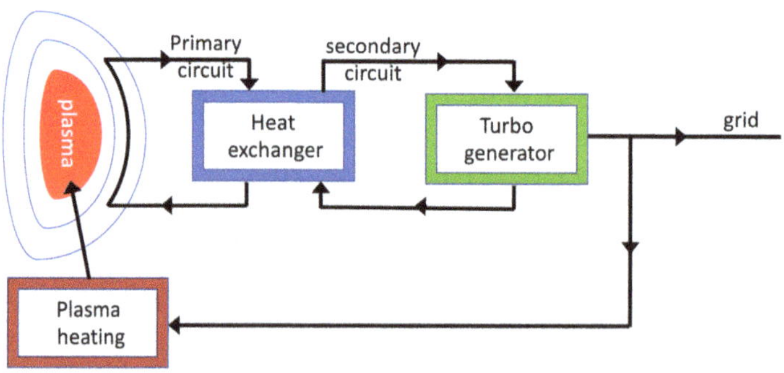

Figure 19.1. Schematic concept of thermonuclear power plant.

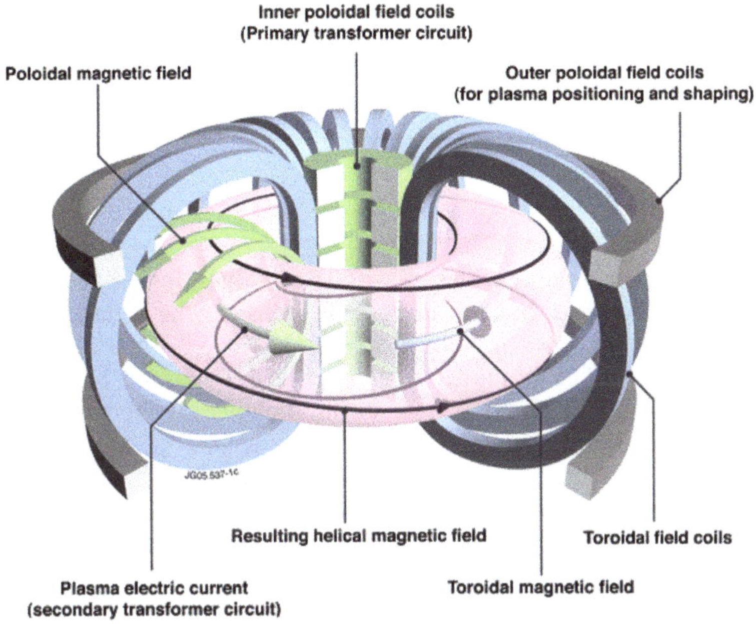

Figure 19.2. Artist's view of the essential elements of a tokamak. Reproduced from reference [8], under CC by 4.0.

In the present text the essential but very complicated topics of stability and of energy and particles losses, for which a vast literature exists, [11, 12, 14–20] are omitted. A variety of perturbations, external to the plasma or internal, associated with the profiles of current and pressure can lead to instabilities of the *ideal* equilibrium, which can destroy the configuration. The instabilities are a time dependent phenomenon; this requires departing from the ideal MHD picture, introducing some typical dissipative and inertial time scales. The subject of ideal and dissipative MHD instabilities requires a separate treatment. The topic of *plasma confinement* should not be confused with that of *plasma insulation* against energy losses, which are due to very complex transport processes related to configuration gradients and at the particles level, involving kinetic mechanisms, turbulence on various scale lengths which cannot be treated here.

Before embarking on the subtle and somewhat thorny development of the physical concepts on which magnetic confinement devices are based, some feeling should be gotten of the absolute magnitudes of the quantities involved. For clarity the requirements for the fusion regime are kept separate from those of actual contemporary devices, and of next fusion grade ones, and the focus is on tokamaks. Although this information and data can be retrieved in hundreds of publications, and most textbooks, we recall here the essential requirements for a thermonuclear fusion device producing net energy, in the form of neutron kinetic energy from a D + T reaction, to be converted into electrical energy supply for the consumer. The values of plasma density, effective impurity number, temperature and Lawson triple product [15] are presented in table 19.1.

Table 19.1. Fusion reactor plasma typical parameters.

n_e [m^{-3}]	Z_{eff}	$T_e \approx T_i$ [keV]	$nT\tau_E$ [m^{-3} keV s]
10^{20}	$\leqslant 1.2$	20	$3 \cdot 10^{21}$

Confining in a finite volume a plasma with these values, for practical times scales $\tau = 1$ s is very challenging, but not impossible; consider for instance that the gas kinetic pressure is of the order of $p \simeq 2 \cdot 10^{21}$ m^{-3} keV = 3.2 atm a counterbalancing magnetic pressure can be obtained with a B field of about 1 Tesla. Detailed engineering studies are required for the optimal choice of the machine size and magnetic field, considering also the constraints imposed on operation by the nuclear physics hazards.

Mathematically the problem is formulated considering the equation expressing the balance between the magnetodynamic force density $\mathbf{J} \times \mathbf{B}$ [N m $^{-3}$], acting on a volume element of plasma assumed to behave as superconducting fluid, and the expansion force density ∇p. The system of equations of magnetostatics and the boundary conditions that enforce thermal and mechanical isolation of the reacting fluid from the outside world complete the basic formulation of the problem.

It should be borne in mind that a fluid description as that just outlined is applicable only when the plasma is considered on length scales much larger than the mean distance between ions. When averaged over such length scales the plasma behaves as electrically neutral and the time scales are much longer than those of any ion or electron oscillation associated with motion in a magnetic field (cyclotron effect) or space charge motion. On these time scales, only the averaged, guiding-centre positions [15–17, 21] of the plasma particles are significant. This is a regime of frequencies and length scales which can be suitably described by a macroscopic, fluid-like model of the plasma. Actually, there is a large variety of possible physically meaningful fluid models that can be developed according to chosen ordering of these scales. The *free choice* of ordering, works as a selection of magnifying lenses to highlight certain problems and help reaching the most economic solution, from the point of view of calculation effort involved.

MHD can be derived systematically from the kinetic theory based on Vlasov–Maxwell equations (see chapter 13 [21]) for a plasma in the limit of large collisionality, and typical scale length larger than the ion Larmor radius, negligible dissipative effect, plus a number of additional assumptions discussed in almost every textbook [15–17, 21]. The applicability of the MHD description to a fusion grade plasma with a macroscopic scale length L, can be tested, according to Freidberg [22], introducing the dimensionless parameters

$$\hat{x} = \left(\frac{m_i}{m_e}\right)^{1/2} \frac{v_{\mathrm{th},i}\tau_D^{i/i}}{L} \tag{19.8}$$

$$\hat{y} = \frac{\rho_{L,i}}{L}.$$

where $v_{\text{th},i} = (k_B T_i/m_i)^{1/2}$ is the ion thermal velocity, $\rho_{L,i} = v_{\text{th},i} m_i/eB$ the ion Larmor (gyro-) radius and τ_D the collisional diffusion time scale.

a) High collisionality requires $\hat{x} \ll 1$;
b) small ion Larmor radius, means $\hat{y} \ll 1$;
c) negligible resistivity requires $\hat{y}^2/\hat{x} \ll 1$.

In the $(\hat{x}, \hat{y})$ plane there is apparently a region of overlap of validity of the three conditions (see figure 19.3). Translating these dimensionless conditions in terms of the physical quantities L [m], T [keV], n [10^{20} m^{-3}] one gets:

$$\begin{aligned} \hat{x} &= 3 \times 10^3 \frac{T^2}{nL} \ll 1 \\ \hat{y} &= 2.3 \times 10^{-2} \left(\frac{\beta}{nL^2}\right)^{1/2} \ll 1 \\ \hat{y}^2 \hat{x}^{-1} &= 1.8 \times 10^{-7} \frac{\beta}{nT^2} \ll 1 \end{aligned} \tag{19.9}$$

Strictly the assumption of high collisionality is not valid in fusion plasmas, but in general the MHD set of equations in the ideal limit gives the same results of other more complex models, even in a range of parameters typical of fusion. This occurs because the high magnetic field tying the motion of particles perpendicular to **B** enforces a fluid-like behaviour.

In any case, there is a certain appeal and advantage in building the MHD model as the simplest possible extension of hydrodynamics. Relatively little knowledge is required about the properties of the constituent substance in order to extract a very subtle and consistent theory, predicting and accounting for fundamental behaviour of the fluid dynamics coupled with a magnetic field. This approach is also more generally applicable because the substance that we will be dealing with need not be gaseous, like plasma, but may include also the physics of liquid metals, and of

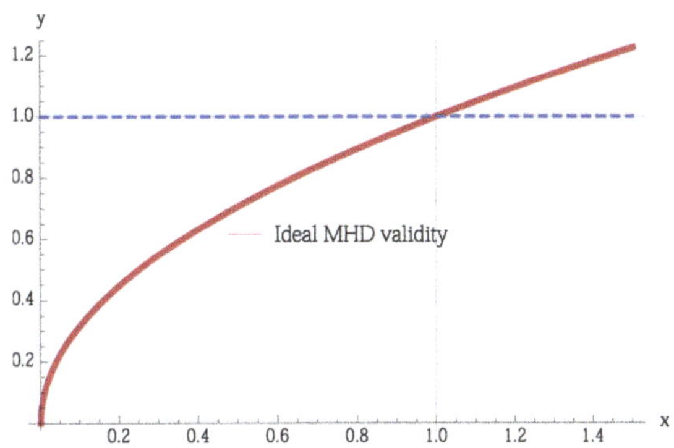

Figure 19.3. Domain of validity of MHD theory, defined by condition equation (19.9) [22].

various conducting solutions, etc. So, the description adopted in the following is based on very basic principles and quantities, which will prove sufficient for most purposes: conservation laws of mass, momentum and energy plus Maxwell's equations for the electric and magnetic fields. The relevant quantities are: ρ_m: mass density, $\mathbf{v}$: flow velocity, p: scalar pressure, ρ: charge density, $\mathbf{J}$: current density, $\mathbf{E}$: electric field $\mathbf{B}$: magnetic field. It is important to be aware of the basic ordering of the MHD model in comparison with the full set of Maxwell equations and constitutive equations [23]. The physical basis of the MHD *model* is the idea that the plasma is strongly tied to the magnetic field lines, with $\delta = \varrho_{L,i}/L_{\mathrm{MHD}} \ll 1$ and possessing an electric drift speed of the order of the thermal velocity $E_{\perp}/B \approx v_{\mathrm{th}}$. Consistent with this ordering is the expression of the fluid velocity as $\mathbf{v}_{\mathrm{MHD}} = \mathbf{v}_{\parallel} + \mathbf{v}_{E\times B}$. The *fluid MHD* version considered here implies in addition the assumption of local thermodynamic equilibrium (LTE) where the particles distribution function is a local Maxwellian $f_{\mathrm{MHD}} = f_M(\mathbf{v} - \mathbf{v}_{\mathrm{MHD}})$. The most important consequence of the LTE condition is that the plasma pressure p is scalar and all the $O(\delta)$ are neglected. Calling L_{MHD}, v_{MHD} the typical MHD macroscopic length and velocity scales, the basic ordering is:

$$\begin{aligned}
&\nabla \times \mathbf{E} = -\frac{\partial \mathbf{B}}{\partial t} \\
&|\nabla \times \mathbf{E}| \approx \omega \mathbf{B} \Rightarrow \frac{E}{B} \simeq \omega L_{\mathrm{MHD}} \simeq v_{\mathrm{MHD}} \\
&\nabla \times \mathbf{B} = \mu_0 \mathbf{J} + \mu_0 \varepsilon_0 \frac{\partial \mathbf{E}}{\partial t} \\
&|\mu_0 \varepsilon_0 \frac{\partial \mathbf{E}}{\partial t}| / |\nabla \times \mathbf{B}| \approx \frac{(\omega L_{\mathrm{MHD}})^2}{c^2} = \frac{v_{\mathrm{MHD}}^2}{c^2}
\end{aligned} \tag{19.10}$$

where $c = 1/\sqrt{\mu_0 \epsilon_0}$ is the speed of light. Therefore

$$\nabla \times \mathbf{B} = \mu_0 \mathbf{J} + O\left(\frac{v_{\mathrm{MHD}}^2}{c^2}\right) \tag{19.11}$$

which leads to the fundamental MHD closure relation, which defines the *MHD quasineutrality* condition [23]:

$$\nabla \cdot \mathbf{J} = 0 \tag{19.12}$$

This condition on moving charges represents a zero rate of change of the net charge density, which is a different concept from that of the static charge density balance condition $\rho_e = n_e - Z_i n_i \approx 0$. The system of pre-Maxwell equations, with the MHD closure equation (19.12), and the one fluid system of equations, constitute the basic dynamic MHD description:

$$\frac{\partial \mathbf{B}}{\partial t} = -\nabla \times \mathbf{E} \tag{19.13}$$

$$\nabla \times \mathbf{B} = \mu_0 \mathbf{J} \tag{19.14}$$

$$\mathbf{E} + \mathbf{v} \times \mathbf{B} = \eta \mathbf{J} \tag{19.15}$$

$$\frac{\partial \rho_m \mathbf{v}}{\partial t} + \mathbf{v} \cdot \nabla \rho_m \mathbf{v} = -\nabla p + \mathbf{J} \times \mathbf{B} \tag{19.16}$$

$$\frac{\partial \rho_m}{\partial t} + \mathbf{v} \cdot \nabla \rho_m + \nabla \cdot \rho_m \mathbf{v} = 0 \tag{19.17}$$

$$\frac{\partial p}{\partial t} + \mathbf{v} \cdot \nabla p + \gamma p \nabla \cdot \mathbf{v} = 0 \tag{19.18}$$

where the fluid inertia is approximated by the ion density and mass $\rho_m \approx n_i m_i$, $\gamma = 5/3$ the adiabatic index. In equation (19.15) for completeness the fluid resistivity is included, $\eta \propto T_e^{-3/2}$ which is omitted further on in the specialized description of the *ideal MHD* physical model. In the next section the attention will be focussed on the fundamental stationary case with isotropic pressure.

19.1 Ideal magnetostatic equilibrium

The simplest case of the steady state condition of a plasma with a scalar isotropic pressure, infinite conductivity, subject to consistent **J** and **B** field helps understanding essential aspects of the problem. The system of equations is:

$$\mathbf{J} \times \mathbf{B} = \nabla p \tag{19.19}$$

$$p \rho_m^{-\gamma} = \text{const} \tag{19.20}$$

$$\nabla \times \mathbf{B} = \mu_0 \mathbf{J} \tag{19.21}$$

$$\nabla \cdot \mathbf{B} = 0 \tag{19.22}$$

$$\nabla \cdot \mathbf{J} = 0 \tag{19.23}$$

The first equation equation (19.19) describes the force balance per unit volume; full mechanical equilibrium requires also a null torque per unit volume, expressed by $\nabla \times (\mathbf{J} \times \mathbf{B}) \equiv \mathbf{B} \cdot \nabla \mathbf{J} - \mathbf{J} \cdot \nabla \mathbf{B} = 0$, which, in equilibrium is automatically satisfied because $\nabla \times \nabla p = 0$. However, the zero torque condition is not as trivial as it seems, since it constrains the components of **J** and **B** also in the presence of perturbations leading to instabilities. The second equation is the polytropic gas law with adiabatic exponent γ. These equations constitute a system of seven scalar partial differential equations, and seven scalar unknowns, and must be considered complemented by suitable boundary conditions (e.g. interface with vacuum, where the r.h.s. of equations (19.19) and (19.21), vanish). Eliminating **J** the system can be reduced to the form:

$$\nabla\left[p + \frac{B^2}{2\mu_0}\right] = \frac{\mathbf{B} \cdot \nabla \mathbf{B}}{\mu_0} \tag{19.24}$$

$$\nabla \cdot \mathbf{B} = 0 \tag{19.25}$$

These are four partial differential equations in four unknowns B_i ($i = 1, \cdots, 3$), p. Equation (19.24) shows the competition between the expansion pressure, modified by the B field, and the tension along the lines of force. This is made apparent introducing the unit vectors parallel, normal and binormal to $\mathbf{B}$,

$$\begin{aligned} \mathbf{b} &= \frac{\mathbf{B}}{B} \\ \mathbf{n} &= \kappa \frac{d\mathbf{b}}{ds} \\ \mathbf{h} &= \mathbf{b} \times \mathbf{n} \end{aligned} \tag{19.26}$$

where $\kappa = |\mathbf{b} \cdot \nabla \mathbf{b}|$ is the field line inverse curvature radius. As a result, the r.h.s. of equation (19.24) is written as:

$$\frac{1}{\mu_0}\mathbf{B} \cdot \nabla \mathbf{B} = \mathbf{b}\frac{1}{2\mu_0}\frac{dB^2}{ds} + \frac{B^2}{\mu_0}\mathbf{n}\kappa \tag{19.27}$$

The first term represents the (mechanical) tension along field lines and the second one describes the compression normal to the curved field lines, illustrated in figure 19.4.

To be definite we refer to the coordinate system pictured in figure 19.7. From the above relations it follows easily that, *in axisymmetry*, a *pure toroidal* magnetic field cannot provide closed isobaric confinement surfaces. Indeed if $\mathbf{B} = B_\phi(R, Z)\mathbf{e}_\phi$ from the condition of zero torque in axisymmetric equilibrium:

$$0 = \nabla \times (\mathbf{J} \times \mathbf{B}) \equiv \mathbf{B} \cdot \nabla \mathbf{J} - \mathbf{J} \cdot \nabla \mathbf{B} = -\boldsymbol{J} \cdot \nabla \boldsymbol{B} \tag{19.28}$$

Then since $\mathbf{J} \cdot \nabla \mathbf{B} = \frac{1}{2\mu_0 R}\frac{\partial B_\phi^2}{\partial Z} = 0$, if B^2 is independent of Z, from equation (19.24) also p does not depend on (R,Z) and cannot define a closed surface. Therefore, just on the basis of the properties of $\mathbf{B}$ it turns out that closed isobaric magnetic surfaces can be realized only with a field with two components, necessarily winding on the surface. The resulting toroidal and poloidal fields have helical field lines, wrapping on the magnetic surfaces. Following a magnetic field line around closed isobaric magnetic surfaces (see figure 19.5) starting from a point a_0, after a turn around the

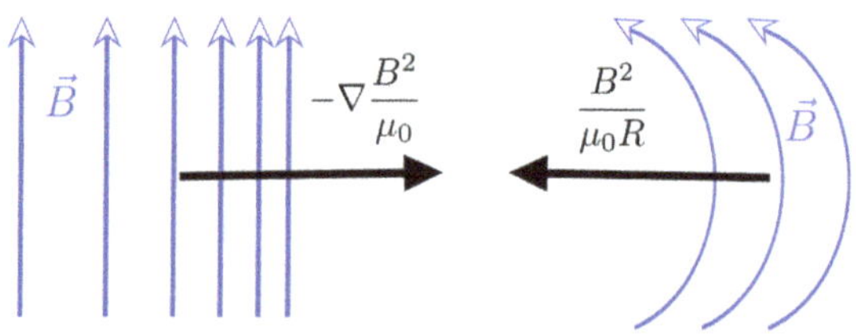

Figure 19.4. Illustration of magnetic pressure normal to **B** field lines and the resultant of the tension along them.

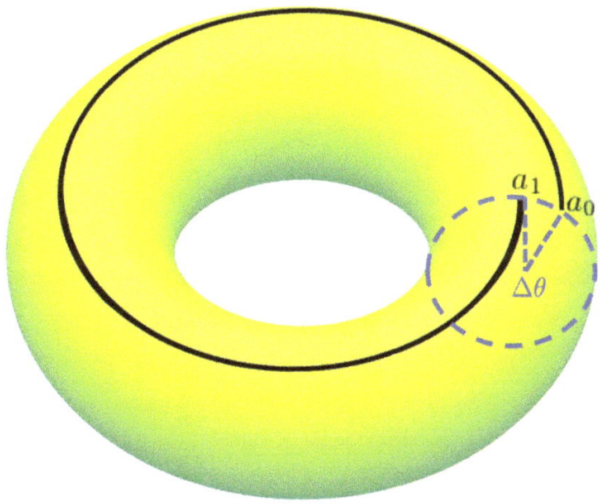

Figure 19.5. Example of magnetic field line around a closed isobaric magnetic surface.

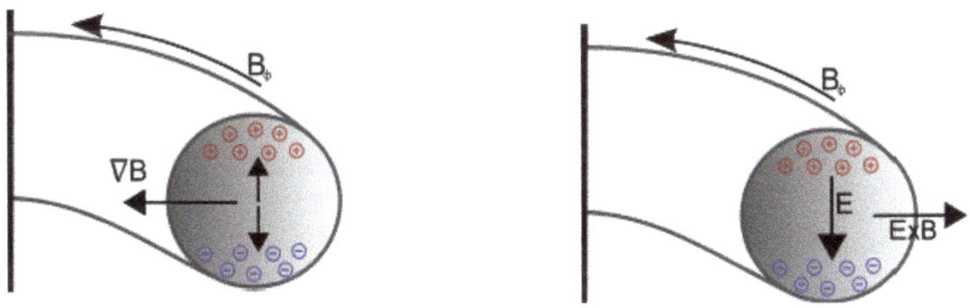

Figure 19.6. Picture of the charge separation induced by (a) $\nabla B \times \mathbf{B}$ and (b) electric $\mathbf{E} \times \mathbf{B}$ outward drift of the plasma, if only a toroidal field is present. An additional (poloidal) component of the **B** field is needed to *short* the electric field **E** on field lines wrapping the isobaric surfaces.

torus, the field line crosses the plane P at a point a_1 at a different angle $\vartheta_0 + \Delta\vartheta$ around the minor axis O. The angle $\Delta\vartheta$ depends on the initial poloidal angle ϑ_0. The average of $\Delta\vartheta$ over an infinite number of transits is called *rotational transform*:

$$\iota = \lim_{n\to\infty} \frac{1}{n} \sum_{r=1}^{n-1} \Delta\vartheta_r \tag{19.29}$$

and is a characteristic label which describes an integral property of a field line on a magnetic surface, discussed in the following paragraphs. If ι is a rational fraction of 2π the field line is closed, otherwise it is ergodic. A fundamental role of the rotational transform is to prevent the $\mathbf{E} \times \mathbf{B}$ plasma drift out of the configuration, shorting-out the electric field due to the separation of charges caused by the $\mathbf{B} \times \nabla B/enB^2$ drifts which depend on charge sign, as shown in figure 19.6.

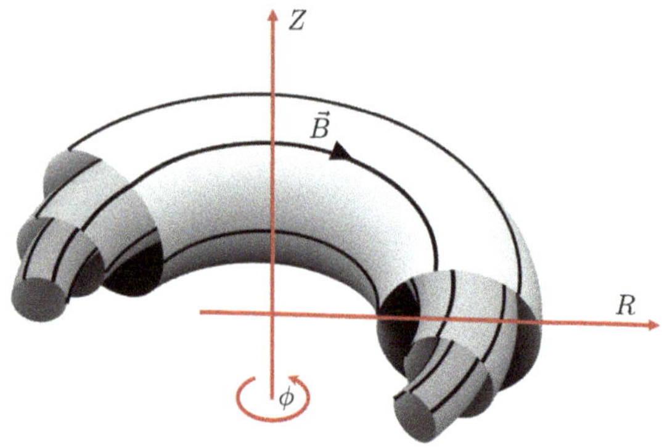

Figure 19.7. Toroidal configuration of (isobaric) magnetic surfaces threaded by field lines of **B** and main right-handed cylindrical coordinate system R, ϕ, Z.

In axisymmetric systems such as the tokamak or the reversed field pinch (RFP) [9, 11, 13–15, 17, 24], where quantities are independent of the coordinate ϕ a configuration with nested magnetic surfaces can be proven to exist if, in addition to the current flowing in external coils and generating a toroidal field, there is current in the plasma or in internal conductors, necessary to generate the other *poloidal* magnetic field B_p components. In a stellarator, which is *not axisymmetric*, the rotational transform is generated by external helical winding, but only mathematically approximate closed magnetic surfaces can be demonstrated.

19.1.1 First principles and topological properties

The ideal MHD system allows deducing important conclusions both qualitative and semi-quantitative on the different types of equilibrium configurations with finite volume, by the simple *translation* of the equations in geometrical and physical terms. The basic question is finding, if at all possible, a class of solutions of the field equations (19.19), (19.22), and (19.21) which has **B** tangent to closed isobaric surfaces, and with **J** sources in a finite volume. A formal treatment of this problem requires working with suitable curvilinear coordinate systems. Topology informs us [25] that a vector field not vanishing anywhere and continuously tangent to a surface, can be tangent *only* to a surface homeomorphic (i.e. topologically equivalent) to a polyhedron with Euler characteristic $\chi_E = N_V - N_E + N_F$ (where N_V, N_E, N_F indicate the number of vertices, edges faces) equal to the sum of the multiplicities (Poincaré index) of the singular points of the field, see [25]. Since $\mathbf{B} \neq 0$ and $\nabla \cdot \mathbf{B} = 0$, **B** has no singular points of the type *knot* or *star* and the Poincaré index is zero. Therefore, also zero is the Euler characteristic χ_E of the surfaces to which the magnetic field can be tangent; therefore, such *magnetic surfaces* are topological tori [17, 26]. An analogous statement holds also for the current density field **J**, which is also solenoidal.

If it is possible to design and construct suitable current distributions within and outside the domain of plasma confinement, the *magnetic surfaces* spanned by the magnetic field lines $\psi = \text{const}$, of toroidal genus turn out to be isobars because

$\mathbf{J} \cdot \nabla p = \mathbf{B} \cdot \nabla p = \mathbf{B} \cdot \nabla \psi = 0$ and consequently $p = p(\psi)$. At this point it is useful to introduce important geometrical and physical properties of the equilibrium configuration, which determine the quantitative scaling of real devices. Concerning the geometry the main controlling parameter is the so-called *inverse aspect ratio* $\epsilon = a/R_0 \ll 1$ where a, the *minor radius*, is the half width of the plasma torus in the equatorial plane $Z = 0$, and R_0 is the distance of the center to the symmetry axis. In some cases it can be assumed that the plasma boundary coincides with a conducting wall surrounding it, so that a and R_0 are those of the vessel. The important physical parameters are the value of the inverse rotational transform at the edge of the plasma, $q(a) = 2\pi/\iota(a) > 1$ related to the total toroidal plasma current, and the ratio of the average plasma kinetic pressure to the magnetic pressure, $\beta = 2\mu_0\langle p\rangle/\langle B^2\rangle \approx 2\mu_0\langle p\rangle/B_0^2 = O(\epsilon^2)$. This ordering of global parameters gives the basic characterization of a tokamak plasma, allowing both the design and the interpretation of experiments.

19.1.2 General representations of the magnetic field

A proper description of the magnetic field properties requires a careful use of general curvilinear coordinate systems. It is desirable to approach the subject in the simplest way possible, nonetheless, some technical details are unavoidable, albeit rather heavy, for a sound understanding of the physics. Let us consider a reference frame cylindrical coordinates system (R, ϕ, Z).

A useful set of curvilinear, right-handed, *non-orthogonal* coordinate systems can be constructed, associated with surfaces of constant fluxes of $\mathbf{B}$

$$\begin{aligned} u^1(R, \phi, Z) &= u \\ u^2(R, \phi, Z) &= \vartheta \\ u^3(R, \phi, Z) &= \phi \end{aligned} \tag{19.30}$$

The unit vectors and the contravariant metric tensor elements are:

$$\mathbf{e}^i = \nabla u^i = (\nabla u, \nabla\vartheta, \nabla\phi),\ g^{ij} = \mathbf{e}^i \cdot \mathbf{e}^j \tag{19.31}$$

with $det[g^{ij}] = 1/g$ and therefore the Jacobian is $\mathfrak{J} = \sqrt{g} = 1/(\nabla u \times \nabla\vartheta \cdot \nabla\phi)$ and volume element $d^3x = \sqrt{g}\, du\, d\vartheta\, d\phi$. The covariant unit vectors and metric tensor elements are

$$\begin{aligned} \mathbf{e}_i &= \frac{\partial \mathbf{r}}{\partial x^i} \\ g_{ij} &= \mathbf{e}_i \cdot \mathbf{e}_j \\ g_{ij}g^{jk} &= \delta_i^k \\ g_{\phi\phi} &= R^2. \end{aligned} \tag{19.32}$$

Many analytical theories and numerical codes use the curvilinear coordinate systems that are constructed with one coordinate surface coinciding with the magnetic surface. In these *magnetic surface coordinates* we need to choose a poloidal coordinate ϑ and a (counterclockwise) toroidal coordinate ϕ. In particular, it is possible and useful to select ϑ and ϕ in such a way that the magnetic field lines become straight lines on the (ϑ, ϕ) plane. Such a coordinate system is also commonly called a *flux coordinate system*. Several kinds of coordinates can be defined, and we address ourselves to the quoted references for a more complete treatment. Here it is sufficient to anticipate the main concepts and fundamental properties necessary for the study of the above equations and of their solutions [11, 14, 24]. Since the magnetic field is solenoidal $\nabla \cdot \mathbf{B} = 0$, it can be expressed in terms of a vector potential $\mathbf{A}$, and a scalar potential Φ_B as $\mathbf{B} = \nabla \times \mathbf{A} - \nabla\Phi_B$. In a generic coordinate system u^i, such as equation (19.30), even with no special symmetry, the vector potential $\mathbf{A}$ can be always written in the form

$$\mathbf{A} = u\nabla\vartheta + \psi\nabla\phi + \nabla G = \psi\nabla\alpha \tag{19.33}$$

in which $\alpha(\mathbf{x})$, $\psi(\mathbf{x})$ are suitable scalar functions of the position vector $\mathbf{x}$ and G is a periodic function of (ϑ, ϕ). Then, momentarily neglecting the scalar potential term for the vacuum field, B can be expressed in the *symplectic* [14] form:

$$\mathbf{B} = \nabla \times \mathbf{A} = \nabla u \times \nabla\vartheta + \nabla\psi \times \nabla\phi = \nabla\psi \times \nabla\alpha \tag{19.34}$$

The physical meaning of the functions $u(\mathbf{x})$, $\psi(\mathbf{x})$, still arbitrary, is obtained from the construction of the *poloidal* and *toroidal* fluxes of $\mathbf{B}$:

$$\Psi_p = \int_{S_p} \mathbf{B} \cdot \mathbf{n}_p \, dS \tag{19.35}$$

$$\Psi_t = \int_{S_t} \mathbf{B} \cdot \mathbf{n}_t \, dS \tag{19.36}$$

The calculation can be performed in terms of the volume integrals related to equations (19.35) and (19.36) applying Gauss theorem and $\nabla \cdot \mathbf{B} = 0$. The contravariant components of the $\mathbf{B}$ field are:

$$\begin{aligned} \mathbf{B} \cdot \nabla\vartheta &= \nabla\psi \times \nabla\phi \cdot \nabla\vartheta = \frac{1}{\sqrt{g}} \frac{d\psi}{du} \\ \mathbf{B} \cdot \nabla\phi &= \nabla u \times \nabla\vartheta \cdot \nabla\phi = \nabla u \cdot \nabla\vartheta \times \nabla\phi = \frac{1}{\sqrt{g}} \end{aligned} \tag{19.37}$$

then the poloidal and toroidal fluxes are:

$$\begin{aligned} 2\pi\Psi_p &= \int_V \mathbf{B} \cdot \nabla\vartheta \, d^3x = \int_V \frac{d\psi}{du} \, dud\vartheta d\phi = 4\pi^2\psi \\ 2\pi\Psi_t &= \int_V \mathbf{B} \cdot \nabla\phi \, d^3x = \int_V dud\vartheta d\phi = 4\pi^2 u \end{aligned} \tag{19.38}$$

Then the final expression of $\mathbf{B}$ in contravariant form becomes

$$\mathbf{B} = \frac{1}{2\pi}\nabla\Psi_t \times \nabla\vartheta + \frac{1}{2\pi}\nabla\Psi_p \times \nabla\phi = \mathbf{B}_t + \mathbf{B}_p \tag{19.39}$$

It must be recalled that the covariant components are $B_i = g_{ij}B^j$ and the *physical components* (e.g., in S.I. units) are:

$$\begin{aligned} B_t &= \sqrt{g_{\phi\phi}}\frac{1}{\sqrt{g}} \\ B_p &= \sqrt{g_{\vartheta\vartheta}}\frac{1}{\sqrt{g}}\frac{d\psi}{du} \end{aligned} \tag{19.40}$$

The local pitch angle of a magnetic field line on an equilibrium surface is defined as the ratio of the contravariant components of the **B** field, and it is constant on a confinement surface:

$$\frac{d\vartheta}{d\phi} = \frac{\mathbf{B}\cdot\nabla\vartheta}{\mathbf{B}\cdot\nabla\phi} = \frac{\iota}{2\pi} \tag{19.41}$$

$\iota/2\pi = 1/q(u)$ is the previously mentioned *rotational transform* (figure 19.5) and is a flux function, therefore constant on a isobaric surface. Since the toroidal and poloidal fluxes equation (19.36), and equation (19.35) are surface functions, $\nabla\Psi_t = \frac{d\Psi_t}{d\Psi_p}\nabla\Psi_p = q\nabla\Psi_p$ and $\nabla(q\vartheta) = \vartheta\frac{dq}{d\Psi_p}\nabla\Psi_p + q\nabla\vartheta$, by setting $\zeta = -\phi$, $\alpha = (q\vartheta - \zeta)$, equations (19.33) and (19.34) can be alternatively written in the Clebsch form

$$\mathbf{A} = -\alpha\nabla\psi = \psi\nabla(q\vartheta - \zeta) \tag{19.42}$$

The second version of expression equation (19.42) follows from the identity $\nabla\times\psi\nabla\alpha = -\nabla\times\alpha\nabla\psi$. In summary, equation (19.22) can be satisfied by a field in the elegant Clebsch form

$$\mathbf{B} = \nabla\psi\times\nabla\alpha \tag{19.43}$$

where α, is a helical coordinate, and ψ, a *radial* coordinate, which are scalar functions of the Cartesian coordinates X, Y, Z corresponding to the integrals of the equations of the **B** field lines of force, expressing the parallelism of a line element $d\ell$ to the field **B**:

$$\begin{aligned} d\ell\times\mathbf{B} &= 0 \\ \psi(X, Y, Z) &= \text{const} \\ \alpha(X, Y, Z) &= \text{const} \end{aligned} \tag{19.44}$$

From equation (19.43) it follows that $\mathbf{B}\cdot\nabla\psi = 0$, $\mathbf{B}\cdot\nabla\alpha = 0$ and $\psi = \text{const}$ and $\alpha = \text{const}$ are the *surfaces* whose intersections are the field lines of **B**:

$$d\ell\times\nabla\alpha\times\nabla\psi = (d\ell\cdot\nabla\psi)\nabla\alpha - (d\ell\cdot\nabla\alpha)\nabla\psi = 0 \tag{19.45}$$

since $\nabla\psi \neq 0$. The surface quantity $q(\psi) = 2\pi/\iota$, called often *safety factor*, because it is associated with stability conditions of the tokamak equilibrium, labels uniquely the magnetic surfaces. If it is an irrational real number, a line of force winds helically on the surface, covering it ergodically, i.e. passing arbitrarily close to any point on the surface. If $q = m/n$ a rational number, the field lines wrapped on this surface close head to tail after n toroidal and m poloidal turns. These surfaces, densely packed among the ergodic ones, are very fragile with respect to magnetic perturbations which alter the topology of the *nested* isobaric magnetic surfaces, breaking them up into layers of unstable (m,n) flux tubes which damage energy confinement [11, 14, 15, 17]. The internal (radial) profile of $q(\psi)$ is related to the surface averaged internal toroidal current. In a tokamak it is generally a monotonic function growing from the magnetic axis to the *magnetic plasma boundary*, defined here as the ψ = const surface where the pressure, and the current density vanish. Physically the plasma extends beyond this boundary, in a shadow region, at least over a space scale of few poloidal ion Larmor radii $\rho_{p,i} = m_i v_{\mathrm{Th},i}/eB_p$, or more precisely over a perpendicular diffusion length $L_\perp = D_\perp^{-1}$, which depends on the composition and collisionality [14, 27] in this scrape-off layer (SOL). These complex problems will be hinted to further on.

The toroidal component of the vector potential equation (19.43), determined up to a gauge transformation [23] can be related to the *poloidal magnetic flux*. It is useful to refer to a surface S in the equatorial plane $Z = 0$, and consider the range:

$$R(\psi_1) < R < R(\psi_2) \tag{19.46}$$

From equation (19.43) the *poloidal field* is $\mathbf{B}_p = \nabla \times (\psi \nabla \phi)$ and the poloidal flux through the surface S is expressed as:

$$\Psi_p = \int_S \mathbf{B} \cdot n \, dS = \oint \psi \nabla \phi \cdot d\ell = 2\pi(\psi_2 - \psi_1) \tag{19.47}$$

In S.I units the poloidal flux in equation (19.47) is measured in Weber = Volt sec. In a similar way we define the poloidal flux in equation (19.48) of current (in Ampere) through a surface such that $\nabla\phi \cdot d\mathbf{S} = 0$:

$$\begin{aligned} I_p &= \int_S \mathbf{J} \cdot dS = \frac{1}{\mu_0} \int_S \nabla \times (T\nabla\phi) \cdot d\mathbf{S} \\ &= \frac{1}{\mu_0} \int_{C_p} T\nabla\phi \cdot d\ell = \frac{2\pi}{\mu_0}(T(\psi_1) - T(\psi_2)) \end{aligned} \tag{19.48}$$

19.1.3 Specific curvilinear flux coordinate system

The scale lengths determining the physical behavior of the plasma in a direction parallel to the magnetic field lines are very different from those across the magnetic surfaces. Therefore, it is necessary, and often inevitable, to employ a coordinate system based on the contours of the magnetic flux function $\psi(X, Y, Z)$, where transport and stability properties are described more naturally. The position vector in rectangular coordinates $\mathbf{r} = X\mathbf{i} + Y\mathbf{j} + Z\mathbf{k}$ is expressed as a function of general curvilinear coordinates $\mathbf{r} = \mathbf{r}(q^1, q^2, q^3)$ where $q^1 = u(\psi)$ as already stated, plays the role of a radial variable, $q^2 = \vartheta$ is a suitably chosen poloidal angle and $q^3 = \phi$ represents a toroidal angle defined in a real space. Here we use the explicit covariant unit vectors and metric tensor through which the line and volume elements are expressed as:

$$d\ell^2 = g_{uu}du^2 + 2g_{u\vartheta}d\vartheta du + g_{\vartheta\vartheta}d\vartheta^2 + g_{\phi\phi}d\phi^2 \tag{19.49}$$

$$dV = \sqrt{g}\, dud\vartheta d\phi \tag{19.50}$$

For these flux coordinates $u = \Psi_t/2\pi$, ϑ, ϕ, the Jacobian $\hat{J}$ is expressed by

$$\hat{J} = (\nabla u \times \nabla\vartheta \cdot \nabla\phi)^{-1} = \frac{R}{|\nabla u|}\frac{d\ell}{d\vartheta} \tag{19.51}$$

With the definition of the coordinates u, ϕ and the Jacobian the polar angle ϑ can be determined from the evaluation of the *polar* arc length element $d\ell_p$:

$$d\ell = \frac{\partial \mathbf{r}}{\partial u}du + \frac{\partial r}{\partial\vartheta}d\vartheta + \frac{\partial \mathbf{r}}{\partial\phi}d\phi \tag{19.52}$$

$$d\ell_p = \frac{\partial \mathbf{r}}{\partial\vartheta}d\vartheta = \hat{J}(\nabla\phi \times \nabla u)d\theta,\ d\vartheta = \frac{R}{\hat{J}|\nabla u|}d\ell_p \tag{19.53}$$

As already stated, any quantity expressed as a function of only a magnetic surface label $\psi(u)$ is called a surface quantity, such as the volume of a magnetic surface $V(\psi)$ and the safety factor $q(\psi)$. The flux surface average of any quantity $X(\psi, \vartheta)$ is defined by:

$$\langle X\rangle = \lim_{\Delta V\to 0}\frac{1}{\Delta V}\int_{\psi}^{\psi+\nabla\psi} XdV = 2\pi\oint\frac{X\sqrt{g}}{dV/d\psi}\cdot d\ell = \frac{\oint X\sqrt{g}d\vartheta}{\oint\sqrt{g}d\vartheta} \tag{19.54}$$

The *average* field lines pitch (*safety factor*) equation (19.41) according to equations (19.54) and (19.51) is:

$$q(\psi) = \frac{1}{2\pi}\oint\frac{d\phi}{d\vartheta}d\vartheta = \frac{1}{2\pi}\oint\frac{B_t}{B_p}\frac{d\ell}{R} = \frac{T}{2\pi}\oint\frac{d\ell}{R|\nabla\psi|} = \frac{T}{2\pi\psi'}\oint\frac{\hat{J}d\vartheta}{R^2} \tag{19.55}$$

For a coordinate system where field lines are straight the Jacobian is

$$\hat{J} = R^2 \tag{19.56}$$

It is important to mention that other systems of curvilinear coordinates are better suited for special studies, such as transport phenomena [24]. Flux surface averages are very important in studies of stability and transport. The first fundamental expression involving flux averages and strictly related to equilibrium and the MHD quasineutrality statement in equation (19.12) is the parallel component of the equilibrium current. The total current density can be written as:

$$\mathbf{J} = J_{\|}\frac{\mathbf{B}}{B} + \mathbf{J}_{\perp} \tag{19.57}$$

For generic non-isodynamic configurations [28–30], $\frac{\partial B}{\partial \vartheta} \neq 0$, the perpendicular current density is not solenoidal

$$\mathbf{J}_\perp = \frac{\mathbf{B} \times \nabla p}{B^2} = \frac{B_p}{B^2} T \frac{dp}{d\psi} \mathbf{e}_\vartheta - \frac{R B_p^2}{B^2} \frac{dp}{d\psi} \mathbf{e}_\phi \tag{19.58}$$

$$\nabla \cdot \mathbf{J}_\perp = -\nabla p \cdot \nabla \times \left(\frac{\mathbf{B}}{B^2} \right) \neq 0 \tag{19.59}$$

The current continuity is then expressed by

$$\mathbf{B} \cdot \nabla \left(\frac{J_\parallel}{B} \right) = -\nabla \cdot \mathbf{J}_\perp \tag{19.60}$$

$$B \frac{d}{d\ell} \left(\frac{J_\parallel}{B} \right) = -\frac{d}{r d\vartheta} \left(\frac{B_p}{B^2} \right) T \frac{dp}{d\psi} \tag{19.61}$$

where $d/d\ell = 1/qR(d/d\vartheta)$. Equation (19.60) for the variation of the parallel current along the field lines is an example of magnetic differential equation:

$$\mathbf{B} \cdot \nabla J_\parallel = S \tag{19.62}$$

Solutions to the magnetic differential equation for $J_\parallel$ are the sum of a particular solution, which varies over the magnetic surface and a homogeneous solution in which $J_\parallel$ is constant along field lines. Both the special and the homogeneous solution can be singular at rational surfaces. The singularity of the particular solution, can arise if the pressure gradient, is non-zero at a rational surface. The solubility condition for the magnetic differential equation is expressed by the Newcomb condition [14]:

$$\oint \frac{S}{B} d\ell = 0 \tag{19.63}$$

on each rational surface. Therefore, integrating equation (19.60) the parallel current density is determined as

$$\frac{J_\parallel}{B} = -\mu_0 T \frac{dp}{d\psi} \frac{B_\vartheta}{B^2} + K(\psi) \tag{19.64}$$

where $K(\psi)$ is an integration constant, which can be determined by averaging. As a result the parallel current is expressed as:

$$J_\parallel = \frac{\mathbf{J} \cdot \mathbf{B}}{\langle B^2 \rangle} B - \frac{T}{B} \frac{dp}{d\psi} \left(1 - \frac{B^2}{\langle B^2 \rangle} \right). \tag{19.65}$$

The second term on the. r.h.s is called the Pfirsch–Schlueter current J_{PS} and represents the parallel *return current* originated by the charge separation effect of toroidal geometry. Since, approximately, $B \approx B_0/(1 + \epsilon \cos \vartheta)$

$$\left(1 - \frac{B^2}{\langle B^2 \rangle}\right) \approx 2\epsilon \cos \vartheta \tag{19.66}$$

and the Pfirsch–Schlueter current shows the explicit dependence:

$$J_{\text{PS}} \approx 2\epsilon \frac{T}{B} \frac{dp}{d\psi} \cos \vartheta \tag{19.67}$$

It is clear that $BJ \cdot \mathbf{B}/\langle B^2 \rangle$ is the *source* of the rotational transform, while J_{PS}, changes sign on the torus axis at $R = R_0$. Although the Pfirsch–Schlueter current is often associated with transport physics, the derivation just summarized shows that it is actually a purely toroidal equilibrium effect.

19.2 Grad–Shafranov equation

In the case of cylindrical symmetry in a system of coordinates (R, ϕ, Z) the longitude ϕ is an ignorable coordinate, and the exact solution of equation (19.43) of the solenoidality conditions equations (19.22) and (19.23) can be written in mixed covariant-contravariant as:

$$\mathbf{B} = T\nabla\phi + \nabla\psi \times \nabla\phi \tag{19.68}$$

$$\mathbf{J} = RJ_\phi \nabla\phi + \nabla T \times \nabla\phi \tag{19.69}$$

With the expressions of the direction vector $\mathbf{e}_\phi = \frac{\nabla\phi}{R}$ the $\mathbf{B}$ field is commonly written in terms of (physical) toroidal and poloidal components

$$\mathbf{B} = B_t \mathbf{e}_\phi + \mathbf{B}_p \tag{19.70}$$

The current density is expressed in terms of the magnetic field by the Ampere law:

$$\begin{aligned} \mu_0 \mathbf{J} &= \nabla \times (T\nabla\phi + \nabla\psi \times \nabla\phi) \\ &= -R\nabla \cdot (R^{-2}\nabla\psi)\nabla\phi + \nabla T \times \nabla\phi \end{aligned} \tag{19.71}$$

It is customary to introduce the elliptic Beltrami operator [11]

$$\Delta^* = R^2 \nabla \cdot (R^{-2}\nabla) = R\frac{\partial}{\partial R}\left(R^{-1}\frac{\partial}{\partial R}\right) + \frac{\partial^2}{\partial Z^2},$$

whereby the toroidal component of the current is written as

$$\mu_0 R J_\phi = -\Delta^* \psi \tag{19.72}$$

and the force balance equation in the direction normal to the magnetic surfaces $\psi = \text{const}$ takes the form of the famous Grad–Shafranov [11, 26] equation:

$$\Delta^* \psi = -\mu_0 R^2 \frac{\partial p}{\partial \psi} - T\frac{\partial T}{\partial \psi} \tag{19.73}$$

The total toroidal current is given by

$$I_t = \frac{1}{2\pi\mu_0}\oint B_p d\ell_p = \frac{1}{2\pi\mu_0}\oint \frac{\psi'}{R} d\vartheta \tag{19.74}$$

and is a constraint to be imposed to obtain a solution of the Grad–Shafranov equation. The current density $J_\phi(R, \psi)$ depends of two *free functions* $p(\psi)$, $T(\psi)$, so that in general the problem is underdetermined and therefore supplementary information or assumptions are necessary, in addition to boundary conditions, to define fully determined problems. The equation (19.72) can be a linear or nonlinear partial differential equation depending on the source $\mu_0 R J_\phi(R, \psi)$. The Grad–Shafranov equation (19.72) associated with the boundary *jump* conditions on the contour Γ, [[]] indicates the finite jump across Γ, at the interface plasma-vacuum

$$[[\mathbf{B} \cdot \mathbf{n}]] = 0, \; [[\mathbf{n} \times \mathbf{B}]] = \mu_0 \mathbf{K}, \tag{19.75}$$

where $\mathbf{n}$ and $\mathbf{t}$ are the unit vectors normal and tangential to the *assigned* or *free* plasma boundary Γ and $\mathbf{K}$ is a surface current.

If the contour Γ coincides with a (super)conducting wall the condition on the tangential components of electric and magnetic fields $\mathbf{E}$, $\mu\mathbf{H} = \mathbf{B}$, $[[\mathbf{n} \times \mathbf{H}]] = 0$, $[[\mathbf{n} \times \mathbf{E}]] = 0$ must be imposed. It then follows from Ohm's law in ideal MHD, $\mathbf{E} + \mathbf{v} \times \mathbf{B} = 0$ that the normal component of velocity also automatically vanishes on the wall $\mathbf{v} \cdot \mathbf{n} = 0$.

The free contour configuration defined by equation (19.72), with given current in equation (19.74) and the boundary conditions just mentioned, represents only the mathematical existence of closed magnetic isobaric magnetic surfaces, but the superposition, to the solution of the Grad–Shafranov equation, of suitably designed external fields (due to currents in conductors outside the plasma) is strictly required to ensure a real equilibrium in the vertical direction and in the major radius direction. The first case *seems* trivial since the Z projection of equation (19.24) prescribes just up–down symmetry $dp/dZ = 0$, with respect to the equatorial plane; actually, most of the more interesting and useful tokamak configurations are not up–down symmetric, and this implies that additional fields for control of the vertical position, and its stability are needed. More crucial is the case of horizontal displacement of the magnetic axis, that is, of expansion of the plasma ring. Since the confining magnetic field behaves as $1/R$ and the torus surface for $R < R_0$ is smaller than for $R > R_0$, a number of body force components, produce a net outward horizontal force, which must be counterbalanced. One is a hoop force which tends to expand the current carrying ring of radius R increasing the (poloidal) flux [23], and another is a *tire tube* force due to the plasma pressure $p(\psi)$ and the non-uniform magnetic pressure $B^2/2\mu_0 \propto 1/R^2$. A detailed force balance evaluation is presented in the literature [11, 13, 22] but in the next paragraphs we present some other derivation of the basic requirements for these external fields, generally multipolar, based on general integral relations, which are a special case of the virial theorem by Chandrasekhar, Fermi *et al* [14, 31]. In reduced form can be presented multiplying equation (19.24) by $R\mathbf{e}_R$ and integrating over an infinite volume:

$$\int_V R\mathbf{e}_R \cdot [\nabla p - \frac{1}{\mu_0}(\nabla \times \mathbf{B}_i) \times \mathbf{B}_i - \mathbf{J} \times \mathbf{B}_e]\, dV = 0 \tag{19.76}$$

where here the magnetic field has been written in terms of the internal *self-consistent* $\mathbf{B}_i$ component and an external field $\mathbf{B}_e$ to be determined. An explicit example is worked out in the section on the large aspect ratio approximation. The two main classes of problems, related to the theoretical and design issues or to the diagnostic ones, are labelled as *direct* or *inverse* and will be presented later on.

19.2.1 Figures of merit of the tokamak equilibria

In the first section, and in equation (19.4) an important parameter was introduced, which characterizes the quality of the magnetic confinement of the plasma in a tokamak, namely the ratio of the thermal pressure $p = nT$ to the magnetic pressure, i.e.

$$\beta = \frac{p}{B^2/2\mu_0} \tag{19.77}$$

This parameter can be rewritten as $\beta = \frac{C_s^2}{V_A^2}$, i.e. the square of the ratio of the sound speed C_s to the Alfvén speed V_A; this gives a basic ordering of the magnetized plasma fluid motion parallel to the magnetic field (sound waves) and perpendicular to it (Alfvén waves). Since the (total) β is $\ll 1$, the latter dominate. Since the pressure in tokamak plasmas is inhomogeneous, the volume-averaged pressure is usually used to define the beta. In tokamak plasmas, the toroidal beta β_t and the poloidal beta β_p are defined, respectively, by

$$\beta_t = \frac{\langle p \rangle}{B_{t0}^2/2\mu_0} \tag{19.78}$$

$$\beta_p = \frac{\langle p \rangle}{B_{\text{pa}}^2/2\mu_0} \tag{19.79}$$

where $\langle \cdots \rangle$ indicates the volume averaging in equation (19.54), B_{t0} is the vacuum toroidal magnetic field at the magnetic axis (or geometrical center of the plasma), B_{pa} is the averaged poloidal magnetic field on the plasma surface. In tokamaks, the toroidal magnetic field is dominant and thus the toroidal beta β_t is generally used to characterize the efficiency of the *magnetic field* in confining plasma energy. The efficiency of the *plasma current* in confining the plasma is monitored by the poloidal β_p, which can be rewritten as

$$\beta_p = \frac{\langle p \rangle}{I_p^2/2/(8\pi^2 a^2)}.$$

The transformation to flux coordinates of the operator

$$\nabla = \frac{1}{\sqrt{g}}\left[\frac{\partial}{\partial u}\nabla u + \frac{\partial}{\partial \vartheta}\nabla \vartheta + \frac{\partial}{\partial \phi}\nabla \phi\right]$$

leads to write the Grad–Shafranov equation in flux coordinates:

$$\frac{R^2}{\sqrt{g}}\frac{\partial \psi}{\partial u}\left[\frac{\partial}{\partial u}\left(\frac{g_{\vartheta\vartheta}}{\sqrt{g}}\frac{\partial \psi}{\partial u}\right) - \frac{\partial}{\partial \vartheta}\left(\frac{g_{u\vartheta}}{\sqrt{g}}\right)\frac{\partial \psi}{\partial u}\right] = -\mu_0 R^2 \frac{\partial p}{\partial u} - T\frac{\partial T}{\partial u} \tag{19.80}$$

Considering the expression of B^{ϑ}, B_{ϑ} and B_t equation (19.80) becomes

$$R^2 B_{\vartheta}\left[\frac{\partial B_{\vartheta}}{\partial u} - B_{\vartheta}\frac{\partial g_{u\vartheta}}{\partial \vartheta}\right] = -\mu_0 R^2 \frac{\partial p}{\partial u} - \frac{1}{2}\frac{\partial}{\partial u}(R^2 B_t^2) \tag{19.81}$$

Applying the averaging operator in equation (19.54), the second term on the l.h.s. of equation (19.80) is suppressed and the resulting first order differential relation

$$\frac{1}{2}\frac{d}{du}\left\langle B_p^2\right\rangle = -\mu_0 \frac{dp}{du} - \frac{1}{2}\frac{d}{du}\langle B_t^2\rangle \tag{19.82}$$

can be integrated, applying the boundary conditions $p(a) = 0$, $B_p(0) = 0$, obtaining:

$$1 + \frac{B_{t,a}^2 - B_{t,0}^2}{B_{p,a}^2} = \beta_p \tag{19.83}$$

The last equation allows understanding the physical meaning of β_p, ratio of the plasma kinetic pressure to the poloidal magnetic pressure, relating it to the para- or diamagnetic behaviour of the plasma. In experimental devices, the typical ordering is $B_p/B_t \ll 1$, $\frac{B_{t,a}^2 - B_{t,0}^2}{B_{p,a}^2} < 1$ and $\beta_p \simeq O(1)$. If there is no poloidal current, affecting the toroidal field, $\beta_p = 1$. When the term $B_{t,a}^2 - B_{t,0}^2 > 0$ the magnetic pressure tends to reduce the plasma pressure and $\beta_p > 1$ and the plasma has *diamagnetic* behavior and the poloidal current contributes to the confinement of the plasma. Otherwise, if $B_{t,a}^2 - B_{t,0}^2 < 0$ of the plasma is *paramagnetic*, with $\beta_p < 1$. This quantity β_p, is related to the geometric shape parameters Δ_{Sh}, κ_a δ_a of the nested magnetic surfaces, and can be limited by equilibrium, stability, or transport conditions [15, 24].

19.2.2 Large aspect ratio limit

An instructive analysis of the general properties of the solution of the Grad–Shafranov equation (19.72) can be done in the linearized and large aspect ratio ϵ^{-1} limit and low β ordering, with a circular meridian cross-section [15]. Several presentations with slightly different notation are available in the literature; here we opt for one of the more streamlined versions available, which still requires some tedious algebra which we propose as an exercise to the reader. A parametric representation which will be more extensively discussed later in the form of equations (19.132) and (19.133) can be written in the leading order form:

$$R = R_0 + \Delta(\rho) + \rho\cos(\theta) \tag{19.84}$$

$$Z = \rho\sin(\theta) \tag{19.85}$$

having substituted $\Delta_{Sh} = \Delta_0 - \Delta(\rho)$, with ρ indicating the flux surface minor radius. The relevant tokamak ordering is recalled

$$\epsilon \ll 1, \rho = O(\epsilon) \tag{19.86}$$

$$B_p/B_t = O(\epsilon), q(a) = O(1) \tag{19.87}$$

$$\beta = O(\epsilon^2), \beta_p = O(1) \tag{19.88}$$

The covariant metric tensor elements specified in these shifted circle coordinates, (ρ, θ, ϕ), which are not flux coordinates, lead to the equation

$$\begin{aligned}(1 - 2\Delta\cos\vartheta)\frac{1}{\rho}\frac{\partial}{\partial\rho}\left(\rho\frac{\partial\psi}{\partial\rho}\right) - \left(\frac{1}{R_0} - \frac{\Delta'}{\rho} + \Delta''\right)\cos\vartheta\frac{\partial\psi}{\partial\rho} + \frac{1}{\rho}\frac{\partial^2\psi}{\partial\vartheta^2}\\ = -\mu_0 R^2 p' - TT'\end{aligned} \tag{19.89}$$

With the tokamak low β ordering the flux function ψ can be expanded as

$$\psi = \psi_0(\rho) + \psi_1(\rho, \theta) + \cdots \tag{19.90}$$

Separating by order in ϵ, indicated by the subscript, and in θ dependence, after some algebra one gets, to $O(1)$ and $O(\epsilon)$

$$\frac{1}{\rho}\frac{\partial}{\partial\rho}\left(\rho\frac{\partial\psi_0}{\partial\rho}\right) = -\frac{d}{d\psi_0}\left(\mu_0 R_0^2 p - \frac{T^2}{2}\right) \tag{19.91}$$

and the $O(\epsilon)$ equation is obtained

$$-2\Delta\frac{\partial^2\psi_0}{\partial\rho^2} - \left(\frac{1}{R_0} + \frac{\Delta'}{\rho} + \Delta''\right)\frac{\partial\psi_0}{\partial\rho} = -2\mu_0 R_0\rho\frac{dp}{d\psi_0} \tag{19.92}$$

Using the expression of the $O(1)$ poloidal field $B_{p0}(\rho) = \frac{1}{R_0}\frac{d\psi_0}{d\rho}$ one obtains:

$$\begin{aligned}B_p(\rho, \theta) &= \frac{\sqrt{g_{\theta\theta}}}{\sqrt{g}}\frac{\partial\psi}{\partial\rho} \approx \frac{1}{R_0}\left[1 - \left(\frac{\rho}{R_0} + \Delta'\right)\cos\theta\right]\frac{\partial\psi}{\partial\rho}\\ &\approx \left[1 - \left(\frac{\rho}{R_0} + \Delta'\right)\cos\theta\right]B_{p0}(\rho)\end{aligned} \tag{19.93}$$

consequently, the equation for the magnetic axis shift $\Delta(\rho)$ is:

$$\frac{d}{d\rho}\left[\rho B_{p0}^2(\rho)\frac{d\Delta}{d\rho}\right] = \frac{\rho}{R_0}(2\rho p' - B_{p0}^2) \tag{19.94}$$

This must be integrated with boundary conditions $\Delta(a) = 0$ leading to the shift of the axis of the nested magnetic surfaces:

$$\Delta' = \frac{\rho}{R_0}\left(\beta_p + \frac{1}{2}\ell_i\right) \tag{19.95}$$

where ℓ_i is the normalized *internal inductance*

$$\ell_i = \frac{2}{I_p^2(a)R_0}\int B_p^2 dV \tag{19.96}$$

The shift of the magnetic axis $\Delta(0) = \Delta_{\rm Sh}$ is called *Shafranov shift*. In presence of a conducting wall of nominal conformal radius b, an approximate expression is $\Delta_{\rm Sh} \approx \frac{b^2}{R_0}\left[\left(\beta_p + \frac{1}{2}\ell_i - \frac{1}{2}\right)\left(1 - \frac{a^2}{b^2}\right) + ln\frac{b}{a}\right]$. It does not depend on the plasma current, but on the particular form of the pressure and current profiles, consistent with the full Grad–Shafranov equation. To gain insight and handy practical estimates it is useful to test the extreme cylindrical limit with profiles of the type suggested by Wesson [15]:

$$p(\rho) = p_0\left[1 - \frac{\rho^2}{a^2}\right] \tag{19.97}$$

$$J_t(\rho) = J_0\left[1 - \frac{\rho^2}{a^2}\right]^\nu \tag{19.98}$$

Then, from Ampere law in the form

$$2\pi\rho B_p(\rho) = \mu_0 \int 0^\rho J_0\left[1 - \frac{\rho^2}{a^2}\right]^\nu \rho d\rho = \mu_0 I(\rho)$$

one obtains

$$B_p = B_{p0}\frac{1 - \left[1 - \frac{\rho^2}{a^2}\right]^{\nu+1}}{\rho/a} \tag{19.99}$$

and from $q(\rho) \approx \rho B_t/RB_p$

$$q = \frac{2(\nu + 1)B_t}{\mu_0 J_0 R}\frac{\rho^2/a^2}{1 - \left[1 - \frac{\rho^2}{a^2}\right]^{\nu+1}} \tag{19.100}$$

and then establish a link with the profiles peakedness ν through the expression [15]

$$\ell_i \approx \ln[1.65 + 0.89\nu] \tag{19.101}$$

and $q(a)/q(0) = \nu + 1$. In practical estimates it is often sufficient to use these cylindrical approximations, and actually to assign the $q(\rho)$ profile and express the current density from it through:

$$\mu_0 J(\rho) = \frac{B_t}{\rho R_0}\frac{d}{d\rho}\left[\frac{\rho^2}{q(\rho)}\right] = \frac{B_t}{R_0}\left[\frac{2q - \rho q'}{q^2}\right] \tag{19.102}$$

Finally, substituting β_p and ℓ_i into equation (19.93) the value of the poloidal field at the plasma boundary turns out to be:

$$B_p(a, \theta) = B_{p0}(a)\left[1 + \frac{a}{R_0}\left(\beta_p + \frac{1}{2}\ell_i - 1\right)\cos\theta\right] \tag{19.103}$$

where $B_{p0}(a) = \frac{\mu_0 I}{4\pi R_0}$. The complete solution requires continuous matching with the vacuum solution, which is a superposition of the magnetic flux generated by the plasma with that created by external currents:

$$\psi_p(\rho, \theta) \approx \psi_0(\rho, \theta) - \Delta\left(\rho\frac{\partial\psi_0}{\partial\rho}\right)\cos\theta \tag{19.104}$$

$$\begin{aligned}\psi_v(\rho, \theta) \approx{}& \frac{\mu_0}{2\pi}I_p R_0\left[\ln\left(\frac{8R_0}{\rho} - 2\right)\right] \\ &+ \frac{\mu_0}{2\pi}I_p\left[\rho\ln\left(\frac{8R_0}{\rho} - 1\right) + \frac{c_1}{\rho} + c_2\rho\right]\cos\theta\end{aligned} \tag{19.105}$$

where the constants c_1, c_2 are determined by the conditions

$$\psi_p(a, \theta) = \psi_v(a, \theta), \quad \psi_p'(a, \theta) = \psi_v'(a, \theta) \tag{19.106}$$

Introducing the quantity $\Lambda = \beta_p + \frac{1}{2}\ell_i - 1$ the constants turn out as

$$c_1 = a^2\left(\Lambda + \frac{1}{2}\right) \tag{19.107}$$

$$c_2 = -\left(\ln\frac{8R_0}{\rho} + \Lambda - 1/2\right) \tag{19.108}$$

From equation (19.104) the external field $B_{e,Z}$ necessary to maintain the equilibrium position in the major radius R direction is

$$B_{e,Z} = \frac{1}{R}\frac{\partial\psi_v}{\partial R} = -\frac{\mu_0 I}{4\pi R_0}\left(\ln\frac{8R_0}{\rho} + \Lambda - \frac{1}{2}\right) \tag{19.109}$$

The external vertical field, on the outer side $R > R_0$ of the torus adds to the plasma poloidal field which goes as $1/R$ and subtracts from it on the inner side $R < R_0$. A limit on the plasma β_p due to equilibrium considerations appears when the vertical field $B_{e,Z}$ compensates exactly the plasma generated poloidal field at some point (see figure 19.8) so that the total poloidal field vanishes on the inside

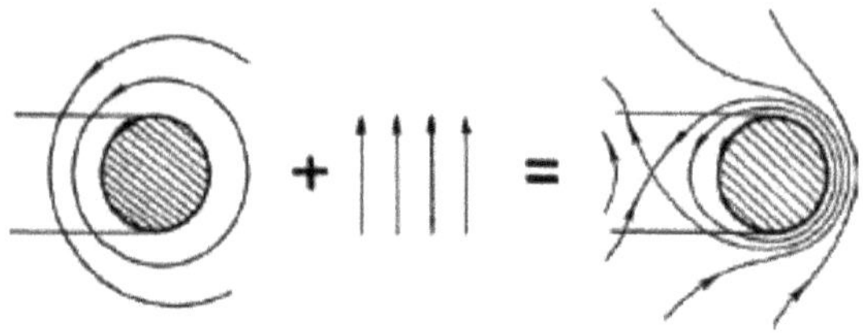

Figure 19.8. Effect of combination of vertical field B_Z and plasma poloidal field. Reprinted from reference [13] with permission. Copyright 1971 IOP, all rights reserved.

plasma boundary, $B_p(a, \pi) = 0$ and $\epsilon\Lambda \approx \epsilon\beta_p = O(1)$, bringing in the vessel the magnetic separatrix between the closed and the open magnetic surfaces [13, 22].

19.2.3 Plasma confined within a conducting shell

Any large displacement of current carrying plasma, magnetically confined within a vessel with conducting wall of thickness d is inhibited by image currents induced in the shell [23]. An equilibrium may last for a time span of the order of the diffusion time $\tau_B \propto L^2\sigma$ of the (poloidal) magnetic field in the metal liner of conductivity σ, where L is a characteristic scale length (thickness) of the conductor. For the estimate of L^2 and of the time constant consider a cylindrical shell of inner radius b and thickness d. An azimuthal (poloidal) image current density j_θ is related to the inductive electric field E_θ by Ohm's law and the latter is obtained from Faraday law:

$$-\frac{\partial B_Z}{\partial t} = (\nabla \times \mathbf{E}) \cdot \mathbf{e}_Z \approx \frac{E_\theta}{b} = \frac{j_\theta}{\sigma b} \tag{19.110}$$

$$\mu_0 j_\theta = (\nabla \times \mathbf{B}) \cdot \mathbf{e}_\theta \approx \frac{B_Z}{d} \tag{19.111}$$

$$-\frac{\partial B_Z}{\partial t} \approx \frac{B_Z}{\mu_0 \sigma b d} \tag{19.112}$$

Therefore, the relevant time scale obtained from equation (19.112) is $\tau_B = \mu_0 \sigma b d$. Such time scales and the associated depth of penetration of the time varying magnetic field in the metal shell modify substantially the boundary conditions for the unstable perturbations that can arise in the plasma. This has important effects for the successful control of the confined plasma. The subject requires a specialized treatment which is beyond the aims of this text.

19.2.4 Radial and vertical equilibrium

The vertical magnetic field equation (19.109) can be obtained also from the virial condition equation (19.76). Actually, in the large aspect ratio limit the latter gives a more precise condition for a stable radial equilibrium. The equation must be rewritten as:

$$\int_V dV\left[\left(p+\frac{B_i^2}{2\mu_0}\right)\nabla\cdot R\mathbf{e}_R-\frac{1}{\mu_0}\mathbf{B}_i\cdot\mathbf{B}_i\nabla R\mathbf{e}_R\right]+\int_V dVR\mathbf{e}_R\cdot\mathbf{J}\times\mathbf{B}_e=0 \quad (19.113)$$

A few elementary steps must be made to transform and combine the two term in the first integral:

$$\mathbf{B}_i\cdot\mathbf{B}_i\nabla(R\mathbf{e}_R)=B_{i,R}^2-B_{i,\phi}^2=B_i^2-B_{i,Z}^2 \quad (19.114)$$

Then, equation (19.113) becomes:

$$\int_V dV\left[2p+\frac{B_{i,Z}^2}{\mu_0}\right]=\int_V dVRJ_\phi B_{e,Z} \quad (19.115)$$

In the spirit of small ϵ and circular cross-section approximation one gets:

$$\int_V dVRJ_\phi B_{e,Z}\approx 2\pi R_0^2B_{e,Z}I \quad (19.116)$$

$$\begin{aligned}\int_V dV\left[2p+\frac{B_{i,Z}^2}{\mu_0}\right]&\approx 2\pi R_0\int_0^{2\pi}d\theta\left[\int_0^a\left(2p+\frac{B_{i,p}^2}{2\mu_0}\right)rdr\right.\\&\left.+\int_a^\infty\frac{B_{i,p}^2}{2\mu_0}\left(1+\frac{r}{R_0}\cos\theta\right)rdr\right]\\&\approx 4\pi^2R_0a^2\langle p\rangle+2\pi^2R_0a^2I^2\left(\frac{L_i}{2}+\frac{L_e}{2}\right)\end{aligned} \quad (19.117)$$

In the last step, the internal and external inductances L_i. L_e have been introduced, in relation to the poloidal magnetic energy. Equating equation (19.116) to equation (19.117) one obtains the expression of $B_{e,Z}$ in equation (19.109), however, in addition, a more precise result can be obtained. If the external field is uniform in R, the equilibrium of the plasma ring, is neutral with respect to displacements along the axis of symmetry. However, if the lines of force are slightly curved, a restoring force $F_Z=\mathbf{J}\times\mathbf{B}_e\cdot\mathbf{e}_Z$ will bring back the plasma ring to its initial position. The necessary field configuration is obtained superimposing a quadrupolar external field to the vertical one, which should decay with R as $(R/R_0)^{-n}$ [13], where $n>0$ is called decay index. If the magnetic axis position R_0 is displaced to $R_0+\delta R$ the equilibrium will be kept if the horizontal restoring force $F_R=\mathbf{J}\times\mathbf{B}_e\cdot\mathbf{e}_R$ increases more than the plasma expansion force related to equation (19.115). During a *virtual displacement* δR the total magnetic flux is assumed to be conserved.

$$\psi_{\text{tot}}=\psi_p+\int_0^{R_0}2\pi RB_{e,Z}dR=\text{const.} \quad (19.118)$$

A prescription on the field decay index follows from equation (19.115) and the constraint equation (19.118). In a $\beta_p=O(\epsilon)$ limit (with no loss of generality) after integration by parts and use of definitions of B_p and J_ϕ one has:

$$\int_V [2p + \frac{B_{Z,\,int}^2}{2\mu_0}]\, dV \approx \int_V \frac{B_p^2}{2\mu_0}\, dV + O(\beta) = -\frac{1}{2\mu_0}\int\int d\theta dr r\left(\frac{1}{r}\frac{\partial r B_p}{\partial r}\right)\psi_p \approx -\frac{1}{2}\psi_p(R_0)I \tag{19.119}$$

Then, combining equations (19.116) and (19.119) and dropping common factors it is convenient to consider the following function, with dimensions of work $[N \cdot m]$:

$$W = 2\pi R_0^2 B_{e,Z} I + \frac{1}{2}\psi_p I = 0 \tag{19.120}$$

A stable displacement δR, *against* flux conservation equation (19.118) implies $\delta\psi_p = -2\pi R B_{e,Z}\delta R_0$ and:

$$\frac{dW}{dR} > 0:\ 2\frac{dR_0^2 B_{e,Z}}{dR_0} - R_0 B_{e,Z} > 0 \tag{19.121}$$

The latter gives an important condition [11, 13] on the external magnetic field *decay index* $n = -\frac{R}{B_{e,Z}}\frac{dB_{e,Z}}{dR_0}$

$$0 < n < \frac{3}{2} \tag{19.122}$$

Furthermore, the field decay index n must be compared with the *critical value*

$$n_{\mathrm{cr}} = 2\left(\frac{\partial M_{\mathrm{Vp}}}{\partial Z}\frac{R_0}{\mu_0 \Lambda L_V}\right) \tag{19.123}$$

where M_{Vp} and L_V are the mutual plasma-vessel inductance and vessel self-inductance. When $|n|/n_{\mathrm{cr}} < 1$, in a perfectly conducting wall the plasma is stable for time scales shorter than the resistive wall time, and can be stabilized for longer time scales using feedback control. When $|n|/n_{\mathrm{cr}} > 1$, the vertical motion approaches the Alfvén velocity [32]. This constitutes a hot vertical disruption (VDE) scenario, where thermal energy is quenched *after* the loss of vertical control. Halo currents shared between the plasma and the vessel wall [33, 34], develop as the plasma squeezes poloidal flux or makes contact with the wall, stressing mechanically the conducting wall structures.

19.2.5 Shape of plasma meridian cross-section

Higher order multipolar fields, solution of the homogeneous Grad–Shafranov equation, control not only the position of the plasma, but also its shape, as shown below, and can be designed to follow in real time the tokamak discharge evolution, within controlled feedback loops. In particular at start-up, in the absence of plasma a multipolar field is required with a high order null on the expected magnetic axis location, to avoid prompt particle drifts to the wall and favor the pinch effect of the seed current channel. The solution of the equilibrium equation (19.72) for an assigned current source profile must be generally evaluated numerically, and several accurate and fast numerical Grad–Shafranov solvers have been produced [35, 36]. However, analytic solutions often afford more insight in the properties of an equilibrium than that provided by numerical codes. For instance analytic methods are particularly useful to derive scalings with the different geometric parameters (aspect ratio, elongation, triangularity) and can help in analytic stability and transport calculations. One historically important analytic solution of equation (19.72) is the Soloveev one, where the pressure and poloidal current profile functions are chosen as $dp/d\psi = -C$ and $TdT/d\psi = A$, equivalent to the straight pinch case [17, 37]. In order to illustrate the relation of the figures of merit with geometry, rather than displaying the standard solution, found in the literature [11] here it is instructive and of sufficient generality to adopt a constructive approach based on the example of reference [37], applied to Soloveev equilibria. With the unrestricting condition $A + C = 1$ equation (19.72) can be rewritten in normalized variables $x = r/R_0$, $z = Z/R_0$, $\bar{\psi} = \psi/\Psi_0$ as

$$x\frac{\partial\bar{\psi}}{\partial x}\left(\frac{1}{x}\frac{\partial\bar{\psi}}{\partial x}\right) + \frac{\partial^2\bar{\psi}}{\partial^2 x} = (1 - A)x^2 + A \tag{19.124}$$

A solution can be expressed as the sum of a particular solution and a solution of the corresponding homogeneous equation.

$$\bar{\psi} = \bar{\psi}_P + \bar{\psi}_H \tag{19.125}$$

The particular solution is

$$\bar{\psi}_P = \frac{x^4}{8} + A\left(\frac{x^2}{2}\ln x - \frac{x^4}{8}\right) \tag{19.126}$$

and the homogeneous part considered in the examples of reference [37] is a linear superposition of a finite number of terms $\bar{\psi}_k$, $k = 1, \cdots, 7$.

$$\begin{aligned}
\bar{\psi}_1 &= 1\\
\bar{\psi}_2 &= x^2\\
\bar{\psi}_3 &= z^2 - x^2\ln x\\
\bar{\psi}_4 &= x^4 - x^2z^2\\
\bar{\psi}_5 &= 2z^4 - 9x^2z^2 + 3x^4\ln x - 12x^2z^2\ln x\\
\bar{\psi}_6 &= x^6 - 12x^4z^2 + 8x^2z^4\\
\bar{\psi}_7 &= 8z^6 - 140x^2z^4 + 75x^4z^2 - 15x^6\ln x + 180x^4z^2\ln x\\
&\quad - 120x^2z^5\ln x
\end{aligned} \tag{19.127}$$

The unknown constants of the linear combination are determined from the boundary constraints on $\bar{\psi}$, as described below.

A magnetic surface in a tokamak is a surface of revolution generated by rotating the projection of a magnetic field line on the (R,Z) plane around the Z axis. The 'midplane' of the configuration is defined as the plane that passes through the magnetic axis and is perpendicular to the symmetric axis (Z axis).

The geometric shape of the projection of a magnetic surface $\psi(R, Z) = \text{const}$ on a meridian plane is conveniently represented, as in figure 19.9 in terms of a small number of parameters. For an up–down symmetric (about the midplane) magnetic surface an intuitive set of parameters is (R_{in}, R_{out}, R_{top}, Z_{top}), indicating the basic extremal points of the poloidal contour. However, it is more convenient to refer to the definitions:

$$R_0 = \frac{R_{\text{out}} + R_{\text{in}}}{2} \tag{19.128}$$

the major radius (machine axis) and

$$a = \frac{R_{\text{out}} - R_{\text{in}}}{2} \tag{19.129}$$

the minor radius of the tokamak. Note that the major radius R_0 of the last closed magnetic surface (LCMS) is usually different from R_{axis}, the R coordinate of the

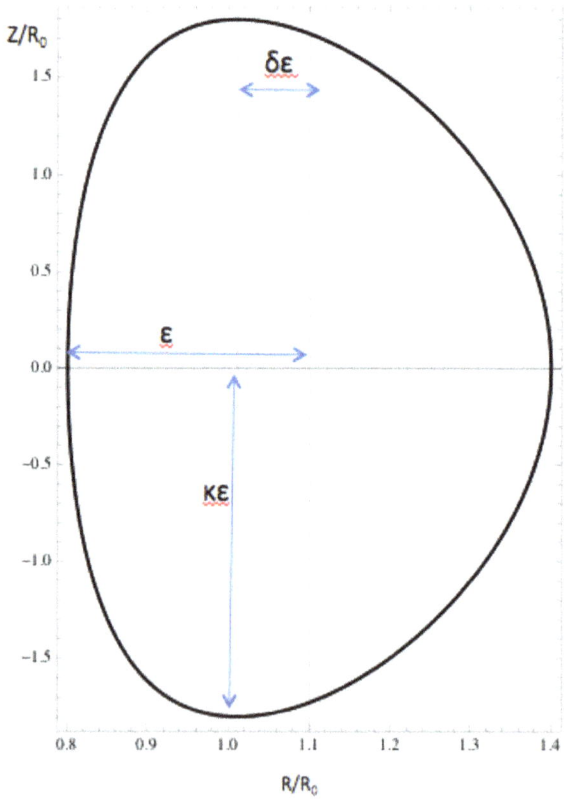

Figure 19.9. Key points of the meridian cross-section of a tokamak confinement surface.

plasma magnetic axis. Generally $R_{\text{axis}} > R_0$ due to the Shafranov shift, Δ_{Sh}. Further significant shape parameters are

$$\kappa_a = \frac{Z_{\text{top}}}{a} \tag{19.130}$$

the ellipticity, related to the elongation $E = \kappa_a - 1$, which vanishes for a circular cross-section, and

$$\delta_a = \frac{R_0 - R_{\text{top}}}{2} \tag{19.131}$$

the triangularity. In this form the parameters typically define the (LCMS), in terms of a very convenient parametric representation of the plasma surface [38]:

$$R = R_0 + \Delta_{\text{Sh}} + a\cos[\vartheta + \delta_a \sin(\vartheta)] \tag{19.132}$$

$$Z = \kappa_a a \sin(\vartheta) \tag{19.133}$$

equations (19.132) and (19.133) are a particular case of a very useful description of generalized magnetic surfaces, provided by a more general, rigorous *representation* [35] in terms of *radial* and *angular* flux variables, akin to a (truncated) Fourier expansion in orders of ϵ.

$$R(r, \vartheta) = R_{\text{axis}} + R_1(r)\cos(\vartheta) + R_2(r)\cos(2\vartheta) - \Delta_{\text{Sh}} + O(\epsilon^3) \tag{19.134}$$

$$Z(r, \vartheta) = Z_1(r)\sin(\vartheta) + O(\epsilon^2) \tag{19.135}$$

The subscripts indicate the ϵ^n order of the coefficients. With a generic dependence on the inner *radial* surface label, they describe the geometric shape of any internal magnetic (isobaric) surface. These geometric parameters are linked to dynamic properties of the equilibrium and through a systematic procedure can be related to properties of the internal current density and pressure profiles. The reader is advised to inspect the original papers for the details, which would be too tedious to be reported here.

The main quantities and the perimeter C_p of the LCMS contour are evaluated as

$$\langle B_{\text{pa}} \rangle = \frac{\oint B_p d\ell_p}{\oint d\ell_p} = \frac{\mu_0 I}{R_0 C_p} \tag{19.136}$$

$$C_p = \frac{\oint d\ell_p}{R_0} = 2\int_{1-\varepsilon}^{1+\varepsilon} [1 + (dz/dx)^2]^{1/2}\, dx \tag{19.137}$$

$$x = R/R_0,\ z = Z/R_0 \tag{19.138}$$

An important link exists between the figure of merit β_p (19.77) with the geometry of the equilibrium configuration, described by the parametric equations (19.132) and (19.133), through equation (19.137) [37, 39]:

$$\beta_p(R_0, a, \kappa, \delta_a) = \frac{2C_p^2}{\mu_0 I^2}\langle p \rangle \tag{19.139}$$

Therefore, equation (19.139) shows that the boundary shaping affects the possible β_p, but further bounds are imposed by general stability conditions. Furthermore, also for β it has been well established, mostly empirically and by numerical experiments, that, under a wide variety of circumstances, the maximum *total* β compatible with ideal MHD stability is bounded by a limit of the form [20]:

$$\beta \leqslant \beta_{\max} = A\frac{I}{aB} \tag{19.140}$$

where the proportionality factor A has been shown to depend on the geometrical characteristics of the plasma cross-section [40]. In other words, the universal character of equation (19.140) scaling of this upper bound is due solely to equilibrium constraints, and only the numerical value of A would be determined by the stability constraints.

In reference [37] it is shown that with the specification at four points of the C_p contour of the values of ψ and at three different points the values of ψ_{xx}, ψ_{zz}, it is possible to express the equilibrium configuration $\psi(R, Z)$ with the desired shape, in terms of the four figures of merit q_a, β, β_t, β_p

$$\begin{array}{ll}
\psi(1+\varepsilon, 0) = 0 & \text{outer equatorial point} \\
\psi(1-\varepsilon, 0) = 0 & \text{inner equatorial point} \\
\psi(1-\delta\varepsilon, \kappa\varepsilon) = 0 & \text{top point} \\
\psi_x(1-\delta\varepsilon, \kappa\varepsilon) = 0 & \text{top point maximum} \\
\psi_{zz}(1+\varepsilon, 0) = -\psi_x(1+\varepsilon, 0)K_1 & \text{outer equatorial point curvature} \\
\psi_{xx}(1-\delta\varepsilon, \kappa\varepsilon) = -\psi_z(1-\delta\varepsilon, \kappa\varepsilon)K_3 & \text{top point curvature}
\end{array} \tag{19.141}$$

Here the curvatures K_j, $j = 1, 2, 3$ are expressed through the geometric parameters

$$\begin{aligned}
K_1 &= -\frac{(1+\alpha_a)^2}{\varepsilon\kappa^2} \\
K_2 &= -\frac{(1-\alpha_a)^2}{\varepsilon\kappa^2} \\
K_3 &= -\frac{\kappa}{\varepsilon\cos^2(\alpha_a)}
\end{aligned} \tag{19.142}$$

with $\alpha_a = \arcsin(\delta_a) \propto \beta_p$, $\delta_a < 0.841$ [37].

A similar procedure can be applied to plasma configurations with a divertor configuration having a separatrix boundary, going through a so-called X-point, which prevents the hot core plasma from contacting the vessel (or *plasmapfacing components* (PFCs)). The cold, not confined plasma layer between the separatrix and the vessel is the SOL, where many processes of recycling occur and have important effects on determining the boundary conditions for thermal energy outward transport, and impurities inflow. At the X-point the second derivative curvature constraint must be replaced by the requirement that both the tangential and normal magnetic field vanish. The conditions at the inboard and outboard equatorial points are left unchanged. The procedure then is to choose and apply the parameters corresponding to the 95% flux surface and replace conditions equation (19.141) by

$$\begin{aligned}
&\psi(1+\varepsilon, 0) = 0 && \text{outer equatorial point}\\
&\psi(1-\varepsilon, 0) = 0 && \text{inner equatorial point}\\
&\psi(R_{\rm sep}, Z_{\rm sep}) = 0 && \text{top point}\\
&\psi_x(R_{\rm sep}, Z_{\rm sep}) = 0 && \\
&\psi_z(R_{\rm sep}, Z_{\rm sep}) = 0 && \\
&\psi_{zz}(1+\varepsilon, 0) = -\psi_x(1+\varepsilon, 0)K_1 && \text{outer equatorial point curvature}\\
&\psi_{zz}(1-\varepsilon, 0) = -\psi_x(1-\varepsilon, 0)K_2 && \text{inner equatorial point curvature}
\end{aligned} \tag{19.143}$$

An illustration is given in figure 19.10.

This procedure is useful for a preliminary approach to the design of a tokamak device, and of various diagnostics issues, subsequently to be carried out with full numerical solution of the MHD equations, possibly coupled with plasma kinetic description. There are some invariance properties of the Grad–Shafranov equation through which it is possible to generate a family of solutions, with the same shape of the LCMS, starting from a known solution. Given an equilibrium with $\psi(R, Z)$, $p(\psi)$, $T(\psi)$, and a constant factor s, positive or negative, it can be verified that a transformation τ, $\psi(R, Z)_\tau = s\psi(R, Z)$, $p_\tau = s^2 p(\psi)$, and $T_\tau = \pm sT(\psi) + C$ is also a solution to the Grad–Shafranov equation. In this case, both the B_p and B_t

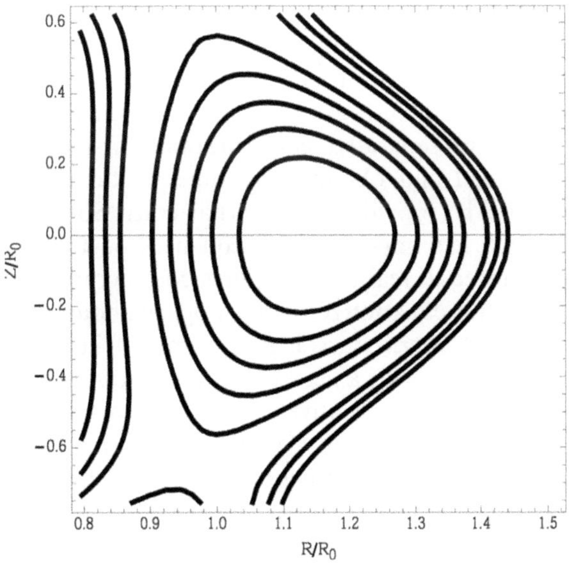

Figure 19.10. Example of equilibrium design, with the procedure of the section on shaping.

magnetic fields are increased by the factor s, and thus the safety factor remains unchanged. The pressure is increased by s^2 factor and thus the value of β remains unchanged. From $T_\tau = \pm sT(\psi)$, it appears that the direction of the toroidal magnetic field can be reversed keeping the equilibrium. Since s can be negative, the direction of the toroidal current can also be reversed without breaking the force balance. Another transformation is obtained by choosing $\psi_\tau = \psi$, $p_\tau = p(\psi) + C$, and $T_\tau^2 = T^2(\psi) + C$. The scaled expression is still a solution to the Grad–Shafranov equation because $T_\tau T_\tau' = TT'$. This scaling keeps the pressure, the poloidal field and β_p unchanged. This transformation modifies the toroidal field and thus can be used to generate a series of equilibria with different profile of safety factor $q(\psi)$.

19.2.6 Shape and boundary conditions

In the design of experimental configurations, boundary conditions must be specified to obtain $\psi = \text{const}$ equilibria that meet the operative requirements. Several alternative concepts must be considered: the simplest, albeit idealized case is the *fixed boundary*: the plasma-vacuum boundary Γ is replaced by the surface of a perfect conductor and only the plasma region is calculated, using as a plasma boundary condition $J_\phi|_\Gamma = 0$ such that at this boundary $\psi - \psi_{\text{vac}} = 0$. A more realistic case is the *free boundary* one: a *nail* value of the poloidal flux ψ is specified on a closed curve, or on a set of points generally in the vacuum region, and corresponding to the material limiter designed to take also the outflowing heat load. The boundary of the solution of the Grad–Shafranov equation for ψ, is forced to touch the limiter beyond which $J_\phi$vanishes and there is a scrape-off plasma very tenuous, but carrying, however, some force-free *halo* current. A further, important class of boundaries comprises the plasma fully detached from the vessel first wall, and bounded by a magnetic separatrix, which is characterized by one or more *hyperbolic X points* where the poloidal field $\mathbf{B}_p$ vanishes. The plasma region beyond the separatrix and the vessel constitutes the divertor region, which must be equipped with additional pumping and fresh gas fuelling systems, as well as important heat load exhaust systems. The separatrix configurations, first tested on the ASDEX device [4–7], are designed and obtained by superposition of particular vacuum multipolar magnetic fields, provided by external coils. So in general the plasma shaping problem, which has important consequences on the device performance, and must be often adjusted in real time, requires consideration also of the vacuum solution of the poloidal flux equation $\Delta^*\psi = 0$. A plasma boundary can be defined by contact with a material *limiter*, consisting of a special heat bearing metallic (or carbon) structure, formed either by a nearly continuous toroidal rail (with insulating gaps) or discreet *mushrooms*. Alternatively, it is defined by a *magnetic separatrix* going through a chosen X point, with a magnetic divertor throat ending on heat bearing target plates. As the development of tokamak devices approaches more closely and reliably the technical conditions required for confinement of a burning plasma, it appears clearly that optimization of the plasma performance requires improved understanding and ultimately better control of the interface conditions between plasma and first material wall. The idealized boundary conditions of the

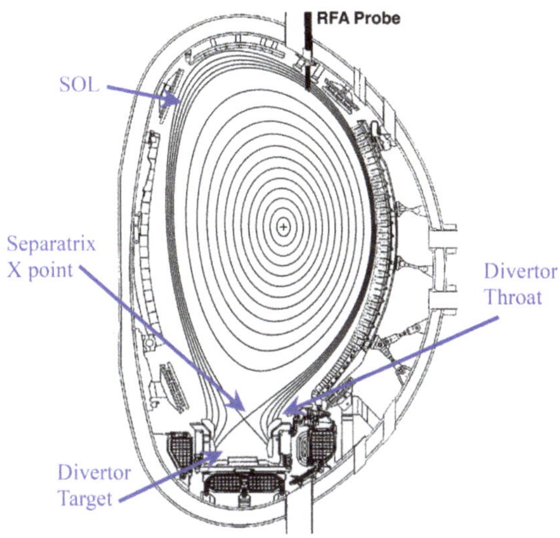

Figure 19.11. Principle of divertor: indication of Plasma Separatrix, X point, SOL, Divertor throat and dumping targets. Reproduced from reference [41] with permission. Copyright 2006 IOP, all rights reserved.

MHD equilibrium configuration and of the more detailed conditions on transport fluxes in the relevant collisionality regimes, should be extended to consider actually as more realistic boundary conditions the diffuse interface with a composite plasma in the SOL, consisting of multiple ion species and heterogeneous meso-particles with variable charge and mass. The complex dynamics of this region is schematized in figure 19.11.

19.3 Direct and inverse problems

Here a concise summary is given of the statement and conceptual steps for the solution of both *direct* (i.e. constructive) and *inverse* (i.e. diagnostic) problems. The subject is too demanding to be included in this context and for detailed information the reader is addressed to the publications suggested. The axisymmetric plasma equilibrium is described by the surfaces $\psi(R, Z) = \text{const}$, solution of the Grad–Shafranov equation:

$$\Delta^{*}\psi = R^2\nabla \cdot \left(\frac{1}{R^2}\nabla\psi\right) = -\mu_0 RJ(R, \psi) \tag{19.144}$$

$$J(R, \psi) = R\frac{dp}{d\psi} + \frac{T}{\mu_0 R}\frac{dT}{d\psi} \tag{19.145}$$

with boundary conditions and specification of undetermined functions $p(\psi)$, $T(\psi)$ This equation, in the absence of a steady flow, is elliptic. The fundamental *direct problem* can be formulated in the form:

For an assigned current density $J(\psi, R)$ and total current I_p find the field $\psi(R, Z)$ solution of equation (19.144), within a fixed or free domain, fulfilling Dirichlet boundary conditions on an external contour $\Gamma(s)$.

Such a direct problem is *well posed* and its solution responds continuously to the boundary data and is valid in the whole domain within $\Gamma(s)$.

One type of inverse problem can be cast in the form:

Given N data $d_i(\psi)$, $i = 1 \ldots N$ corresponding to N functionals $F_i(\psi)$, $i = 1 \ldots N$, find the field $\psi(R, Z)$ solution of equation (19.144), and the current profile $J(\psi, Z)$ within a fixed or free domain, matching the interface and boundary conditions.

The solution is expected, by other diagnosticians, to also be compatible with other available inputs, thought to be functionals of ψ, but not always available and, strictly, not belonging to the Grad–Shafranov equation. This kind of problem requires the solution of the elliptic Grad–Shafranov equation with both Dirichlet and Cauchy boundary conditions and is ill posed [35, 36, 42]. Small variations in the data at the boundary propagate as large variations in the solution, far from the boundary. The functions $p(\psi)$, $T(\psi)$ should fit at best *experimental* $P = nT$, MSE, interferometry, diamagnetic measurements, etc.

The undetermined p and T functions can be expressed in terms of suitable basis functions $p' = \sum_i \alpha_i \phi_i$, $TT' = \sum_i \gamma_i \phi_i$ with a vector of fitting parameters $u_i = (\alpha_i, \gamma_i)$.

Labeling with c and m the calculated and measured quantities, respectively, the square discrepancy of model results $F_i(\psi)$ from data d_i leads to define a constrained cost function:

$$
\begin{aligned}
\chi(\psi, \mathbf{u}) = & \frac{1}{2} \sum_{k=1}^{N} \left| \frac{1}{R} \frac{\partial \psi}{\partial \nu}_k - \tilde{b}_k \right|^2 \\
& + \frac{1}{2} \sum_{k=1}^{M} w_F \left| c_F \int_{L_k} \frac{n}{R} \frac{\partial \psi}{\partial R} d\ell - \alpha_{F,k} \right|^2 \\
& + \frac{1}{2} \sum_{k=1}^{M} w_n \left| \int_{L_k} n d\ell - \bar{n}_{m,k} \right|^2 \\
& + \frac{1}{2} w_p \left| \int_0^1 p_c d\ell - \int_0^1 p_m d\ell \right|^2 + J_\varepsilon(\mathbf{u})
\end{aligned}
\qquad (19.146)
$$

with J_ε a Tikhonov regularizing term, and w_k are weights related to the variances of the measurements. Typically the data $\tilde{b}_i$ are the i_{th} values of the magnetic field, i.e. the values of the normal derivative of ψ measured by N pick-up coils on a contour surrounding the plasma (i.e. the vessel contour); $\bar{n}_{m,k}$ the line integrated density on the kth interferometer channel, and $\alpha_{F,k}$ the line integrated Faraday rotation angles on the kth interferometer channel. The constrained minimization problem then is:

[36, 42] *find* **u** *minimizing* χ^2 *with* $\psi(R, Z)$ *satisfying G-S equation with its boundary conditions*

$$\chi^2(\psi, \mathbf{u}) \rightarrow \text{minimum} \tag{19.147}$$

$$\Delta^*\psi = -\mu_0 RJ(R, \psi|\mathbf{u}) \tag{19.148}$$

with Dirichlet boundary conditions for the flux $\psi_\Gamma = \psi_n$, measured by N toroidal loops coaxial with the vessel. The nonlinear best fit of the data requires defining an *adjoint* problem for a function $\Theta(R, Z|\mathbf{u})$ which plays the role of generalized Lagrange multipliers. For the interested reader, details of the formulation of the inverse problem and the possible numerical implementations, can be found in the specialized literature [35, 36]. The development of ever faster computers has allowed fast and accurate reconstruction of the equilibrium acquiring the periferic magnetic signals during the discharge. Recently, good performance has been reached also in real-time equilibrium reconstructions, which can be coupled to the experiment feedback system. As a historical example, in figure 19.12 the first detection is shown of the natural formation of magnetic separatrix in JET [4–7], at the end of a discharge. The formation of a *double null* magnetic separatrix, occurred naturally as an effect of high $\beta_p \propto I_p^{-2}$ and of the hexapolar leakage magnetic field due to the iron core shape of the JET device.

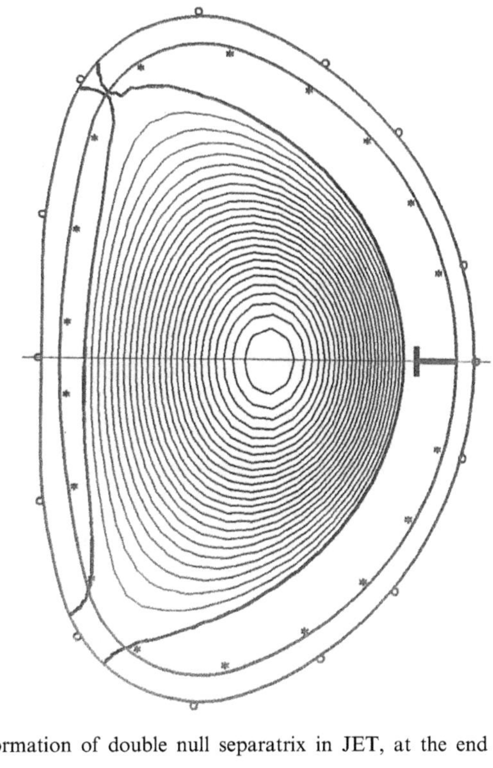

Figure 19.12. Natural formation of double null separatrix in JET, at the end of a discharge, as a result of high β_p.

From this first observation in 1984, the machine was subsequently upgraded to have a divertor configuration, at any I_p, controlled by the poloidal coils system. Since then, access to the H-mode confinement regime and control of impurities influx has become routine.

19.3.1 Tokamak equilibrium with flow

A final extension of the *static equilibrium configuration* is worth quoting. In the presence of a steady equilibrium flow, the system equation (19.24) is extended in the form

$$\rho_m(\mathbf{v} \cdot \nabla\mathbf{v}) = \mathbf{J} \times \mathbf{B} - \nabla p \tag{19.149}$$

$$-\nabla\Phi + \mathbf{v} \times \mathbf{B} = 0 \tag{19.150}$$

$$\nabla \cdot (\rho_m \mathbf{v}) = 0 \tag{19.151}$$

where the second equation represents Ohm's law, with Φ the electrostatic potential. Due to axisymmetry and the incompressibility condition, the **B** field and the mass density flow can be written in the mixed notation as:

$$\mathbf{B} = T\nabla\phi - \nabla\phi \times \nabla\psi \tag{19.152}$$

$$\rho_m \mathbf{v} = \Theta\nabla\phi - \nabla\phi \times \nabla F \tag{19.153}$$

Here T, F, Θ, Φ are all functions of the poloidal flux $\psi(R, Z)$, therefore, are constant on magnetic surfaces. Using expressions equations (19.152) and (19.153) in equations (19.149)–(19.151) with straightforward algebra one obtains:

$$\Phi' = \frac{1}{\rho_m R^2}[TF' - \Theta] \tag{19.154}$$

The projection of the momentum balance equation (19.149) along the $\nabla\phi$ direction, yields another surface quantity, from which a final expression for the flow velocity follows:

$$\begin{aligned} X(\psi) &= T\left[1 - \frac{(F')^2}{\rho_m}\right] + R^2 F'\Phi' \\ \mathbf{v} &= \frac{F'}{\rho_m}\mathbf{B}\nabla\phi - R^2\Phi'\nabla\phi \end{aligned} \tag{19.155}$$

The projection of the momentum balance equation (19.149) along **B** vanishes, and equations (19.154) and (19.155) yield the following relations:

$$\mathbf{B} \cdot \left[\frac{1}{2}(\nabla v^2) + \frac{\Phi'}{\rho_m}(F'T - \Phi'\rho_m R^2) + \frac{\nabla p}{\rho_m}\right] = 0 \tag{19.156}$$

Finally, the last equation (19.156) can be rewritten, using equation (19.154), as a generalized Bernoulli equation

$$\mathbf{B} \cdot \nabla \left[p + \rho_m \left(\frac{v^2}{2} + \Phi' \frac{\Theta}{\rho_m} \right) \right] = 0 \tag{19.157}$$

The quantity $P_s(\psi) = p + \rho_m \left(\frac{v^2}{2} + \Phi' \frac{\Theta}{\rho_m} \right)$ is a surface function; it is convenient to introduce the *poloidal* Mach number $M^2 \equiv \frac{v_p^2}{v_A^2} = \frac{(F')^2}{\rho_m}$ and write the projection of the momentum balance equation in the $\nabla\psi$ direction, obtaining the Grad–Shafranov equation generalized with the presence of a stationary toroidal plasma velocity $v_\phi = \Theta / R\rho_m$:

$$\begin{aligned} &(1 - M^2)\Delta^*\psi - \frac{(M^2)'}{2}|\nabla\psi|^2 + \frac{1}{2}\left(\frac{X^2}{1 - M^2} \right)' \\ &+ R^2 \left(P_s - \frac{XF'\Phi'}{1 - M^2} \right)' + \frac{R^4}{2}\left[\frac{\rho_m(\Phi')^2}{1 - M^2} \right]' = 0 \end{aligned} \tag{19.158}$$

Here Δ^* is the Beltrami operator of equation (19.73). In the subsonic cases $M^2 \ll 1$ equation (19.158) remains elliptic. Comparison of equations (19.158) and (19.73) in the subsonic range, leads to identify the current density in presence of rotation:

$$J_\phi = -\frac{1}{2\mu_0 R}(X^2)' + R[P_s - XF'\Phi')' + \frac{R^3}{2}\left(\rho_m(\Phi')^2\right]' \tag{19.159}$$

The formulation of the equilibrium problems remain unaltered, with the proviso, that the (toroidal) equation produces a centrifugal force effect, akin to the pressure term.

19.4 Principal technical elements of a tokamak

The mathematical properties of the *ideal* axisymmetric equilibrium configurations briefly summarized above, have found a surprising, and in my view, most elegant application in the real tokamak machine. It is worth remarking that the physics of an actual device, while fully embodying the first principles, adds complexity and new phenomena that lead to the need to develop several specialized studies, ranging from high power electrotechnics, radiofrequency high power sources, material science, control theory and techniques to advanced diagnostic science, which will be quoted and referred to when necessary. In a tokamak the *essential* poloidal field is generated by a high current induced in the plasma by external primary circuits, fed by a suitable power supply system, generally operating in a pulsed regime.

In figure 19.5, it is shown that the vessel containing the plasma is surrounded by several large coils wherein a poloidal current flows, producing a static*toroidal* magnetic field B_ϕ, which is the most intense in the tokamak device. The largest amount of electromagnetic energy $E_B = \frac{1}{2}L_c I_c^2$ is stored in this coil system (here the subscript c stands for *coils*), and in the case of technical failure it releases huge, destructive forces on the torus assembly. Around the symmetry axis Z there is the coil of the primary circuit; in the simplest arrangement, for pulsed operation, 'breaking' the primary and crowbarring on resistors loads provokes a variation of the coupled magnetic flux driving an elecromotive force in the toroidal vacuum vessel. More precisely, the vector potential toroidal component generated by the central inductor is $A_\phi = \frac{1}{2\pi R}\psi$ and because of axisymmetry the driven toroidal electromotive force is $V_{\text{loop}} = 2\pi R E_\phi = -d\psi/dt$. This electric field provides the initial gas breakdown ionization, and by driving the current in the ionized gas, generates the poloidal field B_p and the associated pinch effect which initiates the confinement and heating of the plasma. Before a plasma current channel is formed, the magnetic field consists of externally applied toroidal and multipolar components: $\mathbf{B} = \mathbf{B}_\phi + \mathbf{B}_m$. The multipolar field must have a sufficiently high order null on the axis where the current channel formation is desired, as shown in figure 19.13; here, in the (R, ϕ, Z) coordinate frame, with $\rho = R - R_0$, $\theta = \tan^{-1}\frac{Z}{R - R_0}$ as an example, we consider a hexapolar field $\mathbf{B}_m = B_{\text{hex}}\frac{\rho^2}{a^2}(\cos(2\theta)\mathbf{e}_R - \sin(2\theta)\mathbf{e}_Z)$, where R_0 and a are major and minor radius of the tokamak vessel. By surface averaging over an area $S = \pi\rho^2 < \pi a^2$ surrounding the null point, an estimate of the effective L_q length is $L_q \approx 4\frac{B_\phi}{\langle|\nabla B_m|\rangle}$.

The coils mounted concentrically to the plasma vessel produce an additional magnetic field, essentially directed vertically along the Z axis. This is basically the field, expressed by equation (19.109), necessary to ensure and control the radial and vertical position of the plasma loop and to control or modify the shape of the meridian cross-section of the plasma. The shaping is particularly difficult and

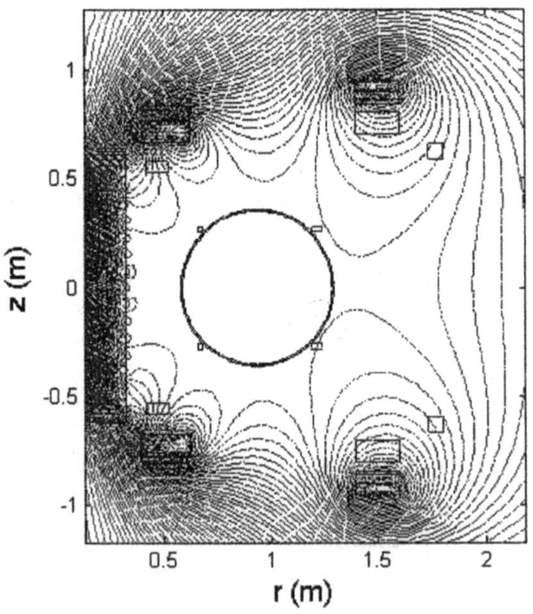

Figure 19.13. Typical start-up multipolar field with a high order null in the center of the vessel.

important in the divertor region, which has to fulfill a variety of tasks. Several years of research on fusion oriented tokamaks has led to selecting an axisymmetric poloidal magnetic X-point divertor as the suitable design for a tokamak plasma-first wall interface. Basically, a null point (actually a line) of the poloidal magnetic field $\mathbf{B}_p$, called conventionally *X-point* is created between the plasma current and a special external coil. A magnetic separatrix boundary is formed, dividing the plasma in the tokamak vessel into a closed field line region (confined plasma) and an open field line SOL region. The SOL field line bundle is *diverted* into a special divertor chamber and terminated on a target plate. This configuration enables energy and particles lost from the confined core plasma, as a result of radial transport and MHD instabilities (e.g., edge localized modes (ELMs)) to flow to the divertor chamber that acts as a separate plasma material interface. Four main tasks are accomplished by the standard poloidal X-point divertor [43]: (1) plasma power exhaust; (2) particle control (D/T and He pumping); (3) control of impurity production (source); (4) impurity shielding by the SOL plasma. A useful design for the divertor must primarily aim at spreading on a sufficient area the plasma heat load passing through the SOL. The amount of conducted power crossing the LCMS and entering the SOL was denoted as P_w in the first section. It is worth recalling that it is the sum of power from auxiliary (P_{aux}), ohmic (P_{ohm}) and alpha heating (P_α) less the core radiated power (P_{rad}). The last one in large part depends on estimates of core plasma impurity content which are based on the existing tokamak database, but have large uncertainties. Therefore, for the design of the divertor targets, an upper limit for P_w must be set, below the target material heat bearing capability. This opens a universe of research in material science, which is still underway. In recent years, a *fashionable* design is that of a *snowflake* [44] configuration with multiple X points and long connection length to the target plates and new experiments are in preparation [2, 3, 45]. In the *experimental* devices built up to now, the operation is pulsed with a duration ranging from few ms to several tens of seconds [4–7, 46], because it is impossible to vary monotonically for an indefinite time the magnetic flux stored in the primary. The unavoidable transient regimes of the real devices produce a variety of important (generally adverse) phenomena not described by the ideal equilibrium scheme outlined above. Furthermore, in an actual device, dissipative effects such as electrical conductivity and fluid viscosity, as well as *finite* Larmor radius effects, impair the ideal *freezing* mechanism tying the plasma on the magnetic field, thereby setting limiting time scales to operationally useful confinement. An uncontrolled progress of these basic and other more sophisticated *imperfections* and non-ideal aspects of the real plasma confinement configurations lead to serious problems for the stability of the system, with losses of energy and particles, up to the eventual collapse and *disruption* of confined plasma.

19.5 Plasma formation

19.5.1 Poynting theorem

The transfer of electromagnetic energy necessary for the start-up and drive of the plasma current from the main external primary coil (and other coupled coils), is governed by the Poynting vector $\mathbf{S} = \mathbf{E} \times \mathbf{B}$ flow. Since the poloidal and toroidal fields are decoupled and the latter is essentially in steady state the relevant equation is [23]:

$$\frac{\partial}{\partial t}\frac{B_p^2}{2\mu_0} = \nabla \cdot \left(\mathbf{E} \times \frac{\mathbf{B}_p}{\mu_0}\right) = J_\phi E_\phi \tag{19.160}$$

Integration of each term of equation (19.160) over the toroidal plasma volume gives:

$$\int_V \frac{\partial}{\partial t}\frac{B_p^2}{2\mu_0}, dV = \frac{\partial}{\partial t}\left(\frac{LI_t^2}{2}\right) \tag{19.161}$$

$$\int_V \nabla \cdot \left(\mathbf{E} \times \frac{\mathbf{B}_p}{\mu_0}\right) dV = \int_S (\mathbf{E} \times \mathbf{B}_p/\mu_0) \cdot \mathbf{dS} = -\frac{d\psi}{dt} \oint \mathbf{B}_p \cdot d\ell \tag{19.162}$$

where L is the total torus inductance. With a basic Ohm's law $E_\phi = \frac{J_\phi}{\sigma_\|}$, where $\sigma_\| = 1.96\eta_\perp^{-1}$, eventually the *circuit equation* is written:

$$\frac{d}{dt}\left(\frac{LI_t^2}{2}\right) + \int_V \frac{J_\phi^2}{\sigma_\|} dV = V_{\mathrm{loop}} I_t \tag{19.163}$$

where V_{loop} is the electromotive force induced by the primary transformer coil.

19.5.2 Start-up and current ramp-up

Here as an example, we consider the principal operation of a tokamak machine with an Ohmic current driven by an *inductive storage* pulsed power supply, which is still the most widely used in experiments, although its intermittency is unsuited for future fusion plant, aiming at continuous energy production. The Ohmic current source is actually a huge transformer, often with an air cored primary, but in some notable examples such as JET with an iron core. The primary is designed to couple its swinging magnetic flux to the secondary, represented by the plasma loop. Before the plasma discharge is initiated in the tokamak vessel, where a high vacuum condition has been prepared, the toroidal field is raised up to its required value and the primary inductor coil (here labelled P_1) is *charged* to its maximum current-turn value $N_1 I_{P_{1,0}}$. As the gas prefilling inlet valve opens, the coil P_1 is disconnected from the flywheel generator and shorted on the crowbar resistor, R the current decays as $I_{P_{1,}} = I_{P_{1,0}} e^{-tL/R}$, where L is the primary self-inductance. The inductively driven electric field E first ionizes the prefilled gas, at a pressure p_g and then drives a current pinch in the region of the multipole field B_m null. The plasma formation phase is *similar* to the Townsend avalanche process and for its success several ionizations must occur before electrons drift to the wall. This implies that the ionization length be much smaller than the connection length

$$\lambda_i \approx \frac{2 \times 10^{-3}}{p_g} e^{-1.3\times 10^4 p_g/E} \ll L_{\mathrm{conn}} \tag{19.164}$$

this shows that high B_t and p_g are favorable for this phase. The rise of the current depends on the type of applied loop voltage wave form, ranging from simple primary circuit breaking to feedback control of the current or the rate of rise: the different operation leads to different matching conditions with the initial, 'spontaneous' avalanche phase. The plasma current evolution in the inductive phase can be described by a single, lumped parameters, nonlinear equation

$$\frac{dI_t}{dt} + \frac{R_p(I_t)}{L}I_t = \frac{V(t)}{L} \tag{19.165}$$

where L is the discharge self-inductance and $R_p = R_p(Z_{\text{eff}}, I_{\text{ind}})$ the plasma resistance, associated with a matching condition with the avalanche current $I_{\text{av}}(t)$ at some time $t = t_m$. The avalanche value, indicated as I_0, can be estimated adapting the Townsend model [47].

As an illustration of the time behavior of the current in the breakdown and inductive phase, the time waveforms of a real FTU discharge are displayed in figure 19.14 illustrating the typical working of an inductively driven tokamak.

19.5.3 Toroidal coils

Contemporary tokamaks, especially those oriented to high performance, and burning plasma regimes [45] have toroidal coils with a D shape. Most of the early machines such as T3, but even important ones such as TFR, TFTR, FTU, Alcator, had circular shaped toroidal coils. The reason for the change of shape, toward a technically more demanding design, is due to electromechanical reasons that are strictly similar to the physical limits on the plasma, concerning the β_p relation with plasma cross-section geometry and stability limits. We summarize here the basic

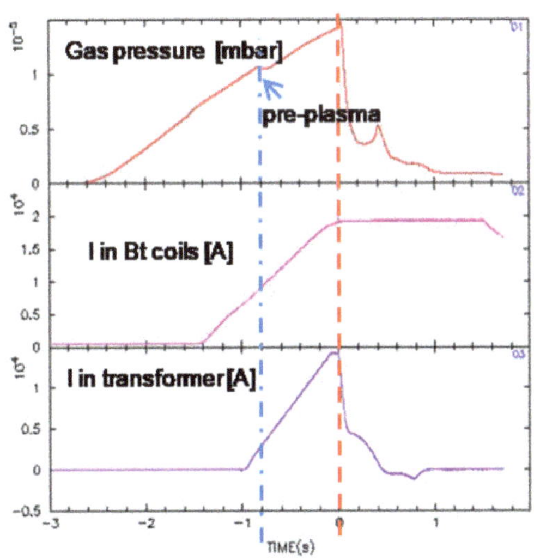

Figure 19.14. Start-up of the FTU tokamak: (top) Wave form of gas pressure inlet; (middle) wave form of current in toroidal B coils; (bottom) wave form of current in transformer (primary) coil.

principle for its instructive and appealing closeness to what happens to the confined plasma. Consider the coil conductor pack as a beam, clamped at either end and subject to a load BI_c normal to it. It will be stable in position if it is not subject to any bending moments, and therefore is under pure constant tension. This condition is specially favorable for superconducting conductors, since large forces will not be transmitted through the thermal and electrical insulation reducing the risk of cracking. Under pure constant tension T, a flexible conductor will take the shape of a curve of radius of curvature, ρ_c, such that the side force per unit length of conductor be $F_n = T/\rho_c \propto BI_c$, where B is the self-generated magnetic field and I_c the current in the conductor. Since the toroidal field, varies inversely as the radius, R, from the axis of symmetry, the coil is subject to a hoop force, similar to the effect on the plasma, which has a net resultant directed toward the axis of symmetry, i.e. the *central pillar*, or better, the central vertical inductor solenoid. This is actually a good feature from the point of view of the statics of the assembly design, as the centering force is taken up (symmetrically) by the inner cylindrical structural element. Thus the toroidal coil contour in contact with it must be tangent and straight. Elsewhere the coil contour should have a radius $\rho_C \propto R$. Therefore, from standard formulae of differential geometry, the contour of the conductor in the (R,Z) plane can be found solving;

$$\begin{aligned} \rho_C &= kR \\ \rho_C &= \pm\left[1 + \left(\frac{dR}{dZ}\right)^2\right]^{3/2} \Big/ \frac{d^2R}{dZ^2} \end{aligned} \tag{19.166}$$

A first integration can be performed and used to impose conditions on the equilibrium curve:

$$\begin{aligned} \wp &= \frac{dR}{dZ} \\ \wp\frac{d\wp}{dR} &= \frac{\pm 1}{kR}[1 + \wp^2]^{3/2} \end{aligned} \tag{19.167}$$

The solution $Z(R)$ which is tangent to the central core for $R = R_c$ must have

$$\frac{dR}{dZ} = \pm\frac{[k^2 - \ln^2(R/R_c)]^{1/2}}{\ln(R/R_c)} \tag{19.168}$$

So $\frac{dR}{dZ} = 0$ at $R_1 = R_c e^k$ and $R_2 = R_c e^{-k}$. Furthermore for chosen R_2 and R_1, $k = \frac{1}{2}\ln(R_2/R_1)$ and the contour shape is described by the formula, which can be evaluated numerically

$$Z(R) = \int_0^{R/R_c} \frac{\pm\ln(R/R_c)}{(k^2 - \ln^2(R/R_c))^{1/2}} dR \tag{19.169}$$

Depending on the parameter k, the result is a family of D shapes (see figure 19.15). In the design of these coils attention must be given to body forces perpendicular to

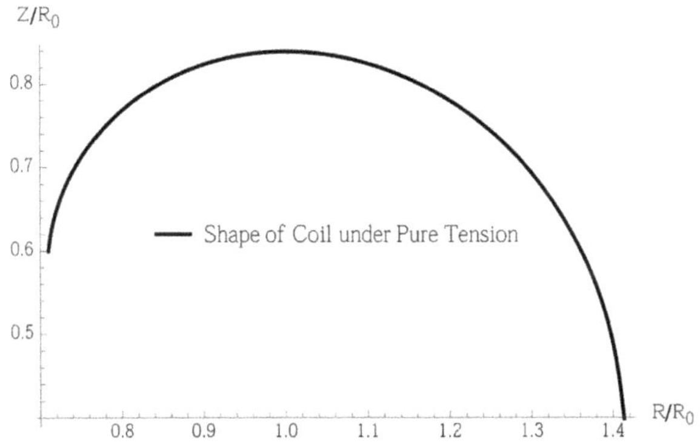

Figure 19.15. Example of calculation of toroidal coil shape from equation (19.169).

the toroidal field coil plane arising from the B_p field interaction with the coil current I_c. These forces tend to move each coil pancake out of its plane. They act cyclically and vary in strength following the discharge evolution. These forces produce overturn moments, M_R, torque, M_Z, and a toroidal force, F_t, to be taken up by the toroidal assembly structure [48]; the bounds on the mechanical stresses have consequences on the operational lower q_a and upper β_p limits.

19.6 Similarity principles applied to tokamaks

A key question for the development of reactor tokamaks is the possibility of scaling from the devices used in the phase of scientific research, up to industrial facilities capable of reaching the objectives of adequate and reliable performance. On the score of long established criteria to characterize physical systems in terms of dimensionless parameters [49], and the assessments of fundamental physics, several studies have outlined a number of criteria of similarity for tokamaks based on scale transformations which leave invariant the governing plasma equations [50–53]. Here a brief summary of the underlying concepts is given. For an idealized tokamak, with no impurities ($Z_{\text{eff}} = 1$) equal ion and electron mean temperatures $T_e = T_i$, a given plasma current I and toroidal field B_t the list of the main dimensional parameters is a [m], R [m], B_t [T], B_p [T], m_e [kg], m_i [kg], n_e [m^{-3}], n_i [m^{-3}], k_{BT} [Joule], e [C]. Applying the formalism of Buckingham's dimensional analysis [49], the state of a tokamak plasma, can be summarized by a *state equation* (here unspecified)

$$f\left(B_t,\, T,\, n,\, a,\, m_i,\, t,\, e,\, \frac{B_p}{B_t},\, \frac{R}{a},\, \frac{m_e}{m_i}\right) = 0 \tag{19.170}$$

where the set of variables a, m_i, t, e can be taken as fundamental units of length, mass, time and charge in a specific system of units (e.g. SI); the other variables, B_t, T, n can be expressed through them and equation (19.170) can be rephrased in a statement involving only dimensionless monomials π_k composed of the fundamental variables [49, 53]:

$$f\left(\pi_1, \pi_2, \pi_3, \pi_4, \frac{B_p}{B_t}, \frac{R}{a}, \frac{m_e}{m_i}\right) = 0 \tag{19.171}$$

For a simple hydrogen plasma one gets:

$$\begin{aligned} \pi_1 &= e m_i^{-1} t B_t \\ \pi_2 &= m_i^{-1} a^{-2} t^2\, k_B T \\ \pi_3 &= a^3 n \\ \pi_4 &= a^{-1} t c \end{aligned} \tag{19.172}$$

In the context of the equations governing quasineutral, collisional, finite- plasmas, other dimensionless parameters appear appropriate to be exploited in a similarity argument [12, 53]:

$$\begin{aligned} \rho_* &= \frac{\pi_2^{1/2}}{\pi_1} \\ \beta &= 16\pi \frac{\pi_2 \pi_3}{\pi_1^2} \\ \nu_* &= \frac{\nu}{\epsilon^{3/2} v_{\text{th}}/Rq} = \nu \frac{aq}{c\epsilon^{3/2}} \frac{\pi_4}{\pi_2^{1/2}} \\ \lambda_* &= \frac{\lambda_D}{a} \end{aligned} \tag{19.173}$$

where $\epsilon = a/R$, $q_a = aB_t/RB_p$. The fundamental goal of both empirical and theoretical scaling laws is reaching some reliable recipe for the global energy confinement time scale τ_E, which pertains to the subject of transport and as such is outside the topics of the present text. However, since the confinement time is so inextricably tied to the structure of the plasma equilibrium and stability, it is worth recalling and summarizing its expected, and sometimes experimentally verified, dependence on the fundamental dimensionless quantities just introduced. A characteristic time scale appearing in the energy confinement issue is the *Bohm time scale* $\tau_B = \rho_*^{-2} \omega_c^{-1}$(where $\omega_c = eB/m$ is the relevant cyclotron pulsation). In terms of this time scale the general form of the energy confinement time is:

$$\tau_E = \tau_B F(\rho_*, \beta, \nu_*, \lambda_*) \propto \tau_B \rho_*^{\kappa} \beta^{\lambda} \nu_*^{\mu} \lambda_*^{\nu} \tag{19.174}$$

Here the exponents in the r.h.s. of the formula, are often found by fitting a large base of experimental data, or partially *fixed* by justified theoretical assumptions. Even if the functional form of $F(\rho_*, \beta, \nu_*, \lambda_*)$ is not known, the concept of similarity helps in predicting or interpreting the performance of a tokamak on the score of the behaviour of other devices, different but having the same values of ρ_*, β, ν_*. It can be shown [15, 51] that the *scale invariance of the equations* governing plasma physics, imposes other constraints on the scaling of the confinement time, such as the Kadomtsev scaling:

$$B\tau_E \propto n^p T^q B^r (\epsilon R)^s \tag{19.175}$$

where p, q, r, s are exponents to be determined by requiring dimensional homogeneity, obtaining:

$$B\tau_E \propto \tau_B F(\rho_*, \beta, \nu_*) \tag{19.176}$$

The applied physics scale invariance constraint removes dependence on λ_*, as phenomena on the Debye scale length are negligible, and the relevant dimensionless variables are (ρ_*, β, ν_*). If small scale turbulence becomes significant, different constraints lead to scalings of the *gyro-Bohm* type $\tau_E \propto \frac{\tau_B}{\rho_\star} F(\beta, \nu_*)$. Tokamaks with identical (ρ_*, β, ν_*) have a confinement $\tau_E \propto B^{-1}$ [4–7, 45]. All experiments aiming at high performance will operate in a limited range of (β, ν_*), around *optimal values*, leaving a scaling $\tau_E = \tau_B F(\rho_*)$. For fixed equilibrium geometry and profiles one can choose one free quantity among B, R, n and T:

$$nR^2,\ BR^{5/4},\ TR^{1/2} \tag{19.177}$$

Different tokamaks having constant the product $BR^{5/4}$ will have the same plasma dynamics; for a burning plasma the requirement of constant $Q = P_\alpha / P_{\text{in}}$ determines a set of values for $\rho_\star$, $\nu_\star$, β_N, which depends on the value of the toroidal field B [45]. The consideration of alpha particle nonlinear dynamics and thermal plasma transport in burning plasmas, where fast ions are produced by auxiliary heating (namely neutral beam injection (NBI) and ion cyclotron range of frequency waves (ICRF)) and in the presence of current drive systems, corresponds to fixing the above parameters. According to [54] the requirement that the alpha particle $\beta_\alpha < \beta$ and the assumption of strong electron–ion equipartition, leads to

$$\beta \approx B^{-1.43} R^{-0.84} \epsilon^{0.37},\ \nu_\star \approx B^{0.67} R^{0.16} \epsilon^{-1.13},\ \tau_{\text{sd}}/\tau_E \approx \beta_H \beta \tag{19.178}$$

where τ_{sd} is the alpha particle slowing down time. The control of edge physics conditions, PWI and core-edge coupling corresponds to fixing T(and P_w/R). Then the heating power is chosen to meet P_w/R and eventually the output requirements of equation (19.4). The complex physics of magnetically confined plasma has to rely both on scaling based on first principles and on empirical evidence, collected from numerous similarity experiments carried over a long span of time. Regimes of enhanced energy confinement have been obtained experimentally in modern tokamaks [4–7] with strong auxiliary plasma heating. The transport physics is too vast and complicated to be presented here, even qualitatively; nonetheless, in the frame of the above scaling arguments it is worthwhile to report the recent validated *empirical* scaling. An important one is the scaling of the thermal energy confinement time *ITERH-97Py* in hydrogen, deuterium, DT and tritium ELMy H-mode discharges, published in [55]:

$$\tau_{\text{th}} = 0.037 I^{0.97} B^{0.08} n^{0.41} P_{\text{aux}}^{-0.63} R^{1.93} \epsilon^{0.23} \kappa^{0.67} A^{0.2} \tag{19.179}$$

This expression shows an important feature emerged from experimental measurements, not foreseen in the similarity arguments presented above for the *design* of the device: the thermal energy confinement degrades with the external power input, presumably because of an increased turbulence level. A fine balance must be found between the requirements of high plasma temperature and sufficiently small losses. Another empirically validated scaling for the existing tokamaks is the so called *ITER-97Py*, illustrated in figure 19.16.

An instructive application of similarity methods to the determination of engineering parameters of ITER was presented in reference [57]. A comprehensive, up-to-date study of the application of similarity techniques to the design of (future) tokamak reactors, is presented in reference [58]. In a steady state magnetically confined fusion plasma the power injected into the plasma through auxiliary heating systems plus that produced by fusion reactions must be extracted from the device. Energy in the neutron channel is isotropically collected in the blanket to be converted into electricity and to drive the tritium breeding process. Energy associated with charged particles, which escape the confinement must be taken up by a structure built to withstand high power loads on a relatively small wetted surface. The SOL width is determined by the competition of parallel and perpendicular transport. While the parallel transport is mainly classical and reasonably well understood, the cross-field transport is expected to be dominated by turbulent transport, in particular, in the reactor relevant high density regimes. Empirical and analytical arguments suggest a scaling with the ion poloidal gyroradius, $\rho_{i,\mathrm{pol}}$. The SOL width decreases with increasing field, but not with size. The value of P_{sep}/R is, therefore, often used as a proxy for the heat flux $q_{\|}$, that has to be handled in the divertor. The power flux on the divertor plasma facing components is $Q \propto 2\pi R P_w \lambda_q$, where P_w is the power leaving the plasma from the non-neutronic channel.

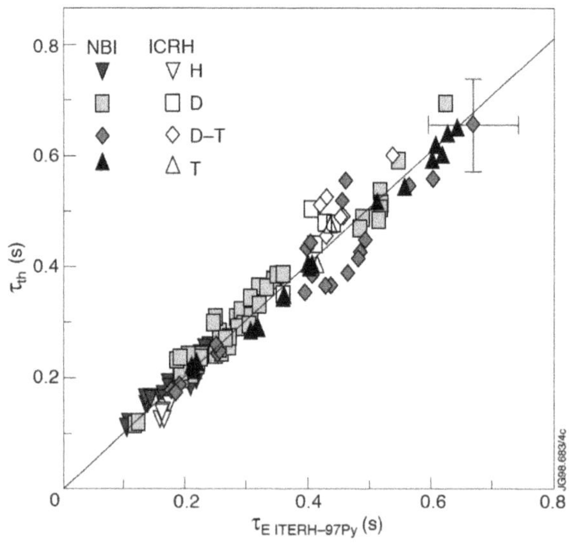

Figure 19.16. Measured thermal energy confinement time τ_{th} in hydrogen (H), deuterium (D), tritium (T) and (D-T) ELMy H-mode discharges plotted against the ITERH-97Py τ_E scaling law, for different auxiliary heating methods: NBI, ICRH. Reproduced from reference [56]. Copyright 1999 IAEA.

The preservation of material integrity of the target presently limits the impinging heat flux to 5 MW m^{-2}. Optimized designs and new radiation resistant materials may double this limit. The constraints on the plasma parameters at the target require a significant reduction of the heat flux via radiative cooling, and a pressure loss via charge exchange, determining eventually, through temperature reduction and recombination, the full detachment of the plasma. These scenarios give strong incentive to the study of alternative configurations which may allow for more radiation in the divertor, and distribute the heat load by charged particles over a larger wall area. The main concepts presently under consideration are obtained by additional multipolar fields created by poloidal coils localized near the original X point. One design foresees an expansion of the flux surfaces near the divertor targets, obtained by decreasing the local poloidal magnetic field. The expansion of this divertor throat near the target leads to an increase of the connection length, $L_{\parallel}$, which decreases the plasma temperature at the target and favors the detachment at lower densities or higher exhaust power. Another advanced configuration is the snowflake divertor [45, 59] which also provides an increased connection length and divertor volume. In this case the corresponding decrease of the poloidal field is obtained by introducing a second order null point. This splits the separatrix around the null into six legs with two enclosing the confined plasma and four divertor legs. Open problems are still to be faced, concerning the increase of cross-field transport bound to occur in the large region of low poloidal field close to the null point, and the MHD stability where the plasma pressure is still relatively high with consequent very large values of local β_p [59], although this is accompanied by a higher q_{95}. So far we have presented the elements of principle for ideal axisymmetric devices. Actual tokamaks are affected (and *plagued* by non-ideal 3-D effects, both configurational (e.g. toroidal field ripples due to finite number of coils) and due to dissipative effects in the plasma and first wall. This is a very complicated physics which is also strictly tied with the actual interface conditions between confined plasma, SOL and first wall, involving operation strategies and choice of materials. This subject is important and must be studied in the relevant literature. Finally, we must recall that in the perspective of a commercially useful power plant, *steady state, stable and efficient operation* must be guaranteed. In this direction the concept of *advanced tokamak* takes a different meaning. While the current design of the largest tokamak reactor under construction (ITER) [4–7, 57], follows the path of the large R, lowish B configurations, recent research has revamped the interest in a high B field, compact ($\epsilon \simeq 1/3$), highly shaped configuration [2, 3, 54, 60–62] as the best to obtain a natural alignment between high β_N operation, favorable stability and transport properties, and a high self-driven (bootstrap) plasma current $J_{\rm bs} \propto -\epsilon \frac{T}{B_p}\frac{dn}{dr}$, which, together with externally driven rf current, sustains the steady state, broadens the total J_t profile and limits disruption hazard. A strong synergy between plasma shaping and broad current and pressure distribution assists the wall stabilization of ideal MHD instability [62]. Plasma shaping (increased elongation and triangularity and Shafranov shift) increases the volume and current carrying capacity, by placing more of the plasma in the high B region. As another consequence of shaping, also the eigenstructure of the modes is modified, strengthening the harmonics that extend beyond the plasma, favoring their passive stabilization by the resistive wall. The full discussion of this subject is beyond the scope of this presentation and the reader should look up the most recent references.

References

[1] Gamow G and Teller E 1938 The rate of selective thermonuclear reactions *Phys. Rev.* **53** 608

[2] Sorbom B *et al* 2015 ARC: A compact, high-field, fusion nuclear science facility and demonstration power plant with demountable magnets *Fusion Eng. Des.* **100** 378–405

[3] Creely A *et al* 2020 Overview of the SPARC tokamak *J. Plasma Phys.* **86** 865860502

[4] The Jet Project. EUR-JET-R7 1975 (http://aei.pitt.edu/94570/)

[5] Rebut P 1992 The JET preliminary tritium experiment *Plasma Phys. Control. Fusion* **34** 1749

[6] Keilhacker M and ASDEX Team *et al* 1985 The ASDEX divertor tokamak *Nucl. Fusion* **25** 1045

[7] Budny R *et al* 1995 Simulations of alpha parameters in a TFTR DT supershot with high fusion power *Nucl. Fusion* **35** 1497

[8] De Tommasi G 2019 Plasma magnetic control in tokamak devices *J. Fusion Energy* **38** 406–36

[9] Tamm I 1961 Theory of a magnetic thermonuclear reactor (Part I) *Plasma Physics and the Problem of Controlled Thermonuclear Reactions* vol 1 (Berlin: Springer)

[10] Sakharov A 1958 Theory of a thermonuclear magnetic reactor (Part II) *Plasma Physics and the Problem of Controlled Thermonuclear Reactions* vol 1 (Berlin: Springer) pp 20–30

[11] Shafranov V 1966 Plasma equilibrium in a magnetic field *Rev. Plasma Phys.* **2** 103

[12] Kadomtsev B and Laing E W 1992 *Tokamak Plasma: A Complex Physical System* (London: Taylor & Francis)

[13] Mukhovatov V and Shafranov V 1971 Plasma equilibrium in a tokamak *Nucl. Fusion* **11** 605

[14] Hazeltine R D and Meiss J D 2003 *Plasma confinement Dover Books on Physics* (New York: Dover)

[15] Wesson J 2004 *Tokamaks* 3rd edn (Oxford: Oxford University Press)

[16] Goedbloed J P, Keppens R and Poedts S 2010 *Advanced Magnetohydrodynamics* (Cambridge: Cambridge University Press)

[17] Freidberg J P 2000 *Plasma Physics and Fusion Energy* (Cambridge: Cambridge University Press)

[18] Kulsrud R 1964 General stability theory *Advanced Plasma Theory* ed M Rosenbluth (New York: Academic)

[19] Laval G, Mercier C and Pellat R 1965 Necessity of the energy principles for magnetostatic stability *Nucl. Fusion* **5** 156

[20] Troyon F, Gruber R, Saurenmann H, Semenzato S and Succi S 1984 MHD-limits to plasma confinement *Plasma Phys. Control. Fusion* **26** 209

[21] Montgomery D C and Tidman D A 1964 *Plasma Kinetic Theory* (New York: McGraw-Hill)

[22] Freidberg J P 1982 Ideal magnetohydrodynamic theory of magnetic fusion systems *Rev. Mod. Phys.* **54** 801

[23] Jackson J D 1965 *Classical Electrodynamics* (New York: Wiley)

[24] Boozer A H 2005 Physics of magnetically confined plasmas *Rev. Mod. Phys.* **76** 1071

[25] Poincaré H 1895 Analysis situs *J. Ec. Polytech* **1** 1

[26] Grad H and Rubin H 1958 Hydromagnetic equilibria and force-free fields *Proc. of the 2nd Conf. on the Peaceful Use of Atomic Energy* vol 31 (Geneva: United Nations)
[27] Golant V E, Zhilinsky A P and Sakharov I E 1980 *Fundamentals of Plasma Physics* (New York: Wiley) (trans. K Z Vedeneyeva and V F Agranat)
[28] Palumbo D 1968 Some considerations on closed configurations of magnetohydrostatic equilibrium *Il Nuovo Cimento* B **53** 507–11
[29] Mercier C 1960 A necessary condition for hydromagnetic stability of plasma with axial symmetry *Nucl. Fusion* **1** 47
[30] Bishop C M and Taylor J B 1986 Degenerate toroidal magnetohydrodynamic equilibria and minimum B *Phys. Fluids* **29** 1144–8
[31] Chandrasekhar S 1981 *Hydrodynamic and Hydromagnetic Stability* (Dover: New York)
[32] Lazarus E, Lister J and Neilson G 1990 Control of the vertical instability in tokamaks *Nucl. Fusion* **30** 111
[33] Strait E, Lao L, Luxon J and Reis E 1991 Observation of poloidal current flow to the vacuum vessel wall during vertical instabilities in the DIII-D tokamak *Nucl. Fusion* **31** 527
[34] Zakharov L E and Li X 2015 Tokamak magneto-hydrodynamics and reference magnetic coordinates for simulations of plasma disruptions *Phys. Plasmas* **22** 062511
[35] Lao L, Hirshman S and Wieland R 1981 Variational moment solutions to the Grad-Shafranov equation *Phys. Fluids* **24** 1431–40
[36] Blum J, Le Foll J and Thooris B 1981 The self-consistent equilibrium and diffusion code SCED *Comput. Phys. Commun.* **24** 235–54
[37] Cerfon A J and Freidberg J P 2010 One size fits all' analytic solutions to the Grad-Shafranov equation *Phys. Plasmas* **17** 032502
[38] Miller R, Chu M, Greene J, Lin-Liu Y and Waltz R 1998 Noncircular, finite aspect ratio, local equilibrium model *Phys. Plasmas* **5** 973–8
[39] Freidberg J P, Cerfon A and Lee J 2015 Tokamak elongation–how much is too much? Part 1. Theory *J. Plasma Phys.* **81** 515810607
[40] Ramos J 1991 Theory of the tokamak beta limit and implications for second stability *Phys. Fluids* B **3** 2247–50
[41] Pitts R A, Fundamenski W, Erents S K, Andrew Y, Loarte A, Silva C and JET-EFDA contributors 2005 Far SOL ELM ion energies in JET *Nucl. Fusion* **46** 82
[42] Blum J, Lazzaro E, O'Rourke J, Keegan B and Stephan Y 1990 Problems and methods of self-consistent reconstruction of tokamak equilibrium profiles from magnetic and polarimetric measurements *Nucl. Fusion* **30** 1475
[43] Pitcher C S and Stangeby P 1997 Experimental divertor physics *Plasma Phys. Control. Fusion* **39** 779
[44] Ryutov D D and Umansky M V 2013 Divertor with a third-order null of the poloidal field *Phys. Plasmas* **20** 092509
[45] Zonca F 2009 The FAST proposal *36th EPS-CPP Conf. (Sofia, Bulgaria, June 30th, 2009)*
[46] Coletti A *et al* 2011 JT-60SA power supply system *Fusion Eng. Des.* **86** 1373–6
[47] Pedersen A 1967 Calculation of spark breakdown or corona starting voltages in nonuniform fields *IEEE Trans. Power Appar. Syst.* **86** 200–6
[48] Alekseev A B 2018 Mechanics of magnetic fusion reactors *Fundamentals of Magnetic Thermonuclear Reactor Design* (Amsterdam: Elsevier) pp 351–86
[49] Buckingham E 1914 On physically similar systems; illustrations of the use of dimensional equations *Phys. Rev.* **4** 345

[50] Kadomtsev B B 1975 Tokamaks and dimensional analysis *Sov. J. Plasma Phys.* **1**
[51] Connor J W and Taylor J B 1977 Scaling laws for plasma confinement *Nucl. Fusion* **17** 1047
[52] Lackner K 1990 Physical equivalence of tokamak discharges *Comments Plasma Phys. Control. Fusion* **13** 163–71
[53] Petty C 2008 Sizing up plasmas using dimensionless parameters *Phys. Plasmas* **15** 781
[54] Romanelli M, Romanelli F and Zonca F 2001 On the optimal choice of the dimensionless parameters of burning plasma physics experiments *28th EPS Conf. on Contr. Fusion and Plasma Phys. (Funchal)* 18–22
[55] Watkins M L and JET Team 1999 High fusion performance from deuterium-tritium plasmas in JET *Nucl. Fusion* **39** 209
[56] The JET Team 1999 Physics of high performance JET plasmas in D-T *Proc. 17th IAEA Fusion Energy Conf. (Yokohama, Japan, October 1998, IAEA-CN-69, OV1/2)* (https://www-pub.iaea.org/MTCD/Publications/PDF/csp_001c/html/node1.htm)
[57] Lackner K 2008 Dimensionless engineering variables for measuring the ITER and reactor relevance of tokamak experiments *Fusion Sci. Technol.* **54** 989–93
[58] Sarazin Y, Hillairet J, Duchateau J-L, Gaudimont K, Varennes R, Garbet X, Ghendrih P, Guirlet R, Pégourié B and Torre A 2019 Impact of scaling laws on tokamak reactor dimensioning *Nucl. Fusion* **60** 016010
[59] Ryutov D, Cohen R, Farmer W, Rognlien T and Umansky M 2014 The 'churning mode' of plasma convection in the tokamak divertor region *Phys. Scr.* **89** 088002
[60] Coppi B, Nassi M and Sugiyama L E 1992 Engineering characteristics of the IGNITOR ULT experiment *Fusion Technol.* **21** 1607–11
[61] Sugiyama K, Schmid K and Jacob W 2016 Sputtering of iron, chromium and tungsten by energetic deuterium ion bombardment *Nucl. Mater. Energy* **8** 1–7
[62] Buttery R J *et al* 2019 DIII-D research to prepare for steady state advanced tokamak power plants *J. Fusion Energy* **38** 72–111

IOP Publishing

Plasma Modeling (Second Edition)

Methods and applications

Gianpiero Colonna and Antonio D'Angola

Chapter 20

Verification and validation in plasma physics

Ivo Furno, Paolo Ricci, Ambrogio Fasoli, Fabio Riva and Christian Theiler

The goal of the verification…validation (V…V) procedure is to assess that the physics models are correctly implemented in numerical codes (code verification), to estimate the numerical errors affecting the simulation results (solution verification), and to establish the consistency of the code results, thus of the physics model, with experimental data (validation). This chapter discusses the methodology to carry out a rigorous V…V exercise and shows, through a practical example, how this can contribute to advancing our physics understanding of plasma turbulence. The method of manufactured solutions is used to assess that the model equations are correctly solved, within the order of accuracy of the numerical scheme. The technique to carry out a solution verification is described to estimate the uncertainty affecting the numerical results. A methodology for plasma turbulence code validation is also discussed, focusing on quantitative assessment of the agreement between experiments and simulations. The V…V methodology is applied to the study of plasma turbulence in the basic plasma physics experiment TORPEX [1], considering both two-dimensional and three-dimensional simulations carried out with the GBS code [2, 3]. The validation procedure allows progress in the understanding of the turbulent dynamics in TORPEX, by pinpointing the presence of a turbulent regime transition, due to the competition between the resistive and ideal interchange instabilities.

20.1 Introduction

Errors affecting the simulations used to improve our understanding and to describe the dynamics of plasmas can have far reaching consequences. To limit these errors, which can be due both to mistakes in the code and to an incomplete, or inaccurate physics model, there is an increasing motivation in plasma physics to use V…V procedures [4–6]. V…V is composed by the *code verification* process, the *solution verification* procedure, and the *validation.*

doi:10.1088/978-0-7503-3559-1ch20

The code verification procedure is a mathematical and numerical procedure targeted to assess that the physics model is correctly implemented in a numerical simulation code, with the accuracy expected by the numerical algorithm. The goal of the code verification procedure is therefore to assess that no mistakes are present in the simulation code.

Due to finite computational power and the related finite precision achievable, the simulation results are always affected by numerical errors, even if the model equations are implemented correctly. Estimating the amplitude of the numerical errors is crucial to ensure the reliability of the numerical results. The estimate of the numerical error affecting the simulations constitutes the solution verification procedure [5, 6, 7, 8].

Finally, validation is used to assess the consistency of the code results, and therefore of the physics model, with experiments. Validation requires the comparison of the simulation results with experimental measurements. The V…V procedure is visualized in figure 20.1.

The purpose of the present chapter is twofold. First, we summarize the methodology for code verification, simulation verification, and validation (section 20.2). Second, in section 20.3, we describe the V…V effort that we have carried out in recent years to study the plasma dynamics in the basic plasma physics experiment TORPEX [1, 9] by using the GBS code [2, 3]. This exemplifies the application of the proposed methodology. Owing to its detailed diagnostics, possibility of parameter scans, and relative simple configuration, TORPEX is an ideal testbed to perform experiment/simulation comparisons and to investigate the corresponding methodological framework. For its simulation three models have been considered, all based on the fluid equations developed by Braginskii for collisional plasmas and simplified by neglecting the fast particle gyration, considering therefore the drift limit of those

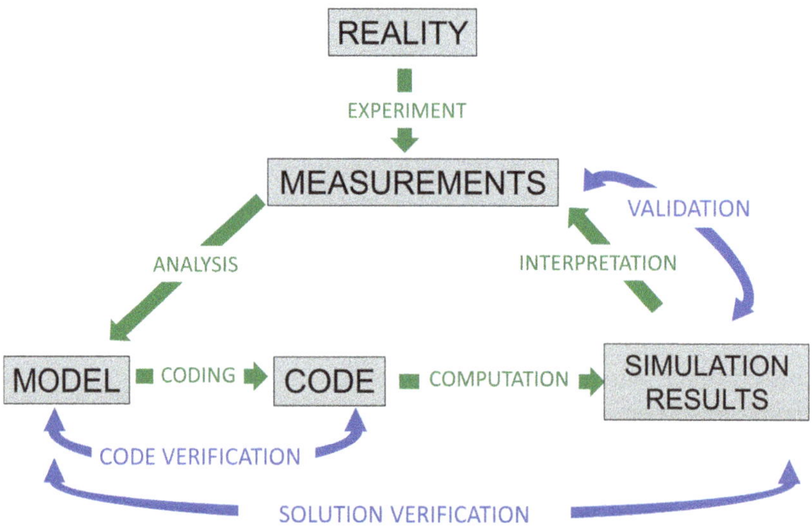

Figure 20.1. Schematic representation of the V…V procedure.

equations: (a) a three-dimensional two-fluid model, able to describe the global evolution of TORPEX plasma [10, 11]; (b) the same three-dimensional model completed by an appropriate first-principle set of boundary conditions that has been recently derived [12]; and (c) a reduced two-dimensional two-fluid model [13, 14], able to describe only the evolution of $k_{\|} = 0$ modes. The three models are implemented in the GBS code. We therefore discuss the verification of the GBS code and of its results. We then compare the simulation results with the experimental measurements, showing that the validation metric is able to point out that the agreement between the two-dimensional model and the experiment is no longer satisfactory when $k_{\|} \neq 0$ modes are present in the experiment [15–17, 18]. The conclusions follow in section 20.4.

20.2 The validation and verification methodology

20.2.1 Code verification methodology

To perform a code verification, one can [5] (a) perform simple tests (e.g., verify that the code conserves energy), (b) compare codes with each other (this is known as code-to-code benchmark), (c) quantify the discretization error with respect to a known solution, (d) check the convergence of the numerical result to a known solution, and (e) prove that the numerical solution converges to the analytical one at the rate expected for the numerical scheme (order-of-accuracy tests). As they do not require an analytical solution, the first two techniques ((a) and (b)) are the simplest to use and, in fact, code-to-code benchmarks are routinely used in plasma physics (see, e.g., references [19–25]). While valuable, this exercise requires that at least one code is fully verified and, generally, it is very difficult to understand if the difference in the code results is due to discretization errors or to a non-correct implementation of the model. Only the three other approaches ((c)–(e)) are rigorous and, in particular, the order-of-accuracy test is the only one that can assess quantitatively the correct implementation of the model equations and of the numerical scheme, verifying that the discretization errors reduce at the rate expected for the numerical scheme, as the spatial mesh and the time step are refined. Order-of-accuracy tests are performed in other scientific domains, such as computational fluid dynamics. The method of manufactured solution (MMS) [26–28] allows order-of-accuracy tests even when an analytical solution is not available for the physics model considered.

Given a theoretical model M with an analytical solution s, such that $M(s) = 0$, and the numerically discretized model of M, M_h, with a numerical solution s_h that satisfies $M_h(s_h) = 0$ (h is a parameter representing the degree of refinement of the mesh), the error affecting the numerical results is expressed as $\epsilon_h = \|s_h - s\|$, where $\|\cdot\|$ denotes a designed norm. The theoretical order of accuracy, p, associated with the numerically discretized operator M_h, represents the rate at which the numerical solution converges to the analytical solution as the mesh is refined. The numerical error satisfies the relation $\epsilon_h = C_p h^p + O(h^{p+1})$, where C_p is independent of h, and p is the order of accuracy of the numerical scheme, typically evaluated through its Taylor expansion [5, 6, 28] with the two numerical solutions of M_h and M_{rh}, i.e., s_h

and s_{rh}, where rh indicates coarsening the h mesh by a factor r, the observed order of accuracy, $\hat{p}$,

$$\hat{p} = \frac{\ln\left(\epsilon_{\mathrm{rh}}/\epsilon_h\right)}{\ln\left(r\right)}. \tag{20.1}$$

If $\hat{p}$ converges to p for $h \to 0$, i.e., when the discretization error is dominated by the lowest order term in the expansion (the so-called asymptotic regime), we can state that the code is verified and the equations are correctly solved, with the order of accuracy expected for the numerical scheme.

The main issue related to the systematic evaluation of $\hat{p}$ is the need for the analytical solution s, to compute the numerical error ϵ_h; in fact, the analytical solution is unknown in most cases. The MMS, developed to overcome this problem [26–28], instead of solving analytically a theoretical model, imposes a solution to the model, the so-called manufactured solution, and modifies the model equations to accommodate the imposed solution. The obtained modified model is then solved to compute the discretized error. More precisely, for a given model M, we choose an analytical function u and we compute a source term, $S = M(u)$, which is subtracted from M to obtain a new analytical model $G = M - S$, whose analytical solution is u [$G(u) = M(u) - S = 0$]. The discretization of the new model is straightforward, i.e. $G_h = M_h - S$. Since the source term S is computed analytically, we do not add any new discretization errors to the numerical model considered. The behavior of the numerical error is preserved. This can be expressed as: $\epsilon_h = \|u_h - u\| = D_p h^p + O(h^{p+1})$. In practice, using the MMS for an order-of-accuracy test implies adding source terms to the discretized equations, performing a simulation scan to obtain the observed order of accuracy, and comparing the observed order of accuracy to the theoretical one to verify the code.

Although the idea behind the MMS is simple, its implementation requires some subtleties. First, the manufactured solution should satisfy the following requirements: (a) be smooth enough and not singular, (b) be general enough to excite all the terms present in the equations, (c) satisfy the code constraints (e.g., positivity for the density or the temperature), and (d) ensure that the magnitude of the different terms composing the equations are of the same order of magnitude. The manufactured solutions are usually built as a combination of trigonometric and/or hyperbolic functions. We note that the MMS cannot be applied to codes used to model singularities, shocks or discontinuities; the verification of these codes is still an open issue [5].

20.2.2 Solution verification methodology

Even when a model is correctly implemented in a code, we note that simulations are affected by numerical errors, which should be evaluated through a solution verification procedure [5, 6, 7, 8]. The sources of numerical error are: (a) round off errors due to the finite number of digits carried over by a computer, (b) statistical sampling errors, (c) errors associated with iterative numerical methods stopped before they reach full convergence, (d) discretization errors due to the finite grid spacing. While all these errors have to be evaluated, if iterative methods are applied using a sufficiently large number of iterations, and simulations are run long enough to decrease the statistical errors, discretization errors typically dominate over the other sources. Focusing on these errors, for grid-based algorithms the Richardson extrapolation can be used as a higher-order estimator of the solution, and the Roache's grid convergence index as a relative numerical uncertainty estimate [29].

In the early 20th century, Richardson developed a method [30, 31], later extended [32, 33], to accelerate the rate of convergence of a numerical sequence. This method is based upon the use of two numerical solutions obtained using two different meshes, s_h and $s_{\rm rh}$, to compute an estimate of the analytical solution that presents a convergence rate one order higher than the original numerical solution. Concretely, the Richardson extrapolation is defined as

$$\bar{s} = s_h + \frac{s_h - s_{\rm rh}}{r^p - 1} \tag{20.2}$$

where p is the formal order of accuracy defined in section 20.2.1. Noting that $\|s_h - s\| = C_p h^p + O(h^{p+1})$, it follows that the extrapolated solution $\bar{s}$ satisfies $\|\bar{s} - s\| = D_p h^{p+1} + O(h^{p+2})$; therefore, for $h \to 0$, $\bar{s} \to s$ faster than the numerical solutions obtained from the simulations. Consequently, we can use $\bar{s}$ as an estimate of the exact solution s, and approximate the numerical error with

$$\epsilon_h \simeq \| s_h - \bar{s} \| = \left\| \frac{s_{\rm rh} - s_h}{r^p - 1} \right\| \tag{20.3}$$

The relative discretization error, *RDE*, is therefore approximated as

$$RDE = \frac{s_h - s}{s} \simeq \frac{s_h - \bar{s}}{\bar{s}} = \frac{s_{\rm rh} - s_h}{s_h r^p - s_{\rm rh}}. \tag{20.4}$$

For $\bar{s}$ to be a reasonable estimate of s, however, several assumptions should be satisfied. First, the Richardson extrapolation method requires the use of uniform mesh spacing, meaning that the degree of the refinement of the meshes can be represented solely by the parameter h discussed before. Therefore, s_h and $s_{\rm rh}$ should be computed over two meshes that are one the uniform systematic refinement of the other. Consequently, the application of Richardson extrapolation to computations involving local mesh refinement or mesh adaptation is not allowed. Second, the simulations used to evaluate $\bar{s}$ should be in the asymptotic regime, meaning that the discretization error is dominated by its lower order term, $C_p h^p$. This requirement could result in computationally very expensive simulations, due to the potential need of very fine meshes. Third, to apply the method presented above for the estimate of the numerical error, the solutions should be smooth enough, with no singularities and/or discontinuities. To allow the expansion of the numerical error in powers of the parameter h, the derivatives of the analytical solution should exist and be continuous. Moreover, we should note that we do not have any guarantee that the Richardson extrapolated solution will meet the same governing equations satisfied by either the numerical solution or the analytical one. Consequently, we use this extrapolation for the computation of the numerical error only.

Usually, it is problematic to satisfy the requirement of being in the asymptotic regime, due to the high computational cost of the simulations. Moreover, it has to be demonstrated that the numerical solutions are in the asymptotic regime, by showing that the observed order of accuracy matches the formal one. This requires at least three simulations, resulting from two subsequent refinements of the coarser mesh of a factor r, from which the observed order of accuracy can be evaluated as

$$\hat{p} = \frac{\ln\left[(s_{r^2h} - s_{\mathrm{rh}})/(s_{\mathrm{rh}} - s_h)\right]}{\ln(r)}. \tag{20.5}$$

If only two simulations are available, or if the observed order of accuracy does not match the formal one, we should substitute the numerical error estimates in equations (20.3) and (20.4) with a numerical uncertainty quantification. In general, the error estimate in equations (20.3) and (20.4) may depend strongly on the refinement factor r and on the precision of the numerical scheme. It is therefore difficult to rely on such an error estimate. To overcome these issues, reference [29] introduces the grid convergence index, *GCI*, defined as

$$GCI = \frac{F_s}{r^{\tilde{p}}-1}\left|\frac{s_{\mathrm{rh}} - s_h}{s_h}\right|, \tag{20.6}$$

which represents an estimate of the relative discretization error affecting the simulation results. The *GCI* value is obtained by approximating in equation (20.4) $s_h r^p - s_{\mathrm{rh}} \simeq (r^{\tilde{p}}-1)s_h$. The factor of safety F_s and $\tilde{p}$ ensures that *GCI* is larger than the numerical discretization error in 95% of the cases. Oberkampf and Roy [5] propose the following: if $| p - \hat{p} | < 0.1p$, we can assume that the simulation is in the asymptotic regime and we use $F_s = 1.25$, as well as $\tilde{p} = p$. If $| p - \hat{p} | > 0.1p$, a more conservative factor of safety, $F_s = 3$, has to be used and $\tilde{p} = \min[\max(0.5, \hat{p}), p]$. If $\hat{p}$ is not evaluated (for example, if only two solutions are available), $F_s = 3$ and $\tilde{p} = p$ are used. We remark that, although these definitions are reasonable, there still is an ongoing discussion in the verification community about their generality.

To conclude our presentation of the error estimate methodology, we have to discuss a few details. First, we draw attention to the fact that the present procedure can be applied not only to point-by-point solution values, but also to solution functionals. This is important to estimate the numerical error affecting the observables used in the validation of the physical model [6]. Second, as s_h and s_{rh} are in general computed on different meshes, the results on the coarser mesh have to be interpolated on the finer grid, using an interpolation scheme whose order is equal to or higher than the order of the numerical scheme used by the code. A complete discussion of this topic is found in reference [32]. Third, using *GCI* as an evaluation of the numerical error requires a non-oscillatory convergence of the numerical solution. If oscillatory convergence is observed, the numerical error has to be evaluated from the difference between the obtained numerical solutions. Finally, we illustrate a useful property of *GCI*, that is, the possibility of computing the overall *GCI* analyzing each coordinate of the problem independently. As it can result numerically very expensive to perform a uniform refinement of the grid along all the coordinates at the same time, it is possible to refine separately each coordinate of the mesh by a factor r_i, where the index i refers to the coordinate under investigation. This allows us to compute a GCI_i and a $\tilde{p}_i$ for the i coordinate, and obtain the overall GCI as $GCI = \sum_i GCI_i$.

20.2.3 Validation methodology

Guidelines to carry out the validation between experiments and simulations have been imported from other domains, from computational fluid dynamics to plasma physics, and these are described in references [34, 35]. Simulations and experiments have to be compared considering a number of physical quantities, common to the experimental measurements and simulation results, analyzed using the same techniques. These physical quantities, denoted as validation observables, should be identified and organized into a hierarchy. This hierarchy is based on the number of model assumptions and combinations of measurements necessary to obtain the observable; i.e., how stringent each observable is for comparison purposes. By combining the result of the comparison of all the observables, while taking into account position in the hierarchy and precision, the agreement between simulations and experiments needs to be quantified by using an appropriate composite metric, χ. This metric χ should be complemented by an index, Q, which assesses the quality of the comparison. Practically, Q is related to the number of observables that are used for the validation and the strength of the constraints they impose. We remark that the validation should take into account the experimental and simulation uncertainties, therefore these have to be accurately evaluated. Sources of experimental uncertainties are the approximations of the models used to interpret the experimental results, the difficulties in the evaluation of the properties of the measuring devices, and the non-perfect reproducibility of the experiments. For the simulations, uncertainties are due to numerical errors (evaluated through the solution verification procedure) and to the uncertainty associated with the non-well known input parameters. The latter can be estimated through a sensitivity analysis.

We point out that the validation procedures should remain simple. The goal is a useful tool that can be easily applied in order to compare different models with experimental results. Through this comparison, it is possible to assess the physics elements that play a role in the dynamics of the system and that should be taken into account for its description. On the other hand, it is very delicate to judge a single model in absolute terms in view of testing its predictive capabilities and this is not a subject of the present chapter.

Simulations and experiments have to be compared considering a number of physical quantities, common to the experimental measurements and simulation results, and analyzed using the same techniques. These validation observables should be independent of each other, and their resolution sufficient to well describe their variation.

Once the observables are defined and evaluated, the agreement between experiments and simulations relative to each observable has to be quantified. We denote with e_j and s_j the values of the jth observable used in the comparison, as coming from the experimental measurement or the simulation results, respectively. Most observables depend on space and time, and are given on a discrete number of points, denoted as N_j. We denote with $e_{j,i}$ and $s_{j,i}$ the values of the jth observable at points $i = 1, 2, \ldots, N_j$ (this notation can therefore be used for zero-, one-, two-, etc, dimensional observables). For the jth observable, we normalize the distance d_j between experiments and simulations to the uncertainty related to these quantities:

$$d_j = \sqrt{\frac{1}{N_j}\sum_{i=1}^{N_j}\frac{(e_{j,i} - s_{j,i})^2}{\Delta e_{j,i}^2 + \Delta s_{j,i}^2}}, \tag{20.7}$$

where $\Delta e_{j,i}$ and $\Delta s_{j,i}$ are the uncertainties related to the evaluation of $e_{j,i}$ and $s_{j,i}$. Since simulations and experiments can be considered to agree if their difference is smaller than their uncertainties, we define the level of agreement between experiments and simulations with respect to observable j as

$$R_j = \frac{\tanh[(d_j - 1/d_j - d_0)/\lambda] + 1}{2} \tag{20.8}$$

with $R_j \lesssim 0.5$ corresponding to agreement, while $R_j \gtrsim 0.5$ denotes disagreement. The function in equation (20.8) is chosen as it describes a transition, whose sharpness can be changed with λ, from agreement to disagreement when $d_j \simeq d_0$. Here, we choose $d_0 = 1$ and $\lambda = 0.5$; our tests show that the conclusions of a validation exercise are not affected by the specific choices of the parameters d_0 and λ, if these parameters are within the reasonable range that point out agreement between experiments and simulations when they fall within their uncertainties. Some authors prefer to normalize the distance between experimental and simulation results to the actual value of the observables, rather than to their uncertainty [5]. We believe that the normalization to the uncertainty is the most appropriate choice in the present case, as we are interested in understanding if the basic physics mechanisms at play in the system are well captured by the model under consideration. The normalization to the actual value of the observable is instead preferable in the case that the predictive capabilities of the code are tested.

In the case of the experiments, we can identify three main uncertainty sources. First, the model of a measuring device provides predictions through which one can infer the physical quantities of interest (e.g., from the I–V curve of a Langmuir probe one can infer n and T_e). Experimental measurements typically do not follow perfectly the model predictions: thus, a fit has to be made in order to evaluate the relevant physical parameters, introducing an uncertainty that we denote with $\Delta e_{j,i}^{\text{fit}}$. Second, a source of uncertainty is due to properties of the measuring device that are often difficult to evaluate accurately (e.g., geometry and surface condition of a Langmuir probe). Thus, measurements should be performed with different tools (e.g., Langmuir probes which differ in dimension, surface condition, and electronics). The quantity $\Delta e_{j,i}^{\text{prb}}$ denotes the uncertainty related to the probe properties. Finally, the plasmas are not perfectly reproducible due to control parameters difficult to set or know precisely (e.g., the vacuum pressure). Experiments should be repeated in order to check the reproducibility of the plasma, while measurements are taken with different measurement devices. The quantity $\Delta e_{j,i}^{\mathrm{rep}}$ is the uncertainty due to the plasma reproducibility, averaged over the different measuring devices. The total experimental uncertainty is given by $\Delta e_{j,i}^2 = (\Delta e_{j,i}^{\mathrm{fit}})^2 + (\Delta e_{j,i}^{\mathrm{prb}})^2 + (\Delta e_{j,i}^{\mathrm{rep}})^2$.

Simulations are also affected by uncertainties resulting from two sources: (i) errors due to the numerics and (ii) errors due to unknown or imprecise input parameters. While errors due to the numerics, $\Delta s_{j,i}^{\mathrm{num}}$, can be estimated through the methodology described in section 20.2.1, the evaluation of the error related to not perfectly known input parameters, $\Delta s_{j,i}^{\mathrm{inp}}$, requires a uncertainty propagation study, i.e. an investigation of how the model results are affected by the input parameter variations. The number of input parameters of a turbulence simulation code is usually quite large and a complete study of the model response is prohibitive. However, the theory can indicate to which input parameters the results are particularly sensitive. The analysis has then to focus on those. We remark that in the literature, a number of useful techniques have been proposed to predict the response of the model to variation of simulation parameters using the smallest possible number of simulations (see, e.g. reference [36]). As in the case of the experimental error bars, the two sources of error should be added, such that $\Delta s_{j,i}^2 = (\Delta s_{j,i}^{\mathrm{num}})^2 + (\Delta s_{j,i}^{\mathrm{inp}})^2$.

We note that the error bars should not take into account the uncertainties related to model assumptions and/or to combinations of measurements, which are often needed to deduce the comparison observables from the simulation results and the raw experimental data [15]. Evaluating rigorously those uncertainties is usually very challenging. The idea is to take them into account approximately through the observables primacy hierarchy. More specifically, the higher the hierarchy level of an observable is, the lower the importance of the observable in the comparison metric.

The overall level of agreement between simulations and experiments can be measured by considering a composite metric, which should take into account the level of agreement of each observable, R_j, and weight it according to how constraining each observable is for comparison purposes. This means that the hierarchy level of each observable and the level of confidence characterizing the measurement or the simulation of each observable have to be considered. The higher the level in the primacy hierarchy and the bigger the error affecting the observable measurement, the smaller the weight of the observable. We thus define the metric χ as

$$\chi = \frac{\sum_j R_j H_j S_j}{\sum_j H_j S_j}, \tag{20.9}$$

where H_j and S_j are functions defining the weight of each observable according to its hierarchy level and the precision of the measurement, respectively. Thanks to the definition of R_j, χ is normalized in such a way that perfect agreement is observed for $\chi = 0$, while simulation and experiment disagree completely for $\chi = 1$.

The definition of H_j and S_j is somewhat arbitrary. H_j should be a decreasing function of the hierarchy level. The definition we adopt is $H_j = 1/h_j$, where h_j is the combined experimental/simulation primacy hierarchy level, which takes into account the number of assumptions or combinations of measurements used in evaluating the observables both from the experiments and from the simulations. In practice, if no assumptions or combinations of measurements are used for obtaining an observable, $h_j = 1$, any assumption or combination of measurement leads h_j to increase of a unity (see reference [15] for more details on the h_j definition). The quantity S_j should be a decreasing function of the experimental and simulation uncertainty. We introduce the following definition:

$$S_j = \exp\left(-\frac{\sum_i \Delta e_{j,i} + \sum_i \Delta s_{j,i}}{\sum_i |e_{j,i}| + \sum_i |s_{j,i}|}\right) \tag{20.10}$$

such that $S_j = 1$ in the case of zero uncertainty.

The validation metric should be complemented by an index, Q, that assesses the 'quality' of the comparison. The idea is that a validation is more reliable with a larger number of independent observables, particularly if they occupy a low level in the primacy hierarchy and the measurement and simulation uncertainties are low. The quality of the comparison Q can thus be defined as

$$Q = \sum_j H_j S_j \tag{20.11}$$

20.3 A practical example of V...V methodology use

We present an example of V...V procedure, to which we apply the methodology described in section 20.2. We discuss the simulations of plasma turbulence in the basic plasma physics experiment TORPEX [1, 9] that have been carried out using the GBS code. The GBS code has been developed in the last few years to simulate plasma turbulence in the open field region of magnetic confinement devices, evolving the drift-reduced Braginskii two-fluid equations [37], without any separation between equilibrium and perturbation quantities [2, 3]. In this development, increasingly complex magnetic configurations have been considered: first, the code was developed to describe basic plasma physics devices, in particular, linear devices such as LAPD [38] and simple magnetized toroidal devices such as TORPEX [11], whose simulations are the focus of the present chapter. GBS was then extended to the tokamak geometry, and it is now able to model the tokamak scrape-off-layer (SOL) region in limited plasmas (see, e.g. references [39–42]).

We first introduce the TORPEX device (section 20.3.1) and then the models used for the simulations of TORPEX and the GBS code (section 20.3.2). GBS has been the subject of a rigorous code verification procedure that is described in references [17, 3] that we briefly summarize in section 20.3.3. The application of the solution verification procedure is the subject of section 20.3.4. Finally, the validation is described in section 20.3.5 We remark that the present example of use of the V…V methodology is reported in reference [18].

20.3.1 The TORPEX device, its diagnostics and ancillary systems

The TORoidal Plasma EXperiment (TORPEX) is a highly flexible toroidal device that allows for full diagnostic access and a variety of magnetic configurations. Plasmas of different gases are produced and sustained by microwaves at 2.45 GHz with typical densities $n_e \approx 10^{15} - 10^{17}$ m^{-3} and temperatures $T_e \approx 5 - 10$ eV. The main TORPEX elements are reviewed in this section. Figure 20.2 shows a drawing of the TORPEX device with the main elements. The circular cross-section toroidal vacuum chamber with major radius $R = 1$ m and minor radius $a = 0.2$ m is made of stainless steel and is divided into twelve sectors, each spanning over 30 degrees. The fourfold symmetrical vessel is divided into four electrically insulated sets, of three sectors each. Four sectors of the vessel, toroidally separated by 90 degrees from each other, are extractable, to facilitate the installation of diagnostics. Spare sectors are available and can be modified for new experiments, without stopping the experimental activity. A total number of 48 port-holes provides access for diagnostics and pumping. Four turbo-pumps (nominal pumping speed of 345 l s^{-1} for nitrogen), backed by four primary pumps, maintain a base pressure $\lesssim 5 \times 10^{-5}$ mbar inside the vessel. Gas or mixture of different gases (argon, hydrogen, deuterium, helium, and neon) are injected at one toroidal position; the injection rate is remotely controlled.

The magnetic field coil system comprises a toroidal, a poloidal and an ohmic system allowing for a variety of magnetic configurations (see reference [43] for details). The toroidal system (figure 20.2) consists of 28 water-cooled coils for the

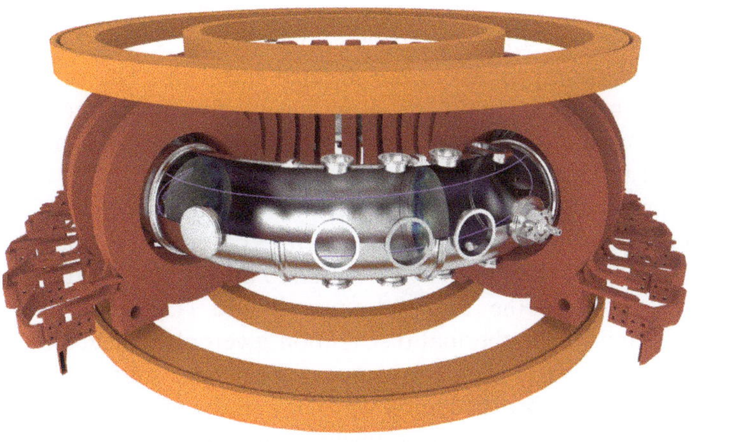

Figure 20.2. View of the TORPEX device with the main elements. A helical magnetic field line of an SMT configuration is shown in violet. Simulated plasma potential profiles at two toroidal positions are shown. These are obtained from numerical simulations using the GBS code.

generation of the toroidal magnetic field up to ~100 mT on axis. The poloidal coil system includes 10 water-cooled coils and allows for a variety of different vertical magnetic field configurations with a maximum vertical field of 5 mT. The design and arrangement of the poloidal coils are optimized to generate also a cusp field up to 100 mT about 10 cm from the centre of the vessel, with a magnetic X-point at the centre of the cross-section. A subset of the poloidal coils can be used to induce in the vessel a loop voltage of the order of 10 V for approximately 30 ms, to drive a plasma current in a tokamak-like configuration [44]. The combination of a toroidal magnetic field B_φ with a vertical magnetic field B_z results in helical open field lines, terminating on the vacuum vessel, as shown in figure 20.2. This magnetic geometry is known as *simple magnetized torus* (SMT) [45], and incorporates the main ingredients for drift and interchange instabilities and turbulence, namely pressure gradients, ∇B and magnetic field curvature. Although it does not possess a rotational transform of the magnetic field lines, the SMT mimics the SOL of magnetic confinement devices for fusion, where magnetic field lines are open and terminate either on the vessel wall or on limiters/divertors.

TORPEX plasmas are produced and sustained by microwaves at 2.45 GHz, corresponding to the electron-cyclotron (EC) range of frequencies. The choice of the frequency is based on the constraints imposed by the maximum available toroidal magnetic field, $B_\varphi \approx 0.1$ T, and by the large availability of sources in this frequency range for industrial applications. The microwave system installed on TORPEX consists of a commercial microwave source and a transmission line, which connects the source to the vacuum chamber. The source can deliver up to 5 kW of microwave power in continuous mode, or up to 50 kW during 100 ms in pulsed mode. The microwave power can be modulated with frequencies up to 100 kHz for a sinusoidal waveform, or 20 kHz for a square waveform. The modulation amplitude is arbitrary, with a lower limit for the injected power of 200 W. The injected and reflected power are measured on the transmission line and provide the value of the power absorbed by the plasma. The microwaves are injected perpendicularly to the magnetic field from the LFS. A truncated waveguide is used as antenna, which introduces finite $k_\perp$ components in the injected spectrum. The ordinary (O-mode) polarization has been chosen to avoid the cut-off for the extraordinary mode at relatively low densities. A complete description of the TORPEX microwave system can be found in reference [46].

The plasma production mechanism associated with the injection of microwaves in the electron-cyclotron range of frequencies in STM plasmas was investigated in great detail [47, 48, 49]. The main contribution to the particle source comes from the electrons accelerated at the EC and upper-hybrid (UH) resonant layers. The electrons in the tail of the thermal distribution give only a negligible contribution to the total ionization rate. The dependence of the plasma source upon experimental parameters, such as the absorbed microwave power, the magnetic field configuration and the neutral gas pressure, was also investigated and an expression for the spatial profile of the particle source is derived from the experimental results and a simple numerical code. These studies are of paramount importance for the code validation program.

TORPEX possesses an extensive set of diagnostics, comprising approximately 200 probes, together with a fast optical imaging system [50–52], which allow for a full reconstruction of plasma profiles and turbulent structures across the entire device. Measurements of time-averaged plasma parameters and electrostatic fluctuations are performed using combinations of arrays of Langmuir probes (LPs) in single, double and triple configuration [53], and Mach probes [54]. The LPs can be operated in ion saturation, floating potential, or swept mode and their signals are acquired at a rate of 250 kHz. The LPs are disposed at different toroidal locations and are installed on remotely controlled 2D positioning systems, which allow covering almost the entire plasma poloidal cross-section.

An array of 85 LPs at one toroidal location, the HEXagonal Turbulence Imaging Probe (HEXTIP), allows characterizing fluctuations and turbulent structures in terms of wavelength, coherence lengths, frequency, speed and statistical properties, in a single plasma discharge [55]. The original HEXTIP design with 86 probes was recently modified to be compatible with the new toroidal conductor system. HEXTIP covers the entire poloidal cross-section with a spatial resolution of 3.5 cm and constitutes the workhorse for a high temporal resolution (250 kHz) real space imaging of the ion saturation current and floating potential dynamical evolution. HEXTIP allows identifying observables related to turbulent structures, such as size, mass, shape, and trajectories from two-dimensional plasma imaging data. The probability distributions of these observables from measurements of many turbulent realizations, can then be used to characterize the nature of turbulence and the related transport. A new version of the HEXTIP array with improved (and variable) spatial resolution has recently been developed. Two new systems are under construction and will be installed in TORPEX at different toroidal locations, thus extending to 3D the measurements available with the previous HEXTIP.

Special arrangements of single and multiple single-sided LPs are used to measure plasma currents collected on a poloidal limiter, in particular for blob studies [56]. The simplest version of these probes consists of a conductor plate (usually made out of tungsten) with a collecting area A_{LP}, which is exposed to the plasma only from one side a few cm in front of the limiter. The plate is oriented perpendicularly to the magnetic field and is kept at the limiter potential, such that the current to the limiter, I_0, is now measured by the probe. The current density is then computed as $J_{\|} = I_0/A_{LP}$. A good agreement is found between single-sided probe and magnetic probe measurements [57] table 20.1.

Magnetic and current probes complement electrostatic probes and provide 3D fluctuating components of the magnetic field with nT resolution. A specially designed magnetic probe, developed in collaboration with Consorzio RFX in Padova (IT), consists of an L-shaped array of three miniaturized three-axial pick up coils (3.5 cm spaced, effective area of 2.3×10^{-3} m^{-2}) [56, 57]. The probe is installed on a 2D movable system and allows mapping of the magnetic field and current density fluctuations associated with turbulent structures [56, 57].

Together with magnetic probes, direct measurements of the current density are obtained using a dedicated miniaturized current probe diagnostic [58]. The probe consists of a Rogowski coil wound around a ferrite core to amplify the magnetic

Table 20.1. Primacy hierarchy for some observables obtained from Langmuir probe measurements. The experimental hierarchy counts the number of assumptions or combinations of measurements used to obtain the observables from experimental data (the first level in the hierarchy denotes no assumptions or combinations of measurements; then, each assumption or combination of measurements increases the hierarchy level of a unity). The same definition is used for the simulation hierarchy. The comparison hierarchy sums the experimental and simulation assumptions.

	Experimental hierarchy	Simulation hierarchy	Comparison hierarchy
Moments, PDF, and PSD of I_{sat} and V_{fl}	1	2	2
$\bar{T}_e$, $\bar{n}$, $\bar{\phi}$	2	1	2
Results from spectral analysis (e.g., k_z)	2	1	2
Statistical analysis of turbulent structures	2	3	4
T_e	3	1	3
Particle flux	3	2	4

flux. Electrical and thermal insulation from the plasma is obtained by placing the detecting coil inside a vacuum safe cylindrical housing, machined as two halves of a container. The toroidal detecting coil consists of a single layer helical winding with an inside return loop to cancel the net single turn formed by the solenoid windings. The probe is operated in current mode, which provides a direct measurement of the total current flowing inside the detection area.

20.3.2 The simulation model

Owing to the low TORPEX plasma temperature, the drift-reduced Braginskii equations (see, e.g., reference [37]) can be used to model the TORPEX plasma dynamics. In the limit of $T_i \ll T_e$ and $\beta \ll 1$, and assuming that $B_z \ll B_\varphi$ so that $B \simeq B_0R/(R + r)$, since the magnetic curvature is constant along a field line and equal to $R + r$, these equations can be written as

$$\frac{\partial n}{\partial t} = \frac{c}{B}[\phi, n] + \frac{2c}{eRB}\left(\frac{\partial p_e}{\partial z} - en\frac{\partial \phi}{\partial z}\right) - \frac{\partial (nV_{\|e})}{\partial x_\|} + S_n \tag{20.12}$$

$$\frac{\partial \nabla_\perp^2 \phi}{\partial t} = \frac{c}{B}[\phi, \nabla_\perp^2 \phi] - V_{\|i}\frac{\partial \nabla_\perp^2 \phi}{\partial x_\|} + \frac{2B}{cm_iRn}\frac{\partial p_e}{\partial z} + \frac{m_i\Omega_{ci}^2}{e^2 n}\frac{\partial j_\|}{\partial x_\|} \tag{20.13}$$

$$\begin{aligned}\frac{\partial T_e}{\partial t} &= \frac{c}{B}[\phi, T_e] + \frac{4c}{3eRB}\left(\frac{7}{2}T_e\frac{\partial T_e}{\partial z} + \frac{T_e^2}{n}\frac{\partial n}{\partial z} - T_e\frac{\partial \phi}{\partial z}\right)\\ &+ \frac{2}{3}\frac{T_e}{en}0.71\frac{\partial j_\|}{\partial x_\|} - \frac{2}{3}T_e\frac{\partial V_{\|e}}{\partial x_\|} - V_{\|e}\frac{\partial T_e}{\partial x_\|} + S_T\end{aligned} \tag{20.14}$$

$$\frac{\partial V_{\parallel i}}{\partial t} = \frac{c}{B}\left[\phi,\ V_{\parallel i}\right] - V_{\parallel i}\frac{\partial V_{\parallel i}}{\partial x_{\parallel}} - \frac{1}{nm_i}\frac{\partial p_e}{\partial x_{\parallel}} \tag{20.15}$$

$$\begin{aligned} m_e n\frac{\partial V_{\parallel e}}{\partial t} &= m_e n\frac{c}{B}\left[\phi,\ V_{\parallel e}\right] - m_e n V_{\parallel e}\frac{\partial V_{\parallel e}}{\partial x_{\parallel}} \\ &\quad - T_e\frac{\partial n}{\partial x_{\parallel}} + en\frac{\partial \phi}{\partial x_{\parallel}} - 1.71n\frac{\partial T_e}{\partial x_{\parallel}} + \frac{en}{\sigma_{\parallel}}j_{\parallel} \end{aligned} \tag{20.16}$$

where $p_e = nT_e$, $[a, b] = \partial_r a\partial_z b - \partial_z a\partial_r b$, $j_{\parallel} = en\left(V_{\parallel i} - V_{\parallel e}\right)$, and S_n and S_T are the density and temperature sources. The r coordinate denotes the radial direction, $x_{\parallel}$ is parallel to B, and z is the direction perpendicular to r and $x_{\parallel}$ (for $B_v \ll B_{\varphi}$ the vertical and z directions are approximately the same).

The computational domain has an annular shape with a cross-section $r = -L_v/2$ to $r = L_v/2$ and $z = 0$ to $z = L_v$. At $r = -L_v/2$ and $r = L_v/2$, Dirichlet boundary conditions are used for n, T_e, ϕ, and $\nabla_{\perp}^2\phi$ and Neumann boundary conditions for $V_{\parallel e}$ and $V_{\parallel i}$.

At the upper and lower walls, at $z = 0$ and $z = L_v$, we consider two sets of boundary conditions. First, we implement an ad hoc set of boundary conditions, where we impose $V_{\parallel i} = \pm c_s$ and $V_{\parallel e} = \pm c_s \exp(\Lambda - e\phi/T_e)$ with $\Lambda = \log\sqrt{m_i/(2\pi m_e)}$ and we explore both Dirichlet and Robin boundary conditions for n and T_e, while for ϕ we use both Dirichlet boundary conditions $e\phi = \Lambda T_e$ (implying $V_{\parallel e} = V_{\parallel i}$) and a boundary condition of the form $\partial_z\phi \propto (e\phi - \Lambda T_e)$. Second, we use the first-principle set of boundary conditions that has been derived in reference [12], that is $V_{\parallel i} = \pm c_s$, $V_{\parallel e} = \pm c_s \exp(\Lambda - e\phi/T_e)$, $\partial_z T_e = k_T\partial_z\phi$, $\partial_z n = \mp(n/c_s)\partial_z V_{\parallel i}$, $\Omega = \nabla_{\perp}^2\phi = -\cos^2\alpha[(\partial_z V_{\parallel i})^2 \pm c_s\partial_z^2 V_{\parallel i})]$, and $\partial_z\phi = \mp c_s\partial_z V_{\parallel i}$.

We have used source profiles that mimic the EC and UH resonance layer in TORPEX, i.e., $S_{n,T} = S_{0;n,T}\{S_{\mathrm{UH}}\exp[-(r - r_{\mathrm{UH}})^2/\lambda_{\mathrm{UH}}^2] + S_{\mathrm{EC}}\exp[-(r - r_{\mathrm{EC}})^2/\lambda_{\mathrm{EC}}^2]\}$, with $S_{\mathrm{UH}} = 1.5$, $S_{\mathrm{EC}} = 1$, $\lambda_{\mathrm{UH}} = 1$ cm, $\lambda_{\mathrm{EC}} = 0.5$ cm, $r_{\mathrm{UH}} = -2$ cm, $r_{\mathrm{EC}} = -6$ cm, and values of the source strength ($S_{0n} = 1.5 \cdot 10^{20}$ m^{-3} s^{-1}, $S_{0T} = 3.5 \cdot 10^4$ eV s^{-1}) estimated experimentally through a global balance of the TORPEX plasma. We remark that the dependence of the UH resonance position on n is neglected in the present model. Other values used are: $R = 1$ m, $L_v = 40$ cm, $m_i/m_e = 200$, $\Lambda = 3$, the resistivity has been varied, ranging from $\nu = 0.01c_s/R$ to $\nu = 1c_s/R$, in order to check the influence of this parameter [$\nu = e^2nR/(m_i\sigma_{\parallel}c_s)$].

GBS solves the equations (20.12)–(20.16) on a field-aligned grid using a second-order finite difference scheme with Runge–Kutta time stepping and small diffusion terms The Poisson brackets are evaluated using the Arakawa scheme [59].

If only $k_{\parallel} \simeq 0$ modes are considered, simple two-dimensional fluid equations that describe the plasma turbulence can be considered. The Braginskii equations are integrated in the parallel direction in order to evolve the line-integrated density, $n(r, z) = \int n(r, z, x_{\parallel})dx_{\parallel}/L_c$, potential, $\phi(r, z) = \int \phi(r, z, x_{\parallel})dx_{\parallel}/L_c$, and temperature, $T_e(r, z) = \int T_e(r, z, x_{\parallel})dx_{\parallel}/L_c$, $L_c = 2\pi NR$ being the magnetic field line length. The magnetic field lines are characterized by the parameter $N = L_v B_{\varphi}/(2\pi RB_z)$, which is the number of field line turns in the device. We use Bohm's boundary conditions to take into account the ion and electron parallel flow at the sheath edge: by assuming that the density at the edge is equal to $n(r, z)/2$, it is possible to approximate the ion and electron flows as $\Gamma_{\|,i} = nc_s/2$ and $\Gamma_{\|,e} = nc_s \exp(-e\phi/T_e + \Lambda)/2$. The evolution equations for n, ϕ, and T_e thus become

$$\begin{aligned}\frac{\partial n}{\partial t} = \left[\frac{c\phi}{B_0}, n\right] + \frac{2c}{eRB_0}\left(n\frac{\partial T_e}{\partial z} + T_e\frac{\partial n}{\partial z} - en\frac{\partial \phi}{\partial z}\right) \\ - \frac{\sigma n c_s}{R}\exp(\Lambda - e\phi/T_e) + S_n,\end{aligned} \tag{20.17}$$

$$\begin{aligned}\frac{\partial \nabla^2\phi}{\partial t} = \left[\frac{c\phi}{B_0}, \nabla^2\phi\right] + \frac{2B_0}{cm_iR}\left(\frac{T_e}{n}\frac{\partial n}{\partial z} + \frac{\partial T_e}{\partial z}\right) \\ + \frac{\sigma c_s m_i \Omega_i^2}{eR}[1 - \exp(\Lambda - e\phi/T_e)],\end{aligned} \tag{20.18}$$

$$\begin{aligned}\frac{\partial T_e}{\partial t} = \left[\frac{c\phi}{B_0}, T_e\right] + \frac{4c}{3eRB_0}\left(\frac{7}{2}T_e\frac{\partial T_e}{\partial z} + \frac{T_e^2}{n}\frac{\partial n}{\partial z} - eT_e\frac{\partial \phi}{\partial z}\right) \\ - \frac{2}{3}\frac{\sigma T_e c_s}{R}[1.71\exp(\Lambda - e\phi/T_e) - 0.71] + S_T,\end{aligned} \tag{20.19}$$

where $\sigma = R/L_c = \Delta/(2\pi L_v)$. We note that a similar system of equations has been used in reference [60]. The system of equations (20.17)–(20.19) has been solved numerically, using the earliest version of GBS that has been developed from the ESEL code [61]. The algorithm used is described in reference [62].

For the 2D simulations, we consider a domain with extension L_r in the radial direction and Δ along z. The boundary conditions are periodic along the vertical direction (due to the flute property of the interchange mode) and we use Dirichlet boundary conditions in the radial direction. In order to study the sensitivity of the results to the parallel boundary conditions, a scan of the σ parameter has been performed.

Both for the 3D and the 2D case, the simulation is started from random noise. Then, the sources introduce plasma and heat, increasing the plasma pressure and triggering the interchange instability. The interchange instability leads to density and particle transport in the radial direction from the source region to the low field side of the machine. At the same time, plasma is removed from the system by parallel losses. The results discussed in this chapter focus on the quasi-steady-state period, established after the initial simulation transient, as a result of a balance between parallel losses, perpendicular transport, and sources.

A detailed analysis of the plasma dynamics described by the 3D model (20.12)–(20.16) has been presented in references [10, 11], while the simulation results obtained from the 2D model (20.17)–(20.19) have been discussed in reference [13].

For the comparison of simulations and experiments, we consider a set of TORPEX configurations with different values of vertical magnetic field, characterized by different properties of plasma turbulence. We analyze four scenarios, characterized by different windings of the magnetic field lines, i.e. $N = 2$, 4, 8, and 16. By using GBS, we carry out simulations of TORPEX plasma turbulence by considering three different models: (a) a global three-dimensional two-fluid model that describes the evolution of the plasma dynamics in the full TORPEX volume, (b) a global three-dimensional two-fluid model that describes the evolution of the plasma dynamics in the full TORPEX volume provided with first-principle boundary conditions recently derived [12], and (c) a two-dimensional model that is able to represent only $k_{\|} = 0$ modes.

20.3.3 Code verification

GBS has been subject to the code verification procedure described in references [17, 3]. Here we recall the main elements of the procedure. To verify the implementation of the drift-reduced Braginskii equations into GBS and to satisfy the requirements given in section 2.1, we choose to manufacture the model solution as the combination of trigonometric functions,

$$s(y, r, \psi, t) = A_s\{B_s + C_s \sin[D_s x_{\|}] \sin(E_s\varphi + F_s) \sin[G_s r + H_s t]\} \qquad (20.20)$$

where A_s, B_s, C_s, D_s, E_s, F_s, G_s, and H_s are arbitrary constants, φ is the toroidal angle, and s represents the scalar fields $\{n, V_{\|i}, V_{\|e}, T_e, \phi\}$ present in the GBS equations. The constant B_s is used to ensure $n > 0$, $T_e > 0$, while the other factors are used to calibrate the numerical error, ensuring that no term dominates the numerical error in the equations. The dependencies imposed through these constants manifest the physical problem of interest, with a term being perfectly aligned to the magnetic field lines, and a perturbation along the vertical z coordinate. Variation along r and a temporal dependence of the manufactured solution are also included.

The computation of the source terms S simply consists in plugging the analytical manufactured solutions into the drift-reduced Braginskii equations. This process is particularly tedious, but it involves only straightforward algebraic manipulations with no conceptual difficulties and can be carried out by using the symbolic manipulation software *Mathematica* [63], which allows the direct translation into the *Fortran* language in which GBS is written.

In figure 20.3 we report the results of the code verification. Six simulations are carried out, with mesh refinement parameters $h = 1, 2, 4, 8, 16, 32$ in each direction. The time step is varied with $\sqrt{h}$, as the Runge–Kutta time advance is fourth order accurate compared to the second order accurate spatial derivatives. The numerical error, evaluated as the norm of the difference between the numerical and the manufactured solutions at each grid point, decreases with the grid spacing linearly on a logarithmic scale, when refining the mesh, with the slope expected from the order of accuracy of the numerical method. The scan we perform shows a reduction of the numerical error by at least three orders of magnitude, giving

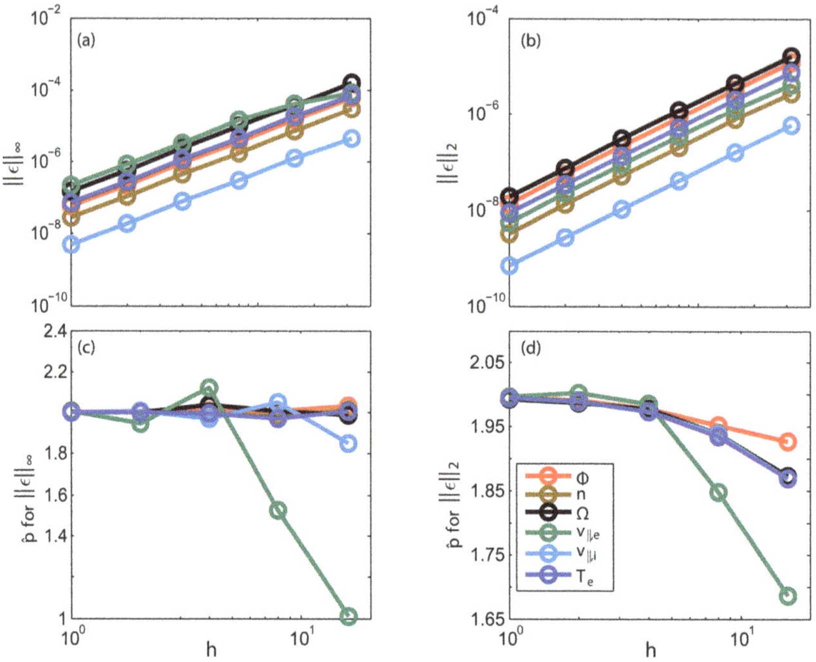

Figure 20.3. $\|\epsilon\|_\infty$ and $\|\epsilon\|_2$ norms of the discretization error ((a) and (b)) and their respective order of accuracy estimates ((c) and (d)) for GBS simulations where the refinement parameter, defined as $h = x/x_0 = y/y_0 = z/z_0 = (t/t_0)^2$, is varied by a factor of 32.

confidence that there are not subdominant errors decreasing at a rate different than the expected one.

20.3.4 Solution verification

A previous solution verification procedure applied to the GBS code [17] shows that the numerical error affecting the simulations is mostly due to the space discretization, while the time discretization leads to a negligible numerical error. We therefore perform three simulations of TORPEX turbulence for $N = 2$ by increasing the spatial grid resolution in all directions, focusing our attention on the numerical error that affects the ten validation observables considered in section 20.3.5. In particular, we consider the equilibrium radial profiles at the vertical midplane of density, n, electron temperature, T_e, electric potential, ϕ, ion saturation current, I_{sat}, normalized I_{sat} fluctuations, $\delta I_{sat}/I_{sat}$, I_{sat} skewness and kurtosis. We also use as validation observables the vertical wave number, k_z, the probability distribution function (PDF), and the power spectrum density (PSD) of I_{sat} at the vertical midplane, at the radial point where I_{sat} is equal to the 3/4 of its peak value (this is the location where we expect to identify more clearly the turbulence properties). We plot the results of the three simulations in figure 20.4. Convergence in some cases shows an oscillatory character: as the necessary condition to use the *GCI* estimate is missing, we estimate the numerical uncertainty as the spatial average of the maximum of the difference

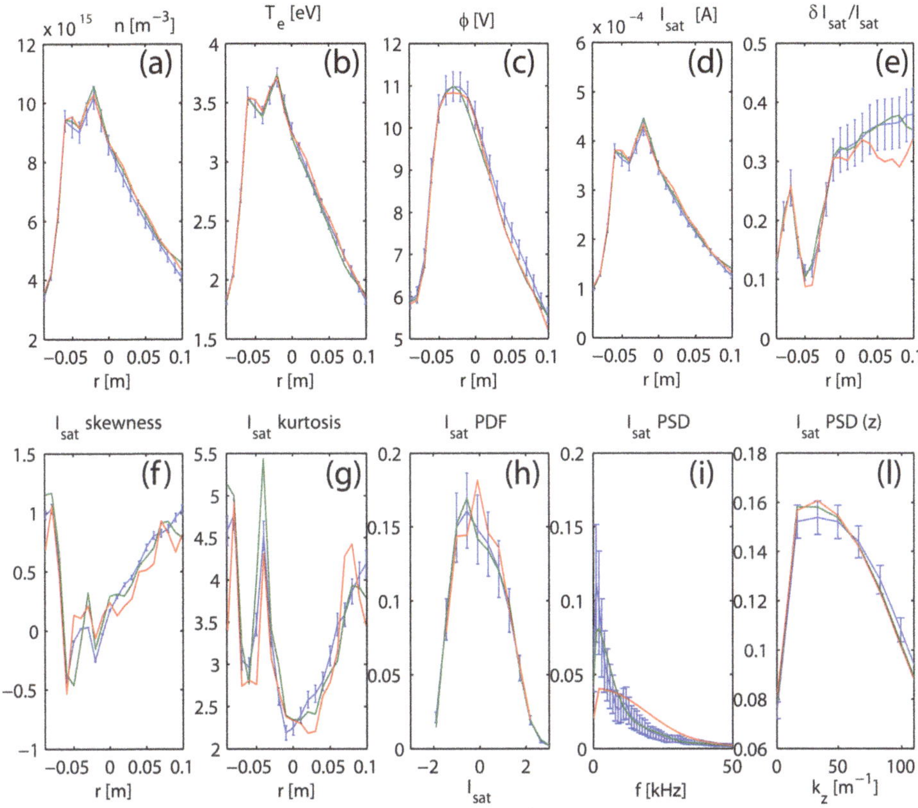

Figure 20.4. Evaluation of the numerical uncertainty due to the numerical discretization. The ten observables considered for the validation procedure are plotted for TORPEX simulations with $N = 2$, considering an increasing refinement of the mesh: $N_r \times N_z \times N_\varphi = 128 \times 128 \times 32$ (blue), $N_r \times N_z \times N_\varphi = 192 \times 192 \times 48$ (green), and $N_r \times N_z \times N_\varphi = 288 \times 288 \times 72$ (red). The resulting estimated uncertainty is represented by the error bars. The I_{sat} PDFs are rescaled to the same values of average, variance, and area. The PSDs are rescaled to the same area.

between the different simulations. The error bar associated with this uncertainty is plotted in figure 20.4.

20.3.5 Validation

To compare experiments and simulations, we use Langmuir probe measurements and the observables [15, 64] listed in table 20.1. Here, the ten observables are all at the second level of the comparison hierarchy ($H = 0.5$ in all cases). For each observable, the experimental uncertainty Δe is evaluated by repeating the experiments a number of times, and comparing the measurements of different probes. The simulation uncertainty, Δs, is evaluated by performing a number of simulations where we vary the input parameters. A complete sensitivity scan would require the analysis of all input parameters; however, due to the high computational cost of the present simulations large parameter scans are prohibitive at the moment, and we have focused our attention on the parameters that are expected to most significantly affect the simulations and that are not well known. For the three-dimensional simulations, these are the plasma resistivity ν and the boundary conditions (in the case of simulations with ad hoc boundary conditions). An example of uncertainty evaluation is shown in figure 20.5, where we show three simulations carried out with different values of ν and the error bar we deduce. For the two-dimensional simulations, instead, we study the sensitivity to the parameter describing the parallel losses, σ.

The values of χ are plotted in figure 20.6, which describes the dependence of the simulation/experiment agreement as a function of the number of field line turns, N, for both the two-dimensional and three-dimensional simulations, with and without the first-principle set of boundary conditions). For the three-dimensional simulations, it is $\chi \simeq 0.5$, showing that the three-dimensional simulations are able to represent equally well low and high N scenarios. A clear trend is instead observed in the case of the

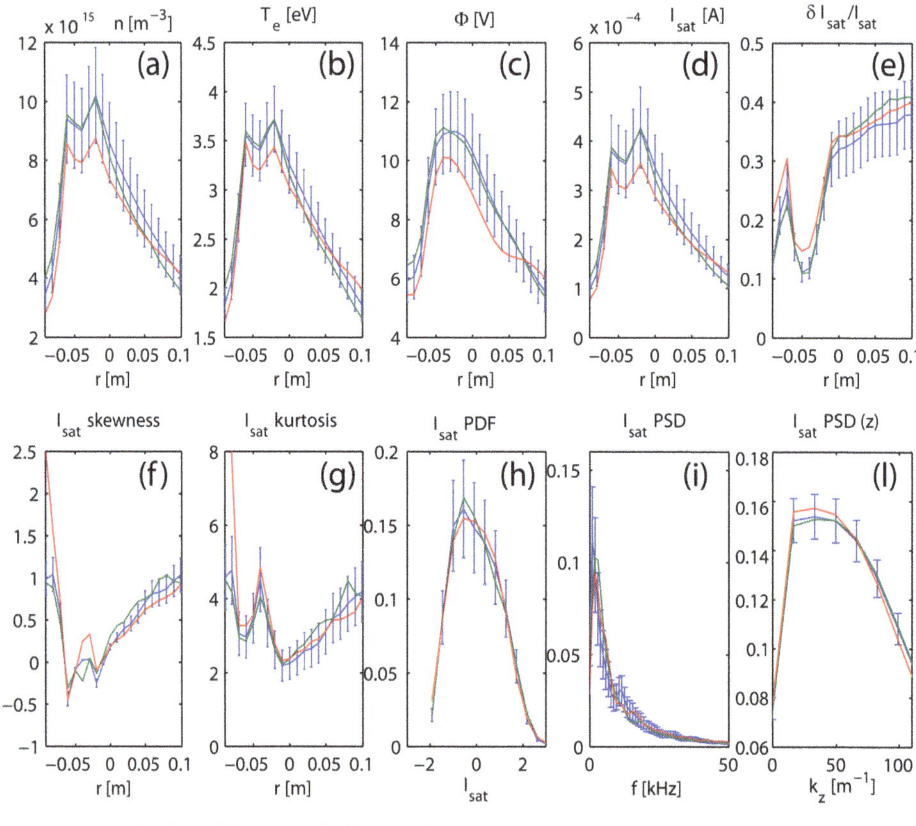

Figure 20.5. Evaluation of the numerical uncertainty due to poorly known input parameters (in this case plasma resistivity). The ten observables considered for the validation procedure are plotted for TORPEX simulations with $N = 2$, considering $\nu = 0.01$ (green), $\nu = 0.1$ (blue), and $\nu = 1$ (red). The resulting estimated uncertainty is represented by the error bars. The I_{sat} PDFs are rescaled to the same values of average, variance, and area. The PSDs are rescaled to the same area.

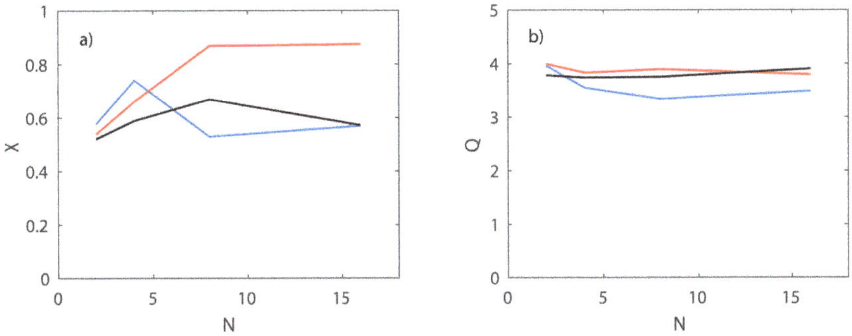

Figure 20.6. Global metric, χ, and quality of the comparison, Q, for the TORPEX simulations as a function of N. The three-dimensional simulations with ad hoc boundary conditions (blue), the three-dimensional simulations with first-principle boundary conditions (black), and two-dimensional simulations (red) are considered.

two-dimensional simulations, where the agreement decreases with N, passing from $\chi \simeq 0.6$ for $N = 2$ to $\chi \simeq 0.9$ (i.e., almost complete disagreement) for $N = 16$. With respect to the observables considered herein, no clear improvement is observed in the simulations carried out with the first-principle set of boundary conditions.

If we look at the origin of the discrepancy between simulations and experiment in more details, we observe that the three-dimensional simulations generally describe reasonably well the equilibrium profiles of n, and I_{sat}. The agreement is worse in the case of the T_e and ϕ profiles. A significant discrepancy between simulations and experiments is revealed by the comparison of the turbulence amplitude levels, as the simulated turbulence amplitude is about a factor of two smaller than the experimental one. The analysis of the PSD reveals that the difference between the experimental and simulated turbulence level is present across all the frequencies.

For two-dimensional simulations, the agreement degrades with increasing N for the majority of the observables. In fact, for $N = 2$ and $N = 4$ the agreement between two-dimensional simulation and experiment is comparable to the one observed in the case of three-dimensional simulations. At $N = 16$, all the observables reveal a complete disagreement.

The quality of the comparison is also plotted in figure 20.6. Since the uncertainties for all the observables are relatively small and all the observables are at the second level of the validation hierarchy, Q is almost independent of N, and in particular $Q \simeq 4$, that is close to the maximum value, $Q = 5$ that can be obtained by using 10 observables at the second level of hierarchy. The Q values reported in the present validation project can be compared with the Q that would be obtained in a validation carried out by comparing exclusively the agreement of the experimental and simulation particle fluxes, that is $Q \leqslant 0.25$.

20.4 Conclusions

In this chapter, we propose a methodology for carrying out a V...V procedure to make progress in the understanding of plasma turbulence. Rigorous techniques have been introduced to assess the correct implementation of the model equations in a simulation code (code verification), and estimate the numerical error affecting the simulation results (solution verification). The assessment of the agreement between experiments and simulations (validation) makes use of a number of validation observables to obtain a global metric χ normalized in order to be equal to 0 in the case of perfect agreement and 1 in the case of complete disagreement. A validation is not concluded until the quality of the comparison, Q, is also provided. The parameter Q is an index that can be used to compare validations among them; it reveals how well a comparison has been made, indicating the number of observables used for the comparison and how constraining they are.

The proposed methodology has been tested on the simulation of the basic plasma physics experiment TORPEX, focusing on measurements from Langmuir probes. We have considered simulations carried out with the GBS code, focusing on two-dimensional and a three-dimensional models. The value of χ and Q are displayed in figure 20.6.

What progress has our V…V exercise allowed? First, by carrying out the verification procedure we have largely increased our confidence on the GBS simulation results. In fact, we have rigorously shown that the drift-reduced model is correctly implemented in GBS and, by establishing a methodology to evaluate the magnitude of the numerical error affecting the simulation results, we have quantified the numerical uncertainty of the TORPEX simulations. Second, by comparing different models, we have made progress in the understanding of plasma turbulence in TORPEX. As discussed in reference [11], there are two turbulent regimes in TORPEX, each primarily driven by a distinct plasma instability: the ideal and the resistive interchange modes. The most obvious difference between the two regimes is the wave number along the magnetic field: $k_{\|} = 0$ in the ideal interchange case, while $k_{\|} \neq 0$ in the resistive case. The main parameter that controls the transition from one instability to the other is the pitch of the field line, expressed in terms of the number of field line turns, N. At low values of N, TORPEX dynamics is dominated by the ideal interchange regime. Theoretical investigations [11] show that the transition to the resistive regime occurs at a value of N that depends on the plasma resistivity and on the vertical size of the device. At the TORPEX resistivity and size, the transition is expected to take place at $N \simeq 10$ as is confirmed experimentally [65] and by the three-dimensional simulations, which show a transition between the two regimes while passing from the $N = 4$ to the $N = 8$ simulations. The validation exercise points out that it is essential to model correctly the transition from the ideal to the resistive interchange mode regime that has $k_{\|} \neq 0$ (not allowed in the two-dimensional simulations), in order to describe the plasma dynamics at high N.

The validation has also pointed out that no significant improvement in the description of the experimental results is made by the implementation of the first-principle set of boundary conditions. However, its implementation eliminates the need of a sensitivity analysis to the boundary conditions, and the possibility of a fortuitous agreement between simulations and experiments. On the other hand, it has been pointed out that the effect of the boundary conditions is particularly important in the parallel velocity profile [66]. We therefore expect that it would be possible to discriminate better the physics of the boundary conditions by considering observables that provide insights on the parallel flow and are based on measurements taken in proximity of the wall. As a matter of fact, the missing elements in the description of the T_e scale length and responsible for the underestimation of the fluctuation amplitude should be searched elsewhere (e.g., plasma source fluctuations, Boussinesq approximation, presence of fast electrons). The importance of these elements can still be assessed through the use of the validation methodology. The example we provide shows that it is relatively easy to use this methodology to discriminate among models and assess whether or not they follow the right trend, pinpointing that the correct physics is described by them. On the other hand, it is much more delicate to judge a single model in absolute terms and to assess its predictive capabilities.

In TORPEX, several advances have been achieved in the basic comprehension of waves, instabilities and turbulence in SMT configurations. To increase the relevance of TORPEX for fusion research and better reproduce the SOL-edge magnetic geometry in tokamak, a new system that creates twisted field line configurations [67] was developed, which is shown in figure 20.7. A toroidal copper conductor (1 cm radius) is suspended inside the vacuum vessel by an electrical coaxial feed-through, figure 20.7(a), and by three vertical stainless steel wires (1 mm diameter). Four horizontal supports are also used to stabilize the conductor during a TORPEX discharge. This set of supports enables different vertical positions of the toroidal conductor, so that the SMT configuration can be recovered by pulling it to the top of the vacuum vessel. A dedicated external power supply drives a current in the toroidal conductor with a maximum flat top current of ≈1 kA. For simplicity, only the feed-through is water-cooled. The flat top current duration is limited by the Ohmic heating of the conductor with almost pure radiative cooling in vacuum. This new

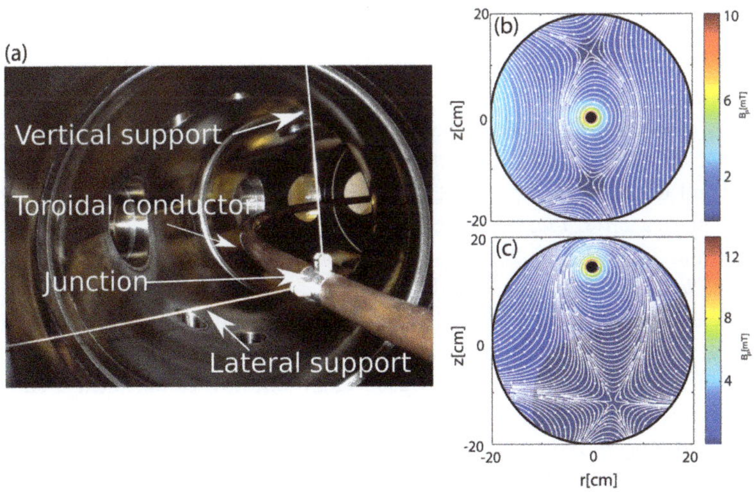

Figure 20.7. (a) The in-vessel toroidal conductor with the main elements installed on TORPEX. Simulation of a (b) double-null configuration and (c) a magnetic second order null-point (magnetic snowflake) on TORPEX.

system opens the possibility of investigating magnetic configurations with both confined regions with closed magnetic flux-surfaces and SOL regions with open field lines. Several new magnetic field configurations are possible, figure 20.7(b, c), from simple plasmas limited on the low-field side (LFS) by the vessel, to more complex magnetic geometries, such as single or double-null X-points, as well as advanced divertor concepts like snowflakes [68, 69]). These more complex magnetic configurations will provide a basis for the validation of the numerical simulations, which is necessary to determine the complexity of the numerical model needed for a realistic description not only of TORPEX data but especially for fusion devices.

References

[1] Fasoli A, Labit B, McGrath M, Müller S H, Plyushchev G, Podestà M and Poli F M 2006 Electrostatic turbulence and transport in a simple magnetized plasma *Phys. Plasmas* **13** 055902

[2] Ricci P, Halpern F D, Jolliet S, Loizu J, Mosetto A, Fasoli A, Furno I and Theiler C 2012 Simulation of plasma turbulence in scrape-off layer conditions: the GBS code, simulation results and code validation *Plasma Phys. Control. Fusion* **54** 124047

[3] Halpern F D, Ricci P, Jolliet S, Loizu J, Morales J, Mosetto A, Musil F, Riva F, Tran T M and Wersal C 2016 The GBS code for tokamak scrape-off layer simulations *J. Comp. Phys. J. Comp. Phys.* **315** 388

[4] Oberkampf W L and Trucano T G 2002 Verification and validation in computational fluid dynamics *Prog. Aerosp. Sci.* **38** 209

[5] Oberkampf W L and Roy C J 2010 *Verification and Validation in Scientific Computing* (New York: Cambridge University Press)

[6] Roache P J 1998 *Verification and Validation in Computational Science and Engineering* (Albuquerque, NM: Hermosa Publishers)

[7] Roache P J 1997 Quantification of uncertainty in computational fluid dynamics *Annu. Rev. Fluid Mech.* **29** 123–60

[8] Stern F, Wilson R V, Coleman H W and Paterson E G 2001 Comprehensive approach to verification and validation of CFD simulations—Part 1: methodology and procedures *J. Fluids Eng.* **123** 793

[9] Fasoli A *et al* 2010 Electrostatic instabilities, turbulence and fast ion interactions in the TORPEX device *Plasma Phys. Control. Fusion* **52** 124020

[10] Ricci P and Rogers B N 2009 Three-dimensional fluid simulations of a simple magnetized toroidal plasma *Phys. Plasmas* **16** 092307

[11] Ricci P and Rogers B N 2010 Turbulence phase space in simple magnetized toroidal plasmas *Phys. Rev. Lett.* **104** 145001

[12] Loizu J, Ricci P, Halpern F D and Jolliet S 2012 Boundary conditions for plasma fluid models at the magnetic presheath entrance *Phys. Plasmas* **19** 122307

[13] Ricci P, Rogers B N and Brunner S 2008 High- and low-confinement modes in simple magnetized toroidal plasmas *Phys. Rev. Lett.* **100** 225002

[14] Ricci P and Rogers B N 2009 High- and low-confinement modes in simple magnetized toroidal plasmas *Phys. Plasmas* **16** 062303

[15] Ricci P, Theiler C, Fasoli A, Furno I, Labit B, Müller S H, Podestà M and Poli F M 2009 Langmuir probe-based observables for plasma-turbulence code validation and application to the TORPEX basic plasma physics experiment *Phys. Plasmas* **16** 055703

[16] Ricci P, Theiler C, Fasoli A, Furno I, Labit B, Gustafson K, Iraji D and Loizu J 2011 Methodology for turbulence code validation: Quantification of simulation-experiment agreement and application to the TORPEX experiment *Phys. Plasmas* **18** 032109

[17] Riva F, Ricci P, Halpern F D, Jolliet S, Loizu J and Mosetto A 2014 Verification methodology for plasma simulations and application to a scrape-off layer turbulence code *Phys. Plasmas* **21** 062301

[18] Ricci P, Riva F, Theiler C, Fasoli A, Furno I, Halpern F D and Loizu J 2015 Approaching the investigation of plasma turbulence through a rigorous verification and validation procedure: A practical example *Phys. Plasmas* **22** 055704

[19] Dimits A M *et al* 2000 Comparisons and physics basis of tokamak transport models and turbulence simulations *Phys. Plasmas* **3** 969

[20] Birn J *et al* 2001 Geospace environmental modeling (GEM) magnetic reconnection challenge *J. Geophys. Res.* **106** 3715

[21] Ricci P, Brackbill J U, Daughton W and Lapenta G 2004 Collisionless magnetic reconnection in the presence of a guide field *Phys. Plasmas* **8** 4102

[22] Umansky V, Day M S and Rognlien T D 2005 On numerical solution of strongly anisotropic diffusion equation on misaligned frids *Num. Hear Transfer* B **47** 533

[23] Myra J R, D'Ippolito D A, Russel D A and Umansky M 2007 *Bull. Am. Phys. Soc.* **52** 533

[24] Falchetto G L *et al* 2008 The European turbulence code benchmarking effort: turbulence driven by thermal gradients in magnetically confined plasmas *Plasma Phys. Control. Fusion* **50** 124015

[25] Bravenec R V, Chen Y, Candy J, Wan W and Parker S 2013 A verification of the gyrokinetic microstability codes GEM, GYRO, and GS2 *Phys. Plasmas* **20** 104506

[26] Seinberg S and Roache P J 1985 Symbolic manipulation and computational fluid dynamics *J. Comput. Phys.* **57** 251

[27] Roache P J 1994 Perspective: a method for uniform reporting of grid refinement studies *J. Fluids Eng.* **116** 405

[28] Roy C J 2005 Review of code and solution verification procedures for computational simulation *J. Comput. Phys.* **205** 131

[29] Roache P J 1994 Perspective: a method for uniform reporting of grid refinement studies *J. Fluids Eng.* **116** 405

[30] Richardson L F, Philos, Trans R and Soc A 1911 *Philos. Trans. R. Soc.* A **210** 307

[31] Richardson L F and Gaunt J A 1927 VIII. The deferred approach to the limit *Philos. Trans. R. Soc.* A **226** 299

[32] Roache P J and Knupp P M 1993 Completed Richardson extrapolation *Commun. Numer. Methods Eng.* **9** 365

[33] Richards S A 1997 Completed Richardson extrapolation in space and time *Commun. Numer. Methods Eng.* **13** 573

[34] Terry P W, Greenwald M, Leboeuf J-N, McKee G R, Mikkelsen D R, Nevins W M, Newman D E and Stotler D P 2008 Task Group on Verification and Validation, U.S. Burning Plasma Organization and U.S. Transport Task Force *Phys. Plasmas* **15** 062503

[35] Greenwald M 2010 Verification and validation for magnetic fusion *Phys. Plasmas* **17** 058101

[36] Myers R H, Montgomery D C and Anderson-Cook C M 2009 *Response Surface Methodology: Process and Product Methodology, using Designed Experiment* 3rd edn (New York: Wiley)

[37] Zeiler A, Drake J F and Rogers B 1997 Nonlinear reduced Braginskii equations with ion thermal dynamics in toroidal plasma *Phys. Plasmas* **4** 2134

[38] Rogers B N and Ricci P 2010 Low-frequency turbulence in a linear magnetized plasma *Phys. Rev. Lett.* **104** 225002

[39] Ricci P and Rogers B N 2013 Plasma turbulence in the scrape-off layer of tokamak devices *Phys. Plasmas* **20** 010702

[40] Mosetto A, Halpern F D, Jolliet S, Loizu J and Ricci P 2013 Turbulent regimes in the tokamak scrape-off layer *Phys. Plasmas* **20** 092308

[41] Halpern F D *et al* 2013 Theory-based scaling of the SOL width in circular limited tokamak plasmas *Nucl. Fusion* **53** 122001

[42] Loizu J, Ricci P, Halpern F D, Jolliet S and Mosetto A 2013 On the electrostatic potential in the scrape-off layer of magnetic confinement devices *Plasma Phys. Control. Fusion* **55** 124019

[43] Müller S 2007 Turbulence in basic toroidal plasmas *PhD Thesis* SB, Lausanne

[44] Furno I, Diallo A, Fasoli A, Labit B, Mueller S H, Plyushchev G, Podestà G and Poli F M 2006 Effect of the magnetic configuration on fluctuations and turbulence in a magnetized toroidal plasma *11th EU-US Transport Task Force Meeting*

[45] Müller S H, Fasoli A, Labit B, McGrath M, Podestà M and Poli F M 2004 Effects of a vertical magnetic field on particle confinement in a magnetized plasma torus *Phys. Rev. Lett.* **93** 165003

[46] Podestà M 2007 Plasma production and transport in a simple magnetised toroidal plasma *PhD Thesis* SB, Lausanne

[47] Podestà M, Diallo A, Fasoli A, Furno I, Labit B, Müller S H and Poli F M 2007 Characterization of the electron distribution function in an electron-cyclotron driven toroidal plasma *Plasma Phys. Control. Fusion* **49** 175

[48] Podestà M, Fasoli A, Labit B, McGrath M, Müller S H and Poli F M 2006 Experimental characterization and modelling of the particle source in an electron-cyclotron wave driven toroidal plasma *Plasma Phys. Control. Fusion* **48** 1053

[49] Podestà M *et al* 2005 Plasma production by low-field side injection of electron cyclotron waves in a simple magnetized torus *Plasma Phys. Control. Fusion* **47** 1989–2002

[50] Iraji D, Diallo A, Fasoli A, Furno I and Shibaev S 2008 Fast visible imaging of turbulent plasma in TORPEX *Rev. Sci. Instrum.* **79** 10F508

[51] Iraji D, Furno I and Fasoli A 2011 Fast imaging of blob motion in TORPEX plasmas *IEEE Trans. Plasma Sci.* **39** 3010–1

[52] Iraji D *et al* 2010 Imaging of turbulent structures and tomographic reconstruction of TORPEX plasma emissivity *Phys. Plasmas* **17** 122304

[53] Theiler C, Furno I, Kuenlin A, Marmillod P and Fasoli A 2011 Practical solutions for reliable triple probe measurements in magnetized plasmas *Rev. Sci. Instrum.* **82** 013504

[54] Labit B, Theiler C, Fasoli A, Furno I and Ricci P 2011 Blob-induced toroidal momentum transport in simple magnetized plasmas *Phys. Plasmas* **18** 032308

[55] Müller S H *et al* 2006 Probabilistic analysis of turbulent structures from two-dimensional plasma imaging *Phys. Plasmas* **13** 100701

[56] Furno I, Spolaore M, Theiler C, Vianello N, Cavazzana R and Fasoli A 2011 Direct two-dimensional measurements of the field-aligned current associated with plasma blobs *Phys. Rev. Lett.* **106** 245001

[57] Furno I, Theiler C, Lançon D, Fasoli A, Iraji D, Ricci P, Spolaore M and Vianello N 2011 Blob current structures in TORPEX plasmas: experimental measurements and numerical simulations *Plasma Phys. Control. Fusion* **53** 124016

[58] Fasoli A *et al* 2013 Basic investigations of electrostatic turbulence and its interaction with plasma and suprathermal ions in a simple magnetized toroidal plasma *Nucl. Fusion* **53** 063013

[59] Arakawa A 1966 Computational design for long-term numerical integration of the equations of fluid motion: Two-dimensional incompressible flow. Part I *J. Comput. Phys.* **1** 119

[60] Bisai N, Das A, Deshpande S, Jha R, Kaw P, Sen A and Singh R 2004 Simulation of plasma transport by coherent structures in scrape-off-layer tokamak plasmas *Phys. Plasmas* **11** 4018

[61] Naulin V, Nycander J and Rasmussen J J 1998 Equipartition and transport in two-dimensional electrostatic turbulence *Phys. Rev. Lett.* **81** 4148
Garcia O E, Naulin V, Nielsen A H and Rasmussen J J 2004 *Phys. Rev. Lett.* **92** 165003

[62] Naulin V and Nielsen A H 2003 Accuracy of spectral and finite difference schemes in 2D advection problems *SIAM J. Sci. Comput.* **25** 104

[63] 2010 *Mathematica* (Champaign, IL: Wolfram Research, Inc.)

[64] Müller S H, Diallo A, Fasoli A, Furno I, Labit B, Plyushchev G, Podestà M and Poli F M 2006 Probabilistic analysis of turbulent structures from two-dimensional plasma imaging *Phys. Plasmas* **13** 100701

[65] Poli F M, Ricci P, Fasoli A and Podestà M 2008 Transition from drift to interchange instabilities in an open magnetic field line configuration *Phys. Plasmas* **15** 032104

[66] Loizu J, Ricci P, Halpern F D, Jolliet S and Mosetto A 2014 Intrinsic toroidal rotation in the scrape-off layer of tokamaks *Phys. Plasmas* **21** 062309

[67] Avino F, Fasoli A and Furno I 2014 The new TORPEX in-vessel toroidal conductor for the generation of a poloidal magnetic field *Rev. Sci. Instrum.* **85** 2014

[68] Piras F *et al* 2009 Snowflake divertor plasmas on TCV *Plasma Phys. Control. Fusion* **51** 055009

[69] Ryutov D D 2007 Geometrical properties of a 'snowflake' divertor *Phys. Plasmas* **14** 064502

[70] Theiler C, Diallo A, Fasoli A, Labit B, Podestà M, Poli F M and Ricci P 2008 The role of the density gradient on intermittent cross-field transport events in a simple magnetized toroidal plasma *Phys. Plasmas* **15** 042303

[71] Bovet A, Furno I, Fasoli A, Gustafson K and Ricci P 2012 Investigation of fast ion transport in TORPEX *Nucl. Fusion* **52** 094017

[72] Bovet A, Furno I, Fasoli A, Gustafson K and Ricci P 2013 Three-dimensional measurements of non-diffusive fast ion transport in TORPEX *Plasma Phys. Control. Fusion* **55** 124021

[73] Bovet A, Gamarino M, Furno I, Ricci P, Fasoli A, Gustafson K, Newman D E and Sanchez R 2014 Transport equation describing fractional Lévy motion of suprathermal ions in TORPEX *Nucl. Fusion* **54** 104009

[74] Diallo A, Fasoli A, Furno I, Labit B, Podestà M and Theiler C 2008 Dynamics of plasma blobs in a shear flow *Phys. Rev. Lett.* **101** 115005

[75] Diallo A, Ricci P, Fasoli A, Furno I, Labit B, Müller S H, Podestà M, Poli F M and Skiff F 2007 Antenna excitation of drift wave in a toroidal plasma *Phys. Plasmas* **14** 102101

[76] D'Ippolito D A, Myra J R and Zweben S J 2011 Convective transport by intermittent blob-filaments: comparison of theory and experiment *Phys. Plasmas* **18** 060501

[77] Fasoli A *et al* 2010 Electrostatic instabilities, turbulence and fast ion interactions in the TORPEX device *Plasma Phys. Control. Fusion* **52** 124020

[78] Fasoli A *et al* 2003 Electrostatic turbulence and transport in a simple magnetized plasma *Bull. Am. Phys. Soc.* **13** 119

[79] Federspiel L, Labit B, Ricci P, Fasoli A, Furno I and Theiler C 2009 Observation of a critical pressure gradient for the stabilization of interchange modes in simple magnetized toroidal plasmas *Phys. Plasmas* **16** 092501

[80] Furno I *et al* 2008 Mechanism for blob generation in the TORPEX toroidal plasma *Phys. Plasmas* **15** 055903

[81] Furno I, Theiler C, Chabloz V, Fasoli A and Loizu J 2014 Pre-sheath density drop induced by ion-neutral friction along plasma blobs and implications for blob velocities *Phys. Plasmas* **21** 012305

[82] Furno I *et al* 2008 Experimental observation of the blob-generation mechanism from interchange waves in a plasma *Phys. Rev. Lett.* **100** 055004

[83] Garcia O E, Naulin V, Nielsen A H and Juul Rasmussen J 2005 Turbulence and intermittent transport at the boundary of magnetized plasmas *Phys. Plasmas* **12** 062309

[84] Garcia O E 2012 Stochastic modeling of intermittent scrape-off layer plasma fluctuations *Phys. Rev. Lett.* **108** 265001

[85] Goodman T P *et al* 2003 An overview of results from the TCV tokamak *Nucl. Fusion* **43** 1619

[86] Greenwald M 2010 Verification and validation for magnetic fusion *Phys. Plasmas* **17** 058101

[87] Gustafson K and Ricci P 2012 Lévy walk description of suprathermal ion transport *Phys. Plasmas* **19** 032304

[88] Gustafson K, Ricci P, Bovet A, Furno I and Fasoli A 2012 Suprathermal ion transport in simple magnetized torus configurations *Phys. Plasmas* **19** 062306

[89] Gustafson K, Ricci P, Furno I and Fasoli A 2012 Nondiffusive suprathermal ion transport in simple magnetized toroidal plasmas *Phys. Rev. Lett.* **108** 035006

[90] Guszejnov D, Lazanyi N, Bencze A and Zoletnik S 2013 On the effect of intermittency of turbulence on the parabolic relation between skewness and kurtosis in magnetized plasmas *Phys. Plasmas* **20** 112305

[91] Halpern F D, Ricci P, Jolliet S, Loizu J and Mosetto A 2014 Theory of the scrape-off layer width in inner-wall limited tokamak plasmas *Nucl. Fusion* **54** 043003

[92] Halpern F D *et al* 2013 Theory-based scaling of the sol width in circular limited tokamak plasmas *Nucl. Fusion* **53** 122001

[93] Heidbrink W W *et al* 2012 Measurements of interactions between waves and energetic ions in basic plasma experiments *Plasma Phys. Control. Fusion* **54** 124007

[94] Henderson M A *et al* 2003 Recent results from the electron cyclotron heated plasmas in tokamak a configuration variable (TCV) *Phys. Plasmas* **10** 1796

[95] Katz N, Egedal J, Fox W, Le A and Porkolab M 2008 Experiments on the propagation of plasma filaments *Phys. Rev. Lett.* **101** 015003

[96] Krasheninnikov S I 2001 On scrape off layer plasma transport *Phys. Lett.* A **283** 368–70

[97] Krommes J A 2008 The remarkable similarity between the scaling of kurtosis with squared skewness for TORPEX density fluctuations and sea-surface temperature fluctuations *Phys. Plasmas* **15** 030703

[98] Labit B *et al* 2007 Statistical properties of electrostatic turbulence in toroidal magnetized plasmas *Plasma Phys. Control. Fusion* **49** B281

[99] Labit B, Furno I, Fasoli A, Diallo A, Müller S H, Plyushchev G, Podestà M and Poli F M 2007 Universal statistical properties of drift-interchange turbulence in TORPEX plasmas *Phys. Rev. Lett.* **98** 255002

[100] Loizu J, Ricci P, Halpern F D, Jolliet S and Mosetto A 2013 On the electrostatic potential in the scrape-off layer of magnetic confinement devices *Plasma Phys. Control. Fusion* **55** 124019

[101] Loizu J, Ricci P, Halpern F D, Jolliet S and Mosetto A 2014 On the electrostatic potential in the scrape-off layer of magnetic confinement devices *Phys. Plasmas* **21** 062309

[102] Mekkaoui A 2013 Derivation of stochastic differential equations for scrape-off layer plasma fluctuations from experimentally measured statistics *Phys. Plasmas* **20** 010701

[103] Mosetto A, Halpern F D, Jolliet S, Loizu J and Ricci P 2013 Turbulent regimes in the tokamak scrape-off layer *Phys. Plasmas* **20** 092308

[104] Müller S H, Theiler C, Fasoli A, Furno I, Labit B, Tynan G R, Xu M, Yan Z and Yu J H 2009 Studies of blob formation, propagation and transport mechanisms in basic experimental plasmas (TORPEX and CSDX) *Plasma Phys. Control. Fusion* **51** 055020

[105] Müller S H *et al* 2008 Plasma blobs in a basic toroidal experiment: origin, dynamics, and induced transport *Phys. Plasmas* **14** 110704

[106] Myra J R and D'Ippolito D A 2005 Edge instability regimes with applications to blob transport and the quasicoherent mode *Phys. Plasmas* **12** 092511

[107] Otsuji K *et al* 2007 Small-scale magnetic-flux emergence observed with Hinode solar optical telescope *Publ. Astron. Soc. Japan* **59** S649–54

[108] Park J, Lühr H, Stolle C, Rother M, Min K W and Michaelis I 2010 Field-aligned current associated with low-latitude plasma blobs as observed by the champ satellite *Ann. Geophys.* **28** 697–703

[109] Perri S and Zimbardo G 2012 Superdiffusive shock acceleration *Astrophys. J.* **750** 87

[110] Perrone D *et al* 2013 Nonclassical transport and particle-field coupling: from laboratory plasmas to the solar wind *Space Sci. Rev.* **178** 233–70

[111] Plyushchev G *et al* 2006 Fast ion source and detector for investigating the interaction of turbulence with suprathermal ions in a low temperature toroidal plasma *Rev. Sci. Instrum.* **77** 10F503

[112] Podestà M, Fasoli A, Labit B, Furno I, Ricci P, Poli F M, Diallo A, Müller S H and Theiler C 2008 Cross-field transport by instabilities and blobs in a magnetized toroidal plasma *Phys. Rev. Lett.* **101** 045001

[113] Poli F M, Brunner S, Diallo A, Fasoli A, Furno I, Labit B, Mueller S H, Plyushchev G and Podestà M 2006 Experimental characterization of drift-interchange instabilities in a simple toroidal plasma *Phys. Plasmas* **13** 102104

[114] Poli F M, Podestà M and Fasoli A 2007 Development of electrostatic turbulence from drift-interchange instabilities in a toroidal plasma *Phys. Plasmas* **14** 052311

[115] Poli F M, Ricci P, Fasoli A and Podestà M 2008 Transition from drift to interchange instabilities in an open magnetic field line configuration *Phys. Plasmas* **15** 032104

[116] Reames D V 1999 Particle acceleration at the Sun and in the heliosphere *Space Sci. Rev.* **90** 413–91

[117] Ricci P, Halpern F D, Jolliet S, Loizu J, Mosetto A, Fasoli A, Furno I and Theiler C 2012 Simulation of plasma turbulence in scrape-off layer conditions: the GBS code, simulation results and code validation *Plasma Phys. Control. Fusion* **54** 124047

[118] Ricci P and Rogers B N 2009 Three-dimensional fluid simulations of a simple magnetized toroidal plasma *Phys. Plasmas* **16** 092307

[119] Ricci P and Rogers B N 2010 Turbulence phase space in simple magnetized toroidal plasmas *Phys. Rev. Lett.* **104** 145001

[120] Ricci P and Rogers B N 2013 Plasma turbulence in the scrape-off layer of tokamak devices *Phys. Plasmas* **20** 010702

[121] Ricci P, Rogers B N and Brunner S 2008 High- and low-confinement modes in simple magnetized toroidal plasmas *Phys. Rev. Lett.* **100** 225002

[122] Ricci P, Theiler C, Fasoli A, Furno I, Gustafson K, Iraji D and Loizu J 2011 Methodology for turbulence code validation: quantification of simulation-experiment agreement and application to the TORPEX experiment *Phys. Plasmas* **18** 032109

[123] Ricci P, Theiler C, Fasoli A, Furno I, Labit B, Mueller S H, Podestà M and Poli F M 2009 Langmuir probe-based observables for plasma-turbulence code validation and application to the TORPEX basic plasma physics experiments *Phys. Plasmas* **16** 055703

[124] Riva F, Ricci P, Halpern F D, Jolliet S, Loizu J and Mosetto A 2014 Verification methodology for plasma simulations and application to a scrape-off layer turbulence code *Phys. Plasmas* **21** 062301

[125] Sandberg I, Benkadda S, Garbet X, Ropokis G, Hizanidis K and del-Castillo-Negrete D 2012 universal probability distribution function for bursty transport in plasma turbulence *Phys. Rev. Lett.* **103** 165001

[126] Schaffner D A, Carter T A, Rossi G D, Guice D S, Maggs J E, Vincena S and Friedman B 2012 Modification of turbulent transport with continuous variation of flow shear in the large plasma device *Phys. Rev. Lett.* **109** 135002

[127] Terry P W, Greenwald M, Leboeuf J-N, McKee G R, Mikkelsen D R, Nevins W M, Newman D E and Stotler D P 2008 Task Group on Verification and Validation, U.S. Burning Plasma Organization and U.S. Transport Task Force 2008 Validation in fusion research: towards guidelines and best practices *Phys. Plasmas* **15** 62503

[128] Theiler C, Diallo A, Fasoli A, Furno I, Labit B, Podestà M, Poli F M and Ricci P 2008 The role of the density gradient on intermittent cross-field transport events in a simple magnetized toroidal plasma *Phys. Plasmas* **15** 042303

[129] Theiler C, Furno I, Fasoli A, Ricci P, Labit B and Iraji D 2011 Blob motion and control in simple magnetized plasmas *Phys. Plasmas* **18** 055901

[130] Theiler C, Furno I, Loizu J and Fasoli A 2012 Convective cells and blob control in a simple magnetized plasma *Phys. Rev. Lett.* **108** 065005

[131] Theiler C, Loizu J, Furno I, Fasoli A and Ricci P 2012 Properties of convective cells generated in magnetized toroidal plasmas *Phys. Plasmas* **19** 082304

[132] Theiler C *et al* 2009 Cross-field motion of plasma blobs in an open magnetic field line configuration *Phys. Rev. Lett.* **103** 065001

[133] Podestà M, Fasoli A, Labit B, McGrath M, Müller S H and Poli F M 2006 xperimental characterization and modelling of the particle source in an Electron-Cyclotron wave driven toroidal plasma *Plasma Phys. Control. Fusion* **48** 1053

[134] Müller S H, Fasoli A, Labit B, McGrath M, Pisaturo O, Plyushchev G, Podestà M and Poli F M 2005 Basic turbulence studies on TORPEX and challenges in the theory-experiment comparison *Phys. Plasmas* **12** 090906
[135] Theiler C, Furno I, Kuenlin A, Marmillod P and Fasoli A 2011 Practical solutions for reliable triple probe measurements in magnetized plasmas *Rev. Sci. Instr.* **82** 013504

Part IV

Data for plasma modelling

IOP Publishing

Plasma Modeling (Second Edition)
Methods and applications
Gianpiero Colonna and Antonio D'Angola

Chapter 21

Thermodynamic and transport properties of complex plasmas

Gianpiero Colonna, Antonio D'Angola and Annarita Laricchiuta

Different systems, such as plasma torches [1], laser induced breakdown spectroscopy (LIBS) [2–6] and high pressure discharges [7, 8], are modeled assuming thermodynamic equilibrium. However, thermodynamic data are necessary also in non-equilibrium conditions to determine the rate of reverse reactions through the detailed balance principle [9].

The thermodynamic properties of a mixture are calculated starting from single species data, that, in the framework of statistical physics, can be derived from the partition functions, which require the knowledge of the internal structure of the considered species.

Another class of data entering plasma simulations are the transport properties, i.e. viscosity, thermal conductivity, diffusion coefficients, and so on. These coefficients are calculated in the framework of the Chapman–Enskog theory [10–12], where the description of the microscopic collisional dynamics requires the collision integrals of binary interactions, whose values can be estimated through quantum, semiclassical or classical approaches from the interaction potential.

Modeling systems in local thermodynamic equilibrium (LTE) would need global properties of the mixture as a function of pressure (P) and temperature (T) [1] to be loaded in the model as lookup tables [13] or fitting functions [14–17].

In this chapter, a general overview of the theoretical foundations is given, presenting the working equations, the rigorous and simplified approaches, the benchmark case of an equilibrium hydrogen plasma and finally the features of the web-tool EquilTheTA.

21.1 Partition functions and thermodynamics

In the framework of the statistical physics, all the thermodynamic properties of a chemical species s in gas (plasma) phase can be determined starting from its partition function Q_s [18]. This section will give a synthesis of statistical thermodynamics of plasmas. For a deeper description readers are referred to specialized books such as reference [13].

doi:10.1088/978-0-7503-3559-1ch21

21.1.1 Single species thermodynamics

The partition function represents the number of states accessible to the system at a given temperature and pressure. Separating the dynamics of particles from their internal structure the partition function can be written as the product of translational and internal contributions

$$\mathcal{Q}_s = \mathcal{Q}_s^{\text{tr}} \mathcal{Q}_s^{\text{int}} \tag{21.1}$$

In the following, temperature and energy are expressed in the same units, to simplify the formalism, so that the Boltzmann constant disappears from the equations. The translational partition function is given by

$$\mathcal{Q}_s^{\text{tr}} = \frac{V}{\lambda_{\text{th},\,s}^3} = \left(\frac{m_s T}{2\pi\hbar^2}\right)^{\frac{3}{2}} \frac{N_s T}{P} \tag{21.2}$$

where m_s and N_s are mass and number of particles of the species s, respectively, V is the gas volume $\hbar$ is the reduced Plank constant, and

$$\lambda_{\text{th},s} = \sqrt{\frac{2\pi\hbar^2}{m_s T}} \tag{21.3}$$

is the thermal de Broglie length. The internal partition function is given by

$$\mathcal{Q}_s^{\text{int}} = \sum_{\ell=0}^{\mathcal{N}_s} g_{s\ell} e^{-\frac{\varepsilon_{s\ell}}{T}} \tag{21.4}$$

where $\mathcal{N}_s$ is the number of levels, $g_{s\ell}$ and $\varepsilon_{s\ell}$, respectively, the statistical weight and the energy of the lth level.

All thermodynamic functions depend on the logarithm of the partition function, therefore the general quantity X_s is calculated as the sum of the translational (X_s^{tr}) and internal (X_s^{int}) contributions. The translational contributions to the internal energy ($\mathcal{U}$), enthalpy ($\mathcal{H}$), constant volume (C_v) and constant pressure (C_v) heat capacities, Gibbs ($\mathcal{G}$) and Helmholtz ($\mathcal{A}$) free energies, and entropy ($\mathcal{S}$) are

$$\begin{aligned}
\mathcal{U}_s^{\text{tr}} &= N_s T \left(\frac{\partial \ln \mathcal{Q}_s^{\text{tr}}}{\partial \ln T}\right)_V = \frac{3}{2} N_s T \\
\mathcal{H}_s^{\text{tr}} &= \mathcal{U}_s^{\text{tr}} + PV = \frac{5}{2} N_s T \\
C_{v,\,s}^{\text{tr}} &= \left(\frac{\partial \mathcal{U}_s^{\text{tr}}}{\partial T}\right)_V = \frac{3}{2} N_s \\
C_{p,\,s}^{\text{tr}} &= \left(\frac{\partial \mathcal{H}_s^{\text{tr}}}{\partial T}\right)_P = \frac{5}{2} N_s \\
\mathcal{G}_s^{\text{tr}} &= -N_s T \ln \frac{\mathcal{Q}_s^{\text{tr}}}{N_s} \\
\mathcal{A}_s^{\text{tr}} &= \mathcal{G}_s^{\text{tr}} - N_s T \\
\mathcal{S}_s^{\text{tr}} &= \frac{\mathcal{U}_s^{\text{tr}} - \mathcal{A}_s^{\text{tr}}}{T} = C_{v,\,s}^{\text{tr}} - \frac{\mathcal{A}_s^{\text{tr}}}{T}
\end{aligned} \tag{21.5}$$

while the internal contributions are given by

$$\begin{aligned}
\mathcal{U}_s^{\text{int}} &= \mathcal{H}_s^{\text{int}} = N_s T \left(\frac{\partial \ln \mathcal{Q}_s^{\text{int}}}{\partial \ln T} \right)_V = N_s \ll \varepsilon \gg_s \\
C_{v,s}^{\text{int}} &= C_{p,s}^{\text{int}} = \frac{\partial \mathcal{U}_s^{\text{int}}}{\partial T} = N_s \frac{\ll \varepsilon^2 \gg_s - \ll \varepsilon \gg_s^2}{T^2} \\
\mathcal{G}_s^{\text{int}} &= \mathcal{A}_s^{\text{int}} = -N_s T \ln \frac{\mathcal{Q}_s^{\text{int}}}{N_s} \\
\mathcal{S}_s^{\text{int}} &= \frac{\mathcal{U}_s^{\text{int}} - \mathcal{A}_s^{\text{int}}}{T} = C_{v,s}^{\text{int}} - \frac{\mathcal{A}_s^{\text{int}}}{T}.
\end{aligned} \tag{21.6}$$

being the statistical mean $\ll x \gg_s$ of a generic properties x

$$\ll x \gg_s = \frac{1}{\mathcal{Q}_s^{\text{int}}} \sum_{\ell=0}^{\mathcal{N}_s} x_{s\ell} g_{s\ell} e^{-\frac{\varepsilon_{s\ell}}{T}} \tag{21.7}$$

Noticing that $C_v^{\text{int}} = C_p^{\text{int}}$, for internal heat capacity the subscript will be removed, using C_s^{int}. It is worth mentioning also the chemical potential μ_s, whose internal and translational contributions are calculated as

$$\mu_s = \frac{\mathcal{G}_s}{N_s} = -T \ln \frac{\mathcal{Q}_s}{N_s} = -T \ln \frac{\mathcal{Q}_s^{\text{tr}}}{N} - T \ln \mathcal{Q}_s^{\text{int}} = \mu_s^{\text{tr}} + \mu_s^{\text{int}}. \tag{21.8}$$

From equation (21.2) we have

$$\mu_s^{\text{tr}} = -T \ln \left(\frac{T}{\lambda_{\text{th}}^3} \right) - T \ln \mathcal{Q}^{\text{int}} = \mu_s^{0,tr} + T \ln P \tag{21.9}$$

As a consequence the reference chemical potential μ_s^0

$$\mu_s^0 = \mu_s^{0,tr} + \mu_s^{\text{int}} \tag{21.10}$$

is only a function of the temperature.

21.1.2 Mixture thermodynamics

The quantities $\mathcal{U}_s$, $\mathcal{H}_s$, $\mathcal{A}_s$, $\mathcal{G}_s$, μ_s have been calculated referring the energy to the species ground state. To calculate global properties of a mixture a common reference system is needed. In practice, neglecting nuclear reactions, a different energy reference must be chosen for each element in the periodic table, determining the formation energy of each chemical species through reactions. For convention, the reference value is that of the pure species in its more stable state at normal conditions. To give some examples, for nitrogen the reference is the molecule N_2, for carbon is the solid carbon. Tables of thermodynamic properties [19–22] reports data using this definition.

As a consequence, the energy of the species should be corrected by adding formation energy ε_s^f to the internal part, i.e. for all the quantities the following substitution should be performed

$$X_s^{\text{int}} \to X_s^{\text{int}} + N_s\varepsilon_s^f \tag{21.11}$$

with the exception of the heat capacity C_s^{int} which remains unchanged and of the chemical potentials

$$\mu_s^{\text{int}} \to \mu_s^{\text{int}} + \varepsilon_s^f \tag{21.12}$$

Including the above corrections, the global thermodynamic properties of a mixture are calculated summing the contributions of each species, i.e., for a generic quantity X

$$X = \sum_s X_s = \sum_s (X_s^{\text{tr}} + X_s^{\text{int}} + N_s\varepsilon_s^f). \tag{21.13}$$

However, there is an exception to equation (21.13) and it is represented by the heat capacities. The constant volume heat capacity is calculated as the derivative with respect of temperature of the mixture internal energy, giving

$$C_v = \frac{\partial \mathcal{U}}{\partial T} = \sum_s N_s \frac{\partial}{\partial T}\frac{\mathcal{U}_s}{N_s} + \sum_s \frac{\mathcal{U}_s}{N_s}\frac{\partial N_s}{\partial T} = C_v^f + C_v^r \tag{21.14}$$

where C_v^f and C_v^r are known as *frozen* and *reactive* heat capacities. The same result can be obtained for the constant pressure specific heat by substituting $\mathcal{U}$ with $\mathcal{H}$ in equation (21.14). When one or more reactions are active, the reactive specific heat is much larger than the frozen one.

As an example, the constant pressure specific heat together with the reactive and frozen contributions, are reported in figure 21.1. The first peak is due to H_2 dissociation and the second one to H ionization. Other processes give a negligible contribution.

In numerical models, specific quantities, normalized per unit volume, per particle or per unit mass are required. Due to the presence of reactions, the total mass is constant while the number of particles and the volume can change, so the thermodynamic functions are generally normalized per unit mass. In the following, the specific quantities per unit mass are indicated with lowercase letters as c_v and c_p for the specific heats.

An important quantity for fluid dynamic modeling is the isentropic coefficient γ defined as [23–27]

$$\gamma = \left(\frac{\partial \ln P}{\partial \ln \rho}\right)_S = \frac{c_p}{c_v}\frac{\rho}{P}\left(\frac{\partial P}{\partial \rho}\right)_T = \gamma_{\text{eq}} z_d \tag{21.15}$$

where ρ is mass density,

$$\gamma_{\text{eq}} = \frac{c_p}{c_v} \tag{21.16}$$

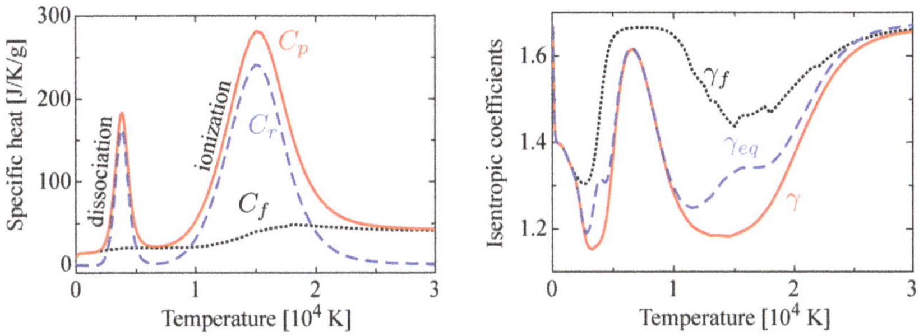

Figure 21.1. Comparison of (left) C_p with frozen (C_p^f) and reactive (C_p^r) contributions and of (right) γ (equation (21.15)) with γ_{eq} (equation (21.16)) and γ_f for the hydrogen plasma at $P = 1$ bar.

is the *equilibrium* isentropic coefficient and

$$z_d = \frac{\rho}{P}\left(\frac{\partial P}{\partial \rho}\right)_T \tag{21.17}$$

is the compressibility factor. It can be shown that for a non reacting ideal gas $z_d = 1$. For this reason, often γ_{eq} is used in place of the isentropic coefficients. The '*frozen*' isentropic coefficient can be calculated using the frozen heat capacities

$$\gamma_f = C_p^f / C_p^v. \tag{21.18}$$

The difference between the three isentropic coefficients can be observed in the right viewgraph of figure 21.1. It is worth noticing that the dumps in the γ are located at the same temperatures as the peaks in the C_p. Quite often, as approximate values, $\gamma_f = \frac{5}{3} = 1.67$, considering only the translational degree of freedom, and $\gamma_f = \frac{7}{5} = 1.4$ for diatomic molecules, adding the contribution due to the molecular rotation. At very low temperature, the rapid variation from 1.67 and 1.4 corresponds to the activation of rotation of H_2. The maximum around $T = 10\,000$ K is the contribution of atomic hydrogen, while at $T = 30\,000$ K it is that of H^+ and electrons.

To conclude this section let us recall the definition of the constant pressure equilibrium constant K_p for a generic chemical reaction

$$\sum_s \nu_s X_s = 0 \tag{21.19}$$

where the sum runs over all the X species and ν_s, the stoichiometric coefficients, are positive for the products, negative for the reactants and null for species not participating to the reaction. Defining

$$\Delta\mu^0 = \sum_s \nu_s \mu_s^0 \tag{21.20}$$

we obtain

$$K_p = e^{-\Delta\mu^0/T} \tag{21.21}$$

21.1.3 Considerations on equilibrium and non-equilibrium plasmas

When modeling non-equilibrium plasmas, some degrees of freedom could be in equilibrium with the gas temperature. In the most common case, translational and rotational degrees of freedom, and sometimes also vibrational ones, are in equilibrium among each other. Reaching the partial equilibrium is a matter of relaxation times: when the characteristic time of the evolution of the system is much larger than the relaxation time of one degree of freedom, it can be treated as in equilibrium with the translational one and can be described by the gas temperature. In contrast, the degree of freedom should be described by an energy relaxation equation[1].

Some questions arise when modeling a plasma in partial equilibrium. An example is represented by the isentropic coefficient, a fundamental quantity in fluid dynamics, determining the speed of sound ($c_{\text{sound}} = \sqrt{\gamma T/\bar{m}}$, where $\bar{m}$ is the mean mass). What value should be used among γ, γ_{eq}, γ_f in numerical simulations? What is the role of the reactive specific heat?

To answer these questions one should start from the following assertion: only those quantities in equilibrium with the translational temperature must enter in the determination of these properties of the gas, all the others are treated by the evolution equation. As an example, let us consider a non-equilibrium flow with a diatomic gas. As a first approximation, assuming the separation of the rotational and vibrational degrees of freedom and that T is much larger than the rotational energy quanta, the contribution to the constant pressure heat capacity is[2] $C_{p,s}/N_s = \frac{7}{2}, \frac{5}{2}$ for the translation and 1 for the rotation. In this case the isentropic coefficient entering the equation of the sound speed is $\gamma_f = \frac{7}{5}$. The contribution of the degrees of freedom in non-equilibrium and of the compressibility factor is accounted for in the continuity equations. The reactive heat capacity must not be considered when chemical non-equilibrium is assumed.

There are other problems related to the calculation of the equilibrium constant. Let us consider a system of diatomic molecules in vibrational non-equilibrium and in dissociation regime [29, 30]. In this case, atoms are considered only in their ground state. Which levels should enter in the calculation of the partition function for the calculation of equilibrium constants? Two possible choices are:

- to include only the levels considered in the simulation;
- to include all the possible levels.

In our opinion the second choice is the most reasonable, because the stationary solution corresponds to the *real* equilibrium composition.

Another issue related to plasma discharges is which temperature should be used when considering the processes involving free electrons in equation (21.9). It should

[1] Two possible models can be used to describe the non-equilibrium of a part of a system, the multi-temperature approach [28], which assumes that such degree of freedom is in internal equilibrium and can be described by a specific temperature, and the state-to-state approach [9], where a kinetic equation for each quantum level is added to determine the distribution of the states.

[2] Remember that the temperature is expressed in units of energy.

be noted that, during a discharge, the electron temperature is determined by the reduced electric field[3] and is not correlated to the gas temperature, therefore, the partition function of electrons should be calculated considering T_e instead of T, while, in the post-discharge the gas temperature should be used for all the species, because electrons rapidly relax to T due to elastic collisions (see chapter 2).

21.1.4 Non-ideal corrections

The results reported in the previous sections refer to an ideal gas, where particles interact only through collisions, the interaction time being negligible. In high pressure and low temperature conditions, this assumption is no longer valid and, to include the effects of non-ideality, the correction terms are estimated through the virial expansion [13, 18]. Typical plasma pressure and temperature are in a range of values such that the virial corrections for interactions between heavy particles are negligible except in very extreme conditions, where quantum effects become effective.

In ionized gases, non-ideal effects are the consequence of the long-range electrostatic interaction, a condition that results in divergency of the viral expansions. In this case a mean field theory should be considered. In fact, while in the ideal thermodynamics the system properties are obtained as the sum of single particle properties, in a plasma the effect of the whole system must be taken into account. A simple approach, that is very accurate, is the Debye–Hückel theory [13].

This theory predicts the corrections for all the thermodynamic quantities considering the characteristic length of the electrostatic interaction in a volume with free charges in condition of quasi-neutrality, the Debye length, λ_D, defined as

$$\lambda_D = \sqrt{\frac{\epsilon_0 T V}{q_s^2 \sum_s N_s}} \tag{21.22}$$

where q_s is the charge of the sth species and ϵ_0 is the vacuum permittivity. The Debye–Hückel corrections to all the thermodynamic quantities are the following

$$\begin{aligned}
\Delta\mathcal{U}_{\mathrm{DH}} &= -\frac{1}{8}T\frac{V}{\pi\lambda_D^3} & \Delta\mathcal{H}_{\mathrm{DH}} &= \frac{4}{3}\Delta\mathcal{U}_{\mathrm{DH}} \\
\Delta\mathcal{A}_{\mathrm{DH}} &= \frac{2}{3}\Delta\mathcal{U}_{\mathrm{DH}} & \Delta\mathcal{G}_{\mathrm{DH}} &= \Delta\mathcal{U}_{\mathrm{DH}} \\
\Delta\mathcal{C}_{\mathrm{vfDH}} &= -\frac{\Delta\mathcal{U}_{\mathrm{DH}}}{2T} & \Delta\mathcal{C}_{\mathrm{pfDH}} &= -\frac{4\Delta\mathcal{U}_{\mathrm{DH}}}{3T} \\
\Delta\mathcal{S}_{\mathrm{DH}} &= \frac{\Delta\mathcal{U}_{\mathrm{DH}}}{3T} & \Delta\mu_s &= -\frac{q_s^2}{8\pi\epsilon_0\lambda_D} \\
\Delta P_{\mathrm{DH}} &= \frac{\Delta\mathcal{U}_{\mathrm{DH}}}{3V} & \frac{\Delta P_{\mathrm{DH}}}{P} &= \frac{\Delta V_{\mathrm{DH}}}{V}
\end{aligned} \tag{21.23}$$

[3] The reduced electric field is E/N, see chapter 2.

It should be noted that, for an isolated atom, the partition function is divergent for $T > 0$ because the number of levels is infinite. However, the presence of other particles limits the number of levels by reducing the ionization potential and the partition function converges. The cutoff energy is influenced and determined by the condition of the system. Two criteria are commonly used, the Fermi and the Griem cutoffs. The Fermi cutoff (Δ_F) excludes all the levels having classical orbital radius larger than the mean particle distance, i.e.

$$\Delta_F = \mathcal{I}_s(z_s + 1)a_0\sqrt[3]{\frac{N}{V}} \tag{21.24}$$

while the Griem cutoff (Δ_G) reduces the ionization energy by the plasma potential

$$\Delta_G = \frac{q_e^2(z_s + 1)}{8\pi\epsilon_0\lambda_D} \tag{21.25}$$

where z_s is the charge number of the sth atomic species, $\mathcal{I}_s$ its ionization energy, a_0 the Bohr radius and q_e the electron charge. These cutoffs act at the same time and the maximum between the two should be chosen.

It is worth mentioning that the plasma modifes the level structure of the atomic species [31, 32] and a quantum approach should be used to determine accurately the internal partition function. By the way, these effects are relevant at very high pressure [33, 34].

As an example, the Debye length, the atomic hydrogen ionization lowering (cutoff) and the Debye–Hückel corrections to pressure and enthalpy for a hydrogen plasma at atmospheric pressure can be observed in figure 21.2.

21.1.5 Fermi–Dirac statistics

To estimate the corrections due to quantum effects, we have to generalize the theory. The basic particle density for a general x statistics (Fermi–Dirac, FD, Bose–Einstein, BE, Boltzmann, B) is calculated as

$$\frac{N_s}{V} = \int_0^\infty g_s f_s^x(\varepsilon)\mathcal{D}_s(\varepsilon)d\varepsilon \tag{21.26}$$

and for any thermodynamic quantity $G_s(\varepsilon)$, its mean value $\mathcal{G}_s$ is obtained as

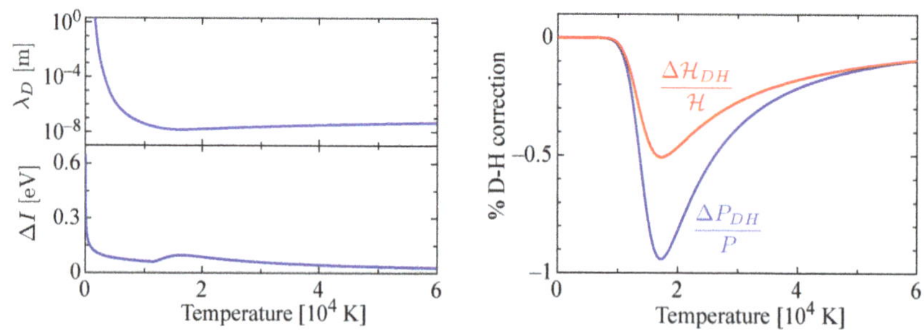

Figure 21.2. Debye length (λ_D) and cutoff of the hydrogen atom (ΔI) (left) and Debye–Hückel correction for pressure and enthalpy (right) for the hydrogen plasma at $P = 1$ bar.

$$\mathcal{G}_s = \int_0^\infty G_s(\varepsilon) g_s f_s^x(\varepsilon) \mathcal{D}_s(\varepsilon) d\varepsilon \tag{21.27}$$

where g_s is the spin multiplicity (2 for electrons) and $\mathcal{D}_s$ is the density of states, that for particles in a box is given by

$$\mathcal{D}_s(\varepsilon) = \frac{\sqrt{(2m_s)^3}}{4\pi^2\hbar^3}\sqrt{\varepsilon} = \frac{2T^{-3/2}}{\lambda_{\text{th},\,s}^3\sqrt{\pi}}\sqrt{\varepsilon}. \tag{21.28}$$

In dense, non-ideal plasmas, free electrons show quantum behaviors following the Fermi–Dirac distribution, f_s^{FD}, [18, 35]

$$f_s^{\text{FD}}(\varepsilon) = [e^{(\varepsilon-\mu_s)/T} + 1]^{-1}. \tag{21.29}$$

In the framework of the Debye–Hückel theory, introducing the reduced energy $x = \varepsilon/T$ and chemical potential $\hat{\mu}_s = \mu_s/T$, we can write the expression for number of particles and energy as

$$\frac{N_s}{V} = g_s\frac{\sqrt{(2m_sT)^3}}{4\pi^2\hbar^3}\int_0^\infty \sqrt{x} f_s^{\text{FD}}(x - \hat{\mu}_s)dx = \frac{g_s}{(\lambda_s^{\text{th}})^3}\frac{2}{\sqrt{\pi}}\int_0^\infty \frac{x^{1/2}}{e^{x-\hat{\mu}_s} + 1}dx \tag{21.30}$$

$$U_s = \frac{g_s}{(\lambda_s^{\text{th}})^3}\frac{2T}{\sqrt{\pi}}\int_0^\infty \frac{x^{3/2}}{e^{x-\hat{\mu}_s} + 1}dx \tag{21.31}$$

The integrals in equations (21.30)–(21.31) are the well known *Fermi integrals* [18, 35]

$$F_\nu(x) = \int_0^\infty \frac{t^\nu}{e^{t-x} + 1}dt \tag{21.32}$$

and have to be numerically calculated, for example using the very efficient fractal integration algorithm [36]. The following transformation [35]

$$t = \exp[s - \exp(-s)] \tag{21.33}$$

of the integration variable allows one to consider a very high upper limit. To calculate thermodynamic properties, we should consider the *complete Fermi integrals*

$$\mathcal{F}_\nu(x) = \frac{F_\nu(x)}{\Gamma(\nu + 1)} \tag{21.34}$$

where Γ is the gamma function, fulfilling the following properties useful for integral calculation

$$\begin{aligned} &\Gamma(n + 1) = n! \quad &&\Gamma(\nu + 1) = \nu\Gamma(\nu) \\ &\Gamma\left(\frac{1}{2}\right) = \sqrt{\pi} \quad &&\Gamma\left(-\frac{1}{2}\right) = -2\sqrt{\pi} \\ &\Gamma\left(\frac{3}{2}\right) = \frac{\sqrt{\pi}}{2} \quad &&\Gamma\left(\frac{5}{2}\right) = \frac{3\sqrt{\pi}}{4} \end{aligned} \tag{21.35}$$

with n a positive integer and ν a real number. The following recursive relation holds for the complete Fermi integrals

$$\frac{d\mathcal{F}_\nu(x)}{dx} = \mathcal{F}_{\nu-1}(x). \tag{21.36}$$

Now we have all the equations to calculate the thermodynamic properties of free electrons in the Fermi statistic. From temperature and density it is possible to determine the chemical potential

$$n_e = \frac{N_e}{V} = \frac{2}{(\lambda_{\mathrm{th},e})^3}\mathcal{F}_{1/2}(\hat{\mu}_e) \Rightarrow \mu_e(n_e, T) = T\mathcal{F}_{1/2}^{-1}\left[\frac{1}{2}n_e(\lambda_{\mathrm{th},e})^3\right] \tag{21.37}$$

There exists an accurate approximation for $\mathcal{F}_{1/2}^{-1}$ [35]

$$\mathcal{F}_{1/2}^{-1}(\eta) = \begin{cases} \ln\eta + 0.353\,6\eta - 0.00495\eta^2 + 0.000125\eta^3 & \eta < 5.5 \\ 1.209\eta^{2/3} - 0.680\,3\eta^{-2/3} - 0.85\eta^{-2} & \eta \geqslant 5.5 \end{cases} \tag{21.38}$$

Finally, the thermodynamic functions are given by [18]

$$\begin{aligned} &U_e = \frac{3}{2}n_eT\frac{\mathcal{F}_{3/2}(\hat{\mu}_e)}{\mathcal{F}_{1/2}(\hat{\mu}_e)} && P_e = n_eT\frac{\mathcal{F}_{3/2}(\hat{\mu}_e)}{\mathcal{F}_{1/2}(\hat{\mu}_e)} && H_e = \frac{5}{2}P_e \\ &G_e = n_e\mu_e(n_e, T) && A_e = \frac{2}{3}U_e + \mu_e n_e && S_e = \frac{5P_e - 2n_e\mu_e}{2T} \\ &\frac{d\hat{\mu}_e}{dT} = -\frac{3}{2T}\frac{\mathcal{F}_{1/2}(\hat{\mu}_e)}{\mathcal{F}_{-1/2}(\hat{\mu}_e)} && C_v = \frac{U_e}{T}\mathcal{Z}(\hat{\mu}_e) && C_p = \frac{H_e}{T}\mathcal{Z}(\hat{\mu}_e) \end{aligned} \tag{21.39}$$

where

$$\mathcal{Z} = \left[1 - \frac{3}{2}\frac{\mathcal{F}_{1/2}(\hat{\mu}_e)}{\mathcal{F}_{-1/2}(\hat{\mu}_e)}\left(\frac{\mathcal{F}_{1/2}(\hat{\mu}_e)}{\mathcal{F}_{3/2}(\hat{\mu}_e)} - \frac{\mathcal{F}_{-1/2}(\hat{\mu}_e)}{\mathcal{F}_{1/2}(\hat{\mu}_e)}\right)\right] \tag{21.40}$$

Also, the Debye length λ_D must be redefined as follows

$$\lambda_{D,s}^{-2} = \frac{n_s q_s^2}{\epsilon_0 T}\frac{\mathcal{F}_{-1/2}(\hat{\mu}_s)}{\mathcal{F}_{1/2}(\hat{\mu}_s)} \tag{21.41}$$

The corrections are calculated subtracting to the Fermi–Dirac expression the Boltzmann one, giving, for single particle dimensionless quantities[4], the expressions

[4] For all the $\mathcal{X}$ thermodynamic quantity $\hat{\mathcal{X}} = \frac{\mathcal{X}}{neT}$ but for entropy and specific heats where $\hat{\mathcal{X}} = \frac{\mathcal{X}}{ne}$.

$$\begin{aligned}
\Delta_{\rm FD}\hat{U}_e &= \frac{3}{2}\Delta_{\rm FD}\hat{P}_e & \Delta_{\rm FD}\hat{H}_e &= \frac{5}{2}\Delta_{\rm FD}\hat{P}_e \\
\Delta_{\rm FD}\hat{C}_v &= \frac{3}{2}\Delta_{\rm FD}\mathcal{Z} & \Delta_{\rm FD}\hat{C}_p &= \frac{5}{2}\Delta_{\rm FD}\mathcal{Z} \\
\Delta_{\rm FD}\hat{A}_e &= \Delta_{\rm FD}\hat{P}_e + \Delta_{\rm FD}\hat{\mu}_e & \Delta_{\rm FD}\hat{S}_e &= \Delta_{\rm FD}\hat{H}_e - \Delta_{\rm FD}\hat{\mu}_e
\end{aligned} \tag{21.42}$$

starting from the correction to the chemical potential and electron pressure

$$\Delta_{\rm FD}\hat{\mu}_e = \mathcal{F}_{1/2}^{-1}\left[\frac{1}{2}n_e(\lambda_{{\rm th},e})^3\right] - \ln\left[\frac{1}{2}n_e(\lambda_{{\rm th},e})^3\right] \tag{21.43}$$

$$\Delta_{\rm FD}\mathcal{Z} = \Delta_{\rm FD}\hat{P}_e - \frac{3}{2}\left(\frac{\mathcal{F}_{1/2}}{\mathcal{F}_{-1/2}} - \frac{\mathcal{F}_{3/2}}{\mathcal{F}_{1/2}}\right) \tag{21.44}$$

$$\Delta_{\rm FD}\hat{P}_e = \frac{\mathcal{F}_{3/2}}{\mathcal{F}_{1/2}} - 1 \tag{21.45}$$

From equation (21.41) the correction for electrons to the Debye length is

$$\Delta_{\rm FD}\lambda_{D,s}^{-2} = \frac{n_s q_s^2}{\epsilon_0 T}\left(\frac{\mathcal{F}_{-1/2}(\hat{\mu}_s)}{\mathcal{F}_{1/2}(\hat{\mu}_s)} - 1\right) \tag{21.46}$$

For a reaction producing or loosing electrons

$$\ldots + c_r \mathrm{e}^- \rightleftharpoons \ldots + c_p \mathrm{e}^- \tag{21.47}$$

the equilibrium constant should be corrected as follows

$$K_{\rm eq} = K_{\rm eq}^0 e^{(c_r - c_p)\Delta_{\rm FD}\hat{\mu}_e} \tag{21.48}$$

To evaluate the effects of the Fermi–Dirac statistic on electron thermodynamic quantities, as an example, figure 21.3 reports the temperature profile of the corrections to pressure and constant pressure specific heats for different values of the electron density. The corrections are relevant only in high-pressure, low-temperature conditions, i.e. when pressure ionization can be important [32, 35].

21.2 Lumped level model

The calculation of partition functions and thermodynamic properties considering all energy levels with cut-off criteria could be computationally very expensive. The strategy firstly conceived to mitigate the problem was the use a piecewise polynomial interpolation of the thermodynamic variables [21, 37–39]. However, this approach is characterized by two main drawbacks:

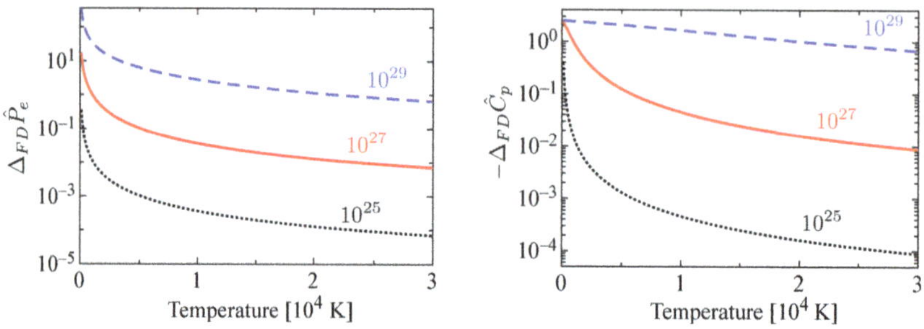

Figure 21.3. Fermi–Dirac corrections to normalized pressure (left) and specific heat (right). The curve labels are the electron particle density n_e in m^{-3}.

- the validity is strictly limited to the interpolation interval (usually 300 K $\leqslant T \leqslant$ 5000 K) and the extrapolation could be highly inaccurate and in many cases returns unphysical behavior of the properties;
- it cannot account that for atomic species thermodynamic properties depend on the cutoff [20–22] which is a function of the thermodynamic state of the system and of the composition [13], as described in detail in section 21.1.4.

An alternative approach has been proposed [40], noticing that the number of levels affects the internal partition function more than their energy. Therefore, lumping levels with a proper criterion will lead to a simplified approach more accurate than the polynomial fitting, using also a limited number of coefficients.

Let us consider a single species with a set of L levels[5] characterized by statistical weight g_ℓ and energy ε_ℓ, ordered by increasing energies. Dividing these levels in $\mathcal{K} + 1$ groups, the partition function can be written as

$$\mathcal{Q}^{\text{int}} = \sum_{k=1}^{\mathcal{K}_s} \sum_{\ell=L_k}^{U_k} g_\ell e^{-\varepsilon_\ell/T} = \sum_{k=1}^{\mathcal{K}_s} \mathcal{Q}_k^{\text{int}} \tag{21.49}$$

where L_k and $U_k = L_{k+1} - 1$ are the index of the first and last levels in the kth group. Assuming

$$\bar{g}_k = \sum_{\ell=L_k}^{U_k} g_\ell \qquad \bar{\varepsilon}_k = \frac{1}{\bar{g}_k} \sum_{\ell=L_k}^{U_k} g_\ell \varepsilon_\ell \qquad \delta_k^2 = \frac{1}{\bar{\varepsilon}_k^2} \sum_{\ell=L_k}^{U_k} g_\ell (\varepsilon_\ell - \bar{\varepsilon}_k)^2 = \frac{\sigma_k^2}{\bar{\varepsilon}_k^2} \tag{21.50}$$

the expansion in Taylor series around the $\bar{\varepsilon}_k$, stopping at the third order the group partition function, and assuming $z_k = \bar{\varepsilon}_k/T$, $\mathcal{Q}_k^{\text{int}}$ can be written as

$$\mathcal{Q}_k^{\text{int}} = \bar{g}_k e^{-z_k}\left(1 + \frac{1}{2}\bar{g}_k \delta_k^2 z_k^2 + \cdots\right) \tag{21.51}$$

It is worth noting that the nth term of the expansion is proportional to the mean of $(\varepsilon_\ell - \bar{\varepsilon}_k)^n z_k^n / \bar{\varepsilon}_k^n$. This method works because the first order term, corresponding to the

[5] Here, for the sake of simplicity, the index of the species has been omitted.

mean error, is null. Moreover, all the coefficients are independent of the temperature and are statistical parameters of the level distribution in the groups. Using the $0th$ order approximation, the second order correction can be used as an estimation of the error.

21.2.1 Lumping levels for atoms

From the theory described above, the lumping method seems suitable for atomic species, whose levels are mostly concentrated in a small energy interval below the ionization limit. For the atomic hydrogen, the error analysis [40] has shown that already the $0th$ order approximation gives an error on the partition function of the order of 1% while the second order correction reduces the error to 0.1%. Another important property is that the error is of the same order for the first and second derivatives, allowing the accurate calculation of the mean energy and the internal specific heat.

A suitable application of the lumped level model is represented by considering the lower levels of an atom separately, grouping all the high energy levels. When the parameters of the lumped levels depend only on the energy levels, the dependence on the cutoff can be easily obtained by fitting the parameters of the group at higher energy as a function of the ionization lowering.

It should be noted that highly excited levels of atoms can be easily approximated by hydrogenoid levels [13, 41, 42]. Energies and statistical weights of hydrogen atom levels are given by

$$\begin{aligned} \varepsilon_n &= \mathcal{I}_H\left(1 - \frac{1}{n^2}\right) \\ g_n &= 2n^2 \end{aligned} \tag{21.52}$$

where n is the principal quantum number. Considering the summation to $n \leqslant \mathcal{N}$ the parameters of the two-level model can be obtained analytically as

$$\begin{aligned} G &= \sum_{n=2}^{\mathcal{N}} 2n^2 = \frac{\mathcal{N}(\mathcal{N}+1)(2\mathcal{N}+1) - 6}{3} \\ \frac{\mathcal{E}}{\mathcal{I}_\mathcal{H}} &= \frac{1}{G}\sum_{n=2}^{\mathcal{N}} 2n^2\left(1 - \frac{1}{n^2}\right) = 1 - \frac{\mathcal{N}-1}{G} \end{aligned} \tag{21.53}$$

$$\begin{aligned} \frac{\varepsilon^2}{\mathcal{I}_H^2} &= \frac{1}{G}\sum_{n=2}^{\mathcal{N}} 2n^2\left(1 - \frac{1}{n^2}\right)^2 = \frac{1}{G}\sum_{n=2}^{\mathcal{N}} 2n^2\left(1 - \frac{2}{n^2} + \frac{1}{n^4}\right) \\ &= 2\frac{\mathcal{E}}{\mathcal{I}_\mathcal{H}} - 1 + \frac{2}{G}\sum_{n=2}^{\mathcal{N}}\frac{1}{n^2} \approx 2\frac{\mathcal{E}}{\mathcal{I}_\mathcal{H}} - 1 + \frac{\mathcal{N}-2}{G\mathcal{N}} \end{aligned} \tag{21.54}$$

where a further approximation (more accurate as $\mathcal{N}$ increases) is obtained substituting the last summation with the integral, giving for the level dispersion

$$\delta_\varepsilon^2 = \frac{\sigma_\varepsilon^2}{\mathcal{E}^2} = \frac{\langle\langle \varepsilon^2 \rangle\rangle - \mathcal{E}^2}{\mathcal{E}^2} \approx \frac{G}{(G - \mathcal{N} + 1)^2}\left[\frac{\mathcal{N}-2}{\mathcal{N}} - \frac{(\mathcal{N}-1)^2}{G}\right] \tag{21.55}$$

Only high energy levels are considered to be hydrogenoid, i.e. those levels such that $\mathcal{N}_{\min} \leqslant n \leqslant \mathcal{N}$, and the generic properties X_g of this group can be calculated as $X_g = X_{\mathcal{N}} - X_{\mathcal{N}_{\min}}$.

21.2.2 Lumping levels for molecules

For molecules, considering the ro-vibrational ladder of electronic states, the parameter δ_k is not small, especially for the ground electronic state, and lumping levels with the approach described above is not feasible. However, the logic of the lumped levels can be of help in finding an alternate strategy. In fact, low-lying levels can be approximated with the rigid rotor and truncated harmonic oscillator, while, for higher levels the lumped level approach is still valid. Differently from atoms, defining the limits of the groups is not straightforward, however, a best fit procedure can be used to optimize the parameters.

For diatomic molecule the expression for the partition function is

$$\mathcal{Q}^{\text{int}} = a_0 T \frac{1 - e^{-a_1/T}}{1 - e^{-a_2/T}} + \sum_k b_k e^{-E_k/T} \tag{21.56}$$

and for polyatomic molecule is

$$\mathcal{Q}^{\text{int}} = c + a_0 T^{p_0} \left(\frac{1 - e^{-a_1/T}}{1 - e^{-a_2/T}} \right)^{p_1} + \sum_k b_k e^{-E_k/T} \tag{21.57}$$

where the power terms p_0 and p_1 have been introduced to account for the variable number of vibrational and rotational modes. The thermodynamic functions can be calculated using the formula in section 21.1.1 where the partition function derivative is calculated analytically.

As an example, figure 21.4 reports the temperature profile of the internal energy, along with the relative error, for N_2 and N_3 calculated with the lumped level model for diatomic (equation (21.56)) and polyatomic (equation (21.57)) molecules, with fitting coefficients reported in table 21.1. It can be observed that the relative error is below 1% except at very low temperature where it remains below 2%.

21.3 Equilibrium calculation

There are plenty of algorithms to calculate equilibrium composition [43], most of them requiring the solution of non-linear systems of algebraic equations. We are presenting here the model implemented in the tool EquilTheTA [44, 45]. This algorithm solves one equation at a time, choosing the reaction ordering in the proper way. This kind of algorithm, named *hierarchical*, was proposed by Villars [46, 47] at the end of 1950s, when the computer power was limited. This algorithm later evolved [48, 49]. However, the method, working perfectly in most of the cases, in some conditions still required too many iterations for the convergence. These problems could be solved by adding ad hoc reactions by hand, and therefore this procedure is unsuitable for automatic calculations. Recently, an accurate analysis of the algorithm [50, 51] has led to improvements allowing the efficient automatization of the equilibrium calculation with the hierarchical approach.

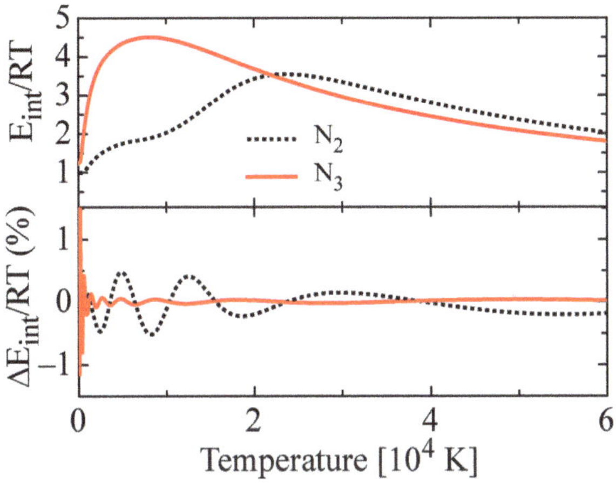

Figure 21.4. Internal energy calculated with equations (21.56) and (21.57) (top) and its percentage relative error with respect to the exact calculation [21] (bottom) of N_2 and N_3 as a function of the temperature. Fitting coefficients are reported in table 21.1.

Table 21.1. Fitting coefficient for N_2 (equation (21.56)) and N_3 (equation (21.57)).

	N_2	N_3		N_2	N_3		N_3
c		2	b_0	7.814 6 10^6	6.219 7 10^3	b_3	1.111 2 10^{10}
a_0	0.175 4	0.630 78	E_0	1.333 9 10^5	2.001 3 10^3	E_3	6.222 8 10^4
a_1	90 641	7 204.4	b_1	5.049 2 10^5	4.300 1 10^7	b_4	6.246 2 10^{10}
a_2	3 308.8	1 977.3	E_1	8.046 3 10^4	2.062 8 10^4	E_4	1.016 10^5
p_0		1.266 7	b_2	2.709 6 10^3	8.816 6 10^8	b_5	8.016 8 10^{10}
p_1		4.563 1	E_2	2.227 1 10^4	3.612 1 10^4	E_5	1.598 3 10^5

The hierarchical method presents the following advantages:

- it does not need the solution of a large system of non-linear equations;
- the set of reactions can be over or under-dimensioned;
- most of the reactions can be chosen to be solved analytically by simply solving second degree polynomials.

In this section, a brief description of the hierarchical algorithm will be given, leaving the details to the cited references.

21.3.1 Reaction equilibrium

Let us write a generic reaction in the following way

$$\sum_{i=1}^{n_s} Cr_{j,i} S_i \Leftrightarrow \sum_{i=1}^{n_s} Cp_{j,i} S_i \tag{21.58}$$

where S_i are the chemical species in the mixture, $Cr_{j,i}$ and $Cp_{j,i}$ are the stoichiometric coefficients, all positive or null, of reactants and products and n_s is the number of species. Given the initial species concentration $[S_i]$, we can define the functions

$$\Pi p_j(x) = \prod_{i=1}^{ns} ([S_i] + Cp_{j,i}x)^{Cp_{j,i}} \tag{21.59}$$

$$\Pi r_j(x) = \prod_{i=1}^{ns} ([S_i] - Cr_{j,i}x)^{Cr_{j,i}} \tag{21.60}$$

where x is the degree of advancement of the reaction. Equilibrium composition can be obtained by the mass action law in which the equilibrium constant K_j is known

$$K_j = \frac{\Pi p_j(x)}{\Pi r_j(x)} \tag{21.61}$$

Equation (21.61) can be solved finding the zeros of a polynomial

$$F_j(x) = K_j \Pi r_j(x) - \Pi p_j(x) = 0 \tag{21.62}$$

where $F_j(x)$ is a polynomial of degree

$$\deg F_j = \max\left(\sum_i Cr_{j,i}, \sum_i Cp_{j,i}\right)$$

It is possible to choose most of the reactions to give a second order equation that can be solved analytically, otherwise a root finding algorithm should be used.

It is straightforward to show that this equation has always at least one solution. The constraint that all the composition must be positive limits the range of x and only one physically meaningful solution is admitted in that interval.

The equilibrium composition is given as

$$[S_i]^{eq} = [S_i] + Cp_{j,i}x - Cr_{j,i}x \tag{21.63}$$

It must be noted that, independently from the quantity used to express the species concentration (partial pressures, moles per unit volume and so on), the equations do not change formally, provided the proper equilibrium constant (K_p, K_c, etc).

Finally, if the equilibrium constant is too large or too small the solution can be smaller than the machine epsilon. In this case one should resort to approximate solutions that can give a more precise value [51].

21.3.2 Reaction ordering

To calculate the mixture equilibrium one must solve equation (21.62) for all the reactions until the convergence criterion is satisfied and including an equation for the constraint (constant pressure, constant volume, etc). By choosing the proper order, the equilibrium can be reached very rapidly. The quantities reaction position ϑ_j and reaction distance $\delta_j = |\vartheta_j|$ (with respect to equilibrium) are introduced with the following properties:

- $\vartheta_j > 0$ if the reaction is shifted toward reactants;
- $\vartheta_j < 0$ if the reaction is shifted toward products;
- $\vartheta_j = 0$ if the reaction is at equilibrium.

The reaction ordering can be implemented by choosing the reaction with larger distance[6]. Therefore, the algorithm can proceed in the following way:

1. calculate ϑ_j for all the reactions;
2. solve the reaction with the largest δ_j;
3. update the solution, applying the constraint equation

and repeating the procedure until the maximum distance is below the required tolerance.

There are different possible definitions of the reaction distance based on the mass action law such as

$$\vartheta_{1j} = F_j(0) \tag{21.64}$$

$$\vartheta_{2j} = \frac{F_j(0)}{\max(K_j \Pi r_j, \Pi p_j)} \tag{21.65}$$

$$\vartheta_{3j} = \frac{F_j(0)}{\max(K_j, 1)} \tag{21.66}$$

Mathematically, these three distances are equivalent, but the choice of the reaction completely changes the ordering. For example ϑ_1 is strongly influenced by equilibrium constant and therefore reactions with relatively small K_j will never be solved. On the other hand, ϑ_2, is limited in the interval $[-1,1]$ and cannot really discriminate the reactions. The distance ϑ_3 is almost the same as ϑ_1 but it is equivalent to changing the reaction direction in order to have the equilibrium constant less than 1. The optimum choice is to reach the convergence firstly with ϑ_1 and then to refine the solution with ϑ_2.

The convergence becomes very slow when a subset of reactions, usually not more than two or three, are solved in sequence, resulting in a small variation of the composition at the end of the sequence. When this condition occurs, the algorithm is caught in a reaction trap. The trap can be easily bypassed by combining the reactions in the trap to make a new *shortcut reaction* that does in one single step what the original reaction set does in hundreds or thousands [50].

[6] It should be noted that the sign of the reaction position ϑ_j is the same as the sign of the solution of the equation.

21.3.3 Vectorial base of chemical reactions

In some cases, the reaction trap is made by a single reaction and it is so deep that the solution is below the machine epsilon. To solve this problem, the vectorial base of chemical reactions has been introduced [52].

Let us consider a mixture with n_s species formed by n_a atoms. A vectorial space of n_a dimension can be introduced, where atoms represent a base. A chemical species is defined by a vector whose elements are the number of the atom in the elements. In this framework, we can exploit the formalism of vector algebra to give a new point of view to the problem. First of all, we are not forced to use the atomic base and the species vector can be expressed with respect to the new base linked to a specified reaction set. The following procedure can avoid all kinds of traps:

1. determine n_a independent species with the largest concentration to form the new base;
2. formulate each reaction in terms of the base species;
3. calculate all the reaction distances;
4. solve the reaction with the highest distance;
5. update the solution, considering also the constraint equation

and repeat until convergence. While in the original model most of the reactions are of second order except the shortcut reactions, now the order of the polynomial can be very high. For these reasons this procedure should be chosen as a third step to further refine the solution.

Let us start considering the trivial base $\mathcal{B}_a = \{A_i\}_{1 \leqslant i \leqslant n_a}$ that includes atomic constituents A_j of the molecular species and the electron to account for charged species. In this base each species S_i is associated to a vector of integer numbers $|S_i\rangle$, whose elements are the number of each atom in the molecule or, equivalently, the stoichiometric coefficients of the reaction of formation of the species from atoms. The procedure[7] to find the new basis set $\mathcal{B}_G$ starts by ordering the species S_i by decreasing concentration. Then the first element of the $\mathcal{B}_G = \{G_i\}_{1 \leqslant i \leqslant n_a}$ is $G_1 \equiv S_1$, the specie with largest concentration. Then one should check if S_2 is independent from G_1 calculating the usual scalar product

$$\langle V|W\rangle = \sum_{j=1}^{n_a} V_j W_j \tag{21.67}$$

to calculate the projections along the parallel (||) and perpendicular ($\perp$) directions with respect to S_1

$$\begin{aligned} \vec{S}_2^{\parallel} &= \frac{\langle S_2|G_1\rangle}{\langle G_1|G_1\rangle}|G_1\rangle \\ \vec{S}_2^{\perp} &= |S_2\rangle - \frac{\langle S_2|G_1\rangle}{\langle G_1|G_1\rangle}|G_1\rangle = |S_2\rangle - |S_2^{\parallel}\rangle \end{aligned} \tag{21.68}$$

[7] The method described here is the classical Gram–Schmidt orthogonalization. [52, 53].

If $|S_2^{\perp}\rangle \neq |0\rangle$, S_2 is independent from $|G_1\rangle$ and we can assume $G_2 = S_2$, otherwise we have to try with the next elements until $|S_i^{\perp}\rangle \neq |0\rangle$ and then $G_2 = S_i$,

Proceeding by induction and supposing to have an incomplete $\mathcal{B}_G$ base with $n_b < n_a$ elements, with $G_{n_b} \equiv S_{i-1}$, $|S_i^{\perp}\rangle$ should be calculated iteratively as

$$|S_i^1\rangle = |S_i\rangle - \frac{\langle S_i|G_1\rangle}{\langle G_1|G_1\rangle}|G_1\rangle \tag{21.69}$$

$$|S_i^h\rangle = |S_i^{h-1}\rangle - \frac{\langle S_i^{h-1}|G_h\rangle}{\langle G_h|G_h\rangle}|G_h\rangle \tag{21.70}$$

until $h = n_b$ giving $|S_i^{\perp}\rangle = |S_i^{n_b}\rangle$. If $|S_i^{\perp}\rangle \neq |0\rangle$ then $G_{n_b+1} = S_i$, otherwise one should try with the next in the order until the condition is fulfilled. This procedure is repeated upon the base is complete.

Limiting this procedure to species with concentration above a given value, we have, as secondary result, the definition of the subspace of relevant species whose dimension can be lower than the number of atoms. For example, this condition is obtained at low temperature, when ion and/or electron concentration is too small and the electron can be excluded by the base.

The vector $|S_i\rangle$ was written with respect to the atomic base. To generalize the notation, we will add an index referring to the base, as $|S_i\rangle_a$ to refer to $\mathcal{B}_a$ and $|S_i\rangle_G$ to refer to $\mathcal{B}_G$. Finding the new base we have defined a matrix $\hat{C}$ so that[8]

$$|G_i\rangle_a = \sum_{j=1}^{n_a} \hat{C}_{ij}|A_j\rangle_a \tag{21.71}$$

The transformation matrix $\hat{T}$ to change the base representation as $|S_i\rangle_G = \hat{T}|S_i\rangle_a$ is $\hat{T} = \hat{C}^{-1}$. The elements of the vector $|S_i\rangle_G$ are the stoichiometric coefficients relating each species to the new base, that now are rational numbers. However, the equilibrium equations in section 21.3.1 are valid also if the coefficients are not integers. Alternately, it is not difficult to transform them into integer numbers.

21.3.4 State of minimal energy

Another trick that improves the convergence is to change the initial composition to reach the state of minimal energy, keeping constant the ratio between atoms. This state, obtained starting from the given initial composition, can be found with the following procedure:

1. find the base following the procedure explained in the section 21.3.3;
2. calculate the maximum advancement[9] $\bar{x}_j$ for each reaction;
3. calculate the energy variation as $\Delta\varepsilon_j = \sum_{i=1}^{\text{ns}}(Cp_{j,i} - Cr_{j,i})\bar{x}_j\varepsilon_i^f$;

[8] Note that $|A\rangle_{a,ij} = \delta_{ij}$ where δ_{ij} is the Kronecker delta.

[9] The maximum advancement is the smallest number such that $\Pi r_j = 0$ (see equation (21.60)), i.e. there exists one i with $Cr_{j,i} > 0$ such that $\bar{x}_j = [S_i]/Cr_{ji}$ and for $k \neq i$ it is $[S_k] - Cr_{j,k}\bar{x}_j \leqslant 0$.

4. choose the reaction m with minimum $\Delta\varepsilon_j$;
5. if $\Delta\varepsilon_m < 0$ advance the reaction with the $\bar{x}_m$ and repeat the procedure;
6. if $\Delta\varepsilon_m \geqslant 0$ the minimum energy state has been found.

Starting from the minimum energy state of the system, will assure that the initial condition is not a local minimum resulting in a reaction trap.

21.4 Transport properties

The transport properties of a mixture are important input data for many applications, as the re-entry space problem and technological applications of thermal plasmas. Calculations can be performed by means of the perturbative Chapman–Enskog method, assuming the plasma in local thermodynamic equilibrium (LTE) and starting from transport cross-sections of the different interactions. The calculations include viscosity, thermal conductivity, electric conductivity and multicomponent diffusion coefficients.

Diffusion represents the transfer of mass from a region to another due to a gradient in the concentration or in the temperature, viscosity represents the transport of momentum due to a gradient in the velocity and thermal conductivity is the transport of thermal energy resulting from the presence of thermal gradients, chemical reactions or internal degrees of freedom in the plasma. Finally, electric conductivity is related to the transport of charged particles due to gradients in concentration, pressure and temperature. In the Chapman–Enskog method, heavy and electron transport coefficients are considered independent. Transport coefficients can be obtained from the expressions of the fluxes of mass, momentum and energy, calculated directly from the Boltzmann equation. In fact, the diffusion velocity, the pressure tensor and the energy flux can be expressed as in reference [10]

$$\bar{\mathbf{V}}_i = \left(\frac{n^2}{n_i\rho}\right)\sum_j m_j D_{ij}\mathbf{d}_j - \frac{1}{n_i m_i}D_i^T\frac{\partial \ln T}{\partial \mathbf{r}} \tag{21.72}$$

$$\overset{\leftrightarrow}{\mathbf{p}} = p\overset{\leftrightarrow}{\mathbf{I}} - 2\eta\overset{\leftrightarrow}{\mathbf{S}} \tag{21.73}$$

$$\mathbf{q} = -\lambda\frac{\partial T}{\partial \mathbf{r}} + \frac{5}{2}kT\sum_i n_i\bar{\mathbf{V}}_i + \frac{kT}{n}\sum_{i,j}\frac{n_j D_i^T}{m_i \mathcal{D}_{ij}(1)}(\bar{\mathbf{V}}_i - \bar{\mathbf{V}}_j) \tag{21.74}$$

where the term

$$\mathbf{d}_i = \frac{\partial}{\partial \mathbf{r}}\left(\frac{n_i}{n}\right) + \left(\frac{n_i}{n} - \frac{n_i m_i}{\rho}\right)\frac{\partial \ln p}{\partial \mathbf{r}} - \frac{n_i m_i}{p\rho}\left(\frac{\rho}{m_i}\mathbf{X}_i - \sum_j n_j\mathbf{X}_j\right) \tag{21.75}$$

contains diffusion forces, due to concentration n_i/n and pressure gradient, and the external force $\mathbf{X}_i$ on the ith species. Here n, p and ρ are the total number density, pressure and mass density of the gas. D_{ij} are the multicomponent diffusion coefficients, D_i^T the multicomponent thermal diffusion coefficients, η the viscosity coefficient, $\overleftrightarrow{\mathbf{S}}$ the rate of shear tensor, λ the thermal conductivity and $\mathcal{D}_{ij}(1)$ is the first approximation to the diffusion coefficient of a binary mixture.

By means of the Sonine polynomial expansions of the perturbation distribution functions, the expressions for the transport coefficients are [10, 54–56]

$$D_{ij}(\xi) = \frac{\rho n_i}{2nm_j}\sqrt{\frac{2kT}{m_i}}\, c_{i0}^{(j,i)}(\xi), \tag{21.76}$$

$$D_i^T(\xi) = \frac{m_i n_i}{2}\sqrt{\frac{2kT}{m_i}}\, a_{i0}(\xi), \tag{21.77}$$

$$\eta(\xi) = \frac{1}{2}kT\sum_j n_j b_{j0}(\xi), \tag{21.78}$$

$$\sigma(\xi) = \frac{e^2 n_e^2}{2kT}\sqrt{\frac{2kT}{m_e}}\, c_{e0}^{(e,e)}(\xi), \tag{21.79}$$

$$\lambda = \lambda'(\xi) - \frac{k}{2n}\sum_{i,j}\frac{n_i n_j}{\mathcal{D}_{ij}(1)}\left(\frac{D_i^T}{n_i m_i} - \frac{D_j^T}{n_j m_j}\right)^2 \tag{21.80}$$

where

$$\lambda'(\xi) = -\frac{5}{4}k\sum_j n_j\sqrt{\frac{2kT}{m_j}}\, a_{j1}(\xi) \tag{21.81}$$

and

$$\mathcal{D}_{ij}(1) = \frac{3(m_i + m_j)}{16nm_i m_j}\frac{kT}{\Omega_{ij}^{(1,1)}} \tag{21.82}$$

where $\Omega_{ij}^{(1,1)}$ is the so-called diffusion type collision integral for the interaction of species i and j [10]. The values of the coefficients $c_{i0}^{(j,i)}(\xi)$, $b_{i0}(\xi)$ and $a_{i0}(\xi)$ are determinated by solving the following linear systems [10]:

$$\sum_j\sum_{m'=0}^{\xi-1}\tilde{Q}_{ij}^{mm'}\, t_{jm'}^{(h,k)}(\xi) = -R_{im}^{(h,k)}(\xi) \tag{21.83}$$

in which the matrix $\tilde{Q}_{ij}^{mm'}$ is defined as

$$\begin{cases} \tilde{Q}_{ij}^{mm'} \quad \text{when } t_{jm'}^{(h,k)} = b_{jm'} \\ \tilde{Q}_{ij}^{mm'} - \dfrac{n_j\sqrt{m_j}}{n_i\sqrt{m_j}}\tilde{Q}_{ii}^{mm'}\delta_{m0}\delta_{m'0} \quad \text{when } t_{jm'}^{(h,k)} = a_{jm'},\ c_{jm'}. \end{cases} \tag{21.84}$$

The dimension of $\tilde{Q}_{ij}^{mm'}$ depends upon the number ξ of terms considered in the expansion and the number of species considered in the plasma. The expansion coefficients $t_{jm'}^{(h,k)}$ and $R_{im}^{(h,k)}$ are linear combinations of collision integrals $\Omega^{(l,s)}$ (see section 21.4.5)

21.4.1 Viscosity

The viscosity coefficient η can be approximated by including only the contribution of heavy particles $\eta \cong \eta_h$ and can be obtained by solving the following linear system

$$\sum_{j=1}^{\nu-1}\left(\frac{Q_{ij}^{00}}{R_{i0}}\right)b_{j0}(1) = -1 \quad i = 1, \ldots, \nu - 1 \tag{21.85}$$

which is equivalent, in the first approximation, to

$$(\eta_h)_1 = \begin{bmatrix} H_{11} & \cdots & H_{1\nu} & x_1 \\ & \vdots & & \\ H_{\nu 1} & \cdots & H_{\nu\nu} & x_\nu \\ x_1 & \cdots & x_\nu & 0 \end{bmatrix} \cdot \begin{bmatrix} H_{11} & \cdots & H_{1\nu} \\ & \vdots & \\ H_{\nu 1} & \cdots & H_{\nu\nu} \end{bmatrix}^{-1} \tag{21.86}$$

where x_i is the molar fraction of the ith species, ν the number of chemical species and the elements of the determinant assume the following form [10]

$$\begin{aligned} H_{ii} &= \frac{x_i^2}{\eta_{ii}} + \sum_{j=1,\, j\neq i}^{\nu} \frac{2x_i x_j}{\eta_{ij}} \frac{M_i M_j}{(M_i + M_j)^2}\left(\frac{5}{3A_{ij}^\star} + \frac{M_j}{M_i}\right) \\ H_{ij} &= -\frac{2x_i x_j}{\eta_{ij}} \frac{M_i M_j}{(M_i + M_j)^2}\left(\frac{5}{3A_{ij}^\star} - 1\right) i \neq j \end{aligned} \tag{21.87}$$

where

$$\eta_{ij} = 2.669\,3 \times 10^{-5} \frac{\sqrt{2(M_i M_j)/(M_i + M_j)T}}{\sigma_{ij}^2 \Omega_{ij}^{(2,2)\star}} [\text{g cm}^{-1}\text{s}^{-1}] \tag{21.88}$$

21.4.2 Diffusion coefficients

Diffusion in plasmas is generated by gradient of particle number densities, gradient of temperature, which can modify a flux of particles, and gradient of concentrations which can cause an energy flux. Moreover, particle fluxes are also influenced by a pressure gradient and by external forces [10]. In a mixture containing N species, the diffusion velocity $\bar{\mathbf{V}}_i$ of the ith species is expressed as [10]

$$\bar{\mathbf{V}}_i = \left(\frac{n^2}{n_i \rho}\right)\sum_j m_j D_{ij} \mathbf{d}_j - \frac{1}{n_i m_i} D_i^T \frac{\partial \ln T}{\partial \mathbf{r}} \tag{21.89}$$

The diffusion coefficients of a multicomponent mixture can be calculated in a sufficiently accurate approximation by considering only one term in the Sonine expansion [10] and the following equations hold

$$\sum_j F_{ij}\{m_h D_{jh}(1) - m_k D_{jk}(1)\} = (\delta_{ih} - \delta_{ik}) \tag{21.90}$$

where

$$F_{ij} = \left\{\frac{n_i}{\rho \mathcal{D}_{ij}(1)} + \sum_{l \neq i} \frac{n_l m_j}{\rho m_i \mathcal{D}_{il}(1)}\right\}(1 - \delta ij) \tag{21.91}$$

The solution of this set of equations can be obtained by defining cofactors F^{ij} of F_{ij} in the determinant $|F|$ of F_{ij} as follows

$$F^{ij} = (-1)^{i+j} \begin{vmatrix} F_{11} & \cdots & F_{1\nu} \\ & \vdots & \\ F_{\nu 1} & \cdots & F_{\nu\nu} \end{vmatrix} \tag{21.92}$$

and finally diffusion coefficients of a multicomponent mixture are calculated by means of the following expression

$$D_{ij}(1) = \frac{F^{ji} - F^{ii}}{m_j |F|} \tag{21.93}$$

where $D_{ii} = 0$.

In a plasma, due to the presence of charges of both positive and negative signs, an electric field may be created and the overall diffusion properties of the charges can be influenced. In fact, diffusion due to concentration, pressure and temperature gradient is affected by the electric field due to charge separation and the ordinary diffusion coefficient D_{ij} and the thermal diffusion coefficient D_i^T must be replaced by the ambipolar diffusion coefficients

$$D_{ij}^a = D_{ij} + \frac{\alpha_i}{\beta} \sum_l Z_l D_{lj} \tag{21.94}$$

$$D_i^{\mathrm{Ta}} = D_i^T + \frac{\alpha_i m_i}{\beta} \sum_l \frac{Z_l D_l^T}{m_l} \tag{21.95}$$

where

$$\alpha_i = \sum_j n_j m_j Z_j D_{ij} \tag{21.96}$$

$$\beta_i = -\sum_{i,j} Z_i Z_j n_j m_j D_{ij} \tag{21.97}$$

21.4.3 Thermal conductivity

The total thermal conductivity of a plasma is calculated assuming as the sum of three contributions:

$$\lambda_{\mathrm{TOT}} = \lambda_{\mathrm{tr}} + \lambda_{\mathrm{int}} + \lambda_r \tag{21.98}$$

in which λ_{tr} is the translational thermal conductivity, calculated assuming that the molecules are spherical and have no internal degrees of freedom, λ_{int} the internal thermal conductivity and λ_r the reactive contribution to the total thermal conductivity. As for the coefficient of viscosity, the translational thermal conductivity λ_{tr} can be considered as the sum of two contributions, due to heavy particles (λ_h) and to the electrons (λ_e).

The second approximation of the Chapman–Enskog method for the translational thermal conductivity of heavy components can be written as [10, 57]

$$(\lambda_h)_2 = 4\begin{bmatrix} L_{11} & \cdots & L_{1\nu} & x_1 \\ & \vdots & & \\ L_{\nu 1} & \cdots & L_{\nu\nu} & x_\nu \\ x_1 & \cdots & x_\nu & 0 \end{bmatrix} \cdot \begin{bmatrix} L_{11} & \cdots & L_{1\nu} \\ & \vdots & \\ L_{\nu 1} & \cdots & L_{\nu\nu} \end{bmatrix}^{-1} \quad (21.99)$$

The elements of the determinant are given by

$$\begin{aligned} L_{ii} &= -\frac{4x_i^2}{\lambda_{ii}} - \sum_{j\neq i} \frac{2x_i x_j[\frac{15}{2}M_i^2 + \frac{25}{4}M_j^2 - 3M_j^2 B_{ij}^\star + 4M_i M_j A_{ij}^\star]}{(M_i + M_j)^2 A_{ij}^\star \lambda_{ij}} \\ L_{ij} &= \frac{2x_i x_j M_i M_j}{(M_i + M_j)^2 A_{ij}^\star \lambda_{ij}}\left[\frac{55}{4} - 3B_{ij}^\star - 4A_{ij}^\star\right] i \neq j \end{aligned} \quad (21.100)$$

where

$$\lambda_{ij} = 1.989\,1 \times 10^{-4} \frac{\sqrt{T(M_1 + M_2)/(2M_1 M_2)}}{\sigma_{ij}^2 \Omega_{ij}^{\star(2,2)}} [\text{cal cm}^{-1}\,\text{s}^{-1}\,\text{K}^{-1}] \quad (21.101)$$

σ_{ij} = collision diameter [Å]

$\Omega_{ij}^{(1,\,1)\star}, \Omega_{ij}^{(2,\,2)\star}$ = reduced diffusion and viscosity-type collision integrals

$$A_{ij}^\star = \frac{\Omega_{ij}^{(2,\,2)\star}}{\Omega_{ij}^{(1,\,1)\star}}$$

$$B_{ij}^\star = \frac{5\Omega_{ij}^{(1,\,2)\star} - 4\Omega_{ij}^{(1,\,3)\star}}{\Omega_{ij}^{(1,\,1)\star}}$$

Collision integrals, $\Omega_{ij}^{(\ell,s)\star}$ are terms related to the dynamics of elastic collisions in the plasma (see section 21.4.5 for details).

The third approximation of the Chapman–Enskog method for the electron contribution is given by

$$(\lambda_e)_3 = \frac{75}{8} n_e^2 k \left(\frac{2\pi k T}{m_e} \right)^{1/2} \frac{q^{22}}{q^{11} q^{22} - (q^{12})^2} \tag{21.102}$$

Expressions for q^{ij} taken from reference [55] are reported below: they are complicated functions of the number density of all species and of the reduced transport cross-sections, $\bar{Q}^{(\ell,\, s)} = \pi\sigma^2 \Omega^{(\ell,\, s)\star}$, of $(\ell,\, s)$ order between electrons (subscript e) and all other species (subscript j).

$$q^{11} = 8\sqrt{2} n_e^2 \bar{Q}_{ee}^{(2,2)} + 8 \sum_{j=1}^{\nu-1} n_j n_e \left[\frac{25}{4} \bar{Q}_{ej}^{(1,1)} - 15 \bar{Q}_{ej}^{(1,2)} + 12 \bar{Q}_{ej}^{(1,3)} \right]$$

$$q^{12} = 8\sqrt{2} n_e^2 \left[\frac{7}{4} \bar{Q}_{ee}^{(2,2)} - 2 \bar{Q}_{ee}^{(2,3)} \right] + 8 \sum_{j=1}^{\nu-1} n_j n_e \left[\frac{175}{16} \bar{Q}_{ej}^{(1,1)} - \frac{315}{8} \bar{Q}_{ej}^{(1,2)} + 57 \bar{Q}_{ej}^{(1,3)} - 30 \bar{Q}_{ej}^{(1,4)} \right]$$

$$q^{22} = 8\sqrt{2} n_e^2 \left[\frac{77}{16} \bar{Q}_{ee}^{(2,2)} - 7 \bar{Q}_{ee}^{(2,3)} + 5 \bar{Q}_{ee}^{(2,4)} \right] + 8 \sum_{j=1}^{\nu-1} n_j n_e \left[\frac{1225}{64} \bar{Q}_{ej}^{(1,1)} - \frac{735}{8} \bar{Q}_{ej}^{(1,2)} + \frac{399}{2} \bar{Q}_{ej}^{(1,3)} - 210 \bar{Q}_{ej}^{(1,4)} + 90 \bar{Q}_{ej}^{(1,5)} \right]$$

Expressions for higher approximations to $(\lambda_e)_n$ can be found in the paper by Devoto and Li [58].

The internal thermal conductivity is calculated according to the Eucken approximation which considers that the presence of internal degrees of freedom affects the heat flux vector. The exact expression for the coefficient of thermal conductivity depends upon the transition probabilities for the transfer of energy among the degrees of freedom of the molecules. However, when the rate of transfer is sufficiently fast so that the internal distributions of atoms and molecules can be considered in equilibrium, the simplified Eucken's expression, including only the ro-vibrational contribution, is given by

$$\lambda_{\text{int}} = \sum_i \frac{\lambda_{\text{int},i}}{\sum_j \frac{\mathcal{D}_{ii} x_j}{\mathcal{D}_{ij} x_i}} \tag{21.103}$$

$$\lambda_{\text{int},i} = \frac{p \mathcal{D}_{ii}}{RT} C_{p,\,i}^{\text{int}} \tag{21.104}$$

Finally, the reactive thermal conductivity includes the contribution of chemical reactions in plasma. In fact, chemical reactions in chemical equilibrium increase the thermal conductivity. When the components of a plasma react, additional heat is transported as formation enthalpy by diffusion of chemical species. By recurring to the method of Butler and Brokaw [59], given a set of μ independent reactions $\{R_i\}$ between ν chemical species S_i

$$R_i\colon\ \sum_{k=1}^{\nu} C_{ik} S_k = 0$$

where C_{ik} is the stoichiometric coefficient of the kth species in the ith reaction, the reactive thermal conductivity is calculated as

$$\lambda_{\rm r} = -\frac{1}{RT^2}\begin{bmatrix} A_{11} & \cdots & A_{1\mu} & \Delta H_1 \\ & \vdots & & \\ A_{\mu 1} & \cdots & A_{\mu\mu} & \Delta H_\mu \\ \Delta H_1 & \cdots & \Delta H_\mu & 0 \end{bmatrix} \cdot \begin{bmatrix} A_{11} & \cdots & A_{1\mu} \\ & \vdots & \\ A_{\mu 1} & \cdots & A_{\mu\mu} \end{bmatrix}^{-1} \tag{21.105}$$

where

$$A_{ij} = A_{ji} = RT\sum_{k=1}^{\nu-1}\sum_{\ell=k+1}^{\nu} \frac{x_k x_\ell}{p\mathcal{D}_{k\ell}}\left[\frac{C_{ik}}{x_k} - \frac{C_{i\ell}}{x_\ell}\right]\left[\frac{C_{jk}}{x_k} - \frac{C_{j\ell}}{x_\ell}\right] \tag{21.106}$$

The set of reactions must be dynamically changed depending on the plasma composition, because if some concentrations are too small, the procedure becomes unstable. To solve the problem, only the subset of species with large enough concentration should be considered. The set of independent reactions can be determined through the Gram–Schmidt orthogonalization, as shown in section 21.3.3.

21.4.4 Electrical conductivity

The electrical conductivity is calculated by using the third approximation for the ordinary diffusion coefficient D_{ee} with the following expression

$$\sigma = \frac{3e^2 n_e^2}{2}\frac{(2\pi/km_e T)^{\frac{1}{2}}}{|\,q\,|}\begin{vmatrix} q^{11} & q^{12} \\ q^{21} & q^{22} \end{vmatrix} \tag{21.107}$$

where

$$|\,q\,| = \begin{vmatrix} q^{00} & q^{01} & q^{02} \\ q^{10} & q^{11} & q^{12} \\ q^{20} & q^{21} & q^{22} \end{vmatrix} \tag{21.108}$$

and

$$q^{00} = 8n_e \sum_{j=1}^{\nu-1} n_j \bar{Q}_{ej}^{(1,1)}$$

$$q^{01} = 8n_e \sum_{j=1}^{\nu-1} n_j\left[\frac{5}{2}\bar{Q}_{ej}^{(1,1)} - 3\bar{Q}_{ej}^{(1,2)}\right]$$

$$q^{02} = 8n_e \sum_{j=1}^{\nu-1} n_j\left[\frac{35}{8}\bar{Q}_{ej}^{(1,1)} - \frac{21}{2}\bar{Q}_{ej}^{(1,2)} + 6\bar{Q}_{ej}^{(1,3)}\right]$$

21.4.5 Collision integrals

The estimation of the coefficients describing the transport of momentum, mass, heat and charge in plasmas relies on the elementary binary interactions entering into the collision integrals in the Chapman–Enskog theory. Classical collision integrals, $\Omega^{(\ell,\, s)}$, entering the working equations, result from a threefold integration: on the inter-particle distance r, on the impact parameter b and on the reduced energy $\gamma^2 = E/kT$ [10, 12].

$$\vartheta(b,\, E) = \pi - 2b \int_{r_c}^{\infty} \left[1 - \frac{b^2}{r^2} - \frac{\varphi(r)}{E}\right]^{-1/2} \frac{dr}{r^2} \tag{21.109}$$

$$Q^{(\ell)}(E) = 2\pi \int_0^{\infty} db\, b(1 - \cos^{\ell} \vartheta) \tag{21.110}$$

$$\Omega^{(\ell,\, s)}(T) = \sqrt{\frac{kT}{2\pi\mu}} \int_0^{\infty} d\gamma\, \gamma^{2s+3} Q^{(\ell)} e^{-\gamma^2} \tag{21.111}$$

Dimensionless *reduced* transport quantities are conveniently introduced, representing the deviation of the actual values of the transport cross-section and of the collision integral from the rigid sphere model

$$Q^{(\ell)\star} = \frac{Q^{(\ell)}}{Q^{(\ell)}_{\mathrm{rs}}} = \frac{4}{\sigma^2}\left[2 - \frac{1 + (-1)^{\ell}}{1 + \ell}\right]^{-1} \int_0^{\infty} db\, b(1 - \cos^{\ell} \vartheta)$$

$$\Omega^{(\ell,\, s)\star} = \frac{\Omega^{(\ell,\, s)}}{\Omega^{(\ell, s)}_{\mathrm{rs}}} = \frac{4(\ell + 1)}{(s + 1)!\, [2\ell + 1 - (-1)^{\ell}]\pi\sigma^2} \int_0^{\infty} d\gamma\, \gamma^{2s+3} Q^{(\ell)} e^{-\gamma^2}$$

While the higher-ℓ orders entails the integration in equation (21.110), the (s+1)-order collision integrals can be straightforwardly obtained through the recurrence relation

$$\Omega^{(\ell,\, s+1)\star} = \Omega^{(\ell,\, s)\star} + \frac{T}{s + 2} \frac{\partial \Omega(\ell,\, s)\star}{\partial T} \tag{21.112}$$

The potential arising in the interaction of collisional-partners, namely neutral–neutral and neutral–charged interactions, is the key element in the derivation of the transport cross-sections and in turn the collision integrals. Assuming binary collisions, classical dynamics, and isotropy of the interaction, the potential is a function of the approaching coordinate, i.e. $\varphi(r)$. In the literature a number of *model potentials* have been proposed. In the following the analytical forms are listed and referenced.

Exponential decay [60, 61] for fully repulsive interactions

$$\varphi(r) = Ae^{-\rho r} \tag{21.113}$$

Inverse power [62], depending on the sign of d for repulsive or attractive interactions

$$\varphi(r) = \frac{d}{r^{\delta}} \tag{21.114}$$

A special case of the general inverse power potential is the attractive **polarization potential**, adopted for example in the interaction of neutral species with bare nuclei ions, with $\delta = 4$, $d = -\frac{ze^2\alpha_{\text{pol}}}{8\pi\varepsilon_0}$, where z is the ion charge and α_{pol} the dipole polarizability of the neutral collider. In this case collision integrals assume a closed form [62]

$$\sigma^2\Omega^{(\ell,\,s)\star} = \frac{4(\ell+1)}{(s+1)![2\ell+1-(-1)^{\ell}]} A^{(\ell)}(\delta)\sqrt{\frac{d\delta}{k_B T}}\,\Gamma\left(s+2-\frac{2}{\delta}\right) \tag{21.115}$$

where Γ is the Gamma function and $A^{(\ell)}(\delta)$ is a temperature-free coefficient correlated to the transport cross-section. The $A^{\ell}(4)$ values have been estimated by Smith [63] for $\ell = 1$, 2 and 3 in the attractive case ($A^1(4) = 0.552\ 3$, $A^2(4) = 0.384\ 6$, $A^3(4) = 0.637\ 7$), leading to simple relations [64]

$$\sigma^2\Omega^{(1,1)\star} = 424.443z\sqrt{\frac{\alpha_{\text{pol}}}{T}};$$
$$\sigma^2\Omega^{(2,2)\star} = 0.870\ \sigma^2\Omega^{(1,1)\star};\ \sigma^2\Omega^{(3,3)\star} = 0.842\ \sigma^2\Omega^{(1,1)\star}$$

Morse [65]

$$\varphi(r) = \varphi_0\left[\exp\left(-\frac{2C}{\sigma}(r-r_e)\right) - 2\exp\left(-\frac{C}{\sigma}(r-r_e)\right)\right] \tag{21.116}$$

Lennard-Jones [66]

$$\varphi(r) = 4\varphi_0\left[\left(\frac{\sigma}{r}\right)^{12} - \left(\frac{\sigma}{r}\right)^{6}\right] \tag{21.117}$$

modified Buckingham [67]

$$\varphi(r) = \frac{\varphi_0}{1-6/\alpha}\left[\frac{6}{\alpha}e^{\alpha(1-r/r_e)} - \left(\frac{r_e}{r}\right)^{6}\right] \tag{21.118}$$

Hulburt–Hirschfelder [68]

$$\begin{aligned}\varphi(r) = \varphi_0\Bigg\{&\exp\left[-2\alpha\left(\frac{r}{r_e}-1\right)\right] - 2\exp\left[-\alpha\left(\frac{r}{r_e}-1\right)\right]\\ &+\beta\left(\frac{r}{r_e}-1\right)^3\left[1+\gamma\left(\frac{r}{r_e}-1\right)\right]\exp\left[-2\alpha\left(\frac{r}{r_e}-1\right)\right]\Bigg\}\end{aligned} \tag{21.119}$$

Tang–Toennies [69]

$$\varphi(r) = A\exp(-br) - \sum_{j=3}^{J} f_{2j}(br)\frac{C_{2j}}{r^{2j}} \tag{21.120}$$

being C_{2j} the dispersion coefficients and

$$f_{2j}(x) = 1 - e^{-x}\sum_{k=0}^{2j}\frac{x^k}{k!} = 1 - \frac{\Gamma(2j+1, x)}{(2j)!} \tag{21.121}$$

Aziz [70]

$$\varphi(r) = \varphi_0 A\exp\left[-\alpha\left(\frac{r}{r_e}\right) + \beta\left(\frac{r}{r_e}\right)^2 - F(r/r_e)\sum_{j=0}^{2}\frac{c_{2j+6}}{(r/r_e)^{2j+6}}\right] \tag{21.122}$$

with

$$F(x) = \begin{cases} \exp\left[-(D/x - 1)^2\right] & x < D \\ 1 & x \geq D \end{cases} \tag{21.123}$$

The Morse and Lennard-Jones model potentials are widely used and reproduce a quite general form including a short-range repulsive part, an attractive long distance part and a more or less deep 'potential well'. For the first four above given potentials, the corresponding reduced collision integrals can be found in the cited references for values of the potential parameters varying in sufficiently large ranges to comprise values to represent the intermolecular interaction for a large variety of systems. Nevertheless, it should be stressed that these potential forms exhibit some inadequacies, for example an excessive short-range repulsion, and poor flexibility to the description of special features of accurate potentials, such as barriers, double minima, shoulders in the repulsive branch. For these reasons analytical expressions characterized by a higher functional complexity and an increasing number of parameters have been also considered in the literature, though in these cases usually collision integrals are not available.

The integrations in equations (21.109)–(21.111) may pose numerical difficulties due to the existence of integrable and non-integrable singularities in the integration of the potential and of rapid oscillations of the cosine function in the integration over the impact parameter and related to the occurrence of *orbiting* trajectories. In the past different solutions were conceived based on the preliminary definition of subdomains in the integration depending on a rigid classification of potential types [68], and the applications of suitable variable transformations to avoid infinities (the Clenshaw–Curtis quadrature [71]), while recently a novel general algorithm has been proposed for the calculation of classical transport collision integrals, able to handle any potential, regardless the number of extrema, relying on an efficient fractal integration method [36].

The modern quantum chemistry techniques allow the derivation of accurate *ab initio* potentials for any collisional pair, however, model potentials are still attractive for different reasons: the models provide a 'full range' description of the interaction, also in the limiting regions of inter-particle distance where theoretical methods could be inaccurate, in some cases the collision integrals can be straightforwardly obtained, but definitely the most important point is the possibility to have a guess of the interaction potential where unknown, through correlation formulas connecting the potential parameters to spectroscopic constants, as in the case of the Hulburt–Hirschfelder potential, or to physical properties of chemical species in the collisional pair (polarizability, charge, …).

This *single-potential* representation of the interaction is almost never the case, especially when open-shell species are involved. In fact, the number of states, arising in the formation of the quasi-molecular system as the colliding partners approach, could be large as predicted by the theory of electronic (angular and spin) momenta composition. Therefore, the rigorous approach prescribes each allowed path of interaction properly accounted for in the so-called **multi-potential approach**, where the effective collision integral of (ℓ, s) order results from each molecular state averaged by its statistical weight [12].

$$\Omega_{\mathrm{av}}^{(\ell,s)} = \frac{\sum_n w_n \Omega_n^{(\ell,s)}}{\sum_n w_n} \tag{21.124}$$

The multi-potential approach still nowadays is not always the choice being more cumbersome and, for systems poorly studied, hindered by the lack of a complete information about the molecular states arising in the partner interaction. These considerations clarify why approaches able to provide potential functions that model the average interaction were explored. In this regard, an improved Lennard-Jones potential, the **phenomenological potential** [72, 73], was proved to predict quite accurately intermolecular interactions in a variety of systems (neutral–neutral and neutral–ion). Fundamental interaction features, i.e. binding energy and equilibrium distance, entering the relevant equations as parameters, are determined on the base of correlation formulas of the physical properties of colliding partners (polarizability, charge, number of electrons effective in polarization) [74–77]

The phenomenological potential [72, 73, 78] has the form

$$\varphi = \varphi_0 \left[\frac{m}{n(x) - m} \left(\frac{1}{x}\right)^{n(x)} - \frac{n(x)}{n(x) - m} \left(\frac{1}{x}\right)^{m} \right] \tag{21.125}$$

where $x = r/r_e$, $n(x) = \beta + 4x^2$. For neutral–neutral and neutral–ion cases the parameter m has the value of 6 and 4, respectively. The value of β parameter, ranging from 6 to 10 depending on the hardness of interacting electronic distribution densities, could be estimated through the following empirical formula [79],

$$\beta = 6 + \frac{5}{(s_1 + s_2)} \tag{21.126}$$

where the subscripts 1 and 2 identify the colliding partners. The *softness* s, entering in equation (21.126), is defined as the cubic root of the polarizability. For open-shell atoms and ions a multiplicative factor, which is the ground state spin multiplicity, should be also considered.

For neutral–neutral systems, the phenomenological method represents the binding energy, φ_0, and the equilibrium distance, r_e, in terms of polarizabilities of the interacting partners, α and the effective number of electrons which contribute to the polarization of the neutral species, N, by the following correlation formulas [74, 75, 80]

$$\begin{gathered} r_e[\text{Å}] = 1.767\frac{\alpha_1^{1/3} + \alpha_2^{1/3}}{(\alpha_1\alpha_2)^{0.095}} \qquad \varphi_0[\text{eV}] = 0.72\frac{C_d}{r_e^6} \\ C_d[\text{eV Å}^6] = 15.7\frac{\alpha_1\alpha_2}{[\sqrt{\alpha_1/N_1} + \sqrt{\alpha_2/N_2}]} \end{gathered} \tag{21.127}$$

where C_d is an effective long-range London coefficient.

Previous formulas have been also extended to neutral–ion systems [76, 77], introducing the parameter ρ, representative of the relative role of dispersion and induction attraction components in proximity to the equilibrium distance.

$$\begin{gathered} r_e[\text{Å}] = 1.767\frac{\alpha_i^{1/3} + \alpha_n^{1/3}}{(\alpha_i\alpha_n[1 + 1/\rho])^{0.095}} \qquad \varphi_0[\text{eV}] = 5.2\frac{z^2\alpha_n}{r_e^4}[1 + \rho] \\ \rho = \frac{\alpha_i}{z^2[1 + (2\alpha_i/\alpha_n)^{2/3}]\sqrt{\alpha_n}} \end{gathered} \tag{21.128}$$

where α_n and α_i are the neutral and ionic polarizability, respectively.

Collision integrals up to high orders have been calculated in a wide range of reduced temperatures considering integer values of β in its range of variability for both neutral–ion and neutral–neutral interactions and fitted with a suitable bi-dimensional function, so that collision integrals can be straightforwardly obtained once known the tuple of parameters (φ_0, r_e, m, β).

$$\begin{aligned} \ln \Omega^{(\ell, s)\star} = {} & [a_1(\beta) + a_2(\beta)x]\frac{e^{(x-a_3(\beta))/a_4(\beta)}}{e^{(x-a_3(\beta))/a_4(\beta)} + e^{(a_3(\beta)-x)/a_4(\beta)}} \\ & + a_5(\beta)\frac{e^{(x-a_6(\beta))/a_7(\beta)}}{e^{(x-a_6(\beta))/a_7(\beta)} + e^{(a_6(\beta)-x)/a_7(\beta)}} \end{aligned} \tag{21.129}$$

where $x = \ln T^\star$ and parameters $a_i(\beta) = \sum_{j=0}^{2} c_j\beta^j$ are polynomial functions of β (parameters c_j are reported in reference [81]).

The approach has been validated through the estimation of benchmark collision-pair interactions relevant to the Earth atmosphere [79], showing how good could be the phenomenological description when compared with results obtained with accurate approaches [82–85].

The phenomenological approach allows the evaluation of consistent collision integral datasets for neutral–neutral and neutral–charged interactions [86], however a *hybrid* approach has been successfully implemented [64, 87, 88], combining the multi-potential collision integrals for atom–atom and atom–atomic ion interactions, data mining the literature for the potential energy curves of the molecular states correlating with the given collision partners, and relying on phenomenological method only for interactions involving polyatomic molecules or where the collisional systems are poorly known.

Obviously, this represents an approximation in the description of the interaction, however, for small and symmetric molecules the anisotropy of the collision geometry should not affect significantly the collision integrals, while for larger chemical systems the phenomenological approach still represents a reliable choice in the lack of exact calculations available only for a few collisional systems and based on the knowledge of the corresponding accurate potential energy surfaces [89–91].

The derivation of collision integrals described so far completely neglects inelastic channels, usually too small to be effective [92, 93]. However, the assumption is not acceptable for resonant charge-transfer in ion–parent-atom interactions

$$\tilde{\mathrm{A}} + \mathrm{A}^{+} \rightarrow \tilde{\mathrm{A}}^{+} + \mathrm{A}$$

In fact, these processes mimic the exchange of momentum in elastic collisions and are characterized by large cross-sections, affecting the odd-order collision integrals [12]. *Effective* collision integrals results by the geometrical mean of elastic and inelastic resonant contributions [94]

$$\Omega^{(\ell,\,s)\star} = \sqrt{(\Omega_{\mathrm{el}}^{(\ell,s)\star})^2 + (\Omega_{\mathrm{ex}}^{(\ell,s)\star})^2}\,(\ell \;\;\text{odd}) \tag{21.130}$$

Though the inelastic contribution to the collision integral results, in principle, by the integration of the transport cross-section, even nowadays it is widely diffuse the use of the closed forms derived in reference [95]. In fact, if the resonant charge-exchange cross-section [Å^2] shows a dependence on the relative velocity v_r [cm s^{-1}] of the type

$$\sqrt{2\sigma_{\mathrm{ex}}} = C - D\ln(v_r) \tag{21.131}$$

the collision integral assumes the form

$$\begin{aligned} \sigma^2\Omega_{\mathrm{ex}}^{(\ell,s)\star} = \frac{1}{\pi}\Bigg\{ & C^2 - CDx + \left[\frac{1}{2}Dx\right]^2 + \frac{D^2}{4}\left\{\frac{\pi^2}{6} - \sum_{n=0}^{s+1}\frac{1}{n^2} + \left[\sum_{n=0}^{s+1}\frac{1}{n} - \bar{\gamma}\right]^2\right\} \\ & + \frac{1}{2}D\xi[Dx - 2C] + \frac{D}{2}[D(x+\xi) - 2C]\ln\left(\frac{T}{M}\right) + \left[\frac{D}{2}\ln\left(\frac{T}{M}\right)\right]^2\Bigg\} \end{aligned} \tag{21.132}$$

with M the molecular weight, $x = \ln(4R)$ con R constant of perfect gas, $\bar{\gamma}$ is the Euler constant, allowing an accurate and straightforward estimation of $\Omega_{\mathrm{ex}}^{(\ell,s)}$ by a simple algebraic formula, once known, through linear regression algorithms, the C and D parameter values. Thus the problem reduces to the estimation of the resonant charge-exchange cross-section either obtained by crossed-beam experiments [96, 97] or through quantum [98] or semiclassical [99, 100] approaches. Among them, the *asymptotic approach* [99, 101–103] has been shown to be predictive and accurate and therefore extensively used for the estimation of resonant charge-exchange cross-sections up to tens of eV in the collision energy. The theoretical development of the method is beyond the scope of the present section, the reader is referred to the literature, and only working equations are here recalled. The cross-section is straightforwardly obtained from estimation of the critical impact parameter $b^\star$ solving the transcendental equation (21.134)

$$\sigma_{\rm ex} = 2\pi \int_0^\infty \sin^2 \zeta b db = \frac{1}{2}\pi b^{\star 2} \tag{21.133}$$

$$\zeta(b^\star) = \frac{1}{\hbar g}\left(\frac{\pi b^\star}{2\alpha}\right)^{\frac{1}{2}} \Delta(b^\star) = \frac{1}{2}e^{-\tilde{\gamma}} = 0.28 \tag{21.134}$$

the phase ζ depending on the exchange interaction potential, $\Delta = \varphi_g - \varphi_u$, defined as the energy separation of the pair states of the quasi-molecule of symmetry gerade and ungerade (*g-u splitting*). Depending on the outer shell configuration of the interacting atoms and the scheme of coupling of momenta, the exchange interaction potential has different formal expressions [101, 102]. For the simplest case of single-electron exchange between bare nuclei

$$\Delta_{\ell_e\mu} = \mathcal{A}^2 R^{\left(\frac{2}{\gamma}-1\right)} e^{\left(-R\gamma-\frac{1}{\gamma}\right)} \frac{(2\ell_e + 1)(\ell_e + |\mu|)!}{(\ell_e - |\mu|)!|\mu|!(R\gamma)^{|\mu|}}, \tag{21.135}$$

where ℓ_e and μ are quantum numbers associated with the electron orbital angular momentum and its axial projection, $-\frac{\gamma^2}{2}$ is the binding energy (in atomic units), and $\mathcal{A}$, the so-called *asymptotic parameter*, connects to the asymptotic representation of the electron wavefunction.

It is worth noticing that the combination of phenomenological and asymptotic approaches, is also a promising tool to tackle one problem still open in the transport characterization of plasmas, i.e. the role of excited states, as demonstrated for the low-lying excited states of oxygen and nitrogen atoms [104–106].

So far neutral–neutral and neutral–charged interactions have been considered, however, the construction of the core transport cross-section database for EquilTheTA tool (see figure 21.5) for any given mixture must include the collision integrals also for charged-particle–charged-particle and electron–neutral interactions, whose derivation requires specific approaches.

The interaction between charged species, determining the transport properties in ionization regimes, is modeled with the **screened Coulomb potential**

$$\varphi(r) = \frac{Z_i Z_j e^2}{r} e^{-r/\lambda_D} = \frac{\varphi_0}{r/\lambda_D} e^{-r/\lambda_D} \tag{21.136}$$

with Z_i and Z_j the charge of the ions i and j, and $\lambda_D = \sqrt{(k_B T)/(4\pi e^2(n_e + \sum_i n_i Z_i^2))}$ the Debye length, calculated including both electron and ion densities [107], though different

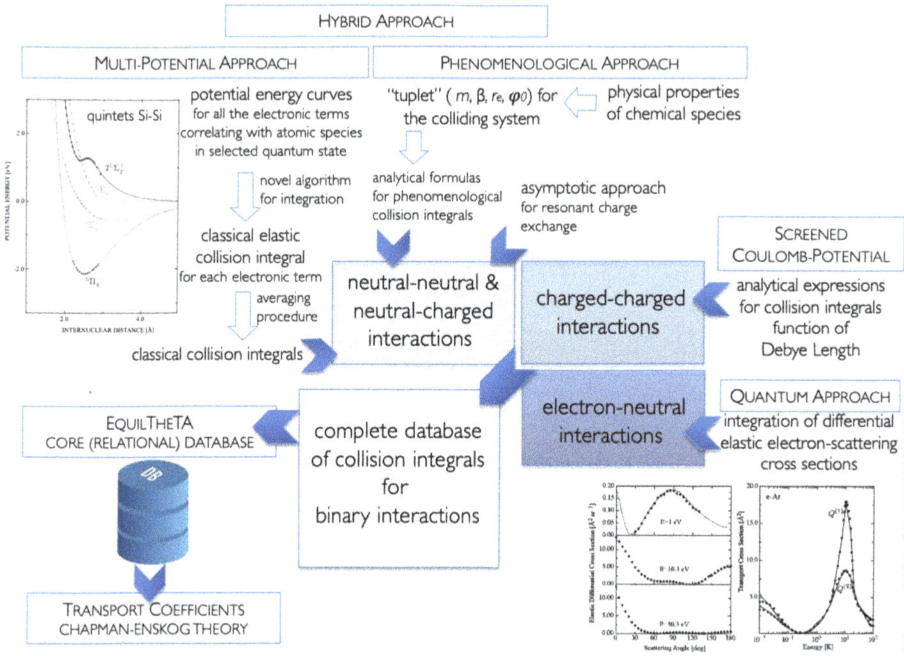

Figure 21.5. Sketch of the general approach in the creation of core collision integral database of EquilTheTA.

authors account only for the screening due to electrons [108–111]. Closed forms for collision integrals were derived by Liboff [112]. Recently accurate collision integrals by Mason [108, 110] have been fitted by the following equation (see reference [15])

$$\ln(\Omega^\star) = \sum_{j=0}^{6} c_j \ln^j T^\star \tag{21.137}$$

The quantum nature (resonances, Ramsauer minimum) of electron–neutral interactions hinders the use of classical approach in the collisional dynamics and collision integrals are calculated by integration of theoretical or experimental differential elastic electron-scattering cross-sections to obtain the different moments and in turn different orders of $\Omega^{(\ell,\, s)}$.

$$Q^{(\ell)}(E) = 2\pi \int_0^\pi d\vartheta \, \sin(\vartheta) \frac{d\sigma}{d\Omega}(E, \vartheta)(1 - \cos^\ell \vartheta) \tag{21.138}$$

21.5 Thermodynamic and transport properties of equilibrium hydrogen plasma

In this section, the thermodynamic and transport properties of an equilibrium hydrogen plasma are reported, discussing the characteristic behaviour of quantities with temperature and pressure, applying the theoretical approaches described in the above sections. The pure H_2 plasma, undergoing reactions, can form the following species: H_2, H_2^+, H_3^+, H, H^+, H^-, *e*. Single species thermodynamic properties are reported in reference [22] and an extensive description of the database of transport cross-sections of binary interactions can be found in reference [64]. Calculations have been performed in the temperature range 300–60000 K for pressures $P = 0.1$, 1, 100 bar.

Figure 21.6 shows the composition of a hydrogen at P = 1 bar (left) and at $P = 100$ bar (right). At $T \approx 3000$ K atomic hydrogen is formed by dissociation, while ionization happens at $T \approx 20\,000$ K. All other species are negligible. Increasing the pressure dissociation and ionization equilibria are shifted to higher temperatures and the maximum fraction of molecular ions and H^- increases with the pressure as a consequence of the Le Châtelier principle.

As an example, specific heat and entropy have been selected as representative of the thermodynamic quantities at different pressures (0.1, 1, 100 bar) and are reported in figure 21.7. The peak corresponds to the activation of a reaction (reactive specific heat, see figure 21.1) corresponding to the flex in the reactant concentration. At the same time the reactions are responsible of the steps in the entropy diagram. The shift of the peaks with the pressure reflects the change observed in the plasma composition.

Also, transport properties reflect the variation of the plasma composition, as can be observed in the profile of the viscosity (see figure 21.8(left)). The peaks in the viscosity are due to the change in dominant interaction between neutral species,

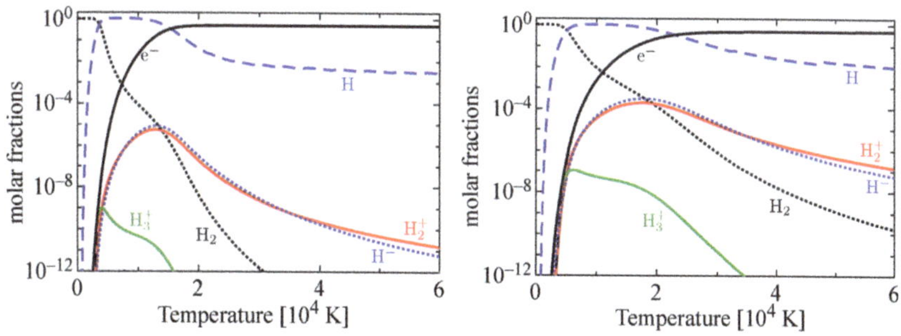

Figure 21.6. Mole fractions of the different species present in a hydrogen plasma at atmospheric pressure and at 100 bar.

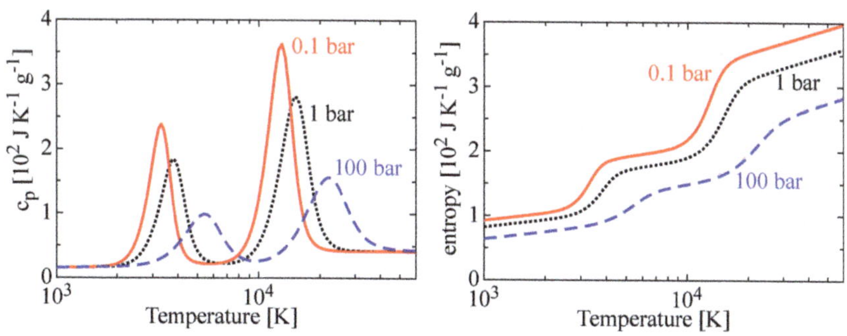

Figure 21.7. Specific heat and entropy at different pressures (0.1, 1, 100 bar).

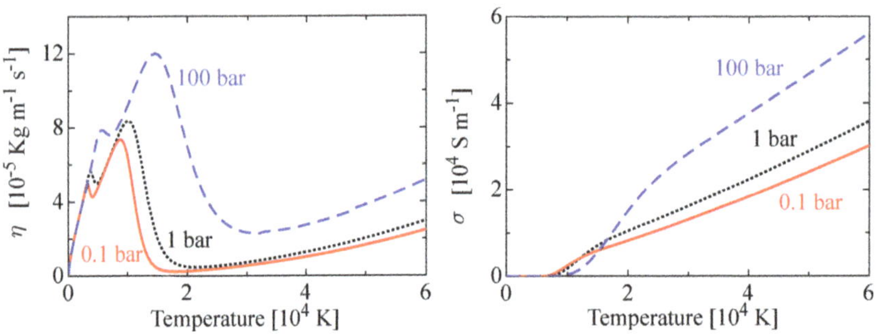

Figure 21.8. Viscosity coefficient and electric conductivity at different pressures (0.1, 1, 100 bar).

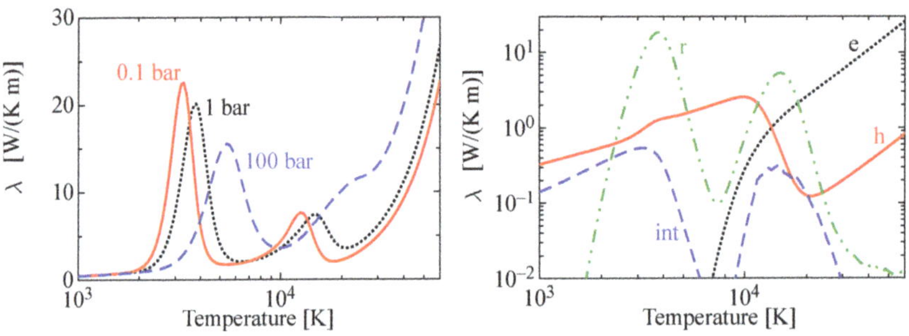

Figure 21.9. Total thermal conductivity at different pressures (0.1, 1, 100 bar) and components of thermal conductivity at atmospheric pressure: reactive (r), translational heavy (h) and electron (e) components, internal (i) contribution.

while abrupt decrease of viscosity is related to the onset of ionization, because electrons are not viscous. Contrarily, the electrical conductivity (see figure 21.8), depending on the electron density, increases only when ionization is active in the plasma. At the onset, the electrical conductivity decreases with the pressure, due to the pressure shifting of ionization equilibrium, while, due to the higher electron density, at high temperature the opposite trend is established.

An interesting behavior is observed in the profile of thermal conductivity (see figure 21.9) showing peaks below $T = 20000$ K and a considerable growth at high temperature. This behavior can be rationalized when looking at the different contributions. The peaks are due to the reactive part, at the temperatures relevant to dissociation and ionization, while at high temperature, electron translational contribution dominates, a consequence of their very low mass and high mobility. In this case internal contribution is practically negligible, hidden by the reactive part.

21.6 EquilTheTA code

The theoretical framework outlined in the chapter represents the foundation of the application EquilTheTA (EQUILibrium for plasma THErmodynamics and Transport Applications), a web-based tool [44] for the calculation of chemical equilibrium and the derivation of thermodynamic and transport properties for complex gas mixtures in wide temperature and pressure ranges [16].

The EquilTheTA code is a C-language software suite integrating different modules for the thermodynamic properties and the equilibrium composition, for the specific heat and for the transport coefficients, the workflow shown in figure 21.10, implementing fast, stable and efficient algorithms based on the approaches above illustrated.

Nested in the core, two My-SQL databases are directly interfaced with the code, providing the fundamental information on the energy levels and properties of chemical species for the statistical thermodynamics and on the collision integrals of binary collisions entering Chapman–Enskog working equations for transport coefficients. These databases, including around 100 chemical species, are characterized by data consistency and a high-level of accuracy, data being collected or derived from the most up-to-date references in the literature, and can be extended to allow the inclusion of new species. The code automatically creates the system of chemical reactions for the hierarchical equilibrium approach, though also allowing the creation of user-defined reaction schemes. Moreover, EquilTheTA is trained to automatically generate, where not available, the collision integrals within the phenomenological approach, provided the physical properties of the colliding species (polarizability, charge, number of equivalent electrons).

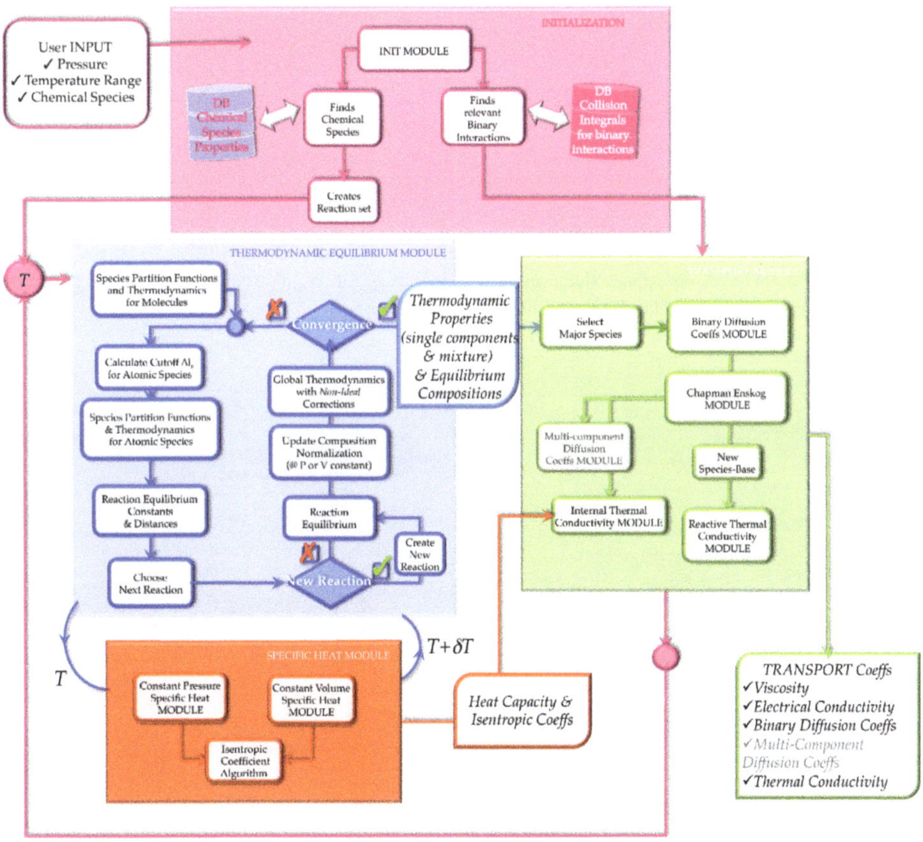

Figure 21.10. EquilTheTA code workflow. Reproduced from EqulTheTA© website with permission.

The simulation requires the minimal configuration data: the chemical model, the pressure and temperature grid for the run performed with different constraints (constant pressure or volume). Models for the mixtures of planetary atmospheres and other chemical systems as well as temperature grids are provided. The interface is designed to accommodate the level of knowledge of the user, therefore advanced options in the configuration of the simulation parameters, connected to the cut-off criterion in the atomic partition functions and the inclusion of non-ideality corrections, can be activated for skilled users or left hidden with the optimized settings predefined by the code developers. The simplified interface for simulation parameters is shown in figure 21.11. The interface has also a friendly builder of specific chemical models, selecting the chemical species to be included and their initial composition in the mixture (see figure 21.11). Results can be either downloaded, as a compressed archive of tab-delimited ASCII files, or graphically rendered for the on-fly analysis (figure 21.12).

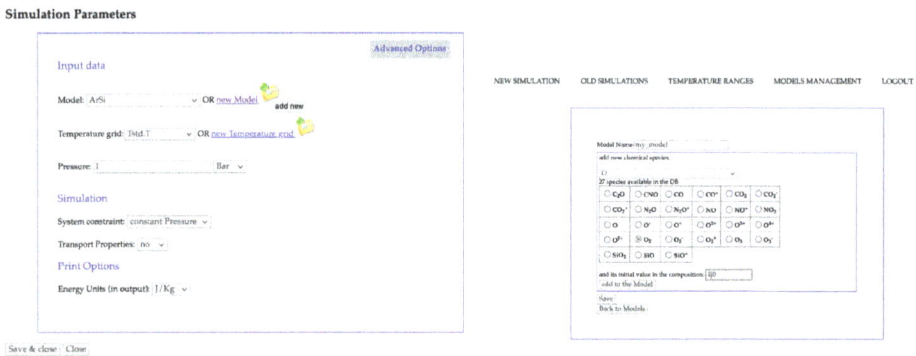

Figure 21.11. EquilTheTA interface: (left) simplified interface for simulation parameters; (right) model builder. Reproduced from EqulTheTA© website with permission.

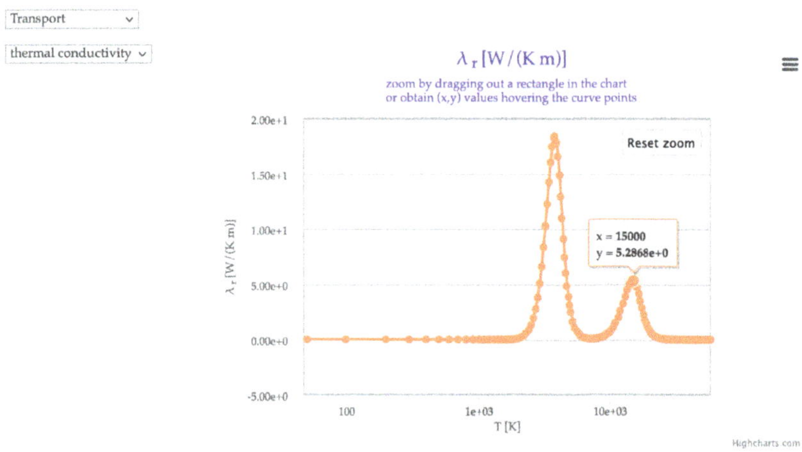

Figure 21.12. EquilTheTA: graphical rendering tool. Reproduced from EqulTheTA© website with permission.

References

[1] Bernardi D, Colombo V, Ghedini E and Mentrelli A 2003 Three-dimensional modelling of inductively coupled plasma torches *Eur. Phys. J.* D **22** 119–25

[2] Casavola A, Colonna G, De Giacomo A and Capitelli M 2004 A combined fluid dynamic and chemical model to investigate the laser induced plasma expansion *Appl. Phys.* A **79** 1315–7

[3] Casavola A, Colonna G, Pesce L and Capitelli M 2005 Laser-induced plasma in a water bubble *Conf. Proc.AIP* vol 762 (New York: AIP) pp 379–84

[4] Casavola A, Colonna G and Capitelli M 2008 Modeling laser-induced plasma expansion under equilibrium conditions *J. Thermophys. Heat Transfer* **22** 407–13

[5] Casavola A, Colonna G and Capitelli M 2009 Kinetic model of titanium laser induced plasma expansion in nitrogen environment *Plasma Sources Sci. Technol.* **18** 025027

[6] Tognoni E, Cristoforetti G, Legnaioli S and Palleschi V 2010 Calibration-free laser-induced breakdown spectroscopy: state of the art *Spectrochim. Acta* **65** 1–14

[7] Curreli D and Chen F F 2011 Equilibrium theory of cylindrical discharges with special application to helicons *Phys. Plasmas* **18** 113501

[8] Lindmayer M 2016 Simulation of switching arcs under transverse magnetic fields for DC interruption *IEEE Trans. Plasma Sci.* **44** 187–94

[9] Capitelli M, Celiberto R, Colonna G, Esposito F, Gorse C, Hassouni K, Laricchiuta A and Longo S 2015 *Fundamental Aspects of Plasma Chemical Physics: Kinetics* (Springer Series on Atomic, Optical, and Plasma Physics vol 85) (Berlin: Springer) https://doi.org/10.1007/978-1-4419-8185-1

[10] Hirschfelder J, Curtiss C and Bird R 1964 *Molecular Theory of Gases and Liquids* (New York: Wiley)

[11] Chapman S and Cowling T G 1970 *The Mathematical Theory of Non-uniform Gases* (Cambridge: Cambridge University Press)

[12] Capitelli M, Bruno D and Laricchiuta A 2013 *Fundamental Aspects of Plasma Chemical Physics: Transport* (Springer Series on Atomic, Optical, and Plasma Physics vol 74) (Berlin: Springer) https://doi.org/10.1007/978-1-4419-8172-1

[13] Capitelli M, Colonna G and D'Angola A 2011 *Fundamental Aspects of Plasma Chemical Physics: Thermodynamics* (Springer Series on Atomic, Optical, and Plasma Physics vol 66) (Berlin: Springer) https://doi.org/10.1007/978-1-4419-8182-0

[14] Capitelli M, Colonna G, Gorse C and D'Angola A 2000 Transport properties of high temperature air in local thermodynamic equilibrium *Eur. Phys. J.* D **11** 279–89

[15] D'Angola A, Colonna G, Gorse C and Capitelli M 2008 Thermodynamic and transport properties in equilibrium air plasmas in a wide pressure and temperature range *Eur. Phys. J.* D **46** 129–50

[16] Colonna G, D'Angola A, Laricchiuta A, Bruno D and Capitelli M 2013 Analytical expressions of thermodynamic and transport properties of the Martian atmosphere in a wide temperature and pressure range *Plasma Chem. Plasma Process.* **33** 401–31

[17] D'Angola A, Colonna G, Gorse C and Capitelli M 2011 Thermodynamic properties of high temperature air in local thermodynamic equilibrium: II Accurate analytical expression for electron molar fractions *Eur. Phys. J.* D **65** 453–7

[18] Landau D and Lifshitz E 1986 *Statistical Physics* (Oxford: Pergamon)

[19] Chase M W J 1998 *NIST-JANAF Thermochemical tables* (New York: AIP) https://doi.org/10.18434/T42S31

[20] Giordano D, Capitelli M and Colonna G 1994 Tables of internal partition functions and thermodynamic properties of high-temperature air species from 50 K to 100000 K *ESA Sci. Tech. Rev.* **237**
[21] Capitelli M, Colonna G, Giordano D, Marraffa L, Casavola A, Minelli P, Pietanza L D and Taccogna F 2005 Tables of Internal Partition Functions and Thermodynamic Properties of High-Temperature Mars-Atmosphere Species from 50 K to 50000 K *ESA Sci. Tech. Rev.* **246**
[22] Pagano D, Casavola A, Pietanza L, Capitelli M, Colonna G, Giordano D and Marraffa L 2009 Internal partition functions and thermodynamic properties of high-temperature Jupiter-atmosphere species from 50 K to 50,000 K *ESA Sci. Tech. Rev.* **257**
[23] Burm K T A L, Goedheer W J and Schram D C 1999 The isentropic exponent in plasmas *Phys. Plasmas* **6** 2622–7
[24] Anderson J D 2000 *Hypersonic and High Temperature Gas Dynamics* (Reston, VA: American Institute of Aeronautics and Astronautics) https://doi.org/10.2514/4.861956
[25] Burm K T A 2005 The isentropic exponent and relation for monoatomic plasmas *Phys. Lett.* A **336** 200–3
[26] Henderson S J and Menart J A 2008 Equilibrium properties of high-temperature air for a number of pressures *J. Thermophys. Heat Transfer* **22** 718–26
[27] Colonna G, D'Angola A and Capitelli M 2012 Electronic excitation and isentropic coefficients of high temperature planetary atmosphere plasmas *Phys. Plasmas* **19** 072115
[28] Prigogine I 1955 *Thermodynamic of Irreversible Processes* (New York: Wiley)
[29] Colonna G, Armenise I, Bruno D and Capitelli M 2006 Reduction of state-to-state kinetic to macroscopic models in hypersonic flows *J. Thermophys. Heat Transfer* **20** 477–86
[30] Colonna G, Pietanza L D and Capitelli M 2008 Recombination assisted nitrogen dissociation rates under nonequilibrium conditions *J. Thermophys. Heat Transfer* **22** 399–406
[31] Capitelli M, Giordano D and Colonna G 2008 The role of Debye-Hückel electronic energy levels on the thermodynamic properties of hydrogen plasmas including isentropic coefficients *Phys. Plasmas* **15** 082115
[32] Colonna G and Laricchiuta A 2020 Thermodynamic and transport properties of equilibrium Debye plasmas *Entropy* **22** 237
[33] Filinov V, Levashov P, Bonitz M and Fortov V 2005 Thermodynamics of hydrogen and hydrogen-helium plasmas: path integral Monte Carlo calculations and chemical picture *Contrib. Plasma Phys.* **45** 258–65
[34] Levashov P R, Filinov V S, Fortov V E and Bonitz M 2002 Thermodynamic properties of nonideal strongly degenerate hydrogen plasma *AIP Conf. Proc.* **620** 119
[35] Kremp D, Schlanges M and Kraeft W-D 2005 *Quantum Statistics of Nonideal Plasmas* (Springer Series on Atomic, Optical and Plasma Physics) (Berlin: Springer) https://doi.org/10.1007/b137422
[36] Colonna G and Laricchiuta A 2008 General numerical algorithm for classical collision integral calculation *Comput. Phys. Commun.* **178** 809–16
[37] Kee R J, Rupley F M and Miller J A 1990 The Chemkin thermodynamic data base *Tech. rep.* Sandia National Labs, Livermore, CA
[38] McBride B J 1993 *Coefficients for Calculating Thermodynamic and Transport Properties of Individual Species* NASA Technical Memorandum 4513 (Washington, DC: NASA)
[39] Gordon S and McBride B J 1994 *Computer Program for Calculation of Complex Chemical Equilibrium* (Washington, DC: NASA) (NASA Reference Publication 1311)

[40] Colonna G and Capitelli M 2009 A few level approach for the electronic partition function of atomic systems *Spectrochim. Acta* **64** 863–73
[41] D'Ammando G, Colonna G and Capitelli M 2013 A simplified approach to calculate atomic partition functions in plasmas *Phys. Plasmas* **20** 032108
[42] Liang R, Liu Y and Li F 2021 Partition functions of atomic and diatomic species in high-temperature atmospheric plasmas *Contrib. Plasma Phys.* **61** e202100036
[43] Smith W R and Missen R W 1982 *Chemical Reaction Equilibrium Analysis: Theory and Algorithms* (New York: Wiley)
[44] 2021 EquilTheTA—EQUILibrium for plasma THErmodynamics and Transport Applications http://phys4entrydb.istp.cnr.it/EquilTHETA/
[45] Colonna G and D'Angola A 2012 EquilTheTA: Thermodynamic and transport properties of complex equilibrium plasmas *Conf. Proc.AIP* **1501** 1110
[46] Villars D S 1959 A method of successive approximations for computing combustion equilibria on high speed digital computer *J. Chem. Phys.* **63** 521
[47] Villars D S 1960 Computation of complicated combustion equilibria on a high-speed digital computer *Proc. of the First Conf. on the Kinetics Equilibria and Performance of High Temperature Systems, Western States section of the Combustion Institute* G Bahn and E Zuckowsky (Washington, DC: Butterworths Scientific Publications) p 18
[48] Cruise D R 1964 Notes on rapid computation of chemical equilibria *J. Phys. Chem.* **68** 3797
[49] Meintjes K and Morgan A P 1985 Performance of algorithms for calculating the equilibrium composition of a mixture of gases *J. Comput. Phys.* **60** 219–34
[50] Colonna G and D'Angola A 2004 A hierarchical approach for fast and accurate equilibrium calculation *Comput. Phys. Commun.* **163** 177–90
[51] Colonna G 2007 Improvements of hierarchical algorithm for equilibrium calculation *Comput. Phys. Commun.* **177** 493–9
[52] Colonna G 2019 Thermodynamic properties of air plasma seeded by SiC molecules Rend. Lincei *Sci. Fis. Nat.* **30** 537–47
[53] Cruise D R 1970 Theoretical Computations of Equilibrium Compositions *Theoretical Computations of Equilibrium Compositions, Thermodynamic Properties, and Performance Characteristics of Propellant Systems* NWC TP 5037, Naval Weapons Center
[54] Devoto R 1966 Transport properties of ionized monatomic gases *Phys. Fluids* **9** 1230–40
[55] Devoto R S 1967 Simplified expressions for the transport properties of ionized monatomic gases *Phys. Fluids* **10** 2105–12
[56] Bonnefoi C and Fauchais P 1976 Fourth approximation to the viscosity of a multi-component ionized gas *Phys. Fluids* **19** 1827–8
[57] Muckenfuss C and Curtiss C F 1958 Thermal conductivity of multicomponent gas mixtures *J. Chem. Phys.* **29** 1273–7
[58] Devoto R S and Li C P 1968 Transport coefficients of partially ionized helium *J. Plasma Phys.* **2** 17
[59] Butler J N and Brokaw R S 1957 Thermal conductivity of gas mixtures in chemical equilibrium *J. Chem. Phys.* **26** 1636
[60] Monchik L 1959 Collision integrals for the exponential repulsive potential *Phys. Fluids* **2** 695
[61] Kalinin A P and Dubrovitskii D Y 2000 The possibility of using repulsive interaction potentials for the calculation of high-temperature integrals of collisions between atoms and molecules *High Temp.* **38** 848

[62] Kihara T, Taylor M H and Hirschfelder J O 1960 Transport properties for gases assuming inverse power intermolecular potentials *Phys. Fluids* **3** 715

[63] Smith F J 1967 High order collision integrals *Final Report on contract* NSR 52-112-001 Washington, DC, NASA

[64] Bruno D *et al* 2010 Transport properties of high-temperature Jupiter atmosphere components *Phys. Plasmas* **17** 112315

[65] Smith F J and Munn R J 1964 Automatic calculation of the transport collision integrals with tables for the Morse potential *J. Chem. Phys.* **41** 3560

[66] Neufeld P D, Janzen A R and Aziz R A 1972 Empirical equations to calculate 16 of the transport collision integrals $\Omega^{(\ell,\ s)\star}$ for the Lennard-Jones (12-6) potential *J. Chem. Phys.* **57** 1100

[67] Mason E A 1954 Transport properties of gases obeying a modified Buckingham (Exp-Six) potential *J. Chem. Phys.* **22** 169

[68] Rainwater J, Holland P and Biolsi L 1982 Binary collision dynamics and numerical evaluation of dilute gas transport properties for potentials with multiple extrema *J. Chem. Phys.* **77** 434–47

[69] Tang K T and Toennies J P 2003 The van der Waals potentials between all the rare gas atoms from He to Rn *J. Chem. Phys.* **118** 4976

[70] Aziz R A and Slaman M J 1990 The repulsive wall of the Ar-Ar interatomic potential reexamined *J. Chem. Phys.* **92** 1030

[71] Clenshaw C W and Curtis A R 1960 A method for numerical integration on an automatic computer *Numer. Math.* **2** 197–205

[72] Pirani F, Albertí M, Castro A, Teixidor M M and Cappelletti D 2004 Atom-bond pairwise additive representation for intermolecular potential energy surfaces *Chem. Phys. Lett.* **394** 37–44

[73] Pirani F, Maciel G, Cappelletti D and Aquilanti V 2006 Experimental benchmarks and phenomenology of interatomic forces: open-shell and electronic anisotropy effects *Int. Rev. Phys. Chem.* **25** 165–99

[74] Liuti G and Pirani F 1985 Regularities in van der Waals forces: correlation between the potential parameters and polarizability *Chem. Phys. Lett.* **122** 245–50

[75] Cambi R, Cappelletti D, Liuti G and Pirani F 1991 Generalized correlations in terms of polarizability for van der Waals interaction potential parameter calculations *J. Chem. Phys.* **95** 1852–61

[76] Cappelletti D, Liuti G and Pirani F 1991 Generalization to ion-neutral systems of the polarizability correlations for interaction potential parameters *Chem. Phys. Lett.* **183** 297–303

[77] Aquilanti V, Cappelletti D and Pirani F 1996 Range and strength of interatomic forces: dispersion and induction contributions to the bonds of dications and of ionic molecules *Chem. Phys.* **209** 299–311

[78] Pirani F, Capitelli M, Colonna G and Laricchiuta A 2019 Transport cross sections from accurate intermolecular forces *Rend. Lincei, Sci. Fis. Nat.* **30** 49–56

[79] Capitelli M, Cappelletti D, Colonna G, Gorse C, Laricchiuta A, Liuti G, Longo S and Pirani F 2007 On the possibility of using model potentials for collision integral calculations of interest for planetary atmospheres *Chem. Phys.* **338** 62–8

[80] Pirani F, Cappelletti D and Liuti G 2001 Range, strength and anisotropy of intermolecular forces in atom-molecule systems: an atom-bond pairwise additivity approach *Chem. Phys. Lett.* **350** 286–96

[81] Laricchiuta A, Colonna G, Bruno D, Celiberto R, Gorse C, Pirani F and Capitelli M 2007 Classical transport collision integrals for a Lennard-Jones like phenomenological model potential *Chem. Phys. Lett.* **445** 133–9

[82] Stallcop J, Partridge H and Levin E 1991 Resonance charge transfer, transport cross sections, and collision integrals for $N^+(^3P)$-$N(^4S)$ and $O^+(^4S)$-$O(^3P)$ interactions *J. Chem. Phys.* **95** 6429–39

[83] Stallcop J, Partridge H, Pradhan A and Levin E 2000 Potential energies and collision integrals for interactions of carbon and nitrogen atoms *J. Thermophys. Heat Transfer* **14** 480–8

[84] Stallcop J, Partridge H and Levin E 2001 Effective potential energies and transport cross sections for atom-molecule interactions of nitrogen and oxygen *Phys. Rev.* A **64** 042722

[85] Levin E and Wright M J 2004 Collision integrals for ion-neutral interactions of nitrogen and oxygen *J. Thermophys. Heat Transfer* **18** 143–7

[86] Laricchiuta A *et al* 2009 High temperature Mars atmosphere. Part I: Transport cross sections *Eur. Phys. J.* D **54** 607–12

[87] Colonna G, D'Angola A, Pietanza L, Capitelli M, Pirani F, Stevanato E and Laricchiuta A 2018 Thermodynamic and transport properties of plasmas including silicon-based compounds *Plasma Sources Sci. Technol.* **27** 015007

[88] Laricchiuta A, D'Angola A, Pirani F, Pietanza L D, Capitelli M and Colonna G 2019 Thermodynamic and transport properties of reacting air including ablated species *Hypersonic Meteoroid Entry Physics IOP Series in Plasma Physics* (Bristol: IOP Publishing) https://doi.org/10.1088/2053-2563/aae894ch14

[89] Subramaniam S, Jaffe R L and Stephani K A 2020 State-resolved transport collision integrals for the O+ O_2 system *Phys. Rev. Fluids* **5** 113402

[90] Luo H, Macheret S and Alexeenko A 2018 First-principle calculations of collision integrals for N_2-O system *2018 Joint Thermophysics and Heat Transfer Conf.* AIAA Paper 2018-3586 https://doi.org/10.2514/6.2018-3586

[91] Dagdigian P J and Alexander M H 2012 Exact quantum scattering calculation of transport properties for free radicals: OH (X $^2\Pi$)-helium *J. Chem. Phys.* **137** 094306

[92] Viehland L A, Dickinson A S and Maclagan R G A 1996 Transport coefficients for NO^+ ions in helium gas: a test of the NO^+-He interaction potential *Chem. Phys.* **211** 1

[93] Maclagan R G A R, Viehland L A and Dickinson A S 1999 Ab initio calculation of the gas phase ion mobility of CO^+ ions in He *J. Phys.* B **32** 4947

[94] Murphy A B 1995 Transport coefficients of air, argon-air, nitrogen-air, and oxygen-air plasmas *Plasma Chem. Plasma Process.* **15** 279

[95] Devoto R S 1967 Transport coefficients of partially ionized argon *Phys. Fluids* **10** 354

[96] Rutherford J A and Vroom D A 1974 The reaction of atomic oxygen with several atmospheric ions *J. Chem. Phys.* **61** 2514–9

[97] Belyaev Y N, Brezhnev B G and Erastov E M 1968 Resonant charge transfer of low energy carbon and nitrogen ions *Sov. Phys—JETP* **27** 924

[98] Krstić P S and Schultz D R 1999 Elastic scattering and charge transfer in slow collisions: isotopes of H and H^+ colliding with isotopes of H and with He *J. Phys. B: At. Mol. Phys.* **32** 3485–509

[99] Nikitin E E and Smirnov B M 1978 Quasiresonant processes in slow collisions *Sov. Phys. - Usp.* **21** 95–116

[100] Bates D R, Massey H S W and Stewart A L 1953 Inelastic collisions between atoms. I. General theoretical considerations *Proc. R. Soc. Lond.* A **216** 437–58

[101] Smirnov B M 2018 *Physics of atoms and ions Atomic Particles and Atom Systems* (Berlin: Springer) https://doi.org/10.1007/978-3-319-75405-5_3

[102] Smirnov B M 1973 *Asymptotic Methods in Theory of Atomic Collisions* (Moscow: Atomizdat) (in Russian)

[103] Firsov O B 1951 Resonance overcharge of ions in slow collisions *JETP* **21** 1001 (in Russian)

[104] Laricchiuta A, Pirani F, Colonna G, Bruno D, Gorse C, Celiberto R and Capitelli M 2009 Collision integrals for interactions involving atoms in electronically excited states *J. Phys. Chem.* A **113** 15250–6

[105] Kosarim A, Smirnov B, Capitelli M, Celiberto R and Laricchiuta A 2006 Resonant charge exchange involving electronically excited states of nitrogen atoms and ions *Phys. Rev.* A **74** 062707

[106] Eletskii A V, Capitelli M, Celiberto R and Laricchiuta A 2004 Resonant charge exchange and relevant transport cross sections for excited states of oxygen and nitrogen atoms *Phys. Rev.* A **69** 042718

[107] André P, Bussière W and Rochette D 2007 Transport coefficients of Ag-SiO_2 plasmas *Plasma Chem. Plasma Process.* **27** 381–403

[108] Mason E A, Munn R J and Smith F J 1967 Transport coefficients of ionized gases *Phys. Fluids* **10** 1827

[109] Devoto R S 1968 Transport coefficients of partially ionized hydrogen *J. Plasma Phys.* **2** 617–31

[110] Hahn H S, Mason E A and Smith F J 1971 Quantum transport cross sections in a completely ionized gas *Phys. Fluids* **14** 278

[111] Murphy A B 2000 Transport coefficients of hydrogen and argon-hydrogen plasmas *Plasma Chem. Plasma Process.* **20** 279–97

[112] Liboff R L 1959 Transport coefficients determined using the shielded Coulomb potential *Phys. Fluids* **2** 40

IOP Publishing

Plasma Modeling (Second Edition)
Methods and applications
Gianpiero Colonna and Antonio D'Angola

Chapter 22

Methods for electron–molecule scattering

**Roberto Celiberto, Jimena D Gorfinkiel,
Annarita Laricchiuta and Jonathan Tennyson**

In low-temperature plasmas the transient and steady conditions are driven by the chemistry and the electron-impact induced processes are key in the kinetics, leading to the build-up of electron density and to the population of excited levels that can be the active species in the relevant mechanisms for the plasma efficiency, depending on the specific application. Therefore, the modeling of technological plasmas needs the construction of complete, accurate and consistent information on the dynamics of elementary processes initiated by electron collision and the collection of cross-sections to be integrated by the energy distribution of free electrons, either Maxwell or non-equilibrium, to derive the actual rate coefficients. This knowledge mainly relies on theoretical approaches [1] to be validated by comparison with electron beam experiments or electron swarm data through Boltzmann analysis and enhancing the predictive capability of kinetic models, allowing one to conduct numerical experiments that not only give a reliable estimation of macroscopic plasma parameters, but also shed light on the collisional mechanisms at microscopic level. In fact, electron–molecule collisions can lead to a number of processes which are important in plasmas and these are hard to capture in a single experiment. Furthermore while the initial (feedstock) plasma gases are usually stable (closed-shell) species, the plasma contains a variety of reactive (open shell) species usually called radicals, as well as molecular ions. Experimental data on electron collisions with radicals are almost non-existent so theory is the prime provider of this vital information.

The modern high-performance computing techniques on multi-processor super-computers design the new frontiers in the investigation of electron–molecule scattering for the derivation of integral and differential cross-sections for elastic, inelastic (excitation) and reactive (dissociation and ionization, …) processes induced in the molecular target (see table 22.1 for a comprehensive list), allowing the exploitation of quantum theoretical approaches characterized by high-level

doi:10.1088/978-0-7503-3559-1ch22

Table 22.1. Processes that can take place in an electron collision with a neutral polyatomic molecule ABC or cation ABC^+. The processes are listed in approximate energy order, that is, the order of their lowest energy threshold. The indices *i* and *j* indicate electronic states, v and v' vibrational states and J and J' rotational states of the target molecule. All processes listed for a neutral target can also take place for cationic ones, except dissociative electron attachment: the process for a cation is dissociative recombination and it is listed at the bottom of the table. If the target is initially in an excited state, superelastic scattering, resulting in its de-excitation and an increase of the kinetic energy of the scattering electron, can also take place.

Process name	Outcome
$e + ABC\ (i,v,J) \longrightarrow$	
Elastic scattering	$ABC\ (i,v,J) + e$
Rotational excitation	$ABC\ (i,v,J') + e$
Vibrational excitation	$ABC\ (i,v',J) + e$
Electronic excitation	$ABC\ (j,v',J') + e$
Impact dissociation/	$A + BC + e$
dissociative excitation	$AB + C + e$
	$A + B + C + e$
Dissociative electron	$A^- + BC$
attachment (DEA)	$AB^- + C$
	etc
(Single) ionization	$ABC^+ + 2e$
Ionization-dissociation/	$A + BC^+ + 2e$
dissociative ionization	$AB^+ + C + 2e$
	etc
Ion pair formation	$A^- + BC^+ + e$
	etc
$e + ABC^+\ (i,v,J) \longrightarrow$	
Dissociative recombination	$A + BC$
(replaces DEA)	$A + B + C$
	etc

accuracy. However, also simplified approaches (classical, semiclassical, phenomenological) still remain attractive tools for the effective estimation of the cross-section with acceptable accuracy and reducing significantly the complexity of the calculations that, however, requiring the knowledge of some quantities characterizing the electronic structure of the molecular target, still need an *ab initio* step.

In this chapter methods for electron–molecule scattering routinely used in the community of data producers are reviewed, illustrating the first principles, reporting the working equations, providing the relevant references, discussing some examples. The chapter can be also regarded as a practical guide to get oriented in the selection of data for the construction of the chemical model. The methods classified as *simplified* are presented in section 22.1 basing on their specificity for selected classes of processes and assessing the level of confidence in using them versus experiments of accurate ones. Section 22.2 illustrates also the presently most reliable methods, at the state-of-the-art level, for electron–molecule scattering, with particular emphasis on the R-matrix approach to cross-section calculations. Examples of the chemical systems investigated with the theoretical approaches described here and relevant to different fields of application are given in section 22.3. The last section 22.4, finally, is devoted to the description of the nowadays available databases that are accessible through the web and allow the direct download of datasets in different formats.

22.1 Simplified approaches

22.1.1 Simplified approaches for excitation and dissociative excitation

In this section some of the methods successfully used in the literature for the derivation of dipole-allowed excitation and dissociative excitation cross-sections in electron–molecule collisions are reviewed. These two-state approaches are largely based on the high-energy Born cross-section, either to set an external parameter as in the semiclassical approach, or as the starting value for rescaling procedures so as to cope with the inaccurate overestimation at low collision energies.

In the following, the state-specific cross-section for the generic vibronic transition from the initial (i, v_i) to the final (j, v_f) levels (see figure 22.1) will be considered.

The *total cross-sections* can be straightforwardly obtained summing up all the excitations

$$\sigma_{v_i}^{i\to j}(E) = \sum_{v_f} \sigma_{v_i,v_f}^{i\to j}(E) + \int_{\epsilon_{\text{th}}}^{\epsilon_{\max}} d\epsilon \frac{d\sigma_{v_i,\epsilon}^{i\to j}(E)}{d\epsilon} \tag{22.1}$$

where the sum extends over all the bound vibrational levels of the j state and the integral accounts for the coupling with the continuum of the j state, leading to dissociation. The continuum integration limits ε_{th}, $\varepsilon_{\max}$ are, respectively, defined as

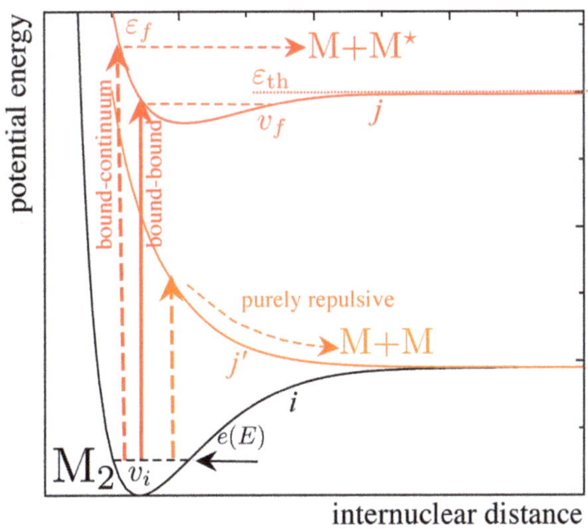

Figure 22.1. Energy diagram for a diatomic molecule. Excitation and dissociative excitation transitions are sketched.

$$\varepsilon_{\text{th}} = V_j(R \to \infty)\; \varepsilon_{\max} = E + \epsilon^i_{v_i} \tag{22.2}$$

where $V_j(R \to \infty)$ is the dissociation threshold for the upper j electronic state.

Impact parameter method (IPM) is a semiclassical approach developed for the non-resonant, spin-conserving, optically allowed transitions (the exchange is neglected) in diatomic molecules with resolution on the ro-vibrational levels of the target. The collisional dynamics considers the electron projectile as a classical particle, while treating quantum mechanically the molecular target undergoing the adiabatic transition from the initial to the final states. The probability is formally derived in the framework of the time-dependent perturbation theory, the interaction of the incident electron with target electrons being the perturbation, whose time dependence results from the classical straight-line motion assumed for the electron. This approach is meant for collisions dominated by the long-range dipole interaction, and this assumption leads to a simplified expression for the perturbation, retaining only one term in the multipole expansion of the interaction.

In the following we give a brief account of the main elements of IPM computational scheme, referring the reader to the literature for a complete description of the method [2–5].

The state-to-state cross-section $\sigma^{i\to j}_{v_i,v_f}(E)$ for an electron-impact induced transition from the v_ith vibrational level of the i electronic state to the v_fth vibrational level of the final j electronic state of a diatomic molecule is expressed as

$$\sigma^{i\to j}_{v_i,v_f}(E) = S^{i,j}_{v_i,v_f} \cdot D^{i,j}_{v_i,v_f}(E) \tag{22.3}$$

where E is the incident kinetic energy. The 'structural factor' $S^{i,j}_{v_i,v_f}$, related to the target structure, is defined by

$$S^{i,j}_{v_i,v_f} = \frac{m^2e^2}{3g_i\hbar^4}(2-\delta_{\Lambda_i,0})(2-\delta_{\Lambda_j,0}) \cdot \left| \int_0^\infty dR \chi^j_{v_f}(R) M^{i,j}_{\Lambda_i,\Lambda_j}(R) \chi^i_{v_i}(R)\right|^2 \tag{22.4}$$

m, e, $\hbar$, and g_i represent, respectively, the mass and charge of electron, Dirac's constant and the degeneracy factor for the i state. $\chi^{i \text{ or } j}_v(R)$ is the vibrational wavefunction that depends on the internuclear distance R, and $M^{i,j}_{\Lambda_i,\Lambda_j}(R)$ is the usual electronic transition dipole moment characterized by the quantum numbers of the projection of electronic angular momentum on the internuclear axis Λ_i and Λ_j.

The 'dynamical factor', $D^{i,j}_{v_i,v_f}(E)$, describing the inelastic scattering effects on the motion of the incident electron, is given by,

$$\begin{aligned} D^{i,j}_{v_i,v_f} = \frac{2\pi\hbar^2}{m^2u_i^2}\Bigg[&\gamma_i(K_0(\gamma_i)K_1(\gamma_i) - \frac{\pi^2}{4}S_0(\gamma_i)S_1(\gamma_i)) + \gamma_f(K_0(\gamma_f)K_1(\gamma_f) \\ &- \frac{\pi^2}{4}S_0(\gamma_f)S_1(\gamma_f)) + \gamma(K_0(\gamma_i)K_1(\gamma_f) + K_0(\gamma_f)K_1(\gamma_i) + \frac{\pi^2}{4}S_0(\gamma_i)S_1(\gamma_f) \\ &+ \frac{\pi^2}{4}S_0(\gamma_f)S_1(\gamma_i)) + \left(\frac{u_i^2 - u_f^2}{u_i^2 + u_f^2}\right) \cdot \left(ln\left(\frac{\gamma_f}{\gamma_i}\right) + \frac{\pi}{2}\int_{\gamma_i}^{\gamma_f} S_0(\gamma)d\gamma\right)\Bigg] \end{aligned} \tag{22.5}$$

where K_i and S_i are the modified Bessel and Struve functions, respectively, and u_i and u_f are the initial and final electron velocities. Moreover:

$$\gamma_i = \frac{\rho^0_{v_i,v_f}|\Delta E^{i,j}_{v_i,v_f}|}{\hbar} \cdot \frac{1}{u_i} \quad \gamma_f = \frac{\rho^0_{v_i,v_f}|\Delta E^{i,j}_{v_i,v_f}|}{\hbar} \cdot \frac{u_i}{u_f^2} \quad \gamma = \frac{\rho^0_{v_i,v_f}|\Delta E^{i,j}_{v_i,v_f}|}{\hbar} \cdot \frac{2u_i}{u_i^2 + u_f^2} \tag{22.6}$$

with $\Delta E^{i,j}_{v_i,v_f}$ being the transition energy defined as:

$$\Delta E^{i,j}_{v_i,v_f} = \epsilon^j_{v_f} - \epsilon^i_{v_i} \tag{22.7}$$

and ϵ_v is the energy eigenvalue of the vth vibrational level of the i or j electronic state. $\rho^0_{v_i,v_f}$ is a cut-off parameter introduced in the impact-parameter method to preserve the unitary of the S-matrix [2–5], whose value is set by equating the impact-parameter and Born approximation cross-sections at high energies.

The existence of an externally adjusted parameter makes this approach not self-consistent, moreover the analytical solution of the dynamical integral introduces a divergence in approaching the threshold energy and produces an unphysical peak in the cross-section energy profile. However, the method has been proven to have a reasonable accuracy, allowing the derivation of complete datasets for bound–bound excitations up to vibrational levels close to the dissociation limits, the estimation of the dissociative channels, and also the derivation of cross-sections for other processes, such as the radiative decay vibrational excitation (EV processes) [6] or predissociation by combining the independent probabilities of the second step following the excitation, i.e. the Einstein coefficient of spontaneous emission and the predissociation to decay lifetime, respectively. In the past the IPM has been successfully used for the creation of the first complete and consistent database for electron-H_2 inelastic processes including hydrogen isotopologues [7, 8] and giving the first example of cross-sections for transitions initiated from excited states [9]. Recently, the quantum convergent close-coupling (CCC) method has been extensively used to re-estimate the cross-sections for these processes [10, 11] and their comparison with IPM results is here reported to assess the accuracy of the semiclassical approach. The transition $X^1\Sigma_g^+ \rightarrow B^1\Sigma_u^+$ in e-H_2 has been reported (figure 22.2), being one of the major components of the emission spectra of the hydrogen molecule. The IPM overestimates up to a factor 1.7 the cross-section in the region of the maximum converging to the CCC method for higher energies. It is worth noticing that the IPM correctly predicts the vibrational dependence.

BEf scaling is a powerful approach for the derivation of accurate excitation cross-sections in dipole-allowed transitions developed in references [12–14]. This approach is based on the plane-wave Born theory and introduces a scaling procedure that brings the cross-section to an excellent agreement with experiments. The BE-scaling is a correction of the low collision energy limit, where the scaling factor depends on the binding energy of the electron excited in the process, B, and on the excitation energy ΔE, as follows

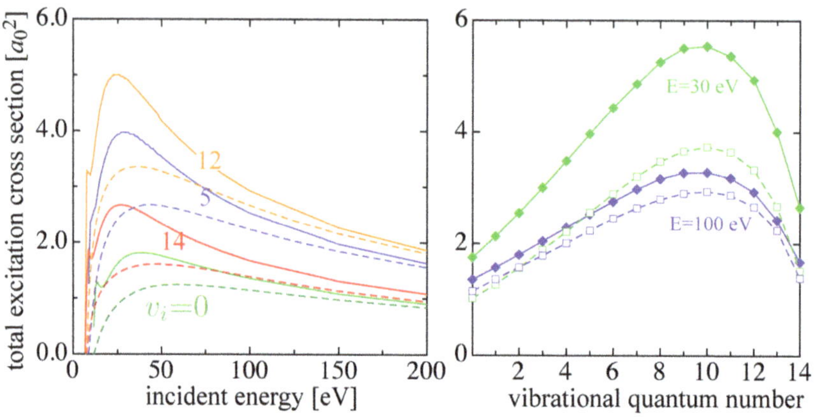

Figure 22.2. Total excitation (including dissociation) cross-sections for the transition to the $B^1\Sigma_u^+$ state of H_2 molecule from vibrationally excited levels, (left) as a function of energy for selected initial vibrational levels and (right) as a function of v_i for two values of the collision energy. (Solid lines) IPM, (dashed lines) CCC method (CCC data have been downloaded from the MCCC database at the URL https://http://www.mccc-db.org/).

$$\begin{aligned}\sigma_{\mathrm{BE}}(E) &= \frac{E}{E+B+\Delta E}\sigma^{\mathrm{Born}}(E)\\ &= \frac{E}{E+B+\Delta E}\frac{2\pi e^2}{a_0 k_i^2}\int_{K_{\mathrm{min}}^2=(ki-kf)^2}^{K_{\mathrm{max}}^2=(ki+kf)^2}\frac{G(K;\,R)}{\Delta E}d\ln(K^2)\end{aligned} \tag{22.8}$$

The Born cross-sections are expressed in terms of the generalized oscillator strength, $G(K;\,R)$, function of the transferred moment $K = |\mathbf{k}_i - \mathbf{k}_f|$, $\mathbf{k}_i$ and $\mathbf{k}_f$ being the initial and final momentum vectors associated with the incident electron, with $k_i^2 = 2\,mE/\hbar^2$ and $k_f^2 = k_i^2 - (2\,m/\hbar^2)\Delta E$. $G(K;\,R)$ is parametrically dependent on the molecular geometry, however, for pure electronic excitation the calculation is usually performed at $R = R_{\mathrm{eq}}$.

The generalized oscillator strength has the form

$$G(K;\,R) = \frac{2m}{\hbar^2 K^2}\,\Delta E_{i,j}(R)\int\frac{d\hat{R}}{4\pi}|\langle j|\sum_q e^{(iK\cdot r_q)}|i\rangle|^2 \tag{22.9}$$

The evaluation of this quantity, corresponding to the optical oscillator strength in the limit $K \to 0$ ($G(K \to 0) = f$), requires in principle *ab initio* electronic structure calculations, preferably at configuration interaction level, for the molecular state wavefunctions. When quantum states are poorly described, a second scaling on the accurate dipole f_{acc}, also obtained from spectroscopy, is performed, further improving the cross-section value

$$\sigma_{\mathrm{BE}f}(E) = \frac{f_{\mathrm{acc}}}{f}\sigma_{\mathrm{BE}}(E) \tag{22.10}$$

This approach, despite its simplicity, is very accurate in the derivation of excitation cross-sections of optically allowed transitions and, in combination with the analytical representation of $G(K)$ proposed by Vriens [15]

$$G(x) = \frac{1}{(1+x)^6}\left[\sum_{m=0}^{\infty} \frac{f_m x^m}{(1+x)^m}\right] \tag{22.11}$$

where the parameters f_m can be optimized to experimental values, could lead to cross-section estimation avoiding the *ab initio* step.

Vibrationally resolved cross-sections can be derived either performing quantum chemistry calculations at different molecular geometries and performing the integration over vibrational wavefunctions, or, where state-specific experimental differential cross-sections are available, deriving an experimental $G_{\text{exp}}(K) = \Delta E k_i/(4a_0^2 k_f) \cdot K\left[\frac{d\sigma(E,\vartheta)}{d\Omega}\right]_{\text{exp}}$ and then integrating to total excitation cross-section from a specific initial state, as demonstrated in reference [13] (figure 22.3).

Dipole-Born approximation and derived variants can be obtained from the Born cross-section, as given in equation (22.8):

$$\sigma^{\text{Born}}(E) = \frac{2\pi e^2}{a_0 k_i^2}\int_{K_{\min}^2}^{K_{\max}^2} \frac{G(K;R)}{\Delta E} d\ln(K^2) \tag{22.12}$$

This can be approximated, for large energies, by expressing the generalized oscillator strength $G(K; R)$ of equation (22.9), as a series of K and stopping the expansion at the first order term (dipole moment approximation). The final expression, referred to in literature as the *dipole-Born approximation*, for a vibro-electronic transition i, $v_i \rightarrow j$, v_f, assumes the form [17]:

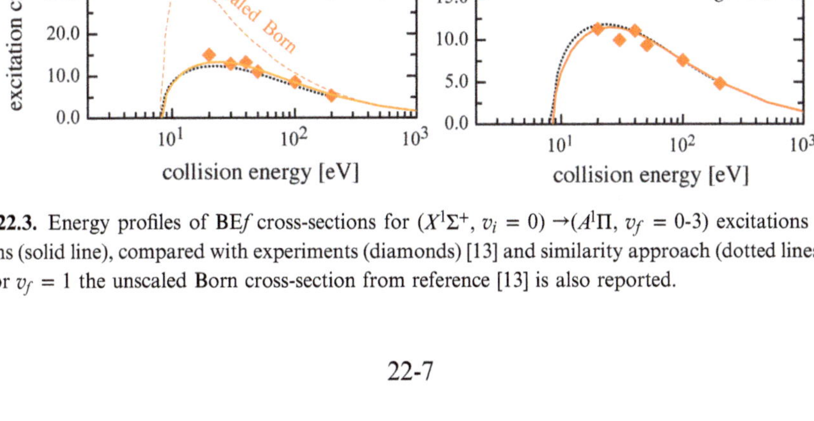

Figure 22.3. Energy profiles of BEf cross-sections for $(X^1\Sigma^+, v_i = 0) \rightarrow (A^1\Pi, v_f = 0\text{-}3)$ excitations in e-CO collisions (solid line), compared with experiments (diamonds) [13] and similarity approach (dotted lines) results [16]. For $v_f = 1$ the unscaled Born cross-section from reference [13] is also reported.

$$\sigma^{\mathrm{Born}}(E) = \frac{8\pi}{3a_0^2 k_i^2} | < v_f | D_{i,j}(R) | v_i > |^2 \ln \frac{k_f + k_i}{|k_f - k_i|} \tag{22.13}$$

where $e \cdot D_{i,j}(R) = M^{i,j}$ is the *vertical transition dipole moment.*

Figure 22.4 shows an example of the dipole-Born approximation cross-sections, as a function of the incident electron energy, compared with the accurate results obtained by the R-matrix method. These calculations were performed for an electron collision inducing a vertical electronic transition, $X^1\Sigma^+ \rightarrow A^1\Sigma^+$, on the LiH molecule initially in its lowest vibrational level $v = 0$ [18]. It can be seen that the dipole-Born calculations largely overestimate the cross-sections with respect the R-matrix results, around the threshold of the excitation process. At very large incident energies, however, the dipole-Born approximation cross-sections show the correct converging behaviour when compared with experimental or accurate theoretical data [19].

In order to mitigate the dipole approximation and make it more rapidly convergent at large energies, equation (22.13) can be further approximated by manipulating the logarithm term [19]. Skipping all the steps, the final expression for the cross-section becomes:

$$\sigma^{\mathrm{Born}}(E) = \frac{4\pi}{3a_0^2 k_i^2} | < v_f | D_{i,j}(R) | v_i > |^2 \ln \frac{4EK_c^2}{\Delta E^2} \frac{\hbar^2}{2m} \tag{22.14}$$

where K_c is an arbitrary parameter to be appropriately set and ΔE the vibro-electronic transition energy. This equation represents the so-called *Born–Bethe approximation* [20]. Different choices can be operated for K_c [20]. One of the possibilities is to express this parameter in terms of the ionization energy, I, for a molecule in its ground state and then put $K_c^2 = 2mI/\hbar^2$. In this case equation (22.14) becomes [21]:

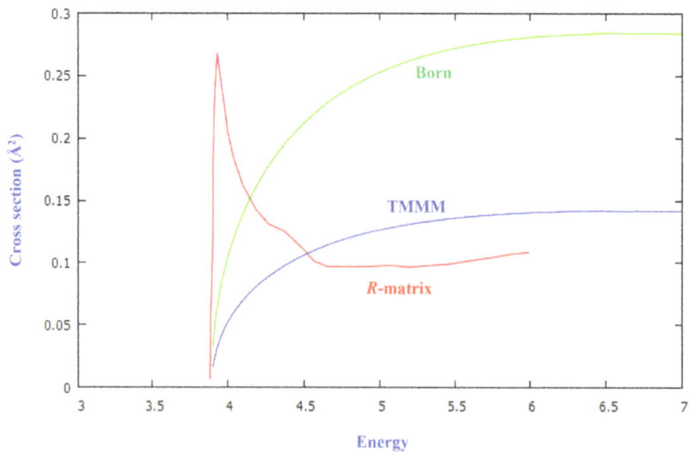

Figure 22.4. Comparison of scattering cross-sections for the electron-impact process involving the lithium hydride molecule and inducing the vertical transition LiH($X^1\Sigma^+$, $v = 0$) → LiH($A^1\Sigma^+$), obtained with different theoretical methods: dipole-Born approximation (green); R-matrix method (red), threshold modified Mott–Massey approximation (blue) [18]. Copyright © IOP Publishing. Reproduced with permission.

$$\sigma^{\mathrm{Born}}(E) = \frac{4\pi}{3a_0^2k_i^2} | < v_f | D_{i,j}(R) | v_i > |^2 \ln \frac{4IE}{\Delta E^2} \tag{22.15}$$

Mott and Massey [22] report a slightly different variant of the above equation, obtained by replacing the ionization potential I with the transition energy ΔE. Equation (22.15) then transforms as:

$$\sigma^{\mathrm{Born}}(E) = \frac{4\pi}{3a_0^2k_i^2} | < v_f | D_{i,j}(R) | v_i > |^2 \ln \frac{4E}{\Delta E} \tag{22.16}$$

Both the above equations, however, diverge at the threshold, as the logarithmic term does not vanish for $E \to \Delta E$. In order to remedy this undesired feature, the Mott and Massey theory can be reformulated following a less approximate treatment. After simple steps (see reference [23]), a threshold-convergent formula is obtained:

$$\sigma_{v, v'}^{X \to A}(E) = \frac{8\pi}{3a_0^2k_i^2} | < v_f | D_{i,j}(R) | v_i > |^2 \ln \frac{\sqrt{\Delta E}}{\sqrt{E} - \sqrt{E - \Delta E}} \tag{22.17}$$

which is known as the *threshold modified Mott–Massey* (TMMM) approximation. The comparison in figure 22.4 includes also the calculated cross-section in the TMMM approximation. A further example is provided in figure 22.5. Here the cross-sections, calculated within different theories, for the electronic excitation $X^2\Sigma^+ \to A^2\Pi$ induced by electron-impact and involving the BeH molecule in its $v = 0$ vibrational level, are shown.

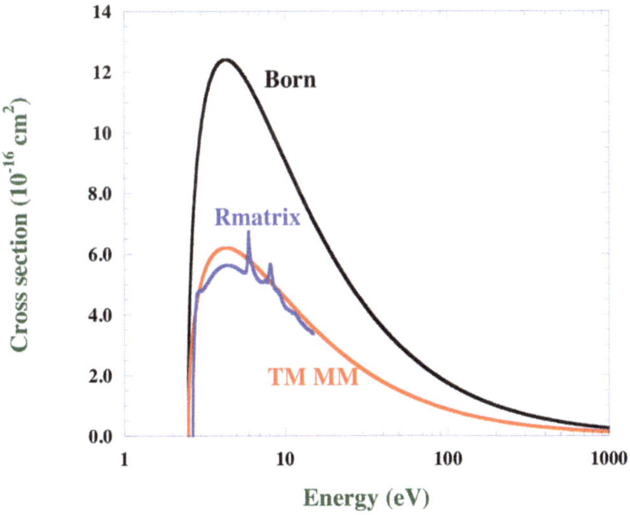

Figure 22.5. As in figure 22.4 for the vertical electron-impact excitation $X^2\Sigma^+ \to A^2\Pi$, occurring in a $v = 0$ BeH molecule: dipole-Born approximation (black); *R*-matrix method (blue), threshold modified Mott–Massey approximation (red) [23]. Copyright © IOP Publishing. Reproduced with permission.

Similarity approach is a method that represents the cross-section through a universal dimensionless function of the reduced energy of the incident electron and the oscillator strength associated with the excitation [24]. The state-specific cross-section has the form

$$\sigma_{v_i,v_f} = \frac{2\pi e^4}{(\Delta E_{v_i,v_f})^2} f_{v_i,v_f}\,\varphi(x), \tag{22.18}$$

where φ is the *similarity function*, describing the collision dynamics,

$$\varphi(x) = \frac{\ln[1 + a\sqrt{x-1}\,]}{x+b} \qquad x = E/\Delta E_{v'v} \tag{22.19}$$

The values of $a \sim 0.5$ and $b \sim 3$ have been optimized for the best agreement with a large number of atomic and molecular systems, however, they could be considered as adjustable parameters.

This method is super-simplified. However, it is worth mentioning because it allowed the inclusion of non-adiabatic effects due to the vibronic coupling in excited states [25, 26] by a mere re-definition of the oscillator strength.

$$\begin{aligned} f_{v_i,v_f} = \frac{2}{3}\frac{g_e}{g_g}\frac{m}{e^2\hbar^2}\Delta E_{v_i,v_f} |\langle\chi_a^d|M_a^d(R)|\chi_{\text{ground}}\rangle + \langle\chi_b^d|M_b^d(R)|\chi_{\text{ground}}\rangle \\ + \langle\chi_c^d|M_c^d(R)|\chi_{\text{ground}}\rangle|^2, \end{aligned} \tag{22.20}$$

where g_e and g_g are the statistical weights of the excited and ground electronic terms, respectively. $\chi_{\text{ground}}(R)$ and $\chi^d_{a,b,c}(R)$ are the ground and excited state vibrational wavefunctions in the *diabatic representation*, these last solutions of a system of coupled radial Schrödinger equations in the form of a normalized vector of mixed-character wave function. $M^d_{a,b,c}(R)$ are the transition dipole moments coupling the ground and the components of the excited coupled states.

The vibronic coupling is due to the Rydberg-valence character-mixing in excited states, markedly affecting the vibrational progression in the vacuum and extreme ultraviolet emission spectra of many diatomic systems, e.g. the singlet states of N_2 [29, 30] (see figure 22.6), the Schumann–Runge system in O_2 [31], the $\beta-$ and $\delta-$ bands of NO [32], therefore the vibrationally-specific cross-sections are expected to exhibit perturbation as well.

22.1.2 Classical and binary-encounter approaches for ionization

Different simplified (semi-empirical, classical and semi-semiclassical) methods [33–37] have been proposed in the literature for the derivation of ionization cross-sections, most of them avoiding the intrinsic theoretical complexity of the physics of ionization and reducing to the derivation of analytical expressions containing few parameters usually related to the structure of the molecular target.

Most of them descend from the classical Thomson cross-section [38]

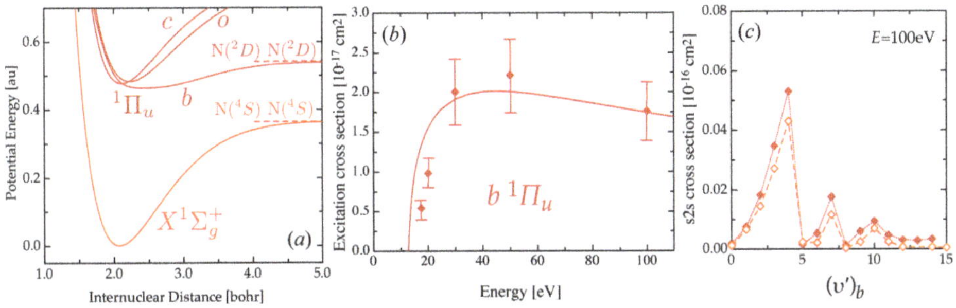

Figure 22.6. (*a*) Diabatic representation of potential energy curves of valence and Rydberg states of $^1\Pi_u$ symmetry in N_2. (*b*) Electron-impact-induced excitation cross-section of $N_2(X^1\Sigma_g^+, v_i = 0)$ to $b^1\Pi_u$ state (Birge–Hopfield system), (solid line) similarity method from reference [26], (circles) experimental results from reference [27]. (*c*) State-specific cross-sections from $v_i = 0$ to final states attributed to the b state, (close diamonds) similarity method from reference [26], (open diamonds) experimental results from reference [28].

$$\sigma^{\text{ion}}(E) = \sum_i \sigma_i(E) = \sum_i 4\pi a_0^2 N_i \left(\frac{Ry}{B}\right)^2 f(E/B) \text{ with } f(E/B) = \frac{E/B - 1}{(E/B)^2} \quad (22.21)$$

where the index i runs over the electrons of molecular state configuration, thus adding the contributions for non-equivalent electrons of the valence shells (partial cross-sections), the core–shells assumed to have negligible contribution and the orbital electron assumed at rest. Ry is the Rydberg constant, B is the electron binding energy in the ith orbital participating in the ionization process, N_i its occupation number.

Gryzinski method results from a classical theory of inelastic collisions in atomic target and has been developed to treat a number of processes, namely excitation, excitation with exchange and ionization [37, 38].

The general expression differs in the function $f(E/B)$ with respect to the Thomson cross-section as follows:

$$\sigma_{\text{Gry}}^{\text{ion}}(E) = \sum_i \frac{4\pi a_0^2 Ry^2 N_i}{BE} \left[\frac{E/B - 1}{E/B + 1}\right]^{3/2} \left[1 + \frac{2}{3}(1 - B/(2E)) \ln\left(\exp(1) + \sqrt{\frac{E}{B} - 1}\right)\right] \quad (22.22)$$

and it has been used in the past for the derivation of the dissociative ionization in electron-H_2 collisions, through the channels $^2\Sigma_g^+$ and $^2\Sigma_u^+$ of H_2^+, accounting for the vibrational excitation of the molecular target that is described by the Franck–Condon density [7, 39].

Recently, those results for H_2 have been extended in the framework of the same approach [40] deriving the non-dissociative and dissociative ionization channels considering also the ionization from excited electronic states, the results showing a remarkable agreement with the accurate CCC cross-sections [41] (see figure 22.7).

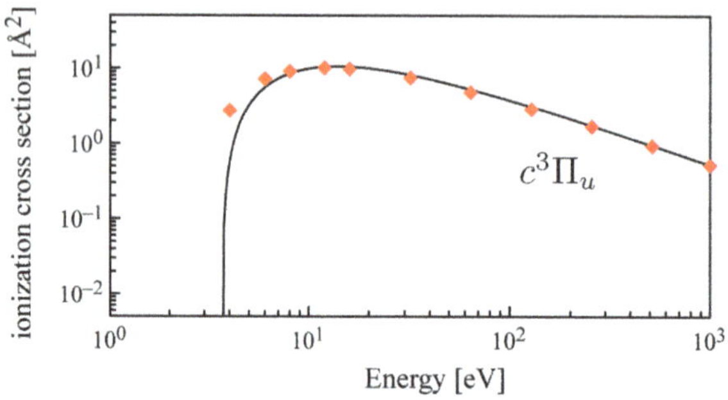

Figure 22.7. Non-dissociative ionization induced by electron in $H_2(c^3\Pi_u)$ state. (Solid line) Gryzinski method [40] (markers) CCC method [41].

Universal function method can be classified as a semi-empirical approach, in fact, starting from the classical Thomson ionization cross-section for atoms, it has been based on the observation that for a number of atomic and molecular targets the experimentally derived function $f(E/B)$ can be described by a *universal* analytical form [35, 42].

$$f(x) = \frac{10(x - 1)}{\pi(x + 9)(x + 1.5)} \tag{22.23}$$

The partial σ_i cross-section to a specific final electronic state of the molecular ion can be estimated assuming electrons of a single shell actually involved in the process. This assumption is physically sound when, given the reference configuration of the initial (ground) state, the states of the molecular ion are dominated by 1-hole configurations, e.g. for the N_2 molecule at the equilibrium internuclear distance, R_{eq},

$$N_2(^1\Sigma_g^+)(1\sigma_g^2 1\sigma_u^2 2\sigma_g^2 2\sigma_u^2 3\sigma_g^2 1\pi_u^4) = \begin{array}{l} N_2^+(X^2\Sigma_g^+) - e(3\sigma_g) \\ N_2^+(A^2\Pi_u) - e(1\pi_u) \\ N_2^+(B^2\Sigma_u^+) - e(2\sigma_u) \end{array} \tag{22.24}$$

The method has been used for the derivation of vibrationally resolved cross-sections for partial non-dissociative ionization cross-sections, assuming the one-electron model described above, for N_2 and O_2 [35, 42, 43].

$$\sigma_{i,\,v}^{\text{ion}} = \sum_{v'} N_i 4\pi a_0^2 \frac{Ry^2}{I_{v,v'}^2} f\left(\frac{E}{I_{v,v'}}\right) S_{v,v'} \tag{22.25}$$

where v' runs over the vibrational levels of the molecular ion in the electronic state of the considered ionization channel and the binding energy is replaced by the energy transitions between levels of the neutral and ionic states, corresponding to the state-specific ionization threshold, $I_{v,v'}$. The analysis of the Franck–Condon factors suggested that the main contributions to the sum in equation (22.25) come from the transitions to few v' levels, those connected to the initial vibrational level by vertical transitions from the classical inner and outer turning points, thus leading to the simplified expression

$$\sigma_i^{\text{ion}} = \frac{N_i 4\pi a_0^2 Ry^2}{2}\left[\frac{1}{I^2_{v_i,v_i'}} f\left(\frac{E}{I_{v_i,v_i'}}\right) + \frac{1}{I^2_{v_i,v_o'}} f\left(\frac{E}{I_{v_i,v_o'}}\right)\right] \tag{22.26}$$

The universal function parameters have been slightly adjusted to better fit the total ionization cross-section from the molecule ground state and then used for the derivation of partial channel-specific vibrationally-resolved ionization cross-sections and for the derivation of ionization from excited molecular targets. The method showed good accuracy and it correctly predict the branching ratios of the different channels (figures 22.8 and 22.9).

Binary-encounter approaches have been developed by Kim and co-workers [36, 46], combining the Mott cross-section and the Bethe theory, for low and high incident energies, respectively. The working equations are briefly reported in the following.

The total ionization cross-section in the *binary-encounter dipole* (BED) approximation is expressed as

$$\sigma_{\text{BED}}^{\text{ion}}(E) = \sum_i \sigma_i(E) = \sum_i \frac{4\pi a_0^2 N_i}{(t + u + 1)}\left(\frac{Ry}{B}\right)^2 \left[D(t)\ln t + (2 - N_i'/N_i)\left(1 - \frac{1}{t} - \frac{\ln t}{t + 1}\right)\right] \tag{22.27}$$

where $t = E/B$ and $u = U/B$, $U = \langle \mathbf{p}^2/2m \rangle$ being the average kinetic energy in the orbital. The first logarithmic term in the square parentheses in equation (22.27) accounts for long-range dipole interaction, while the other terms are related to the close collisions, including interference of direct and exchange scattering.

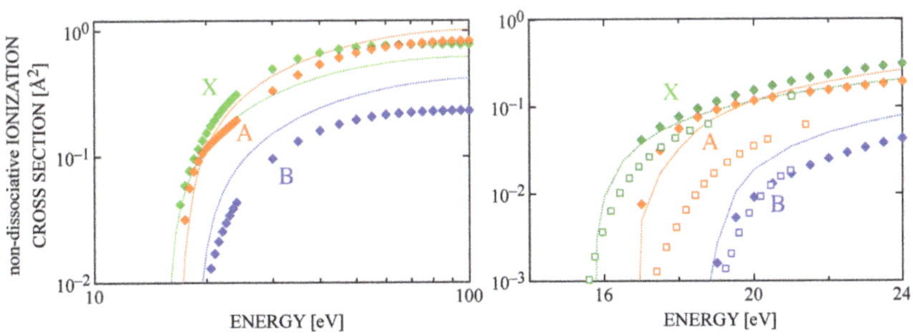

Figure 22.8. (left) Non-dissociative partial ionization cross-sections as a function of collision energy from $N_2(v = 0)$ vibrational level. (right) Threshold region magnified. (dotted line) universal function [35, 43] with $f(x) = 9.39(x - 1)/(\pi(x + 10)(x + 2))$; (close diamonds) derived from experimental integrated electric dipole photoionization oscillator strengths [44]. (open squares) Gryzinski approach [45]. Reproduced from Phy4Entry Database© website with permission.

Comparison of theoretical and experimental branching ratios at E = 100 eV.

Electronic term	[Laricchiuta et al (2006)]	[Goembel et al. (1994)]	[Van Zyl & Pendleton (1995)]	[Doering & Yang (1997)]	[Shemansky & Liu (2005)]
$X^2\Sigma_g^+$	30.1	0.56 ± 0.05	0.320 ± 0.147	0.448 ± 0.03	0.423
$A^2\Pi_u$	49.4	0.35 ± 0.03	0.535 ± 0.112	0.452 ± 0.033	0.450
$B^2\Sigma_u^+$	20.4	0.09±0.02	0.145±0.017	0.099±0.01	0.127

Figure 22.9. Reproduced from Phy4Entry Database© website with permission.

$$D(t) = \frac{1}{N}\int_0^{(t-1)/2} \frac{1}{w+1}\frac{df(w)}{dw}dw \tag{22.28}$$

$$N_i' = \int_0^{\infty} \frac{df(w)}{dw}dw \tag{22.29}$$

$\frac{df(w)}{dw}$ is the continuum dipole oscillator strength and $w = W/B$, with $W = E - B$ the energy of the ejected electron.

The average kinetic energy of the electron in the orbital can be obtained by energy structure calculations, the hardly accessible quantity is instead represented by $\frac{df(w)}{dw}$ entailing the coupling of bound and continuum electron wavefunctions and requiring in principle a fully quantum theoretical approach [47].

However, for many molecular systems this limit can be overcome deducing the continuum dipole oscillator strength from experimental photoionization cross-sections and, where partial photoionizations are available, also the single channel ionizations can be calculated.

A further simplification, proposed to model systems of unknown $\frac{df(w)}{dw}$, consists of assuming for this quantity the analytical form

$$\frac{df(w)}{dw} = \frac{N}{(w+1)^2} \tag{22.30}$$

deriving from equation (22.27) a new expression for the total ionization cross-section in the so-called *binary-encounter-Bethe* (BEB) model.

$$\sigma_{\mathrm{BEB}}^{\mathrm{ion}}(E) = \sum_i \frac{4\pi a_0^2 N_i}{(t+u+1)}\left(\frac{Ry}{B}\right)^2\left[\frac{\ln t}{2}\left(1 - \frac{1}{t^2}\right) + \left(1 - \frac{1}{t} - \frac{\ln t}{t+1}\right)\right] \tag{22.31}$$

The binary-encounter approach is nowadays widely used for the derivation of total ionization cross-section also for polyatomic systems, and it has been proved to be characterized by significant accuracy, the error being within 10% in the entire energy range. The accuracy of the threshold energy is connected to the value of the binding energy of the HOMO electron and therefore the experimental ionization threshold is sometimes used to reach a better agreement. The undoubtable merit of this approach relies on a simple formulation leading to full-range expressions, depending only on a limited number of parameters, associated to the properties of the orbital structure of the molecular target.

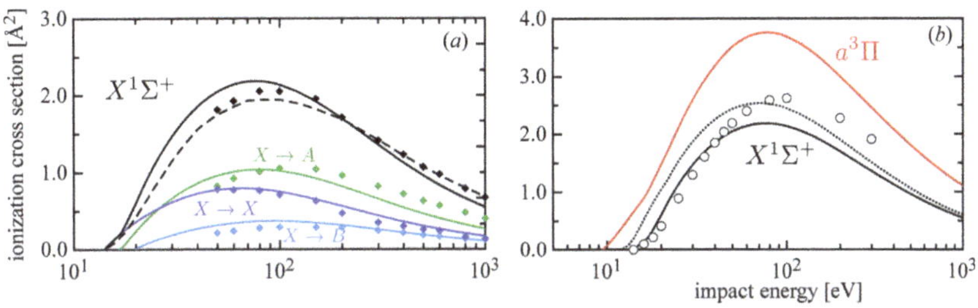

Figure 22.10. (*a*) Total (sum of three channels : X, A and B states of CO^+) and partial BED non-dissociative ionization cross-section in e-CO($X^1\Sigma^+$, $v = 0$) collision (solid lines) [16] compared with experimental results in the literature. (diamonds) reference [48], (dashed line) recommended data [49]. (*b*) BEB cross-sections for total ionization (solid lines) of $a^3\Pi$ metastable [16] and of $X^1\Sigma_g^+$ ground (circles) state [46] of CO by electron-impact. The ground state ($X^1\Sigma^+$) ionization cross-section to $CO^+(X, A, B)$ from $v = 0$ (dashed line) and from $v = 40$ (dotted line) are also reported.

The derivation of vibrationally-specific ionization cross-sections in the BED approach, has been recently performed [16] for the ionization of CO molecule, resorting to the R-centroid approximation in consideration of the vertical nature of the transition

$$\sigma_{v,\,v'}^{\text{ion}} = (\sigma^{\text{ion}}(E)_{\text{BED}})_{R=R_c} S_{v,v'} R_c = \langle \chi_{v'} | R | \chi_v \rangle / \langle \chi_{v'} | \chi_v \rangle \tag{22.32}$$

assuming the continuum dipole oscillator strengths for the channel-specific ionization geometry independent and describing the vibrational dependence by the Franck–Condon factors $S_{v,v'}$ and for the dissociative ionization transitions by the Franck–Condon density, assuming the transition occurring at the inner turning point to the repulsive branch of the ionized state. The BEB approach has been also applied to the estimation of ionization cross-section from the metastable states of CO molecule [16] (see figure 22.10).

22.2 Accurate approaches

Open shell species, and indeed electronically excited states of closed-shell species, have complicated electronic structures, so to obtain reliable results requires the use of sophisticated quantum mechanical theoretical methods capable of giving an accurate treatment of both the target molecule and the collision process. The most successful methods for doing these combine the use of quantum chemistry program suites to represent the electronic structure of the target plus detailed, wavefunction-based, treatments of the scattering processes. While more approximate methods, described elsewhere in this chapter, can give useful results in special cases and in particular at higher energies, collisions of electrons with molecules at a few eV requires consideration of the full problem. In this section we briefly describe the theoretical methods available for such studies. We start by considering the R-matrix method, which is both the most widely used theoretical method for studying electron collisions with molecules (atoms and ions) and our personal method of choice. We then consider other methods, give examples of results generated for plasma models and end with a summary of data sources.

All scattering methods applied to the study of electron collisions from a variety of targets solve the time-independent Schrödinger equation for the electronic degrees of freedom. The methods usually employ the fixed-nuclei (FN) approximation: during the time of the collision the nuclei are assumed to be fixed in space. Calculations are therefore performed for a specific geometry and the electronic part of the problem is solved. Nuclear motion (rotational, vibrational or dissociative) is treated beyond the FN approximation, in most cases employing FN electronic quantities (scattering amplitudes or matrices) obtained for a range of geometries.

In this section, we use atomic units (so that the mass and charge of the electron and $\hbar$ are set to 1). Within the FN approximation, the Hamiltonian operator included in the Schrödinger equation can be written as the sum of three terms:

$$H = H^N + K^{\mathrm{proj}} + V \tag{22.33}$$

where H^N is the electronic target Hamiltonian operator, K^{proj} is the operator for the kinetic energy of the projectile and V is the projectile–target interaction potential operator.

22.2.1 The *R*-matrix method

The *R*-matrix approach [50, 51], and its computational implementations in the UKRmol [52] and UKRmol+ [53] suites are well established for the study of electron collisions with molecules. The method is normally applied within the FN approximation. The approach is based in the division of space into an inner and outer region separated by a sphere of radius a, usually called the *R*-matrix sphere. When the projectile electron is inside the *R*-matrix sphere, it is indistinguishable from the N target electrons: an accurate description of electron exchange interactions and electron correlation is essential. In this region, the target plus scattering electron, $(N+1)$-electron, system is described by wavefunctions of the form:

$$\psi_k^{N+1} = \mathcal{A}\sum_{i,j}^{n} c_{ijk}\Phi_i^N(\mathbf{x}_1, \ldots, \mathbf{x}_N)\eta_{ij}(\mathbf{x}_{N+1}) + \sum_m b_{mk}\chi_m^{N+1}(\mathbf{x}_1, \ldots, \mathbf{x}_{N+1}). \tag{22.34}$$

where $\mathbf{x}_i$ stands for the space-spin coordinate of electron i, $\Phi_i^N(\mathbf{x}_1, \ldots, \mathbf{x}_N)$ is the wavefunction of the target electronic state i with energy ε_i and n is the number of target electronic states considered. The single-particle functions $\eta_{ij}(\mathbf{x}_{N+1})$ are known as continuum orbitals and describe the (unbound) projectile in the inner region; $\chi_m^{N+1}(\mathbf{x}_1, \ldots, \mathbf{x}_{N+1})$, known as L^2 functions, describe the short-range correlation-polarization and are built from (bound) target orbitals. The coefficients c_{ijk} and b_{mk} are obtained from diagonalizing the Hamiltonian matrix, which is made Hermitian by the addition of the Bloch operator [51] to the Hamiltonian operator H (Equation (22.33)), and $\mathcal{A}$ is the antisymmetrization operator. The eigenvalues E_k associated with the ψ_k^{N+1} are known as *R*-matrix poles.

Outside the *R*-matrix sphere the projectile is distinguishable from the target electrons. In this region, the problem can be reduced to a single centre one, with the projectile–target interaction described by means of a multipole potential expansion. The wavefunctions describing the $N + 1$-electron system thus have a simpler form:

$$\psi_k(\mathbf{x}_1, \mathbf{x}_2 \ldots \mathbf{x}_{N+1}) = \sum_i^{n'} \sum_j \Phi_i^N(\mathbf{x}_1 \ldots \mathbf{x}_N) \frac{F_j(r_{N+1})}{r_{N+1}} Y_{l_j m_j}(\hat{\mathbf{r}}_{N+1}) \tag{22.35}$$

where $n' \leqslant n$, so that the number of target states included in this expression can be smaller than that used in the inner region (equation (22.34)), r_{N+1} and $\hat{\mathbf{r}}_{N+1}$ indicate the radial and angular coordinates of the scattering electron, respectively, and $Y_{l_j m_j}(\hat{\mathbf{r}}_{N+1})$ are spherical harmonics. It is from the reduced radial functions $F_j(r_{N+1})$ that the scattering data is obtained. These functions are determined by solving a set of coupled differential equations, subject to boundary conditions expressed in terms of the R-matrix built entirely from inner region information,

$$R_{ij}(E) = \frac{1}{2a} \sum_k \frac{f_{ik}(a) f_{jk}(a)}{E_k - E}, \tag{22.36}$$

and using a propagation method [50]. The $f_{ik}(a)$, known as boundary amplitudes, give the amplitude of the inner region wavefunction at the R-matrix sphere and are defined as

$$f_{ik}(a) = \left\langle \frac{(\Phi_j)_i Y_{l_i m_i}}{r_{N+1}} \middle| \Psi_k \right\rangle_{\text{inner}} \tag{22.37}$$

where *inner* indicates that the integration implied by the bra and ket is performed in the inner region only. The product of the target state and the spherical harmonics, $(\Phi_j)_i Y_{l_i m_i}$, (with the correct total spin) defines collision channel i.

The coupling term in the differential equations that are solved to obtain the $F_j(r_{N+1})$ is proportional to interaction ('coupling') long-range potentials given by the multipole, single centre, expansion whose coefficients are obtained from the permanent and transition multipole moments of the target electronic states included in the calculation; calculations usually include dipole and quadrupole moments. Matching of the reduced radial functions to known asymptotic forms:

$$F_{ij} \overset{r \to \infty}{\longrightarrow} \delta_{ij} \delta_{l_i l_j} \delta_{m_i m_j} R_i^- + K_{ij} R_j^+, \tag{22.38}$$

allows for the determination of the scattering energy-dependent elements K_{ij} of the K-matrix, $\mathbf{K}(E)$, from which all the scattering data can be obtained. In the above equation, the $R^\pm$ can be seen as free and scattered radial waves of the form:

$$R_i^\pm = \frac{1}{\sqrt{k_i}} \begin{cases} \sin \\ \cos \end{cases} \left(k_i r - l_i \frac{\pi}{2} + \frac{Z}{k_i} \ln 2k r_i + \Gamma\left(l_i + 1 - i\frac{Z}{k_i} \right) \right) \tag{22.39}$$

where Z is the total charge of the target, Γ is the Gamma function and $k_i = \sqrt{2E - \varepsilon_i}$ and l_i correspond to the momentum of the projectile and its partial wave, respectively, in channel i. The 'transition' T-matrices and 'scattering' S-matrices can be trivially obtained from the K-matrix:

$$\mathbf{T}(E) = \mathbf{S}(E) + 1 = \frac{2i\mathbf{K}(E)}{1 - i\mathbf{K}(E)} \tag{22.40}$$

and, from them, electronically elastic and inelastic cross-sections:

$$\sigma_{ij} = \frac{\pi}{k_i^2}\sum\frac{2S + 1}{2(2S_i + 1)}|T_{ij}(E)|^2 \tag{22.41}$$

where S is the spin of the $N + 1$-electron system and S_i the spin of the N-electron target in channel i. Note that since i and j again indicate an initial and final channel, to obtain a cross-section for a transition between two target states, one needs to sum over all the channels associated with those specific states. Angle differential cross-sections can be calculated using external software that uses as input T- or K-matrices, for example, POLYDCS [54] (see below).

In addition, the energies of quasibound states, known as resonances, and their lifetimes can be obtained either from:

- fitting a Breit–Wigner profile to the eigenphase sums which are obtained trivially from the diagonalized K-matrices [55];
- analysis of the eigenvalues of the time-delay matrix $\mathbf{Q}(E)$ that is calculated from the S-matrices via the expression $\mathbf{Q}(E) = i\hbar\mathbf{S}(E)\frac{d\mathbf{S}(E)}{dE}$ [56].

These resonance data can then be employed to perform dissociative electron attachment (DEA, e.g. [57]) or dissociative recombination (DR, e.g. [58]) calculations.

In practical terms, performing an R-matrix calculation requires three steps: (1) the calculation of the target wavefunctions and energies, $\Phi_i^N(\mathbf{x}_1...\mathbf{x}_N)$ and ε_i, and the associated target permanent and transition moments; (2) the calculation of the inner region $N + 1$ wavefunctions ψ_k^{N+1} and associated eigenvalues E_k; (3) the solution of the outer region part of the calculation and determination of scattering quantities.

R-matrix calculations can employ fairly sophisticated approaches to constrict the required wavefunctions. The choice of approach determines how the target and $N + 1$ wavefunctions are constructed. Standard ones, used by other methods as well, are the relatively simple static-exchange (SE) and static-exchange-plus-polarization (SEP) models [51, 53]. In SE and SEP calculations only the ground state of the target molecule is included in the close-coupling expansion (equation (22.34)), described at Hartree–Fock (HF) level (i.e. by a single, antisymmetrized, product of bound target orbitals). Hartree–Fock self-consistent-field orbitals are normally used to build the target and L^2 functions. These models enable the determination of integral and differential (and momentum transfer) elastic cross-sections and resonance parameters for shape resonances.

When a better representation of the ground state is needed or, especially, when excited states are included in the close-coupling expansion, a configuration interaction (CI) description of the target states is preferred. For very small targets, a full-CI description is possible (see the example of H_2 below). It is more usual, however, to use a complete active space (CAS) approach in which a number of electrons are frozen in the orbitals they occupy in the ground state configuration and excitations to and from a set of occupied and virtual target orbitals (the active space) are allowed for the remaining electrons. In these cases, more sophisticated orbitals, obtained from standard quantum chemistry software are used: MOLPRO [59] is regularly employed to determine state-averaged complete active space self-consistent field (CASSCF) orbitals. When excited states are included in the close-coupling expansion, the model is known as close-coupling (CC) and both electronic excitation cross-sections and core-excited resonance parameters can be determined.

The target orbitals mentioned above are expanded in standard bases of Gaussian type orbitals (GTOs) which are readily available. One should always take into account the need for these orbitals to be contained inside the inner region: basis sets that include very diffuse functions are likely to require the use of large R-matrix radii. The continuum orbitals $\eta_{ij}(\mathbf{x}_{N+1})$ can be expanded in terms of GTOs centred on the centre of mass of the system, B-spline type orbitals (BTOs) or both. The older UKRmol suite only allowed for the use of GTOs, their exponents determined by fitting Coulomb or Bessel functions [60]. When only GTOs are employed, the majority of integrals required for the evaluation of the Hamiltonian matrices and other quantities can be determined analytically (and hence, computationally cheaply and with high accuracy).

The use of GTOs for the continuum description limits the size of R-matrix sphere to those with radii $\lesssim$ 18 a_0, therefore also limiting the target states that can be accurately described. Use of BTOs, on the other hand, enables much larger radii to be used, and therefore more diffuse target states to be included in the close-coupling expansion. The drawback is that integrals involving these functions need to be evaluated numerically, significantly increasing the computational cost while reducing the numerical accuracy. Nonetheless, use of BTOs has, for example, enabled more accurate description of the electronic continuum for higher kinetic energies of the projectile [53]. An intermediate alternative, helpful to reduce computational cost, is to use GTOs in the radial coordinate region close to the atomic nuclei and BTOs from there up to the R-matrix boundary. We note that the UKRmol/UKRmol+ R-matrix computational implementations require all (target and continuum) orbitals to be orthogonal. Poor orthogonalization can introduce non-physical features in the scattering quantities. 'Over' orthogonalizing (by removing too many of the continuum orbitals) leads to a poor description of the continuum; use of BTOs virtually eliminates this problem for low scattering energies. Whether using GTOs or BTOs, one must choose the maximum angular momentum, $l_{\max}$, of the continuum orbitals. This will determine the maximum angular momentum employed in the partial wave expansion implicit in the expression for the outer region wavefunctions (equation (22.35)), since $l_j \leqslant l_{\max}$.

A final consideration, when performing the inner region calculations, is that of the description of polarization effects, which is crucial to accurately model features that may be present in electron scattering, e.g. Ramsauer–Townsend minima and virtual states. The description of resonances is also quite sensitive to the amount of polarization being modelled. Polarization effects are described by including appropriate L^2 functions in the close-coupling expansion (equation (22.34)). These L^2 functions also allow for the relaxation of this orthogonalization and are therefore needed even in the SE model. The appropriate L^2 functions to be included are dictated by the need to ensure that the calculation is 'balanced', i.e. that the target (N-electron) and the $N + 1$ wavefunctions are described to the same level of accuracy or quality. A number of ad hoc rules have been developed that guide the choice of these functions (detailed discussion of these and the different models can be found in [51] and [53]).

An approach that has yielded good results for small targets (e.g. H_3^+, Li_2, CH_4) in terms of fully describing polarization effects is the molecular version of the R-matrix with pseudostates approach (RMPS) [61, 62], in which the close-coupling expansion is extended to include eigenvectors of the target Hamiltonian that do not describe true states of the target. These pseudostates are built to give a complete representation of all the target bound and continuum electronic states (up to some energy threshold) *inside* the R-matrix sphere. In practice, this requires the inclusion of an extra, even-tempered, GTO basis set at the centre of mass of the system [61], to generate 'pseudocontinuum' orbitals. Use of pseudostates has enabled, for example, the calculation of near-threshold ionization cross-sections for two-electron systems [61, 62] and the study of electron collisions with C_2^- [63].

The calculations for the outer region are usually simpler (though they can be computationally demanding if many channels are included) and less sensitive to the specifics of the target molecule. Usually, one need only make sure that the propagation is performed to a sufficiently large radial distance to ensure results are converged: around 100 a_0 is normally sufficient. The calculation of integral elastic and electronically inelastic cross-sections is straightforward using equation (22.41), but identifying and characterizing resonances requires more care, particularly if core-excited ones are of interest.

The modelling of processes involving nuclear motion requires going beyond the FN approximation and thus further methodology and, sometimes, software is needed. The above-mentioned POLYDCS can be used to calculate state-to-state rotational excitation cross-sections (for electronically elastic scattering). The program performs a frame transformation of the body-fixed K-matrices[1] into laboratory-frame ones [64] and can be applied to spherical, symmetric and asymmetric rotors. In addition, POLYDCS determines 'Born-corrected' integral and differential elastic cross-section. This 'correction' overcomes the non-convergence of these cross-sections for polar molecules in the FN approximation [65]. The program calculates the cross-sections using the Born approximation, then subtracts the partial Born cross-sections for $l < l_{max}$ and finally adds the (*ab initio*) R-matrix cross-sections that are calculated including partial waves up to l_{max}; the procedure can be seen as correcting the low partial-wave contributions to the Born cross-section by replacing them with the *ab initio* (R-matrix in this case) ones. A similar approached, also based on the Born approximation, can be used to add a correction to the R-matrix inelastic

[1] A modification of the published version of the code allows for the input of T-matrices, and therefore an accurate calculation of elastic cross-section when electronically inelastic channels are open.

cross-sections for dipole-allowed transitions using one of the programs in the UKRmol/UKRmol+ suites that implements the approach of Chu and Dalgarno [66].

The evaluation of both electronically elastic and inelastic vibrational resolved cross-sections for diatomic molecules is implemented within the UKRmol+ suite [67]. The approach is based on the adiabatic nuclei (AN) approximation [68] and involves the use of vibrational wavefunctions to average the FN T-matrices over a range of bond lengths. For simplicity, the approach assumes the electron kinetic energy to be geometry-independent, something that holds if the potential energy curves associated with the initial and final electronic states are approximately parallel. (For vibronic excitation, an even simpler approach is the use of Franck–Condon factors to multiply the FN cross-section, usually evaluated at the ground state equilibrium geometry.) The vibrational wavefunction (or Franck–Condon factors) required are determined using external software (e.g. [69]). The approach can model non-Franck–Condon effects, but it is unable to deal with resonances; however, if these are very narrow, they will not affect the vibrational excitation significantly. If they are broader, other approaches do exist to describe resonance-driven vibrational excitation [70].

Both UKRmol and UKRmol+ suites are available for download from Zenodo [71], where updates and new releases of UKRmol+ are also made available. Performing R-matrix calculations can be complex and require a reasonable understanding of quantum chemical concepts as well as some fluency in the use of the software. Two initiatives, one much better established, are aimed at lowering the barriers to perform R-matrix calculation. QEC (Quantemol electron collisions) [72] provides a graphical interface system for running the UKRmol+ codes that greatly simplifies the input generation and harvesting of results. Currently, an initiative is developing a platform for computational atomic and molecular physics that includes a web interface to a number of atomic and molecular codes [73, 74] of which the UKRmol+ suite is one.

22.2.2 Schwinger multichannel method

The Schwinger multichannel method (SMC) is a variational approach to the scattering amplitude based on the Lippmann–Schwinger integral equation. The method is applied within the FN and is restricted to closed-shell molecular targets. A detailed description of the SMC method, as applied with inclusion of pseudopotentials (see below), can be found in [75].

In the SMC approach, the scattering state $\Psi_{\mathbf{k}}^{(\pm)}(\mathbf{x}_1, \cdots, \mathbf{x}_{N+1})$ associated with wave vector $\mathbf{k}$ (related to the momentum, and therefore the energy, of the scattering electron) is expanded in the basis ϕ_j:

$$|\Psi_{\mathbf{k}}^{(\pm)}\rangle = \sum_j c_j^{(\pm)}(\mathbf{k})\, |\phi_j\rangle = \sum_j c_j^{(\pm)}(\mathbf{k})(\mathcal{A}|\Phi_{i_j}^N\rangle|\varphi_{r_j}\rangle) \tag{22.42}$$

where Φ_i^N again describes a target electronic state and $\mathcal{A}$ is the antisymmetrization operator; φ_r is a single-electron wavefunction, in practice an L^2 integrable orbital. One can write the scattering amplitude as

$$f_{\mathbf{k}_i,\mathbf{k}_f} = -\frac{1}{2\pi}\sum_{jn}\langle S_{\mathbf{k}_f}|V|\phi_j\rangle(d^{-1})_{jn}\,\langle\phi_n|V|S_{\mathbf{k}_i}\rangle, \tag{22.43}$$

where i and f indicate initial and final channels, respectively, V is the electron–molecule interaction potential (see equation (22.33)), the $S_{\mathbf{k}}$ describe asymptotic states and are the product of a target electronic wavefunction and $\exp(i\mathbf{k}\cdot\mathbf{r}_{N+1})$ and $d_{jn} = \langle\phi_j|A^{(+)}|\phi_n\rangle$ with

$$A^{(+)} = V - VG_0^{(+)}V; \tag{22.44}$$

$G_0^{(+)}$ is the free-particle Green's operator. In principle, representing the Green's operator would require a sum over bound and integration over (infinite) target continuum states. This requirement is overcome by projecting the Lippmann–Schwinger equation onto the energetically open electronic states of the target molecule by means of a projection operator P; this enables the use of a projected Green's operator $G_P^{(\pm)} = PG_0^{(\pm)}$.

The key point of the SMC method is obtaining an expression for $A^{(\pm)}$ that incorporates the projected Green's operator while satisfying $A^{(+)\dagger} = A^{(-)}$. This is achieved [76] by writing the Schrödinger equation to be solved as

$$(E - H)\,[(1 - aP) + aP]\,|\Psi_{\mathbf{k}_i}^{(\pm)}\rangle = 0, \tag{22.45}$$

which is exact if a is real. Here E is the total energy and H is once again the Hamiltonian for the $N + 1$ system defined by equation (22.33). From this equation, one obtains:

$$A^{(\pm)} = \frac{1}{2}(PV + VP) - VG_P^{(\pm)}V + \frac{1}{a}\left[\hat{H} - \frac{a}{2}(\hat{H}P + P\hat{H})\right], \tag{22.46}$$

where $\hat{H} = (E - H)$. Setting $a = N + 1$ ensures $A^{(+)\dagger} = A^{(-)}$. Using this expression for $A^{(+)}$, the d_{jn} can be determined and, from them, the scattering amplitudes (equation (22.43)). Differential cross-sections are calculated from these and their integration produces the integral cross-section.

In practice, SMC calculations use some of the strategies also employed in the R-matrix approach. GTOs are used to expand the target (bound) but also the scattering orbitals and a Hartree–Fock description is employed for the target ground state. Calculations for elastic scattering and resonance parameters are performed at the SE and SEP levels. In the latter, the expansion of the scattering state given by equation (22.42) includes both the ground and 'excited' states of the target, the latter described by single excitations from a hole orbital into a particle orbital. The choice of these states and scattering orbitals is key for the description of polarization effects [77]. For polar molecules, a Born-closure procedure (similar to, but not the same as that employed in the R-matrix approach) is used [75] both for elastic and inelastic scattering.

In order to reduce the computational effort related to the integral evaluation, analytic pseudopotentials can be used to describe the nuclei and core electrons. This implementation of the method [78] is known as SMCPP and it facilitates the study of targets containing electron-rich atoms (see, for example, a recent study on elastic scattering on halomethanes [79]).

Electronic excitation calculations with this method are rarer (although becoming more frequent); when performed, they involve (formally) the inclusion of electronically excited states in the summation that defines the projection operator P. In practice, the excited states were often represented by one or two configurations (more accurately Slater determinants) built from improved virtual orbitals (see [75] for details) and more recently using a single-excitation CI treatment. The former approach leads to a poor description of states of different spin associated with occupation of the same molecular orbital; the latter introduces large number of pseudostates in the calculation and therefore significantly increases the computational cost. A recent example of this type of calculations is the study of electronically inelastic scattering from ethanol [80], that included more than 400 target states.

22.2.3 Molecular convergent close-coupling approach

The molecular implementation of the convergent close-coupling approach (MCCC) [81, 82] enables the calculation of elastic, vibrationally inelastic and electronically inelastic (integral and differential), ionization and neutral dissociation cross-sections for quasi-one and two-electron targets. So far, it has been used extensively to determine cross-sections for H_2 and its isotopologues [83].

MCCC is an *ab initio* method based on a close-coupling expansion, like the R-matrix approach:

$$\psi_k^{N+1} = \mathcal{A} \sum_i^n f_i^{(+)}(\mathbf{x}_{N+1})\Phi_i^N(\mathbf{x}_1, \ldots, \mathbf{x}_N). \tag{22.47}$$

where the notation is the same as for equation (22.34) and the $f_i^{(+)}$ describe the scattering electron; all calculations performed so far are for $N \leqslant 2$. The target wavefunctions Φ_i^N for $N = 2$ are obtained from configurations generated as products of single-particle orbitals. These orbitals are built from Laguerre functions (the product of associated Laguerre polynomials and an exponential of the type $\exp(-\alpha r)$ and spherical harmonics).

The method relies on the completeness of the Laguerre basis: the FN scattering wave function (equation (22.47)) is expanded over a near-complete set of (positive- and negative-energy) states and pseudostates of the target. The use of pseudostates that represent ionization enable accurate calculations well above the ionization threshold (i.e. the hard-to-describe intermediate energy regime) and the determination of ionization cross-sections.

The MCCC method solves the time-independent Schrödinger equation by using expansion (22.47) and formulating a set of close-coupling equations in momentum-space for the T-matrix:

$$\langle \mathbf{k}_f^- \Phi_f^N | T | \Phi_i^N \mathbf{k}_i^- \rangle = \langle \mathbf{k}_f^- \Phi_f^N | V | \sum_i^n f_i^{(+)} \Phi_i^N \rangle \tag{22.48}$$

where T is T-matrix operator and V, as usual, the projectile–target interaction potential operator. The distorted waves $|\mathbf{k}^{\pm}\rangle$, normalized in momentum space, are solutions of

$$\left(\frac{k^2}{2} - K^{\text{proj}} - U_0\right)|\mathbf{k}^{\pm}\rangle = 0. \tag{22.49}$$

Here, U_0 is a short-ranged central distorting potential, chosen so that it cancels the spherical part of the projectile–nuclei interaction potential and is included to save computational resources and K^{proj} is the kinetic operator for the projectile.

A partial wave expansion of the $|\mathbf{k}^{\pm}\rangle$ is used to solve the coupled equations given by equation (22.48). This results in a set of Lippmann–Schwinger equations for the partial-wave T-matrix. These body-frame T-matrices are transformed to the laboratory frame and orientation averaged to obtain cross-sections. Similarly to the methods just described, a Born-approximation based approach is used to top-up the partial wave expansion for dipole-allowed transitions at intermediate and large scattering energies.

The method can go beyond the FN approximation by using the AN approximation [68] to provide vibrationally resolved cross-sections [41] in an analogous way to the R-matrix approach, as well as dissociative excitation cross-sections. When this is the case, spheroidal coordinates are used to solve the electronic (FN) part of the problem, as these coordinates are more appropriate for large internuclear diatomic distances. The vibrational wavefunctions are obtained diagonalizing the (Born–Oppenheimer) vibrational Hamiltonian in a basis of Laguerre functions similar to those used for the electronic wavefunctions.

In addition to many studies of electron scattering from diatomics in the ground electron state, the method has been applied to collisions with the target initially in an electronic excited state: calculations have been performed of non ro-vibrationally resolved cross-sections for scattering from triplet states of H_2 [84].

22.2.4 Other methods

We briefly mention a couple of methods that have been applied to the study of electron scattering from molecules.

Kohn variational method: in its most used implementation, the complex Kohn variational method [85, 86], is also based on a close-coupling expansion of the type employed by the R-matrix approach (see equation (22.34)) and a description of the target states at the CAS or multi-reference CI levels with GTO basis sets used to expand the bound molecular orbitals and Bessel functions for the continuum. The method has been employed to calculate (within the FN approximation) elastic and electronically inelastic cross-section. The method is based on the application of the Kohn variational principle to the T-matrix, that is expressed as the stationary value of a functional that is determined using the close-coupling expansion wavefunctions. Its current implementation is usually applied to relatively small molecules.

Recent applications of the method to electron scattering are scarce and have focused on dissociative recombination (e.g. [88]) and the characterization of resonances that lead to DEA: for example, DEA to MgCN has been studied using resonance data from Kohn variational calculations as input to form the Hamiltonian relevant to the nuclear dynamics; the associated multidimensional time-dependent Schrödinger equation is then solved using the multi-configuration time-dependent Hartree approach [87].

Single centre expansion (SCE) approach is a one-electron method in which V, the interaction between projectile and target, is expressed in terms of three contributions describing: (i) the electrostatic interaction between projectile and target; (ii) the non-local exchange interaction, sometimes described by means of a local exchange model, particularly for larger targets; (iii) correlation-polarization, typically described using density functional theory. Similar to the R-matrix approach, SCE uses a partial wave expansion of the wavefunction describing the free electron: the radial contribution, from which the scattering data is obtained, is calculated solving a set of coupled radial differential equations, where the coupling is given by the interaction potential just described.

The relatively lower cost of treating the scattering problem effectively as single-electron means this approach can be more easily applied to larger (more electron-rich) targets. In addition, the use of explicit potentials to describe specific contributions to the projectile–molecule interaction makes it easier to modify these in a systematic manner. However, this characteristic also limits the accuracy with which these interactions can be described. The method, as implemented in the VOLSCAT suite [89, 90] has been applied to the study of electronically elastic scattering from a number of small and medium size (mainly biological) molecules and to the analysis of their shape resonances.

22.3 Some examples

Plasma chemistry sets: the provision of complete chemistry sets is important for performing plasma models. This means that data are not only required for the initial gas but also for the many fragments that are created in the plasma environment. Take for example the molecule nitrogen trifluoride, NF_3. NF_3 is used in remote source plasmas, silicon etching and chemical vapour deposition processes. In the plasma, NF_3 fragments to give fluorine atoms and ions which provide the active plasma component; this also gives the radical species NF_2 and NF. While there are some experimental measurements of electron collision processes with NF_3 [91], the only process which has been studied in the laboratory for NF and NF_2 is electron impact ionization. This situation is typical.

Hamilton *et al* [92] undertook a systematic study of electron collision processes involving NF_3, NF_2 and NF using the R-matrix method. Table 22.2 lists the

Table 22.2. Electron collision processes considered by Hamilton *et al* [92] for NF_3. (Species or electronic states given in brackets represent different possible starting points or outcomes.)

Elastic scattering and momentum transfer: $AB + e \rightarrow AB + e$

$NF_3 + e \rightarrow NF_3 + e$

$NF_2 + e \rightarrow NF_2 + e$

$NF + e \rightarrow NF + e$

Inelastic scattering: $AB + e \rightarrow AB^* + e$

$NF_3(X\,^1A' + e \rightarrow NF_3(a^3A', b\,^3A'', c\,^3A', A\,^1A', B\,^1A'', C\,^1A'', d\,^3A'', e\,^3A') + e$

$NF_2(X\,^2A_1 + e \rightarrow NF_2(A\,^2A_1, B\,^2B_2, C\,^2A_1, a\,^4A_2, D\,^2B_2, b\,^4B_1, E\,^2A_2, F\,^2A_2) + e$

$NF(X\,^3\Sigma^-) + e \rightarrow NF(a\,^1\Delta, b\,^1\Sigma^+, A\,^3\Pi, B\,^3\Sigma^+, d\,^1\Pi, E\,^3\Pi, e\,^1\Pi, F\,^3\Pi) + e$

Superelastic scattering: $AB^* + e \rightarrow AB + e$

$NF(a\,^1\Delta, b\,^1\Sigma^+) + e \rightarrow NF_3(X\,^3\Sigma^-) + e$

Dissociative electron attachment: $AB + e \rightarrow A + B^-$

$NF_3 + e \rightarrow NF_2 + F^-$

$NF_2 + e \rightarrow NF + F^-$

$NF + e \rightarrow N + F^-$

Electron impact dissociation: $AB + e \rightarrow A + B + e$

$NF_3 + e \rightarrow$ fragments $+ e$

$NF_2 + e \rightarrow$ fragments $+ e$

$NF + e \rightarrow N + F + e$

Electron impact dissociation (state resolved): $AB + e \rightarrow A^* + B + e$

$NF_3 + e \rightarrow NF(b\,^1\Sigma^+, B\,^3\Sigma^+) + F_2 + e$

$NF_2 + e \rightarrow NF_2(A\,^2A_1, a\,^4A_2, B\,^2B_2) + F + e$

Metastable electron impact dissociation: $AB^* + e \rightarrow A + B + e$

$NF(a\,^1\Delta, b\,^1\Sigma^+) + e \rightarrow N + F + e$

Electron impact ionization: $AB + e \rightarrow AB^+ + 2e$

$NF_3 + e \rightarrow NF_3^+ + 2e$

$NF_2 + e \rightarrow NF_2^+ + 2e$

$NF + e \rightarrow NF^+ + 2e$

Electron impact dissociative ionization: $AB + e \rightarrow A + B^+ + 2e$

$NF_3 + e \rightarrow (NF_3^+, NF_2^+ + F, NF^+ + 2F, F^+ + NF_2, N^+ + 3F) + 2e$

$NF_2 + e \rightarrow (NF_2^+\ NF^+ + F, F^+ + NF, N^+ + 2F) + 2e$

$NF + e \rightarrow (NF^+, N + F^+, N^+ + F) + 2e$

processes considered in this study. Detailed cross-sections as a function of electron impact energy given in the supplementary data of their paper and can be obtained from the Quantemol Database (QDB) [93]. Table 22.2 shows that a whole variety of processes were considered but these still do not represent a comprehensive dataset for electron collisions with NF_x ($x = 1, 2, 3$). For example, the table does not include electron impact vibrational excitation about which there is little information for these systems [94].

These datasets are often collected into sets of recommended cross-sections which combine available experimental and theoretical data to give the most reliable cross-sections for each process of importance [94].

Fusion: The centre of fusion plasmas is too hot to support molecules so electron–molecule collisions are only important at the edge of these where the plasma temperature is only a few eV. However, despite the limited region of interest, theory has a particularly important role in providing electron–molecule collision data for fusion studies because of the near impossibility of performing experiments on molecules containing tritium. Conversely, adapting a theoretical model developed, and possibly tested against experiment, for H containing species to consider T instead is straightforward.

The most important molecule in fusion plasmas is H_2 (or its deuterated and triterated isotopologues). Recently, Scarlett *et al* [95, 96] have used a the molecular convergent close-coupling (MCCC) approach to produce comprehensive sets of cross-sections for electron collisions with H_2, HD, HT, D_2, DT and T_2. The MCCC approach is particularly well suited for this as the methodology is designed for treating few electron systems in a very comprehensive manner which can cover a very extended range of energies. In this case, Scarlett *et al* considered electron collision energies up to 500 eV and excitation to 17 electronically excited states of each isotopologue. Consideration of vibrational motion, within the adiabatic nuclei approximation, led to about 5000 reactions for each isotopologue. A prior benchmark study had shown that the MCCC approach gives very similar results for electron collisions with H_2 as an *R*-matrix study [97].

Another role played by molecules in the study of fusion is as diagnostics. For example, erosion of beryllium walls can be monitored by emissions of beryllium hydride; emissions from the BeD molecule have been observed from the Joint European Torus (JET) [67, 98]. As these emissions are non-thermal, electron collision rates are required to model them. An *R*-matrix study has provided cross-sections for vibrationally-resolved electron impact excitation of the A $^2\Pi$ excited electronic state, which is the one observed to emit, for BeH, BeD and BeT [67].

Re-entry physics: the (re-)entry of a spacecraft into a planetary atmosphere causes the spacecraft to experience intense heat flux with a corresponding sharp rise in the temperature of the shock region. This results in plasma conditions involving the atmospheric gases, N_2 and O_2 on Earth and other gas mixtures for Mars, Venus and Jupiter. Modelling these requires a wealth of reaction rates or cross-sections with a particular emphasis on vibrationally resolved processes including vibrational excitation, dissociative electron attachment (DEA) and electron impact dissociation.

Electron collisions with vibrationally excited molecules can lead to processes that do not normally occur with molecules in their ground state. The obvious one of these is electron impact dissociation with electrons whose energy is less than the dissociation energy of the molecule, which is usually the case in these plasmas. Another interesting example is DEA to the oxygen molecule. O_2 has a low-lying resonance of $^2\Pi$ symmetry; when O_2 is in its vibrational ground state ($v = 0$), excitation of this resonance to create $O_2^-(v = 0)$ does not provide the system with enough energy to undergo DEA to $O + O^-$. The process can only occur through higher-lying resonances which are accessed by higher energy electrons. As a result, occupation of the $^2\Pi$ resonance leads to strong resonance-enhanced vibrational excitation. However, if the initial oxygen molecule is vibrationally excited then there can be enough energy in the system to undergo DEA via the low-lying $^2\Pi$ resonance. This leads to DEA becoming the dominant process and vibrational excitation being largely suppressed. Figure 22.11 compares the key cross-sections for $O_2(v)$ with $v = 0$ and $v = 20$.

Astrophysics: the interstellar medium (ISM) contains giant clouds of molecules which are in fact so diffuse that they are almost a collision-free environment. These clouds are weakly ionized and thus form a very low-temperature plasma environment which is far from thermal equilibrium. Despite this, the molecular clouds are often assigned a temperature, typically in the 10–100 K region. This temperature is usually quoted as the electron temperature which, of course, cannot actually be measured.

Over 200 molecules have now been identified in the ISM, largely by the use of radio astronomy. These long wavelengths are sensitive to the rotational spectrum of the constituent molecules. However, given the low density environment, it is largely generated by photon emission following collisional excitation of the species being observed. Hydrogen (H_2) is the dominant species in these clouds and it is usually assumed that rotational excitation due to collisions with H_2 is the dominant excitation process. However, the rotational excitation rates for electron collisions with molecular ions are vastly bigger (maybe a factor of 10^5) than those given by collisions with H_2, which is actually a rather inefficient exciter of rotational motion. This means that in interstellar environments where the electron density is 10^{-5} that of hydrogen or greater, electron impact rotational excitation becomes an important process.

However, there is a problem: the spacing between the rotational energy levels in molecules is so small that thus far it has proved impossible to measure the electron impact rotational excitation cross-section for any molecular ion. The best that has been obtained experimentally was an electron collision rotational cooling rate for HD^+ ions which proved to be consistent with theoretical predictions [100]. This has

Figure 22.11. Calculated cross-sections for electron-oxygen molecule collisions with the O_2 in initial vibrational states $v = 0$ (left) and $v = 20$ (right). The processes considered are vibrational excitation (VE) for excitation by one vibrational quanta ($\Delta v = 1$), dissociative electron attachment (DeA), and electron impact dissociation. The cross-sections, which are due to Laporta *et al* [99], were generated by considering a sum over the processes arising from the four lowest O_2^- resonance states. Reprinted with permission from [99], Copyright (2015) by the American Physical Society.

meant that all electron impact rotational excitation rates have thus far been obtained from theory.

At low electron collision energies, rotational excitations are largely driven by the permanent dipole moment of the molecule. If a molecular ion has a large dipole moment then a simple dipole theory based on the Born approximation [66] can be used. This theory only allows changes in the rotational quantum J of one, i.e. transitions with $\Delta J = 1$. However, astronomical studies implied that a much larger set of transitions occur, up to $\Delta J = 6$ in some cases. For this reason an R-matrix methodology has been developed for solving this problem [101]. This again uses a Born approximation treatment to allow for the long-range effect of the dipole interaction but also explicitly allows for the short-range effects which are responsible for the other transitions. This methodology has been extensively used to give electron impact rotational excitation rates for a range of astronomically important molecular ions; these rates can be found in the BASECOL database [102], see next section.

Another process important in astrophysics, and indeed elsewhere, is dissociative recombination (DR). The DR of H_3^+

$$H_3^+ + e \rightarrow H + H + H \text{ or } H_2 + H \tag{22.50}$$

is particularly important in cold hydrogen plasmas such as the ones in the ISM. This process is important as H_3^+, which formed following cosmic ray ionization, seeds much of the interstellar chemistry and DR is the main competing destruction mechanism. Initial theory suggested that DR of H_3^+ was an inefficient process which only took place slowly in ISM conditions. However, it is now accepted on the basis of both theory and experiments that this DR rate is actually rapid [103].

To get a theory of rapid DR for H_3^+ it is necessary to abandon the standard Born–Oppenheimer approximation, which separates the motion of the electrons from that of the nuclei, and to explicitly consider couplings that allow for interactions between these motions. With such a theory, it has proved possible to calculate accurate rates which have been found to depend on the initial state of the H_3^+ [104]; this has led to realization that full state dependent chemistry is required to model the processes correctly.

22.4 Databases

Access to good data is often the key to generating useful plasma models and, in the past, the construction of the chemical kinetic model proceeded through the laborious process of literature data-mining, retrieving data published either in tabular format or plotted in figures. Various data compilations have been undertaken to provide easy access to interested users. For key molecules these compilations are often themselves in the form of papers considering all the available measurements and calculations on a system and then making recommendations on which cross-sections are deemed to be the most reliable with, in many cases, an estimated uncertainty included [94]. Examples of this approach include the recent compilation of electron collision cross-sections for water [105] or the compilation of largely experimental data on electron collisions with NF_3 [91] mentioned above.

An alternative source of data is through a number of web-access specialist databases [83, 106–111], that allow the direct download of datasets for cross-sections or rate coefficients of different classes of processes relevant to many atomic and molecular systems, with customizable queries for the efficient database exploration. These databases, exploiting modern web technologies, represent the most valuable aid to share the dynamical information for plasma modelling, collecting and classifying data, offering graphical tools for the on-the-fly rendering of selected datasets or their critical comparison and providing the complete and correct references in the literature.

The efforts put in by the scientific community are extraordinary and stimulated the debate on some issues relevant to the impact of the chemistry on the predictivity of the kinetic simulations: the accuracy of data and the completeness of the chemical model. The assessment of theoretical data accuracy in principle requires the comparison with electron-energy loss or emission spectra, not always available, especially when transitions from excited states are considered. However, it has been stated the importance of quantifying the *uncertainty* related to the intrinsic approximations of the theoretical approach adopted [112, 113], this aspect is mainly related to the possibility of characterizing the dataset for a specific process with an index of reliability. Unfortunately, the uncertainty quantification is still very far from being completed, committing the community of data producers for the coming years. Presently, different databases are built with different perspectives with respect to this problem, some of them collecting multiple datasets others recommending data. As for the completeness issue, it is usually related to the inclusion of the relevant channels and to the range of collision energy, from threshold to high-energy limits connected to the model applications. This is the case of chemical models in LXCat [106], where the consistency of the datasets is validated by comparison with swarm data, or in the Quantemol database [93, 108], ranking the chemistries on the base of the validation level achieved in comparing the results obtained with plasma modeling softwares, e.g. GlobalKIN and ChemKin, to experiments.

When moving to state-to-state kinetics the databases need a structure able to classify state-specific cross-sections, to handle large dynamical datasets and to include the dependence of process probability on the electronic, vibrational and rotational degrees of freedom accounting for the complete manifold of levels, thus redefining the concept of *completeness.* The Phys4EntryDB database [111], mainly intended for the modeling of non-equilibrium conditions during entry in planetary atmospheres [114], and, more recently, the LXCat [106] and the database maintained at Perth University [83], distributing the vibrationally-resolved cross-sections for the electron-impact induced processes in H_2 and its isotopologues, are the only examples of this sort of database. The validation of these data relies, in many cases, on the comparison with experimental cross-sections for processes initiated from the fundamental vibrational level of the ground electronic state. It is worth noticing that in dealing with state-to-state vibrational kinetics the chemical scheme needs to integrate also the collisional transfer processes redistributing the energy among the translational and the internal degrees of freedom or convert vibrational energy into dissociation. These processes, thoroughly treated in the chapter 23, are available for

Table 22.3. Currently active databases providing electron–molecule collision data.

Database	Reference	Electron-collision data	Target field	Other data included
BASECOL	[102]	Rotational excitation	Astrophysics	Heavy particle inelastic cross-sections
KIDA	[117]	Dissociative recombination	Astrophysics	Reaction rates
LXCat	[118]	Excitation processes	Plasma physics	Atomic cross-sections
NIFS	[119]	Excitation processes	Fusion	Reactions rates
NFRI	[120]	Excitation processes	Fusion	Reactions rates
ALADDIN	[121]	Excitation processes	Fusion	Reactions rates
Phys4EntryDB	[114]	Vibrational excitation	Atmospheric re-entry	Heavy particle inelastic cross-sections
QDB	[93]	Excitation processes	Technological plasmas	Reaction rates
UDfA	[122]	Dissociative recombination	Astrophysics	Reaction rates

some collisional systems in the Phys4EntryDB and also in the dedicated database STELLAR [115].

The main currently active databases which provide cross-sections and/or rates for electron–molecule collisions are listed in table 22.3; these databases provide a mixture of theoretically and experimentally determined results. Several of these databases are accessible simultaneously using the portal provided by the Virtual Atomic and Molecular Data Centre (VAMDC) [110, 116].

References

[1] Bartschat K and Kushner M J 2016 Electron collisions with atoms, ions, molecules, and surfaces: fundamental science empowering advances in technology *Proc. Natl Acad. Sci.* **113** 7026–34

[2] Hazi A U 1981 Impact-parameter method for electronic excitation of molecules by electron impact *Phys. Rev.* A **23** 2232–40

[3] Redmon M J, Garrett B C, Redmon L T and McCurdy C W 1985 Improved impact-parameter method for electronic excitation and dissociation of diatomic molecules by electron impact *Phys. Rev.* A **32** 3354–65

[4] Garrett B C, Redmon L T, McCurdy C and Redmon M J 1985 Electronic excitation and dissociation of O_2 and S_2 by electron impact *Phys. Rev.* A **32** 3366

[5] Celiberto R and Rescigno T N 1993 Dependence of electron-impact excitation cross sections on the initial vibrational quantum number in H_2 and D_2 molecules: $X^1\Sigma_g^+ \rightarrow B^1\Sigma_u^+$ and $^1\Sigma_g^+ \rightarrow C^1\Pi_u$ transitions *Phys. Rev.* A **47** 1939–45

[6] Celiberto R, Capitelli M and Lamanna U T 1994 Vibrational excitation of $H_2(X^1\Sigma_g^+, v)$/$D_2(X^1\Sigma_g^+, v)$ through excitation of electronically excited singlet states and radiative cascade *Chem. Phys.* **183** 101–06

[7] Celiberto R, Janev R K, Laricchiuta A, Capitelli M, Wadehra J M and Atems D E 2001 Cross section data for electron-impact inelastic processes of vibrationally excited molecules of hydrogen and its isotopes *At. Data Nucl. Data tables* **77** 161–213

[8] Laricchiuta A, Celiberto R, Esposito F and Capitelli M 2006 State-to-state cross sections for H_2 and its isotopic variants *Plasma Sources Sci. Technol.* **15** S62

[9] Laricchiuta A, Celiberto R and Janev R 2004 Electron-impact-induced allowed transitions between triplet states of H_2 *Phys. Rev.* A **69** 22706

[10] Tapley J K, Scarlett L H, Savage J S, Zammit M C, Fursa D V and Bray I 2018 Vibrationally resolved electron-impact excitation cross sections for singlet states of molecular hydrogen *J. Phys. B: At. Mol. Opt. Phys.* **51** 144007

[11] Tapley J K, Scarlett L H, Savage J S, Fursa D V, Zammit M C and Bray I 2018 Electron-impact dissociative excitation cross sections for singlet states of molecular hydrogen *Phys. Rev.* A **98** 032701

[12] Kim Y-K 2007 Scaled Born cross sections for excitations of H_2 by electron impact *J. Chem. Phys.* **126** 064305

[13] Kato H, Kawahara H, Hoshino M, Tanaka H, Brunger M and Kim Y-K 2007 Cross sections for electron impact excitation of the vibrationally resolved $A^1\Pi$ electronic state of carbon monoxide *J. Chem. Phys.* **126** 064307

[14] Tanaka H, Brunger M, Campbell L, Kato H, Hoshino M and Rau A 2016 Scaled plane-wave Born cross sections for atoms and molecules *Rev. Mod. Phys.* **88** 025004

[15] Vriens L 1967 Excitation of helium by electrons and protons *Phys. Rev.* **160** 100

[16] Laricchiuta A, Pietanza L D, Capitelli M and Colonna G 2018 Electron-CO excitation and ionization cross sections for plasma modeling *Plasma Phys. Control. Fusion* **61** 014009

[17] Itikawa Y 2007 *Molecular Processes in Plasma* (Berlin: Springer)

[18] Celiberto R, Janev R and Laricchiuta A 2020 Electron-impact state-to-state cross sections and rate coefficients for $X(v) \rightarrow A(v')$ excitation of LiH molecule *Plasma Sources Sci. Technol.* **29** 035008

[19] Geltman S 1969 *Topics in Atomic Collision Theory* (New York: Academic)

[20] Inokuti M 1971 Inelastic collision of fast charged particles with atoms and molecules–The Bethe theory revisited *Rev. Mod. Phys.* **43** 297

[21] Celiberto R, Janev R K and Reiter D 2009 Basic molecular processes for hydrocarbon spectroscopy in fusion edge plasmas: vibrationally state-selective excitation of $A^2\Delta$, $B^2\Sigma^-$ and $C^2\Sigma^+$ states of CH by electron impact *Plasma Phys. Control. Fusion* **51** 085012

[22] Mott N F and Massey H S W 1965 *The Theory of Atomic Collision* 3rd edn (Oxford: Oxford University Press) p 497

[23] Celiberto R, Baluja K L and Janev R K 2013 Electron-impact state-to-state resolved cross sections and rate coefficients for the $X(v) \rightarrow A(v')$ excitation in BeH molecules *Plasma Sources Sci. Technol.* **22** 015008

[24] Adamson S, Astapenko V, Deminskii M, Eletskii A, Potapkin B, Sukhanov L and Zaitsevskii A 2007 Electron impact excitation of molecules: calculation of the cross section using the similarity function method and *ab initio* data for electronic structure *Chem. Phys. Lett.* **436** 308–13

[25] Laricchiuta A, Celiberto R, Capitelli M and Colonna G 2017 Calculation of electron-scattering cross sections relevant for hypersonic plasma modeling *Plasma Process. Polym.* **14** 1600131

[26] Celiberto R, Laporta V, Laricchiuta A, Tennyson J and Wadehra J 2014 Molecular physics of elementary processes relevant to hypersonics: electron–molecule collisions *Open Plasma Phys. J.* **7** 33–47

[27] Malone C P, Johnson P V, Liu X, Ajdari B, Kanik I and Khakoo M A 2012 Integral cross sections for the electron-impact excitation of the $b^1\Pi_u$, $c_3^1\Pi_u$, $o_3^1\Pi_u$, $b'^1\Sigma_u^+$, $c_4'^1\Sigma_u^+$, $G^3\Pi_u$, and $F^3\Pi_u$ states of N_2 *Phys. Rev.* A **85** 062704

[28] James G K, Ajello J M, Franklin B and Shemansky D E 1990 Medium resolution studies of extreme ultraviolet emission from N_2 by electron impact: the effect of predissociation on the emission cross section of the $b^1\Pi_u$ state *J. Phys. B: At. Mol. Opt. Phys.* **23** 2055

[29] Stahel D, Leoni M and Dressler K 1983 Nonadiabatic representations of the $^1\Sigma_u^+$ and $^1\Pi_u$ states of the N_2 molecule *J. Chem. Phys.* **79** 2541–58

[30] Spelsberg D and Meyer W 2001 Dipole-allowed excited states of N_2: Potential energy curves, vibrational analysis, and absorption intensities *J. Chem. Phys.* **115** 6438–49

[31] Lewis B R, England J P, Gibson S T, Brunger M J and Allan M 2001 Electron energy-loss spectra of coupled electronic states: effects of Rydberg-valence interactions in O_2 *Phys. Rev.* A **63** 022707

[32] Langhoff S R, Partridge H, Bauschlicher C W Jr and Komornicki A 1991 Theoretical study of the NO β system *J. Chem. Phys.* **94** 6638–43

[33] Deutsch H, Scheier P, Matt-Leubner S, Becker K and Märk T 2005 A detailed comparison of calculated and measured electron-impact ionization cross sections of atoms using the Deutsch-Märk (DM) formalism *Int. J. Mass Spectrom.* **243** 215–21

[34] Deutsch H, Scheier P, Matt-Leubner S and Becker K *et al* 2005 Erratum: A detailed comparison of calculated and measured electron-impact ionization cross sections of atoms using the Deutsch–Märk (DM) formalism (*Int. J. Mass Spectrom.* **243** 215–21) *Int. J. Mass Spectrom.* **246** 113

[35] Kosarim A, Smirnov B, Capitelli M, Celiberto R, Petrella G and Laricchiuta A 2005 Ionization of excited nitrogen molecules by electron impact *Chem. Phys. Lett.* **414** 215–21

[36] Kim Y-K and Rudd M E 1994 Binary-encounter-dipole model for electron-impact ionization *Phys. Rev.* A **50** 3954

[37] Gryziński M 1965 Classical theory of atomic collisions. I. Theory of inelastic collisions *Phys. Rev.* **138** A336–58

[38] Bauer E and Bartky C D 1965 Calculation of inelastic electron-molecule collision cross sections by classical methods *J. Chem. Phys.* **43** 2466–76

[39] Celiberto R, Capitelli M and Cacciatore M 1990 Electron impact direct dissociative-ionization cross sections from vibrationally excited H_2 molecules and translational energy distribution functions of protons *Chem. Phys.* **140** 209–15

[40] Wünderlich D 2021 Vibrationally resolved ionization cross sections for the ground state and electronically excited states of the hydrogen molecule and its isotopomeres *At. Data Nucl. Data tables* **140** 101424

[41] Scarlett L H, Savage J S, Fursa D V, Bray I, Zammit M C and Schneider B I 2021 Convergent close-coupling calculations of electrons scattering on electronically excited molecular hydrogen *Phys. Rev.* A **103** 032802

[42] Kosarim A, Smirnov B, Capitelli M, Laricchiuta A and Paniccia F 2006 Electron impact ionization cross sections of vibrationally and electronically excited oxygen molecules *Chem. Phys. Lett.* **422** 513–7

[43] Laricchiuta A, Capitelli M, Celiberto R and Colonna G 2006 *Dissociation and ionization cross sections and rate coefficients of air molecules by electron impact: the role of vibrational energy* AIAA Paper 2006-2898

[44] Shemansky D E and Liu X 2005 Evaluation of electron impact excitation of $N_2X^1\Sigma_g^+(0)$ into the $N_2^+X^2\Sigma_g^+(v)$, $A^2\Pi_u(v)$, and $B^2\Sigma_u^+(v)$ states *J. Geophys. Res. Space Phys.* **110** A07307

[45] Cacciatore M, Capitelli M and Gorse C 1982 Non-equilibrium dissociation and ionization of nitrogen in electrical discharges: the role of electronic collisions from vibrationally excited molecules *Chem. Phys.* **66** 141–51

[46] Hwang W, Kim Y-K and Rudd M E 1996 New model for electron-impact ionization cross sections of molecules *J. Chem. Phys.* **104** 2956–66

[47] Huo W M 2001 Convergent series representation for the generalized oscillator strength of electron-impact ionization and an improved binary-encounter-dipole model *Phys. Rev.* A **64** 042719

[48] Aarts J F M and De Heer F J 1970 Emission cross sections of the $A^2\Pi$ and $B^2\Sigma^+$ states of CO^+ *Physica* **49** 425–40

[49] Itikawa Y 2015 Cross sections for electron collisions with carbon monoxide *J. Phys. Chem. Ref. Data* **44** 013105

[50] Burke P G 2011 *R-Matrix Theory of Atomic Collisions: Application to Atomic, Molecular and Optical Processes* (Berlin: Springer)

[51] Tennyson J 2010 Electron–molecule collision calculations using the R-matrix method *Phys. Rep.* **491** 29–76

[52] Carr J, Galiatsatos P, Gorfinkiel J D, Harvey A, Lysaght M, Madden D, Mašín Z, Plummer M, Tennyson J and Varambhia H 2012 UKRmol: a low-energy electron- and positron-molecule scattering suite *Eur. Phys. J.* D **66** 58

[53] Mašín Z, Benda J, Gorfinkiel J D, Harvey A G and Tennyson J 2020 UKRmol+: A suite for modelling electronic processes in molecules interacting with electrons, positrons and photons using the R-matrix method *Comput. Phys. Commun.* **249** 107092

[54] Sanna N and Gianturco F A 1998 Differential cross sections for electron/positron scattering for polyatomic molecules *Comput. Phys. Commun.* **114** 142–67

[55] Tennyson J and Noble C J 1984 RESON–A program for the detection and fitting of Breit-Wigner resonances *Comput. Phys. Commun.* **33** 421–4

[56] Little D A, Tennyson J, Plummer M, Noble C J and Sunderland A G 2017 TIMEDELn: A programme for the detection and parametrization of overlapping resonances using the time-delay method *Comput. Phys. Commun.* **215** 137–48

[57] Laporta V, Tennyson J and Celiberto R 2016 Carbon monoxide dissociative attachment and resonant dissociation by electron-impact *Plasma Sources Sci. Technol.* **25** 01LT04

[58] Laporta V, Chakrabarti K, Celiberto R, Janev R K, Mezei J Z, Niyonzima S, Tennyson J and Schneider I F 2017 Theoretical resonant electron-impact vibrational excitation, dissociative recombination and dissociative excitation cross sections of ro-vibrationally excited BeH^+ ion *Plasma Phys. Control. Fusion* **59** 045008

[59] Werner H-J, Knowles P J, Lindh R, Manby F R and Schütz M *et al* 2009 MOLPRO, version 2 009.1, a package of *ab initio* programs

[60] Faure A, Gorfinkiel J D, Morgan L A and Tennyson J 2002 GTOBAS: fitting continuum functions with Gaussian-type orbitals *Comput. Phys. Commun.* **144** 224–41

[61] Gorfinkiel J D and Tennyson J 2004 Electron-H_3^+ collisions at intermediate energies *J. Phys. B: At. Mol. Opt. Phys.* **37** L343–50

[62] Gorfinkiel J D and Tennyson J 2005 Electron impact ionization of small molecules at intermediate energies: the molecular R-matrix with pseudostates method *J. Phys. B: At. Mol. Opt. Phys.* **38** 1607–22

[63] Halmová G, Gorfinkiel J D and Tennyson J 2008 Low and intermediate energy electron collisions with the C_2^- molecular anion *J. Phys. B: At. Mol. Opt. Phys.* **41** 155201

[64] Gianturco F A and Jain A 1986 The theory of electron scattering from polyatomic molecules *Phys. Rep.* **143** 347–425

[65] Zhang R, Faure A and Tennyson J 2009 Electron and positron collisions with polar molecules: studies with the benchmark water molecule *Phys. Scr.* **80** 015301

[66] Chu S-I and Dalgarno A 1974 Rotational Excitation of CH^+ by Electron Impact *Phys. Rev.* A **10** 788–92

[67] Darby-Lewis D, Tennyson J, Yurchenko S N and Lawson K D 2020 Vibrational resolved electron impact electronic excitation of BeH *J. Phys. B: At. Mol. Opt. Phys.* **53** 135202

[68] Lane N F 1980 The theory of electron-molecule collisions *Rev. Mod. Phys.* **52** 29–119

[69] Yurchenko S N, Lodi L, Tennyson J and Stolyarov A V 2016 A general program for calculating spectra of diatomic molecules *Comput. Phys. Commun.* **202** 262–75

[70] Laporta V, Cassidy C M, Tennyson J and Celiberto R 2012 Electron-impact resonant vibration excitation cross sections and rate coefficients for carbon monoxide *Plasma Sources Sci. Technol.* **21** 045005

[71] European Organization For Nuclear Research, OpenAIRE, Zenodo 2013 (https://zenodo.org/)

[72] Cooper B *et al* 2019 Quantemol Electron Collisions (QEC): an enhanced expert system for performing electron molecule collision calculations using the R-matrix method *Atoms* **7** 97

[73] 'AMOS Gateway', A Portal for Research and Education in Atomic and Molecular Physics 2021 (https://amosgateway.org/)

[74] Schneider B I *et al* 2020 Atomic and molecular scattering allocations in an Apache Airavata science gateway *Practice and Experience in Advanced Research Computing, PEARC '20* (New York: Association for Computing Machinery) pp 270–7

[75] da Costa R F, Varella M T d N, Bettega M H F and Lima M A P 2015 Recent advances in the application of the Schwinger multichannel method with pseudopotentials to electron-molecule collisions *Eur. Phys. J.* D **69** 159

[76] Lima M A P and McKoy V 1988 Aspects of the Schwinger multichannel variational formulation *Phys. Rev.* A **38** 501–3

[77] Kossoski F and Bettega M H F 2013 Low-energy electron scattering from the aza-derivatives of pyrrole, furan, and thiophene *J. Chem. Phys.* **138** 234311

[78] Bettega M H F, Natalense A P P, Lima M A P and Ferreira L G 1996 Note on the generation of Gaussian bases for pseudopotential calculations *Int. J. Quantum Chem.* **60** 821–24

[79] Kiataki M B, Varella M T d N, Bettega M H F and Kossoski F 2020 Shape resonances and elastic cross sections in electron scattering by CF_3Br and CF_3I *J. Phys. Chem.* A **124** 8660–7

[80] Falkowski A G, Lima M A P and Kossoski F 2020 Electronic excitation of ethanol by low-energy electron impact *J. Chem. Phys.* **152** 244302

[81] Zammit M C, Fursa D V, Savage J S and Bray I 2017 Electron–and positron–molecule scattering: development of the molecular convergent close-coupling method *J. Phys. B: At. Mol. Opt. Phys.* **50** 123001

[82] Scarlett L H, Savage J S, Fursa D V, Bray I and Zammit M C 2020 Electron-scattering on molecular hydrogen: convergent close-coupling approach *Eur. Phys. J.* D **74** 36

[83] MCCC, Molecular Convergent Close-Coupling Database 2021 (https://http://www.mccc-db.org/)

[84] Wünderlich D, Scarlett L H, Briefi S, Fantz U, Zammit M C, Fursa D V and Bray I 2021 Application of molecular convergent close-coupling cross sections in a collisional radiative model for the triplet system of molecular hydrogen *J. Phys. D: Appl. Phys.* **54** 115201

[85] Rescigno T, McCurdy C, Orel A and Lengsfield B III 1995 The complex Kohn variational method *Computational Methods for Electron Molecule Collisions* ed W M Huo and F A Gianturco (New York: Plenum) pp 1–44

[86] McCurdy C W and Rescigno T N 1989 Collisions of electrons with polyatomic molecules: electron-methane scattering by the complex Kohn variational method *Phys. Rev.* A **39** 4487–93

[87] Orel A and Larson Å 2020 Dissociative electron attachment to MgCN *Eur. Phys. J.* D **74** 15

[88] dos Santos S F, Ngassam V, Orel A E and Larson A 2016 Dissociative recombination of N_2H^+ *Phys. Rev.* A **94** 022702

[89] Gianturco F A, Thompson D G and Jain A 1995 Electron-scattering from polyatomic molecules using a single-center-expansion formulation *Computational Methods for Electron Molecule Collisions* ed W M Huo and F A Gianturco (New York: Plenum) pp 75–118

[90] Sanna N, Morelli G, Orlandini S and Baccarelli I 2019 VOLSCAT2.0: The new version of the package for electron and positron scattering off molecular targets *Comput. Phys. Commun.* **235** 366–77

[91] Song M-Y, Yoon J-S, Cho H, Karwasz G P, Kokoouline V, Nakamura Y, Hamilton J R and Tennyson J 2017 Cross sections for electron collisions with NF_3 *J. Phys. Chem. Ref. Data* **46** 043104

[92] Hamilton J R, Tennyson J, Huang S and Kushner M J 2017 Calculated cross sections for electron collisions with NF_3, NF_2 and NF with applications to remote plasma sources *Plasma Sources Sci. Technol.* **26** 065010

[93] Tennyson J *et al* 2017 QDB: a new database of plasma chemistries and reactions *Plasma Sources Sci. Technol.* **26** 055014

[94] Song M-Y, Yoon J-S, Cho H, Karwasz G P, Kokoouline V, Nakamura Y and Tennyson J 2020 'Recommended' cross sections for electron collisions with molecules *Eur. J. Phys.* D **74** 60

[95] Scarlett L H, Fursa D V, Zammit M C, Bray I and Ralchenko Y 2021 Complete collision data set for electrons scattering on molecular hydrogen and its isotopologues: II. Fully vibrationally-resolved electronic excitation of the isotopologues of H_2 $(X^1\Sigma_g^+)$ *At. Data Nucl. Data tables* **139** 101403

[96] Scarlett L H, Fursa D V, Zammit M C, Bray I, Ralchenko Y and Davie K D 2020 Complete collision data set for electrons scattering on molecular hydrogen and its isotopologues: I. Fully vibrationally-resolved electronic excitation of $H_2(X^1\Sigma_g^+)$ *At. Data Nucl. Data tables* **137** 101361

[97] Meltzer T, Tennyson J, Mašín Z, Zammit M C, Scarlett L H, Fursa D V and Bray I 2020 Benchmark calculations of electron impact electronic excitation of the hydrogen molecule *J. Phys. B: At. Mol. Opt. Phys.* **53** 145204
[98] Duxbury G, Stamp M F and Summers H P 1998 Observations and modelling of diatomic molecular spectra from JET *Plasma Phys. Control. Fusion* **40** 361–70
[99] Laporta V, Celiberto R and Tennyson J 2015 Dissociative electron attachment and electron-impact resonant dissociation of vibrationally excited O_2 molecules *Phys. Rev.* A **91** 012701
[100] Shafir D *et al* 2009 Rotational cooling of HD^+ molecular ions by superelastic collisions with electrons *Phys. Rev. Lett.* **102** 223202
[101] Faure A and Tennyson J 2001 Electron-impact rotational excitation of linear molecular ions *Mon. Not. R. Astron. Soc.* **325** 443–8
[102] Dubernet M-L *et al* BASECOL2012: A collisional database repository and web service within the Virtual Atomic and Molecular Data Centre (VAMDC) *Astron. Astrophys.* **553** A50
[103] Johnsen R and Guberman S L 2010 Dissociative recombination of H_3^+ ions with electrons: theory and experiment *Advances in Atomic, Molecular, and Optical Physics* vol 59 (New York: Elsevier) ch 3
[104] dos Santos S F, Kokoouline V and Greene C H 2007 Dissociative recombination of H_3^+ in the ground and excited vibrational states *J. Chem. Phys.* **127** 124309
[105] Song M-Y, Cho H, Karwasz G P, Kokoulinea V, Nakamura Y, Tennyson J, Faure A, Mason N J and Itikawa Y 2021 Cross sections for electron collisions with H_2O *J. Phys. Chem. Ref. Data* **50** 023103
[106] LXcat 2021 Plasma Data Exchange Project database (https://us.lxcat.net/)
[107] ALADDIN 2021 Numerical database maintained by the IAEA Nuclear Data section A + M Data Unit (https://amdis.iaea.org/databases/)
[108] QDB 2021 Quantemol-Database (https://http://www.quantemoldb.com/)
[109] NIFS 2021 National Institute for Fusion Science database (http://dbshino.nifs.ac.jp/)
[110] Dubernet M-L *et al* 2016 The Virtual Atomic and Molecular Data Centre (VAMDC) consortium for astrophysics *J. Phys. B: At. Mol. Opt. Phys.* **49** 074003
[111] Phys4Entry 2021 (http://phys4entrydb.istp.cnr.it/Phys4EntryDB/)
[112] Drake G W 2020 Accuracy in atomic and molecular data *J. Phys. B: At. Mol. Opt. Phys.* **53** 223001
[113] Chung H, Braams B, Bartschat K, Császár A, Drake G, Kirchner T, Kokoouline V and Tennyson J 2016 Uncertainty estimates for theoretical atomic and molecular data *J. Phys. D: Appl. Phys.* **49** 363002
[114] Celiberto R *et al* 2016 Atomic and molecular data for spacecraft re-entry plasmas *Plasma Sources Sci. Technol.* **25** 033004
[115] STELLAR 2021 (http://esther.ist.utl.pt/pages/stellar.html)
[116] Albert D *et al* 2020 A decade with VAMDC: results and ambitions *Atoms* **8** 76
[117] Wakelam V *et al* 2014 The 2014 KIDA network for interstellar chemistry *Astrophys. J. Suppl.* **217** 20
[118] Pitchford L C *et al* 2017 LXCat: an open-access, web-based platform for data needed for modeling low temperature plasmas *Plasma Proc. Polym.* **14** 1600098
[119] Emoto M, Murakami I, Kato D, Yoshida M, Kato M and Imazu S 2019 Improvement of the NIFS Atom and Molecular Database *Atoms* **7** 91

[120] Park J-H, Choi H, Chang W-S, Chung S Y, Kwon D-C, Song M-Y and Yoon J-S 2020 A new version of the plasma database for plasma physics in the data center for plasma properties *Appl. Sci. Convergence Technol.* **29** 5–9

[121] Hulse R A 1990 The ALADDIN atomic physics database system *AIP Conf. Proc.* **206** 63–72

[122] McElroy D, Walsh C, Markwick A J, Cordiner M A, Smith K and Millar T J 2013 The UMIST database for astrochemistry 2012 *Astron. Astrophys.* **550** A36

IOP Publishing

Plasma Modeling (Second Edition)
Methods and applications
Gianpiero Colonna and Antonio D'Angola

Chapter 23

Rate coefficients for energy transfer and chemical reactions in heavy particle collisions

Igor V Adamovich, Fabrizio Esposito and Sergey O Macheret

Weakly ionized low temperature plasmas, especially at low or moderate gas pressures, are in thermodynamic non-equilibrium. Not only is the electron temperature much higher than that of the neutral molecules and atoms, but different molecular degrees of freedom are not in equilibrium with each other. Most commonly, the populations of vibrational states of molecules greatly exceed those in thermodynamic equilibrium, which can be approximately expressed as the vibrational temperature being much higher than the translational temperature. The high vibrational populations or temperatures are due to the large cross-sections of vibrational excitation by electron impact and to the slow vibrational–translational energy transfer. In such non-equilibrium plasmas, chemical reactions of neutral heavy particles (molecules and atoms) are significantly affected by the vibrational–translational non-equilibrium. The redistribution of vibrational energy in vibration–vibration (V–V), vibration–translation (V–T), and other processes affects the populations of excited states and, through that, the chemical reaction rates. Although quantum-mechanical (QM) calculations of chemical reaction cross-sections have progressed significantly, they are still limited to a few very simple systems and/or collinear collisions. At the same time, over a wide range of conditions of interest, the motion of particles during nonreactive and reactive collisions can be described either classically, or as a combination of classical relative motion of colliding particles and quantum vibrations of the molecules. Combination of the classical approach with the *potential energy surfaces* (PES) increasingly available from *ab initio* quantum chemistry computations make the *quasiclassical trajectory* (QCT) modeling of energy transfer and chemical reactions an increasingly powerful tool. Alternative approaches to the energy transfer and reactions in non-equilibrium include analytical and semi-analytical techniques, such as those based on the semiclassical *forced harmonic oscillator* (FHO) theory, on classical impulsive approximation, or on the vibronic term model. All these approaches are reviewed in this chapter.

doi:10.1088/978-0-7503-3559-1ch23

23.1 Semiclassical analytic theory of vibrational energy transfer in molecular collisions

The theory of state-to-state vibrational energy transfer in nonreactive collisions of small molecules and atoms have been covered in a number of authoritative reviews in the past [1–4]. Briefly, the methods used for the analysis of V–T, vibrational–rotational (V–R), and V–V energy transfer in molecular collisions can be separated in three groups. Purely quantum calculations (e.g. [5]) are rigorous but computationally expensive, due to a large number of coupled vibrational and rotational states, such that their use remains rather limited.

In semiclassical trajectory calculations, the vibrational energy mode is treated quantum mechanically, and the vibrational energy transfer is uncoupled from the classical rotational/translation trajectory [4]. This approximation is justified for temperatures much higher compared to the characteristic rotational temperature. It violates the energy conservation, which is enforced using the *symmetrization* of the collision kinetic energy between the initial and the final values. The validity of this procedure has been demonstrated by comparing the semiclassical and purely quantum calculations for several molecular systems [6]. In the semiclassical method, the vibrational eigenfunctions of the colliding molecules are used as the expansion basis for the vibrational wavefunctions before and after collisions, such that reactive collisions are not considered. An extensive database of the state-specific V–T/R and V–V rate coefficients for diatomic molecules such as N_2, O_2, and CO, ranging from cryogenic temperatures to several thousand degrees K, has been accumulated over the last three decades [4, 7–11].

*Q*CT calculations incorporate full coupling between all three energy modes, and may also include reactive and electronically nonadiabatic collisions, if accurate *potential energy surfaces* (PES) are available. In this method, the state-specific vibrational energy transfer cross-sections are evaluated from the ratio of the number of trajectories resulting in the energy transfer into a specific discrete vibrational energy *bin* to the total number of trajectories considered. This limits the accuracy of the modeling predictions at low collision energies, when the energy transfer cross-sections may be very low. Recently, the QCT method, based on accurate *ab initio* PES has been used for predictions of the state-to-state energy transfer rates in N_2–N [12, 13], O_2–O [14–16], O_2–Ar [17], O_2–N [18], N_2–O [19], and N_2–N_2 [20, 21] systems at high collision energies typical for high-enthalpy hypersonic flows.

Analytic theories of vibrational energy transfer, either classical or semiclassical [1] provide closed-form expressions for the energy transfer cross probabilities. Most of these models assume a collinear collision model of harmonic oscillators and a simplified (e.g. repulsive exponential) intermolecular potential, tractable analytically. Both the classical and semiclassical models are based on the perturbation theory and valid at the low values of the vibrational energy transferred per collision, and low-probability single-quantum transitions, respectively. Much work has been done to incorporate the effects of the oscillator anharmonicity, non-collinear orientation of the colliding partners, non-zero impact parameter collisions, and repulsive-attractive potentials, including the long-range dipole–dipole interaction of polar molecules, such as CO and CO_2. The semiclassical analytic theory has been extended to incorporate close coupling of multiple vibrational states of molecules [22–24], which made it applicable for high-energy collisions and multi-quantum transitions. The development of semiclassical and quasiclassical trajectory numerical methods reduced the demand for the approximate theory predictions. The strongest limitation of analytic models is the use of simplified potentials, compared to the more accurate PES employed in the trajectory calculations. However, the availability of the higher fidelity methods, e.g. [25, 26] also made possible the critical analysis of the analytic theory, to identify its deficiencies and isolate the dominant collision dynamics features controlling the energy transfer cross-sections. Some of these features are discussed below.

The critical factor affecting the predictive capability of a semiclassical theory of vibrational energy transfer over a wide range of temperatures is the close coupling of multiple vibrational states. This is essential for analyzing multi-quantum transitions, which become significant at high collision energies and vibrational quantum numbers. This has been done in [27, 28], based on the *forced harmonic oscillator* (FHO) theory developed in [22, 23] for V–T energy transfer and in [24] for V–V–T processes. For diatom–atom collisions,

$$AB(i) + M \rightarrow AB(f) + M$$

where i, f are the initial and final vibrational quantum numbers, the FHO transition probabilities are expressed in terms of a single trajectory integral ε, the average vibrational energy transferred per collision, $P_{i \rightarrow f}(\varepsilon)$ [23].

For collisions of diatomic molecules,

$$AB(i_1) + CD(i_2) \rightarrow AB(f_1) + CD(f_2)$$

the energy transfer probabilities depend on two trajectory integrals, ε and ρ, $P_{i_1 \rightarrow f_1}^{i_2 \rightarrow f_2}(\varepsilon, \rho)$ [24]. In the limit of low collision energies, $\varepsilon = P_{1 \rightarrow 0}^{0 \rightarrow 0}$. For head-on collinear collisions governed by a repulsive exponential potential, $U(R) = Ae^{-\alpha R}$, and repulsive-attractive Morse potential, $U(R) = A(e^{-\alpha R} - 2e^{-\alpha R/2})$, where R is the center-of-mass distance, the closed-form expressions for the trajectory integrals ε and ρ are well known [1].

In references [27, 28], this theory was applied to frequency-corrected harmonic oscillators (with harmonic wavefunctions but anharmonic energy spectrum). The infinite series expressions for the energy transfer probabilities, $P(i \rightarrow i - n)$ and $P(i_1, i_2 \rightarrow i_1 - n, i_2 + n)$, where n is the number of quanta transferred per collision, were replaced by the more tractable asymptotic analytic forms. The adjustable *steric factors*, $0 < S_{VT}, S_{VV} < 1$, were used to account for the effect of non-collinear collisions resulting in V–T and V–V energy transfer, respectively. These corrections yielded surprisingly good agreement with the higher fidelity 3D semiclassical trajectory calculations for N_2–N_2 and O_2–O_2 [9, 29, 30], both for V–T and V–V processes. However, this encouraging result left room for skepticism, since the use of steric factors indicated lack of understanding of collision dynamics, and therefore little predictive capability. Also, the FHO rates and the 3D trajectory rates were calculated for a simplified Morse potential, such that the comparison was not entirely adequate. The state-specific FHO V–T and V–V rate coefficients, including multi-quantum jumps, were used in computational fluid dynamics (CFD) modeling of state-specific vibrational relaxation and dissociation in a non-equilibrium hypersonic shock layer [31]. State-specific modeling of vibrational kinetics was also employed in CFD simulations of a hypersonic shock layer in reference [32] (see also chapter 10).

The second critical advance of this theory, which demonstrated its accuracy without the use of adjustable parameters, was accounting for the molecular rotation [33, 34]. Briefly, this effect results in the periodic modulation of the time-dependent interaction potential during the collision. For a diatom–diatom collision governed by a purely repulsive intermolecular potential, this yields

$$U[R(t),\, r_i,\, \theta_i(t)] = 4Ae^{-\alpha R(t)}\cosh\left[\frac{\alpha r_1}{2}\cos\theta_1(t)\right]\cosh\left[\frac{\alpha r_2}{2}\cos\theta_2(t)\right] \tag{23.1}$$

where r_1, r_2 are the distances between the nuclei, and θ_1, θ_2 are the rotation angles relative to the center-of mass radius vector R (see figure 23.1). Because of this, for some values of the rotational energy and phase, the time-dependent perturbation of the molecules during the collision becomes more *sudden*, increasing the vibrational energy transition probability. The modulation can be calculated easily if the rotation is assumed to be free, i.e. the rotational angular momenta remain constant during the collision. This approximation is similar to uncoupling of the vibrational energy transfer from the classical translational–rotational trajectory, used in semiclassical scattering theories. With the rotational modulation taken into account, the time-dependent perturbation becomes $U(t) = \frac{E\gamma^2}{\cosh^2(\gamma\alpha u t)}$, where E is the total collision energy, $u = \sqrt{E/2m}$ (m is the collision reduced mass), and γ is a closed-form analytic factor which depends on the rotational energies, $E_{\text{rot},i}$, impact parameter, b, and molecular orientation angles, θ_i and ϑ_i (see figure 23.1). The trajectory integral ε, which controls the entire set of V–T probabilities, becomes

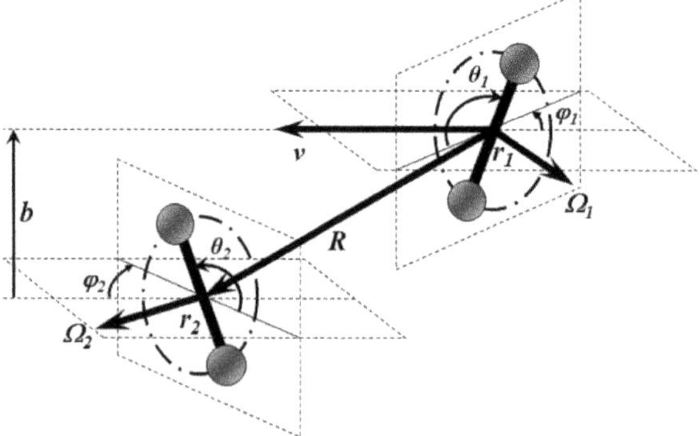

Figure 23.1. Schematic of diatom–diatom collision geometry.

$$\varepsilon(E, E_{\mathrm{rot},i}, b, \theta_i, \phi_i) = \frac{\theta'}{4\theta}\frac{\cos^2\theta\cos^2\varphi}{\sinh^2\left(\dfrac{\pi\omega}{\gamma\alpha u}\right)} \tag{23.2}$$

where ω is the vibrational frequency, $\theta = \hbar\omega/k$ is the characteristic vibrational temperature, and $\theta' = (4\pi^2\omega^2 m)/(\alpha^2 k)$. For head-on collinear collisions, $\gamma = 1$, and equation (23.2) is reduced to the well-known one-dimensional theory result. Thus, the effect of molecular rotation is expressed in terms of a single parameter γ, obtained from the analysis of collision dynamics.

Comparison of the *forced harmonic oscillator—free rotation* (FHO-FR) model with the semiclassical trajectory calculations for N_2–N_2 [25, 26], using the same intermolecular potential, exhibits good agreement [34]. Note that at low collision energies, the probability of V–T relaxation $P(1, 0 \rightarrow 0, 0) \approx \varepsilon$ reaches maximum for a non-collinear collision orientation of two rotating molecules, $\theta_1 = \theta_2 = \pi/4$, $\phi_1 = \phi_2 = 0$, $E_{\mathrm{rot},1}/E = E_{\mathrm{rot},2}/E = 1/6$, and $b = 0$ (at $\gamma_{\max} = 3/2$), when the time-dependent perturbation during the collision becomes the most sudden. The major effect of the potential modulation by rotation dominates the collision dynamics, greatly exceeding the effect of orientations of nonrotating molecules. As shown in [34], for trajectories near the *optimum* rotational energy values, the V–T probability can exceed that for the collinear trajectory by over an order of magnitude. Therefore, the use of a *steric factor* to correct the predictions of one-dimensional collision models is misleading.

Another major effect of collision three-dimensionality, which is sometimes overlooked, is the removal of non-monotonous dependence of the rates on the collision energy, predicted by the one-dimensional theory. For example, the one-dimensional FHO V–T rates in diatom–diatom collisions decrease at high collision energies, e.g. $P(1, 0 \rightarrow 0, 0) = \varepsilon e^{-2\varepsilon}$. For higher vibrational quantum numbers, the V–T rates begin to decrease at progressively lower collision energies. In fact, this does not occur in reality. As the total collision energy E increases, the optimum configurations of the phase space variables $(\theta_i, \varphi_i, E_{\mathrm{rot},i}, b)$ for which the transition probability is maximized, simply shift toward higher impact parameters. Therefore the V–T energy transfer probability continues to increase with the total collision energy. For this reason, accurate predictions of the energy transfer at high temperatures and vibrational quantum numbers (e.g. vibrational relaxation and dissociation behind hypersonic shock waves) require the use of a 3D collision model, such as FHO-FR [35]. The 3D FHO-FR theory is in good agreement with the *exact* semiclassical trajectory calculations in nitrogen, including the cross-section dependence on collision angles, impact parameter, rotational energy, and total collision energy [33, 34]. Comparison with such *exact* numerical solutions are essential for validation of approximate analytic theories, before they can be compared with the experimental data. This theory has also been applied to predict the state-specific rates of molecular dissociation [36], although in this case no head-to-head comparison with the higher fidelity trajectory calculations have been made.

The third critical element of a 3D semiclassical analytic theory is predicting the rotational energy transfer, which is ignored in the FHO-FR model assuming free rotation. This is essential for the high-fidelity simulations of flows in a hypersonic shock layer, where rotational non-equilibrium may be significant and rotational–translational (R–T) processes may well contribute to dissociation. Classical trajectory analysis of R–T energy transfer in 3D molecular collisions appears to be analytically intractable. However, there is an approximate *stochastic* analytic solution for the R–T probabilities [37],

$$P(j_0 \to j) = \frac{1}{\sqrt{4\pi S}} e^{-(j_0-j)^2/4S} \tag{23.3}$$

In equation (23.3), j_0 and j are the initial and the final rotational quantum numbers, and S is an integral over the collision trajectory, evaluated numerically in [37]. For the repulsive interaction potential, the trajectory integral S is controlled by a short part of the trajectory near the point of the closest approach, $R_0 \gg 1/\alpha$, and is proportional to the energy of the relative translational motion in the radial direction [38],

$$S(j_0 \to j, E, B) = \frac{1}{8}\left(\sqrt{\frac{E}{B}} - j_0\right)^2 \left[1 - \left(\frac{b}{R_0}\right)^2\right] \tag{23.4}$$

Here, B is the rotational constant. This greatly simplifies the R–T energy transfer model, which is in good agreement with the 3D trajectory calculations over a wide range of collision parameters (e.g. see figure 23.2). The V–R–T probabilities, approximated as follows, $P_{\mathrm{VRT}}(v, j_1 \to w, j_2) = P_{\mathrm{RT}}(j_1 \to j_2)P_{\mathrm{VT}}(v \to w)$, and

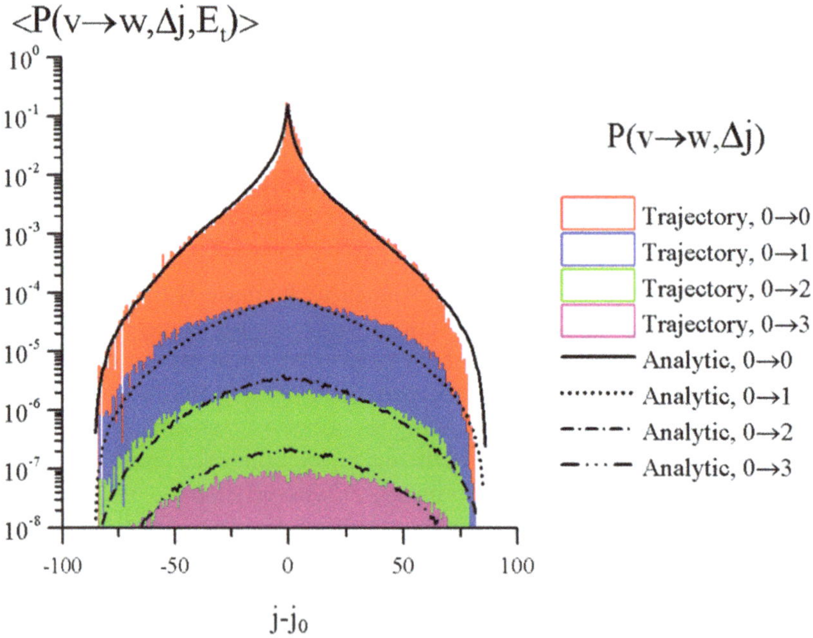

Figure 23.2. Analytic V–R–T transition probabilities for N_2–N_2 ($v = 0, j_0 \to w = 0 - 3, j$), versus 3D semiclassical trajectory calculations. Total collision energy $E = 3 \times 10^4$ cm^{-1}. Reproduced with permission from [38], copyright 2014 AIP.

averaged over the initial and final rotational quantum numbers, also exhibit fairly good agreement with the trajectory calculations, as illustrated in figure 23.2. Note that the *factorization* of the V–R–T probabilities does not account for the full coupling between the rotational and vibrational energy modes, the same way all semiclassical theories decouple the classical trajectory from the quantized vibrational energy transfer.

The wide use of the semiclassical energy transfer theory discussed above is hindered by several issues requiring additional analysis. The most significant limitation is not accounting for reactive collisions (such as atom exchange reactions for N_2–N and O_2–O), which are shown to contribute significantly into the overall V–T probabilities [16, 21]. Although the rigorous analysis of the atom exchange collisions is likely to be analytically intractable, there may well be a stochastic solution similar to the one in the R–T energy transfer. Second, the theory predicts the analytic energy transfer cross-sections as functions of individual collision parameters (orientation angles, rotational energies, impact parameter, and total energy). At high collision energies and vibrational quantum numbers, multiple combinations of these parameters result in effective energy transfer, instead of a single *optimum* configuration discussed above. This makes the analytic averaging over the phase space, to obtain the closed-form analytic cross-sections versus total collision energy, challenging. At this time, the analytic cross-sections are averaged over the phase space numerically, using a Monte Carlo method [33, 34]. Third, predicting the energy transfer in collisions affected by the long-range (dipole–dipole and dipole–quadrupole) molecular attraction, which may result in complex *orbiting trajectories* at low collision energies, remains an unresolved problem and requires additional work. This issue is mostly important for collisions of polar molecules such as CO and CO_2 at relatively low temperatures, up to several hundred degrees. In high-enthalpy hypersonic flows, this effect is overshadowed by the strong repulsion and becomes relatively minor.

Finally, an obvious question is whether there is a need to know all state-specific V–R–T energy transfer rates, including multi-quantum jumps. Even at R–T equilibrium, the number of these rates for diatom–diatom collisions can be up to $\approx 10^7$. Incorporating the state-to-state vibrational–rotational energy transfer makes the use of the full rate matrix completely unfeasible. In recently developed reduced-order *coarse grain* models [20, 39], some of the quantum states are grouped together, a method that greatly reduces the computational cost but somewhat lacks rigor. An alternative approach would be to obtain closed-form, *self-similar* analytic expressions for the integral parameters such as the energy transfer moments, playing the role of the diffusion coefficients in the vibrational energy space [40]. Then the vibrational energy distribution could be predicted by solving a single Fokker–Planck type diffusion equation, instead of a state-specific master equation. Previously, the accuracy of this approach was shown to be limited by the multi-quantum relaxation processes [41]. This makes the energy transfer nonlocal, such that the diffusion approximation is no longer applicable. The use of higher order expansion of the distribution function, and higher order energy transfer moments, may circumvent this problem and expand the applicability of the quasi-continuous diffusion method.

In summary, the approximate 3D analytic theory of vibrational energy exchange incorporates the major features of collision dynamics controlling the energy transfer cross-sections. Although the fidelity of this theory is lower compared to QCT calculations using an accurate PES, at high collision energies the *exponential repulsive wall* in the intermolecular potential is likely to be the dominant factor. This theory also has greater flexibility, does not require an extensive computational effort for the PES development, and is straightforward to use in non-equilibrium flow codes (CFD and DSMC). The development of the collision dynamics based formalism for the reactive collisions may well make this approach competitive with the state-of-the-art QCT calculations.

23.2 Chemical reaction rates for thermally non-equilibrium plasmas

The focus of this section is on chemical processes that occur at the conditions of vibrational non-equilibrium, when the vibrational temperature of molecules is greater (as occurs in electric discharges and in high-enthalpy expansion flows) or smaller (e.g. in hypersonic shock layers) than the translational–rotational one. In general, though, the following types of chemical processes in non-equilibrium plasmas with participation of only heavy particles (molecules, atoms, some possibly in their excited states), but not electrons, can be distinguished:

1. Associative ionization and Penning ionization:

$$A + B \rightarrow AB^{+} + e^{-}$$

$$\left.\begin{matrix} A^{\star} + B \\ A^{\star} + B^{\star} \end{matrix}\right\} \rightarrow AB^{+} + e^{-}$$

$$A^{\star} + B \rightarrow A + B^{+} + e^{-}$$

   Here A and B are atoms, molecules, or radicals in their ground electronic states, and $A^{\star}$ and $B^{\star}$ are the corresponding electronically excited states. These processes are important for ionization kinetics in many electric discharges and high-enthalpy flows (e.g. behind strong shock waves). However, in this chapter, we will not consider such processes.
2. Conventional exothermic chemical reactions of radicals, atoms, and molecules, and ion–molecule reactions. The rate coefficients of such reactions are determined mostly by the R–T temperature of the reactants. Thus, the rate coefficients can be taken as those at thermal equilibrium, and therefore we will not consider such processes in this chapter.
3. Dissociation of molecules in heavy particle collisions:

$$AB + C \rightarrow A + B + C$$
$$AB + CD \rightarrow A + B + CD$$

4. Exchange reactions:

$$AB + C \rightarrow A + BC$$

Both dissociation and exchange reactions are significantly affected by the vibrational non-equilibrium, and thus are considered in this section.

The purpose of a theoretical model of dissociation or exchange reaction is to find vibrational state-specific or vibrational energy-specific rate coefficient $k(E_v, T)$, where E_v is the vibrational energy viewed as a continuous variable. The two-temperature rate coefficient $k(T_v, T)$ can then be found as

$$k(T_v, T) = \int_0^{\mathcal{D}} k(E_v, T)f(E_v, T)dE_v \tag{23.5}$$

where $f(E_v, T)$ is the vibrational energy distribution function normalized as

$$\int_0^{\mathcal{D}} f(E_v, T)dE_v = 1 \tag{23.6}$$

It is important to note that the chemical reaction itself can affect $f(E_v, T)$, and thus the calculation of $k(T_v, T)$ should be performed self-consistently. This is particularly true for dissociation that depletes the upper vibrational states (and, at very high temperatures, even the lower states), thus reducing its own rate; this effect is mitigated at the late stages of the dissociation process by the reverse reaction of recombination which populates the upper vibrational states. In contrast, exchange reactions, as a rule, do not significantly disturb the vibrational distribution, justifying uncoupled calculation of the two-temperature rate.

23.2.1 Non-equilibrium dissociation: classical impulsive theory

In most simulations of non-equilibrium hypersonic flows, the models used for dissociation probabilities in DSMC (direct simulation Monte Carlo) and two-temperature rate coefficients in CFD are empirical. Examples include Park model for CFD [42] and *total collision energy* (TCE) model for DSMC [43]. These models are simple and therefore computationally efficient, but they are not based on the physics of reactive collisions and can be quite inaccurate. On the other hand, with advances in quantum chemistry that can now generate potential energy surfaces (PES) for simple colliding molecules and the rapidly growing computational power, QCT simulations are becoming more accessible, but incorporation of QCT-generated data into CFD and DSMC still poses challenges.

However, in addition to the two extremes-empirical models and QCT-based approximations, there is a third approach: theoretical (physical) models based on reasonable and physically justified assumptions and approximations. Algebraic, i.e. closed-form, formulas based on such physical models can be almost as simple as Park and TCE models and can be conveniently used in both DSMC and CFD, while also ensuring accuracy of the modeling and providing insight into the underlying physics.

In this chapter, we review one such approach: the classical impulsive mode of dissociation, known also as the Macheret–Fridman model [44–49]. The theoretical model was originally proposed by Macheret *et al* [44, 45] based on the classical impulsive limit. The Macheret–Fridman (MF) model [44–49] assumes that translational, rotational, and vibrational motion can be described by classical mechanics. Additionally, because of the high velocity required for dissociation especially from low vibrational states, the Massey parameter $= \Delta UR/\hbar V \ll 1$, where ΔU is the energy transferred into the vibrational mode, R is the characteristic radius of interaction, V is the relative velocity of the colliding particles, and $\hbar$ is the reduced Planck constant. Therefore, the collision can be viewed as instantaneous, i.e. close to the impulsive limit. The details of PES are not important in such an approximation; additionally, the molecules do not experience three-body interactions before and after the collision. The last approximation is that the reaction rates are only determined by the energy threshold and the behavior of the cross-sections just above the threshold. That follows from the exponential dependence of the distribution function on energy.

In the classical impulsive limit, the energy transfer from kinetic energy to vibrational energy of dissociating molecule is determined by the masses of colliding particles and by the set of angles and phases in the instant when the chemical rearrangement occurs. Detailed derivations of closed-form reaction probabilities and rates can be found in [45, 46].

By making post-collision vibrational energy of molecule AB equal to the dissociation energy $\mathcal{D}$, we can obtain the threshold function $F = E_{t,\min}$, which is the minimum collisional energy for the dissociation reaction $AB + CD \rightarrow A + B + CD$ to occur. The rotational energy of dissociating molecule E_r contributes to the dissociation by its component along the direction of collision. However, rotational energy also increases the effective dissociation energy due to the centrifugal barrier, which arises as the rotational energy decreases when the molecule is stretched during dissociation. The strategy proposed in [45] is to use an effective dissociation energy $\mathcal{D}_{\text{ef}}$, that includes both effects. $\mathcal{D}_{\text{ef}}$ is evaluated with a long-range potential energy curve $V(r) = -2\mathcal{D}(R_{\text{eq}}/r)^6$, where R_{eq} is the equilibrium distance of molecular bond and r is the internuclear distance,

$$\mathcal{D}_{\text{ef}} = \mathcal{D} - E_r + \frac{2E_r^{3/2}}{3\sqrt{6\mathcal{D}}} \tag{23.7}$$

In the MF model for CFD, the first step is to determine the optimum configuration of the reactants with the minimum threshold kinetic energy. The probability of dissociation is then obtained based on Taylor expansion of the threshold function near the optimum configuration [45, 46]. Multi-temperature reaction rates coefficients can then be calculated by integrating the probability with distribution functions of translational, rotational and vibrational energy. To be consistent with basic assumptions in CFD, Maxwell–Boltzmann distribution functions are used as a first approximation. However, it should be noted that dissociation reaction favors high vibrational levels and thus depletes their population, making the vibrational energy distribution (VDF) deviate from Boltzmann distribution. The prediction of reaction rates with non-Boltzmann distribution requires state-to-state calculations, which is possible but so far has not been done for the MF model.

It is important to note that the MF model analyzes the energy transformation in reactions and calculates the energy threshold and the probability of dissociation assuming that the collision has occurred. Thus, in order to obtain the rate coefficient or the cross-section of dissociation, the dissociation probability calculated by the model should be multiplied by the *total* rate coefficients or total cross-sections of collisions; such total cross-sections or collisional rate coefficients should be taken from sources other than the model itself. In references [44–49], instead of attempting to calculate the total cross-sections and the collisional rates $k_{\text{coll}}(T)$, the focus was on the non-equilibrium factor $Z = k(T, T_v)/k_a(T)$, i.e. on the behavior of the dissociation rate in thermally non-equilibrium conditions normalized by its thermally equilibrium (Arrhenius) value,

$$k_a(T) = AT^n e^{-\mathcal{D}/T} \tag{23.8}$$

where the dissociation energy $\mathcal{D}$ is in Kelvin. Accordingly, we use the condition $k(T = T_r = T_v) = k_a(T)$. In such context, the multi-temperature rates are calculated by integrating the reaction probability multiplied by the translational, rotational, and vibrational distribution functions over the translational, rotational, and vibrational energies.

The extreme complexity of the reaction probability and the integration itself precludes the direct analytical integration, and rough approximation was employed in [44, 45]. The two-temperature rate coefficient was presented as the sum of contributions from low, $E_v \leqslant \alpha\mathcal{D}_{\text{ef}}$, and high, $E_v > \alpha\mathcal{D}_{\text{ef}}$, vibrational levels:

$$k(T, T_v) = k_l + k_h \tag{23.9}$$

Here α is the mass factor that, for the case $m_A = m_B = m$ and $m_C = m_D = M$, is

$$\alpha = \left(\frac{m}{m + M}\right)^2 \tag{23.10}$$

A simple version of the steepest descend method can be applied to calculate the rates for $E_v \leqslant \alpha\mathcal{D}_{\text{ef}}$. The method approximates the pre-exponential factor by its value in the vicinity of the exponent's maxima. The two-temperature rate coefficient for dissociation from low vibrational states is:

$$k_l = AT^n L \exp\left[\frac{-\mathcal{D}}{T} + \Delta\mathcal{D}\left(\frac{1}{T_a} - \frac{1}{T}\right)\right] \tag{23.11}$$

and the contribution from the high vibrational levels is roughly

$$k_h = AT^n(1 - L)\frac{1 - e^{h\nu/T_v}}{1 - e^{h\nu/T}} e^{-\mathcal{D}/T} \tag{23.12}$$

Here ν is the vibrational frequency of the dissociating molecule. The formulas for L in both diatom–atom and diatom–diatom cases can be found in references [46, 49]. In equations equations (23.11) and (23.12), T_a is the *average* temperature:

$$T_a = \alpha T_v + (1 - \alpha)T \tag{23.13}$$

and ΔD is the approximate value of the reduction of effective dissociation energy due to rotational motion, $\Delta \mathcal{D} = 3b\alpha^2\mathcal{D}$, where b is a numerical factor that can be calculated [45]. Within this simple version of the MF model, the average vibrational energy removed by dissociation reaction is:

$$E_{v,\mathrm{rem}} = \frac{\left[\alpha(\mathcal{D} - \Delta\mathcal{D})\left(\frac{T_v}{T_a}\right)^2\right]k_l + \mathcal{D}k_h}{k_l + k_h} \tag{23.14}$$

The non-equilibrium factor calculated with the MF model shows a good agreement, within a factor of 3 or better [46, 49], with those computed by QCT and inferred from shock tube experiments [50–54]. The average vibrational energy removed by dissociation is somewhat overpredicted by the MF model [46, 49]. Relaxing some of the simplifying assumptions made in the integration can potentially yield substantially more accurate results. It is worth noting that the version of MF model for DSMC [47, 48] deals directly with the threshold kinetic energy for dissociation and reaction probabilities without the need for approximate integration to obtain the multi-temperature rates. This MF-DSMC model is in much better agreement with QCT and experimental data [47, 48], confirming the fundamental accuracy the MF approach.

23.2.2 Exchange reactions

The key question in the development of models for exchange reactions in thermodynamic non-equilibrium is the efficiency of vibrational energy in lowering the activation barrier for translational energy. In general, vibrational energy is quite efficient in endothermic reactions and much less so in exothermic processes.

In a limited number of cases, QCT computations of exchange reactions are available. Notable examples are the key reactions in air chemistry [55–57]:

$$N_2 + O \rightarrow NO + N \tag{23.15}$$

$$O_2 + N \rightarrow NO + O \tag{23.16}$$

However, one must be cautious when applying such QCT results to low-temperature plasmas where the vibrational temperature is high but the translational–rotational one is low. Indeed, as shown in [58–60], the very mechanism of reaction (equation (23.15)) changes from electronically-adiabatic at high temperatures to electronically-nonadiabatic (curve-crossing) one at low temperatures ($T < 1000$ K or so). Theoretical formulas for the rate coefficient of equation (23.15) as a function of vibrational energy or temperature [58, 59], derived for the electronically-non-adiabatic mechanism, are consistent with the room-temperature experiments [61].

If no theory or QCT computations exist for an exchange reaction, use of the so-called α-model, also known as the Fridman–Macheret model [42, 60, 62], can be recommended. The quantity α is defined so that the reaction activation energy with respect to translational–rotational energy is reduced by αE_v, where E_v is the vibrational energy of the reactant. The parameter α reflects the efficiency of vibrational energy in reducing the activation barrier and should not be confused with the mass factor α used in the MF model for dissociation.

The Fridman–Macheret model is based on analysis of transitions between vibronic terms of the reactants and products and yield the following expression for α:

$$\alpha = \frac{\gamma_+ E_{a+}}{\gamma_+ E_{a+} + \gamma_- E_{a-}}$$

Here E_{a+} and E_{a-} are the activation energies of the forward and reverse reactions, and γ_+ and γ_- are the slopes of vibronic terms for the reactants and products. Assuming further that $\gamma_+ \approx \gamma_-$ yields a simplified formula:

$$\alpha = \frac{E_{a+}}{E_{a+} + E_{a-}}$$

that can be recommended in the absence of more detailed information.

23.3 Quasiclassical trajectory method

The quasiclassical trajectory method (QCT) is a popular method for performing calculations of cross-sections and rate coefficients involving molecular collisions. The whole collision is treated in the framework of classical mechanics, using the potential energy surface (PES) suited for the collisional system studied. The PES is obtained with semi-empirical [63] or *ab initio* methods [64], and it describes the electronic interaction among all the nuclei involved in the collision, following this interaction from reactants to products. This picture neglects the possibility of a change in electronic state (it is *adiabatic*), which is realistic when all the other possible electronic states are sufficiently far in energy from the one considered in the PES used, at least in the range of energies explored by the dynamical simulation. If this not the case, multiple PES and couplings among them should be obtained for the studied system, and a suitable method for implementing *nonadiabatic* dynamics should be used [65]. QCT can be used even in this case, provided the suited modifications are applied, consisting essentially in a method for choosing the most probable PES during time evolution, on the basis of a quantum-mechanical (QM) framework [66].

The QCT procedure includes a method for defining reactants, one for propagating the collisional system in time, one for checking the end of the collision and one for analyzing products. It is a time-dependent method, so it is possible to follow the evolution of a collision and to study where it takes place on the PES in time. The essentials of the method are present in the pioneering paper of Karplus *et al* [67] about $H + H_2$ exchange reaction on a semi-empirical PES, but many other authors have then refined the method over the years. Some well-known reviews are available in the literature [68, 69]. Here we only sketch out the main features and provide some observations about the relations with other semiclassical methods, in order to understand the capabilities and limits, particularly important in applications in the context of plasma modelling.

23.3.1 Main features

In QCT reactants are classical particles with the masses of the bodies in the collision that start with the dynamical conditions dictated by the problem under study. Considering the simplest case of atom–diatom collision, reactants relative velocity is obtained from the collision energy that one wants to study. Initial and final conditions for the dynamics are chosen so that the atom–molecule interaction is negligible. As a consequence, quantized quantities corresponding to these conditions, such as vibrational action and molecular angular momentum, correspond to classical constants of motion, always associated with angular quantities (action-angle variables [70]). All these angular quantities are QM indeterminate (for the uncertainty principle, being the respective constants of motion completely determined), and can be uniformly distributed classically, because they are cyclic coordinates conjugated to the constants of motion. As a consequence, in QCT the angles corresponding to initial conditions (in particular relative to vibrational actions and angular momenta) are always uniformly distributed in a random way. The vibrational action and angular momentum modulus of reactants are quantized by imposing the quantization conditions (Langer relations):

$$v_c = \left(v + \frac{1}{2}\right)h \tag{23.17}$$

$$j_c = \sqrt{j(j+1)}\,\hbar \tag{23.18}$$

where v and j are the (integer) quantum vibrational and rotational quantum numbers, v_c and j_c the corresponding classical quantities (vibrational action and angular momentum modulus), respectively[1]. Imposing the classical angular momentum value j_c as initial condition is not particularly difficult and will not be described here. The initial orientation of the molecular axis of rotation, w.r.t. a given space fixed reference frame xyz, is randomly distributed in more common QCT calculations, using uniform distributions of the azimuthal angle and of the cosine of the polar angle. Alternatively, the j_z component of angular momentum along a given fixed z axis can be quantized using $j_z = m\hbar$, imposing in this way a quantization on the polar angle relative to the z axis. Even the orbital angular momentum, relative to the atom–molecule system, has to be considered for the initial conditions. It is commonly considered taking into account a suited distribution of the impact parameter of the collision, i.e. the distance of the initial relative velocity vector of the atom w.r.t. the parallel line passing through the center of mass of the molecule (in this case it is not quantized). But it can be taken into account even as a quantized vector [71], and summed to the molecular angular momentum with the same quantum-mechanical rules of angular momenta composition. This can be extremely useful when directly comparing partial results with QM calculations.

[1] h is Planck constant and $\hbar = h/2\pi$.

Concerning the vibrational action, from its classical definition:

$$v_c = \oint p dr = 2\int_{r_-}^{r_+} \left\{ 2\mu_{BC} \left[\varepsilon_{vj} - V_{BC}(r) - \frac{j_c^2}{2\mu_{BC} r^2} \right] \right\}^{1/2} dr \tag{23.19}$$

(r = diatomic internuclear distance, p = linear momentum of vibration) one can obtain the result provided the classical extrema r_-, r_+ of vibration are known, which are readily available once the value of (ro)vibrational energy ε_{vj} and the diatomic potential $V_{BC}(r)$ are known, being μ_{BC} the reduced mass of BC molecule. Therefore, it is necessary to invert equation (23.19) in order to obtain a given value of classical action. This can be done iteratively by adjusting the initial vibrational energy level by successive approximations within a given error in order to reproduce the v_c value obtained by equation (23.17). The reverse process, i.e. to find the final vibrational quantum number v' in products, is straightforward from equation (23.19) and then inverting equation (23.17). It is worth noting that the final value of v' is still a real number, and a procedure for quantizing it must be used (see below). The WKB procedure just described can be performed quite accurately including also the angular momentum j effect without approximations (for example, in [72] the energy discrepancy with accurate QM calculations is less than 6 cm^{-1}). The outcome of this procedure, which is more complex for polyatomic aggregates, can be tabulated once and for all the possible reactant states, allowing a fast initialization of trajectories, including in the tables also vibrational extrema, vibrational period, etc for given values of (v, j). As a consequence of all these distributions of initial quantities associated with the collision, results are extracted in QCT method from averages over initial conditions on the trajectories (Monte Carlo averages).

During collision evolution in time, obtained integrating the classical equations of motion relative to all the bodies involved, final conditions for atom–diatom collisions are met when two of the three atom–atom distances are greater than a *large* value $R_{\max}$, which means that one atom is sufficiently far from the other two atoms, so that the interaction is too weak to significantly change the final conditions. This distance $R_{\max}$ has to be found empirically by a preliminary study of convergence for increasing values of $R_{\max}$. In QCT there is no distinction of treatment if the final molecular arrangement is the same as in the initial conditions or different. In the first case the collision results in a rovibrationally inelastic process[2]

$$A + BC(v, j) \to A + BC(v'j') \tag{23.20}$$

in the second one it is reactive

$$A + BC(v, j) \to \begin{cases} B + AC(v'j') \\ C + AB(v'j') \end{cases} \tag{23.21}$$

Of course, the reactive final rearrangement can be found only if the PES used for the dynamics allows that, i.e. if the final channel of AC and/or AB have been

[2] if $v' = v$ and $j' = j$ the collision is elastic.

implemented in the PES. Even dissociation (by some authors called atomization) events can easily be managed by the QCT method:

$$A + BC(v, j) \rightarrow A + B + C \tag{23.22}$$

For monitoring this kind of event, the first check is the internal molecular energy E_{int} of product (i.e. of the couple out of the three atoms whose internuclear distance is the lowest at final analysis). If $E_{\text{int}} > 0$, than the further check is if it is greater than E_b, the top of the rotational barrier, which depends on the final rotational angular momentum, or if the current internuclear distance is greater than the position of the rotational barrier top. In both cases, obviously, the system is classically unbound. If this is not the case but $E_{\text{int}} > 0$, the final molecule is trapped classically by the rotational barrier in a metastable state called quasibound. Quasibound states have a probability different from zero of escaping the barrier quantum mechanically, that can easily be calculated by WKB approximation [73, 74]. From the probability one can easily obtain the lifetime of the state. As a consequence, dissociation can also come from quasibound states, provided their lifetime is sufficiently small in comparison with the characteristic collision time. This means that the problem of quasibound states and dissociation cannot be treated within only a dynamic method, even a kinetic scheme is needed for accurate results. Recombination can be obtained as the reverse of dissociation by detailed balance (for many other details see reference [74]). The level of accuracy that can be attained with QCT concerning dissociation/recombination processes is high, considering also that accurate QM methods are practically unfeasible for high-lying rovibrational states and that the necessary approximations ordinarily introduced in QM methods can have an effect even worse than those ones implicit in QCT [75].

The main interest in QCT concerning cold plasma modelling is of course the determination of cross-sections of a given process taking place in the collision. In order to perform calculations, the probability of the process is determined by the frequency of occurrence, i.e. dividing the number N_{ev} of trajectories in which the event of interest has taken place over the total number N_{tot} of trajectories used in the calculation:

$$P(\text{event}) = \frac{N_{\text{ev}}}{N_{\text{tot}}} \tag{23.23}$$

This probability can then be used in the calculation of cross-sections, using the distribution of impact parameter b:

$$\sigma = 2\pi \int_0^{b_{\max}} P(b) b db \tag{23.24}$$

Alternatively, one can implement the dynamics using the complete quantization of angular momenta, so the cross-section is obtained as [71]:

$$\sigma = \frac{\pi}{k^2} \frac{1}{2j + 1} \sum_{J=0}^{J_{\max}} (2J + 1)[2 \min(J, j) + 1] P(J) \tag{23.25}$$

being $k^2 = 2\mu E_{coll}/\hbar^2$, with μ = reduced mass of $A + BC$ system, E_{coll} = collisional energy and:

$$P(J) = \frac{1}{2\min(j, J) + 1} \sum_{\ell=|J-j|}^{J+j} P(J, \ell) \tag{23.26}$$

with $P(J, \ell) = N_{ev}(J, \ell)/N_{tot}(J, \ell)$ (quasiclassical probability) and ℓ = orbital angular momentum quantum number. The last equation for σ is the one used even in quantum-mechanical calculations, if $P(J)$ is obtained with those methods. b_{max} or J_{max} are empirically found by a convergence study. They strongly depend on the kind of event studied, even for a given collisional system.

The Monte Carlo averages in QCT are of importance of course, because the statistics about the initial conditions sampled should be sufficient to obtain a small statistical error on the quantities of interest. It is important to stress, however, that this error is only a necessary condition for good quality results, but not sufficient. The problem is to understand if the approximation operated in the QCT method is adequate for the problem studied. This will be treated in the following sections.

For product analysis, the easiest and intuitive way to *pseudoquantize* v' is to round it to the nearest integer. This method is named *histogram binning*, because it is based on a series of bins centered on quantized values v' and with integer width (from $v' - 0.5$ to $v' + 0.5$), collecting the probability values (see below) coming from the uniform distributions of angle variables. In the following sections a better justification of this procedure, as well as the possibility of improving it, will be presented. Final molecular angular momentum is classically evaluated and then pseudoquantized following the same criterion.

23.3.2 The classical S-matrix theory and its relation with QCT

The theoretical justification of the pseudoquantization operated in histogram binning comes as a corollary of the classical S-matrix theory of W H Miller, in which a rigorous treatment allows to use classical mechanics for approximating quantum mechanics of molecular collisions, with a fundamental difference from the quasiclassical treatment, that will be briefly presented in the following. Time propagation of a quantum system can be accurately described by using the Feynman's interpretation of quantum mechanics, based on the concept of path integral expressing the quantum-mechanical exact probability amplitude [76]:

$$\langle r_2 | e^{-iH(t_2-t_1)/\hbar} | r_1 \rangle \approx \sum_{\text{all paths}} e^{-i\varphi(r_1, r_2)/\hbar} \tag{23.27}$$

related to the evolution from position r_1 at time t_1 to position r_2 at time t_2 (H is the Hamiltonian of the system, while the symbol $\approx$ indicates that the relation written here neglects the presence of a normalization, that will appear in the following). $\varphi(r_1, r_2)$ is the classical action integral along a given path. The semiclassical idea is to reduce the sum over all possible paths (as operated in the Feynman's path integral) to the classical paths only. The classical S-matrix theory proceeds then by substituting ordinary Cartesian coordinates and momenta with the action-angle variables. The amplitude in equation (23.27) is then rewritten by changing the coordinates from the Cartesian to action-angle ones, quantizing the actions and leaving *free* the conjugated variables (this is exactly what is done even in QCT). With the passage to the action-angle variables, and considering here for simplicity only the vibrational degree of freedom, the transition amplitude (the S-matrix) can be rewritten as:

$$S_{21} = \sum_{\text{paths}} \left[2\pi \left(\frac{\partial \tilde{n}_2}{\partial q_1} \right) \right]^{-1/2} e^{-i\varphi(q_1,\, n_1)/\hbar} \tag{23.28}$$

where $\varphi(q_1, n_1)$ is the time integral of the Lagrangian of the system, while q_1 is the initial angle variable (the vibrational phase angle in this case), n_1/n_2 are, respectively, the initial/final vibrational quantum numbers for which the transition probability is calculated, and $\tilde{n}_2$ is the classical (unquantized) equivalent of the final quantum vibrational number (practically v' of QCT) for a given initial n_1, taken when equal to n_2, i.e. for which the equation

$$\tilde{n}_2(q_1, n_1) = n_2 \tag{23.29}$$

holds (it is a search over q_1 values). In figure 23.3 is shown an example of $\tilde{n}_2$ in the asymptotic region of the dynamics, as a function of the initial vibrational phase[3] and for a given n_1. It is periodic by construction, and in the example spans a range of quantized values from 0 to *almost* 4. Following figure 23.3, equation (23.28) can be evaluated calculating $\frac{\partial n_2}{\partial q_1}$ at $v' = 0, 1, 2, 3$, and the phase $\varphi(q_1, n_1)$ should be calculated in the same conditions. As an example, in the figure there is the line at

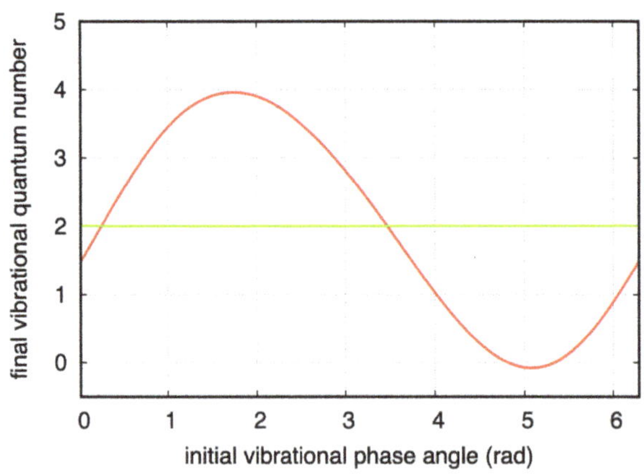

Figure 23.3. Model of final vibration $\tilde{n}_2$ as a function of initial vibrational phase for explaining some features of the classical S-matrix theory (see text).

[3] The actual calculation is based on a slightly different definition of vibrational phase [77], inessential for the discussion here.

$v' = 2$. There are two intersections with the final vibration curve, at about $q_1 = 0.25$ rad and $q_1 = 3.5$ rad, so two trajectories are contributing to the *semi-classical* sum in equation (23.27). The semiclassical character of equation (23.28) consists in the approximation of the result using only classical paths, i.e. classical trajectories, instead of the more general paths used in a Feynman's integral. The physical meaning of equation (23.28) can appear difficult to appreciate, but it is possible to show [78] that the term $[2\pi(\partial n_2/\partial q_1)]^{-1}$ (without the square root) is the classical probability of the event following a given trajectory, while $e^{-i\varphi(q_1, n_1)/\hbar}$ is the quantum phase associated. By neglecting the phase term attached to each classical path, one can sum purely *classical probabilities*:

$$P_{21} = \sum_{\text{all classical path}} \left\{ 2\pi \left[\frac{\partial \tilde{n}_2}{\partial q_1} \right] \right\}^{-1} \tag{23.30}$$

This is reminiscent of the ideal experiment in the famous Feynman's paper [76] about his alternative formulation of quantum mechanics. Summing probabilities as in equation (23.30) means neglecting the phases of interfering paths, which is equivalent to including the possible paths taken one at a time, individually: in this way we measure where the system has passed through, going from initial to final conditions (the classical picture). Summing amplitudes as in equation (23.27) means to include the interference among all the considered paths, which is equivalent to saying that the system has passed somewhere using every possible path simultaneously (the quantum picture). If all the possible paths are included, an exact quantum picture is recovered (equation (23.27)), if all the classical trajectories are considered satisfying equation (23.29), the semiclassical picture is obtained, which then allows the calculation of the transition probabilities from the square modulus of the transition amplitudes:

$$P_{21} = |S_{21}|^2 \tag{23.31}$$

Even quasiclassical trajectories are smoothly explained in this picture: it is sufficient [79] to average the purely classical probability in equation (23.30) in the interval $(n_2 - 1/2, n_2 + 1/2)$ for obtaining the Monte Carlo average of QCT, i.e. the histogram binning cited above. Furthermore, it is possible to manipulate [80] e: molmol:eq30 to substitute it with an integral over the Dirac distribution:

$$P_{21} = \frac{1}{2\pi} \int dq_1 \delta(\tilde{n}_2(q_1) + n_2)) \tag{23.32}$$

Then, by approximating the Dirac distribution with a Gaussian of suited width, one can obtain an approximation to P_{21}, starting not from a root search as in the discussion for figure 23.3, which becomes practically intractable when considering further degrees of freedom, but from a cheaper Monte Carlo average as in the standard procedure of QCT. This is the Gaussian binning, that has shown good performance in reactive scattering calculations compared with accurate QM calculations. From a practical point of view, it consists in weighting the final vibration (or even rotation) with a Gaussian function centered in the quantized value of interest. The Gaussian decreases very rapidly as a function of distance between the final action and the quantized value, so it is a bit more expensive than standard QCT (many trajectories far from the quantized values are practically discarded), but of very practical use.

Concluding this section, it is necessary to stress that it is definitely *not* easy to numerically apply the classical S-matrix theory in general [81, 82], and in particular to the intensive calculations required when building databases for plasma chemistry models, which span large total energy ranges. The problem is relatively solvable when applying equation (23.28) to the conditions $v' = 0, 1, 2, 3$ of figure 23.3 (the *classically allowed* transitions), but it becomes even harder for $v' \geqslant 4$ (the *classically forbidden* transitions). Theoretically, it is possible to analytically continue the $\tilde{n}_2$ in the complex plane [83], where the intersections with $v' \geqslant 4$ can be found. It is feasible to do that for $v' = 4$, but it becomes more and more numerically difficult to find accurate results for higher v' values not reached by the real curve of the classical action (*highly* classically forbidden cases). The greatest interest, on the other side, is just in the missed real intersections. In those cases, the use of binning in QCT allows finding some non-zero result for $v' = 4$, because some trajectories can show values of final vibration near $v' = 4$ and are so included in the relative bin, but for more classically forbidden transitions surely QCT will give a zero result, erroneously. Of course, any kind of QCT binning will be unable to extract good results in this case. So just where QCT fails, classical S-matrix theory is numerically weaker. Another point to stress is the possibility of conditions of *classicality* in the dynamics, when trajectories tend to show a complex behaviour, with random jumps of both $\tilde{n}_2$ and phases when q_1 is slightly changed (see in particular reference [84]). In this case it is numerically impossible to apply equation (23.28), but fortunately equation (23.30) or equivalently QCT can be successfully used.

However, the great strength of the classical S-matrix theory is in providing the logical path which brings from the ideally accurate Feynman's path integral to the semiclassical and classical trajectories calculations. It is a precious guide to understand when and why QCT can correctly work and where it cannot, and as a consequence how to get the best of quasi/semiclassical methods in real calculations.

23.3.3 Some observations about QCT binning

It is clear from the preceding sections that the QCT method is based on the simple binning of final actions to classify products. How reliable is this tool? It depends on the distribution of final actions to treat. If the distribution is significantly wider than one bin and without rapid variations, the binning *per se* is expected to be accurate. If the distribution is narrow and rapidly changing from one-bin limits to the next, the result is much more prone to errors due the *rough* binning. Of course, binning cannot be made finer, because it depends on the quantization. It is the method of detecting probability of final states that should be changed in that case (see below). The binning becomes *rough* when the system is close to a classically forbidden condition (as $v' = 4$ of figure 23.3), while it does not give any result for a highly classically forbidden condition (as $v' = 5$ of figure 23.3). This fundamental aspect of QCT is particularly clear and important for inelastic processes. Suppose the reactant molecule is in vibrational state $v = 1$, and a low-energy inelastic collision takes place with an atom. The interaction is supposed to be weak, as a consequence there will be a low probability of a change of vibrational state to, for example, $v' = 0$. But in classical terms the effect will be different: we will see a small variation of final vibration w.r.t. the initial one, for example one trajectory will show $v' = 0.99$ and another 0.95. Of course, $v' = 0$ bin is too far to be reached in these conditions, and QCT will count the process as vibrationally *elastic* ($v' = v$), even if a contribution of non-elastic transitions is surely present. The moments method [85, 86] tries to solve this problem, considering that the moments of vibrational energy transfer should be equal in the QM and classical treatments:

$$\langle \Delta E_{v^n} \rangle = \sum_{v'=0}^{v'_{\max}} (E^q_{v'} - E^q_v)^n P_{vv'} \cong \frac{1}{N_{\text{tot}}} \sum_{v'=0}^{v'_{\max}} (E^c_{v'} - E^c_v)^n \tag{23.33}$$

where E^q_v, $E^q_{v'}$ are the initial and final vibrational state energy quantum values, while E^c_v, $E^c_{v'}$ are the corresponding classical values (E^c_v is equal to E^q_v because of the initial exact quantization, while $E^q_{v'}$ is the unquantized final value obtained from the classical dynamics). The last sum is over all the trajectories (N_{tot}) employed in the calculation, while the first sum is over all the open vibrational channels up to $v'_{\max}$ (which depends on the total energy available). The moments can be calculated with n index values from 1 to $v'_{\max}$. In principle, from a system of $v'_{\max}$ equations like equation (23.30), one can obtain the quantum $P_{vv'}$ transition probabilities. The method can also be applied to rotational transitions [69]. In practical calculations it is possible to obtain good results only for the lowest states [86] and with a limited number of open channels (i.e. low total energy). However, the method is physically useful to understand that in the inelastic processes the molecular classical dynamics should be considered as a sort of average of many quantum transitions, rather than an exact reproduction of a single transition, especially for low-lying states. This is exactly what can be observed in other methods for treating inelastic processes, all based on the exact correspondence between the quantum and the classical forced harmonic oscillator (FHO) dynamics. We will distinguish here the general physical model by calling it *FHO model* from the specific *FHO method* presented in section 23.1 [28, 34] of this chapter, which is based on the same correspondence but is an analytical model based on a potential model and does not include a trajectory dynamics. The FHO model is well described in [22, 87, 88], and applied in [23, 89]. Kerner [22] has shown that if an initially stationary harmonic oscillator is subject to a force $f(t)$ depending only on time, then the exact solution of the Schrödinger equation is an oscillating wave packet which maintains the original shape and whose center follows a classical evolution on the same potential. The final analysis of this oscillating wave packet, obtained as usual in quantum dynamics by superposition to vibrational eigenstates, is exact. So, for atom–diatom inelastic collisions the idea [89] is to perform a classical dynamics similar to QCT, but for the vibrational initial condition, which corresponds to the bottom of the vibrational potential (and not to the energy of any initial state v). Classical dynamics provides the final vibrational energy of the wave packet, which is sufficient to calculate the full transition probability matrix relative to the harmonic potential [23]. If only vibration is considered in the collisional system (typically in a collinear atom–diatom collision), one trajectory is able to provide the full matrix at a fixed energy. In cases more useful for plasma modelling, a Monte Carlo average over rotation, similarly to QCT, is able to provide a quite good result for low lying states (i.e. those states of the harmonic oscillator very close to vibrational states of the actual potential). The method is used with any PES available for the system and is significantly cheaper than QCT when applied to many transitions (one run covers theoretically the whole vibrational matrix, including all possible initial states, at variance with QCT). In comparison with QCT, it can be much more accurate for inelastic processes relative to low-lying states. However, it is intrinsically inelastic, it does not treat reaction, dissociation, and quasi-reactive inelastic processes, that will be introduced in the following sections. The atom–diatom interaction is modelled with the force $f(t)$, which physically means that the atom interaction enters the FHO model only by its temporal dependence, and it is not affected by the quantum vibrational state. All these conditions can realistically be satisfied if the interactions of the atomic projectile A with each atom B and C in the diatom are similar and not very strong, which in turn can reasonably be correct if R, the atom–diatom distance, is always greater than r, the diatomic internuclear distance, along the whole collision. Therefore, the best condition for FHO model application is in collisions in which the atom passes by the diatom at a certain minimum distance R larger than the internuclear distance r along the whole trajectory. This condition can easily be verified by monitoring the *collision remoteness measure* (*CRM*), defined as [19, 90–92]:

$$CRM = \min_t \left\{ \frac{\min(R_{AB}, R_{AC})}{R_{BC}} \right\} \tag{23.34}$$

in which R_{AB} is the $A-B$ distance, and so on for the other terms. The first minimum is calculated over each whole trajectory, while at each time step the inner minimum is the smallest distance of the atomic projectile with B and C. CRM is a value useful to classify the strength of the interaction for the trajectory associated: a high value means that the atom remains quite far from the diatom[4] and the interaction cannot be very high, a low value means that the interaction is strong at least somewhere during trajectory evolution. If the collisional system is potentially reactive, the internuclear distances can be normalized to the respective equilibrium distances $R_{AB}^{\text{eq}}, R_{AC}^{\text{eq}}, R_{BC}^{\text{eq}}$. In this case a CRM value less than 1 corresponds to the condition of traversing the reaction barrier, which undoubtedly implies a strong interaction. If the trajectory is inelastic and its $CRM < 1$, then it can be classified as *quasi-reactive* (QR), while with $CRM > 1$ it can be called *purely nonreactive* (PNR) or purely inelastic. This distinction is quite interesting because of the completely different features of the two inelastic processes, as also found experimentally [93]. QR events are essentially frustrated reactive trajectories, which explore the reactive PES regions

[4] Hence the name *collision remoteness*.

and terminate in the initial channel, but with a reactive dynamics (see also [94]). This means that there is a break (or at least a weakening) of the initial bond and in the end a reconstruction of it, but the initial internal energy of the molecule has mixed with the relative translational energy of the collisional system, so the final arrangement, even if identical to the initial one, shows states not strictly related to its initial state. This is a fundamental difference from the PNR events, which are characterized by a clear memory of the initial state: the initial vibration is only slightly perturbed by the atomic projectile, so the final state in a given classical trajectory is only slightly different from the initial one. If the *CRM* is presented as a function of the final vibration $\tilde{n}_2$ already presented in preceding sections, the result is quite clear and meaningful, as will be briefly presented for the case of $O + N_2(v = 1, j = 0) \rightarrow O + N_2(v' = 1, j' = \text{all})$ inelastic process, extracted here from calculations performed in reference [19]. The $O + N_2$ collisional system is reactive with a high threshold, about 3.3 eV. Under the threshold it behaves inelastically, with a QCT cross-section that at relatively low collision energy (up to six times the vibrational quantum) is zero, while still unpublished accurate QM and semiclassical calculations give a low but never zero result, even at one quantum of collision energy.

In figure 23.4, *CRM* is reported for a bunch of trajectories for this system as a function of classically obtained final vibrational quantum number. It is clear the presence of two distinct regions, one of PNR events ($CRM > 1$) and another of QR events for $CRM \leqslant 1$. The difference is plain: in the upper part there is a distribution of trajectories that is highly peaked in the initial vibrational state and widening for v' values on both sides of the initial $v = 1$, with a quite compact distribution overall. In the lower part there is a wide, almost flat distribution with a much weaker relation with the initial vibrational state. The distributions are shown at three different and increasing collision energy ranges. For the first one (E_{col} = 1.6–1.7 eV) only PNR events are found, and the distribution is unable to reach the boundaries of the two adjacent bins of $v' = 0$ and 2. For the intermediate range (2.9–3.0 eV) even the QR

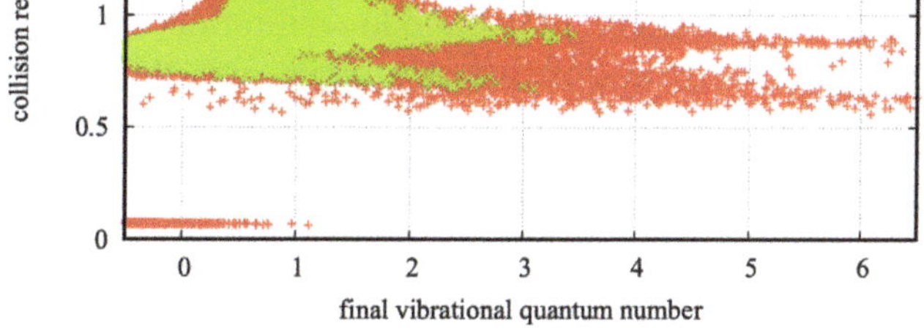

Figure 23.4. Collision remoteness as a function of classically obtained final vibrational quantum number for three different collision energy ranges. See text for comments.

distribution is present and is significantly larger than the PNR one, with a completely different shape of the distribution. For the third range (3.4–3.5 eV) QR distribution is much wider and even the reaction appears, as a distinct low-lying group of trajectories at *CRM* < 0.1. Logically, the QR distribution is ideal for the binning treatment, while the PNR distribution is the worst possible for that. For increasing collision energy, the PNR distribution tends to widen, being captured in a more accurate way by the binning. Summarizing, what QCT *sees* in inelastic transitions is a sum of a PNR distribution, difficult to be captured at low energy, and for sufficient collision energy, a QR distribution that is correctly binned. This means that one can think of separating the QCT *accurate* partial result from the less accurate one by using the *CRM*. It happens that concerning inelastic processes, QCT and FHO model dynamics are complementary, because QR contribution is correctly binned by QCT, while the FHO model captures the PNR contribution with accuracy. Even the reverse is true: QCT is inaccurate in PNR events, and QR processes should not be treated with the FHO model, because of the considerations already presented. Work is in progress [92] to develop a merging of QCT with another simple semiclassical method following the idea just exposed in order to extend the accuracy of QCT to inelastic collisions, which are of paramount importance in cold plasmas of technological interest. From this discussion it should be clear that in the case of reaction, the detection of final vibration by binning in QCT can be expected to be accurate, because there is a strong interaction that produces an almost flat vibrational distribution, as in the QR case. Indeed, in reactive processes a severe limitation to QCT accuracy can be the presence of tunnelling in the translational coordinate of the reaction. If this effect is important in the conditions studied, QCT cannot give an accurate result. However, in the comparisons with the accurate QM calculations concerning complete cross-sections (excluding partial results, like reaction probability limited to total angular momentum $J = 0$, or low-dimensionality models, scarcely useful for plasma modelling), the differences seem to be limited to the lightest species around room temperature [92, 95], but it seems really a minor problem for air species starting from 300 K [19, 91], of paramount importance in practical applications. While for low-temperature, low-masses species, an accurate QM calculation is feasible [75, 96], the analogous calculations for air species are in many cases beyond the current limits of computations, and this is the important reason for studying the level of approximation of semi- and quasiclassical methods. Another source of error in reaction processes treated with QCT is the zero point issue, which can be seen as another aspect of a binning issue related in this case to low vibrational energy available after the reaction event [92]. Even in this case one might also think to model the final vibration as an oscillating wave packet instead of a purely classical object, with the complication of the strong interaction generally associated with the reactive process.

23.3.4 Cold plasmas and state-to-state data

Typically, in a cold plasma the initial excitation can be an electrical discharge or a microwave flux, able to excite just the first vibrational levels of the molecules involved. By inelastic collisions, very high values of vibration can be reached, up to the dissociation limit. As a consequence, the inelastic processes, both PNR and QR, play a fundamental role. Semiclassical methods are expected to perform well on PNR collisions, while QCT is expected to be accurate on QR, and it can manage even reactivity and dissociation. The problem of an efficient nitrogen fixation using a cold plasma, whose accurate modelling relies on reliable state-to-state data [97], is of great technological value due to the industrial, agricultural and even clean transportation [98] applications. The current model [60, 99] for the CO_2 plasma reactor in order to obtain renewable fuels directly from CO_2 and water using low-temperature plasmas is an example of a fundamental model that needs accurate data. There is enormous interest for this process, due to the urgent problems linked to the constant increase of human-produced CO_2 presence in the atmosphere that in turn increases the Earth's average temperature. To be reliable, models need accurate state-to-state rate coefficients, that are often unfeasible to obtain with highly accurate QM methods, hence the interest for continuing the development of approximate methods for molecular dynamics.

References

[1] Rapp D and Kassal T 1969 Theory of vibrational energy transfer between simple molecules in nonreactive collisions *Chem. Rev.* **69** 61–102

[2] Ormonde S 1975 Vibrational relaxation theories and measurements *Rev. Mod. Phys.* **47** 193

[3] Shin H K 1976 Vibrational energy transfer *Dynamics of Molecular Collisions* ed W H Miller (Berlin: Springer) pp 131–210

[4] Billing G 1986 Vibration–vibration and vibration-translation energy transfer, including multiquantum transitions in atom-diatom and diatom-diatom collisions *Nonequilibrium Vibrational Kinetics* ed M Capitelli (Berlin: Springer) pp 85–112

[5] Hernández R, Toumi R and Clary D C 1995 State-selected vibrational relaxation rates for highly vibrationally excited oxygen molecules *J. Chem. Phys.* **102** 9544–56

[6] Billing G D and Jolicard G 1982 Comparison of quantum mechanical and semiclassical (classical path) probabilities for vibrational transitions in diatom-diatom collisions *Chem. Phys.* **65** 323–33

[7] Cacciatore M, Kurnosov A and Napartovich A 2005 Vibrational energy transfer in N_2-N_2 collisions: a new semiclassical study *J. Chem. Phys.* **123** 174315

[8] Coletti C and Billing G D 2002 Vibrational energy transfer in molecular oxygen collisions *Chem. Phys. Lett.* **356** 14–22

[9] Billing G D 1994 VV and VT rates in N_2-O_2 collisions *Chem. Phys.* **179** 463–7

[10] Cacciatore M and Billing G D 1981 Semiclassical calculation of VV and VT rate coefficients in CO *Chem. Phys.* **58** 395–407

[11] Kurnosov A, Cacciatore M and Billing G D 2003 State-to-state rate constant calculations for V-V energy transfer in CO-N_2 collisions *J. Phys. Chem.* A **107** 2403–10

[12] Esposito F and Capitelli M 2006 QCT calculations for the process $N_2(v) + N \rightarrow N_2(v') + N$ in the whole vibrational range *Chem. Phys. Lett.* **418** 581–5

[13] Panesi M, Jaffe R L, Schwenke D W and Magin T E 2013 Rovibrational internal energy transfer and dissociation of N_2 $(^1\Sigma_g^+)$-$N(^4S_u)$ system in hypersonic flows *J. Chem. Phys.* **138** 044312

[14] Esposito F, Armenise I, Capitta G and Capitelli M 2008 O-O_2 state to state vibrational-relaxation and dissociation rates based on quasiclassical calculations *Chem. Phys.* **351** 91–8
[15] Andrienko D A and Boyd I D 2016 Rovibrational energy transfer and dissociation in O_2-O collisions *J. Chem. Phys.* **144** 104301
[16] Grover M S, Schwartzentruber T E, Varga Z and Truhlar D G 2019 Vibrational energy transfer and collision-induced dissociation in O + O_2 collisions *J. Thermophys. Heat Transfer* **33** 797–807
[17] Kim J G and Boyd I D 2015 Thermochemical nonequilibrium analysis of O_2 + Ar based on state-resolved kinetics *Chem. Phys.* **446** 76–85
[18] Andrienko D A and Boyd I D 2016 Thermal relaxation of molecular oxygen in collisions with nitrogen atoms *J. Chem. Phys.* **145** 014309
[19] Esposito F and Armenise I 2017 Reactive, inelastic, and dissociation processes in collisions of atomic oxygen with molecular nitrogen *J. Phys. Chem.* A **121** 6211–9
[20] Macdonald R, Jaffe R, Schwenke D and Panesi M 2018 Construction of a coarse-grain quasi-classical trajectory method. I. Theory and application to N_2-N_2 system *J. Chem. Phys.* **148** 054309
[21] Grover M S, Valentini P, Josyula E and Chaudhry R S 2020 Vibrational state-to-state and multiquantum effects for N_2 + N_2 interactions at high temperatures for aerothermodynamic applications *AIAA Scitech 2020 Forum* AIAA Paper 2020-1227
[22] Kerner E H 1958 Note on the forced and damped oscillator in quantum mechanics *Can. J. Phys.* **36** 371–7
[23] Treanor C E 1965 Vibrational energy transfer in high-energy collisions *J. Chem. Phys.* **43** 532–8
[24] Zelechow A, Rapp D and Sharp T E 1968 Vibrational-vibrational-translational energy transfer between two diatomic molecules *J. Chem. Phys.* **49** 286–99
[25] Billing G 1984 Rate constants and cross sections for vibrational transitions in atom-diatom and diatom-diatom collisions *Comput. Phys. Commun.* **32** 45–62
[26] Billing G 1987 Rate constants for vibrational transitions in diatom-diatom collisions *Comput. Phys. Commun.* **44** 121–36
[27] Adamovich I V, Macheret S O, Rich J W and Treanor C E 1995 Vibrational relaxation and dissociation behind shock waves. Part 1-kinetic rate models *AIAA J.* **33** 1064–9
[28] Adamovich I V, Macheret S O, Rich J W and Treanor C E 1998 Vibrational energy transfer rates using a forced harmonic oscillator model *J. Thermophys. Heat Transfer* **12** 57–65
[29] Billing G D and Fisher E 1979 VV and VT rate coefficients in N_2 by a quantum-classical model *Chem. Phys.* **43** 395–401
[30] Billing G D and Kolesnick R 1992 Vibrational relaxation of oxygen. State to state rate constants *Chem. Phys. Lett.* **200** 382–6
[31] Candler G V, Olejniczak J and Harrold B 1997 Detailed simulation of nitrogen dissociation in stagnation regions *Phys. Fluids* **9** 2108–17
[32] Giordano D, Bellucci V, Colonna G, Capitelli M, Armenise I and Bruno C 1997 Vibrationally relaxing flow of N_2 past an infinite cylinder *J. Thermophys. Heat Transfer* **11** 27–35
[33] Adamovich I V and Rich J W 1998 Three-dimensional nonperturbative analytic model of vibrational energy transfer in atom-molecule collisions *J. Chem. Phys.* **109** 7711–24
[34] Adamovich I V 2001 Three-dimensional analytic model of vibrational energy transfer in molecule-molecule collisions *AIAA J.* **39** 1916–25

[35] Gimelshein S F, Wysong I J and Adamovich I V 2018 Application of the 3D forced harmonic oscillator model in the DSMC method *J. Thermophys. Heat Transfer* **32** 882–91

[36] Macheret S O and Adamovich I V 2000 Semiclassical modeling of state-specific dissociation rates in diatomic gases *J. Chem. Phys.* **113** 7351–61

[37] Nyeland C and Billing G D 1979 Approximative treatments of rotational relaxation *Chem. Phys.* **40** 103–10

[38] Adamovich I V 2014 Three-dimensional analytic probabilities of coupled vibrational-rotational-translational energy transfer for DSMC modeling of nonequilibrium flows *Phys. Fluids* **26** 046102

[39] Magin T E, Panesi M, Bourdon A, Jaffe R L and Schwenke D W 2012 Coarse-grain model for internal energy excitation and dissociation of molecular nitrogen *Chem. Phys.* **398** 90–5

[40] Skrebkov O 1995 Diffusional description of vibrational relaxation in a binary mixture of diatomic molecules–quantum oscillators *Chem. Phys.* **191** 87–99

[41] Adamovich I V, Rich J W and Macheret S O 1997 Existence of the bottleneck in vibrational relaxation of diatomic molecules *J. Thermophys. Heat Transfer* **11** 261–5

[42] Macheret S, Losev S, Chernyi G and Potapkin B 2002 *Physical and Chemical Processes in Gas Dynamics: Cross sections and Rate Constants* vol I (Reston, VA: American Institute of Aeronautics and Astronautics)

[43] Bird G A 1994 *Molecular Gas Dynamics and the Direct Simulation of Gas Flows* (Oxford: Clarendon)

[44] Macheret S O and Rich J W 1993 Nonequilibrium dissociation rates behind strong shock waves: classical model *Chem. Phys.* **174** 25–43

[45] Macheret S O, Fridman A A, Adamovich I V, Rich J W and Treanor C E 1994 Mechanisms of nonequilibrium dissociation of diatomic molecules *6th Joint Thermophysics and Heat Transfer Conf.* AIAA Paper 1994-1984

[46] Luo H, Alexeenko A A and Macheret S O 2018 Assessment of classical impulsive models of dissociation in thermochemical nonequilibrium *J. Thermophys. Heat Transfer* **32** 861–8

[47] Luo H, Sebastiãgo I B, Alexeenko A A and Macheret S O 2018 Classical impulsive model for dissociation of diatomic molecules in direct simulation Monte Carlo *Phys. Rev. Fluids* **3** 113401

[48] Luo H, Alexeenko A A and Macheret S O 2019 Development of an impulsive model of dissociation in direct simulation Monte Carlo *Phys. Fluids* **31** 087105

[49] Macheret S O, Luo H and Alexeenko A A 2019 Nonempirical analytical model of nonequilibrium dissociation in high-temperature air *Hypersonic Meteoroid Entry Physics* ed G Colonna, M Capitelli and A Laricchiuta (Bristol: IOP Publishing)

[50] Chaudhry R S, Grover M S, Bender J D and Schwartzentruber T E 2018 Quasiclassical trajectory analysis of oxygen dissociation via O_2, O, and N_2 *2018 AIAA Aerospace Sciences Meeting* AIAA Paper 2018-0237

[51] Bender J D, Valentini P, Nompelis I, Paukku Y, Varga Z, Truhlar D G and Schwartzentruber T 2015 An improved potential energy surface and multi-temperature quasiclassical trajectory calculations of $N_2 + N_2$ dissociation reactions *J. Chem. Phys.* **143** 054304

[52] Singh N and Schwartzentruber T E 2017 Coupled vibration-rotation dissociation model for nitrogen from direct molecular simulations *47th AIAA Thermophysics Conf.* AIAA Paper 2017-3490

[53] Losev S, Makarov V and Nikolsky V 1994 Thermochemical nonequilibrium kinetic models in strong shock waves on air *6th Joint Thermophysics and Heat Transfer Conf.* AIAA Paper 1994-1990

[54] Yalovik M and Losev S 1972 Vibrational kinetics and dissociation of nitrogen molecules at high temperature *Sci. Proc.* **18** 3–34

[55] Bose D 1996 Kinetics of the N_2 + O yields NO + N reaction under thermodynamic nonequilibrium *J. Thermophys. Heat Transfer* **10** 148–54

[56] Bose D, Candler G, Bose D and Candler G 1997 Thermal nonequilibrium rates of the Zeldovich reactions *35th Aerospace Sciences Meeting and Exhibit.* AIAA Paper 1997-133

[57] Bose D and Candler G V 1997 Simulation of hypersonic flows using a detailed nitric oxide formation model *Phys. Fluids* **9** 1171–81

[58] Macheret S O, Rusanov V D, Fridman A A and Sholin G V 1978 Synthesis of nitrogen oxides in a nonequilibrium plasma Sov *Tech. Phys. Lett.* **4** 140

[59] Macheret S O, Rusanov V D, Fridman A A and Sholin G V 1980 Nonequilibrium plasmachemical synthesis of nitrogen oxides *Sov. Phys. - Tech. Phys.* **25** 421

[60] Fridman A 2008 *Plasma Chemistry* (Cambridge: Cambridge University Press)

[61] Ricard A, Oh S-g and Guerra V 2013 Line-ratio determination of atomic oxygen and metastable absolute densities in an rf nitrogen late afterglow *Plasma Sources Sci. Technol.* **22** 035009

[62] Macheret S O, Rusanov V D and Fridman A A 1984 Reactivity of vibrationally excited molecules *Sov. Phys. - Dokl.* **276** 522

[63] Kuntz P 1976 Features of potential energy surfaces and their effect on collisions *Dynamics of Molecular Collisions* ed W H Miller (Boston, MA: Springer) pp 53–120

[64] Hollebeek T, Ho T-S and Rabitz H 1999 Constructing multidimensional molecular potential energy surfaces from *ab initio* data *Annu. Rev. Phys. Chem.* **50** 537–70

[65] Prigogine I and Rice S A 2003 *The Role of Degenerate States in Chemistry* vol 150 (New York: Wiley)

[66] Tully J C 1990 Molecular dynamics with electronic transitions *J. Chem. Phys.* **93** 1061–71

[67] Karplus M, Porter R N and Sharma R D 1965 Exchange reactions with activation energy. I. Simple barrier potential for (H, H_2) *J. Chem. Phys.* **43** 3259–87

[68] Truhlar D G and Muckerman J T 1979 Reactive scattering cross sections III: quasiclassical and semiclassical methods *Atom–Molecule Collision Theory* ed R B Bernstein (Berlin: Springer) pp 505–66

[69] Pattengill M 1979 *Rotational excitation III: classical trajectory methods Atom-Molecule Collision Theory* (Berlin: Springer) pp 359–75

[70] Goldstein H, Poole C and Safko J 2000 *Classical Mechanics* 3rd edn (San Francisco, CA: Addison-Wesley)

[71] Aoiz F, Sáez-Rábanos V, Martínez-Haya B and González-Lezana T 2005 Quasiclassical determination of reaction probabilities as a function of the total angular momentum *J. Chem. Phys.* **123** 094101

[72] Blais N C and Truhlar D G 1976 Monte Carlo trajectory study of Ar.H_2 collisions. I. Potential energy surface and cross sections for dissociation, recombination, and inelastic scattering *J. Chem. Phys.* **65** 5335–56

[73] Kuntz P 1979 Collision-induced dissociation II: Trajectories and models *Atom–Molecule Collision Theory* ed R B Bernstein (Berlin: Springer) pp 669–92

[74] Esposito F and Capitelli M 2009 Selective vibrational pumping of molecular hydrogen via gas phase atomic recombination *J. Phys. Chem.* A **113** 15307–14

[75] Esposito F, Coppola C M and De Fazio D 2015 Complementarity between quantum and classical mechanics in chemical modeling. The $H + HeH^+ \rightarrow H_2^+ + He$ reaction: a rigorous test for reaction dynamics methods *J. Phys. Chem.* A **119** 12615–26

[76] Feynman R P 1948 Space-time approach to non-relativistic quantum mechanics *Rev. Mod. Phys.* **20** 367–87

[77] Miller W H 1970 The classical S-matrix: a more detailed study of classically forbidden transitions in inelastic collisions *Chem. Phys. Lett.* **7** 431–5

[78] Miller W H 1971 Semiclassical nature of atomic and molecular collisions *Acc. Chem. Res.* **4** 161–7

[79] Miller W H 1971 Classical S matrix for rotational excitation; quenching of quantum effects in molecular collisions *J. Chem. Phys.* **54** 5386–97

[80] Bonnet L 2008 The method of Gaussian weighted trajectories. III. An adiabaticity correction proposal *J. Chem. Phys.* **128** 044109

[81] Duff J W and Truhlar D G 1974 Classical S matrix: numerical applications to classically allowed chemical reactions *Chem. Phys.* **4** 1–23

[82] Duff J W and Truhlar D G 1975 Use of semiclassical collision theory to compare analytic fits to the interaction potential for vibrational excitation of H_2 by He *J. Chem. Phys.* **63** 4418–29

[83] Miller W H 1970 Classical S matrix: numerical application to inelastic collisions *J. Chem. Phys.* **53** 3578–87

[84] Rankin C and Miller W H 1971 Classical S matrix for linear reactive collisions of $H + Cl_2$ *J. Chem. Phys.* **55** 3150–6

[85] Truhlar D G and Duff J W 1975 Classical probability matrix: prediction of quantum-state distributions by a moment analysis of classical trajectories *Chem. Phys. Lett.* **36** 551–4

[86] Gordon R J 1976 A comparison of exact classical and quantum mechanical calculations of vibrational energy transfer *J. Chem. Phys.* **65** 4945–57

[87] Husimi K 1953 Miscellanea in elementary quantum mechanics, II *Prog. Theor. Phys.* **9** 381–402

[88] Takayanagi K 1963 Vibrational and rotational transitions in molecular collisions *Prog. Theor. Phys. Suppl.* **25** 1–98

[89] Giese C F and Gentry W R 1974 Classical trajectory treatment of inelastic scattering in collisions of H^+ with H_2, HD, and D_2 *Phys. Rev.* A **10** 2156

[90] Capitelli M, Colonna G, Esposito F, Hassouni K, Laricchiuta A and Longo S 2015 Fundamental aspects of plasma *Chemical Physics: Kinetics* (Berlin: Springer)

[91] Esposito F and Armenise I 2021 Reactive, inelastic, and dissociation processes in collisions of atomic nitrogen with molecular oxygen *J. Phys. Chem.* A **125** 3953–64

[92] Esposito F 2019 Reactivity, relaxation and dissociation of vibrationally excited molecules in low-temperature plasma modeling *Rend. Lincei, Sci. Fis. Nat.* **30** 57–66

[93] Greaves S J, Wrede E, Goldberg N T, Zhang J, Miller D J and Zare R N 2008 Vibrational excitation through tug-of-war inelastic collisions *Nature* **454** 88–91

[94] Jankunas J, Sneha M, Zare R N, Bouakline F, Althorpe S C, Herráez-Aguilar D and Aoiz F J 2014 Is the simplest chemical reaction really so simple? *Proc. Natl Acad. Sci.* **111** 15–20

[95] Bell R 1980 *The Tunnel Effect In Chemistry* (London: Chapman and Hall)

[96] Lique F 2020 Theoretical study of the $T + D_2$ and $D + T_2$ collisions in low-temperature hydrogen plasmas *Phys. Rev.* A **102** 012806

[97] Van Alphen S, Vermeiren V, Butterworth T, van den Bekerom D C, Van Rooij G and Bogaerts A 2020 Power pulsing to maximize vibrational excitation efficiency in N_2 microwave plasma: a combined experimental and computational study *J. Phys. Chem.* C **124** 1765–79

[98] Wan Z, Tao Y, Shao J, Zhang Y and You H 2021 Ammonia as an effective hydrogen carrier and a clean fuel for solid oxide fuel cells *Energy Convers. Manage.* **228** 113729

[99] Pietanza L D *et al* 2021 Advances in non-equilibrium CO_2 plasma kinetics: a theoretical and experimental review *Eur. Phys. J.* D **75** 1–55

www.ingramcontent.com/pod-product-compliance
Ingram Content Group UK Ltd.
Pitfield, Milton Keynes, MK11 3LW, UK
UKHW051043120926
8556IPUK00030B/849